FUNDAMENTALS OF MATHEMATICS

VOLUME II

Geometry

Fundamentals of Mathematics

Volume I

Foundations of Mathematics
The Real Number System and Algebra

Volume II

Geometry

Volume III

Analysis

FUNDAMENTALS OF MATHEMATICS

VOLUME II

Geometry

Edited by
H. Behnke
F. Bachmann
K. Fladt
H. Kunle

with the assistance of
H. Gerike
F. Hohenberg
G. Pickert
H. Rau

Translated by
S. H. Gould

The MIT Press   Cambridge, Massachusetts, and London, England

Originally published by Vandenhoeck & Ruprecht, Göttingen, Germany, under the title *Grundzüge der Mathematik*. The publication was sponsored by the German section of the International Commission for Mathematical Instruction. The present translation of this volume is based upon the second German editions of 1967 and 1971.

Fifth printing, 1988

First MIT Press paperback edition, 1983

ISBN 0-262-02069-6 (hardcover)
       0-262-52094-X (paperback)
Library of Congress catalog card number: 68-14446

Contents

Translator's Foreword

The pleasant task of translating this unique work has now extended over several years, in the course of which I have received invaluable assistance from many sources. Fortunately I had the opportunity, in personal conversation or in correspondence, of discussing the entire translation with the original authors, many of whom suggested improvements, supplied exercises, or made changes and additions in the German text, wherever they seemed desirable to bring the discussion up to date, for example, on the continuum hypothesis, Zorn's lemma, or groups of odd order. To all these authors I express my gratitude.

For technical and clerical help I am especially indebted to Linda Shepard, of the Law School at the University of Utah, for her expert typing and discriminating knowledge of English; to Diane Houle, supervisor of the Varitype Section of the American Mathematical Society, for her unrivaled skill and experience in the typing of mathematical translations; to Linda Rinaldi and Ingeborg Menz, secretaries, respectively, of the Translations Department of the Society and the firm Vandenhoeck and Ruprecht, for keeping straight a long and complicated correspondence; to the staff of The MIT Press for their customary technical expertness; and to my wife, Katherine Gould, for help too varied and too substantial to be readily described.

S. H. Gould
Institute of Mathematics
Academia Sinica
Taipei, Taiwan
Republic of China

September 1973

Preface

Euclid's geometry has always been the model for a deductive, axiomatic theory, but the attitude toward the axioms themselves has undergone constant change up to the present day. Although, in 1882, Pasch made important progress toward the explicit statement of all assumptions of geometry not of a purely logical nature, the work of basic importance for modern axiomatics is Hilbert's *Foundations of Geometry* (1899). Pasch still regarded the "facts" of geometry as given realities and set himself the task of selecting from them certain "basic theorems," as few as possible, from which all the others could be logically derived. But with Hilbert all questions of the meaning of the fundamental concepts or the sense in which the axioms are true were excluded altogether from geometry, a step that at first may seem superficial but has, in fact, given rise to the whole modern, far-reaching theory of axiomatics. The habit of thinking in terms of axiomatically defined structures has produced a fundamental change in our conception of the nature of mathematics and the interrelation of its various parts. Structures like group, field, vector space, and lattice, which in themselves are algebraic, provide geometric forms of thought basic to the study of such geometric structures as affine, projective, or metric planes and projective or topological spaces. In contrast to Hilbert's system of axioms, concerned only with the classical Euclidean geometry, each of these structures subsumes many nonisomorphic geometries. Argument in terms of structures brings together what is common to hitherto diverse parts of mathematics, thereby unraveling complicated sets of facts, giving increased precision to geometry, and greatly strengthening its bonds with algebra in such a fruitful way that the term "foundations of geometry" (though objections may be raised) is now customarily used for the study of the relationship between geometric structures and algebraic forms.

Let us examine the individual chapters in the light of these remarks. Since the source of all geometry is intuition, the basic problem is to define its intuitive data in terms of concepts that can be discussed scientifically.

ix

Thus the first chapter deals with the phenomenological aspect of geometry.

Chapter 2, also introductory but from a different point of view, presents a particularly simple system of axioms, to show just how a deductive theory is built up. The resulting theory, a part of Euclidean geometry, yields a rich harvest of theorems and figures, although it has very few axioms.

Chapters 3 through 6 are central to this presentation of the foundations of geometry. The first edition of Hilbert's *Foundations* started from the classical geometry of Euclid and derived the other systems by changing its axioms. But here we proceed from the conceptually simpler to the more complex. We begin with the structures of pure incidence, namely, the affine and the projective planes, and end in Chapter 6 with a system of axioms for Euclidean geometry, appropriately modified from Hilbert's system.

Chapter 3 deals chiefly with closure theorems and collineation groups, extensively investigated in recent years, and with their role in the algebraization of affine and projective planes.

Chapter 4 introduces a metric concept into the affine plane, namely, line reflection, thereby producing the plane geometry of Euclid, except for order, continuity, and free mobility. These Euclidean planes are seen to be planes over an (almost) arbitrary field, including finite planes. In spite of their generality, they already have many of the metric and algebraic properties of the real Euclidean plane.

Chapter 5 emphasizes the purely group-theoretic construction of geometry, independently of the earlier chapters. The basic concept here is line reflection, and proofs consist of computations with reflections. "Absolute geometry," in a very general sense, is built up by this method, which applies to many other metric planes, including the general Euclidean planes of Chapter 4, and the non-Euclidean planes.

Chapter 6 axiomatizes the classical Euclidean and hyperbolic geometry, again independently of earlier chapters.

Chapters 7 and 8 deal with the construction of geometric figures with given instruments, such as the ruler and the compass, and with the systematic modern theory of regular polygons and polyhedra, topics particularly suitable for the enrichment and enlivenment of introductory instruction.

In Chapter 9 the analytic treatment of Euclidean and affine geometry is based, as is now customary, on the concept of a vector space, with suitable attention to the growing importance of linear algebra. The properties of affine and Euclidean *n*-dimensional point spaces are discussed, together with the corresponding mappings and configurations, especially for two and three dimensions, that is, for plane and spherical trigonometry.

Similarly, in Chapter 10 projective geometry, with its various mappings (collineations and correlations) and its quadratic configurations (conics and quadrics), is developed from linear algebra. By distinguishing suitable "improper" configurations, we can then obtain the Euclidean and affine geometries from projective geometry.

Chapter 11 gives a brief introduction to the concepts and methods of algebraic geometry.

Chapter 12 deals with Euclidean and affine geometry from the point of view of Klein's Erlanger Program (1872), namely, the ordering of the corresponding groups of transformations as subgroups of the projective group. Other examples of geometries that can be subsumed in this way under projective geometry are circle geometry, sphere geometry, line geometry, and particularly the non-Euclidean geometries.

Chapter 13 deals with other aspects of the theory of groups as an ordering principle for geometries. The discussion of transformation groups and group representations leads to recent results on the space problem in its various forms.

Chapter 14 presents the basic concepts of the differential geometry of curves and surfaces, including the fundamental ideas of Riemannian geometry, important for our conception of space in physics.

Chapter 15 deals with convex figures, which can be treated by differential geometry but also by more general methods not requiring a highly developed theory and therefore suitable, like the topics in Chapters 7 and 8, for lively discussion in introductory classes.

The subject of Chapter 16, topology, has had a profound effect on the mathematics of the present century. Originating in close association with geometry, it has now developed into an independent discipline, of great importance for other branches of mathematics. Its problems concern not only the analysis of geometric figures by combinatorial and algebraic methods but also set-theoretic topology. By giving some description of both types of questions, this final chapter* takes account of the connection between topology and geometry, laying at the same time a foundation for the basic concepts of analysis, to be dealt with in Volume III.

---

* A chapter on descriptive geometry has been omitted in the translation.

PART A

Foundations of Geometry

Geometry—A Phenomenological Discussion

1. The axiomatic geometer demands nothing of his reader except the ability to draw a logical conclusion. He sets up a number of axioms, containing words that sound like geometry, and then from these axioms he undertakes to derive theorems of many different kinds. On the other hand, the analytic (or better, the algebraic) geometer attaches geometric names to certain algebraic objects and then proves by algebraic methods that they have certain properties. But in both cases some sort of groundwork should be laid; there should be some discussion of the particular choice of axioms and of the geometric names for the given algebraic objects.

Since the concepts of geometry have been taken from the space of our everyday experience and visualization, and since conversely they often find applications there, we can proceed a surprisingly long way with a purely phenomenological analysis of this empirical space before making any start on a more or less clean-cut axiomatic or analytical treatment. In school the intuitive approach is never entirely abandoned, and Euclid himself, in spite of all his rigor, did not set up an unobjectionable system of axioms. Thus, in dealing with any particular part of geometry, the teacher must clearly realize why and how far he is willing, or compelled, to base his instruction on the intuitive powers of his students; he must know what further steps, and what choice of axioms, would be necessary to make his instruction entirely independent of intuition. In short, both for his own knowledge of the subject and for his instruction of others, he must undertake an analysis of our intuition of space. Only then can he teach with a good conscience; only then will he be able to lead his pupils, who at the beginning of the journey are at the mercy of their intuition, across its treacherous shoals onto higher ground.

The discussion in the present chapter is entirely phenomenological, although we assume that the reader knows his geometry. In our analysis of space we unhesitatingly make use of concepts analyzed in later chapters and do not give any proofs, often assuming that the reader can easily prove certain simple statements for himself.

Order

2.   In our intuition of space the concept of a segment precedes that of a straight line. In fact, we arrive at the concept of a straight line by continually extending a segment in both directions. The straight line contains many segments, each of which is determined by its two endpoints. Every segment is an *infinite* set but can be determined by two data, namely, its endpoints.

By the segment AB we obviously mean the points of the straight line AB lying *between A* and *B* (exclusive). The *relation of betweenness,* which underlies the concept of a segment, is a relation "*C* lies between *A* and *B*" among three (arbitrary) points of a (fixed) line. Euclid, giving free rein to intuition, paid no attention to a relation of this sort, and Pasch was the first to recognize its importance. In the early stages of geometry, recognizing the similarity of two figures such as la and lb, which differ only in their order properties, represented a difficult feat of abstraction, so difficult indeed that even today many beginners are confused by it.

In the time of Pasch, on the other hand, it was a bold deed to free oneself from the Euclidean tradition and recognize the mathematical importance of these neglected questions of *order.*

Betweenness is one of the concepts of order. With its help, for example, we can describe the intuitive order which is imposed on the set of points in a straight line when we traverse the line in one direction; in a passage in the direction $A \to B$ the points *between A* and *B* are those which come *before B.*

But this relation of betweenness is quite inconvenient, since it is a relation among three things (a three-place relation), so that any nontrivial statement about it must take at least four things into account; for example, one of Hilbert's axioms runs as follows: "If four points are given on a line, they can

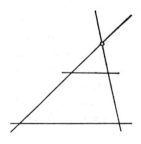

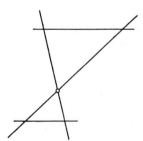

Fig. 1a   Fig. 1b

always be denoted by *A, B, C, D* in such a way that *B* lies between *A* and *C* and also between *A* and *D*, and *C* lies between *A* and *D* and also between *B* and *D*."[1]

It is much more satisfactory if we proceed not from our intuition of betweenness but from the idea of passage along a straight line. Here the relation "*A* before *B*," which we shall also write as $A < B$, is meaningful and, fortunately, is a two-place relation, so that we can make nontrivial statements about it by considering only three things.

A set on which the relation "before" is defined is thereby made into a *totally ordered set*. More precisely, a set is said to be ordered if for every pair of distinct elements *A* and *B* exactly one of the relations

$$A < B, \quad B < A$$

is satisfied in such way that for every three elements *A*, *B*, *C* it follows from

$$A < B, \quad B < C$$

that

$$A < C.$$

Instead of $A < B$ we also write $B > A$.

Of course, a set can be ordered in many different ways. But on a straight line we intuitively distinguish two special orders, one of them being the *opposite* of the other; i.e., if $A < B$ in one of them, then $B < A$ in the other. Instead of the axioms of betweenness, as they are to be found in Hilbert, we can postulate: on every straight line (i.e., *oriented, or directed straight line*) two (opposite) orders are distinguished.

Every point *A* on a line determines two *halflines*, the set of points $B < A$ and the set of points $B > A$, and it does not matter which of the two orders is adopted. Two points *A* and *B* on a line determine four halflines and then, if $A < B$, the segment *AB* is defined as the intersection of the sets $C < B$ and $C > A$.

If a halfline is distinguished, the line is thereby oriented; for if *A* is the point determining the halfline and *B* is any point belonging to it, we may distinguish the order in which $A < B$.

3. There is not much more of importance to say about order on a line. But there is also a certain natural order in the *plane*.

Every line divides the plane into two parts, namely, two *halfplanes*; every point of the plane that does not lie on the line lies in exactly one of its two halfplanes. A halfplane has the property that two arbitrary points in it can be joined by a segment lying entirely in the halfplane. On the other hand, two points in different halfplanes determined by the same line *l* cannot be joined by a line segment that does not cross *l*.

[1]This axiom was subsequently derived from axioms of order in the *plane*.

This situation can be described in another way, in terms of convexity. A set is said to be *convex* if with every pair of points *A, B* in it the whole segment *AB* belongs to the set. Thus a line, a halfline, a segment, a disk, and the surface of a triangle are convex sets.

Then the above property of the two halfplanes of a line *l* can be described by saying that each of the two halfplanes is convex, but if to either of them we add a single point not on *l* from the other halfplane, the resulting set is no longer convex.

4.   Like the line, a plane α can also be oriented. For let us choose an oriented line *l* in α and decide which of the two resulting halfplanes is to be called the *left side* of *l* (in α). Then we shall say that the plane α has been *oriented*, or *directed*, since we have now distinguished between the two sides of it as a plane in space; for when we are looking along the directed line *l*, our choice for its left-hand side will obviously depend on which side of the plane we are on in space. A plane in space has exactly two sides.

But the concept of an oriented plane can also be understood intuitively without any reference to space. For we need only consider, in addition to the oriented line *l*, a second line *m*, crossing *l* from right to left; i.e., the orientation of *m* is such that on it the points of the right halfplane of *l* come *before* those of the left halfplane. Or conversely, if we begin with the pair *l,m* of oriented lines in α, we know which side of *l* is to be regarded as the left; it is the side into which the line *m* points. (Automobiles on *m* have the right-of-way over those on *l*.)

Thus the choice of two intersecting oriented lines *l,m* in α orients the plane α. Let us note the importance here of the order in which lines *l,m* are taken. If the order is reversed, the plane α is given the opposite orientation; for if *m* crosses *l* from right to left, then *l* crosses *m* from left to right. Thus an ordered pair of intersecting oriented lines *l,m* in a plane α, or alternatively an ordered pair of intersecting halflines (such a pair will be called a *bilateral*), orients the plane α. This orientation is reversed if *l* and *m* are interchanged, or if either *l* or *m* is reversed in direction. The orientation of a plane can also be described by means of an *oriented triangle ABC*, where *B* is the intersection of *l* and *m* and $B < A, B < C$ on *l* and *m* respectively. The same orientation is determined by the triangles *ABC, BCA, CAB* and the opposite one by the triangles *ACB, CBA, BAC*, so that in an oriented triangle we are interested only in the *sense* in which the triangle is traversed. In an oriented plane the area of a triangle can be given a sign, which is positive or negative according to whether or not the triangle determines the given orientation of the plane.

If for an oriented line *l* in a plane α we have determined which is its left side, then from the above discussion we also know which is the left side of any oriented line *m* intersecting *l* (Fig. 2); for if *m* crosses *l* from right to left, then *l* will cross *m* from left to right. The manner in which the left

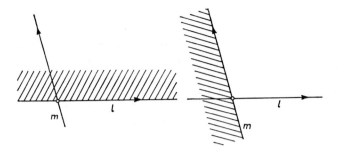

Fig. 2

side of m is determined by the left side of l is clear from the two sketches in Fig. 2. Also, it is intuitively clear that if the oriented line l moves continuously into the position of the oriented line m, its left side is "carried along with it"; i.e., its left side remains its left side in any continuous motion. (If m is parallel to l, we can determine the left side of m either by means of a third line cutting both l and m or by a continuous motion.)

Instead of an oriented line l and its left side, we may continuously transport a pair of oriented lines l,m (a bilateral), which will then constantly determine the same orientation of the plane.

Thus an affine transformation with which the identical transformation is continuously connected within the set of (nondegenerate) affine transformations takes a bilateral into another bilateral determining the same orientation of the plane. But there also exist affine transformations of the plane into itself (for example, reflections) that reverse its orientation. A given bilateral cannot be transported continuously in the plane into a bilateral determining the opposite orientation; at some stage the two lines of the moving bilateral must coincide, but then it ceases to be a bilateral.

5. From the algebraic point of view the situation is as follows; in the oriented plane let us choose an orienting bilateral, whose oriented lines can now be taken as the x-axis and the y-axis. The equation of a straight line is

$$l \equiv ax + by + c = 0.$$

Taking l as a symbol for the oriented line, we let ρl denote the same oriented line, for all $\rho > 0$, or the oppositely directed line, for all $\rho < 0$. We then take the left side of l to be the set of points (x, y) with $l > 0$, and note that under multiplication with $\rho > 0$ or $\rho < 0$ the sides are in fact preserved or interchanged.

The reader may verify that the line $l \equiv -x + y$ points from the lower halfplane into the upper, and the line $l \equiv x - y$ from the upper, and the line $l \equiv x - y$ from the upper into the lower.

Instead of operating with ordered pairs of real numbers we can also coordinatize the plane by means of the complex numbers. After choice of an oriented line as the "real axis" and assignment of its points in the usual way to the real numbers, we choose another line, perpendicular to the first, as the "imaginary axis," whose points correspond to the pure imaginary numbers. If the points 0 and 1 have been chosen, the real axis and its co-ordinatization are thereby determined. But now we must fix the position of i. Here there are the two possibilities that i may lie on the left or right of the real axis. If the given plane is already oriented, we take i on the *left* side of the real axis, traversed in the sense of increasing numbers; or conversely, we orient the plane in such a way that i lies on the left side of the real axis.

A circle centered on the origin consists of the set of points $re^{i\varphi}$ ($r > 0$ fixed, φ a real variable). If φ traverses the real axis in the positive sense (i.e., if φ increases), then $re^{i\varphi}$ traverses the circle in the sense $1, i, -1, -i$, which we agree to call the *positive sense*, where it is to be noted that the positive sense depends on the orientation of the plane. Or conversely, we may orient the plane by stating which is the positive sense of traversal on the circumference of the circle.

If the circle is traversed in the positive sense, the origin (together with the whole interior of the circle) lies to the left of the direction of traversal, i.e., to the left of the tangent directed at each point in the sense of the traversal. We have already spoken about the sense of traversal of a triangle. Here again the interior of the triangle lies in each case to the left of the positive direction of traversal. More generally, we can define a positive traversal for arbitrary convex curves; the interior must always lie to the left of the direction of traversal.

6. The situation in space is analogous. A plane divides the space into two halfspaces. Each of these two halfspaces is convex and becomes nonconvex when a single point (not on the plane) of the other halfspace is added to it. The space becomes *oriented* (left- and right-handed screws are distinguished) if for an oriented plane α we state which is its left side. Or we may choose an oriented plane α and an intersecting oriented line. Or again we may orient the space by means of a *trilateral,* i.e., an ordered triple of distinct oriented lines (for example, all of them through the same point) or of halflines. Interchange of two elements of the trilateral produces the opposite orientation, but cyclic permutation of its three elements leaves the orientation unchanged. Again, in place of all these methods, we may take an *oriented tetrahedron ABCD* (where A is the intersection of the three lines and in each case $A < B, A < C, A < D$). An even permutation of the vertices preserves the orientation of the space, and an odd permutation reverses it.

It is a remarkable fact that the space can be oriented by means of an ordered pair of oriented lines $l, m, provided\ l, m\ are\ skew.$ For we have only to draw a third oriented line n intersecting l and m and pointing from l to m.

Of course, the orientation of space obtained in this way is independent of the choice of n.

A continuous rotation about an oriented line l in an oriented space can take place in either the positive or the negative sense; if we construct a plane α perpendicular to l in such a way that l passes through α from right to left, the given rotation will take place in the positive sense if it moves a point of α in the positive sense (see 5 above).

If we combine a rotation about l with a steady motion along l, we obtain a *screw*, which will be positive if the rotation about l takes place in the same sense (for example, in the positive sense) as the motion along l. The points of the space then describe helical lines like the thread on a screw. The ordinary screws of everyday life are right-handed. In the space of physics the right-handed sense is called positive.

The above discussion for the plane can be repeated here, and we can proceed analogously in higher dimensions. The n-dimensional space is oriented by an ordered set of n-oriented lines (an n-lateral), the even or odd permutations of which preserve or reverse the orientation of the space.

Cyclic Order

7. In the oriented plane it is obvious that there also exists an order among the halflines issuing from a given point (*pencil of halflines*) and that this order is different in character from the order of the points on a straight line (see 2 above). The order among the halflines is said to be *cyclic,* and the same sort of order is to be found on the face of a clock or in the cycle of months in a year. On the oriented line we were able to ask whether A comes before B or not, but we cannot ask whether noon comes before midnight or summer before winter. Of course, we can say that the sequence "morning, noon, evening," or "summer, autumn, winter" is correct and the reverse sequence is wrong; but the sequence "noon, evening, morning," for example, or "winter, summer, autumn" is also correct.

The n objects $a_1, a_2, \ldots, a_n$ can be arranged in $n!$ ways. Two arrangements such as

$$a_{i_1}, a_{i_2}, \ldots, a_{i_p}, a_{i_{p+1}}, \ldots, a_{i_n}$$

and

$$a_{i_p}, a_{i_{p+1}}, \ldots, a_{i_n}, a_{i_1}, a_{i_2}, \ldots, a_{i_{p-1}}$$

are said to be cyclically equivalent and are assigned to the same (cyclic) equivalence class.[2] To provide a cyclic order for a finite set means simply to

[2]In the symmetric group S_n this equivalence class is a left coset with respect to the cyclic subgroup generated by $\{i \rightarrow i + 1\}$.

distinguish one equivalence class among all its (cyclic) equivalence classes. If V and W are arbitrary sets with $W \subset V$, each cyclic order in V generates the cyclic order in W obtained by simply discarding all the elements not in W. To provide a cyclic order for an arbitrary set Z means providing a cyclic order for all its finite subsets in such a way that for $W \subset V \subset Z$ (where V is finite) the cyclic order in W is the one determined by V (in the subset W).

A triple a, b, c admits two cyclic orders: $abc = bca = cab$ and $acb = cba = bac$, and it can be shown that the cyclic order of any set is already determined by the cyclic order of each of its triples.

By omitting a fixed element a we can interpret a cyclically ordered set Z as an ordered set Z'; we have only to write $x < y$ if axy is a triple in the cyclic order of Z. If we do this, the transitive law does in fact hold; for if $x < y$, $y < z$, then the triples axy and ayz correspond to the cyclic order of Z, and this result admits only the cyclic order $axyz$ for the quadruple, so that $x < z$ as desired.

A cyclically ordered set Z admits an n-fold "covering," as follows. For every $z \in Z$ we define a set of elements z_i (where i is an integer mod n) and agree, for example, that for $x < y$, $z \neq a$ the order

$$a_i x_i y_i a_{i+1} z_{i+1} a_{i+2}$$

is to be cyclic, where $x < y$ is defined as just above by means of a fixed element a.

The set Z can also be ∞-times covered, but then the result is an ordered set (i.e., not cyclically ordered). To do this we define, for every $z \in Z$, a sequence of elements z_i (where i is an integer) and agree that (for $x < y$, $z \neq a$)

$$a_i < x_i < y_i < a_{i+1} < z_{i+1}.$$

These "coverings" are essentially independent of the choice of a.

8. The lines through a point in the oriented plane can be so ordered as to form a cyclically ordered set (a cyclically ordered pencil of lines); for let us orient one of these lines a arbitrarily and then orient the others in such a way that they cross a from right to left. For two such lines x, y let us set $x < y$ if and only if x is crossed by y from right to left, and then regard $axyz \ldots$ as a cyclic order if $x < y < z < \ldots$. This order is independent of the choice of the line a and of the orientation given to it, but is reversed by a reversal of the orientation of the plane.

The oriented lines or halflines through a point in the oriented plane can also be cyclically ordered, and in fact as a double covering of the cyclic order of the pencil of (unoriented) lines described above. It is easy to see how this is done.

The cyclic order of the pencil of lines or of halflines can also be called a *sense of rotation*. Orienting a plane is thus equivalent to determining a sense of rotation.

Magnitude

9.   The basic statements in Euclid fall into two classes: postulates
($\alpha i\tau \eta \mu \alpha \tau \alpha$) and axioms ($\varkappa o\iota \nu \alpha i$ $\xi \nu \nu o\iota \alpha \iota$, common notions). The postulates
are geometric in nature, whereas the axioms refer to magnitudes in general.[3]
The first of these statements is: "Things that are equal to the same thing are
equal to each other." Nowadays we would say: equality is a two-place relation $a = b$ with the property of comparativity; namely, from $a = c$ and
$b = c$ it follows that $a = b$. The words "equal to each other" imply that this
relation is also symmetric; i.e., from $a = b$ it follows that $b = a$. We also
assume that the relation is reflexive; i.e., every magnitude is equal to itself.
(The axiom of symmetry is then superfluous.)

A relation with these properties is nowadays called an *equivalence*. Examples of such relations are: equally long, equally heavy, equally old. An
equivalence relation in a set generates a partition into classes. A definite
length, weight, or age is an example of equivalence class (a class of equally
long, equally heavy, equally old things). But in this respect present-day
language is usually somewhat careless. Concerning a segment AB, for
example, people say that $AB = 3$ cm. But "3 cm" is not a segment; it is an
equivalence class of segments (which are 3 cm long). A segment is not equal
to an equivalence class of segments but is at most contained in it. When AB
denotes a segment, we should say something like $AB \in 3$ cm.

Things can be compared not only with respect to equality but also with
respect to "greater and smaller," whereupon the equivalence classes become an ordered set. But we arrive at the concept of magnitude only when
we are able to add and subtract (the smaller from the greater). In general, we
cannot add segments but only their lengths, i.e., we can only add equivalence
classes. A system of magnitudes is thus an ordered set with an addition that
has certain properties (such as commutativity). The exact definition is rather
complicated, and it is easier to begin in the first place with an ordered
Abelian group (IB1, §§2.5 and 2.3). Its positive elements constitute exactly
what is meant by a system of magnitudes.

10.   We can also take multiples of magnitudes: if x is a magnitude and n
is a natural number, then $nx = x + \cdots + x$ (with n summands). Given
two magnitudes, it may happen that neither of them is a multiple of the
other; in fact, they do not even need to have a common multiple; for example, the diagonal and side of a square are *incommensurable*, i.e., they have
no common measure and thus no common multiple.

This situation becomes quite unpleasant when we wish, for example, to
prove that the areas of the rectangles $ABB'A'$ and $ACC'A'$ (with equal altitudes) are to each other as their bases AB and AC (Fig. 3); or again (Fig. 4)

[3]But the tradition on this division into postulates and axioms is by no means consistent.

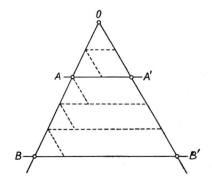

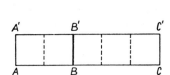

Fig. 3   Fig. 4

that $OA:OB = OA':OB'$. If the segments are proportional to integers (i.e., if they are commensurable), it is easy to give a proof by subdividing the two rectangles and using the theorems on congruence. But how are we to proceed in general?

Eudoxus (in the Fifth Book of Euclid's Elements) avoids this difficulty in a very ingenious way. He simply states that by definition

$$a:b = a':b'$$

means that for all positive integers m and n the two relations in each of the three pairs of relations

$$ma > nb \quad \text{and} \quad ma' > nb',$$
$$ma = nb \quad \text{and} \quad ma' = nb',$$
$$ma < nb \quad \text{and} \quad ma' < nb'$$

are either both correct or both incorrect, whereupon the proof of the desired proportions follows at once.

Eudoxus continues in the natural way by defining

$$a:b > a':b'$$

to mean the existence of a pair m, n such that

$$ma > nb, \quad \text{but} \quad ma' \leqq nb'.$$

However, we are now involved in a new difficulty. If we wish to show, for example, that for

$$a < b$$

we have

$$a:a > a:b,$$

we must find m, n such that

$$ma > na, \quad \text{but} \quad ma \leqq nb.$$

If we try to do this by setting $m = n + 1$, we have

$$(n + 1)a \leqq nb$$

or

$$a \leqq n(b - a).$$

In other words, for the magnitudes a and d (with $d = b - a$) we must find a positive integer n such that

$$nd \geqq a.$$

The requirement that "for two magnitudes a and d there exists an n such that $nd \geqq a$" is called the *Axiom of Archimedes*,[4] although, disguised as a definition, it was already formulated by Eudoxus. A system of magnitudes satisfying this axiom is called an *Archimedean system*.

The concepts of Eudoxus are closely related to those of Dedekind. The ratio $a:b$ of two magnitudes determines two sets of rational numbers m/n such that $ma > nb$ if m/n is in the first set and $ma \leqq nb$ if m/n is in the second set. These sets have the properties that Dedekind requires for the upper and lower classes of a cut. The definition given by Eudoxus for the equality of two ratios means that a cut determines *at most* one (real) number. For Dedekind a cut must also, by definition, determine at least one number. Dedekind is seeking to define the real numbers in terms of the rational numbers. On the other hand, for Eudoxus, magnitudes are already given geometrically. Unlike Dedekind, he has no need to provide a definition for $\sqrt{2}$, for example; for him this magnitude already exists as the ratio of the diagonal to the side of a square.

An Archimedean system of magnitudes is isomorphic to a subset of the system of real numbers. An Archimedean system that satisfies Dedekind's postulate is isomorphic to the system of real numbers.

11. In one respect the concept of a magnitude, formulated in this way, is still too restrictive. Angles are an example of a cyclic magnitude, at least if we count up to $360°$ and identify $360°$ with $0°$ (i.e., if we calculate mod $360°$), so they do not form an ordered set but, like the halflines in a pencil, a cyclically ordered set (and furthermore an Abelian group). Let us look at what we mean by an angle.

Angle

12. In elementary geometry the concept of an angle is ambiguous and hazy. Euclid defines it as an inclination of lines (including curved lines) to

[4]Compare here also IB1, §3.4.

each other, where he is obviously thinking of halflines, since otherwise he could not distinguish an angle from its adjacent angle. But in the next definition he goes on to speak of the lines (straight lines) enclosing an angle, where it is clear that he is (also) thinking of a part of the plane.

Euclid does not recognize zero angles, straight angles, or reflex angles. But this procedure is often inconvenient; for example, given an obtuse angle at the circumference of a circle, what becomes of the theorem that it is half as great as the angle at the center; or what about the sum of a set of angles that add up to more than 180°?

In the theory of the measurement of angles (i.e., in goniometry) angles are considered as being at the center of a circle, say with unit diameter, and are related to the corresponding arcs (Fig. 5); in fact, the angle is even measured by the arc, with the result that, unlike line segments, angles have a natural unit of measure (the complete circumference, corresponding to 360° or 2π). Thus angle magnitudes are dimensionless.

In goniometry angles are measured, not up to 180°, but up to 360°. Then we can either go on or else neglect multiples of 360°; in other words, we can calculate mod 360° (mod 2π). But even this latter procedure, though it is the most satisfactory one from a logical point of view, does not get us out of all our difficulties; in the statement that the sum of the angles in a quadrilateral is equal to 360° it is *not* convenient to replace 360° by 0°.

Moreover, the goniometric definition of an angle deals with arcs of a circumference and not with angles between halflines. For if we are given only two halflines (with common endpoint) we cannot say which of the two circular arcs α and $2\pi - \alpha$ should be regarded as measuring the angle between them; nothing in the appearance of the halflines themselves will settle this question. Of course, we mean the arc that lies "between" the sides of the angle. But what is meant by the word "between"? Again the answer depends on which of the two sides is taken as the first. The goniometric angle is a function of the *ordered* pair of halflines, and the corresponding arc is the one that begins on the left of the first halfline and ends on the right of the second.

But again we must be cautious. The left side and the right side of a straight line are meaningful only in an oriented plane, and it is only in such a plane that angles are defined at all in goniometry.

Fig. 5

The concept of an ordered pair of halflines is reminiscent of the bilateral but is more inclusive. For we now allow the halflines to coincide or to point in opposite directions, with corresponding angles 0 and π whereas in §4 matters were so arranged that a bilaterial agreeing with the given orientation of the plane corresponds to an angle between 0 and π. But if we reverse the orientation, the angle α becomes the angle $2\pi - \alpha$, which means that in an unoriented plane we cannot distinguish between these two angles. In this case it is better to deal with angles only from 0 to π, so that Euclid was quite right in not admitting greater angles. For then he would have had to begin by orienting the plane, a procedure quite foreign to his way of thought, since the choice of orientation is arbitrary. This inability to distinguish between α and $2\pi - \alpha$ can also be interpreted as meaning that an angle is no longer a function of an ordered pair of halflines but of an unordered pair, since interchange of the order of the halflines takes α into $2\pi - \alpha$.

The formula for calculating the angle α between two unit vectors x and y is

(1) $$\cos \alpha = x\mathfrak{y}.$$

This formula is symmetric in x and $\mathfrak{y}$. So we are dealing here with the angle between an unordered pair of vectors, in agreement with the fact that the value of the cosine does not indicate whether it comes from α or from $2\pi - \alpha$. To be sure, we have another formula

(2) $$\sin \alpha = x_1 y_2 - x_2 y_1,$$

which seems to help us if we are trying to decide between α and $2\pi - \alpha$. But this help is only apparent. For the right side of (1) does not depend on our choice of (rectangular) coordinate system, whereas the right side of (2) changes sign if we replace one of the axes by the oppositely oriented line (say $x_2 \to -x_2$, $y_2 \to -y_2$). The choice of axes has oriented the plane, and the angle α calculated from (2) depends on this orientation. Thus formulas (1) and (2) together determine the goniometric angle in the oriented plane.

Confusion about the concept of an angle is particularly troublesome in plane analytic geometry, where it is customary to talk about the angle between two lines (instead of two halflines), so that apparently we cannot even distinguish between an angle and its adjacent angle. But again things are not so bad as they seem. In analytic geometry an angle between two lines l and m is determined by its (trigonometric) tangent, which is of period π, so the angle is determined only mod π. Let us look at this more closely.

If in the $x_1 x_2$-plane we choose the one line l as the x_1-axis and describe the other line m by the equation $x_2 = \mu x_1$, the angle α between m and the x_1-axis is given by the formula

$$\operatorname{tg} \alpha = \mu.$$

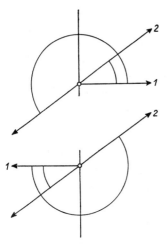

Fig. 6

We are dealing here with an ordered pair of halflines, the first of which lies on x_1-axis and the second on m. Two of the four goniometric angles thus obtained are equal to each other and the other two (also equal to each other) differ from them by π (Fig. 6).

More generally, the angle between an ordered pair of lines l,m in the sense of analytic geometry is the goniometric angle (in the oriented plane) of an ordered pair of halflines, the first of which lies on l and the second on m. This angle is determined mod π. Here the plane is oriented by the choice of coordinate system (the bilateral consisting of the positive x_1-axis and the positive x_2-axis).

In addition to these three concepts for an angle there is still a fourth, commonly used in elementary solid geometry, where it is remarkable that we speak not of the angle between a pair of (unoriented) halflines but of a pair of (unoriented) lines. The lines may be skew, but in order to determine the angle they are translated into the same plane.

Let us set up a table for these four concepts of an angle (Fig. 7).

The reader should not assume from this table that in analytic geometry, for example, it would be impossible to consider any other concept of an angle. On the contrary, we have already seen in §5 that an oriented line can be defined in analytic geometry, so that the goniometric concept is quite possible there. Similarly, in solid geometry we could very well consider the angle between halflines (see 14 below); the table (Fig. 7) merely represents the usual procedure in elementary instruction.

13.   The angles of elementary geometry form an ordered set in which

Goniometry	The angle in		
	Elementary Geometry	Analytic Geometry	Solid Geometry
	between an		
ordered	unordered	ordered	unordered
	pair of		
halflines	halflines	lines	lines
	in an		
oriented	unoriented	oriented	unoriented
	plane is defined		
mod 2π	between 0° and 180°	mod π	between 0° and 90°

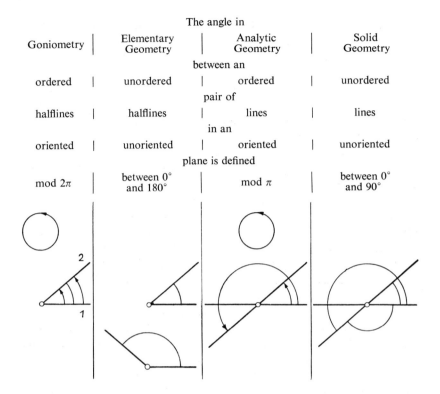

Fig. 7

addition is not yet unrestrictedly possible, and calculation with angles greater than 180° is rather hard to justify. The system of angles mod 360° is a double covering of the system mod 180° and is cyclically ordered in the same way as the lines or halflines in a pencil. The ∞-fold covering produces exactly the kind of angles needed in the statement that the sum of the angles in n-gon is $(n - 2)\pi$.

14. If a space is already oriented, its planes are not necessarily oriented thereby. On the contrary, by a rotation in space it is possible to take an oriented plane into the plane with opposite orientation. Thus we cannot meaningfully define an angle mod 2π for an ordered pair of halflines of a plane in space; the angles α and $2\pi - \alpha$ are necessarily indistinguishable. Similarly, we cannot define the angle mod π between two coplanar lines in space; i.e., we cannot distinguish between an angle and its adjacent angle.

The situation is quite different if we confine our attention to skew lines or halflines l,m in oriented space. A plane ε parallel to l and m can be oriented by postulating that an oriented line n (intersecting l and m and pointing from l to m) crosses ε from left to right (see also §6.) The lines l and m can then be translated into this oriented plane so as to determine the various types of angles; in particular, the "goniometric" angle is the angle by which *l* must be twisted, in the direction prescribed by the orientation of the space, in order to reach *m*.

In oriented space the angle φ between the line *l* and an oriented plane α can be reduced by definition to the angle ψ between *l* and the normal to α oriented toward the left side of α, and the angle between two oriented planes can be defined analogously.

Spaces of higher dimension give rise to complications. The relative position of two nonparallel planes in four-dimensional space can no longer be described in terms of *one* angle.

15.   Up to now we have considered an angle as a magnitude, in agreement with Euclid's first notion of it as the "inclination of two lines." But other procedures are possible. Compare, for example, our treatment of a line segment, not as a length, but as the set of points between its two endpoints, which completely determine the segment. Similar possibilities are available for angles, if we wish to avoid considerations of magnitude altogether.

Thus two intersecting lines *l,m* determine four angles (*sets* of points) in the plane, all of them logically on a equal footing. But if in an oriented plane we consider *l,m* as an ordered pair of oriented lines we can assign a *unique* angle to this pair, namely, the set of points to the left of *l* and (at the same time) to the right of *m*. All the angles defined in this way are convex, since they are the intersections of two (convex) halfplanes, but it is easy to see how we may introduce nonconvex angles (Fig. 8).

Area and Volume

16.   In elementary instruction, areas and volumes are introduced numerically, i.e., as numbers for which certain rules of calculation are pre-

Fig. 8

scribed. But in Euclid, and to some extent in the schools today, the interest lies not in calculating areas and volumes but in comparing them. Among Euclid's axioms (more precisely, his κοιναὶ ἔννοιαι—common notions), in addition to those concerning the general notion of magnitude there is one that reads: "Things that can be superposed on each other are equal."

Thus we consider congruent figures to be equal in area or volume. But we also make use of Euclid's axiom: "If equals are added to equals, the wholes are equal," and are thereby led to Hilbert's concept of *decomposable equality* (Zerlegungsgleichheit): two figures that can be decomposed into pairwise congruent figures are said to be equal, like the rectangle and rhombus in Fig. 9. But this concept is not yet adequate if we wish to prove, for example, that parallelograms with equal bases and altitudes are equal in area. The method is successful for the parallelograms $ABDC$ and $ABD'C'$ in Fig. 10, each of which is the sum of the same trapezoid and of congruent triangles, but it will no longer work for Fig. 11, where we must argue differently: the parallelograms can be obtained by subtraction from the trapezoid $ABD'C$; in the first case by subtraction of the triangle BDD' and in the second of the congruent triangle ACC'. Here we are making use of Euclid's axiom "If equals are subtracted from equals, the remainders are equal," and are thereby led to Hilbert's concept of *supplementwise equality* (Ergänzungsgleichheit): two figures are said to be supplementwise equal if by the adjunction of decomposably equal figures they can be supplemented in such a way as to become decomposably equal. (The case of Fig. 11 could also be dealt with by reducing it to Fig. 10 by means of a step-by-step insertion between the two parallelograms of a sequence of parallelograms each of which is in the same position with respect to the next as the two parallelograms in Fig. 10; but then it would be necessary to make use of the Axiom of Archimedes, without which the concept of supplementwise equality is actually more inclusive than that of decomposable equality.)

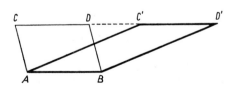

Fig. 10

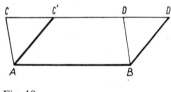

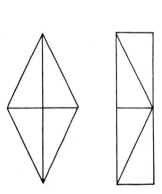

Fig. 9

Fig. 11

Since the relation of supplementwise equality has the properties of an equivalence, we can combine supplementwise equal figures into a class, which we shall call the *area* (or in three dimensions, the *volume*) of these figures. For these areas (equivalence classes) we can now define an order relation and an addition, whereupon they will form a system of magnitudes. But this system is not yet altogether satisfactory, since up to now we do not even know whether or not all geometric figures have the same area. In fact, some extremely pathological cases can arise. For example, it is possible to decompose the surface of a sphere into three congruent sets such that two of them, rearranged in a suitable way, again produce the whole surface of the sphere.[5]

But if as "figures" we admit only planar polygons we can show that all such figures are supplementwise equal to rectangles with a fixed altitude and that two such rectangles can be supplementwise equal to each other only if they have the same base. The base of the corresponding rectangle can then be taken as a measure for the area of the figure, whereby we return to the elementary notion of area.

But in space, with its polyhedral surfaces, this method is no longer successful. Two pyramids with bases of equal area and with equal altitudes are no longer necessarily supplementwise equal to each other.[6] In order to establish a theory for the volumes of polyhedra it is customary in the schools to refer to a principle usually named after Cavalieri but already to be found in Democritus and Archimedes; namely, if two three-dimensional figures are such that their intersections with any plane parallel to a given plane are equal in area (supplementwise equal), then the figures themselves are equal in volume (*Cavalieri equal*). This concept, together with the concept of supplementwise equality in space, is sufficient for the theory of volumes of polyhedra in space.[7]

The *Cavalieri principle* can in its turn be based on a passage to the limit, and such limiting processes are necessary if we wish to consider figures bounded by curved lines.

The problem of showing that the areas of circles C, C' are to each other as the squares of their radii r, r' is hardly more difficult than the proof that the areas of rectangles with the same height are to each other as their bases. For example, if we had $C:C' > r^2:r'^2$, there would exist positive integers m,n such that

(3) $$mC > nC', \text{ but } mr^2 \leqq nr'^2$$

(see 10), and we could find a regular polygon inscribed in C with an area V

[5]F. Hausdorff, Grundzüge der Mengenlehre, 1. Aufl., Berlin-Leipzig 1914, S. 469. J. von Neumann, Fundamenta Math. (1929), 73–116.
[6]M. Dehn, Math. Ann. 55 (1902).
[7]W. Süss, Math. Ann. 82 (1921), 297–305.

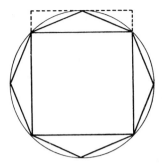

Fig. 12

such that also

(4) $$mV > nC'.$$

Such a polygon can be found in the following way: let V_i be the area of the inscribed regular polygon with 2^i-sides. Then it is easy to see (Fig. 12) that

$$C - V_{i+1} < \tfrac{1}{2}(C - V_i),$$

and thus

$$C - V_p < 2^{-(p-2)}(C - V_2).$$

The axiom of Archimedes now guarantees the existence of a q such that

$$2^{q-2}(mC - nC') \geqq m(C - V_2).$$

For this q,

$$(mC - nC') > m(C - V_q),$$

and thus

$$mV_q > nC',$$

so that (4) is satisfied with $V = V_q$. Then if V' is the area of the regular 2^q-gon inscribed in C', we have

$$mV > nV',$$

and, on the other hand,

$$V: V' = r^2: r'^2,$$

in contradiction to the second half of (3). The assumption $C:C' > r^2:r'^2$ is thus refuted, and similarly for the opposite inequality.

Here we have made use of continuity only in the sense of the axiom of Archimedes, but if we wish to show that for a given circle, for example, there exists a rectangle, with prescribed altitude, that is equal in area to the circle, we must make use of Dedekind continuity,[8] and in fact in analysis, the area of arbitrary figures is treated systematically from a modern point of view.

17. Up to now we have based the concept of area on congruence ("Things that can be superposed on each other are equal"). But this is not the only possibility, and from the algebraic standpoint it is not even the most convenient. The concept of area is less closely related to congruence than to affine transformations, under which the ratio of areas is invariant.

We first replace the requirement of invariance of area under congruence by *invariance of area under translation,* since translation is itself an affine-invariant concept. The area of a parallelogram $ABCD$ in the plane is then completely determined by the vectors AB and AD; it is a function $f(a, b)$ of the two vectors a, b forming the sides. We next require, as is natural, that if one side is multiplied by a factor c (multiplication of a vector by a number is also an affine-invariant concept) the area is thereby multiplied by the same number, and then it is desirable to admit negative factors, which lead to the concept of negative areas, removing the difficulty that the side of a parallelogram determines not one vector but rather two (opposite) vectors. But what does it mean intuitively that the parallelograms $ABCD$ and $ABC'D'$ (Fig. 13) have opposite areas? We see that in the plane they determine opposite orientations, so that the area must be a function of an *ordered* pair of vectors and must change sign with interchange of the vectors:

(5) $$f(a, b) = -f(b, a),$$

and we must also have

(6) $$f(ca, b) = f(a, cb) = cf(a, b).$$

Furthermore, the parallelograms $ABDC$ and $CDFE$ taken together (Fig. 14) have the same area as $ABFE$; for we may subtract the triangle BDF, move it to the position ACE and then add it again. Now the three parallelograms have one side ($AB = CD$) in common, and the fourth side of $ABFE$ is the

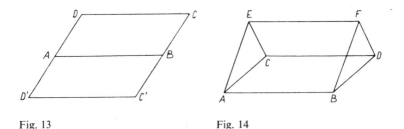

Fig. 13   Fig. 14

[8]See also H. Hasse and H. Scholz, Die Grundlagenkrisis der Mathematik. Pan-Bücherei, Gruppe: Philosophie, Nr. 3, 1928.

sum of the corresponding sides of $ABDC$ and $CDFE$; i.e., for the vector AE we have $AE = AC + CE$. So we must also require:

$$(7) \qquad\qquad f(a, b + b') = f(a, b) + f(a, b')$$

and

$$(8) \qquad\qquad f(a + a', b) = f(a, b) + f(a', b).$$

The equations (5) through (8) can be summed up as follows: area, regarded as a function of two vectors, is antisymmetric and linear in each of its arguments.

But this function is not completely determined until we have chosen a unit of area. To do this, we take an arbitrary parallelogram, with the vectors e and f for sides, and set and

$$(9) \qquad\qquad f(e, f) = 1,$$

by definition. The unique existence of such a function f is shown in algebra (IB3, §3.4).

18. The procedure in three-dimensional space (or in n-dimensional) is entirely analogous, the analogue of a parallelogram being a parallelepiped (in n dimensions, a parallelotope). By definition, the volume is an antisymmetric function of three (or n) vectors, linear in each of its arguments, and uniquely determined by means of a standard figure.

Groups

19. How many of the elements of a triangle are necessary to determine it completely? In elementary geometry the answer is three, but in plane analytic geometry it is the six coordinates of the three vertices. How does the contradiction arise?

With the six coordinates of the vertices we determine the triangle not only in shape and size but also in position (with respect to the coordinate system). But in elementary geometry we often regard a triangle as already constructed if, in a class of congruent triangles, we have found one triangle that satisfies the requirements of a given problem. But even in elementary geometry the usage varies. Consider the two theorems:

1. A triangle is completely determined by the lengths of its three sides.
2. Two triangles ABC and $A'B'C'$ are congruent if $AB = A'B'$, $BC = B'C'$, $CA = C'A'$.

If the word "triangle" is to have precisely the same meaning in these two sentences, then the second sentence, while not in contradiction to the first, is trivial and superfluous; for by the first theorem the two triangles ABC and $A'B'C'$ are already identical with each other.

But in fact the word "triangle" means something quite different in the two theorems. In the first theorem (and more generally in many construction problems) congruent triangles are regarded as not essentially different, or, as it is usually expressed nowadays, the word "triangle" means a *class* of congruent triangles, in the sense of the second theorem.

Construction of classes is a logical process of widespread usefulness. Usually it rests on an equivalence relation, i.e., on a two-place relation ($\cdots \sim \cdots$) with the properties that $a \sim a$ and that $a \sim c$, $b \sim c$ implies $a \sim b$. In a set in which such a relation has been established among the pairs of elements, we can combine equivalent elements into a "class." Since congruence is an equivalence, congruent figures in the plane can be combined into classes. A congruence class of segments is simply a length, and a congruence class of triangles is a "triangle" in the sense of (most) construction problems. Similarity and equality of area are other concepts often regarded as equivalences.

20.   From the logical point of view a geometry is a system of elements (i.e., the elements of a set) and relations among them. The elements may (intuitively) be points, lines, circles, angles, distances, and so forth. The relations may be one-place (X is a point), two-place (X is incident with Y), three-place (Y lies between X and Z), or four-place (the distance from X to Y is the same as from Z to U), and so forth.

To every geometry there belongs a group, its *automorphism group,* i.e., the totality of all mappings of the set of elements onto itself under which all the relations are preserved.

For example, if we regard the plane as a set of points and for every ρ consider the relation "X and Y are at a distance ρ from each other," we obtain the group of *rigid mappings* (direct and opposite isometries), in which two points at a distance ρ are taken into two points at the same distance ρ. But if for our relation we take "X and Y are the same distance apart as Z and U," we obtain the group of *similarities,* namely, the transformations that leave invariant the ratios of distances.

Conversely, a geometric concept of equivalence (see 19) often depends on a *group of transformations G.* Two figures Φ and Φ' are said to be equivalent if there exists a transformation f in G taking one of them into the other; thus

$$\Phi \sim \Phi' \text{ if and only if } f\Phi = \Phi'$$

for a suitably chosen $f \in G$. From the axioms for a group it follows that this relation is actually an equivalence.

21.   We have already mentioned the group B of rigid mappings obtained by transforming the plane (or space) as a rigid body, i.e., by requiring that distances remain invariant. If for a triangle ABC (or tetrahedron $ABCD$) we prescribe the position of its (congruent) image $A'B'C'$ ($A'B'C'D'$), the transformation is completely determined, but if only the segment AB (the triangle ABC) has a prescribed image $A'B'$ ($A'B'C'$), then for the image

$C'(D')$ of a point $C(D)$ not on the line AB (the plane ABC) two positions are still possible, one on each side of AB (ABC). However, if we require that in addition to distances the transformation f must leave invariant the orientation of the plane (or space), there is only one possibility for $C'(D')$; by prescribing the image of AB (ABC) we have already completely determined the transformation f. The set of rigid mappings that preserve orientation forms a subgroup B_0 of B. (For the definition of B_0 it is of no importance *how* we orient the plane (or space); all that matters is the fact that we *can* orient it.)

22. "This segment is 3 cm long" and "This parallelogram has an area of 12 cm^2" are statements invariant under the group B. Nevertheless, such statements are hardly regarded as being part of geometry (but rather of geodesy). When they occur in geometry, they are regarded as references to a certain unit of measurement (1 cm, 1 cm^2), which may in fact be chosen in a completely arbitrary way (on a blackboard usually about 10 times as large as in a notebook). Consequently the group B of rigid mappings is much less important than the group A of *similarities,* i.e., transformations that leave invariant the *ratios* of distances (and therefore angles, and ratios of areas). If to a triangle ABC (tetrahedron $ABCD$) we assign its (similar) image $A'B'C'$ $(A'B'C'D')$, it is again true that the corresponding $f \in A$ is completely determined. Here also we can add the requirement of invariance of the orientation of the plane (or space) and thus arrive at a subgroup A_0 of A.

23. By parallel projection we can map a plane onto another plane. A pair of parallel lines is then taken into a pair of parallel lines, and parallel segments are multiplied by the same factor. If by a further sequence of parallel projections we finally bring the images back into the original plane, we obtain a mapping in the plane which preserves parallelism and the ratios of parallel segments. Such a mapping is said to be *affine.* The affine mappings of the plane onto itself form a group F. If we prescribe the image $A'B'C'$ of a triangle ABC, the corresponding affine mapping $f \in F$ is thereby completely determined. For it follows from the invariance of ratios of segments that every point on the lines AB and AC has a predetermined image on $A'B'$ and $A'C'$, while an arbitrary point X can be regarded as the vertex of a parallelogram AB_1XC_1 (B_1 on AB, C_1 on AC) and its image X' as a vertex of the corresponding parallelogram $A'B_1'X'C_1'$ (Fig. 15).

The group A of similarities is a subgroup of the group F of affine transformations and A is certainly a proper subgroup of F, since under the mappings of A all ratios of segments remain invariant, whereas under F (in general) only the ratios of parallel segments remain invariant. The group F has another subgroup F_0 consisting of those affine transformations that preserve orientation. Ratios of areas of parallelograms remain invariant under all the transformations of F, since their definition depends only on parallelism and on the ratios of intervals, both of which are affine-invariant.

Affine transformations in space are defined in exactly the same way as in

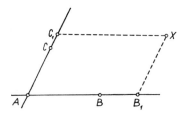

Fig. 15

the plane; a tetrahedron and its image completely determine an affine transformation. Ratios of volumes of parallelepipeds are invariant, but this statement is not necessarily true for areas of parallelograms in nonparallel planes
or for nonparallel segments.

24.   If a plane is mapped by a central projection (i.e., a perspectivity)
onto another plane, straight lines will be mapped into straight lines, as will
again be the case if we project back (from another center) onto the original
plane. Yet there are difficulties here. Even for one projection there will be
points that have no images, and points in the image plane that are not
images of any point, i.e., when the projecting ray is parallel to one of the
planes. In order to avoid these "exceptions," we supplement the plane by
ideal points (parallel lines being considered to pass through the same ideal
point) and thereby obtain the *projective plane.*

Let us take four points *A, B, C, D* in the plane, no three of them on the
same straight line, draw the six lines joining them in pairs, construct the
intersections of these lines, and then proceed by successively joining points
and taking the intersections of lines. The configuration thus obtained is
called the *Möbius net* (for *A, B, C, D*). It does not contain every point in the
plane; for example, if *A, B, C, D* have rational coordinates, then only points
with rational coordinates can be obtained. But the points of the net come
arbitrarily close to every point of the plane.

In a mapping φ of the projective plane that takes lines into lines in such a
way that the points *A, B, C, D* have prescribed images *A', B', C', D'* (no three
of them on the same straight line), every point of the Möbius net for *A, B,
C, D,* will have a unique image in the net for *A', B', C', D'.* If φ is continuous,
the φ-image of every point in the plane is uniquely determined.

A continuous mapping of the projective plane onto itself that takes lines
into lines is called a *projectivity.* If we prescribe the images of four points in
general position, the corresponding projectivity is completely determined.

In space the situation is precisely analogous; a projectivity is completely
determined by prescribing the images of five points in general position.

Adjunction of the ideal elements to the projective plane (space) destroys
the order properties of the ordinary plane (space). Order on the projective
line is the same as in a pencil of lines; in other words, it is cyclic. A point does

not divide the projective line into two parts, a line does not divide the projective plane, and a plane does not divide projective space.

It is possible to construct a model of the projective plane that lies entirely in a bounded part of the Euclidean plane, in the following way. From the center M of a sphere let us project the projective plane onto the sphere, delete the northern hemisphere as being superfluous, identify diametrically opposite points of the equator (since they correspond to the same point of the projective plane) and then project vertically onto the equatorial plane, thereby providing a model of the projective plane in the form of a circular disk with identification of opposite points on its boundary.

In the same way the ball (solid sphere) with identification on its boundary is a model of projective space.

25. From Euclid up to the end of the 19th century, every textbook in geometry began with definitions like "A point is that which has no part," and these definitions were followed by axioms (or postulates) like "A straight line may be drawn from any point to any point" and "If equals are added to equals, the wholes are equal." In general, such so-called definitions played no role in any proof, and many more axioms were used than were actually stated. A reasonably complete system of axioms for Euclidean geometry is to be found for the first time in Pasch (1882),[9] who explicitly requires that "the process of deductive proof must everywhere be independent, not only of the figures, but of the meaning of the geometric concepts. Poincaré (1902), in his description of Hilbert's "Grundlagen der Geometrie," expresses himself more brusquely: we must be able to insert the geometric axioms into a machine, which will then produce the whole of geometry.

The question "What is space?" was much debated in the 19th century and was finally settled by Hilbert. The answer is that this question is of no concern to the geometer. What are points, lines, and circles? The answer is that the meaning of these words is implicitly determined by the axioms in which the words occur. Whether there exists anything in nature that satisfies the axioms and what it looks like are questions for the physicist, not the mathematician. The mathematician requires from his system of axioms only that it shall not produce contradictions; but the physicist requires that it shall have useful applications to the external world.

The axiomatic method goes further; it investigates the mutual relationship of the axioms. For example, can we omit Axiom 7, i.e., can we deduce it from the other axioms? Or is it independent of the others? To prove its independence, we must construct a consistent system in which all the axioms except Axiom 7 are valid and Axiom 7 itself is false. It was in this way that the independence of the parallel axiom was proved by means of non-

---

[9]Vorlesungen über neuere Geometrie. Leipzig 1882. 2. Aufl. Berlin 1926.

Euclidean geometry, and the method has given rise to a great number of new geometries.

The axiomatic method has proved valuable in many fields; it brings into clear relief exactly what is necessary in the proof of a theorem and often allows us to combine many branches of mathematics into one.

But we must note that the natural approach does not begin with axioms. We discover certain relations in the physical world, or in some already mathematicized system, and register their properties. When we have carried this process sufficiently far, we select as "axioms" some of the properties that appear to be of fundamental importance, and then we operate with this system of axioms in a purely deductive way; that is, we draw conclusions from it. The resulting mathematical system can then be applied, either in the situation from which it was originally derived, or elsewhere. Discrepancies between the consequences of the mathematical theory and the observations of the physicists do not demonstrate any error, either in mathematics or in physics; they only show that the mathematical theory does not fit the observable world.

CHAPTER 2

Points, Vectors, and Reflections

Visualization is the source of all geometry. But geometry attains the rank of a science only when it is set up as a deductive theory, only when a system of concepts and abstract processes is developed in such a way that conclusions follow from axioms with all the certainty of logic.

The teacher of geometry must constantly strive to do equal justice to the various interacting elements of geometric thought: figures with their visual immediacy, the orderly march of abstract thought, experimentation with conjectures, the challenge to the mind that lies in the question "How can this be proved?" and the insight that comes from a complete proof.

But precisely this question "How are we to give the students some idea of geometry as a deductive theory?" gives rise to problems of the most varied kind. One reason for the difficulty lies in the fact that the Hilbert system of axioms for Euclidean geometry[1] is an extremely complicated structure. Of course, it is easy to say that the system provides a firm logical basis for the synthetic geometry that is taught in secondary schools; yet even the instructor does not find it easy to see exactly how all the details are to be carried out in a precise way.

However, it is by no means necessary to study the whole of classical Euclidean geometry in order to answer the question "What is a deductive geometric theory?" Various parts of the subject will suffice, and it is the aim of this chapter to present such a *partial theory*, based on a clear-cut system of axioms, in which we can prove a reasonably large number of theorems in an attractive, concise, and yet logically complete way. Our system of axioms is satisfied by numerous mathematical systems, including finite ones. At any rate, all the theorems hold in affine spaces of arbitrary dimension, and therefore in the Euclidean plane and in three-dimensional Euclidean space.

---

[1]D. Hilbert, Grundlagen der Geometrie. 9. Aufl., Stuttgart 1962.

For simplicity, our only geometric objects will be *points* (no lines, planes, and so forth), and the ideas will be carried only so far as they remain easily understood. Our theorems will have less to do with triangles than with parallelograms. As starting point and as model throughout we take the theorem : *in every quadrangle the midpoints of the sides form a parallelogram ;* in other words, by a certain geometric process (construction of the midpoint-quadrangle) every quadrangle is associated with a quadrangle of greater regularity.

Of course, when we decrease the number of axioms, we also decrease the number of theorems that can be stated and proved; but in return we gain generality. We shall see that in our theory we can formulate questions of a very elementary nature, almost without prerequisites of any kind at all, and can then prove theorems that will nevertheless provide insight into the inherent nature of geometry.

From the very beginning we shall take the point of view of group theory; that is, we shall first define a certain group and then formulate all our theorems in this group.[2] The use of group-theoretic language, first developed by G. Thomsen,[3] is not in itself of any great importance, but it enables us to proceed in a more uniform way. However, it would also be quite possible to take the earlier[4] (at first sight perhaps more natural) attitude in which points, etc., are regarded as given geometric entities (not as elements of an abstract group).

We thus arrive at a calculus with points which is both a *vector calculus* and a *reflection calculus* and therefore leads directly to two methods of modern geometric analysis which in recent decades have played an increasing role in instruction in geometry.

Although this chapter is built up axiomatically, it is to be understood as a chapter not only in axiomatics but also in *geometry*. The accompanying figures are an essential part of the text. The reader is urged to draw them for himself and also to construct other figures, since otherwise he may remain insensitive to the charming interplay of visualization and proof, and to the astonishing simplicity with which a whole ensemble of geometric facts can sometimes be expressed in the notation of this calculus.

---

[2]This procedure does not mean that in order to operate with our calculus one must have a knowledge of the theory of groups. All that is necessary is the ability to write down capital letters one after another according to very few rules.

[3]G. Thomsen, Grundlagen der Elementargeometrie in gruppenalgebraischer Behandlung. Leipzig-Berlin 1933.

[4]In II5, §1, the example of a Euclidean plane is used to show how we can proceed from the earlier point of view to the group-theoretic. For the theory developed in this chapter, the passage from one point of view to the other is much simpler, since here we consider only points, point reflections, and the products of point reflections, and not lines, incidence, orthogonality, line reflections, and arbitrary products of line reflections.

Remarks. 1. It is well known that certain important geometrical statements are valid only under assumptions of distinctness, nonincidence, and so forth (for example, the line joining two points is uniquely determined only if the points are distinct), a fact which can give rise to certain complications, especially in closure theorems. One of the chief reasons why the theorems proved in this chapter are so simple is that they do not depend at all on questions of distinctness. If two or more points coincide, the theorems remain valid. In the proofs we do not operate with negation, so that the logical pitfalls associated with that principle are altogether avoided.

2. In a recent article[5] the same subject was treated in such a way that each of the various theorems was proved without reference to preceding theorems. This chapter is constructed somewhat more abstractly, so that the relationships among the various theorems are more immediately clear and various further questions are suggested. The two treatments of the subject serve as complements to each other.

The elementary group-theoretic concepts and facts needed throughout the chapter are briefly as follows. Let there be given a group with a multiplicatively written law of composition and with unit element 1; for the time being we shall use lower case Greek letters to denote the elements of the group.

An element σ is said to be *involutory* if $\sigma^2 = \sigma\sigma = 1$ and $\sigma \neq 1$; of course, $\sigma\sigma = 1$ is equivalent to $\sigma = \sigma^{-1}$. If $\sigma_1, \ldots, \sigma_n$ are involutory, then $(\sigma_1\sigma_2\cdots\sigma_n)^{-1} = \sigma_n\cdots\sigma_2\sigma_1$: *the inverse of a product of involutory elements is obtained by inverting the order of the terms in the product.* If the product $\sigma_1\sigma_2\cdots\sigma_n$ is also involutory, then $\sigma_1\sigma_2\cdots\sigma_n = \sigma_n\cdots\sigma_2\sigma_1$.

The element $\gamma^{-1}\alpha\gamma$ is called *the transform of α by γ.* For brevity, we shall denote the element $\gamma^{-1}\alpha\gamma$ by α^γ. Then $(\alpha\beta)^\gamma = \alpha^\gamma\beta^\gamma$ and $(\alpha^\delta)^\delta = \alpha^{\gamma\delta}$. If σ is involutory, then σ^γ is also involutory.

As the geometrical source of the system of axioms to be introduced in §1 the reader should visualize the following situation.

Let E be a Euclidean plane, and in it consider the group of motions of E into itself (the group "multiplication" of two motions consists of carrying them out successively). For every point A of E consider the reflection in A: that is, the motion for which A is the midpoint of every pair of points consisting of a point and its image. Reflection in A (rotation through $180°$ around A) is involutory. The product of a reflection in A and a reflection in B is a translation, namely through twice the vector determined by the points A, B. The product of reflections in A, in B, and in C is equal to the reflection in a point D, where A, B, C, D form a parallelogram. Thus every product of an odd number of point reflections is a point reflection, and every product of an even number of point reflections is a translation. If we let $\mathfrak{P}_E$ denote the set of point reflections in E, then the set of all products of finitely many elements of $\mathfrak{P}_E$ is certainly a group, namely, the subgroup $\mathfrak{G}_E$ of the group of motions of E generated by $\mathfrak{P}_E$. Thus $\mathfrak{G}_E$ is the group of translations and point reflections in E.

The fundamental assumption and the axioms of §1 are valid for the group $\mathfrak{G}_E$ and the set $\mathfrak{P}_E$.

[5]F. Bachmann, Punkte, Vektoren, Spiegelungen. Der math. und naturwiss. Unterricht 18 (1965), Heft 4.

I. Foundations

Fundamental assumption: *Let there be given a group* $\mathfrak{G}$ *and a nonempty system* $\mathfrak{P}$ *of generators of* $\mathfrak{G}$; *let all the elements of* $\mathfrak{P}$ *be involutory.*

The elements of P will be called *points* and will be denoted by uppercase Latin letters.

Axiom 1:   *The product of three points is a point.*

This axiom, our basic one, is an existence axiom, stating that for three given points A, B, C there always exists a point D such that $ABC = D$. Then D is uniquely determined. We shall call D the "fourth vertex of the parallelogram determined by A, B, C," or the *fourth* parallelogram point for A, B, C (Fig. 1).

Since by Axiom 1 every product ABC is involutory, we have

Rule 1: $ABC = CBA$.

This rule *"Every product of three points can be inverted"* is the next most important rule in our calculus after the rule $AA = 1$. It is equivalent to

Rule 2: $(AB)^C = CABC = BA$.

By repeated application of Axiom 1: Every product of an odd number of points is a point (and can thus be inverted); every product of an even number of points is equal to a product of two points. Those elements of $\mathfrak{G}$ that can be represented as a product of two points form a *subgroup* $\mathfrak{B}$ *of* $\mathfrak{G}$. *The subgroup* $\mathfrak{B}$ *is commutative:*

Rule 3: $AB \cdot CD = CD \cdot AB$.

Proof. $AB \cdot CD = ABC \cdot D = CBA \cdot D = C \cdot BAD = C \cdot DAB = CD \cdot AB$.

The elements of the subgroup $\mathfrak{B}$ will be called *vectors.*[6] For vectors we

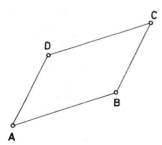

Fig. 1

---

[6]The composition of vectors is here written multiplicatively and not, as is usual, additively. This fact should not give the reader any trouble; he need only remember that we are dealing here with vectors in precisely the usual way.

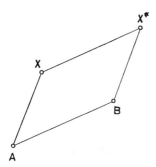

Fig. 2

have the *relation of Chasles:*[7] $AB \cdot BC = AC$, and more generally $A_1 A_2 \cdot A_2 A_3 \cdots A_{n-1} A_n = A_1 A_n$, as follows from the rule $AA = 1$ applied to B and to $A_2, \ldots, A_{n-1}$. *Every element of* $\mathfrak{G}$ *is either a vector or a point.* An equation like $AB = C$, and thus $ABC = 1$, is impossible by Axiom 1. The subgroup $\mathfrak{B}$ is of index 2 in $\mathfrak{G}$.

Two ordered pairs of points A, B and D, C are said to be *parallel-equal* if $AB = DC$. Parallel-equality is an equivalence relation on the set of ordered pairs. Axiom 1 has the same meaning as the statement: for every vector AB and every point X there exists a point X^* such that A, B is parallel-equal to X, X^* (Fig. 2). In short, *any desired vector can be marked off from any given point;* also, $X^* = XAB$, so that this process of marking off a vector is unique.

Thus for every vector AB we can consider the mapping

(1) $$X \to XAB.$$

This one-to-one mapping of $\mathfrak{P}$ onto itself is called *translation by the vector AB;* every pair of points consisting of a preimage and its image is parallel-equal to the pair A, B (since $X \cdot XAB = AB$). The translations defined in this way form a *commutative group* $\mathfrak{F}$. The correspondence which to a vector AB assigns the translation (1) is an isomorphism of the vector group $\mathfrak{B}$ onto the translation group $\mathfrak{F}$. The set $\mathfrak{P}$ of points, together with the translation group $\mathfrak{F}$ operating on it, forms a *homogeneous space.*[8] A translation which leaves a point fixed must be the identity (since $PAB = P$ implies $AB = 1$).

It follows from Axiom 1 that $X^A = AXA$ is always a point (Fig. 3). Thus

(2) $$X \to X^A$$

is a mapping of $\mathfrak{P}$ into itself. The twofold application of this mapping is the identity (since $X^{AA} = X$); thus (2) is a one-to-one mapping of $\mathfrak{P}$ onto itself. Also, A is a fixed point of this mapping (since $A^A = A$). The mapping (2)

[7]G. Choquet, L'Enseignement de la Géométrie. Paris 1964, p. 37.
[8]N. Bourbaki, Algèbre. Chap. 1, Structures algébriques. § 7, 6. Paris 1951.

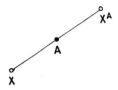

Fig. 3

is called *reflection in the point A*, and the point X^A is called the (*reflected*) *image of X with respect to A.*

In dealing with these mappings it is often useful to employ the following rule (for its geometric interpretation see §3.1).

Rule 4: $PP^{AB} = (AB)^2$, $P^A P^B = (AB)^2$.

Proof. $PP^{AB} = PBAP \cdot AB = AB \cdot AB$, $P^A P^B = A \cdot PABP \cdot B = A \cdot BA \cdot B$, by Rule 2 in each case.

The mapping

(3) $X \rightarrow X^{AB}$

that consists of reflection in a point A followed by reflection in a point B is the *translation by the vector* $(AB)^2$, or in other words by "twice" the vector AB; for by Rule 4 we have $X^{AB} = X(AB)^2$.

Every one-to-one mapping $X \rightarrow X^*$ of $\mathfrak{P}$ onto itself that takes parallel-equal pairs of points into parallel-equal pairs of points induces a one-to-one mapping of $\mathfrak{V}$ onto itself, defined by $XY \rightarrow X^* Y^*$. Every translation induces on $\mathfrak{V}$ the identical mapping (since $XAB \cdot YAB = XAB \cdot BAY = XY$); every point reflection induces on $\mathfrak{V}$ the mapping which to each vector associates its inverse (since $X^A Y^A = (XY)^A = YX$, by Rule 2). In other words: *translations take every vector into itself, and point reflections reverse every vector.*

The Rules 1 through 4 are special cases of the *general rule:* a product $A_1 A_2 \cdots A_n$ is not changed by an arbitrary permutation of the elements in odd-numbered positions only, or in even-numbered only. Application of the general rule merely means iterated application of Rule 1. The general rule facilitates simplification of the longer products in §8.

Axiom 2:   *From $P^A = P$ follows $P = A$; that is, A is the only fixed point of the reflection in A.*

Many of the following theorems are independent of Axiom 2, but its importance is to be seen from its consequences (§1.1 through 1.3), which, more precisely, are equivalent to it, provided Axiom 1 is valid.

1.1.   *From $(AB)^2 = 1$ it follows that $AB = 1$ (and conversely); in other words, $\mathfrak{V}$ contains no involutory element.*

Proof. The equation $(AB)^2 = 1$ can be written in the form $B^A = B$. Then from Axiom 2 it follows that $B = A$, or in other words, $AB = 1$.

If $P^A = Q$, we give to A the name *midpoint* of P, Q.

1.2. *From $P^A = Q$ and $P^B = Q$ it follows that $A = B$; in other words, two points have at most one midpoint.*

Proof. By multiplying the given equations we obtain $P^A P^B = 1$. By Rule 4 this result is equivalent to $(AB)^2 = 1$, and thus by §1.1 equivalent to $AB = 1$.

1.3. *From $P^{AB} = P$ it follows that $AB = 1$ (and conversely).*

Proof. By hypothesis we have $PP^{AB} = 1$. By Rule 4 this equation is equivalent to $(AB)^2 = 1$, and thus by §1.1 equivalent to $AB = 1$.

Axiom 3: *For given P, Q there exists a point A such that $P^A = Q$; in other words, every pair of points has a midpoint.*

In fact, Axiom 3 is of little importance for our calculus, but we shall not consider the question of what can be asserted independently of Axiom 3 (or of Axiom 2); we already make use of Axiom 3 in the second exercise below.

Axioms 2 and 3 state that every vector can be uniquely bisected. It would be natural to require further that every vector can be uniquely trisected, or uniquely divided into n equal parts. Unique trisection would be provided for by two additional axioms: a) *from $(AB)^3 = 1$ follows $AB = 1$*; b) *for every P, Q there exists an X such that $(PX)^3 = PQ$*. The solution of those exercises that are stated here in the form of questions would be simpler if we were to adopt these axioms, and the special features that would then be lost are of no great significance.

In the geometric interpretation of the consequences of our axioms we shall use the ordinary language of three-dimensional Euclidean geometry, because it is so easily visualized.

Exercises. 1 (Independent of Axiom 3). If $\mathfrak{P}$ contains only one element A, then $\mathfrak{G} = \{1, A\}$ is a cyclic group of two elements, and therefore Abelian. If $\mathfrak{P}$ contains at least two elements, then the center of $\mathfrak{G}$ consists of the unit element alone (so that in particular $\mathfrak{G}$ is non-Abelian).

2 (Use Axiom 3). Every translation is a product of two point reflections and is therefore representable in the form (3). The point reflections and the translations form a group, which is the group of inner automorphisms of $\mathfrak{G}$ restricted to $\mathfrak{P}$.

In what follows we shall usually, though not always, be dealing with n-gons, in particular with triangles, quadrangles, and hexagons, so that we now define:

An n-gon is an ordered n-tuple of points $A_1, A_2, \ldots, A_n$, whose order may be permuted cyclically and may also be inverted. The individual points are called *vertices,* and pairs of consecutive vertices are *sides.* An *n*-gon all of whose vertices coincide will be called *trivial.*

In a 2*n*-gon $A_1, \ldots, A_{2n}$ we shall say that any two vertices A_i and A_{i+n} ($i = 1, 2, \ldots, 2n$; indices to be taken mod 2*n*) are *opposite* to each other. If to every vertex there is assigned its opposite vertex, then to every ordered pair of consecutive vertices there is assigned an ordered pair of consecutive vertices. Two ordered pairs of consecutive vertices assigned to each other in this way will be called *opposite sides.*

By selecting *n* alternate vertices in a 2*n*-gon, namely, $A_1, A_3, \ldots, A_{2n-1}$ and $A_2, A_4, \ldots, A_{2n}$, we can form two *part-n-gons:* these *n*-gons will be called the *alternate part-n-gons of the 2n-gon.*

2. Parallelograms

Two ordered pairs of points *A, B* and *C, D* are said to be *parallel-equal* (see §1) if the vectors *AB* and *CD* are equal, $AB = CD$. This parallel-equality is an equivalence relation on the set of ordered pairs of points. The defining equation $AB = CD$ is equivalent to $AC = BD$ and also to $BA = DC$: If *A, B* and *C, D* are parallel-equal, then so also are *A, C* and *B, D* (interchangeability of the inner terms), and also *B, A* and *D, C* (inversion of each of the pairs).

2.1.   *The pairs A, B and C, D are parallel-equal if and only if there exists a translation taking A, B into C, D* (Fig. 4).

Proof.   Let us express the parallel-equality of *A, B* and *C, D* by $AC = BD$ (interchange of the inner terms). A translation taking *A* into *C* and *B* into *D* will exist if and only if there exists a vector α such that $A\alpha = C$ and $B\alpha = D$, or in other words if $\alpha = AC = BD$, so that $AC = BD$.

The pairs of points *A, B* and *C, D* will be said to be *antiparallel-equal,* if the vectors *AB* and *CD* are opposite, in other words if $AB = (CD)^{-1}$. The defining equation is equivalent to $AB = DC$ and also to $ABCD = 1$. Anti-parallel-equality is a symmetric relation on the set of ordered pairs of points. If *A, B* and *C, D* are antiparallel-equal, then so also are *A, D* and *C, B*

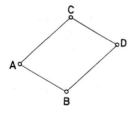

Fig. 4

(interchangeability of the second terms), and also B, A and D, C (invertibility of the pairs).

2.2.   *The pairs A, B and C, D are antiparallel-equal if and only if there exists a point reflection taking A, B into C, D; that is, if the pairs A, C and B, D have a common midpoint* (Fig. 5).

Proof.   The pairs A, C and B, D have a common midpoint if and only if the midpoint of A, C is also the midpoint of B, D. Let M be the midpoint of A, C; in other words, $A^M = C$. By Rule 2 we have $AB = (BA)^M = B^M A^M$. Thus $AB = B^M C$. Consequently, $B^M = D$ if and only if $AB = DC$.

A quadrangle will be called a *parallelogram* if its opposite sides are antiparallel-equal. A quadrangle A, B, C, D has this property if and only if $AB = DC$ or in other words $ABCD = 1$. From §2.2 we have:

2.3.   *A quadrangle is a parallelogram if and only if its pairs of opposite vertices have a common midpoint.*

More generally, we may consider the n-tuples $A_1, \ldots, A_n$ and $B_1, \ldots, B_n$ together with a correspondence between their points (unless otherwise stated, points with the same subscript will correspond to each other). The two n-tuples will be called *parallel-equal* if every two corresponding pairs of points are parallel-equal; thus $A_i A_k = B_i B_k$ for $i, k = 1, \ldots, n$; the n-tuples will be called *antiparallel-equal* if every two corresponding pairs of points are antiparallel-equal; thus $A_i A_k = (B_i B_k)^{-1}$. The results in §2.1 and 2.2 for $n = 2$ can be generalized. We shall restrict our attention here to the case $n = 3$.

Let there be given two point triples A_1, A_2, A_3 and B_1, B_2, B_3, and let

(1) $A_1 A_2 = B_1 B_2, A_2 A_3 = B_2 B_3.$

2.4.   *From* (1) *follows* $A_1 A_3 = B_1 B_3$.

Proof. $A_1 A_3 = A_1 A_2 \cdot A_2 A_3 = B_1 B_2 \cdot B_2 B_3 = B_1 B_3$.
Interpreting §2.4 geometrically gives the special affine theorem of Desargues (Fig. 6) (cf. II3, §1.2).

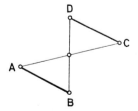

Fig. 5

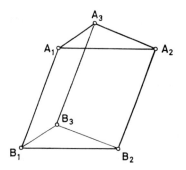

Fig. 6

The point triples A_1, A_2, A_3 and B_1, B_2, B_3 are parallel-equal if and only if (1) holds. Interchange of the inner terms shows that the equations (1) are equivalent to

(2) $$A_1B_1 = A_2B_2 = A_3B_3.$$

2.5.   *The point triples A_1, A_2, A_3 and B_1, B_2, B_3 are parallel-equal if and only if there exists a translation taking A_1, A_2, A_3 into B_1, B_2, B_3.*

Proof.   The proof follows from (2) and is like the proof of §2.1.
Now let A_1, A_2, A_3 and B_1, B_2, B_3 be point triples for which

(3) $$A_1A_2 = (B_1B_2)^{-1}, \quad A_2A_3 = (B_2B_3)^{-1}.$$

2.6.   *From (3) follows $A_1A_3 = (B_1B_3)^{-1}$.*

Proof. $A_1A_3 = A_1A_2 \cdot A_2A_3 = (B_1B_2)^{-1} \cdot (B_2B_3)^{-1} = B_2B_1B_3B_2 = (B_1B_3)^{-1}$, the last equality by Rule 2.
Interpreting §2.6 geometrically gives a generalization of the special affine theorem of Pappus (Fig. 7).[9]

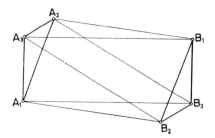

Fig. 7

[9]The figure for the special affine theorem of Pappus arises if A_1, A_2, A_3 and B_1, B_2, B_3 are respectively collinear (cf. II3, § 1.2).

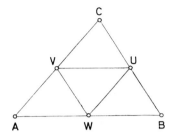

Fig. 8

The point triples A_1, A_2, A_3 and B_1, B_2, B_3 are antiparallel-equal if and only if (3) holds. From §2.2 we have

2.7.   *The triples A_1, A_2, A_3 and B_1, B_2, B_3 are antiparallel-equal if and only if there exists a point reflection taking A_1, A_2, A_3 into B_1, B_2, B_3; that is, if the three pairs A_i, B_i ($i = 1, 2, 3$) have a common midpoint.*

2.8.   For a given point triple U, V, W we can construct the fourth parallelogram point for each permutation of U, V, W. By Rule 1 the six resulting points reduce to three (e.g., $UVW = WVU$). Let us set

(4) $A = WUV$, $B = UVW$, $C = VWU$.

The point triple (4) is called the *derived triple* (the *derivative*) of U, V, W. We have the vector equations

(5) $CU = VW = UB$, $BW = UV = WA$, $AV = WU = VC$.

These equations show in particular that U is the midpoint of C, B, and so forth. Every point triple is the midpoint triple of its derived triple (Fig. 8).

Exercises.   1.   Let A_1, A_2 and B_1, B_2 be antiparallel-equal. If C_1, C_2 is antiparallel-equal to A_1, A_2, then C_1, C_2 is parallel-equal to B_1, B_2, and conversely. The corresponding remarks hold for n-tuples.
2.   The two point triples A_1, A_2, A_3 and A_3, $A_3 A_2 A_1$, A_1 are antiparallel-equal.
3.   If two triples are parallel-equal or antiparallel-equal, their derived triples have the same property.

3.   Iterated Reflection

3.1.   If we reflect a point P first in A and then in B, we obtain the point P^{AB}. By Rule 4 we have $PP^{AB} = (AB)^2$, so that the vector PP^{AB} is the "double" of the vector AB. The mapping $X \to X^{AB}$ is a translation by the vector $(AB)^2$. The midpoint of PP^{AB} is the point PAB, since $P^{PAB} = P^{AB}$

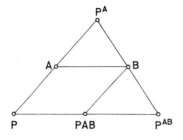 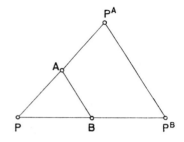

Fig. 9   Fig. 10

(Fig. 9). This point PAB is the image of P in the translation $X \to XAB$; for fixed A, B, it varies with P.

If we reflect P in A and also in B, we obtain the images P^A and P^B. Since $P^A P^B = (AB)^2$, the vector $P^A P^B$ is also the "double" of the vector AB (Fig. 10).

3.2.   If we reflect P first in A, then in B, and then in C, we obtain the point P^{ABC}. The mapping $X \to X^{ABC}$ is the reflection in the fourth parallelogram point ABC of A, B, C. The midpoint of PP^{ABC} is the point ABC; for given A, B, C, this point remains the same for all P (Fig. 11). If an arbitrary point P is reflected successively in the vertices of a parallelogram $A, \ldots, D$, the point P returns to its original position: $P^{ABCD} = P$ (since $ABCD = 1$).

3.3.   Let A, B, C be given. If we reflect P first in A, then in B, then in C, and then once more in A, in B, and in C, the figure "closes": $P^{ABCABC} = P$. For by Axiom 1, ABC is a point D, so that $ABCABC = DD = 1$. The six points

(1) $P, P^A, P^{AB}, P^{ABC}, P^{ABCA} = P^{CB}, P^{ABCAB} = P^C$

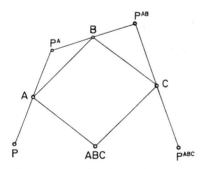

Fig. 11

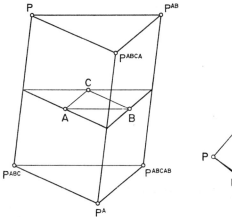

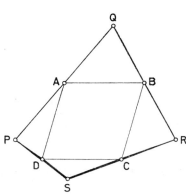

Fig. 12   Fig. 13

form a "prism" (Fig. 12). The translation $X \to X^{PABC}$ takes the points P, P^{AB}, P^{CB} into the points P^{ABC}, P^C, P^A (for example, we have $(P^{AB})^{PABC} = P^{ABP \cdot ABC} = P^{PBA \cdot ABC} = P^{PC} = P^C$); these two point triples are parallel-equal (cf. §2.5).

3.4. *The midpoints of the sides of an arbitrary quadrangle form a paral-lelogram.* Let $P, \ldots, S$ be a quadrangle, and let $A, \ldots, D$ ($P^A = Q$, etc.) be the midpoints of its sides (Fig. 13).

Proof 1. $P^{ABC} = Q^{BC} = R^C = S$ and $P^D = S$. Since ABC is a point by Axiom 1, it follows that ABC and D are midpoints of P, S. Thus by §1.2 we have $ABC = D$, so that $ABCD = 1$.

Proof 2.   By hypothesis we have $P^{ABCD} = Q^{BCD} = R^{CD} = S^D = P$. Thus it follows from §1.3 that $ABCD = 1$.

3.5.   From §2.3 and §3.4 we thus have: *in an arbitrary quadrangle the pairs of midpoints of opposite sides have a common midpoint.*

3.6.   Let U, V, W be the triple of midpoints of a point triple (triangle) A, B, C; thus

(2) $A^W = B, B^U = C, C^V = A.$

Then we have $A^{WUV} = B^{UV} = C^V = A$, so that Axiom 2 gives $WUV = A$ (Fig. 14). Taken together, these results lead to the three equations §2 (4). *Thus if U, V, W is the midpoint triple of A, B, C, it follows that A, B, C is the derived triple of U, V, W* (the converse holds by §2.8). *Consequently every point triple is the midpoint triple of only one triple, namely, its derived triple.*

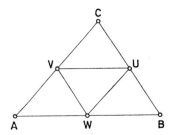

Fig. 14

4. Midpoint *n*-gon and Derivative of an *n*-gon

The midpoints of the sides of an *n*-gon $\mathscr{A}_n$, taken in the cyclic order induced by the cyclic order of the vertices of $\mathscr{A}_n$, form an *n*-gon $\mathscr{A}_n^0$, called the midpoint *n*-gon of $\mathscr{A}_n$. If $\mathscr{A}_n = \{A_1, \ldots, A_n\}$, we shall denote the midpoint of A_i, A_{i+1} by A_i^0. Then $A_i^{A_i{}^0} = A_{i+1}$ and $\mathscr{A}_n^0 = \{A_1^0, \ldots, A_n^0\}$.

4.1. For even *n* we have: *The product of the vertices of any midpoint n-gon is 1.*

Proof (cf. the second proof of §3.4). In the notation $A_1^{A_1{}^0 A_2{}^0 \cdots A_n{}^0} = A_2^{A_2{}^0 \cdots A_n{}^0} = A_n^{A_n{}^0} = A_1$. Thus it follows from §1.3, since *n* is even and therefore $A_1^0 A_2^0 \cdots A_n^0$ is a vector, that $A_1^0 A_2^0 \cdots A_n^0 = 1$.

4.2. Problem (Choquet l.c., p. 53). *Let there be given an n-gon A_1, $A_2, \ldots, A_n$. Determine all the n-gons for which $A_1, A_2, \ldots, A_n$ is the midpoint n-gon.*

Solution. If *P* is an arbitrary point, then

(1) $$P, \; P^{A_1}, \; P^{A_1 A_2}, \ldots, \; P^{A_1 A_2 \cdots A_{n-1}}$$

is an *n*-gon in which A_1 is the midpoint of the first side, A_2 of the second, and $\ldots, A_{n-1}$ of the $(n-1)$-th side, and every *n*-gon with this property can be written in the form (1). Thus the desired *n*-gons are exactly those *n*-gons (1) in which A_n is the midpoint of the *n*-th side; that is, exactly those *n*-gons (1) containing a point *P* for which

(2) $$P^{A_1 A_2 \cdots A_{n-1} A_n} = P.$$

For the discussion of equation (2) we distinguish two cases.

Case 1: *n even.* Then $A_1 A_2 \cdots A_n$ is a vector. By §4.1 the problem has a solution only if $A_1 A_2 \cdots A_n = 1$. If $A_1 A_2 \cdots A_n = 1$, *then* (2) *holds for every point P and $A_1, A_2, \ldots, A_n$ is the midpoint n-gon of every n-gon* (1)

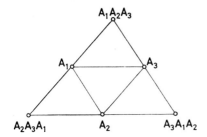

Fig. 15

with arbitrary P. For example, a parallelogram $A, \ldots, D$, since $ABCD = 1$, is the midpoint quadrangle of every quadrangle P, P^A, P^{AB}, P^{ABC}.

Case 2: *n odd.* In this case $A_1 A_2 \cdots A_n$ is a point; there exists a point P and by Axiom 2 only one such point, with the property (2), namely, $P = A_1 A_2 \cdots A_n$. Thus $A_1, A_2, \ldots, A_n$ *is the midpoint n-gon of exactly one n-gon, namely,*

(3) $\qquad A_1 A_2 \cdots A_n, A_2 \cdots A_n A_1, \ldots, A_n A_1 \cdots A_{n-1}.$

For $n = 3$ (Fig. 15) cf. §§2.8., 3.6.

For every triple of consecutive vertices of an n-gon $\mathscr{A}_n$, let us construct the fourth parallelogram point. In the cyclic order induced by the cyclic order of the vertices of $\mathscr{A}_n$, these points form an n-gon $\mathscr{A}'_n$, the *derivative of* $\mathscr{A}_n$. If $\mathscr{A}_n = \{A_1, \ldots, A_n\}$, then every vertex A_i is the middle point of a triple of consecutive vertices; we denote the fourth parallelogram point of this triple by $A'_i : A_{i-1} A_i A_{i+1} = A'_i$. Then $\mathscr{A}'_n = \{A'_1, \ldots, A'_n\}$.

(For even n the product of the vertices of $\mathscr{A}'_n$ and the product of the vertices of $\mathscr{A}_n$ are related by $A'_1 A'_2 \cdots A'_n = (A_1 A_2 \cdots A_n)^{-3}$.)

The correspondences

(4) $\qquad\qquad\qquad\qquad \mathscr{A}_n \to \mathscr{A}_n^0$

(5) $\qquad\qquad\qquad\qquad \mathscr{A}_n \to \mathscr{A}'_n$

are mappings of the set of all n-gons (for each fixed n) into itself. These mappings commute, as follows.

4.3. *In every case* $\mathscr{A}_n^{0'} = \mathscr{A}_n^{'0}$; *in other words, the derivative of the midpoint n-gon of* $\mathscr{A}_n$ *is equal to the midpoint n-gon of the derivative of* $\mathscr{A}_n$.

Proof. E.g., we have $(A_n A_1 A_2)^{A_0{}^0 A_1{}^0 A_2{}^0} = A_n^{A_n{}^0 A_1{}^0 A_2{}^0}(A_1 A_2)^{A_n{}^0 A_1{}^0 A_2{}^0}$ $= A_3 A_2 A_1 = A_1 A_2 A_3$, as follows from $A_n^{A_n{}^0 A_1{}^0 A_2{}^0} = A_1^{A_1{}^0 A_2{}^0} = A_2^{A_2{}^0} = A_3$ and Rule 2. This equation states that $(A'_1)^{A_1{}^{0'}} = \mathscr{A}'_2$, i.e., $A_1^{0'}$ is the midpoint of $A'_1 A'_2$, so that $A_1^{0'} = A_1^{'0}$.

By the definitions of this section, every n-gon $\mathscr{A}_n$ is associated with three

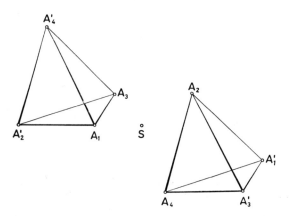

Fig. 16

n-gons: the midpoint n-gon $\mathscr{A}_n^0$, the derivative $\mathscr{A}_n'$, and the *"mid-derivative"* $\mathscr{A}_n^{0'} = \mathscr{A}_n'^0$.

By §§2.8 and 3.6 we always have $\mathscr{A}_3'^0 = \mathscr{A}_3^{0'} = \mathscr{A}_3$; thus in the case $n = 3$, the mappings (4) and (5) of the set of all triangles into itself are one-to-one and inverse to each other.

Figure 16 shows the quadrangle $\mathscr{A}_4$ and its derivative $\mathscr{A}_4'$. For $n = 4$ we have: the translation $X \rightarrow XA_1A_2A_3A_4$ takes the points A_1, A_2', A_3, A_4' into the points A_3', A_4, A_1', A_2. *Every parallelogram is equal to its derivative.* The derivative of hexagons will be studied in the next section.

Exercises. 1. A quadrangle and its derivative have the same midpoint-quadrangle.

2. $\mathscr{A}_4 \rightarrow \mathscr{A}_4^0$ maps the set of all parallelograms one-to-one onto itself.

3. Is $\mathscr{A}_4 \rightarrow \mathscr{A}_4''$ a one-to-one mapping of the set of quadrangles onto itself?

4. Let each vertex of a pentagon $\mathscr{A}_5 = \{A_1, \ldots, A_5\}$ be reflected in the midpoint of the opposite side; we thus obtain the pentagon $\mathscr{A}_5^{\ddagger} = \{A_3A_1A_4, A_4A_2A_5, A_5A_3A_1, A_1A_4A_2, A_2A_5A_3\}$. Here $\mathscr{A}_5^{\ddagger'} = \mathscr{A}_5'^{\ddagger} = \mathscr{A}_5$. Thus $\mathscr{A}_5 \rightarrow \mathscr{A}_5'$ is a one-to-one mapping of the set of all pentagons onto itself, and $\mathscr{A}_5 \rightarrow \mathscr{A}_5^{\ddagger}$ is the inverse mapping.

5. Prisms

In this section and the next we shall be dealing with hexagons. We distinguish special classes of hexagons (prisms, A-hexagons, C-hexagons, AC-hexagons) and investigate such hexagons in detail. Although they will not be defined in terms of midpoint hexagons and derivatives, it will nevertheless turn out that these special classes of hexagons can be characterized by the fact that for the hexagons in a given class the midpoint hexagons or the derivatives are in some sense degenerate. Here the following types of degeneration of a hexagon $\mathscr{A}_6 = \{A_1, \ldots, A_6\}$ will occur:

Let $\mathscr{A}_6$ be called a *doubly covered triangle* if the opposite vertices of $\mathscr{A}_6$ coincide, that is, if $A_1 = A_4$, $A_2 = A_5$, $A_3 = A_6$. Let $\mathscr{A}_6$ be called a *triply covered segment* if both the alternate part-triangles of $\mathscr{A}_6$ are trivial, that is, if $A_1 = A_3 = A_5$ and $A_2 = A_4 = A_6$. Then $\mathscr{A}_6$ is *trivial* if and only if it is both a doubly covered triangle and a triply covered segment.

We now define the first of the above-mentioned classes: $\mathscr{A}_6$ is called a prismatic hexagon, or more briefly a *prism*, if (Fig. 17)

(1) $$A_1 A_4 = A_3 A_6 = A_5 A_2.$$

Then translation by the vector (1) takes A_1, A_3, A_5 into A_4, A_6, A_2. The equations (1) state that the two alternate part-triangles of $\mathscr{A}_6$, related to each other by associating vertices with opposite vertices, are parallel-equal (cf. §2).

Degenerate hexagons (doubly covered triangles, triply covered segments, and trivial hexagons) are prisms. The doubly covered triangles are exactly those prisms for which the translation vector (1) is equal to 1.

5.1. $\mathscr{A}_6$ *is a prism if and only if every pair of opposite sides of $\mathscr{A}_6$ has a common midpoint; that is, if $\mathscr{A}_6^0$ is a doubly covered triangle.*

Proof.   By §2.2 the equation $A_1 A_4 = A_5 A_2$ is equivalent to the condition that A_1, A_2 and A_4, A_5 have a common midpoint, so that $A_1^0 = A_4^0$.

If A, B, C is a triangle and P is an arbitrary point, then §3(1) is a prism with midpoint hexagon A, B, C, A, B, C. Every prism can be written in this form.

Remark. $\mathscr{A}_6$ *is a triply covered segment if and only if $\mathscr{A}_6^0$ is trivial.*
The derivatives of hexagons may be described as follows:

5.2.   a)   *The derivative of a hexagon is a prism.*
     b)   *Every prism is the derivative of a hexagon.*

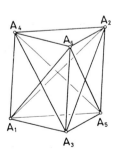 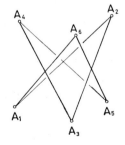

Fig. 17

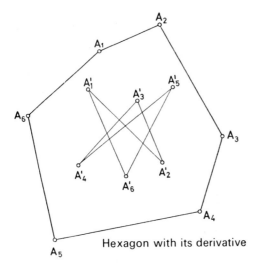

Hexagon with its derivative

Fig. 18

Proof of a).   Let $A_1, \ldots, A_6$ be an arbitrary hexagon (Fig. 18). For its derivative $A_1', \ldots, A_6'$ we have: $A_1'A_4' = A_6A_1A_2 \cdot A_3A_4A_5 = (A_1 \cdots A_6)^{A_6} = (A_1 \cdots A_6)^{-1}$, by Rule 2; $A_3'A_6' = A_2A_3A_4 \cdot A_5A_6A_1 = (A_1 \cdots A_6)^{A_1} = (A_1 \cdots A_6)^{-1}$; $A_5'A_2' = A_4A_5A_6 \cdot A_1A_2A_3 = A_6A_5A_4 \cdot A_3A_2A_1 = (A_1 \cdots A_6)^{-1}$. Thus the derivative $A_1, \ldots, A_6$ is a prism, and we also see that the "translation vector" of this prism is equal to $(A_1 \cdots A_6)^{-1}$.

Remark.   For an arbitrary set of four points $A_1, \ldots, A_4$ there exists exactly one pair of points A_5, A_6 such that $A_1, \ldots, A_4, A_5, A_6$ is a prism, namely, $A_1A_4A_2 = A_5$, $A_3A_1A_4 = A_6$, as follows from (1).

Proof of b).   Every prism can be written in the form $A_1, \ldots, A_4$, $A_2A_4A_1$, $A_4A_1A_3$. Thus it is the derivative of the hexagon A_3, A_3, A_2, A_2, A_4, A_1.

From §5.2 a) and §5.1 we have the following corollary.

5.3.   *For every hexagon the mid-derivative is a doubly covered triangle.*

The vertices of a doubly covered triangle $\mathscr{A}_6^{0'} = \mathscr{A}_6^{'0}$, as we remark in passing, are the midpoints of the pairs of opposite vertices of the given hexagon $\mathscr{A}_6$. For example, $A_1^{A_1^0 A_2^0 A_3^0} = A_2^{A_2^0 A_3^0} = A_3^{A_3^0} = A_4$, so that the point $A_1^0 A_2^0 A_3^0 = A_2^{0'}$ is the midpoint of the pair of opposite vertices A_1, A_4 of the hexagon $\mathscr{A}_6$.

Exercises.   1.   Draw a prism, its derivative, and its mid-derivative.
2.   A prism $A_1, \ldots, A_6$ is the derivative of an arbitrary hexagon P_1, P_2, $P_2P_1A_2, A_2P_1A_3, A_3P_2A_4, A_1P_2P_1$ (with arbitrary P_1, P_2).

3.   Characterize the dodecagons which are the derivatives of dodecagons.

4.   Apart from the parallelograms and the trivial n-gons, are there any other n-gons that are equal to their derivatives? In particular, is a hexagon necessarily trivial if it is equal to its derivative?

6. A-Hexagons and C-Hexagons

As we have seen, parallelograms are characterized in the set of all quadrangles by the following properties:

(A)   *the opposite sides are antiparallel-equal,*

(B)   *the pairs of opposite vertices have a common midpoint,*

(C)   *the product of the vertices is 1,*

(D$_4$)   *they are the midpoint quadrangles of arbitrary quadrangles.*

Let us now consider hexagons possessing at least one of the properties (A), (B), (C), (D$_6$), where by (D$_6$) we mean the property analogous to (D$_4$) (that is, in (D$_4$) the word "quadrangles" is to be replaced by "hexagons"). Thus we are studying hexagons analogous in some sense to parallelograms.

Hexagons with the property (A) will be called A-hexagons (Fig. 19a, b). As an example in the Euclidean plane, consider the outline (Fig. 19a) of the parallel projection of a parallelepiped (for example, a cube or a brick); the three-dimensional preimage of such an outline (that is, a set of six edges of the parallelepiped) is also an A-hexagon. The defining property of an A-hexagon $A_1, \ldots, A_6$ is

(1) $$A_1 A_2 = A_5 A_4, \quad A_2 A_3 = A_6 A_5, \quad A_3 A_4 = A_1 A_6,$$

any two of which equations imply the third; cf. §2.6. The defining property states that every triple of consecutive vertices is antiparallel-equal to the triple of opposite vertices. Thus from §2.7 we have

6.1.   *For hexagons the properties* (A) *and* (B) *are equivalent.*

Given four arbitrary points $A_1, \ldots, A_4$, there exists exactly one pair of

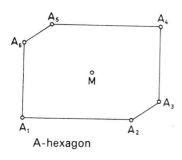

A-hexagon

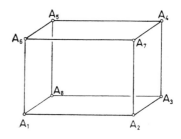

Fig. 19a   Fig. 19b

points A_5, A_6 such that $A_1, \ldots, A_4, A_5, A_6$ is an A-hexagon, namely, $A_1 A_2 A_4 = A_5$, $A_1 A_3 A_4 = A_6$.

Hexagons with the property (C) will be called C-hexagons. The defining property of a C-hexagon $A_1, \ldots, A_6$ is

(2) $$A_1 A_2 A_3 A_4 A_5 A_6 = 1.$$

This equation states that the three vectors $A_1 A_2$, $A_3 A_4$, $A_5 A_6$ form a closed vector polygon (for the significance of this condition see §9.4). The C-hexagons are of a quite general kind. For if we choose five arbitrary points $A_1, \ldots, A_5$ there always exists exactly one point A_6 such that $A_1, \ldots, A_5$, A_6 is a C-hexagon, namely, the point $A_1 A_2 \cdots A_5 = A_6$. In n-dimensional Euclidean space the C-hexagons can be four-dimensional figures, whereas the A-hexagons and prisms are at most three-dimensional.

The C-hexagons are precisely those hexagons that have the following property (D_6):

6.2. *The midpoint hexagon of every hexagon is a C-hexagon. Every C-hexagon $A_1, \ldots, A_6$ is the midpoint hexagon of all hexagons §4(1) with $n = 6$ and arbitrary P, as follows from §§4.1 and 4.2.*

The properties (A) and (C), although equivalent for quadrangles, are distinct for hexagons: an A-hexagon is not necessarily a C-hexagon, and conversely. Those hexagons that the both A-hexagons and C-hexagons will be called AC-*hexagons* (Fig. 20).

6.3 (Construction of the AC-hexagons). By §6.1 the A-hexagons are those hexagons for which the pairs of opposite vertices have a common midpoint M; in other words, they are the hexagons $A_1, A_2, A_3, A_1^M, A_2^M, A_3^M$. The product of the vertices of such a hexagon is $A_1 A_2 A_3 (A_1 A_2 A_3)^M$; thus it is equal to 1 if and only if $A_1 A_2 A_3 = (A_1 A_2 A_3)^M$, or in other words, if $A_1 A_2 A_3 = M$ (Axiom 2). Thus the AC-hexagons are precisely those of the

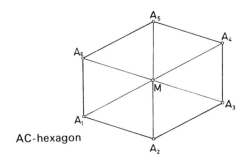

AC-hexagon

Fig. 20

form

(3) $$A_1, A_2, A_3, A_1^{A_1A_2A_3}, A_2^{A_1A_2A_3}, A_3^{A_1A_2A_3}.$$

So we obtain an AC-hexagon if we take three arbitrary points A_1, A_2, A_3 and reflect these points in their fourth parallelogram point $A_1A_2A_3$; every AC-hexagon can be constructed in this way. Thus for every three points $A_1, A_2,$ A_3 there exists exactly one point triple A_4, A_5, A_6 such that $A_1A_2, A_3, A_4, A_5,$ A_6 is an AC-hexagon. In a Euclidean space, an AC-hexagon is always a plane figure. AC-hexagons are well-known as the affine images of regular hexagons. We may call them the *affine-regular hexagons*.

The A-hexagons, C-hexagons, and AC-hexagons can be characterized in the following way as hexagons with degenerate derivatives.

6.4.   *Let* $\mathcal{A}_6 = \{A_1, \ldots, A_6\}$ *be a hexagon, and let* $\mathcal{A}_6' = \{A_1', \ldots, A_6'\}$ *be its derivative. Then*
   a) $\mathcal{A}_6$ *is an A-hexagon if and only if* $\mathcal{A}_6'$ *is a triply covered segment.*
   b) $\mathcal{A}_6$ *is a C-hexagon if and only if* $\mathcal{A}_6'$ *is a doubly covered triangle.*
   c) $\mathcal{A}_6$ *is an AC-hexagon if and only if* $\mathcal{A}_6'$ *is trivial.*

Proof.   The defining equations (1) of an A-hexagon can be written in the form

(4) $$A_1' = A_3' = A_5', \quad A_2' = A_4' = A_6'.$$

Thus a) is true. The defining equation (2) of a C-hexagon is equivalent to each of the equations

(5) $$A_1' = A_4', \quad A_2' = A_5', \quad A_3' = A_6'.$$

Thus b) is true. By a) and b) the hexagon $\mathcal{A}_6$ is an AC-hexagon if and only if the equations (4) and (5) hold, in other words, if $A_1' = A_2' = A_3' = A_4' = A_5' = A_6'$.

By examining their derivatives we can discover geometric properties of these hexagons, as follows.

6.5.   The characteristic equations (4) for an A-hexagon $\mathcal{A}_6$ make a statement about the triples of consecutive vertices of $\mathcal{A}_6$ and the alternate part-triangles A_1, A_3, A_5 and A_2, A_4, A_6, of $\mathcal{A}_6$. Equation (4) states that the triples of consecutive vertices whose middle point belongs to the same alternate part-triangle have the same fourth parallelogram point. Consequently, just as in the interpretation in Euclidean space, we refer to the two points (4) for a given A-hexagon $\mathcal{A}_6$ as its *seventh* and *eighth parallelepiped* points and write $A_1' = A_3' = A_5' = A_7$, $A_2' = A_4' = A_6' = A_8$. Then we also have $A_1A_2 \cdots A_6A_7A_8 = 1$ and thus $A_7A_8 = (A_1A_2 \cdots A_6)^{-1}$. Let M be the common midpoint of the pairs of opposite vertices of the A-

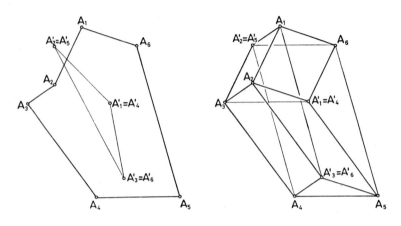

Fig. 21   Fig. 22

hexagon (cf. §6.1). Reflection in M interchanges the two alternate part-triangles, and thus also interchanges the triples of consecutive vertices associated with them and the points A_7, A_8: $A_7^M = A_8$. Here $\mathscr{A}_6' = \{A_7, A_8, A_7, A_8, A_7, A_8\}$ and $A_6'^0 = \{M, M, M, M, M, M\}$.

The characteristic equations (5) for a C-hexagon $\mathscr{A}_6$ state that triples of consecutive vertices whose middle points are opposite vertices of $\mathscr{A}_6$ have the same fourth parallelogram point (Figs. 21 and 22).

An A-hexagon $\mathscr{A}_6$ with M as the common midpoint for its pairs of opposite vertices is an AC-hexagon if and only if $A_7 = A_8 = M$;[10] that is, if M is the fourth parallelogram point of all triples of consecutive vertices, from which it already follows that $\mathscr{A}_6' = \{M, M, M, M, M, M\}$.

6.6.   The AC-hexagons are the midpoint hexagons of the A-hexagons (Fig. 23).

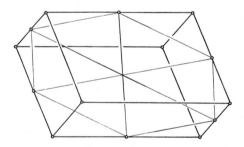

Fig. 23

----

[10]In view of $A_2^M = A_8$, this equation is equivalent to each of the equations $A_7 = A_8$, $A_7 = M$, $A_8 = M$.

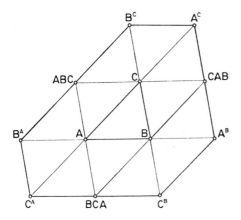

Fig. 24

By §6.2 an AC-hexagon is in any case the midpoint hexagon of some hexagons, since it is certainly a C-hexagon. Thus it is enough to show:

$\mathscr{A}_6$ is an A-hexagon if and only if $\mathscr{A}_6$ is an AC-hexagon.

Proof.   The following statements are equivalent to one another: $\mathscr{A}_6$ is an A-hexagon; $\mathscr{A}'_6$ is a triply covered segment (§6.4,a)); $\mathscr{A}'_6$ is trivial (Remark in §5.1); $\mathscr{A}'_6$ is trivial (§4.3); $\mathscr{A}'_6$ is an AC-hexagon (§6.4,c)).

6.7 (special type of C-hexagons). *Let each vertex of a triangle A, B, C be reflected in the other two vertices.* The resulting hexagon $B^C, A^C, A^B, C^B, C^A, B^A$ (Fig. 24) is a C-hexagon (since by Rule 2 and Axiom 1 we have $B^C A^C A^B C^B C^A B^A = AB \cdot CA \cdot BC = (ABC)^2 = 1$); its derivative is the doubly covered triangle A, B, C (e.g., $B^A B^C A^C = B^A \cdot AB = ABAAB = A$). The hexagon is always "planar" (cf. §6.3), but it is an AC-hexagon only if the nine points all coincide.[11]

Exercises.   1.   A hexagon is an A-hexagon if and only if its alternate part-triangles are antiparallel-equal when each vertex corresponds to its opposite vertex.

2.   The mappings $\mathscr{A}_6 \to \mathscr{A}_6^0$ and $\mathscr{A}_6 \to \mathscr{A}'_6$ map the set of prisms, the set of A-hexagons, the set of C-hexagons, and the set of AC-hexagons respectively into itself.

3.   If four consecutive points of the derivative of a hexagon coincide, the derivative is trivial, so that the hexagon is an AC-hexagon.

4.   Apart from the AC-hexagons and the trivial n-gons, do there exist other n-gons with trivial derivatives?

[11]This hexagon is a special *"Thomsen hexagon."* For *Thomsen hexagons* see K. Reidemeister, Grundlagen der Geometrie. Berlin 1930 and 1968, Fig. 3, or W. Blaschke and G. Bol, Geometrie der Gewebe. Berlin 1938, Fig. 28, 42. Thomsen hexagons are affine specializations of the figure for the projective theorem of Pappus-Pascal-Brianchon. They are C-hexagons.

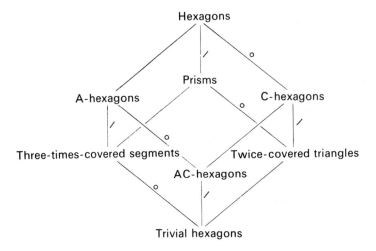

Fig. 25.   Diagram of the eight classes of hexagons. A line joining two classes means
that the one higher on the page includes the lower one. The marks ° and ′
mean that the former is mapped onto the latter by §4(4) and §4(5), respectively.

5.   Two hexagons have the same derivative if and only if they differ from each
other by a hexagon with trivial derivative (an AC-hexagon). More precisely: 1) if
$C_1, \ldots, C_6$ is a hexagon with trivial derivative $M, \ldots, M$, then $A_1, \ldots, A_6$ and
$A_1 M C_1, \ldots, A_6 M C_6$ have the same derivative; 2) if $A_1, \ldots, A_6$ and $B_1, \ldots, B_6$
have the same derivative (without loss of generality we may take $A'_i = B'_i$ for
$i = 1, \ldots, 6$), then for arbitrary M the hexagon $M A_1 B_1, \ldots, M A_6 B_6$ has a
trivial derivative (namely, $M, \ldots, M$).

6.   Does $\mathscr{A}_6 \to \mathscr{A}'_6$ map the set of prisms one-to-one onto itself?

7.   Two $2n$-gons have the same midpoint $2n$-gon if and only if they differ from
each other by a $2n$-gon with trivial midpoint $2n$-gon, namely, by an n-fold covered
segment. For a more precise formulation compare Problem 5.

8.   For $2n$-gons property (A) is equivalent to (B), and (C) is equivalent to
(D_{2n}). Moreover, for $4n$-gons (B) implies (C).

9.   In the diagram for the eight classes of hexagons, there are four lines not
marked by either of the symbols °, ′; in other words there are four pairs of classes
joined by an unmarked line. Does there exist a mapping of the set of all hexagons
onto itself which in all these four cases maps the upper class onto the lower?

7.   Hexagons and Tetrahedra

Let us now present a method for the construction of A-hexagons leading
to a connection between A-hexagons and tetrahedra.

7.1.   Let A, B, C be a point triple, and let P be an arbitrary point. Form the fourth parallelogram points APB, BPC, CPA. The hexagon

(1) $$A, APB, B, BPC, C, CPA$$

is an A-*hexagon* (for example, we have $A \cdot APB = PB = C \cdot CPB$). The point P is the eighth parallelepiped point of the A-hexagon as defined in §6, and the point $APBPC$ is the seventh. Let us denote by $\mathscr{A}_6(A, B, C; P)$ the hexagon (1) associated with a quadruple $A, B, C; P$ (Fig. 26).

Every A-hexagon $A_1, \ldots, A_6$ is equal to $\mathscr{A}_6(A_1, A_3, A_5; A_8)$ with $A_8 = A_1 A_2 A_3$, as can be verified on the basis of §6.5.

By a *tetrahedron* we mean an arbitrary point quadruple $A_1, \ldots, A_4$; the pairs of points A_i, A_j $(i \neq j; i, j = 1, 2, 3, 4)$ are called the *edges of the tetrahedron,* and it is clear what is meant by opposite edges.

7.2.   Given a point triple A, B, C and a point P, we can form the tetrahedron

(2) $$P, P^A, P^B, P^C.$$

In this tetrahedron the points A, B, C are the midpoints of the edges containing P, and the six points (1) are the midpoints of the six edges (for example, APB is the midpoint of $P^A P^B$, since $P^{A \cdot ABB} = P^{BB} = P^B$). Thus the midpoints of the edges of the tetrahedron (2), taken in suitable order, form an A-hexagon (Fig. 27), namely, $\mathscr{A}_6 (A, B, C; P)$.

7.3.   *The* A-*hexagons are the midpoint figures of the tetrahedra.*

In view of Axiom 3, every tetrahedron can be written in the form (2); by §7.2, the midpoints of its edges, when suitably enumerated, form the

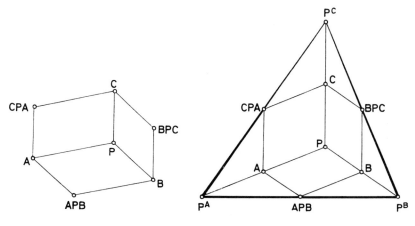

Fig. 26   Fig. 27

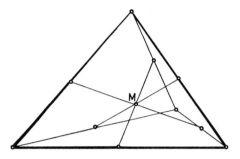

Fig. 28

A-hexagon $\mathscr{A}_6(A, B, C; P)$. Every A-hexagon can be written in the form $\mathscr{A}_6(A, B, C; P)$; by §7.2, its vertices are the midpoints of the edges of the tetrahedron (2).

Since the midpoints of the opposite edges of the tetrahedron (2) are the opposite vertices of the A-hexagon $A_6(A, B, C; P)$, we obtain from §6.1 the following result.

7.4.   *The three pairs of midpoints of opposite edges of a tetrahedron have a common midpoint* (Fig. 28).

Exercises.   1.   The mapping $X \to P^X$ maps $\mathfrak{P}$ one-to-one onto itself (dilation with center P and magnification factor 2). Verify that this mapping takes parallelograms into parallelograms, prisms into prisms, and so forth, and is permutable with the mappings §4 (4), (5).

2.   Determine the prisms and translation-equal tetrahedra contained in the ten-point figure consisting of a tetrahedron and the midpoints of its edges (Euclid, Book 12, Theorem 3).

8.   Isobaric *n*-tuples, Center of Gravity

Definition.   Let two *n*-tuples of points $C_1, \ldots, C_n$ and $D_1, \ldots, D_n$ be called *isobaric* if

(1) $C_1 D_1 \cdots C_n D_n = 1;$

in other words, if the vectors $C_1 D_1, \ldots, C_n D_n$ form a closed vector-polygon (have the "sum zero").

The relation of isobarism is a relation between unordered *n*-tuples; that is, it is preserved if the points $C_1, \ldots, C_n$ or the points $D_1, \ldots, D_n$ are permuted among themselves. This statement follows from Rule 1.

Example.   Let $A_1, \ldots, A_n$ be arbitrary points. The *n*-tuple $A_1, \ldots, A_n$ is then isobaric with itself. The proof is trivial (for $A_1 A_1 \cdots A_n A_n = 1$, on the basis alone of the rule $AA = 1$). But this implies the less trivial result: if π is an arbitrary permutation of the digits $1, \ldots, n$, then $A_1 A_{1\pi} \cdots A_n A_{n\pi} = 1$. The *Chasles* relation, written in the form $A_1 A_2 \cdot A_2 A_3 \cdots A_{n-1} A_n \cdot A_n A_1 = 1$, is the special case arising when π is the cyclic permutation $i \to i + 1 \pmod{n}$.

Remark. Every equation between two products of points can be written in the form $A_1 \cdots A_m = 1$. Since by Axiom 1 the product of an odd number of points is never equal to 1, the only interesting case is that m is even. Every equation $A_1 \cdots A_{2n} = 1$ states that two n-tuples are isobaric.

8.1. *The n-tuples $C_1, \ldots, C_n$ and $D_1, \ldots, D_n$ are isobaric if and only if*

(2) $$OC_1 \cdots OC_n = OD_1 \cdots OD_n$$

for all points O. If there exists a point for which (2) holds, then (2) holds for all points O.

Equation (2) states that the vectors $OC_1, \ldots, OC_n$ and $OD_1, \ldots, OD_n$ have the same "sum." Thus the equivalent definition of isobaric provided by §8.1 is also easily visualized; however, since a "subsidiary point" O has been used, this definition is not quite so basic as the other one.

Proof of §8.1. By Rule 3, we have the identity

(3) $$(OC_1 \cdots OC_n)(C_1 D_1 \cdots C_n D_n) = OD_1 \cdots OD_n.$$

From this identity (3) we see that if (2) holds for a point O, then (1) holds. If (1) holds, then (2) holds for every point O.

Let us now state certain formal properties of the isobarism relation; these properties are easily verified on the basis of the definition, or in some cases even more easily by §8.1:

(i) *If $C_1, \ldots, C_n$ is isobaric to $D_1, \ldots, D_n$, then every permutation of $C_1, \ldots, C_n$ is isobaric to every permutation of $D_1, \ldots, D_n$.*

(ii) *For every fixed n, the isobarism relation is an equivalence relation on the set of n-tuples of points.* Thus instead of (1) we also write

(4) $$C_1, \ldots, C_n \sim D_1, \ldots, D_n.$$

For $n = 1$ isobaric is the same as equal.

(iii) *Let (4) hold. Then if $C_{n+1}, \ldots, C_{n+m} \sim D_{n+1}, \ldots, D_{n+m}$, then $C_1, \ldots, C_n, C_{n+1}, \ldots, C_{n+m} \sim D_1, \ldots, D_n, D_{n+1}, \ldots, D_{n+m}$, and conversely.*

Thus if isobaric n-tuples are adjoined to isobaric m-tuples, the resulting $(n + m)$-tuples are isobaric, and the same remark holds for subtraction. In particular, we have the substitution rule: if in a given n-tuple an m-tuple $(m \leq n)$ is replaced by an m-tuple isobaric to it, then the resulting n-tuple is isobaric to the original n-tuple.

Example. From (4) it follows that

(5) $\qquad C_1, \ldots, C_n, C_1, \ldots, C_n \sim D_1, \ldots, D_n, D_1, \ldots, D_n.$

The foregoing results depend only on the fundamental assumption and on Axiom 1. If Axiom 2 is also used, we have the bisection rule:

(iv) *From* (5) *follows* (4).

Proof. From $(C_1 D_1 \cdots C_n D_n)^2 = 1$ it follows by §1.1 that $C_1 D_1 \cdots C_n D_n = 1$.

Remark. Let $B_1, \ldots, B_n, C_1, \ldots, C_n \sim D_1, \ldots, D_n, D_1, \ldots, D_n$. Then it follows that if two of the n-tuples $B_1, \ldots, B_n$; $C_1, \ldots, C_n$; $D_1, \ldots, D_n$ are isobaric, all three are isobaric to one another.

Proof. If the first and the second n-tuples are isobaric, (5) follows by the substitution rule; but then by (iv) the second and third n-tuples, and thus by (ii) all three n-tuples are isobaric. If one of the first two n-tuples is isobaric to the third, then so is the other, by (iii); but then by (ii) all three n-tuples are isobaric.

Definition. A point S is called a *center of gravity* of an n-tuple of points $C_1, \ldots, C_n$ if

(6) $\qquad\qquad\qquad S, \ldots, S \sim C_1, \ldots, C_n$

(on the left-hand side S is written n times, since otherwise (6) would be un-defined), or in other words if $SC_1 \cdots SC_n = 1$; that is, if the vectors $SC_1, \ldots, SC_n$ form a closed vector polygon.

Since the isobarism relation is an equivalence, we may at once make the following assertions:

(v) *If* $C_1, \ldots, C_n$ *and* $D_1, \ldots, D_n$ *have a common center of gravity, they are isobaric.*

(vi) *If* $C_1, \ldots, C_n$ *and* $D_1, \ldots, D_n$ *are isobaric, then for an arbitrary point S we have: S is a center of gravity of* $C_1, \ldots, C_n$ *if and only if S is a center of gravity of* $D_1, \ldots, D_n$.

Of course, it may happen, for example, that a pair of points and a quad-ruple of points have the same center of gravity, but such a case is not included in the isobarism relation, since this relation is defined only for sets contain-ing the same number of points. However, we may assert, for example:

(vii) *If* $C_1, \ldots, C_n, C_1, \ldots, C_n \sim D_1, \ldots, D_{2n}$, *then S is a center of gravity of* $C_1, \ldots, C_n$ *if and only if S is a center of gravity of* $D_1, \ldots, D_{2n}$.

Proof. By the bisection rule (iv) the relation (6) is equivalent to $S, \ldots, S \sim C_1, \ldots, C_n, C_1, \ldots, C_n$ (where S is written $2n$ times) and thus equivalent by hypothesis to $S, \ldots, S \sim D_1, \ldots, D_{2n}$.

8.2. *S is a center of gravity of A, B if and only if S is a midpoint of A, B.*

Proof. $S, S \sim A, B$ is equivalent to $SASB = 1$, or in other words, to $A^S = B$.

8.3.   *Let there be given n points and let them be divided into two sets*
$C_1, \ldots, C_k$ *and* $D_1, \ldots, D_l$ $(k + l = n)$. *Let A be a center of gravity of the*
first set and B a center of gravity of the second, so $A, \ldots, A \sim C_1, \ldots, C_k$
and $B, \ldots, B \sim D_1, \ldots, D_l$. *Then it follows that S is a center of gravity of*
$C_1, \ldots, C_k D_1, \ldots, D_l$ *if and only if*

(7) $S, \ldots, S \sim A, \ldots, A, B, \ldots, B$ (*A is written k times, and B is written*
l times);

that is, if $(SA)^k(SB)^l = 1$, *or in other words if S is a center of gravity of the*
k-fold covered point A and the l-fold covered point B.

Proof.   By the hypothesis it follows from (iii) that $A, \ldots, A, B, \ldots, B$
$\sim C_1, \ldots, C_k, D_1, \ldots, D_l$ (*A is written k times and B is written l times*).
Examples.   a)   *Let A, B, C be a triangle* (Fig. 29) *and U be a midpoint of B, C.*
Then S is a center of gravity of A, B, C if and only if $SA(SU)^2 = 1$, *so that* $AS =$
$(SU)^2$; *that is, if S divides the segment AU in the ratio 2:1.* (The first set consists
of *A*, and the second set consists of *B, C*.)
b)   *Let there be given a tetrahedron* (Fig. 30) *in which P is a vertex and T is a*
center of gravity of the three other vertices. Then S is a center of gravity of the
tetrahedron if and only if $SP(ST)^3 = 1$, *so that* $PS = (ST)^3$; *that is, if S divides the*
segment PT in the ratio 3:1.
Exercises.   1.   The relation (6) is equivalent to $S^{C_1}, \ldots, S^{C_n} \sim C_1, \ldots, C_n.$
2.   Every 2^m points have exactly one center of gravity.
3.   If $C_1, \ldots, C_n$ are given, then for every point X there exists exactly one
point X^* such that $C_1, \ldots, C_n \sim X, \ldots, X, X^*$; thus $X \to C_1 X C_2 X C_3 \cdots$
$C_{n-1} X C_n$ is a mapping of $\mathfrak{P}$ onto itself which remains unchanged if the given
n-tuple is replaced by another *n*-tuple isobaric to it. Investigate the properties of
this mapping.

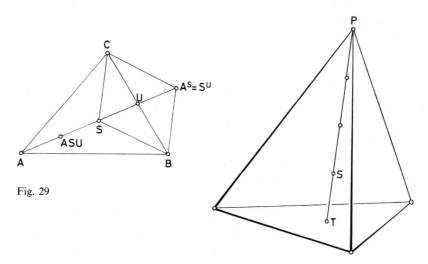

Fig. 29

Fig. 30

9.   Theorems on Centers of Gravity

9.1.   *M is a center of gravity of $C_1, \ldots, C_n, C_1^M, \ldots, C_n^M$.*

Proof.   We have $M, M \sim C_i, C_i^M$ (cf. §8.2) and thus it follows from (iii) and (i) that $M, \ldots, M \sim C_1, \ldots, C_n, C_1^M, \ldots, C_n^M$.

Example.   For every $2n$-gon whose pairs of opposite vertices have a common midpoint (for example, a parallelogram or an A-hexagon), this common midpoint is a center of gravity.

9.2.   *Let there be given $2n$ points. Arrange these $2n$ points in n pairs of points, and for each pair construct the center of gravity (midpoint). Then S is a center of gravity of the given points if and only if S is a center of gravity of the centers of gravity (midpoints) of the pairs.*

Proof.   Let the given points be denoted by $C_1, \ldots, C_{2n}$ in such a way that $(C_1, C_2), \ldots, (C_{2n-1}, C_{2n})$ are the pairs in question. Let $A_1, \ldots, A_n$ be the midpoints of the pairs. Then (cf. §8.2) we have $A_1, A_1 \sim C_1, C_2; \ldots;$ $A_n, A_n \sim C_{2n-1}, C_{2n}$, and thus it follows from (iii) and (i) that $A_1, \ldots, A_n$, $A_1, \ldots, A_n \sim C_1, C_2, \ldots, C_{2n-1}, C_{2n}$. The assertion then follows from (vii).

§9.1 follows from §9.2. Further examples for §9.2 are
   a)   *For a quadrangle*: S is a center of gravity if and only if S is the midpoint of midpoints of opposite sides.
   b)   *For a tetrahedron*: S is a center of gravity if and only if S is the midpoint of midpoints of opposite edges.
   c)   *For a hexagon*: S is a center of gravity if and only if S is a center of gravity of the triangle consisting of the midpoints of the pairs of opposite vertices (for this triangle cf. §5.3).

Let us now consider n-gons.

9.3.   a)   *An n-gon and its midpoint n-gon are isobaric.*
   b)   *An n-gon and its derivative are isobaric.* In the notation of §4 we thus have for every n-gon $\mathscr{A}_n$: $\mathscr{A}_n \sim \mathscr{A}_n^0 \sim \mathscr{A}_n' \sim \mathscr{A}_n^{0\prime}$.

Proof of a).   Let $A_1^0, A_2^0, \ldots, A_n^0$ be the midpoint n-gon of $A_1, A_2, \ldots,$ A_n. By §8.2 we have $A_1^0, A_1^0 \sim A_1, A_2; A_2^0, A_2^0 \sim A_2, A_3; \cdots; A_n^0, A_n^0 \sim A_n,$ A_1, and thus it follows from (iii) and (i) that $A_1^0, \ldots, A_n^0, A_1^0, \ldots, A_n^0 \sim$ $A_1, \ldots, A_n, A_1, \ldots, A_n$. Consequently, by the bisection rule (iv) we have $A_1^0, \ldots, A_n^0 \sim A_1, \ldots, A_n$.

Proof of b).   Let $A_1', A_2', \ldots, A_n'$ be the derivative of $A_1, A_2, \ldots, A_n$. Then $A_1 A_1' A_2 A_2' \cdots A_n A_n' = A_n A_1' A_1 A_2' \cdots A_{n-1} A_n' = A_n A_n A_1 A_2 \cdot$ $A_1 A_1 A_2 A_3 \cdots A_{n-1} A_{n-1} A_n A_1 = A_1 A_2 \cdot A_2 A_3 \cdots A_n A_1 = 1$ (Relation of Chasles).

Example. The fact, already mentioned, that for an A-hexagon $\mathscr{A}_6$ the common midpoint M of the pairs of opposite vertices is a center of gravity can be proved from §9.3 in the following way: by §6.5 we have $\mathscr{A}_8^{0'} = \{M, M, M, M, M, M\}$. Thus it follows from $\mathscr{A}_6 \sim \mathscr{A}_8^{0'}$ that $A_1, \ldots, A_6 \sim M, \ldots, M$.

9.4. We have seen that the $2n$-gons with the property

(C) *the product of the vertices is equal to* 1

constitute precisely the midpoint $2n$-gons (cf. §§4.1 and 4.2). The property (C) now admits a new interpretation: it is equivalent to

(C′) *the two alternate partial n-gons are isobaric.*

For $n = 2$, cf. §2.3. Thus if we take an arbitrary n-gon $A_1, \ldots, A_n$ and construct for it an isobaric n-gon $B_1, \ldots, B_n$ (for example, by §9.3), then $A_1, B_1, A_2, B_2, \ldots, A_n, B_n$ is a $2n$-gon with the property (C). Moreover, if we permute the points $A_1, \ldots, A_n$, or the points $B_1, \ldots, B_n$, arbitrarily among themselves, we obtain a $2n$-gon with the property (C).

For a given $2n$-gon $\mathscr{A}_{2n} = \{A_1, \ldots, A_{2n}\}$ we may consider not only the two alternate partial n-gons $A_1, A_3, \ldots, A_{2n-1}$ and $A_2, A_4, \ldots, A_{2n}$, but also a further n-gon, namely, the *n-gon of the midpoints* $M_1, M_2, \ldots, M_n$ *of the pairs of opposite vertices* (for $n = 3$, cf. §5.3). The property

(C″), *one of the partial n-gons and the n-gon consisting of the midpoints of opposite vertices are isobaric to each other,*

is equivalent to (C′), and thus also to (C).

Proof. Let M_i be the midpoint of A_i, A_{i+n}, so that $A_i, A_{i+n} \sim M_i, M_i$ (cf. §8.2). Then from (iii) and (i) it follows that

$A_1, A_3, \ldots, A_{2n-1}, A_2, A_4, \ldots, A_{2n} \sim M_1, \ldots, M_n, M_1, \ldots, M_n$.

Consequently, from the remark in (iv) it follows from (C′) or also from (C″) that all three of the n-gons in question are isobaric. Thus (C′) and (C″) are equivalent.

9.5. A refinement of §9.3, b), which we shall here formulate only for a hexagon $\mathscr{A}_6$, is given by the following two relations between the alternate part-triangles of $\mathscr{A}_6$ and $\mathscr{A}_6$:

(1) $A_1 A_1' A_3 A_3' A_5 A_5' = (A_2 A_2' A_4 A_4' A_6 A_6')^{-1} = (A_1 A_2 A_3 A_4 A_5 A_6)^2$

(2) $A_1 A_2' A_3 A_4' A_5 A_6' = A_1' A_2 A_3' A_4 A_5' A_6 = (A_1 A_2 A_3 A_4 A_5 A_6)^{-1}$.

Now let $\mathscr{A}_6$ be an A-hexagon and let A_8 be its eighth parallelepiped point. Then $A_2' = A_4' = A_6' = A_8$ (cf. §6.5) and thus it follows from (2) and (1) that

(3) $(A_8 A_1 A_8 A_3 A_8 A_5)^2 = (A_1 A_2 A_3 A_4 A_5 A_6)^2 = A_8 A_2 A_8 A_4 A_8 A_6$.

Thus (from §1.1) we see that: *if A_8 is a center of gravity of an alternate part-triangle of $\mathscr{A}_6$, then A_8 is also the center of gravity of the other alternate part-triangle; this will be the case if and only if the A-hexagon $\mathscr{A}_6$ is an AC-hexagon.*

9.6. We now give a supplement to §7 and a characterization of a special case of the geometric situation considered there.

As in §7 we discuss the three entities: a quadruple of points A, B, C, P, the tetrahedron P, P^A, P^B, P^C, and the A-hexagon $\mathscr{A}_6 (A, B, C; P)$ defined by

§7.1, (1), whose vertices are the midpoints of the edges of the tetrahedron (cf. §7.2). The opposite vertices of the A-hexagon are the midpoints of opposite edges of the tetrahedron, and their midpoint M is the center of gravity of the tetrahedron (Example for §9.2). We investigate the special case where the vertex P of the tetrahedron coincides[12] with the center of gravity M of the tetrahedron, and show that for arbitrary points A, B, C, P the following statements are equivalent to one another:

1) *P is a center of gravity of A, B, C;*

2) $\mathscr{A}_6(A, B, C; P)$ *is an AC-hexagon;*

3) *P is a common midpoint of the pairs of opposite vertices of* $\mathscr{A}_6(A, B, C; P)$;

4) *P is a center of gravity of the tetrahedron P, P^A, P^B, P^C* (Fig. 31).

Proof 1.   The eighth parallelepiped point of $\mathscr{A}_6(A, B, C; P)$ is P (cf. §7.1). Thus by §9.5 the statements 1) and 2) are equivalent, and by §6.5 the statements 2), 3) are equivalent. Since 3) and 4) both state that $P = M$, the two statements 3) and 4) are equivalent.

In order to encourage the reader to prove theorems of our theory without reference to preceding theorems but merely on the basis of the definitions and by explicit calculation with the calculus defined in §1, we now give a second proof.

Proof 2.   1) states by definition that

(4) $$PAPBPC = 1.$$

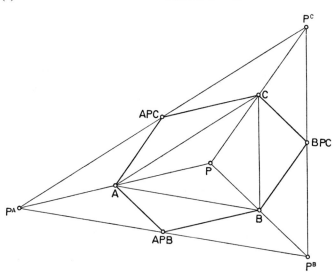

Fig. 31

[12]In Euclidean space this figure is then necessarily plane; it will be recognized as the figure for the proof of the median theorem.

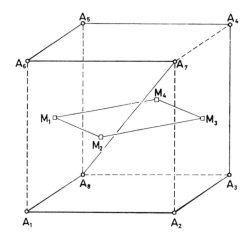

Fig. 32

Let us show that 2), 3), and 4) are equivalent to (4). As for 2), we have that the product of the vertices of $\mathcal{A}_6$ $(A, B, C; P)$ is $A \cdot APB \cdot B \cdot BPC \cdot C \cdot CPA = PBPCPA = PAPBPC$, as follows from Rule 3; this product is equal to 1 if and only if (4) holds. As for 3), A, BPC is a pair of opposite vertices of $\mathcal{A}_6$ $(A, B, C; P)$; the point P is the midpoint of this pair of points if and only if $A^P = BPC$, or in other words if (4) holds. As for 4), we have: 4) states by definition that $PPPP^A PP^B PP^C = 1$; this equation, since by Rule 3 the product on the left gives $(PAPBPC)^2$, is equivalent to $(PAPBPC)^2 = 1$, and thus by §1.1 it is equivalent to (4).

Exercises.   1.   The product of the vertices of an octagon is equal to 1 if and only if the midpoints of opposite vertices form a parallelogram. The center of gravity of this parallelogram is then the center of gravity of the octagon (Fig. 32). For examples see §§6.5 and 9.4.

2.   Let a triangle A, B, C be given $X \rightarrow AXBXC$ be a one-to-one mapping of **P** onto itself. By this mapping A, B, C is mapped onto the derived triangle. The image of X is the seventh parallelepiped point of the A-hexagon $\mathcal{A}_6$ $(A, B, C; X)$; the preimage of X is the center of gravity of the tetrahedron X, X^A, X^B, X^C. Every triple isobaric to A, B, C defines the same mapping.

3. (Completion of §8, Problem 2).   a) $\mathfrak{U}_n = \{\alpha \in \mathfrak{V}, \alpha^n = 1\}$ is a subgroup of $\mathfrak{V}$.

b)   For the set $\mathfrak{S}$ of centers of gravity of an n-gon we have in general: If $S \in \mathfrak{S}$, then $\mathfrak{S} = S\mathfrak{U}_n$. If $\mathfrak{S}$ is not empty, then, in the notation of §10, $(\mathfrak{U}_n \cup \mathfrak{S}, \mathfrak{S})$ is a model of the axiom system.

10.   Models

The axiom system in §1 refers to a *"generated group"*; that is, to a pair $(\mathfrak{G}, \mathfrak{P})$, where $\mathfrak{G}$ is a group and $\mathfrak{P}$ is a system of generators of $\mathfrak{G}$. Every generated group $(\mathfrak{G}, \mathfrak{P})$ satisfying the fundamental assumption (that is, the condition: $\mathfrak{P}$ is not empty and every element of $\mathfrak{P}$ is involutory) and also satisfying Axioms 1 through 3, will be called a *model of the axiom system*.

Let us first consider a generated group $(\mathfrak{G}, \mathfrak{P})$ satisfying the fundamental assumption and Axiom 1. Let $\mathfrak{B}$ be the vector group defined in §1; that is, the group of products of two elements of $\mathfrak{P}$. Then $\mathfrak{B}$ is an Abelian subgroup of $\mathfrak{G}$ of index 2 and $\mathfrak{P}$ is the coset of $\mathfrak{B}$. We have $(\mathfrak{G}, \mathfrak{P}) = (\mathfrak{B} \cup \mathfrak{P}, \mathfrak{P})$. If A is an arbitrary element of $\mathfrak{P}$, then $\mathfrak{P} = A\mathfrak{B}$ and A satisfies the relations

(1) $$AA = 1 \text{ and } \alpha^A = \alpha^{-1} \text{ for all } \alpha \in \mathfrak{B}.$$

Conversely, let $\mathfrak{B}$ be an arbitrary Abelian group. Let us imagine a larger group including $\mathfrak{B}$ and write it, in the form $\mathfrak{B} \cup \mathfrak{P}$, where the set of elements not in $\mathfrak{B}$ has been denoted by $\mathfrak{P}$. The group $\mathfrak{B} \cup \mathfrak{P}$ will be called a D-*extension* of $\mathfrak{B}$ if 1) the group $\mathfrak{B}$ *is a subgroup of index 2 in the group* $\mathfrak{B} \cup \mathfrak{P}$ (so that $\mathfrak{P}$ is the coset of $\mathfrak{B}$) and 2) *every element of* $\mathfrak{P}$ *is involutory*. These conditions, 1) and 2), imply that every element $A \in \mathfrak{P}$ satisfies the relations (1).

For every Abelian group $\mathfrak{B}$ *a* D-*extension can be constructed*, for we may take any new element A and multiply the elements of the group $\mathfrak{B}$ and those of the set $A\mathfrak{B}$ of the formal products $A\alpha(\alpha \in \mathfrak{B})$ in accordance with (1) by the rules

(2) $$\alpha \cdot \beta = \alpha\beta, \quad \alpha \cdot A\beta = A(\alpha^{-1}\beta), \quad A\alpha \cdot \beta = A(\alpha\beta),$$
$$A\alpha \cdot A\beta = \alpha^{-1}\beta \ (\alpha, \beta \in \mathfrak{B}).$$

This multiplication is associative, as is proved from the fact that $\mathfrak{B}$ is commutative. Every element of $A\mathfrak{B}$ is inverse to itself: $A\alpha \cdot A\alpha = \alpha^{-1}\alpha = 1$. Also, $\mathfrak{B} \cup A\mathfrak{B}$ is a group and, more precisely, is a D-extension of $\mathfrak{B}$. *Every two* D-*extensions of a fixed Abelian group* $\mathfrak{B}$ *are isomorphic over* $\mathfrak{B}$. If $\mathfrak{B} \cup \mathfrak{P}$ and $\mathfrak{B} \cup \mathfrak{P}'$ are D-extensions of $\mathfrak{B}$ and $A \in \mathfrak{P}$, $A' \in \mathfrak{P}'$, then the mapping $\alpha \to \alpha$, $A\alpha \to A'\alpha$ ($\alpha \in \mathfrak{B}$) is an isomorphism of the D-extensions.

For every D-extension $\mathfrak{B} \cup \mathfrak{P}$ of an Abelian group $\mathfrak{B}$, the set $\mathfrak{P}$ is a system of generators (every element of $\mathfrak{B}$ can be represented as the product of two elements of $\mathfrak{P}$). We shall can $\mathfrak{P}$ the *canonical system of generators of the* D-*extension*. Then the canonically generated D-extension $(\mathfrak{B} \cup \mathfrak{P}, \mathfrak{P})$ satisfies the fundamental assumption and Axiom 1.

Thus we have the result: *the generated groups which satisfy the fundamental assumption and Axiom 1 constitute precisely the canonically generated* D-*extensions of the Abelian groups*. Up to this point therefore, our theory has exactly the same generality as the theory of Abelian groups and can be called the *geometry of Abelian groups*. It is only Axioms 2 and 3 that produce a restriction.

If $\mathfrak{B}$ is an arbitrary Abelian group, the mapping

(3) $$\xi \to \xi^2$$

is a homomorphism of $\mathfrak{B}$ into $\mathfrak{B}$. A canonically generated D-extension of

$\mathfrak{B}$ satisfies Axiom 2 if and only if the mapping (3) of $\mathfrak{B}$ into itself is one-to-one, and satisfies Axiom 3 if and only if (3) is a mapping of $\mathfrak{B}$ onto $\mathfrak{B}$; thus the D-extension satisfies both the Axioms 2 and 3 if and only if the mapping (3) is an automorphism of $\mathfrak{B}$. We then have the following theorem.

Theorem: *The models of the axiom system are precisely the canonically generated* D-*extensions of those Abelian groups for which* (3) *is an automorphism.*

Example. Let $\mathfrak{B} = \mathfrak{B}_n(K)$ be the n-dimensional vector space over a field K of characteristic $\neq 2$. The canonically generated D-extension of $\mathfrak{B}$ is a model of the axiom system. The canonical system $\mathfrak{P}$ of generators, with the translation group defined on it in accordance with §1 and isomorphic to $\mathfrak{B}$, is an n-dimensional affine space[13] corresponding to the vector space $\mathfrak{B}$.

It is clear that we can also approach the theory of n-gons by starting from a vector space of arbitrary dimension over a field and defining an n-gon as an n-tuple of vectors. A further treatment of the theory of n-gons, from this point of view, will be published elsewhere.[14]

---

[13]In the sense of N. Bourbaki, Algèbre, Chap. 2: Algèbre linéaire. Paris 1962, § 9 Def. 1.

[14]F. Bachmann and E. Schmidt, n-Ecke (BI-Hochschultaschenbuch). In preparation.

CHAPTER 3

Affine and Projective Planes

Introduction

The next four chapters deal with the foundations of geometry and are thus concerned with the same range of ideas as the famous work of David Hilbert. For centuries geometry had served as the model of an exact science, since the time of Euclid, who had tried to lay a rigorously logical foundation. From our modern point of view his attempt was marked by many imperfections and omissions; nevertheless the realization that the inherent nature of the subject matter calls for such an altogether abstract treatment remains one of the immense intellectual achievements of the Greeks. In modern times Hilbert has given a complete system of axioms for Euclidean geometry and thus has put the subject on a firm basis. Imperfections and omissions have been removed, at least so far as we can judge at the present time. Of course, it remains an open question whether the future will not demand a still higher degree of precision and logical exactness.

The axiomatization of Euclidean geometry by David Hilbert not only represents the final stage of a geometric tradition that had lasted for millennia, but has also opened up new problems in the investigation of geometry. For if we weaken the axioms, as is suggested by the discovery of non-Euclidean geometry, we obtain many new geometries, whose individual existence is perhaps less important than the immense fund of attractive and interesting interrelationships among them. The process of weakening the axioms leads to a careful study of the actual significance of each of them, with a resulting gain in clarity for many of the theorems. The various questions here, some of which were already raised by Hilbert, have been investigated and extended to such an extent that the study of the foundations of geometry is now divided into two very extensive parts. The first part deals with incidence structures, in particular the affine and projective planes, and the second part with metric structures, for example, the metric planes, of which the Euclidean and non-Euclidean planes are special cases.

This chapter, which is to be regarded as an introduction to the foundations of affine and projective geometry, deals with the theory of affine and projective planes, and includes a section on projective spaces. The subject matter of affine and projective planes forms, as it were, the core of plane geometry since, apart from the axioms guaranteeing a sufficient number of points, it requires only the existence and uniqueness of lines joining two distinct points, together with the condition that there exists, in the affine case, a parallel to a given straight line through a given point and, in the projective case, a unique intersection of two distinct lines.

The chapter begins with the axiomatic construction of the affine planes and then proceeds to prove some of the characteristic propositions in the richly developed theory of closure theorems and collineation groups, both of which are particularly important for the structure of such planes. Within the very general class of affine planes as a whole, the validity of certain closure theorems (for example, the theorem of Desargues) and, on the other hand, the existence of certain collineations (for example, translations) determine in each case a definite class of affine planes (Desargues planes, translation planes).

The second section turns to the problem of algebraization of the affine planes and provides an answer to the fundamental question of the connection between a purely synthetic construction of geometry and an analytic construction. Analytic geometry is based on the concept of number and defines the objects of geometry by numbers and equations; for example, a point in the plane is defined as an ordered pair of numbers and a line is defined by a linear equation, so that algebraic calculation with numbers corresponds to proof of geometric theorems. On the other hand, in synthetic geometry we deal only with points, lines, incidence, parallelism, congruence and at most order, none of which are directly related to numbers or calculation. Of course, it has long been a habit to associate each line segment with a number, namely its length, but we must not forget that from an abstract logical point of view the two concepts are quite distinct. However, we will see in §2 that the connection between synthetic and analytic geometry is surprisingly simple and can be described in an esthetically satisfactory way.

In §3 the concept of an affine plane will be extended to that of a projective plane, with a short discussion of certain special features of the theory of projective planes.

After a short axiomatization of three-dimensional projective space in §4, we turn in §5 to the concept of a projective-metric plane, in preparation for the next chapter.

I. Affine Planes

1.1. *A System of Axioms for an Affine Plane*

If in ordinary Euclidean geometry we consider only those theorems that deal with incidence and parallelism of points and lines, we are led to the concept of the affine plane. The attempt to define this class of theorems more precisely leads to the following system of axioms.

Let $\mathfrak{P}, \mathfrak{G}$ be two sets, and let the elements of $\mathfrak{P}$ be called points, and those of $\mathfrak{G}$ be called lines. Also, let there be given a relation I between points and lines expressed as follows: *"The point $\mathfrak{P}$ is incident with the line g,"* in symbols P I g (to be read, P incident with g). The notation g I P has exactly the same meaning as P I g.

Two lines g, h are said to be *parallel*, in symbols $g \parallel h$, if either $g = h$ or g, h are not both incident with any point.

The triple $\mathfrak{A} = (\mathfrak{P}, \mathfrak{G}, I)$ is called an affine plane if the following axioms hold:

A1: *For two distinct points P, Q there exists exactly one line incident with P and Q.*

A2: *For every line g and every point P there exists exactly one line incident with P and parallel to g.*

A3: *There exist three points not incident with the same line.*

This system of axioms is of such great generality that no interesting theorems have been deduced from it alone. So we introduce suitable additional axioms and then study the affine planes that satisfy the new axioms as well. In what follows we shall present some of the many interesting results.

But first we require certain simple consequences of the Axioms A1 through A3, which we shall now deduce. If P I g, we shall say for variety of expression that P lies on g, or that g goes through P; and if g I P, Q (i.e., if g I P and g I Q), with $P \neq Q$, then g will be called the *line joining P, Q* and will be denoted by (P, Q). Similarly, if P I g, h, $g \neq h$, then P is called the *intersection* of the lines g, h and is denoted by (g, h). Three points are *collinear* if they are incident with the same line, and three lines are *copunctual* if they are incident with the same point.

Lemma 1: *Two nonparallel lines g, h have exactly one intersection.*

Proof.   By hypothesis we have $g \neq h$, so that there exists at least one point P with P I g, h. If there existed a point P', with $P' \neq P$ and P' I g, h, then P, P' would be joined by two distinct lines g and h, in contradiction to Axiom A1.

Lemma 2: *Parallelism is an equivalence relation.*

Proof.   Reflexivity and symmetry follow immediately from the definition. As for transitivity: let $g \parallel h$ and $h \parallel l$, where we may assume that $g \neq l$. From the assumption that g and l have a common point P it follows that through P there pass two parallels g and l to h. But then Axiom A2 states $g = l$, in contradiction to the hypothesis. Thus $g \parallel l$.

The equivalence classes defined by parallelism as an equivalence relation are called *parallel pencils*. The parallel pencil containing the line g will be denoted by Π_g, so that $\Pi_g = \{h \mid g \parallel h\}$. Clearly,

(i)   *from $h \in \Pi_g$ follows $\Pi_h = \Pi_g$.*
(ii) *from $r \in \Pi_g$, Π_h follows $\Pi_g = \Pi_h$.*

If $g \nparallel h$ (g is nonparallel to h), we shall say that g and h *intersect*.

Lemma 3:   *If the line g intersects the line h then every line r parallel to g also intersects h.*

Proof.   By hypothesis, $g \nparallel h$. If we had $h \parallel r$, then since $r \parallel g$ it would follow from Lemma 2 that $g \parallel h$. Thus $h \nparallel r$.

A *triangle* is a set of three noncollinear points; the lines joining each pair of points are called the *sides* of the triangle. A *quadrangle* is a set of four points such that no three are collinear; the lines joining pairs of points are called the sides of the quadrangle.

Lemma 4:   *On every line there lie at least two points.*

Proof.   By Axiom A3 there exist at least three noncollinear points A, B, C (Fig. 1). The sides of the triangle ABC are pairwise nonparallel. A line g distinct from the sides of the triangle is nonparallel to at least two of its sides, say $(A, C) = a$ and $(B, C) = b$ and therefore intersects them at points S and T. If $S \neq T$, the desired assertion is already proved. But since $S, C \ \mathrm{I} \ a, b$, it follows from $S = T$ by Lemma 1 that $S = C$ and thus $C \ \mathrm{I} \ g$. Then by Lemmas 3 and 1 the parallel d to a through B cuts the line g in exactly one point R. From $R = C$ it would follow from $C \ \mathrm{I} \ d, a$ by Axiom A2 that $d = a$ and therefore $A, B, C \ \mathrm{I} \ a$, in contradiction to the assumption on A, B, C. Thus the two distinct points C and R both lie on g.

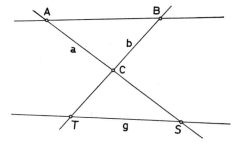

Fig. 1

Consequence.   *Two lines g, h intersect if and only if they have exactly one point in common.*

By means of two examples we shall now show that affine planes actually exist:

a)   It is well known that in the (ordinary) Euclidean plane the affine incidence Axioms A1 through A3 are valid if the concepts point, line, incidence, and parallel are defined in the usual way.

b)   We can construct a model of an affine plane $\mathfrak{A} = (\mathfrak{P}, \mathfrak{G}, I)$ which consists of only finitely many points and lines and is in fact the smallest possible model: let $\mathfrak{P}$ be a set of four elements denoted by A, B, C, D, and let $\mathfrak{G}$ be a set of six elements denoted by a, b, c, d, e, f. The crosses in the accompanying table indicate which elements are incident.

	a	b	c	d	e	f
A	×			×	×	
B	×	×				×
C		×	×		×	
D			×	×		×

It is easy to verify the incidence Axioms A1 through A3. Figure 2 gives an illustration of the model.

As stated above this model $\mathfrak{A}$ is the smallest model of an affine plane.

1.2.   *Closure Theorems*

Among the theorems in ordinary Euclidean geometry involving only points, lines, incidence, and parallelism, the "closure theorems," asserting that a given figure closes on itself, are of particular importance for the theory of affine planes. Let us mention the following examples:

1°   Theorem (P) of Pappus-Pascal (affine form):   *If in a hexagon $P_1Q_2P_3Q_1P_2Q_3$ the vertices lie alternately on two lines g, h, but no vertex*

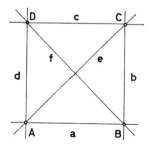

Fig. 2

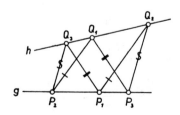

Fig. 3

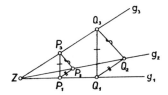

Fig. 4

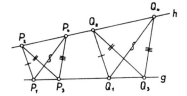

Fig. 5

*lies on both these lines, and if two pairs of opposite sides are parallel, then
the remaining pair is also parallel: From* P_i I g, I h; Q_i I h, I g *for* $i = 1, 2, 3$
and $(P_1, Q_2) \parallel (P_2, Q_1); (P_1, Q_3) \parallel (P_3, Q_1)$ *follows* $(P_2, Q_3) \parallel (P_3, Q_2)$ (Fig.
3). We shall call g, h the *supporting lines* of the configuration.

2° Theorem (D) of Desargues (affine form): *If corresponding vertices of
two triangles lie in each case on exactly one of three copunctual lines and if two
pairs of corresponding sides are parallel, then the third pair of corresponding
sides is also parallel: From* Z I g_i; P_i, Q_i I g_i, I g_4 *for* $i, k = 1, 2, 3$ *and*
$i \neq k$ *and* $(P_1, P_2) \parallel (Q_1, Q_2), (P_1, P_3) \parallel (Q_1, Q_3)$ *follows* $(P_2, P_3) \parallel (Q_2, Q_3)$
(Fig. 4). The point Z is called the *center of the Desargues configuration*, and
the lines g_1, g_2, g_3 are its supporting lines.

3° Theorem (S) (scissors theorem): *If the vertices of two quadrangles lie
alternately on two lines, no vertex lying on both lines, and if three pairs of
corresponding sides are parallel, then the fourth pair is also parallel: From*
P_i, Q_i I g, I h *for* $i = 1, 3$ *and* P_j, Q_j I h, I g *for* $j = 2, 4$ *and* $(P_1, P_2) \parallel$
$(Q_1, Q_2), (P_2, P_3) \parallel (Q_2, Q_3), (P_3, P_4) \parallel (Q_3, Q_4)$ *follows* $(P_1, P_4) \parallel (Q_1, Q_4)$
(Fig. 5). Here g, h are the supporting lines of the configuration.

   If the supporting lines in these theorems are parallel, we speak of the
"little theorem (p) of Pappus-Pascal," of the "little theorem (d) of Desargues," and of the "little scissors theorem (s)." The theorems (p) and (s) are
special cases of (P) and (S). It will be shown below that (d) follows from (D).

4° Theorem (D*): *If the vertices of two triangles* $P_1 P_2 P_3$, $Q_1 Q_2 Q_3$ *lie on three
distinct parallel lines* $g_1 g_2 g_3$, *then either there exists a line* a, *the axis of the
figure, such that each pair of corresponding sides belongs to the same pencil of
lines as* a, *or else corresponding sides of the triangles are parallel* (Fig. 6).

   The question now arises whether any of these theorems is valid in an arbitrary affine plane. The answer is in the negative. For each of the theorems
(P), (D), (S), (p), (d), (s), (D*) we can construct an affine plane in which the
given theorem is false. But it remains an important problem to investigate the
logical interdependence of these theorems. In this connection we shall prove
the following theorems.

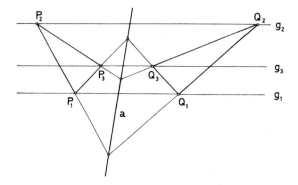

Fig. 6

Theorem 1 :   *From* (d) *follows* (p).[1]

Proof.   Let the conditions of (p) be satisfied and let $P_1 \neq P_2$ and $P_2 \neq P_3$ (for $P_1 = P_2$ or $P_2 = P_3$ the desired assertion follows at once) (Fig. 7). Let S be the intersection of the parallel to (Q_1, P_2) through Q_3 and the parallel to (Q_1, P_3) through Q_2. Since $(Q_1, P_2) \nparallel (Q_1, P_3)$ the point S exists by Lemma 3. By applying (d) to the triangles $P_2 P_1 Q_3$, $Q_1 Q_2 S$ we see that $(P_2, Q_3) \parallel (S, Q_1)$, and by applying (d) to $Q_3 Q_1 S$ and $P_1 P_3 Q_2$ we obtain $(P_3, Q_2) \parallel (Q_1, S)$. Thus $(P_2, Q_3) \parallel (P_3, Q_2)$.

In contrast to theorems (p) and (s), which are special cases of (P) and (S), the statement of (d) is not included in (D). However, we have

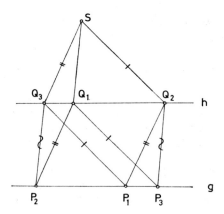

Fig. 7

[1]Whether, conversely, (p) implies (d) is still an unsolved problem.

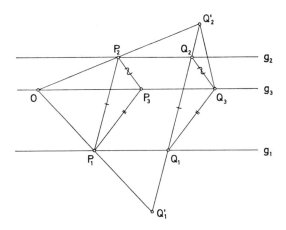

Fig. 8

Theorem 2: *From* (D) *follows* (d).

Proof. Let the assumptions of (d) be satisfied and let $P_1 \neq Q_1$ (otherwise the assertion follows at once from the affine axioms). We must show that $(P_2, P_3) \parallel (Q_2, Q_3)$. By Lemma 3 the line parallel to (P_2, P_3) through Q_3 cuts (Q_1, Q_2) in a point Q_2' (Fig. 8). If we now assume $Q_2 \neq Q_2'$, then by Lemma 3 there would exist a point of intersection O of (P_2, Q_2') and g_3 and also, by the same lemma, a point of intersection Q_1' of (O, P_1) and (Q_1, Q_2) with $Q_1' \neq Q_1$. From Theorem (D) applied to the triangles $P_1P_2P_3$ and $Q_1'Q_2'Q_3'$ it would follow that $(Q_1', Q_3) \parallel (P_1, P_3)$, so that there would exist two distinct parallels to (P_1, P_3) through Q_3, in contradiction to Axiom A2. Thus the assumption is false, and the assertion is proved.

Theorem 3: *The theorems* (D) *and* (S) *are equivalent.*

Proof. A) From (D) follows (S). Now let the hypotheses of (S) be satisfied (Fig. 9). Let us first assume that $g \nparallel h$ and that O is the intersection of g and h. By §1, Lemma 2, the assertion will follow if $(P_2, P_3) \parallel (P_1, P_4)$ and $(Q_2, Q_3) \parallel (Q_1, Q_4)$. Without loss of generality, we may therefore assume that $(P_2, P_3) \nparallel (P_1, P_4)$, since if $(Q_2, Q_3) \nparallel (Q_1, Q_4)$ we may simply interchange the quadrangles.

Let S be the intersection of (P_2, P_3) and (P_1, P_4). Also, let T be the intersection (which exists by Lemma 3) of (O, S) and (Q_2, Q_3). Applying (D) to the triangles SP_3P_4, TQ_3Q_4 we obtain $(S, P_4) \parallel (T, Q_4)$. For the triangles SP_2P_1, TQ_2Q_1 it follows from (D) that $(S, P_1) \parallel (T, Q_1)$; that is, $(T, Q_1) = (T, Q_4) = (Q_1, Q_4)$, which proves the assertion.

On the other hand, if $g \parallel h$, we may apply the argument of (d), as is possible by Theorem 2.

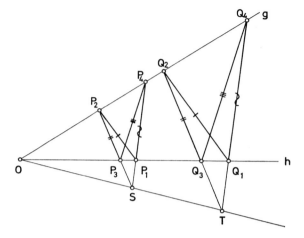

Fig. 9

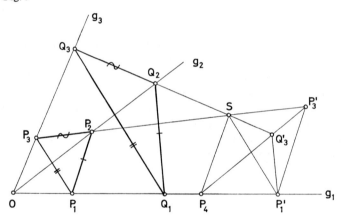

Fig. 10

B)   From (S) follows (D). For let the hypothesis of (D) be satisfied, where we may obviously assume that $P_3 \neq Q_3$. We first show:

(*) Either $(P_2, P_3) \parallel (Q_2, Q_3)$ or their intersection lies on g_1 (Fig. 10).

Let us assume that $(P_2, P_3) \not\parallel (Q_2, Q_3)$ and that the intersection S of (P_2, P_3) and (Q_2, Q_3) does not lie on g_1. Then let P_4 be the intersection (which exists by §1, Lemma 3) of g_1 and the line parallel to g_3 through S, let P_1' be the intersection of g_1 and the line parallel to (P_1, P_3) through S, let P_3' be the intersection of (P_2, P_3) and the line parallel to g_2 through P_4, and finally let Q_3' be the intersection of (Q_2, Q_3) and (P_4, P_3'). Applying (S) to the quadrangles $OP_2P_1P_3$, $P_4P_3'P_1'S$ and to $OQ_2Q_1Q_3$, $P_4Q_3'P_1'S$ we obtain $(P_1, P_2) \parallel (P_1', P_3')$ and $(Q_1, Q_2) \parallel (P_1', Q_3')$. Since $(Q_1, Q_2) \parallel (P_1, P_2)$

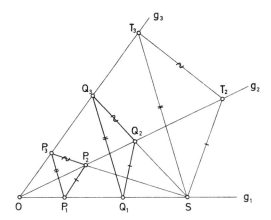

Fig. 11

it follows that $(P_1', P_3') \parallel (P_1', Q_3')$, so that $(P_1', P_3') = (P_1', Q_3')$ and therefore $P_3' = Q_3'$ and $(Q_2, Q_3) = (P_2, P_3)$, in contradiction to the assumption that $(Q_2, Q_3) \nparallel (P_2, P_3)$. Thus S lies on g_1.

Let us now show that the case $S \mathrel{I} g_1$ is not possible: for if $S \mathrel{I} g_1$, we may draw through S the parallels to (P_1, P_3) and to (P_1, P_2), which cut g_3 in T_3 and g_2 in T_2 (Fig. 11). If we now apply (*) to the triangle $P_1 P_2 P_3$ and $S T_2 T_3$, it follows that $(P_2, P_3) \parallel (T_2, T_3)$, since an intersection of (P_2, P_3) and (T_2, T_3) obviously cannot lie on g_1. Similarly for the triangles $Q_1 Q_2 Q_3$ and $S T_2 T_3$ it follows from (*) that $(Q_2, Q_3) \parallel (T_2, T_3)$. From Lemma 2 we then have $(P_2, P_3) \parallel (Q_2, Q_3)$, in contradiction to the assumption that (P_2, P_3) and (Q_2, Q_3) intersect.

Thus the assertion of (D) follows from (*).

In the same way as in Part A of the preceding proof we can show that Theorem (d) implies (s): for if $g \parallel h$, then instead of the line (O, S) we may consider the parallels to g through S, whereupon the desired assertion follows from Theorem (d) in the same way as above.

Theorem 4: *From* (P) *follows* (S).

Proof. Let the assumptions of (S) be satisfied, let R be the intersection with g of the line r parallel to (P_3, P_4) through Q_2, and let S be the intersection with h of the line s parallel to (P_3, P_4) through P_1 (Fig. 12). Then $R \mathrel{I} h$; for from $R \mathrel{I} h$, or in other words from Q_2, $R \mathrel{I} h, r$, it would follow, in view of $Q_2 \neq R$ (since $Q_2 \mathrel{\not I} g$) that $h = r$, so that $(P_3, P_4) \parallel h$ and, by Axiom A2, $(P_3, P_4) = h$, so that $P_3 \mathrel{I} h$, contrary to assumption. In the same way we can show that $S \mathrel{\not I} g$.

Thus P_1, R, Q_1 do not lie on h, and P_2, Q_2, S do not lie on g; and also $(P_2, P_1) \parallel (Q_2, Q_1)$ and $(R, Q_2) \parallel (P_1, S)$. So it follows from (P) for the

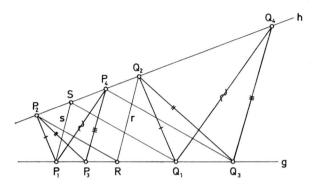

Fig. 12

hexagon $P_2RQ_2Q_1SP_1$ that $(P_2, R) \parallel (S, Q_1)$. Correspondingly, from (P) applied to the hexagon $P_3P_2RQ_2Q_3P_4$ it follows that $(P_2, R) \parallel (P_4, Q_3)$, and finally a third application (P) to the hexagon $P_4P_1SQ_1Q_4Q_3$ shows that $(P_1, P_4) \parallel (Q_1, Q_4)$. Thus (S) is valid.

Consequence. *In an affine plane the theorem of Desargues follows from the theorem of Pappus-Pascal.*

We note that the converse of this assertion is not true. There exist affine planes in which the Desargues theorem holds but not the theorem of Pappus-Pascal.

Theorem 5: *The theorems* (D) *and* (D*) *are equivalent.*

Proof. A)   From (D) follows (D*). Now let the assumptions of (D*) be satisfied. If $(P_1, P_2) \parallel (Q_1, Q_2)$ and $(P_1, P_3) \parallel (Q_1, Q_3)$ the desired assertion follows from (d), which by Theorem 2 is implied by (D). Without loss of generality we may assume that $(P_1, P_3) \nparallel (Q_1, Q_3)$. Also we may assume that $P_i \neq Q_i$ (since otherwise the assertion follows at once). If $(P_1, P_2) \parallel (Q_1, Q_2)$, let a be the parallel to (P_1, P_2) through the intersection R_2 of (P_1, P_3) with (Q_1, Q_3); and if $(P_1, P_2) \nparallel (Q_1, Q_2)$, let a be the line joining R_2 with the intersection R_3 of (P_1, P_2) with (Q_1, Q_2) (Fig. 13). Since (D*) is obviously valid if $a \parallel (P_2, P_3), (Q_2, Q_3)$, we may assume $a \nparallel (P_2, P_3)$. Let R_1 be the intersection of a with (P_2, P_3), and let R be the intersection of a with the parallel to (P_1, P_3) through P_2. Then $(R_2, Q_1) \parallel (R, Q_2)$ follows from (D) or (d) for the triangles $P_1Q_1R_2$ and P_2Q_2R. If Q_3' is the intersection of (R_1, Q_2) with g_3, then $(R_2, Q_3') \parallel (R, Q_2)$ follows from (D) for the triangles RP_2Q_2, $R_2P_3Q_3'$. Since $(R_2, Q_3) = (R_2, Q_1) \parallel (R, Q_2)$, we then have $(R_2, Q_3) \parallel (R_2, Q_3')$; that is, $Q_3 = Q_3'$. Thus the lines $a, (P_2, P_3), (Q_2, Q_3)$ are copunctual.

B)   From (D*) follows (D). Let the assumptions of (D) be satisfied. We may also assume $P_1 \neq Q_1$ (since otherwise the assertion of the theorem

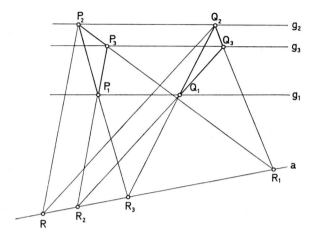

Fig. 13

is trivial). Now let us assume $(P_2, P_3) \nparallel (Q_2, Q_3)$ (Fig. 14), and let S be their intersection. Then S does not lie on (P_1, P_3), (Q_1, Q_3). Let h be the parallel to (P_1, P_3) through S, and let T be the intersection (which exists by Lemma 3) of h and (P_1, P_2). Let us apply (D*) to the triangles SP_3Q_3, TP_1Q_1; it follows that (S, Q_3) and (T, Q_1) intersect on g_2, and thus intersect at the point Q_2. So $(T, Q_1) = (Q_1, Q_2)$. Since $(Q_1, Q_2) \parallel (P_1, P_2)$ and $T \mathrel{I} (P_1, P_2)$, (Q_1, Q_2) it follows that $(Q_1, Q_2) = (P_1, P_2)$, in contradiction to the assumption. Thus we have $(P_2, P_3) \parallel (Q_2, Q_3)$, and therefore the assertion of (D).

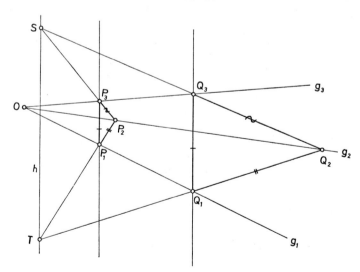

Fig. 14

The results stated in Theorems 1–5 on the interdependence of the closure theorems can be summarized in the accompanying diagram. In this diagram we recognize three levels or ranks, corresponding to three classes of affine planes. The first rank consists of the Pappus planes, characterized by the theorem of Pappus-Pascal, the second of Desargues planes, characterized by the the Desargues theorem, and the third of those affine planes in which the little theorem of Desargues is valid. In §1.4 below, these latter planes will be seen to be the translation planes. In the Pappus planes all the above closure theorems are valid, and in the Desargues planes all of them except the Pappus-Pascal theorem.

$$(P)$$
$$\downarrow$$
$$(D^*) \leftrightarrow (D) \leftrightarrow (S)$$
$$\downarrow$$
$$(p) \leftarrow (d) \rightarrow (s)$$

1.3.   *Collineations*

Collineations are another subject of interest in the affine plane. As with closure, we give a few outstanding theorems illustrative of many others. We first define:

A one-to-one incidence-preserving mapping α of the set of points and the set of lines on an affine plane $\mathfrak{A}$, each onto itself, is called a *collineation* of $\mathfrak{A}$. It is obvious that a collineation takes collinear points into collinear points, and conversely it is easy to show that noncollinear points are mapped into noncollinear points.

In what follows we denote by $P\alpha$ the image point of P under α, and by $g\alpha$ the image line of g.

The inverse α^{-1} of a collineation α is also a collineation. Furthermore,

Theorem 6   *The set of collineations of an affine plane forms a group with respect to successive application of the mapping.*

Proof.   The set of one-to-one mappings of a set onto itself is a group with respect to successive application of the mapping (IB2, §1.2.5). The collineations form a subgroup, since with α, β the mapping $\alpha\beta^{-1}$ is also a collineation.[2]

Lemma 5:   *A collineation of an affine plane takes parallel lines into parallel lines.*

Proof.   Let us assume that there exist two parallel lines g, h with $g\alpha \nparallel h\alpha$.

---

[2]If α and β are mappings, then $\alpha\beta$ denotes the mapping that consists of first α and then β.

Then there exists a point P with $P \text{ I } g\alpha, h\alpha$, and thus $P\alpha^{-1} \text{ I } g, h$, or in other words, $g = h, g\alpha = h\alpha, g\alpha \parallel h\alpha$, in contradiction to the assumption.

Let α be a collineation of the affine plane $\mathfrak{A}$. Then a point F is said to be a *fixed point* of α if $F\alpha = F$, and a line g is a *fixed line* of α if $g\alpha = g$. The set of all fixed points and fixed lines of α is called the *fixed configuration* of α and is denoted by $\mathfrak{F}(\alpha)$. The triple corresponding to $\mathfrak{F}(\alpha)$, namely, the set of fixed points, the set of fixed lines, and the incidence relation, is also denoted by $\mathfrak{F}(\alpha)$.

Let us now show that $\mathfrak{F}(\alpha)$ is a subplane belonging to $\mathfrak{A}$. Here we call a triple $\mathfrak{T} = (\mathfrak{P}', \mathfrak{G}', \text{I})$ a *subplane* of the affine plane $\mathfrak{A} = (\mathfrak{P}, \mathfrak{G}, \text{I})$ if $\mathfrak{P}' \subset \mathfrak{P}$, $\mathfrak{G}' \subset \mathfrak{G}$, and the Axioms A1, A2 hold for $\mathfrak{T}$.

Lemma 6: *The fixed configuration $\mathfrak{F}(\alpha)$ of a collineation of the affine plane $\mathfrak{A}$ is a subplane of $\mathfrak{A}$.*

Proof. In the first place we obviously have

(i) *The line joining two distinct fixed points is a fixed line ; the intersection of two distinct fixed lines is a fixed point.*

From (i) it follows that the Axiom A1 holds for $\mathfrak{F}(\alpha)$. In order to verify A2 let us choose F, g from $\mathfrak{F}(\alpha)$. If h is the line (in $\mathfrak{A}$) through F parallel to g, then $h\alpha \text{ I } F\alpha = F$ and $h\alpha \parallel g\alpha = g$ by Lemma 5, so that by A2 (for $\mathfrak{A}$) we have $h\alpha = h$. Thus h is a parallel (in $\mathfrak{F}(\alpha)$) through F to the line g.

Now if in $\mathfrak{F}(\alpha)$ there were a second line $h' \neq h$ through F parallel to g, the lines h', g would have a point of intersection by A2. By (i) this point would also belong to $\mathfrak{F}(\alpha)$, so that in $\mathfrak{F}(\alpha)$ the lines h', g would also not be parallel, in contradiction to the assumption. Thus A2 also holds for $\mathfrak{F}(\alpha)$, and therefore $\mathfrak{F}(\alpha)$ is a subplane of $\mathfrak{A}$.

Let us now turn to the study of certain special collineations of great importance: namely, the translations, dilations, and axial affinities. As we shall see, these collineations resemble one another, the reason being that from the projective point of view they are all of the same type (cf. §3.1). We begin with the translations.

Definition. A collineation τ in an affine plane is called a *translation* if $g \parallel g\tau$ for all lines g and either no point or all points are fixed under τ.

A line that contains at least one pair $P, P\tau$ is called a *trace of P under τ*, or more briefly a *trace* of τ.

Lemma 7: *The traces of a translation τ are simply all its fixed lines.*

Proof. By definition every fixed line is a trace. Conversely, let g be a trace of P with respect to τ; then we have $g \parallel g\tau$ and $P\tau \text{ I } g, g\tau$, so that $g = g\tau$.

Lemma 8: *The traces of a translation $\tau \neq 1$ form a parallel pencil.*

Proof. From the definition of a translation and from Lemma 7 it follows that $\mathfrak{F}(\tau)$ contains only lines, namely, all the traces of τ. Since at least one trace must pass through every point of $\mathfrak{A}$, it follows from Lemma 6 that $\mathfrak{F}(\tau)$ is a parallel pencil.

If Π_g is the parallel pencil of the traces of the translation τ with $\tau \neq 1$, then τ is called a *translation in the g-direction*. The identity is a translation in the g-direction for every g.

Lemma 9: *If A, B are two points of an affine plane, there exists at most one translation τ with $A\tau = B$.*

Proof. If $A = B$, then $\tau = 1$ is the only translation with $A\tau = B$. But if $A \neq B$ and if C is an arbitrary point with $C \nmid (A, B)$, then for every translation τ with $A\tau = B$ the point $C\tau$ is uniquely determined as the intersection of the parallel to (A, C) through $B = A\tau$ (by the definition of a translation) and the parallel to (A, B) through C, since τ must be a translation in the (A, B)-direction. For the case $C \text{ I}(A, B)$ we can argue correspondingly from a point and its image not on (A, B).

Theorem 7: *The set T of all translations of an affine plane forms a group, and so does the set $T(g)$ of all translations in the g-direction.*

Proof. We must show that T is a subgroup of the group of collineations of the affine plane, or in other words that with τ', τ the mapping $\tau'\tau^{-1}$ is also a translation (cf. IB2, §3). Since $g \parallel g\tau'$ and since by Lemma 5 we have $g \parallel g\tau^{-1}$ and $g\tau^{-1} \parallel g\tau'\tau^{-1}$, therefore $g \parallel g\tau'\tau^{-1}$ for all lines g. If there exists a fixed point F of $\tau'\tau^{-1}$ we have $F = F\tau'\tau^{-1}$, or in other words $F\tau = F\tau'$, and thus $\tau' = \tau$ by Lemma 9. Thus $\tau'\tau^{-1} = 1$, so that $\tau'\tau^{-1}$ is in every case a translation. Consequently, T is a group.

For the proof that $T(g)$ is a group, we must show (cf. IB2, §3) that with τ, τ' the mapping $\tau' \tau^{-1}$ also belongs to $T(g)$. But by Lemmas 7 and 8 it suffices to show that $g\tau'\tau^{-1} = g$ for τ', $\tau \in T(g)$, which is clearly true, since $g\tau' = g$ and $g\tau = g$, so that $g = g\tau^{-1}$ and therefore $g\tau'\tau^{-1} = g\tau^{-1} = g$.

There exist affine planes in which for a fixed preassigned line g and an arbitrary pair of distinct points A, B with $(A, B) \parallel g$ there does not always exist a translation in the g-direction taking A into B. Let us make the definition: the group $T(g)$ of translations in the g-direction is said to be *linearly transitive* if for every two points A, B on a line parallel to g there exists a translation τ in $T(g)$ with $A\tau = B$.

Lemma 10: *If τ is a translation in the g-direction and α is a collineation, then $\alpha^{-1}\tau\alpha$ is a translation in the $(g\alpha)$-direction.*[3]

Proof. From $b \parallel b\tau$ for all lines b it follows that $a\alpha^{-1} \parallel (a\alpha^{-1})\tau$ for all lines a and thus by Lemma 5 that $a = a\alpha^{-1}\alpha \parallel ((a\alpha^{-1})\tau)\alpha = a\alpha^{-1}\tau\alpha$. If

---

[3]The translation $\alpha^{-1} \tau\alpha$ is called the *transform* of τ by α.

there exists a fixed point F for $\alpha^{-1}\tau\alpha$, we have $F = F\alpha^{-1}\tau\alpha$ and thus $F\alpha^{-1} = F\alpha^{-1}\tau$, so that $\tau = 1$ by Lemma 9 and $\alpha^{-1}\tau\alpha = 1$ is a translation. Thus $\alpha^{-1}\tau\alpha$ is in every case a translation. Since $(g\alpha)\alpha^{-1}\tau\alpha = g\tau\alpha = g\alpha$, it follows that $\alpha^{-1}\tau\alpha \in T(g\alpha)$.

Theorem 8:　*If there exist translations $\neq 1$ in distinct directions, then the group of translations is Abelian.*

Proof. a)　If τ, τ' are translations in distinct directions, then $\tau\tau' = \tau'\tau$; for let $\tau \in T(g)$ and $\tau' \in T(g')$ with $g \nparallel g'$, so that we see by Lemma 10 that $\tau'^{-1}\tau\tau' \in T(g)$ and by Theorem 7 that $\tau^{-1}(\tau'^{-1}\tau\tau') \in T(g)$. Correspondingly we have $(\tau^{-1}\tau'^{-1}\tau)\tau' \in T(g')$, so that $\tau^{-1}\tau'^{-1}\tau\tau' \in T(g)$, $T(g')$ with $g \nparallel g'$ and thus $\tau^{-1}\tau'^{-1}\tau\tau' = 1$, so that as desired $\tau\tau' = \tau'\tau$.

b)　If τ, τ' are translations in the same direction, then again $\tau\tau' = \tau'\tau$. For let τ, $\tau' \in T(g)$ and let τ'' be a translation, existent by hypothesis, such that $\tau'' \notin T(g)$; then also $\tau'\tau'' \in T(g)$ (otherwise by Theorem 7 we would have $\tau'\tau^{-1}\tau'\tau'' \in T(g)$). Thus by a) we have $(\tau\tau')\tau'' = \tau(\tau'\tau'') = (\tau'\tau'')\tau = \tau'(\tau''\tau) = \tau'(\tau\tau'') = (\tau'\tau)\tau''$, and thus $\tau\tau' = \tau'\tau$.

Let us now turn to the dilations. A collineation δ of the affine plane is called a *dilation* if $g \parallel g\delta$ for all lines g and if δ has at least one fixed point. As with translations, every line containing at least one pair P, $P\delta$ is called a *trace* of δ, and we have:

Lemma 11:　*The traces of a dilation consist of all its fixed lines.*

Lemma 12:　*A dilation $\delta \neq 1$ has exactly one fixed point.*

Proof.　Every line through a fixed point is a trace of δ and thus by Lemma 11 is a fixed line. If δ had two fixed points F, F', then obviously every point P with $P \not\mathrel{I} (F, F')$, being the intersection of the fixed lines (P, F) and (P, F'), would be a fixed point, and thus every point on (F, F') would be fixed, so that $\delta = 1$, in contradiction to the assumption.

By Lemma 12 the fixed configuration $\mathfrak{F}(\delta)$ of a dilation $\delta \neq 1$ consists of a point and of the pencil of lines through this point.

Consequence.　*The traces of a dilation $\delta \neq 1$ comprise the pencil of lines through the fixed point of δ.*

Every line through the single fixed point O of δ is a trace, and thus a fixed line. If there were a trace l with $l \not\mathrel{I} O$, then δ would have fixed points $\neq O$.

Theorem 9:　*The set $D(O)$ of dilations with the same fixed point O form a group.*

Proof.　We must show (cf. IB2, §3) that with δ', δ also $\delta'\delta^{-1}$ is in $D(O)$. Obviously (as in the case of translations) we have $g \parallel g\,\delta'\delta^{-1}$ for all lines g. Also, $O\delta'\delta^{-1} = O\delta^{-1} = O$, since O is a fixed point of δ' and δ, and thus also of δ^{-1}. Consequently $\delta'\delta^{-1} \in D(O)$.

Lemma 13: *If A, B are two points $\neq O$, there exists at most one dilation δ in $D(O)$ with $A\delta = B$.*

Proof. From δ, $\delta' \in D(O)$ and $A\delta = B = A\delta'$ it follows that $A\delta'\delta^{-1} = A$ and $\delta'\delta^{-1} \in D(O)$. But since $A \neq O$, Lemma 12 shows that $\delta'\delta^{-1} = 1$ and thus $\delta = \delta'$.

As with translations, the group $D(O)$ does not necessarily contain, for given points A, $B \neq O$ collinear with O, a dilation with $A\delta = B$. But if for every pair of points A, $B \neq O$ collinear with O the group $D(O)$ does contain such a dilation, then it is said to be *linearly transitive*.

Finally, we now consider the axial affinities. Here a collineation α of an affine plane is said to be an affinity (affine transformation) with axis a, or more briefly an *axial affinity*, if a is pointwise fixed. If P is a point such that $P \ \mathrm{I} \ a$, then any line containing P and $P\alpha$ is called a *trace* of P with respect to α.

Lemma 14: *The traces of an axial affinity α, together with the axis, comprise all its fixed lines.*

Proof. Let a be the axis of α and let g be the trace of P with respect to α. If $g \nparallel a$, then (g, a) is fixed and $g, g\alpha \ \mathrm{I} \ P\alpha, (g, a)$, and thus since $(g, a) \neq P\alpha$, we have $g = g\alpha$. On the other hand, if $g \parallel a$, then also $g\alpha \parallel a\alpha = a$ and, since $g, g\alpha \ \mathrm{I} \ P\alpha$, it follows that $g = g\alpha$.

Lemma 15: *If α is an affinity $\neq 1$ with axis a, then the fixed configuration $\mathfrak{F}(\alpha)$ consists of all points on a, of a parallel pencil and of the line a.*

Proof. Since $\alpha \neq 1$, it follows from Lemma 6 that $\mathfrak{F}(\alpha)$ is a proper subplane of $\mathfrak{A}$. Then obviously

(i)   *If $\mathfrak{F}(\alpha)$ contains a line g not parallel to a, then $\Pi_g \subset \mathfrak{F}(\alpha)$.*
For through every point on a the parallel to g is in $\mathfrak{F}(\alpha)$.

Thus $\mathfrak{F}(\alpha)$ does not contain any fixed point F not on a, since otherwise $(F, G) = g$ would be a fixed line for every point G on a, so that we would have $\Pi_g \subset \mathfrak{F}(\alpha)$ for all g with $g \nparallel a$, and every point, being the intersection of two lines not parallel to g, would be fixed.

Thus (i) implies the assertion of Lemma 15, provided $\mathfrak{F}(\alpha)$ contains a line not parallel to a. If $\mathfrak{F}(\alpha)$ contains a line parallel to a, then $\Pi_a \subset \mathfrak{F}(\alpha)$, since through every point of $\mathfrak{A}$ there passes a trace and therefore a fixed line.

If Π_z denotes the parallel pencil in Lemma 15 for the axial affinity α, then α is called an *axial affinity in the z-direction*.

Theorem 10: *The set $A(a)$ of affinities with a as axis forms a group, and so does the set $A(z, a)$ of axial affinities in $A(a)$ in the z-direction.*

As before, the group $A(z, a)$ of affinities in the z-direction with a as axis is said to be *linearly transitive* if for every pair P, Q with P, $Q \ \mathrm{I} \ a$ and P, $Q \ \mathrm{I} \ l$, where $l \parallel z$ there exists an axial affinity α in $A(z, a)$ with $P\alpha = Q$.

Lemma 16: *An axial affinity α is uniquely determined by any pair consisting of a point and its image not on the axis.*

Proof. If there were two affinities α, α' with axis a and $A\alpha = A\alpha'$ for $A \nmid a$, it would follow that $A\alpha'\alpha^{-1} = A$ and thus, from Theorem 10 and Lemma 15, that $\alpha'\alpha^{-1} = 1$, so that $\alpha' = \alpha$.

The axial affinities are particularly important, because all affinities of the affine plane can be constructed from them. Here we shall first give a purely synthetic definition of the *affinity*[4] *of an affine plane* $\mathfrak{A}$, namely as all the collineations of $\mathfrak{A}$ that are products of axial affinities. Then later (in §2) we shall show that the affinities in a Pappus plane coincide with the mappings defined in the usual way in analytic geometry. The affinities of an affine plane obviously form a group, which we shall denote by $\bar{A}$.

Theorem 11: *In an affine plane in which every group $A(z, a)$ is linearly transitive, there always exists, for any two nondegenerate triangles $P_1 P_2 P_3$ and $Q_1 Q_2 Q_3$ an affinity α taking one triangle into the other (for which $P_i\alpha = Q_i$ $(i = 1, 2, 3)$).*

Proof. If a_1 is a line not passing through P_1 or Q_1, the linear transitivity of the group $A(z, a_1)$ (here z denotes a line parallel to some line through P_1, Q_1) implies the existence of an affinity α_1, with a_1 as axis, taking P_1 into Q_1 (Fig. 15). Now if a_2 is a line incident with $P_1\alpha_1 = Q_1$ but not with $P_2\alpha_1$ or with Q_2, then as before there exists an affinity α_2, with a_2 as axis, taking $P_2\alpha_1$ into Q_2. Finally if $a_3 = (Q_1, Q_2)$, then by assumption $Q_3 \nmid a_3$ and also

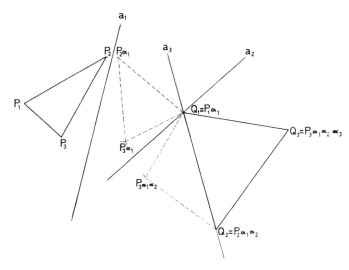

Fig. 15

[4]Cf. also II9 and 10.

$P_3\alpha_1\alpha_2 \nmid a_3$, since otherwise we would have $Q_1 = P_1\alpha_1\alpha_2$, $Q_2 = P_2\alpha_1\alpha_2$, $P_3\alpha_1\alpha_2$, so that P_1, P_2, P_3 would be collinear. Consequently there exists an affinity α_3 with α_3 as axis taking $P_3\alpha_1\alpha_2$ into Q_3. But then $\alpha = \alpha_1\alpha_2\alpha_3$ is an affinity with $P_i\alpha = Q_i$ $(i = 1, 2, 3)$.

1.4. *Relations between Collineations and Closure Theorems*

In the preceding subsections we have pointed out that in an arbitrary affine plane it is not necessary that any of the closure theorems should hold or that any collineations $\neq 1$ should exist at all. We now turn to the question whether and to what extent certain closure theorems are connected with the existence or particular properties of collineations. For example, we shall find that the affine planes in which (d) holds are exactly those affine planes in which for every pair of points A, B there exists a translation τ with $A\tau = B$. These affine planes will also be called *translation planes*. More generally, we shall show that

Theorem 12: *In an affine plane* (D) *is valid if and only if for all configurations with O as center the group D(O) is linearly transitive. Also,* (d) *is valid if and only if for all configurations with supporting lines parallel to g the group T_g is linearly transitive.*

Proof.   We give only the proof of the first part of Theorem 12, since the proof of the second is quite similar.

A)   Let $D(O)$ be linearly transitive, and let the conditions of Theorem (D) be satisfied (Fig. 16). Then we must show that $(P_2, P_3) \parallel (Q_2, Q_3)$. By hypothesis there exists a dilation $\delta \in D(O)$ with $P_1\delta = Q_1$. Then (P_1, P_2) is mapped onto (Q_1, Q_2) and, since $g_2\delta = g_2$, therefore $P_2\delta = Q_2$. In exactly the same way, for the lines (P_1, P_3) and (Q_1, Q_3) we have $P_3\delta = Q_3$. Since δ is a dilation, it follows that $(P_2, P_3) \parallel (P_2\delta, P_3\delta) = (Q_2, Q_3)$.

B)   Now let (D) hold. Then we must show that for every pair of points

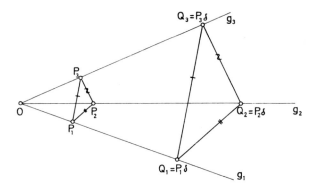

Fig. 16

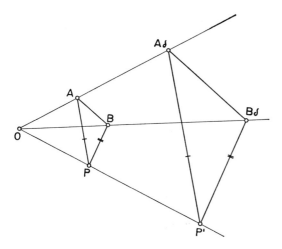

Fig. 17

P, P' collinear with O there exists a δ in $D(O)$ with $P\delta = P'$. For this purpose we construct a point mapping, for simplicity also denoted by δ, in the following way (here R is a point with $R \mathbin{\text{I}} (P, Q)$, chosen once for all).

For $X \mathbin{\text{I}} (P, Q)$ let $X\delta = X'$ with $(P, X) \parallel (P', X')$ and $X' \mathbin{\text{I}} (X, O)$.

For $X \mathbin{\text{I}} (P, O)$ and $X \neq O$ let $X\delta = X'$ with $(R, X) \parallel (R\delta, X')$ and $X' \mathbin{\text{I}} (P, O)$.

For $X = O$ let $X\delta = O$.

It is clear that δ is a one-to-one mapping of $\mathfrak{P}$ onto itself, taking P into P'. Also

(*)   If $A \neq B$, then $(A, B) \parallel (A\delta, B\delta)$.

Then we have $A, B \mathbin{\text{I}} (P, O)$ and $(A, B) \mathbin{\text{I}} O$, so that (*) follows immediately from (D) (Fig. 17). In the other cases we obtain figures from which (D) allows us to conclude in the same way that (*) is valid, but we omit the details here for lack of space.

From (*) it follows that the transformation δ sends collinear points into collinear points, so that δ induces a collineation of $\mathfrak{A}$, which we shall also denote by δ. Under δ, again by (*), every line goes into a line parallel to it. Since O is a fixed point of δ, we have $\delta \in D(O)$. Thus the proof of Theorem 12 is complete.

Theorem 13:   *A linear transitive group $D(O)$ is Abelian if and only if the theorem* (P) *holds for all configurations with supporting lines through O. A linearly transitive group $T(g)$ is Abelian if and only if the theorem* (p) *is valid for all configurations with supporting lines parallel to g.*

Proof.   Again we prove only the first part of the theorem.

A)   Let $D(O)$ be Abelian, and let the assumptions of the theorem (P) be

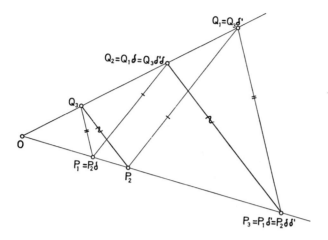

Fig. 18

satisfied (Fig. 18). We must show that $(P_1, Q_2) \parallel (P_2, Q_1)$ and $(P_1, Q_3) \parallel (P_3, Q_1)$ imply that (P_2, Q_3) and (P_3, Q_2) are parallel. Since $D(O)$ is linearly transitive, there exist dilations δ, $\delta' \in D(O)$ with $P_2\delta = P_1$, $P_1\delta' = P_3$. Then $Q_1\delta = Q_2$, $Q_3\delta' = Q_1$. Thus it follows that $P_2\delta\,\delta' = P_3$, $Q_3\delta'\delta = Q_2$ and, since $D(O)$ is Abelian, also $Q_3\delta\delta' = Q_2$. But $\delta\delta'$ is a dilation taking (P_2, Q_3) into (P_3, Q_2), so that $(P_2, Q_3) \parallel (P_3, Q_2)$.

B) Under the assumption that (P) holds for the given configurations, we must show that $D(O)$ is Abelian. Let δ, $\delta' \in D(O)$, and choose two arbitrary points P_2, Q_3 such that O, P_2, Q_3 are collinear. Also, let $P_2\delta = P_1$, $Q_3\delta' = Q_1$, $P_2\delta\delta' = P_3$, $Q_3\delta'\delta = Q_2$. It is easy to verify the assumptions of (P) with respect to the hexagon $P_1Q_2P_3Q_1P_2Q_3$ with supporting lines $(O, P_2), (O, Q_3)$. From $(P_1, Q_2) \parallel (P_2, Q_1)$ and $(P_1, Q_3) \parallel (P_3, Q_1)$ it follows that $(P_2, Q_3) \parallel (P_3, Q_2)$. Thus $Q_2 = Q_3\delta\delta'$, which means $Q_3\delta'\delta = Q_3\delta\delta'$ and by Lemma 12, $\delta\delta' = \delta'\delta$.

The following theorem is proved in the same way as Theorem 10.

Theorem 14:   *In an affine plane $\mathfrak{A}$ the theorem (D^*) with a as axis and with supporting line parallel to z is valid if and only if the group $A(z, a)$ is linearly transitive.*

Let us state the following consequence of Theorems 5 and 14.

Theorem 15:   *An affine plane is a Desargues plane if and only if every group $A(z, a)$ is linearly transitive.*

The statement about transitivity proved in Theorem 12 for the affinities of an affine plane holds, by Theorem 15, for all Desargues planes.

2. Introduction of Coordinates

In the preceding section, our discussion of the fundamental concepts and facts of the theory of affine planes was carried out synthetically, on an axiomatic basis. But we may also consider the analytic construction of geometry, which has been developed in recent years from linear algebra and the theory of vector spaces over a field.

Then it is natural to ask how the axiomatic-synthetic construction is is related to the analytic-algebraic; in particular, in the synthetic treatment how is it possible to introduce coordinates?

A first answer to this question is given by a surprisingly simple and attractive theorem: the Pappus (Desargues) planes are identical with the affine-coordinate planes defined by a two-dimensional vector space over a field (a skew field). This theorem is the basis for the geometry of vector spaces.

But, as we have seen in §1.1, non-Desarguesian planes also exist, and while it is true that coordinates can be introduced into such planes, they will not have all the properties of a skew field, so that we shall not discuss them here. Let us only note that it has been possible to build up a stepwise structure in which each successive step consists of adding, in the synthetic case, another axiom and in the analytic, another of the laws for addition and multiplication in a field. In each case, the second-to-last step produces the Desargues planes and the last produces the Pappus planes.

2.1. *Affine-Coordinate Planes over a Skew Field*

Let K be a skew field and let $U = K^2$ be the two-dimensional right vector space over K. The one-dimensional subspace of U generated by the vector A will be denoted by $\langle A \rangle$.

Each element of U is called a *point,* and each residue class of a one-dimensional subspace of U is called a *line.* A point P is incident with a line g if $P \in g$. If G is the set of all lines, then the triple $(U, G, \in)$ is called the *affine-coordinate plane* $\mathfrak{A}(K)$ over the skew field K.

Thus by definition every pair $P = (x, y)$ with $x, y \in K$ is a point. A line is a set of points; the point $X = (x, y)$ is incident with the line $g = A + \langle B \rangle$ if and only if $X = A + Bu$ for some u in K, or in components, if

(1)
$$x = a_1 + b_1 u$$
$$y = a_2 + b_2 u$$

with $A = (a_1, a_2)$ and $B = (b_1, b_2)$. Here (1) is the usual parametric representation of the line; u runs through all the elements of K (Fig. 19).

Theorem 16: *The plane $\mathfrak{A}(K)$ is Desarguesian.*

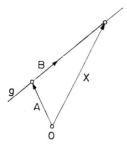

Fig. 19

For the proof we make use of simple familiar facts from the theory of vector spaces over a skew field (cf. IB2 and 3).

We first show that $\mathfrak{A}(K)$ is an affine plane, i.e., that Axioms A1 through A3 are valid. As a preliminary we prove

(i) *The two lines* $A + \langle B \rangle$, $A' + \langle B' \rangle$ *are parallel if and only if* $\langle B \rangle = \langle B' \rangle$.

If $\langle B \rangle = \langle B' \rangle$, the two lines are residue classes of the same subspace and thus are either identical or without common element, and therefore are parallel.

But if $\langle B \rangle \neq \langle B' \rangle$, then the residue classes $A + \langle B \rangle$, $A' + \langle B' \rangle$ are also distinct. Moreover, B and B' span the vector space U, so that there exist elements $u, u' \in K$ with $A' - A = Bu - B'u'$. But then $A + Bu = A' + B'u' \in A + \langle B \rangle$, $A' + \langle B' \rangle$; i.e., the two lines have a point in common and thus are not parallel.

Verification of Axioms A1 through A3:

As for A1: If P, Q are two distinct points in $\mathfrak{A}(K)$, then $P - Q \neq O$ (where O denotes the zero vector in U), from which it follows, since $P = Q + (P - Q) \cdot 1$ and $Q = Q + (P - Q) \cdot 0$, that $P, Q \in Q + \langle P - Q \rangle$. Thus $Q + \langle P - Q \rangle$ is the line joining P, Q. Now if these points were joined by a second line $A + \langle B \rangle$, we would have $P + \langle B \rangle = A + \langle B \rangle = Q + \langle B \rangle$, since $P, Q \in A + \langle B \rangle$. But then $P - Q \in \langle B \rangle$, so that $\langle P - Q \rangle = \langle B \rangle$ and thus $A + \langle B \rangle = Q + \langle P - Q \rangle$.

As for A2: Let P be a point and $A + \langle B \rangle$ a line; then the line $P + \langle B \rangle$ contains the point P and thus by (i) is parallel to $A + \langle B \rangle$. If there existed a second line $A' + \langle B' \rangle$ through P parallel to $A + \langle B \rangle$, then by (i) we would have $\langle B \rangle = \langle B' \rangle$ and $A' + \langle B' \rangle = P + \langle B' \rangle$, in view of the fact that $P \in A' + \langle B' \rangle$. Thus $A' + \langle B' \rangle = P + \langle B \rangle$.

As for A3: The points $(0, 0)$, $(1, 0)$, $(0, 1)$ are obviously not collinear.

Verification of Axiom (D): By Theorem 12 it is enough to show that every group $D(F)$ $(F \in U)$ is linearly transitive.

Let F be chosen arbitrarily. The mapping δ with

$$(2) \qquad\qquad X\delta = Xa + F(1 - a) \qquad\qquad (a \neq 0)$$

is a one-to-one mapping of U onto itself. We now show that δ induces a dilation of $\mathfrak{A}(K)$ from the group $D(F)$.

If $A + \langle B \rangle$ is a line, then $X \in A + \langle B \rangle$ is equivalent to $X\delta \in A\delta + \langle B \rangle$, since from $X = A + Bu$ it follows that $X\delta = Xa + F(1 - a) = Aa + F(1 - a) + (Bu)a = A\delta + B(ua)$. Thus δ takes the line $A + \langle B \rangle$ into the line $A\delta + \langle B \rangle$ parallel to $A + \langle B \rangle$, so that δ induces a dilation. Since $F\delta = Fa + F(1 - a) = F$, we have $\delta \in D(F)$. If P, Q are two points $\neq F$ collinear with F, there exists a δ in $D(F)$ with $P\delta = Q$, since there is an element $a \neq 0$ in K with $Q - F = (P - F)a$. For the mapping δ with (2) we then have $P\delta = Pa + F(1 - a) = Pa - Fa + F = Q$. Thus the group $D(F)$ is linearly transitive.

Corollary: *If K is a field, then $\mathfrak{A}(K)$ is a Pappus plane.*

Proof.   Let us assume that K is a field and show that every group $D(F)$ is Abelian. Then §1, Theorem 13, shows that (P) holds in $\mathfrak{A}(K)$. Let δ, $\delta' \in D(F)$; then there exist a, a' in K such that $X\delta = Xa + F(1 - a)$ and $X\delta' = Xa' + F(1 - a')$ for all X in U, as is shown by the above proof of (D) in $\mathfrak{A}(K)$. Thus by the commutativity of multiplication in K we have: $X\delta\delta' = (X\delta)a' + F(1 - a') = (Xa + F(1 - a))a' + F(1 - a') = Xaa' + F(1 - aa') = Xa'a + F(1 - a'a) = (Xa' + F(1 - a'))a + F(1 - a) = (X\delta')a + F(1 - a) = X\delta'\delta$ for all X in U, so that $\delta\delta' = \delta'\delta$ for all δ, $\delta' \in D(F)$, as asserted.

2.2.   Desargues Planes and Affine-Coordinate Planes over a Skew Field

Theorem 17:   *Every Desargues plane $\mathfrak{A} = (\mathfrak{P}, \mathfrak{G}, I)$ is isomorphic to the affine-coordinate plane $\mathfrak{A}(K) = (U, G, \in)$ over a skew field K.*

Here the word "isomorphic" means that there exists a one-to-one mapping φ of $\mathfrak{P}$ onto U and of $\mathfrak{G}$ onto G such that $P \mathrel{I} g(P \in \mathfrak{P}, g \in \mathfrak{G})$ if and only if $P\varphi \in g\varphi$.

Proof.   Let $\mathfrak{A}$ be a Desargues plane. Also let O and E be two distinct points in $\mathfrak{A}$, and let K be the set of points incident with the line $x = (O, E)$. Then, by Theorems 2 and 12, for every A in $\mathfrak{A}$ there exists exactly one translation taking O into A; we denote this translation by τ_A. The dilation in $D(O)$ (which exists by Theorem 12 and is uniquely determined by Theorem 13) taking E into the point C with $C \neq O$ in K will be denoted by δ_C.

Now by means of the definition (*) below, we introduce two operations in the set K, an addition and a multiplication:

$$A + B = O\tau_A\tau_B = A\tau_B \qquad \text{for all } A, B \in K,$$

$$(*) \qquad A \cdot B = E\delta_A\delta_B = A\delta_B \qquad \text{for all } A, B \neq O \text{ in } K,$$

$$A \cdot O = O \cdot A = O \qquad \text{for all } A \in K.$$

These definitions mean that $A + B$ is the point in K into which the point O is taken by successive application of the translations τ_A and τ_B, and $A \cdot B$ with $A, B \neq O$ is the point in K into which E is taken by successive application of the mappings δ_A, δ_B. Then by (*) we have $A \cdot B = A\delta_B$ for all A and all $B \neq O$.

From the definition (*) it follows at once that

$$(1) \qquad \tau_{A+B} = \tau_A\tau_B \qquad \text{for } A, B \text{ in } K,$$

$$(2) \qquad \delta_{A \cdot B} = \delta_A\delta_B \qquad \text{for } A, B \text{ in } K \text{ with } A, B \neq O.$$

We first show

(i) *The set K of points on the line x forms a skew field with respect to the operations defined by* (*). *Notation*: $K = K(\mathfrak{A})$.

By (1) we see that, with respect to addition, K is isomorphic to the Abelian group $T(x)$ and by (2) that, with respect to multiplication, $K \setminus \{O\}$ is isomorphic to the group $D(O)$. In order to show that K is a skew field we must still prove the two distributive laws.

Let us first remark that if y is an arbitrary line through O distinct from x and if α is an affinity in $A(x, y)$, then

$$(3) \qquad \alpha^{-1}\tau_A\alpha = \tau_{A\alpha} \qquad \text{for } A \text{ in } K.$$

For Lemma 10 shows that $\alpha^{-1}\tau\alpha$ is a translation from $T(x)$ and also that $O\alpha^{-1}\tau_A\alpha = O\tau_A\alpha = A\alpha = O\tau_{A\alpha}$.

By Theorems 5 and 14 and Lemma 16, for every $A \neq O$ in K there exists exactly one α in $A(x, y)$ with $E\alpha = A$; let us denote this affinity by α_A. Then

$$(4) \qquad \alpha\delta = \delta\alpha$$

for all $\alpha \in A(x, y)$ and all $\delta \in D(O)$. For it is clear that $\alpha^{-1}\delta\alpha \in D(O)$ since $O\alpha^{-1}\delta\alpha = O$ and, for a point $E' \neq O$ on y, that $E'\alpha^{-1}\delta\alpha = E'\delta\alpha = E'\delta$ since $E', E'\delta$ I y. By Lemma 13, we have $\alpha^{-1}\delta\alpha = \delta$ and therefore $\alpha\delta = \delta\alpha$. But then by (*) and (4), $A\alpha_B = B\delta_A = B \cdot A$ for $A, B \neq 0$, and thus for $A = 0$ it follows at once that $A\alpha_B = B \cdot A$.

$$(5) \qquad \alpha_B^{-1}\tau_A\alpha_B = \tau_{BA} \qquad \text{with } B \neq O.$$

If δ_B is an arbitrary dilation in $D(O)$, then

$$(6) \qquad \delta_B^{-1}\tau_A\delta_B = \tau_{AB} \qquad B \neq O \text{ and } A \in K,$$

as before.

Now by (5) and (6) the distributive laws follow for $C \neq O$:

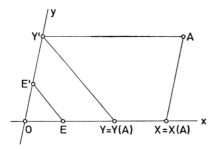

Fig. 20

$C(A + B) = (A + B)\alpha_C = (A\tau_B)\alpha_C = A\alpha_C\tau_{CB} = (CA)\tau_{CB} = CA + CB;$
$(A + B)C = (A + B)\delta_C = (A\tau_B)\delta_C = A\delta_C\tau_{BC} = (AC)\tau_{BC} = AC + BC.$
Obviously, also $O(A + B) = OA + OB$ and $(A + B)O = AO + BO$, so that (i) is valid.

To every point A in the plane we now assign an element from the two-dimensional vector space U over $K(\mathfrak{A})$, by means of the following construction (Fig. 20). Let E' be a point of $\mathfrak{A}$ that is not collinear with O, E, and let $y = (O, E')$. Then corresponding to the point A we construct, $X, Y \in K(\mathfrak{A})$ as follows: let X be the point of intersection of x with the parallel to y through A. If Y' is the point of intersection of y with the parallel to x through A, let Y be the intersection of x with the parallel to (E, E') through Y'. Then set $X = X(A)$, $Y = Y(A)$ and

$$A\varphi = [X, Y].$$

In order to avoid confusion with the line (X, Y), in the present section we denote the pair (X, Y) by $[X, Y]$. Then φ is a one-to-one mapping of the set of points in $\mathfrak{A}$ onto U.

The coordinates of the point of a line g satisfy a linear equation; for if $g \nparallel x$ and if g' is the parallel to g through E', and if furthermore A, B are the points of intersection of g, g' with x and if P is an arbitrary point $\neq B$ on g (Fig. 21), then for $X = X(P)$, $Y = Y(P)$

(7) $A\delta_Y\tau_X = B,$

and thus by (*)

(8) $AY + X = B.$

But if $g \parallel x$, a point P lies on g if and only if $Y(P) = C$ and C is the y-coordinate of the intersection of g with y. Thus the line g contains all the points P with

(9) $Y = C,$

if $Y = Y(P)$.

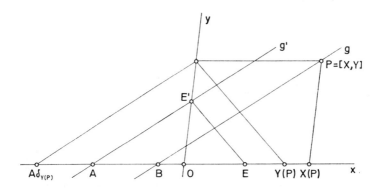

Fig. 21

We now extend φ in the following way. If g is a line in $\mathfrak{A}$, we set $g\varphi = \bar{A} + \langle \bar{B} \rangle$, if $\bar{A} = [B, O]$ in the case (8) and $\bar{A} = [O, C]$ in the case (9), and if $\bar{B} = [-A, E]$ in case (8) and $\bar{B} = [E, O]$ in case (9). We then have:

$$P \text{ I } g \quad \text{if and only if} \quad P\varphi \in g\varphi.$$

Thus φ is an isomorphism of $\mathfrak{A}$ onto $\mathfrak{A}(K)$, and Theorem 17 is proved.

Corollary: *If $\mathfrak{A}$ is a Pappus plane, then $K(\mathfrak{A})$ is a field.*

For it follows from (P) that the group $D(O)$ is Abelian. But since $K \setminus \{O\}$ is isomorphic to the group $D(O)$, the multiplication in K is commutative.

2.3. *The Hilbert Calculus of Segments*

There exists still another method of introducing coordinates, namely, the Hilbert calculus of segments (Hilbert 1899), which we now sketch in brief.

As before, we choose three noncollinear points O, E, E' in $\mathfrak{A}$ and set $g = (O, E)$ and $g' = (O, E')$. Each pair $[O, A]$ with A I g is called a *segment* (Fig. 22). The segments will be denoted by lowercase Latin letters, except that the particular segment O, O is denoted by 0 (zero) and the particular segment O, E by 1. If A' is a point on g' and $(E, E') \parallel (A, A')$, we also call O, A' a segment, namely, the segment a' on g' corresponding to a (Fig. 22.)

For the segments on g we then introduce an addition and a multiplication by means of the constructions illustrated in Figs. 23 and 24. Here g'' is the parallel to g through E'.

In the ordinary Euclidean plane, addition obviously corresponds to the juxtaposition of segments, and multiplication to the ray theorem. But it must be noted that the segments thus defined do not necessarily have a "length."

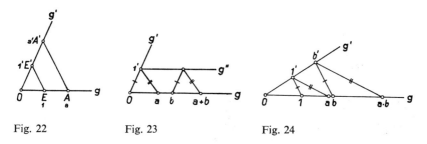

Fig. 22   Fig. 23   Fig. 24

We now wish to show that the segments on the line g of a Desargues plane form a skew field. In the first place it is clear that $a + 0 = a$ and $a \cdot 1 = a$ and $0 \cdot a = 0$ for all segments a. Furthermore, from the affine incidence axioms alone it follows that for every segment a there exists a segment $-a$ with $a + (-a) = 0$ and also a segment a^{-1} with $a \cdot a^{-1} = 1$.

The associative, commutative, and distributive laws then correspond to certain closure theorems, and in fact it is easy to see from Figs. 25 through 27 that (s) corresponds to the associativity of addition, (S) to the associativity of multiplication, and (p) to the commutativity of addition. By Theorems 1 through 3 we see that (p), (s), (S) are valid in a Desargues plane, so that the corresponding rules for calculation are also valid. Figure 28 illustrates the closure theorem (with an auxiliary line h and auxiliary points S_1, S_2) corresponding to the right-distributive law $a(b + c) = ab + ac$, with the inessential restrictions $a, b, c \neq 0$ and $a \neq 1$. This closure theorem is easily proved from the little theorem of Desargues by drawing through P_1 the parallel h to g, determining the intersections S_1, S_2 of h with the paral-

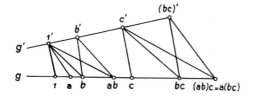

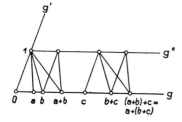

Fig. 25

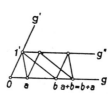

Fig. 26

Fig. 27

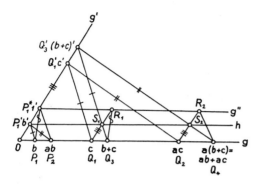

Fig. 28

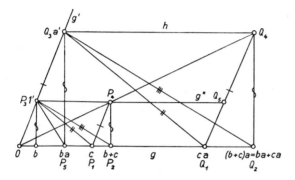

Fig. 29

lels to g' through Q_1 and Q_2, and then applying (d) three times, first to the triangles $P_1P_1'P_1^*$, $Q_3A_1R_1$, then to the triangles $Q_1Q_1'Q_2$, $S_1Q_3'S_2$, and finally to the triangles $P_1'P^*P_2$, $S_2R_2Q_4$.

Figure 29 illustrates the closure theorem (with the auxiliary line h and the auxiliary point Q_4) corresponding to the left-distributive law $(b + c)a = ba + ca$ (again with the restrictions $a, b, c \neq 0$ and $a \neq 1$). This closure theorem follows from (D) if we draw through Q_3 the parallel h to g and then determine the intersection Q_4 of h with the line joining O and P_4. If we then apply (D) to the triangles $P_1P_3P_4$ and $Q_1Q_3Q_4$, and subsequently also to the triangles $P_2P_3P_4$ and $Q_2Q_3Q_4$, we obtain the result that $(P_4, P_1) \parallel (Q_4, Q_1) \parallel g'$ and $(P_2, P_4) \parallel (Q_2, Q_4)$, so that $(Q_2, Q_4) \parallel (P_5, Q_3)$. From (d) for the triangles $P_3Q_3P_5$, $Q_2Q_4Q_5$, it then follows that $(P_3, P_5) \parallel (Q_2, Q_5)$, so that $(b + c)a = ba + ca$.

Thus we have the following theorem.

Theorem 18: *The segments on a line x of a Desargues plane $\mathfrak{A}$ form a skew field K.*

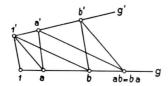

Fig. 30

If $\mathfrak{A}$ is a Pappus plane, the multiplication of segments on x is commutative, as can be seen from Fig. 30, so that the segments on a line x of a Pappus plane form a field.

For the proof that the Desargues plane is isomorphic with the affine coordinate plane over the skew field K we may proceed as before. Equation (1) follows here from the validity of (d).

2.4.   Result of the Process of Algebraization

We can sum up the results of §§2.2 and 2.3 as follows.

Proposition 1.   *The concept of a Desargues (Pappus) plane determines the same geometric entity as an affine-coordinate plane over a skew field (a field).*

Since this theorem states that for the Desargues planes the synthetic and analytic point of view lead to exactly the same results, any given problem can be dealt with either synthetically or analytically, as may be more convenient in each case. All the above results, obtained synthetically for the Desargues plane, could just as well be proved analytically. The analytic apparatus will be developed in Chapter 7, so that at this point we can bring to an end our study of coordinate planes. Let us merely describe, as an illustration of Proposition 1, the analytic representation of affinities.

2.5.   Representation of Affinities

Theorem 19:   *Let $\mathfrak{A}(K)$ be the coordinate plane over a field K, let U be the set of $\mathfrak{A}(K)$, let λ be a linear mapping of U onto itself, and finally let C be an element of U. Then the mapping α with*

$$X\alpha = X\lambda + C \qquad X \text{ in } U$$
(1)
$$(A + \langle B\rangle)\alpha = A\alpha + \langle B\lambda\rangle \qquad B \neq O, \text{ and } A \text{ in } U$$

is an affinity of $\mathfrak{A}(K)$. For every affinity α of $\mathfrak{A}(K)$ there exists a linear mapping λ of U onto itself and a C in U such that (1) holds.

Proof.   We first show
(i)   *The mapping α with the properties (1) is a collineation of $\mathfrak{A}(K)$.*
Obviously α is a one-to-one mapping of U and of G (the set of lines of

$\mathfrak{A}(K)$), each onto itself, such that: from $X \in A + \langle B \rangle$ it follows that $X\alpha \in A\alpha + \langle B\lambda \rangle = (A + \langle B \rangle)\alpha$, in view of the fact that $X = A + Bu$ implies $X\alpha = X\lambda + C = (A + Bu)\lambda + C = A\lambda + (B\lambda)u + C = A\alpha + (B\lambda)u$. Furthermore,

(ii)   *The set $\mathfrak{L}$ of collineations α of the form* (1) *constitutes a group.*

Given two mappings α, α' of the form (1) the mapping $\alpha'\alpha^{-1}$ is also of the form (1): moreover, $X(\alpha'\alpha^{-1}) = X(\lambda'\lambda^{-1}) + (C' - C)\lambda^{-1}$, and similarly for the lines. With α, α' we also have $\alpha'\alpha^{-1} \in \mathfrak{L}$, so that $\mathfrak{L}$ is a group by the criterion for subgroups.

(iii)   *If $P_1 P_2 P_3$ and $Q_1 Q_2 Q_3$ are two nondegenerate triangles, there exists at most one collineation α in $\mathfrak{L}$ with $P_i\alpha = Q_i$ ($i = 1, 2, 3$).*

If there were two mappings of α, α' in $\mathfrak{L}$ with $P_i\alpha = Q_i$ and $P_i\alpha' = Q_i$ ($i = 1, 2, 3$), then for the corresponding linear mappings λ, λ' of U onto itself we would have $(P_2 - P_1)\lambda = P_2\lambda - P_1\lambda = P_2\alpha - P_1\alpha = P_2\alpha' - P_1\alpha' = P_2\lambda' - P_1\lambda' = (P_2 - P_1)\lambda'$ and correspondingly $(P_3 - P_1)\lambda = (P_3 - P_1)\lambda'$. Since $P_2 - P_1$, $P_3 - P_1$ are linearly independent (for otherwise P_1, P_2, P_3 would be collinear) we have $\lambda = \lambda'$. Furthermore, for the corresponding points C, C' we then have $C = P_1\alpha - P_1\lambda = P_1\alpha' - P_1\lambda' = C'$, so that $\alpha = \alpha'$.

(iv)   *Every axial affinity α of $\mathfrak{A}(K)$ is an element of $\mathfrak{L}$.*

Let $\bar{A} + \langle \bar{B} \rangle$ be the axis of α, and let P, Q be a point and its image not belonging to the axis. Then $P - \bar{A}, \bar{B}$ are linearly independent vectors in U, and so also are $Q - \bar{A}, \bar{B}$, and there exists a linear mapping λ of U onto itself with $(P - \bar{A})\lambda = Q - \bar{A}$ and $\bar{B}\lambda = \bar{B}$. If we set $C = \bar{A} - \bar{A}\lambda$, then for the collineation $\bar{\alpha}$ constructed with λ, C in accordance with (1) we have $X\bar{\alpha} = X$ for all X in $\bar{A} + \langle \bar{B} \rangle$ (since $\bar{A}\bar{\alpha} = \bar{A}\lambda + C = \bar{A}$ and $\bar{B}\lambda = \bar{B}$) and also $P\bar{\alpha} = Q$. Thus $\bar{\alpha}$ is an affinity with axis $\bar{A} + \langle \bar{B} \rangle$ taking P into Q, and is therefore equal to α by Lemma 16. Consequently $\alpha \in \mathfrak{L}$.

From (ii) and (iv) we have

(v)   *Every affinity of $\mathfrak{A}(K)$ is an element of $\mathfrak{L}$.*

Now if α is an element of $\mathfrak{L}$ and $P_1 P_2 P_3$ is a nondegenerate triangle, then the triangle $P_1\alpha P_2\alpha P_3\alpha$ is also nondegenerate, so that by Lemma 14 there exists an affinity α' in $\mathfrak{A}(K)$ with $P_i\alpha' = P_i\alpha$ ($i = 1, 2, 3$). By (v) we have $\alpha' \in \mathfrak{L}$ and by (iii) also $\alpha' = \alpha$, so that the group of affinities of $\mathfrak{A}(K)$ is identical with the group $\mathfrak{L}$. Thus the proof of Theorem 19 is complete.

Let us now present the usual representation of affinities for the coordinates of points,[5] which follows immediately from (1):

(1')
$$x^* = a_1 x + b_1 y + c_1$$
$$a_1 b_2 - b_1 a_2 \neq 0.$$
$$y^* = a_2 x + b_2 y + c_2$$

[5]Cf. II10, § 2.3.

Here we have set $X = (x, y)$, $X\alpha = (x^*, y^*)$, $C = (c_1, c_2)$ and have described the linear mapping λ by means of the matrix $\begin{pmatrix} a_1 & b_1 \\ a_2 & b_2 \end{pmatrix}$. In particular, an affinity with axis defined by the equation $ax + by + c = 0$ is represented by

$$
\begin{aligned}
x^* &= x + u(ax + by + c) \\
(2) &\qquad\qquad\qquad\qquad\qquad\qquad ua + vb + 1 \neq 0. \\
y^* &= y + v(ax + by + c)
\end{aligned}
$$

3. Projective Planes

3.1. *The System of Axioms for a Projective Plane*

Given two distinct points in an affine plane, it is always possible to find a line incident with both points, whereas for two distinct lines there does not always exist a point incident with both lines. Thus the affine axioms of incidence are not symmetric with respect to points and lines, and then it is natural to think of defining a "plane" in such a way that points and lines occur symmetrically. In this way we arrive at the concept of a projective plane.

A geometric triple $\Pi = (\mathfrak{P}, \mathfrak{G}, I)$ consisting of a set $\mathfrak{P}$ of points, a set $\mathfrak{G}$ of lines and an incidence relation I is called a projective plane if the following axioms hold:

P1: *For two points P, Q there exists at least one line incident with P, Q. For two lines g, h there exists at least one point incident with g, h.*

P2: *If P, Q are two points incident with two lines g, h, then P = Q or g = h.*

P3: *There exist at least four points such that no three of them are collinear.*

Axioms P1 and P2 imply the existence and uniqueness of the intersection of two distinct lines and of the line joining two distinct points. Thus the symmetry of the statements in these axioms with respect to point and line is immediately evident. On the other hand, the statement dual to Axiom P3 can be deduced from the three given axioms. For it requires the existence of four lines, no three of which are copunctual, and if A, B, C, D are the four points guaranteed by Axiom P3 such that no three of them are collinear, then by Axiom P1 there exist the four lines (A, B), (B, C), (C, D), and (D, A) which by Axiom P2 have the desired property. Thus in a projective plane we have the following basic principle.

Principle of duality: *If in a projective plane a theorem, stated in terms of the basic concepts of point, line, and incidence, is valid and follows solely from*

Axioms P1 *through* P3, *then the dual theorem arising from it by interchange of the words point and line is also valid.*

Thus projective planes and affine planes are equivalent in the following sense:

Theorem 20: *Every affine plane can be extended to a projective plane by the adjunction of the "points at infinity" and the "line at infinity." Every projective plane is an affine plane with respect to a fixed line regarded as the line at infinity.*

Proof. a) Let $\mathfrak{A} = (\mathfrak{P}, \mathfrak{G}, I)$ be an affine plane, let every pencil of parallel lines be called a *point at infinity,* and let the set g_∞ of these infinitely distant points be called the *line at infinity.* Then we can extend the given affine plane to a projective plane $(\mathfrak{P}', \mathfrak{G}', I')$, where $\mathfrak{P}'$ is the set $\mathfrak{P}$ together with all points at infinity and $\mathfrak{G}'$ is the set $\mathfrak{G}$ together with g_∞. The incidence relation I is extended to I' by stipulating that a line g in $\mathfrak{G}$ is incident with a point at infinity Π_h if $g \in \Pi_h$, and g_∞ is incident with all the points at infinity and with no other points. Then it is easy to show that $(\mathfrak{P}', \mathfrak{G}', I')$ is a projective plane.

b) Let $(\mathfrak{P}, \mathfrak{G}, I)$ be a projective plane, and let u be a fixed line. Then if $\mathfrak{P}'$ is the set of all points excluding the points on u, and $\mathfrak{G}$ is the set of all lines excluding the line u, and if I' is the restriction of I to $\mathfrak{P}'$ and $\mathfrak{G}'$, it is easy to show that $(\mathfrak{P}', \mathfrak{G}', I')$ is an affine plane.

This theorem shows that the geometric entities determined by the two systems of axioms are the same. For a given problem we may take either the affine or the projective point of view.

We may also formulate closure theorems for the projective plane. Let us here give only the projective form of the Pappus-Pascal theorem.

Six given points $P_1, Q_2, P_3, Q_1, P_2, Q_3$ with the property that no two cyclically successive points are collinear with any of the other points will be called a *simple hexagon.*

$1^{\circ\circ}$ Theorem of Pappus-Pascal (projective form): *If the vertices of a simple hexagon $P_1Q_2P_3Q_1P_2Q_3$ lie alternately on two lines,* then the intersections of the opposite sides $(P_1, Q_2), (P_2, Q_1); (P_1, Q_3), (P_3, Q_1); (P_2, Q_3), (P_3, Q_2)$ *lie on a line p* (the Pascal line) (Fig. 31).

Two triangles $P_1P_2P_3$ and $Q_1Q_2Q_3$ with sides $p_1 = (P_2, P_3)$, $p_2 = (P_3, P_1)$, $p_3 = (P_1, P_2)$, $q_1 = (Q_2, Q_3)$, $q_2 = (Q_3, Q_1)$, $q_3 = (Q_1, Q_2)$ are said to be *perspective with respect to the point Z* (the center) if for each fixed $i = 1, 2, 3$ the three points Z, P_i, Q_i are collinear; and the two triangles are *perspective with respect to the line a* (the axis) if for each fixed $i = 1, 2, 3$ the lines a, p_i, q_i are copunctual.

$2^{\circ\circ}$ Theorem of Desargues (projective form): *If two triangles are in perspective with respect to a point, they are also in perspective with respect to a line* (Fig. 32).

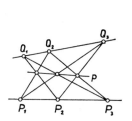

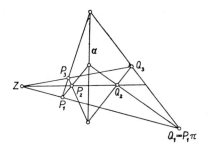

Fig. 31   Fig. 32

In the projective extension Π of a Pappus plane the validity of (P) means that Theorem 1°° holds for all configurations for which the line at infinity is the Pascal line. It is easy to show that in Π the projective theorem of Pappus-Pascal holds generally. A projective plane in which 1°° holds is likewise called a Pappus plane.

3.2. *Mappings in the Projective Plane*

Certain important observations can now be made about mappings in a projective plane.

Every projective mapping consists of one or more perspectivities of the following three types: first of all, a perspectivity of a range of points[6] onto another range of points (Fig. 33). In such a transformation the lines joining corresponding points pass through a certain point, the *center Z of the perspectivity*. In the second place, we define a perspectivity of a pencil of lines onto another pencil as a mapping in which corresponding lines intersect on a fixed line *a*, the *axis of the perspectivity* (Fig. 34). In the third place, a mapping of a pencil of lines onto a range of points in which corresponding elements are incident is called a *perspectivity of the pencil of lines onto the range of points*. A mapping of a one-dimensional fundamental

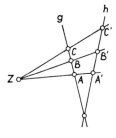

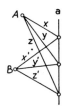

Fig. 33   Fig. 34

[6]A range of points is defined as the set of all points on a line.

configuration (by which we mean a pencil of lines or a range of points) onto another such configuration is called a *projective mapping* of the one configuration onto the other if the mapping consists of a finite sequence of perspectivities.

For the projective mappings of a range of points onto another range of points the fundamental theorem of projective geometry is as follows.

There exists at most one projective mapping of a range of points onto itself, or onto another range, taking three given distinct points into three given distinct points.

In an arbitrary projective incidence plane the fundamental theorem does not necessarily hold. It has been shown that the fundamental theorem in a projective incidence plane is equivalent to the theorem of Pappus-Pascal.[7]

A collineation in a projective plane is defined in exactly the same way as in an affine plane. Among the collineations of a projective plane the most important are the *perspective collineations*, which leave a line a, called the axis, pointwise fixed and a point Z, called the center, linewise fixed. If a I Z, such collineations are called *elations,* and otherwise *homologies.*

By §1, Lemma 5, the collineations of an affine plane $\mathfrak{A}$ induce collineations of the projective extension Π of $\mathfrak{A}$; in particular, the translations induce elations (for which reason elations are often called translations), the dilations induce homologies with g_∞ as axis, and the affinities with a as axis induce elations or homologies with a as axis, depending on whether the traces are parallel to a or not. Thus translations, dilations, and axial affinities of an affine plane correspond in the projective extension to perspective collineations, which is the type common to all of them, and the various theorems requiring separate proofs in affine geometry are seen to be special cases of one and the same projective theorem. As examples let us mention

(i) *A projective collineation has at most one fixed point not incident with the axis.*

(ii) *There exists at most one perspective collineation, with a as axis and Z as center, taking a point P into a point Q that is collinear with P and Z (P, Q $\neq$ Z; P, Q Ι a).*

(iii) *The set P(Z, a) of perspective collineations with a as axis and Z as center forms a group.*

In view of Theorem 20, the proof of (i) through (iii) can be reduced to the proofs in §1. As before, a group $P(Z, a)$ is said to be linearly transitive if for every two points P, Q [P, Q Ι a; P, $Q \neq Z$] collinear with Z there exists a perspective collineation π in $P(Z, a)$ with $P\pi = Q$. The same relation

---

[7]See, for example, F. Schur, Grundlagen der Geometrie; Veblen-Young, Projective geometry.

between the linear transitivity of the group $P(Z, a)$, or its commutativity, and the validity of the (projective) closure theorems (Pappus-Pascal) holds here as in §1.4.

Every product of perspective collineations is called a projective collineation[8] of the projective plane. In the same way as for affinities in a Desargues plane, we can show that for every two quadruples of points $P_1P_2P_3P_4$, $Q_1Q_2Q_3Q_4$ such that no three points are collinear there always exists a projective collineation π with $P_i\pi = Q_i$ for $i = 1, 2, 3, 4$, if every group $P(Z, a)$ is linearly transitive.

The existence of collineations $\neq 1$, to say nothing of projective collineations, is by no means guaranteed for an arbitrary projective plane.

On a projective plane we may also consider *correlations,* which in a certain sense are dual to the collineations. Thus a *correlation* of the projective plane is a one-to-one mapping of the set of points of the projective plane onto the set of lines, and of the set of lines onto the set of points, such that incidence is preserved. A correlation is said to be projective if a range of points is mapped projectively onto the corresponding pencil of lines. Not every correlation is projective. In a projective plane the existence of correlations (to say nothing of projective correlations) is as little guaranteed as the existence of collineations $\neq 1$.

In the construction of metric geometry a central role is played by the involutory correlations, which are called *polarities.*[9] The elements corresponding to each other in a polarity are called *pole* and *polar.* If a point lies on its polar, it is called *self-polar.* Depending upon whether it contains self-polar points or not, a given polarity is said to be *hyperbolic* or *elliptic,* a terminology that will be further explained in Chapter 5.

As an example of a hyperbolic polarity let us mention the well-known relationship between pole and polar with respect to a conic section in the Euclidean (analytic) plane. For example, in the polarity corresponding to the circle $-x^2 - y^2 + 1 = 0$ we have the following pairs of corresponding elements: (1) the point $O = (0, 0)$ and the line at infinity; (2) the point $(x_0, y_0) \neq (0, 0)$ and the line $-x_0x - y_0y + 1 = 0$; and (3) the point on the line at infinity defined by the pencil of lines parallel to $y_0x - x_0y = 0$ (which thus has the homogeneous coordinates $(x_0, y_0, 0)$) and the line $-x_0x - y_0y = 0$. Here the self-polar points are exactly the points on the circle.

In this example the polar p_0 of a point $P_0 = (x_0, y_0) \neq (0, 0)$ can also be constructed in the following way: if P_1 is the mirror image of P_0 under inversion with respect to the given circle, then the line through P_1 perpendicular to the line

---

[8]A projective collineation (and correspondingly a projective correlation) may be defined as a collineation in which every line is mapped projectively onto its image, and the reader is also referred to the definition of projective collineation in II10, §2.2. In a projective plane all these definitions are equivalent to one another.

[9]It should be noted that here, though not in n-dimensional space with $n > 2$, every involutory correlation is called a polarity.

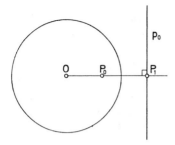

Fig. 35

(O, P_1) (on which P_0 lies) is the polar of P_0 (Fig. 35). For if $P_1 = (x_1, y_1)$, then

$$x_1 = \frac{x_0}{x_0^2 + y_0^2}, y_1 = \frac{y_0}{x_0^2 + y_0^2}$$

and the line (O, P_1) has the equation $y_1 x - x_1 y = 0$, so that the equation for the line perpendicular to it at P_1 is $-x_1 x - y_1 y + x_1^2 + y_1^2 = 0$, so that $-x_0 x - y_0 y + 1 = 0$.

Let us here observe that this relationship between a conic section and a hyperbolic polarity can be made the basis of a purely synthetic definition of a conic section in a projective incidence plane Π (v. Staudt):

Given a hyperbolic projective polarity, the set of self-polar points is called a conic section.

If the theorem of Pappus-Pascal holds in the projective incidence plane, we can deduce the usual facts about conic sections from the properties of a polarity (see II10, §6): (1) a line cuts a conic section in exactly two points or, if it is the polar of a self-polar point (and thus a "tangent"), in exactly one point or else in no point at all; (2) the theorem of Pascal holds for a hexagon inscribed in a conic section.

3.3.   The Projective Coordinate Plane over a Field

Let V be the vector space of triples over the field K. Every one-dimensional subspace T can then be represented in the form

$$T = K\mathfrak{x} = \{\lambda\mathfrak{x} | \lambda \in K\}$$

with $\mathfrak{x} \neq 0$ in T. The vector $\mathfrak{x}$ is called a representative of T. For $\mathfrak{a}, \mathfrak{b} \in V$ let $\mathfrak{a}\mathfrak{b}$ be the ordinary scalar product of a and b (i.e., $\mathfrak{a}\mathfrak{b} = a_1 b_1 + a_2 b_2 + a_3 b_3$ for $\mathfrak{a} = (a_1, a_2, a_3)$ and $\mathfrak{b} = (b_1, b_2, b_3)$) and let $\mathfrak{a} \times \mathfrak{b}$ be the vector product. The mapping $(\mathfrak{a}, \mathfrak{b}) \to \mathfrak{a}\mathfrak{b}$ is a symmetric bilinear form. If U is a subspace, the set of vectors $\mathfrak{v}$ with $\mathfrak{u}\mathfrak{v} = 0$ for all $\mathfrak{u} \in U$ forms a subspace, which we shall denote by U^0. Then dim U + dim $U^0 = 3$.

Every one-dimensional subspace of V is called a *point* and also a *line*. A point X is said to be *incident* with a line u if $X \subset u^0$ (and therefore $u \subset X^0$).

It is clear that a point $X = K\mathfrak{x}$ is incident with a line $u = K\mathfrak{u}$ if and only if $\mathfrak{x}\mathfrak{u} = 0$. The triple consisting of the set of points, the set of lines, and the incidence relation is then called a *projective coordinate plane* over K and is denoted by $\Pi\,(K)$.

Thus by definition a point of $\Pi(K)$ consists of the set of all triples in K proportional to a triple (x_1, x_2, x_3) with $(x_1, x_2, x_3) \neq (0, 0, 0)$, and a line consists of the set of all triples in K proportional to a triple (u_1, u_2, u_3) with $(u_1, u_2, u_3) \neq (0, 0, 0)$; incidence is defined by the equation $x_1u_1 + x_2u_2 + x_3u_3 = 0$.

Since the correspondence $u \to u^0$ is a one-to-one mapping of the set of lines $\Pi(K)$ onto the set of two-dimensional subspaces of V, we could also define the projective coordinate plane over K by saying that the one-dimensional subspaces are points and the two-dimensional subspaces are lines and by defining incidence of point and line by inclusion. Correspondingly, we could say that all one-dimensional subspaces are lines and all two-dimensional subspaces are points (pencils of lines).

Theorem 21:   *The projective coordinate plane $\Pi(K)$ is a projective plane.*

Proof.   Let us consider the validity of the Axioms P1 through P3 successively for $\Pi(K)$.

As for P1:   Two one-dimensional subspaces T, T' are always included in a two-dimensional subspace U. If T, T' are points, then U^0 is a line incident with T, T'. If T, T' are lines, then U^0 is a point incident with T, T'.

Let us note that if $X = K\mathfrak{x}, Y = K\mathfrak{y}$ are two distinct points, then $\mathfrak{x} \times \mathfrak{y}$ is the line joining X and Y. If $u = K\mathfrak{u}, v = K\mathfrak{v}$ are two distinct lines, then $\mathfrak{u} \times \mathfrak{v}$ is the point of intersection of u and v.

As for P2:   Let X, Y be two points each of which is incident with both lines at u, v. Then $X, Y \subset u^0, v^0$, so that $X = Y$ or $X \neq Y$ and $u^0 = v^0$, and thus $u = v$, since there exists only one two-dimensional subspace containing two distinct one-dimensional subspaces.

As for P3:   The vectors $\mathfrak{e}_1 = (1, 0, 0)$, $\mathfrak{e}_2 = (0, 1, 0)$, $\mathfrak{e}_3 = (0, 0, 1)$, $\mathfrak{e}_4 = (1, 1, 1)$ represent four one-dimensional subspaces, no three of which lie in the same two-dimensional subspace. So no three of the points $K\mathfrak{e}_i$ $(i = 1, 2, 3, 4)$ are collinear.

Thus, as can be proved from Proposition 1, we have:

Proposition 2:   *The concepts of a Pappus (projective) plane and a projective coordinate plane over a field characterize the same geometric entity.*

With respect to Proposition 2 let us make a brief remark about the representation of projective collineations in the projective coordinate plane over a field K. It is clear that every linear mapping of V onto itself induces a collineation in $\Pi(K)$, since one-dimensional subspaces are mapped into one-dimensional subspaces, and two-dimensional subspaces into two-dimensional. Conversely we can show, e.g., on the basis of §2.5, that every projective collineation is induced by a mapping of V onto itself. If the linear mapping is represented by the matrix $\mathfrak{A}$, the corresponding projective collineation will be given by

$$x^* = \mathfrak{A}x \qquad \det \mathfrak{A} \neq 0.$$

Correspondingly, for every projective correlation κ there exists a linear mapping of V onto itself which induces κ.

4. Projective Geometry in Three-Dimensional Space

In our study of axiomatic foundations let us now make some remarks about the synthetic geometry of projective spaces, particularly the embedding of projective incidence planes in a projective space. Up to now we have said nothing about whether the planes under discussion are embedded in a surrounding space or not, but it seems reasonable that certain additional properties can be derived for a plane that is embeddable in a space.

4.1. *System of Axioms for Three-Dimensional Projective Space* [10]

Let there be given a set $\mathfrak{R}$ of elements divided into two subsets, called *points* and *planes*. Between the points and the planes let there be defined a concept of incidence, with the following axioms:

PR1: *For any three given points there is a plane incident with them.*

PR2: *For any three given planes there is a point incident with them.*

PR3: *If each of two distinct points is incident with each of two distinct planes, then every point incident with each of the planes is also incident with every plane that is incident with each of the points.*

PR4 (Fundamental figure): *There exist five points P_i in $\mathfrak{R}$ and five planes ε_i in $\mathfrak{R}$ ($i = 1, 2, 3, 4, 5$) such that the point P_i is not incident with the plane ε_k if and only if $i - k \equiv \pm 2 \pmod 5$.*

The meaning of Axiom PR4 is that no four points in the fundamental figure are coplanar and no four planes are copunctual. From Axiom PR3 it readily follows that every set of three points in the fundamental figure is incident with exactly one of the planes, and every set of three planes is incident with exactly one of the points.

A set $\mathfrak{R}$ of points and planes satisfying the Axioms PR1 through PR4 will be called a *projective space*.

[10] As formulated by A. Winternitz: Zur Begründung der projektiven Geometrie: Einführung idealer Elemente unabhängig von der Anordnung. Ann. of Math. 41 (1940), pp. 365–390.

By a line in the space $\Re$ we shall mean, on the one hand, any set of points incident with two distinct planes in $\Re$ and, on the other hand, every set of planes incident with two distinct points. Two lines are said to be equal if either they agree in their sets of points or in their sets of planes or else every point of one line is incident with every plane of the other.

From this definition it might appear that there are two distinct kinds of lines in $\Re$. But this is not the case:

Theorem 22: *Every line g defined as a set of points is equal to a line defined as a set of planes, and every line g' defined as a set of planes is equal to a line defined as a set of points.*

Proof. Let g be defined as the set of points incident with two distinct planes α, β. By PR2 there exists, corresponding to the planes α, β, a point P incident with α, β, and by PR4 there exists a plane γ not incident with P. Then by PR2 the planes α, β, γ have a point Q in common, with $P \neq Q$ (since $P \, \mathrm{I} \, \gamma$), and the set of planes incident with P and Q is by definition a line g'. But now by PR3 every point of g is incident with every plane containing the two points P, Q, so that g and g' are identical by definition. The second part of the assertion is proved in exactly the same way from PR1, PR3 and PR4.

Thus a line g is defined either by a set of points $\mathrm{p}(g)$ or by a set of planes $\mathrm{e}(g)$. Every point in $\mathrm{p}(g)$ and every plane in $\mathrm{e}(g)$ is said to be *incident* with g, and every other point and every other plane is not incident with g. By PR3, the set of points $\mathrm{p}(g)$ consists precisely of all the points incident with two distinct but arbitrarily chosen planes $\in \mathrm{e}(g)$, and $\mathrm{e}(g)$ consists precisely of all the planes incident with two distinct but arbitrarily chosen points $\in \mathrm{p}(g)$. It follows that every line is incident with at least two points and at least two planes.

As can be seen at once from the axioms, the following duality principle holds in projective space, in complete analogy with the plane:

If in a theorem in projective geometry in space, whose proof depends only on PR1 through PR4, we replace points by planes and planes by points, we again have a valid theorem.

For it is clear that the axioms are dual to one another: PR1 is dual to PR2, and both PR3 and PR4 are self-dual. Since the definition of a straight line in projective space is self-dual, in the dualization of a theorem the lines are left unchanged.

From the system of axioms we at once obtain the following table of incidences for points, lines, and planes in the projective space $\Re$:

Ia. For two distinct points P and Q there exists exactly one line g incident with P and Q.

Ib. For two distinct planes ε and ε' there exists exactly one line g incident with ε and ε'.

IIa. For two distinct lines g and h incident with a plane ε there exists ex-

IIb. For two distinct lines g and h incident with a point P there exists exactly one point P incident with g and h.

IIIa. For three points not incident with the same line there exists exactly one plane incident with all three points.

IVa. If a point P is not incident with a line g, there exists exactly one plane incident with P and g.

Va. If a point P is incident with a line g and g is incident with a plane ε, then P is also incident with ε.

VIa. If a line g is incident with two distinct points P and Q, both of which are incident with a plane ε, then g is also incident with ε.

actly one plane incident with g and h.

IIIb. For three planes not incident with the same line there exists exactly one point incident with all three points.

IVb. If a plane ε is not incident with a line g, there exists exactly one point incident with ε and g.

Vb. If a plane ε is incident with a line g and g is incident with a point P, then ε is also incident with P. (Identical with Va.)

VIb. If a line g is incident with two distinct planes ε and ε', both of which are incident with a point P, then g is also incident with P.

VII. On every line there lie at least three distinct points. In every plane there lie at least four points, no three of which are collinear. There exist at least five points, no four of which lie in a plane.

To prove all the statements in the table it suffices, on account of duality, to prove one of the two statements a and b in each case. The statements Va and VIa are at once clear from the definitions and from what has already been proved.

As for Ia: The set $\mathfrak{e}$ of planes incident with P and Q forms by definition a line g. The set $\mathfrak{e}(g)$ contains two distinct planes. By Theorem 22 the line g can also be represented as a set of points $\mathfrak{p}(g)$, where $\mathfrak{p}(g)$ contains all those points incident with every plane $\in \mathfrak{e}(g)$. Thus P and Q belong to the set $\mathfrak{p}(g)$ and are therefore incident with g. Now if g' were a second line incident with P and Q, so that P and Q would belong to the set of points $\mathfrak{p}(g')$ defining g', there would exist a set $\mathfrak{e}(g')$ of planes which would also represent g'. This set $\mathfrak{e}(g')$ consists precisely of all those planes that are incident with two distinct points $\in \mathfrak{p}(g')$, and thus consists of all planes incident with P and Q; consequently it is equal to $\mathfrak{e}(g)$, which means that $g = g'$.

As for IIa: In view of Ia we need only prove the existence of a point common to g and h. By what has already been proved, g and h are incident with two distinct planes, one of which is the plane ε and the other is, say, α (for g) and β (for h). Then by PR2 the three planes α, β, ε have a point P in common, which by definition is incident with g and h.

As for IIIa: If g, g' are the lines joining P to Q and R (by Ia they exist and are uniquely defined), then by IIb there exists a plane ε incident with g and g' and therefore with P, Q, R (by Va). If ε' were a second plane incident with P, Q, R, then by VIa the plane ε' would also be incident with g and g', from which it would follow from IIb (since P, Q, R cannot be incident with the same line) that $\varepsilon = \varepsilon'$ and therefore $g \neq g'$.

As for IVa: On g there must exist two distinct points Q and R not collinear with P, since otherwise it would follow from Ia that P lies on g. Thus by IIIa there exists exactly one plane incident with P, Q, R and therefore, by VIa, with P and g.

The statements of VII can easily be derived from the existence of the fundamental figure in PR4.

4.2.   *The Theorem of Desargues and Perspectivities in Projective Space*

From Ia, IIa, and VII it follows that every plane ε of the projective space $\mathfrak{R}$ is a projective incidence plane, but the fact that plane ε is embedded in a space $\mathfrak{R}$ has the additional consequence that the Desargues Theorem must hold in ε, as we shall now prove from Theorem 12 and §3.2, by showing that the group of perspective collineations with fixed center and fixed axis is linearly transitive in ε. Since the perspective collineations can be compounded from projective mappings in $\mathfrak{R}$, we first define, in analogy with the plane:

A one-to-one mapping of the points of a plane ε onto another plane ε' (Fig. 36) such that every point and its image lie on a straight line through a fixed point O [O Ɪ ε, ε'] is called a *perspectivity* π of the plane ε onto the plane ε' with center O. Collinear points in ε are thereby mapped onto collinear points in ε'; for if P_1, P_2, P_3 are three points on a line g with g Ɪ ε, then in view of the fact that O Ɪ ε and therefore O Ɪ g, there exists by IVa exactly one plane α incident with O and g. Then by IVa the lines (O, P_i) $(i = 1, 2, 3)$ lie in α and consequently, by Va, the points $P_i\pi$ $(i = 1, 2, 3)$ also lie in α. On the other hand, we also have $P_i\pi$ Ɪ ε', so that the points $P_i\pi$ must be incident with the lines of intersection of α and ε'. By Ib these lines exist and are uniquely defined, and also $\alpha \neq \varepsilon'$, since O Ɪ α, O Ɪ ε'.

The product of a perspectivity π of ε onto ε' with a perspectivity π' of ε' onto ε is thus a collineation in the plane ε, and in fact, since the line a of intersection of ε and ε' remains pointwise fixed, by §3.1 this product is a perspective collineation in ε with a as axis. If the centers O, O' of π, π' are

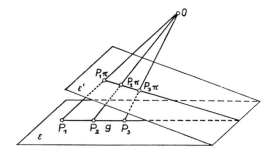

Fig. 36

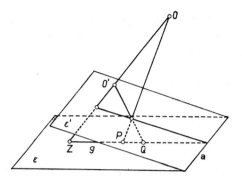

Fig. 37

distinct, the center Z of the collineation $\pi\pi'$ in ε is given by the point of intersection of (O, O') with ε (which exists by IVb).

The linear transitivity of the group of perspective collineations with a as axis and Z as center for arbitrary a and Z in ε is proved as follows (Fig. 37). Let ε' be an arbitrary plane $\neq \varepsilon$ through a and let O be an arbitrary point with $O \mathrel{\rlap{\,/}{\mathrm{I}}} \varepsilon, \varepsilon'$, and finally let π be the perspectivity of ε onto ε' with center O. Then if P, Q are two arbitrary points on the line g through $Z[P, Q \neq Z;$ $P, Q \mathrel{\rlap{\,/}{\mathrm{I}}} a]$, it follows from IIa that the line joining Q with $P\pi$ intersects the line (Z, O) in a point O', since Q and $P\pi$ lie in the plane α spanned by g and O, so that by VIa the lines $(Q, P\pi)$ and (O, Z) also lie in this plane. If π' is the perspectivity of ε' onto ε with center O', it is clear that $\pi\pi'$ is a perspective collineation with Z as center and a as axis, taking the point P into the point Q. Thus the proof of our assertion is complete and with it, by Theorem 12 and §3.2, the proof of the theorem of Desargues in the plane ε.

Of course, this proof of the Desargues theorem could also be stated in terms of the configuration involved, since the application of a perspectivity corresponds to the argument that corresponding sides of perspective (non-coplanar) triangles must intersect along a straight line (namely, the intersection of the two planes in which the triangles lie).

Thus we have:

Theorem 23: *The Desargues Theorem holds in every plane that can be embedded in a projective space.*

Finally, let us note that the converse is also true; every projective plane Π in which the Desargues Theorem is valid can be embedded in a projective space. Namely, Π can be represented as the projective coordinate plane over a skew field K; for if we distinguish a line in Π as the line at infinity, the resulting affine plane can be represented, by Theorem 17, as the affine coordinate plane over a skew field K. If we then define the projective coordinate plane over the skew field K in the same way as in §3.1, we can show that Π is isomorphic to the projective coordinate plane over K. Then by means of quadruples of elements from K we can define a projective coordinate space $P_3(K)$ in which the coordinate plane corresponding to Π is included.

5. Projective-Metric Geometry

It has been shown by Cayley and Klein that all metric properties can also be interpreted as projective-metric properties. Let us consider how we can define a metric in a projective Pappus plane. Here we take into account only Pappus planes in which the Fano axiom holds:

Axiom P4 (Fano):   *The diagonal points of a complete quadrilateral are not collinear* (Fig. 38).

By a complete quadrilateral we here mean any set of four points (no three collinear) together with the set of six lines joining them in pairs. If A, B, C, D are the points of a complete quadrilateral, the intersections of the three pairs of opposite sides $(A, B), (C, D)$; $(A, C), (B, D)$; $(B, C), (A, D)$ are the *diagonal points* of the complete quadrilateral.

If P4 holds, then for every point Z in the projective plane and every line a not incident with Z there exists exactly one involutory homology with Z as center and a as axis. On the other hand, no elation can be involutory. Conversely, if P4 does not hold, then all elations are involutory, but no homology can be involutory.

For the corresponding coordinate field K the Fano axiom implies that the characteristic[11] of the field K is different from 2.

In order to introduce a (projective) metric in a projective Pappus plane Π (in which the Fano axiom is valid) it is natural, from the synthetic point of view, to begin by introducing a definition of orthogonality in Π, since the concept of distance requires us to introduce numbers or, more generally, elements of a field. Orthogonality can be defined as follows:

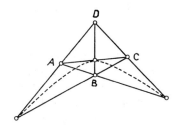

Fig. 38

[11]Cf. IB5, § 1.11.

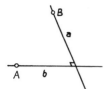

Fig. 39

In Π let there be given a projective polarity π. We then define an orthogonality in Π with respect to this polarity in the following way:

Two lines a and b are said to be *orthogonal* (perpendicular) to each other if a is incident with the pole B of b (and therefore b is incident with the pole A of a) (Fig. 39).

Orthogonality is symmetric, since the polarity π is involutory.

The pair (Π, π) consisting of the Pappus plane Π (with the Fano axiom) and the projective polarity π, will be called an *ordinary projective-metric plane*.

If σ is a harmonic homology with a nonincident pole-polar pair as center Z and axis a, then $\pi\sigma\pi = \sigma$, since $\pi\sigma\pi$ is again an involutory homology with $Z\pi = a$ as axis and $a\pi = Z$ as center, and since there can exist only one involutory homology, namely σ, with this axis and this center. Since $\sigma\pi\sigma = \pi$, the transformation σ takes a pole-polar pair into a pole-polar pair, and thus takes perpendicular lines into perpendicular lines. Thus the group $\mathfrak{B}$ of projective collineations generated by the harmonic homologies with nonincident pole-polar pair as center and axis will be called the *group of motions* of the ordinary projective-metric plane. Every element of $\mathfrak{B}$ leaves orthogonality invariant.

By §3.3, Theorem 2, an ordinary projective-metric plane can be represented as the projective coordinate plane over a field K of characteristic $\neq 2$.

To π there corresponds a symmetric bilinear form of rank 3[12] over the three-dimensional vector space V over K by means of which a projective coordinate plane over K is described. For, since π is a projective correlation, it is induced, as remarked in §3.3, by a linear mapping α. If we then set

$$f(\mathfrak{x}, \mathfrak{y}) = \mathfrak{x}(\mathfrak{y}\alpha) \qquad \mathfrak{x}, \mathfrak{y} \in V,$$

it follows that f is a bilinear form of rank 3, since $f(\mathfrak{a}, \mathfrak{x}) = 0$ for all $\mathfrak{x}$ in V implies $\mathfrak{a} = 0$. Then mutually perpendicular lines will be represented by vectors $\mathfrak{x}$, $\mathfrak{y}$ for which $f(\mathfrak{x}, \mathfrak{y}) = 0$ and the corresponding quadratic form $f(\mathfrak{x}, \mathfrak{x})$ can be considered as the square of the length of the vector $\mathfrak{x}$. In this way we have given a natural definition of distance that agrees very well with

[12]A symmetric bilinear form f over the vector space V (over a field K of characteristic $\neq 2$) of dimension n is said to be of rank k if $k = \dim V - \dim V^{\perp}$ and $V^{\perp} = \{\mathfrak{x} \mid f(\mathfrak{x}, \mathfrak{y}) = 0 \text{ for all } \mathfrak{y} \in V\}$.

everyday ideas. In a basis e_1, e_2, e_3 of V the bilinear form is represented as follows:

$$f(\mathfrak{x}, \mathfrak{y}) = \sum_{i,k=1}^{3} f_{ik} x_i y_k,$$

where we have set $f_{ik} = f(e_i, e_k)$ and $\mathfrak{x} = \sum_{i=1}^{3} x_i e_i$

$$\text{and } \mathfrak{y} = \sum_{i=1}^{3} y_i e_i.$$

To say that f is of rank 3 is the same as to say that det $(f_{ik}) \neq 0$.

Then by §3.3 the motions of the ordinary projective-metric plane are represented by linear transformations $\mathfrak{x}^* = \mathfrak{A}\mathfrak{x}$. Since motions in the ordinary projective-metric plane leave the polarity invariant, the corresponding linear transformations must take the bilinear form into itself up to a constant factor c; that is, we must have $f(\mathfrak{A}\mathfrak{x}, \mathfrak{A}\mathfrak{y}) = cf(\mathfrak{x}, \mathfrak{y})$.

In order to subsume Euclidean geometry under the concept of a projective-metric plane, we now also define a *singular projective-metric plane*:

In a projective Pappus plane Π (with the Fano axiom) let there be given a projective involution ρ on a line u without fixed points. The pairs of points on u corresponding to each other under ρ will be said to be *polar*. Two lines $a, b \neq u$ in Π are then said to be orthogonal (perpendicular) to each other if a and b intersect the line u in two polar points B and A (Fig. 40). Moreover, the line u is orthogonal to all lines and to itself.

Here also orthogonality is symmetric. The pair (Π, ρ), consisting of the projective plane Π (with the Fano axiom) and the projective involution ρ without fixed points, is called a *singular projective-metric plane*.

In the singular case also, we can show that a harmonic homology whose center Z lies on the line u and whose axis cuts the line u in the point on u polar to Z takes a pair of polar points on u into a pair of polar points on u, and thus takes perpendicular lines into perpendicular lines. The group of projective collineations generated by all such harmonic homologies will also be called the *group of motions* of the singular projective-metric plane.

A singular projective-metric plane can likewise be represented as the co-

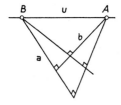

Fig. 40

ordinate plane over a field K of characteristic $\neq 2$, namely, as the projective coordinate plane with u as the line at infinity. To the projective involution ρ on u without fixed points there then corresponds a degenerate symmetric bilinear form of rank 2. In a given basis the form f can be represented as follows:

$$f(x, \mathfrak{y}) = \sum_{i,k=1}^{3} f_{ik} x_i y_k \quad \text{with det}(f_{ik}) = 0.$$

Then the motions in the singular projective-metric plane are the projective collineations in Π induced by linear mappings that take the form f into itself.

Bibliography

[1]   ARTIN, E.: Geometric algebra. New York 1957.
[2]   BACHMANN, F.: Aufbau der Geometrie aus dem Spiegelungsbegriff. Berlin-Göttingen-Heidelberg 1959.
[3]   CHOQUET, G.: L'enseignement de la géométrie. Paris 1964.
[4]   HESSENBERG, G.: Grundlagen der Geometrie. Berlin and Leipzig 1930.
[5]   HILBERT, D.: The foundations of geometry. Translated from the German by E. J. Townsend. Reprint edition. The Open Court Publishing Co., La Salle, Ill., 1959.
[6]   PICKERT, G.: Projektive Ebenen. Berlin-Göttingen-Heidelberg 1955.
[7]   VEBLEN, O., and YOUNG, W.: Projective geometry. 2 vols. Boston 1910–1916.

Bibliography Added in Translation

BAER, R.: Linear algebra and projective geometry. Academic Press, New York 1952.

COXETER, H. S. M.: Projective geometry. Blaisdell Publishing Co., New York 1964.

COXETER, H. S. M.: The real projective plane. McGraw-Hill, New York 1949.

DORWART, H. L.: The geometry of incidence. Prentice-Hall, Englewood Cliffs, N. J. 1966.

HOPKINS, E. J., and HAILS, J. S.: An introduction to plane projective geometry. Oxford 1953.

JEGER, M.: Transformation geometry (tr. from the German by A. W. Deicke and A. G. Howson). George Allen and Unwin, London 1966.

LEVY, H.: Projective and related geometries. Macmillan Co., New York 1964.

MODENOV, P. S., and PARHOMENKO, A. S.: Geometric transformations (translated from the Russian by M. B. P. Slater). 2 vols. Academic Press, New York and London 1965.

PEDOE, D.: An introduction to projective geometry. Pergamon Press, Oxford-New York-Paris 1963.

ROBINSON, G. DE B.: The foundations of geometry. University of Toronto Press, Toronto 1940.

SEIDENBERG, A.: Lectures in projective geometry. D. Van Nostrand Co. Inc., Princeton, N. J. 1962.

VEBLEN, O.: The foundations of geometry. Chapter One of Monographs in topics of modern mathematics. New York 1955.

YALE, P. B.: Geometry and symmetry. Holden-Day, San Francisco 1968.

Euclidean Planes

Introduction

In general, instruction in mathematics in the secondary schools takes two very different forms: first comes synthetic or elementary geometry in the earlier years, and then analytic geometry. The transition from one type of geometry to the other, consisting essentially in the introduction of co-ordinates, is usually motivated only by pointing out that, given a fixed unit segment, every positive number corresponds to a certain length, i.e., to a class of congruent segments (cf. III1, §9), so that not only lengths but also numbers can be marked off on suitably chosen lines, subsequently called the coordinate axes.

But an important question is thereby passed over in silence: what have numbers to do with geometry? More precisely, what geometric laws enable us to assign numbers to the points of a line in such a way that expressions involving these numbers have a geometric interpretation? In the axioms and theorems of elementary geometry as stated by Euclid or Hilbert there is no mention of numbers. So we must ask ourselves how the axioms in geom-etry can have any relation whatever to the laws for addition and multiplica-tion in algebra. The algebraization of an axiomatically defined geometry consists in a demonstration that the objects and relations of the geometry can all be expressed in a domain of numbers, or more generally in an algebraic structure, formed from the geometry itself.

For certain geometries, i.e., for the Desargues and Pappus affine (and projective) geometries, which do not involve any unit of length, this demon-stration has been given in II3 (cf. §2.4, Proposition 1), where, in addition to the axioms of incidence, we assumed certain closure theorems, or equivalently, the transitivity (and commutativity) of certain collineation groups. But we now ask: can these assumptions in turn be derived from others, e.g., from "metric" laws that are more easily visualized, and so perhaps more natural, than the rather complicated closure theorems. To this end we introduce the concept of line reflection, and take certain of its fundamental properties as "reflection axioms." The result, based on the axioms of reflection and incidence, is a very general *Euclidean geometry,* making no use of order or free mobility, which admits other models as well as the real Euclidean plane and yet includes many of the familiar theorems on parallelograms, or transversals of triangles, and so on, and is thus perhaps easier to visualize than the general affine or projective geometries.

In this chapter we make use of some of concepts and results of the preceding chapter, namely, §§1.1, 1.3, 2.2 and the definition of the affine coordinate plane in §2.1; but §§1.2 and 1.4, on closure theorems as related to collineations, will not be used; instead, we will give a direct proof of the transitivity of the groups of translations and dilations. In §2.2 of the preceding chapter the expression "Desargues plane" is only an abbreviation for an affine plane with these transitive collineation groups.

In our present type of Euclidean geometry the translations can be characterized (§3) very simply as certain (rigid) motions, but the dilations must be constructed (§4) as products of halfrotations which are not motions. However, we can then give an algebraic description not only of incidence (§5) but of perpendicularity, as defined by line reflections (§6), and also, though we omit it in this chapter, of the line reflections themselves.

I. A System of Axioms for Euclidean Planes

In addition to the basic concepts *point, line,* and *incidence* of plane affine geometry (cf. II3, §1.1) and the derived notions of *parallel, pencil, collineation,* etc., we now deal with a fourth fundamental concept, namely, *line reflections.* These reflections are mappings of the plane which take points into points and lines into lines. As usual, the image of an element x (point or line) under a mapping α will be denoted by $x\alpha$ and the product of two mappings α, β (first α and then β) by $\alpha\beta$. Thus $x(\alpha\beta) = (x\alpha)\beta$.

A mapping σ for which $\sigma^2 = 1$ is the identity, but $\sigma \neq 1$, is called *involutory* or an *involution,* and it is in terms of these concepts that the following axioms are stated.

Reflection Axioms

Re1: *In every line g there exists exactly one reflection σ_g.*

Re2: *The reflection σ_g is an involutory collineation which leaves g pointwise fixed.*

Re3: *If the line h is fixed under the reflection σ_g, then g and h have at least one point in common.*

Re4:   *If and only if three lines a, b, c lie in a pencil,*[1] *the product* σ_a, σ_b, σ_c *of their reflections is again a line reflection* (*the three-reflections theorem*).

These axioms, taken together with the affine incidence axioms A1 through A3 (II3, §1.1), form an axiom system for our Euclidean geometry, and by a *Euclidean plane* we shall mean any model of the system, i.e., any affine plane for which line reflections are so defined as to satisfy the axioms Re1 through Re4.

Line reflections will play a central role in later chapters, in view of the fact that the axioms of absolute geometry can be expressed entirely in terms of the group-theoretic properties of line reflections regarded as elements in the group of all motions (cf. II5).

In general absolute geometry, if the axioms are stated in our present (non-group-theoretic) language of points and lines regarded as geometric entities, the line reflections are not enough to serve as the only metric concept; it is necessary to have some additional relation like orthogonality. But in our special case of Euclidean geometry, where pencils of lines are already defined by affine incidence, adequate reflection axioms can be formulated without reference to orthogonality. In fact, the chief purpose of the rest of the chapter is to show that our present systems of axioms is sufficient for the following complete algebraization of Euclidean planes.

Theorem: *Every Euclidean plane can be represented as an affine coordinate plane over a commutative field* $\Re$ *with characteristic* $\neq 2$, *in which perpendicularity* (*orthogonality*) *is defined as follows:*

1.   *The lines* $x = c$ *are perpendicular to the lines* $y = b$.
2.   *The lines* $y = ax + b$ *and* $y = a'x + b'$ *with* a, $a' \neq 0$ *are perpendicular to each other if* $1 + kaa' = 0$, *where* $-k$ *is a fixed element in* $\Re$ *that is not the square of any element in* $\Re$.

2.   Line Reflections and Point Reflections

As our starting point we take the definitions of translation, dilation, and axial affinity in §1.3 of the preceding chapter, together with the theorems proved there (from Theorem 6 to Lemma 15). In view of Axiom Re2 we will give special attention to the involutory collineations. By definition, the involutory mappings of a set $\mathfrak{M}$ are those mappings σ of $\mathfrak{M}$ onto itself for which $\sigma = \sigma^{-1} \neq 1$. The following rules for the involutions σ of $\mathfrak{M}$ will be in constant use. For arbitrary mappings α, β of $\mathfrak{M}$ the relation $\sigma\alpha = \beta$ is equivalent to $\alpha = \sigma\beta$. If α is a one-to-one mapping of $\mathfrak{M}$ onto itself, then for any involutory mapping σ the "transformed" mapping $\alpha^{-1}\sigma\alpha$ is also involutory. The inverse of a product of involutions $\sigma_1, \sigma_2, \ldots, \sigma_n$ is ob-

---

[1]We recall the definition in II3, §1.1: lines lie in a pencil if they are all either co-punctual or parallel.

tained by inverting the order of the factors: $(\sigma_1\sigma_2 \cdots \sigma_n)^{-1} = \sigma_n \cdots \sigma_2\sigma_1$.

Turning now to the simplest properties of line reflections and their products, we make the following definitions.

Definition 1. A (finite) product of line reflections is called a *motion*.[2] Let g, h be two lines. The product $\sigma_g\sigma_h$ is called a *rotation about the point P* if g, h I P, and, in particular, is called a *point reflection* in P if $\sigma_g\sigma_h$ is involutory.

Definition 2. A line h is said to be *perpendicular* to a line g, in symbols $h \perp g$, if $h\sigma_g = h \neq g$, or in other words if h is a fixed line of σ_g distinct from g (Fig. 1).

Thus Re3 states that perpendicular lines have exactly one point of intersection. The three-reflections theorem (Re4) can be sharpened as follows.

Supplement to Re4: *If $\sigma_a\sigma_b\sigma_c = \sigma_d$ for four lines a, b, c, d, then the four lines all lie in a pencil.*

Proof. The given equation implies $\sigma_d\sigma_a\sigma_b = \sigma_c$, so that from Re4 we see that a, b, c, and also a, b, d, lie in a pencil. Thus if $a \neq b$, all four lines belong to the same pencil, and if $a = b$, there is nothing to prove (Fig. 2).

The three-reflections theorem can be expressed somewhat as follows. The representation of a rotation, or of a motion $\sigma_g\sigma_h$ with $g \parallel h$, in the form of a product $\sigma_a\sigma_b$ of line reflections is not unique, since for a given $a \neq b$ we can choose an arbitrary line c in the pencil determined by a, b and then determine a line d in the pencil in such a way that $\sigma_a\sigma_b = \sigma_d\sigma_c$; in the same way we can choose at will the axis of the first line reflection and then determine the second axis.

Lemma 1: *The fixed points of the reflection σ_g are precisely the points of the line g, and the fixed lines of σ_g consist, apart from g itself, of a pencil of parallels not including g.*

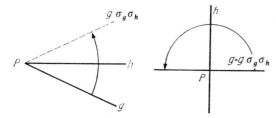

Fig. 1

[2]Here we must point out an inconsistency in terminology, apparently arbitrary but in fact justifiable from various points of view. One speaks either of "motions" or of "rigid mappings" (i.e., "isometries"). The first term corresponds to the point of view adopted here and in other chapters (II3 and 5) on the foundations of geometry; the second is introduced in III, §20, where rigid mappings are subdivided into "motions" (direct isometries) and "indirect isometries," and is retained in most of the chapters in this volume.

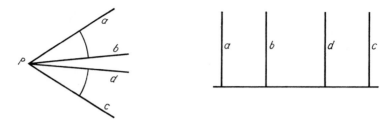

Fig. 2

Proof.   From Re2 it follows that σ_g is a collineation $\neq 1$ leaving g point-wise fixed. Thus by Lemma 15 in §1.3 of the preceding chapter the fixed points of σ_g are precisely the points of g, and the fixed lines consist of g itself and a pencil B of parallels. Since the lines in B that are distinct from g must by Re3 intersect g, the line g does not belong to B. Thus by the parallel axiom (Axiom A2 in II3, §1.1) we have at once the following lemma:

Lemma 2:   *Through any given point a unique perpendicular can be drawn to a given line. In other words, for every point P and every line g there exists exactly one line l with P I l $\perp$ g.*

Lemma 3:   *If α is a motion and g is a line, then $\alpha^{-1}\sigma_g\alpha = \sigma_{g\alpha}$.*

Proof.   By Definition 1 the motion α is a product of line reflections, say $\alpha = \sigma_1 \ldots \sigma_n$. But by Re4 the motion $\sigma_1\sigma_g\sigma_1$ is also a line reflection, and this reflection leaves $g\sigma_1$ pointwise fixed (by Lemma 1 it leaves no other point fixed). Consequently $\sigma_1\sigma_g\sigma_1 = \sigma_{g\sigma_1}$. Since $\alpha^{-1}\sigma_g\alpha = \sigma_n \ldots \sigma_1\sigma_g\sigma_1 \ldots \sigma_n$, an n-fold application of this result completes the proof.

In the rest of this section g, h will always be two lines.

Lemma 4:   *The motion $\sigma_g\sigma_h$ is involutory if and only if $h \perp g$.*

Proof.   By the above remarks on involutions, the motion $\sigma_g\sigma_h$ is involutory if and only if $\sigma_g\sigma_h = \sigma_h\sigma_g \neq g$, or in other words if $\sigma_g\sigma_h\sigma_g = \sigma_h$ and $g \neq h$. But by Lemma 3 this statement is equivalent to $h\sigma_g = h \neq g$, which implies the desired result $h \perp g$.

Applying this result twice we immediately obtain

Lemma 5:   *From $h \perp g$ follows $g \perp h$: perpendicularity is symmetric.*

From Lemma 1, we now have

Lemma 6:   *Lines with a common perpendicular are parallel. Parallels have all their perpendiculars in common.*

Proof.   For lines a, g, h it follows by Lemma 5 from $a \perp g$, h that $g, h \perp a$, and thus by Lemma 1 that $g \parallel h$. Conversely, by Lemma 5 it follows from $g \parallel h$ and $a \perp g$ that $g \perp a$ and then by Lemma 1 that $h \perp a$ and therefore $a \perp h$, again by Lemma 5.

Lemma 7: *In any point P there is exactly one point reflection. The fixed elements of this reflection are precisely P and the lines through P.*

Proof. There exists a line g' through P; and for the line h' perpendicular to g' at P the mapping $\sigma_{g'} \sigma_{h'}$ is involutory by Lemma 4, and is thus a point reflection in P. Now assume $g, h \mathrel{I} P$ and let $\sigma_g \sigma_h$ be another reflection in P. By the complement to Re4, the mapping $\sigma_g \sigma_{g'} \sigma_{h'} = \sigma_l$ is the reflection in a line $l \mathrel{I} P$, and $\sigma_g \sigma_l = \sigma_{g'} \sigma_{h'}$ is involutory. Thus $h = l$ is the line, uniquely determined by Lemma 2, perpendicular to g at P, and therefore the point reflection $\sigma_g \sigma_h = \sigma_{g'} \sigma_{h'}$ in P is uniquely determined. Also, since

$$g \sigma_{g'} \sigma_{h'} = g \sigma_g \sigma_h = g \sigma_h = g,$$

the point reflection in P is a dilation with center P, so that by Lemmas 11 and 12 in §1.3 of the preceding chapter its fixed elements are as stated.

The point reflection in the point P, uniquely determined in this way, will from now on be denoted by σ_P. Also, the perpendicular, uniquely determined by Lemma 2, to a line g at or from a point P will often be denoted by perp (P, g). Our use of the notation perp (P, g) and σ_P will tacitly depend on Lemmas 2 and 7.

Lemma 8: *The conditions $g \perp h$ and $P = (g, h)$ taken together are equivalent to $\sigma_P = \sigma_g \sigma_h$.*

Proof. Lemma 4 shows that $g \perp h$ is equivalent to the fact that $\sigma_g \sigma_h$ is a point reflection. Obviously (g, h) is a fixed point of $\sigma_g \sigma_h$, and by Lemma 7 it is the only fixed point, so that the assertion follows.

Lemma 9: *For a motion α and a point P we always have*

$$\alpha^{-1} \sigma_P \alpha = \sigma_{P\alpha}.$$

Proof. We have $\sigma_P = \sigma_g \sigma_h$ for suitably chosen lines g, h, so that Lemma 3 gives $\alpha^{-1} \sigma_P \alpha = \alpha^{-1} \sigma_g \alpha \alpha^{-1} \sigma_h \alpha = \sigma_{g\alpha} \sigma_{h\alpha}$. By the above remarks on involutions this mapping is again involutory and is thus the reflection in $(g\alpha, h\alpha) = (g, h)\alpha = P\alpha$.

3. Translations

Now that some of the usual properties of perpendicularity and reflection have been deduced from our system of axioms, we turn our attention in the next two sections to the translations and dilations introduced in §1.3 of Chapter 3 and utilized in §2.2 of that chapter to define coordinates in an affine plane.

Fig. 3

Lemma 10: *If a, b are the perpendiculars to a line v through two points A and B, then $\sigma_A \sigma_B = \sigma_a \sigma_b$.*

For by Lemma 8 we have $\sigma_A \sigma_B = \sigma_a \sigma_v \sigma_v \sigma_b = \sigma_a \sigma_b$.

Lemma 11: *The motions $\sigma_A \sigma_B$ comprise all the motions $\sigma_a \sigma_b$ with $a \parallel b$.*

Proof. If two points A, B are given, let v be a line through A and B, and let a, b be the perpendiculars to v through A and B (Fig. 3). Then $a \parallel b$ by Lemma 1 and $\sigma_A \sigma_B = \sigma_a \sigma_b$. Conversely, if two parallel lines a, b are given, choose a point A on a, draw the perpendicular v from A to b and set $B = (b, v)$. Then by Lemma 6 we have $a \perp v$, so that $\sigma_a \sigma_b = \sigma_A \sigma_B$ by Lemma 10.

Lemma 12: *Every motion $\sigma_A \sigma_B$ is a translation, and if $A \neq B$ this translation is in the direction of (A, B).*

Proof. Since the identity is itself a translation, we need to prove the lemma only for $A \neq B$. Since σ_A and σ_B are dilations, for every line g we have

$$g \sigma_A \sigma_B \parallel g \sigma_A \parallel g,$$

so that $\sigma_A \sigma_B$ is necessarily either a translation or a dilation, in view of the fact that only such mappings take every line into a parallel line. Now by Lemma 11 there exist parallel lines a, b for which $\sigma_A \sigma_B = \sigma_a \sigma_b$, an equation which shows that the mapping $\sigma_A \sigma_B$ leaves linewise fixed the pencil of lines consisting of the common perpendiculars to a and b. So this mapping is not a dilation but is a translation in the (A, B)-direction.

For $A \neq B$ it is to be noted that the translation $\sigma_A \sigma_B$ does not take A into B but takes A into $A \sigma_B$. So we make the following definition.

Definition 3. *A point M is called the midpoint of the points A, B if $A \sigma_M = B$. A line m is called the right bisector of A, B if $A \sigma_m = B$.*

For a nondegenerate triangle ABC with sides $a = (B, C)$, $b = (C, A)$, $c = (A, B)$ we have

Lemma 13: *If N is the midpoint of A, C and l is the parallel to a through N, then the point of intersection $(l, c) = M$ is the midpoint of A, B.*

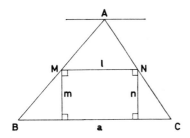

Fig. 4

Proof.   Let n and m be the perpendiculars to l at N and M, respectively (Fig. 4). Since $l \parallel a$ it follows from Lemma 6 that $n, m \perp a$ and thus

$$a\sigma_N = a\sigma_n\sigma_l = a\sigma_l = a\sigma_m\sigma_l = a\sigma_M.$$

Moreover, this line $a\sigma_N = a\sigma_M \neq c$ is incident with A, since from $a \mathrel{\mathrm{I}} C = A\sigma_N$ it follows that $a\sigma_N \mathrel{\mathrm{I}} A$. Consequently, $A = (a\sigma_M, c)$ and

$$A\sigma_M = (a\sigma_M, c)\sigma_M = (a, c\sigma_M) = (a, c) = B.$$

Thus we have at once the first part of

Lemma 14:   *Two points A, B have exactly one midpoint M and for $A \neq B$ exactly one right bisector m. Here m is the perpendicular to (A, B) at M.*

Proof of the existence of a midpoint. For $A = B$, the point A is a midpoint of A and B. If $A \neq B$, take an arbitrary point N outside $c = (A, B)$ and apply Lemma 13 to the triangle $ABA\sigma_N$.

Proof of the uniqueness of the midpoint. If M, M' are two midpoints of A, B, we have $A\sigma_M = A\sigma_{M'}$, so that the translation $\sigma_M\sigma_{M'}$ has the fixed point A and is therefore the identity. Then $\sigma_M = \sigma_{M'}$ so that $M = M'$.

Proof of the existence and uniqueness of the right bisector for $A \neq B$. Again, let $c = (A, B)$. The midpoint M of A and B lies on c, so that for the perpendicular m to c at M it follows from $\sigma_M = \sigma_m\sigma_c$ that

$$A\sigma_m = A\sigma_M\sigma_c = B\sigma_c = B.$$

Conversely, for every right bisector m of $A \neq B$ we have $m \mathrel{\mathrm{I}} A, B$, so that $m \neq c$ and

$$c\sigma_m = (A\sigma_m, B\sigma_m) = (B, A) = c,$$

and thus $c \perp m$, so that from $A\sigma_c\sigma_m = A\sigma_m = B$ it follows that (c, m) is the midpoint of A, B. Thus m is uniquely determined as the perpendicular to c at the point M.

Theorem 1:   *For every two points A, B there is exactly one translation* *taking A into B. The set of translations coincides with the set of products of two point reflections and forms a commutative group* $\mathfrak{T}$.

Proof. By Lemma 14 the points A, B have exactly one midpoint M, for which it is obvious that $A\sigma_A\sigma_M = B$, and by Lemma 12 the mapping $\sigma_A\sigma_M$ is a translation. Since such a mapping is determined by one point and its image (Lemma 9 in §1.3 of Chapter 3), $\sigma_A\sigma_M$ is the only translation taking A into B; thus all translations are of the form described in Theorem 1 and therefore, by Theorem 8 in §1.3 of Chapter 3, they form a commutative group.

From Theorem 1 it is easy to deduce the following lemmas:

Lemma 15: *The product of three point reflections is always a point reflection.*

If, given the three points A, B, C, we denote the midpoint of A and $A\sigma_B\sigma_C$ by D (Fig. 5), we obviously have

$$A\sigma_B\sigma_C = A\sigma_D = A\sigma_A\sigma_D,$$

and since the two translations $\sigma_B\sigma_C$ and $\sigma_A\sigma_D$ are thereby already determined, we must have

$$\sigma_B\sigma_C = \sigma_A\sigma_D \quad \text{or} \quad \sigma_A\sigma_B\sigma_C = \sigma_D.$$

Lemma 16: *If $A \neq B$, D and $\sigma_A\sigma_B = \sigma_D\sigma_C$, then also $C \neq B$, D and not only $(A, B) \parallel (C, D)$ but also $(A, D) \parallel (B, C)$: the quadrilateral $ABCD$ is a parallelogram.*

For by Lemma 12 the translation $\sigma_A\sigma_B = \sigma_D\sigma_C \neq 1$ has not only the direction (A, B) but also the direction (C, D), and similarly the translation $\sigma_B\sigma_C = \sigma_A\sigma_D \neq 1$ has both the direction (B, C) and also the direction (A, D).

Lemma 17: *Translations are never involutory.*

For if $\sigma_A\sigma_B$ is a translation whose square is the identity, so that $\sigma_A\sigma_B\sigma_A =$

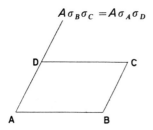

Fig. 5

σ_B, it follows from Lemma 9 that $B\sigma_A = B$ is a fixed point of σ_A, so that $A = B$ and $\sigma_A\sigma_B = 1$ is not involutory.

On the basis of Theorem 1 we may again regard translations as ordinary vectors and a product of translations as a sum of vectors. But up to now the multiplication of a vector by a scalar does not correspond to any combination of mappings. For this purpose we must construct the dilations of the Euclidean plane by means of halfrotations.

4. Halfrotations and Dilations

The theory of halfrotations makes essential use of a theorem that is also of fundamental importance for absolute geometry, namely, the theorem of perpendiculars, which we shall obtain from the following lemma.

Lemma 18: *For two points A, C and a line b (Fig. 6) the mapping $\sigma_A\sigma_b\sigma_C$ is a line reflection if and only if* perp $(A, b) =$ perp (C, b).

Proof. Let us set $v =$ perp (A, b), $v' =$ perp (C, b), $a =$ Perp (A, v) and $c =$ Perp (C, v'). By Lemmas 1 and 6 we have $v \parallel v'$ and $a \parallel b \parallel c$, from which by Lemma 8 and Re4 it follows that $\sigma_A\sigma_b\sigma_C = \sigma_v\sigma_a\sigma_b\sigma_c\sigma_{v'} = \sigma_v\sigma_d\sigma_{v'}$ for any a parallel to b and therefore, by Lemma 6, perpendicular to v and v'. Thus by Re4 this motion is a line reflection if and only if $v = v'$.

Theorem 2 (theorem of perpendiculars): *Let $\sigma_a\sigma_b\sigma_c = \sigma_d$ for four lines a, b, c, d, and let A and C be points on a and c respectively; also let $a' =$ perp (A, a) and let $c' =$ perp (C, c). Then a', b, c' lie in a pencil if and only if* perp $(A, d) =$ perp (C, d). (See Fig. 7.)

Proof. By Re4 and the above result, the assertion is an immediate consequence of the following equation for reflections, valid by Lemma 8,

$$\alpha_{a'}\sigma_b\sigma_{c'} = \sigma_{a'}\sigma_a\sigma_a\sigma_b\sigma_c\sigma_c\sigma_{c'} = \sigma_A\sigma_d\sigma_C.$$

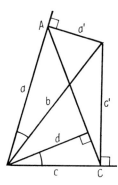

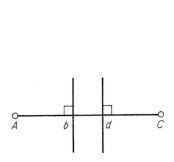

Fig. 6   Fig. 7

If $A \neq C$ in this theorem, then the assertion perp $(A, d) = $ perp (C, d) can be replaced by $(A, C) \perp d$, and the corresponding remark holds for Lemma 18.

Definition 4.   *Let u, v be two lines through a point O not perpendicular to each other. Then for points and lines we define a mapping (which we shall denote by a *) as follows:*

* 1)   If $g \mathrel{I} O$, let g^* be the line for which $\sigma_g \sigma_{g^*} = \sigma_u \sigma_v$.
   2)   If $g \mathrel{\not I} O$, let $l = $ perp (O, g) and let $P = (g, l)$. Then l^* is defined by 1) and we set $g^* = $ perp (P, l^*).
   3)   Let $O^* = O$.
   4)   If $P \neq O$, let $l = (O, P)$. Here again l^* is defined by 1) and we set $P^* = (l^*, $ perp $(P, l^*))$.

The mapping * (see Fig. 8) is called the *halfrotation about O defined by the rotation $\sigma_u \sigma_v$* and is denoted by η_{uv}.

The name "halfrotation" comes from the fact that the mapping * takes every P into the midpoint of P and $P \sigma_u \sigma_v$, as can easily be verified.

Theorem 3:   *The halfrotations about a fixed point O and their inverses are collineations that are permutable with one another.*

Let us successively prove the following partial assertions in this theorem for halfrotations $\eta_{ur} \neq 1$ (i.e., $u \neq v$) about O, which for brevity we shall again denote by *:

a)   ** is a one-to-one mapping of the set of points onto itself and of the set of lines onto itself.*
b)   ** preserves incidence.*
c)   *If η, θ are two halfrotations about O, then $\eta\theta = \theta\eta$, and also $\theta^{-1}\eta = \eta\theta^{-1}$ and $\eta^{-1}\theta^{-1} = \theta^{-1}\eta^{-1}$.*

Proof of a).   The definition of η_{uv} provides for each point and each line a unique rule for constructing the image point and image line. Conversely, let us consider the following mapping $^\circ$:

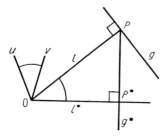

Fig. 8

1°) If g I O, let $g°$ be the line for which $\sigma_{g°}\sigma_g = \sigma_u\sigma_v$.

3°) Let $O° = O$.

4°) If $P \neq O$, let $l = (O, P)$ and $g = \text{perp}(P, l)$. Then $l°$ is defined by 1°), and since $u \not\perp v$ we have $l° \not\perp l \perp g$, so that $l° \not\parallel g$, and we set $P° = (g, l°)$.

2°) If $g \nmid O$, let $l = \text{perp}(O, g)$ and let $P = (l, g)$. Then $P°$ is defined by 4°), and we set $g° = \text{perp}(P°, l°)$.

The mapping $°$ also sets every element x of the plane in unique correspondence to an element $x°$, with $x°* = x*° = x$, so that $°$ is the inverse of $*$, which immediately implies a).

Proof of b). Let P be a point on a line g (Fig. 9). For $P = O$ the assertion $P*$ I $g*$ is trivial. For $P \neq O$ let $k = (O, P)$, $l = \text{perp}(O, g)$ and $F = (l, g)$. Then $\sigma_u\sigma_v = \sigma_k\sigma_{k*} = \sigma_l\sigma_{l*}$, and also $(P, P*) \perp k*$, $g* \perp l*$, $g*$ I F.

Thus the theorem of perpendiculars implies that $\text{perp}(P*, l*) = \text{perp}(F, l*) = g*$, so that $P*$ I $g*$.

Proof of c). From a) and b) it follows that η and θ are collineations, so that we need only prove that $\eta\theta = \theta\eta$ for points $P \neq O$.

In the notation of b) let $P\eta = P*$, $P\theta = F$ and $P\theta\eta = F*$, and also let $k\eta = k*$, $k\theta = l$ and $k\theta\eta = l\eta = l*$. Then $\sigma_k\sigma_l = \sigma_{k*}\sigma_{l*}$ and by the theorem of perpendiculars we again have $\text{perp}(P*, l*) = \text{perp}(F, l*)$. Thus $P\eta\theta = P*\theta = F* = P\theta\eta$. Since P was an arbitrary point distinct from O, we thus have $\eta\theta = \theta\eta$ in complete generality. But then it follows that $\theta^{-1}\eta = \theta^{-1}\eta\theta\theta^{-1} = \theta^{-1}\theta\eta\theta^{-1} = \eta\theta^{-1}$ and also $\eta^{-1}\theta^{-1} = \theta^{-1}\eta^{-1}$. Thus the proof of Theorem 3 is complete.

We now formulate and prove a theorem analogous to Theorem 1:

Theorem 4: *For every three collinear points O, A, B (see Fig. 10) with $O \neq A$, B there exists exactly one dilation with center O taking A into B. The*

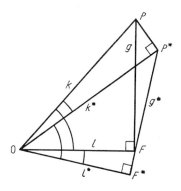

Fig. 9

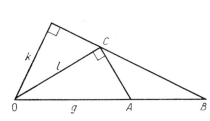

Fig. 10

dilations with center O are precisely the products of halfrotations about O and of inverses of such halfrotations which leave the lines through O fixed. These products form a commutative group $\mathfrak{D}_O$.

Proof.   There exists a line l through O which is neither equal nor perpendicular to $g = (O, A)$. The halfrotation η_{gl} takes A into a point $C \neq O$ on l. Let $k = \text{perp } (O, (B, C))$. Since (B, C) intersects the lines g and l and since $k, g, l \text{ I } O$, we have $k \not\perp g, l$. Thus the halfrotations η_{lk} and η_{gk} exist, and the product $\delta = \eta_{gl}\eta_{lk}\eta_{gk}^{-1}$ takes A into B. For every line $h \text{ I } O$ it follows from the definitions 4, 1), and 1°) that

$$\sigma_{h\delta} = \sigma_h(\sigma_g\sigma_l)(\sigma_l\sigma_k)(\sigma_g\sigma_h)^{-1} = \sigma_h,$$

so that $h\delta = h$ and thus δ is a dilation with center O, taking A into B. Since a dilation is thereby uniquely determined (Lemma 13 in §1.3 of Chapter 3), the dilation δ is the only one with this property, and all dilations with center O have the form described in Theorem 4. Finally, by Theorem 3 the dilations with center O generated in this way are permutable with one another, so that they form a commutative group.

5.   Euclidean Planes as Affine Coordinate Planes

With Theorems 1 and 4 we have now made the connection with §2.2 of the preceding chapter.

Theorem 5 (Algebraization of Euclidean planes):   *Every Euclidean plane is representable, apart from its metric properties (i.e., in the present context, its reflection properties), as an affine coordinate plane over a (commutative) field of characteristic $\neq 2$.*

For we see that all the assumptions of §2.2 of the preceding chapter are satisfied: the (commutative) group $\mathfrak{T}$ of translations is transitive on the whole plane and the group $\mathfrak{D}_O$ of dilations with center O is linearly transitive and also commutative. Moreover, it is unnecessary in this case to refer to theorems on axial affinities, since the commutativity of $\mathfrak{D}_O$ means that one of the distributive laws of the coordinate field is already implied by the other, and the latter is proved by means of dilations and translations alone:

$$C(A + B) = (A + B)C = AC + BC = CA + CB.$$

Finally, if char $\mathfrak{K} = 2$, it would follow that $E + E = O$; thus the translation taking O into E would be involutory, in contradiction to Lemma 17.

6.   Perpendicularity in the Coordinate Plane

The problem that now remains is to give an algebraic characterization of perpendicularity in the coordinate plane. Since for this purpose we shall be making use of the theorem of altitudes, which in Euclidean geometry follows at once from the theorem of right bisectors, we first prove these two theorems. In our proofs the perpendicular from a vertex of a (nondegenerate) triangle to the (nonincident) side will, as usual, be called an *altitude* of the triangle.

Theorem 6 (Theorem of right bisectors): *The right bisectors (of the pairs of vertices) of a triangle intersect in one point; i.e., if A, B, C are three non-collinear points, there exist three uniquely determined lines u, v, w such that $A\sigma_w = B$, $B\sigma_u = C$, and $C\sigma_\delta = A$, and u, v, w intersect in one point.*

Proof.   The existence and uniqueness of u, v, w are part of the assertion of Lemma 14. Since $w \perp (A,B) \not\Vert (B, C) \perp u$, the lines w and u intersect in a point M. We set $v' = (B, M)$ and by Re4 define a line v'' through M by means of $\sigma_{v''} = \sigma_u\sigma_{v'}\sigma_w$. From

$$C\sigma_{v''} = C\sigma_u\sigma_{v'}\sigma_w = B\sigma_{v'}\sigma_w = B\sigma_w = A$$

it follows by Lemma 14 that $v'' = v \text{ I } M$.

Theorem 7 (Theorem of altitudes): *The altitudes* (see Fig. 11) *of a triangle intersect in one point; namely, if A, B, C are three noncollinear points, the lines $u = \operatorname{perp}(A, (B, C))$, $v = \operatorname{perp}(B, (C, A))$ and $w = \operatorname{perp}(C, (A, B))$ intersect in one point.*

The proof consists in the classical reduction of this theorem to the theorem of right bisectors. If by Lemma 15 we determine three points A', B', C' by means of $\sigma_{A'} = \sigma_C\sigma_A\sigma_B$, $\sigma_{B'} = \sigma_A\sigma_B\sigma_C$ and $\sigma_{C'} = \sigma_B\sigma_C\sigma_A$, then by Lemma 9 we have $A'\sigma_C = B'$, so that A', C, B' are collinear, and from Lemma 16 it follows that $(A', B') \parallel (A, B)$. Thus by Lemma 6, $w = \operatorname{perp}(C, (A, B)) = \operatorname{perp}(C, (A', B'))$ and finally $A'\sigma_w = A'\sigma_C = B'$. The same remarks hold for u and v: the altitudes of the triangle ABC are the right bisectors of the triangle $A'B'C'$, which by the preceding theorem intersect in one point.

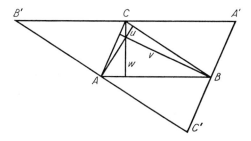

Fig. 11

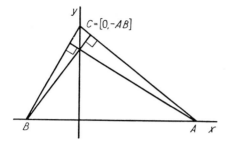

Fig. 12

If for the coordinate axes we now choose two mutually perpendicular lines x, y through the origin of coordinates O, we can determine how the perpendicularity of two lines is expressed in general in terms of coordinates. Since $y \perp x$, it follows from Lemma 6 that the lines $Y = C$ are perpendicular to the lines $X = D$. So let us now consider the lines that are not parallel to the axes, with equations of the form $Y = AX + C$, $A \neq O$. A pencil of parallel lines will then be defined by its slope $A \neq O$, and if we let $f(A)$ denote the slope of the pencil of parallel lines perpendicular to the given pencil, f is a function defined for all elements of the field distinct from O, with values $f(A)$ that are again nonzero elements of the field, and with $f(A) \neq A$, since perpendicular lines are distinct. We now determine the function f by means of the theorem of altitudes.

Let us choose $B \neq O$, A: then in the triangle (see Fig. 12) with the vertices

$$A = [A, O], \quad B = [B, O], \quad C = [O, -AB]$$

and the sides

$$(B, C): Y = A(X - B), \quad (A, C): Y = B(X - A), \quad (A, B): Y = 0$$

the altitudes are

$$Y = f(A)(X - A), \quad Y = f(B)(X - B), \quad X = O.$$

Since the altitudes intersect in one point, the first two of them must cut the y-axis $X = O$ in the same point; i.e.,

$$f(A)A = f(B)B.$$

Since $B \neq O$, A was chosen arbitrarily, $f(A)A$ is a universal constant $\neq O$, which we denote by $-1/K$, where K is called the orthogonality constant. The function to be defined is then given by

$$f(A) = \frac{-1}{KA},$$

and since $f(A) \neq A$, it follows that $-K$ cannot be the square of an element in the coordinate field $\Re$. Thus we have proved the following theorem.

Theorem 8: *In the coordinate plane with perpendicular axes x, y we have:* 1) *the lines* $Y = C$ *are perpendicular to the lines* $X = D$; 2) *the lines* $Y = AX + C$ *and* $Y = A'X + C'$ *are perpendicular to each other if and only if*

$$1 + KAA' = O,$$

where $-K$ *is a fixed nonsquare in* $\Re$.

With Theorems 5 and 8 the proof of the theorem in §1 is complete.

In ordinary analytic geometry $f(A) = -1/A$, so that $K = 1$. This simplification is based on a special choice of the unit point E' on the y-axis. Here it is necessary to find a line w through O such that reflection in it interchanges the x- and y-axes, i.e., such that $y = x\sigma_w$, and if we then set $E' = E\sigma_w$, we obtain $K = 1$, since w is represented by $Y = X$ and $w\sigma_x \perp w$ by $Y = -X$, so that $f(1) = -1$. Of course, our assumptions Re1 through Re4 are not enough to prove the existence of a line w with $x\sigma_w = y$, so that $K = 1$ can be proved only under a suitable additional assumption.

Conclusion

With this theorem the Euclidean planes, as defined by our system of axioms, have been completely algebraized. For the converse is also true: our system of axioms is satisfied by every coordinate plane of the above kind. More precisely:

Converse. *Let* $\Re$ *be a commutative field of characteristic* $\neq 2$ *and let* $-K$ *be a nonsquare in* $\Re$. *Then the affine coordinate plane over* $\Re$, *with perpendicularity defined by* 1) *and* 2) *in Theorem 8, satisfies not only the affine axioms of incidence but also the reflection axioms* Re1 *through* Re4, *where reflection in a line g is defined as the involutory collineation leaving fixed all points on g and all perpendiculars to g.*

The proof of this converse is simpler than the proof of the theorem itself, since it consists of computation (omitted here) in a commutative field, on the basis of §2.5 in Chapter 3.

We have now shown that Euclidean geometry can be constructed without reference to the real numbers. Admittedly, the real number field is a coordinate domain for a Euclidean plane in accordance with our definition, but so is the field of rational numbers and, more generally, every field of characteristic $\neq 2$ in which not every element is a square. On the other hand, it is impossible to define a Euclidean plane over the complex field, where every number is a square.

Again, the minimal model of affine geometry, i.e., the affine coordinate

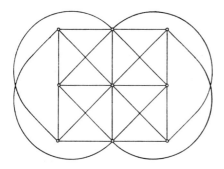

Fig. 13

plane over the field with two elements, cannot be made into a Euclidean plane, since the field is of characteristic 2 and every element in it is a square. However, we have the following theorem.

Over every finite field of characteristic $\neq$ 2 there exists a Euclidean plane, uniquely determined up to isomorphism.

Apart from the zero element all the elements of such a field fall into the class of squares or the class of nonsquares, and since a given Euclidean plane is transformed isomorphically when its orthogonality constant is multiplied by a square factor, all the elements of the field that are admissible as orthogonality constants lead to planes that are isomorphic to one another. If the given field contains n elements, the corresponding Euclidean plane, as follows from the definition of a coordinate plane, contains n points on each line, $n + 1$ lines through each point, n^2 points altogether, $n^2 + n$ lines, $n^3 + n^2$ incidences, and half as many pairs of perpendicular lines.

The Euclidean plane over the prime field of characteristic 3 (see Fig. 13) with the orthogonality constant 1 is thus the smallest Euclidean plane altogether, and the plane over the prime field of characteristic 5 with orthogonality constant 2 is the smallest Euclidean plane in which a right angle cannot be bisected.

These two models show not only that our system of axioms can be satisfied in a finite plane, but also that in a finite plane even the simplest metric theorems like *"All right angles can be bisected"* or the equivalent *"Some square exists"* may or may not be valid. On the other hand, the theorem *"Every angle can be bisected"* is false in any finite plane.

But in spite of these possibilities, our theorem and its converse show that the synthetic Euclidean geometry we have discussed is equivalent to a generalized analytic Cartesian geometry in which many theorems, even including many that arise in metric problems, can be dealt with by the usual algebraic methods.

Absolute Geometry

The subject matter of so-called *elementary geometry,* namely the Euclidean plane, has a complicated structure, as is shown, for example, by the Hilbert system of axioms (see [5]), involving such diverse concepts as incidence, order, metric, and continuity. We shall not try to answer the question, exactly which of the concepts definable in the Euclidean plane are of interest in elementary geometry, nor whether, indeed, the everyday use of the phrase "elementary geometry" is consistent enough to allow any agreement on its meaning. Our purpose here is to give an axiomatic discussion of certain geometric concepts forming the core of elementary geometry in the sense that they include a great part of what usually goes by that name and, on the other hand, would themselves be included in any interpretation of elementary geometry, however restricted.

One natural possibility for setting up such a core is presented by the *group of motions.* Certain geometric objects and relations, in particular, points, straight lines, the incidence relation, and such metric relations as perpendicularity and congruence, can be represented in a natural way in the group of motions, the words "in a natural way" meaning that the representation depends *only on the given multiplication for elements of the group* (in this case the successive application of motions) without reference to any additional structure such as order, neighborhood, or the like.

The necessary means for setting up this representation are provided by the totality of *reflections* in points and lines (i.e., a subset of the set of motions), since the points and lines are in one-to-one correspondence with the reflections in them, and geometric relations among points and lines correspond to group-theoretic equations among the reflections. In the structure of its subset of point and line reflections, the group of motions contains an image of the properties of the plane, so that we can now do *geometry in the group of motions;* i.e., we can formulate geometric theorems about elements of the group of motions and can then prove these theorems by group-theoretic calculation. The geometric objects can now be dealt with directly by the characteristic method of modern mathematics, namely, the study of a given structure by investigation of its group of automorphisms, without the sort of detour that occurs, for example, in analytic geometry, where, instead of working with the objects themselves, we use numbers and sets of numbers that describe them.

Reflections have already occurred in some of the axioms and theorems of the preceding chapter, but there our purpose was not so broad. At that time we wished to construct an axiomatic foundation for *Euclidean* geometry up to the theorem of algebraization (II4, §1), whereas the subject matter of the present chapter, namely, *absolute* geometry, is more inclusive: the term "absolute," first used by J. Bolyai, refers to a geometry that includes both Euclidean and non-Euclidean geometry as special cases, which means that we now wish to set up a system of axioms not yet implying any decision about parallelism, since only then will the full scope of the calculus of reflections become clear. The method will enable us, as we shall see, to set up absolute geometry without any hypothesis concerning order, continuity, or free mobility.[1]

A further difference from the preceding chapter is that now the whole system of axioms is formulated in group-theoretic terms alone, so that geometric concepts like point, line, and incidence no longer are primary but are derived, and the whole calculus of reflections is immediately at our disposal (a simpler example of this method is to be found in II2).

The systematic development of the theory is given in §§2, 4, and 5; in §2 the system of axioms is stated, together with its interpretation in the group plane; in §4 some of the theorems of absolute geometry are proved, and in §5 the various special geometries (Euclidean, elliptic, hyperbolic) are classified as subdivisions of absolute geometry. These sections, in which the proofs are given in full, are logically independent of the other sections and of the preceding chapters.

Sections 1, 3, and 6 are written in freer style, without complete detail and to some extent in the form of a review. In §1, as a preparation for the system of axioms in §2, we discuss certain theorems on reflections in the real Euclidean plane. In §3, the great generality of our axioms is illustrated by some examples, namely by algebraic models. Like the Euclidean planes in the preceding chapter, the present axioms for absolute geometry allow an algebraization over a commutative field (i.e., the introduction of coordinates from such a field). In §6, we give a general survey of this process.

---

[1]For a first introduction to calculation with reflections in the Euclidean plane, see Bachmann [1], §1.

The method of proving geometric theorems by calculation with reflections was used by Hessenberg,[2] and more systematically by Hjelmslev,[3] but a completely consistent treatment, emphasizing the group-theoretic point of view (for Euclidean geometry), was first given in a book by Thomsen,[4] in which, however, the author had certain purposes of no concern to us here; the corresponding treatment for absolute geometry was first given by Arnold Schmidt.[5] A more complete account of the theory, as discussed in the present chapter, is to be found in Bachmann [1], with further biblio- graphical references.

1. Introduction

1.1. As our starting point for the development of a plane metric geometry we take the concepts of reflection in a line and successive application of such reflections, by the system of axioms in §2. As a preparation for those axioms we now present a series of statements about reflections which are valid in every Euclidean plane (in the sense of II4, §1).

Since we are interested only in motivating the axioms in §2, we shall make the present introduction independent of Chapter II4 by dealing only with the continuous Euclidean plane, considered as a coordinate plane over the field of real numbers. For the same reason, we shall not give proofs for any intuitively obvious statements.

No logical gaps will arise in the systematic development beginning in §2 if this introduction (apart from §1.8, which is independent of the other sections) is re- garded simply as an argument from plausibility.

1.2. *Reflections and Motions*

Points of the (continuous, Euclidean) plane will be denoted by A, B, $C, \ldots$, and lines by $a, b, c, \ldots$. We write $P \mathrm{I} g$ (or $g \mathrm{I} P$) for "the point P is incident with (lies on) the line g" and $g \perp h$ for "the lines g, h are mutu- ally orthogonal (perpendicular)." Also, $P \mathrm{I} g, h$ stands for "$P \mathrm{I} g$ and $P \mathrm{I} h$"; $P, Q \mathrm{I} g, h$ stands for "$P \mathrm{I} g, h$ and $Q \mathrm{I} g, h$," and so forth.

We now consider mappings of the plane. The image of a point A or a line

[2]G. Hessenberg, Neue Begründung der Sphärik. Sitzungsberichte der Berliner math. Ges. 4 (1905), 69–77. Hessenberg [4].

[3]J. Hjelmslev, Neue Begründung der ebenen Geometrie. Math. Ann. 64 (1907), 449– 474. J. Hjelmslev, Einleitung in die ebene Kongruenzlehre. Matematisk-fysiske Meddelelser. Copenhagen 1929–1949.

[4]G. Thomsen, Grundlagen der Elementargeometrie in gruppenalgebraischer Behand- lung. Hamburger Math. Einzelschriften 15. Leipzig 1933.

[5]Arnold Schmidt, Die Dualität von Inzidenz und Senkrechtstehen in der absoluten Geometrie. Math. Ann. 118 (1943), 608–625.

a under a mapping α will be denoted by $A\alpha$ or $a\alpha$. If α and β are mappings, then $\alpha\beta$ denotes the mapping "first α, then β."

By a reflection in a line g, or in a point P, we shall mean a mapping which to every point A of the plane assigns the point symmetric to A with respect to the line g or the point P (the mirror image of A in g or P), and which to every line a assigns the line symmetric to a with respect to g, or P. It is helpful to visualize a movable copy of the (fixed) plane, which transports the elements into their images: for the reflection in a line g, the copy is turned over along the line g; and for the reflection in a point P, it is turned through $180°$ around the point P.

We note

(i)   *In every line and in every point there is exactly one reflection.*

The reflection in the line g will be denoted by σ_g, and the reflection in the point P by σ_p, and for brevity we shall use the expressions *line reflection* and *point reflection*.

A mapping α of the set of points and the set of lines (in the plane) each onto itself is called an *orthogonal collineation* if it is consistent with the relations of incidence and orthogonality, i.e., if it has the following two properties:

$A \text{ I } b$ *implies* $A\alpha \text{ I } b\alpha$, *and conversely.*

$a \perp b$ *implies* $a\alpha \perp b\alpha$, *and conversely.*

The orthogonal collineations form a group.

An element σ of a group is said to be *involutory* if it is of order 2, i.e., if $\sigma^2 = 1$ but $\sigma \neq 1$, or in other words, if $\sigma = \sigma^{-1}$ and $\sigma \neq 1$. Constant use will be made of the following rule: *if $\sigma_1, \sigma_2, \ldots, \sigma_n$ are involutory elements of a group, then*

(1) $(\sigma_1\sigma_2\cdots\sigma_n)^{-1} = \sigma_n\sigma_{n-1}\cdots\sigma_2\sigma_1.$

Then clearly,

(ii)   *The line reflections and the point reflections are involutory orthogonal collineations,* and a simple argument from symmetry shows

(iii)   *If $P \text{ I } g, h$ and $g \perp h$, then $\sigma_P = \sigma_g\sigma_h = \sigma_h\sigma_g$.*

Of particular importance are those orthogonal collineations that can be obtained by successive line reflections. Such a mapping will be called a *motion,*[6] since it is easy to visualize how the movable copy of the plane can be brought into any prescribed position by a sequence of turns about suitably chosen lines.

Thus a motion is a product $\sigma_{g_1}\sigma_{g_2} \cdots \sigma_{g_n}$ of line reflections. The mapping inverse to a motion is also a motion (see (1)). The motions form a group,

[6]On our use of the term "motion," see fn. 2 in Chapter 4.

namely, the subgroup of orthogonal collineations generated by the line reflections. By (iii) the point reflections are motions.

In a Euclidean plane the orthogonal collineations also include all similarity transformations.

1.3.   *The Movement (Transformation) of Motions*

Let α be an element of a group. By *transformation* with α we mean the mapping which to each element of the group γ assigns the element $\gamma^\alpha = \alpha^{-1}\gamma\alpha$. For this transformation we have the following rules:

$$(2) \qquad (\gamma_1\gamma_2)^\alpha = \gamma_1^\alpha\gamma_2^\alpha, \quad (\gamma^{-1})^\alpha = (\gamma^\alpha)^{-1}, \quad (\gamma^{\alpha_1})^{\alpha_2} = \gamma^{\alpha_1\alpha_2}.$$

The first rule states that transformation with α is consistent with the multiplication of the group. Since transformation with α has an inverse (namely transformation with α^{-1}) every transformation is an automorphism of the group.

In a group of one-to-one mappings of a set $\mathfrak{M}$ onto itself (e.g., motions), we can apply a group element α not only to the elements of $\mathfrak{M}$ but also to the elements of the group, namely, in the sense of transforming them with α. Now it is particularly important to consider these two processes side by side. For every group element γ and every pair $A, B \in \mathfrak{M}$ we have

$$(3) \qquad A\gamma = B \text{ implies } (A\alpha)\gamma^\alpha = B\alpha, \text{ and conversely,}$$

since in fact $(A\alpha)\gamma^\alpha = A\alpha\alpha^{-1}\gamma\alpha = A\gamma\alpha = B\alpha$, and the converse follows by application of α^{-1}. The rule (3) can be interpreted thus: application of α to elements of $\mathfrak{M}$ and to elements of the group are both consistent with the relation "B is the image of A under γ," regarded as a three-place relation in A, B, γ (Fig. 1).

This result holds in particular for the group of motions: let γ be a motion; if to pairs like "point A, image B" or "line a, image b" (with respect to γ) we apply a motion α, and then look for the mapping which takes $A\alpha$ into $B\alpha$ and $a\alpha$ into $b\alpha$, the result is precisely the motion γ^α. Thus we may say that α "moves the motion γ into the motion γ^α."

Let us now consider the special case that the motion γ to be transformed is a line reflection σ_g. We wish to make visually clear the result that σ_g^α is

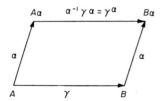

Fig. 1

again a line reflection; or more precisely, that $\sigma_g = \sigma_{g\alpha}$. If $A\sigma_g = B$, then by (3) we have $(A\alpha)\sigma_g^\alpha = B\alpha$. But also $(A\alpha)\sigma_{g\alpha} = B\alpha$, since A, B are mirror images of each other with respect to the line g and therefore $A\alpha$, $B\alpha$ must be mirror images with respect to the line $g\alpha$. So altogether: $(A\alpha)\sigma_g^\alpha = (A\alpha)\sigma_{g\alpha}$ for all points A. Since α is one-to-one, this means that $A\sigma_g^\alpha = A\sigma_{g\alpha}$ for all points A. Similarly $a\sigma_g^\alpha = a\sigma_{g\alpha}$ for all lines a. From the last two relations it follows that $\sigma_g^\alpha = \sigma_{g\alpha}$. From the corresponding remark about point reflections (proved in exactly the same way) we have

(iv) $$\sigma_g^\alpha = \sigma_{g\alpha}, \quad \sigma_P^\alpha = \sigma_{P\alpha}.$$

In general, a set of elements of a group is said to be an *invariant complex* if it is mapped into itself (and thus onto itself) by every transformation with an element of the group. From (iv): *the line reflections and the point reflections form an invariant complex in the group of motions.* Thus the totality of line reflections is an invariant system of generators for the group of motions.

1.4. *Fixed Elements*

If α is a motion, a point P or a line g is called a *fixed point* or a *fixed line* of α if $P\alpha = P$ or $g\alpha = g$, respectively. For fixed elements under reflection the following statements are immediately evident:

(v) *The fixed points of a line reflection σ_g are the points incident with g. The fixed lines of σ_g consist of the line g and the lines perpendicular to g.*

(vi) *The only fixed point of a point reflection σ_P is the point P. The fixed lines of σ_P are the lines incident with P.*

Let us note that the concept of a metric plane, defined below in our axiomatic construction, also includes certain planes (namely, the elliptic planes, cf. §§3 and 5.1) in which statements (v) and (vi) do not hold in the above form. In an elliptic plane the fixed elements also include the points and lines "polar" to g or P.

1.5. *Formulation of Geometry in the Group of Motions*

Let us consider the two mappings

(4) $$g \to \sigma_g \quad \text{and} \quad P \to \sigma_P,$$

the first one from the set of lines onto the set of line reflections, and the second from the set of points onto the set of point reflections. These mappings are one-to-one, since reflections in two distinct points have distinct sets of fixed points (by vi), and similarly for lines and line reflections.

It is now our purpose to replace the original geometric objects, i.e., points and lines, by the reflections assigned to them in the mappings (4), and to express geometric relations among points and lines by relations among the corresponding reflections, the latter relations being defined solely in terms of the multiplication in the group of motions.

Let us first see how a statement about a motion of points and lines (i.e., about an element of the group of motions) is to be "translated" (into a statement about reflections). The statement $P\alpha = Q$ (where α is a motion, and P, Q are points) is equivalent, in view of the fact that the mappings (4) are one-to-one, to the statement $\sigma_{P\alpha} = \sigma_Q$. But by (iv) we have $\sigma_{P\alpha} = \sigma_P^\alpha$. Thus:

(vii) $P\alpha = Q$ if and only if $\sigma_P^a = \sigma_Q$,

and correspondingly for lines and line reflections:

(viii) $g\alpha = h$ if and only if $\sigma_g^\alpha = \sigma_h$.

So in our replacement procedure a motion of a point or a line is to be replaced by a transformation of the corresponding reflection. This transformation is in fact defined in terms of the group multiplication alone, and the circumstance that the elements of the group (namely, the motions) happen themselves to be mappings no longer plays any role.

The translation of the relations of incidence and orthogonality into our new language is given by the following statements:

(ix) $P \mathrel{I} g$ if and only if $\sigma_P\sigma_g$ is involutory.

(x) $g \perp h$ if and only if $\sigma_g\sigma_h$ is involutory.

On (ix) let us remark that if $P \mathrel{I} g$ and if h is the perpendicular to g at P, then by (iii) we have $\sigma_P\sigma_g = \sigma_h$, so that $\sigma_P\sigma_g$ is involutory. Conversely, if for a given point P and line g the mapping $\sigma_P\sigma_g$ is involutory, then $(\sigma_P\sigma_g)^2 = 1$, so that $\sigma_P^{\sigma_g} = \sigma_P$, and thus by (vii) the point P is a fixed point of σ_g. Then by (v) we have $P \mathrel{I} g$.

With respect to (x): the statement $g \perp h$ is equivalent by (v) to $h\sigma_g = h$ and $g \neq h$, and thus to $\sigma_h^{\sigma_g} = \sigma_h$ and $\sigma_g \neq \sigma_h$, and therefore to $(\sigma_g\sigma_h)^2 = 1$ and $\sigma_g\sigma_h \neq 1$.

Since through every point there exist pairs of mutually orthogonal lines, by (iii) every point reflection is an involutory product of two line reflections. Conversely, if $\sigma_g\sigma_h$ is involutory, then by (x) we have $g \perp h$, so that by (iii) the mapping $\sigma_g\sigma_h$ is equal to the reflection in the point of intersection of g and h. Consequently:

(xi)   *The point reflections are the involutory products of two line reflections.*

If now, instead of points and lines as heretofore, we consider the reflections as the object of primary interest, statements such as (vii) through (x), by being read from right to left, may be regarded as geometric interpretations of group-theoretic relations among reflections. Later, in our axiomatic construction of the theory, we will take this point of view exclusively, so

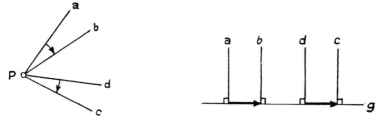

Fig. 2   Fig. 3

that, for example, the relations of incidence and orthogonality will be *defined* by the group-theoretic relations on the right-hand side of statements (ix) and (x).

As a further example of geometric interpretation of a relation among reflections, let us consider

(5) $\sigma_a\sigma_b = \sigma_d\sigma_c.$

We shall first assume that the lines a, b, c, d are incident with a point P. Then $\sigma_a\sigma_b$ is a rotation about P, as may be visualized on a movable copy of the plane, as mentioned above (Fig. 2). The line a is transformed by this rotation to the line $a\sigma_a\sigma_b = a\sigma_b$. Thus we are dealing with a rotation through an angle twice as great as the angle formed by a, b in the direction from a to b. So on the assumption a, b, c, d I P we may interpret (5) as follows: the directed angle from a to b is equal to the directed angle from d to c.

Let us now consider the case that the lines a, b, c, d are parallel to one another (Fig. 3). Then $\sigma_a\sigma_b$ is a translation (a parallel shift), perpendicular to a and b, in the direction from a to b through a distance equal to twice the distance from a to b. In this case the relation (5) is to be interpreted as follows: the directed distance from a to b is equal to the directed distance from d to c.

1.6.   *The Three-Reflections Theorem*

This theorem is one of the statements about line reflections that will be taken as an axiom in the next section (see also II4, §1). It reads

(xii)   *If the three lines a, b, c lie in a pencil (i.e., if they are either co-punctual or parallel), then $\sigma_a\sigma_b\sigma_c$ is a line reflection.*

This theorem may be visualized from the above remark on relation (5), since it asserts that a line d exists such that $\sigma_a\sigma_b\sigma_c = \sigma_d$, which is equivalent to relation (5). In the case a, b, c I P the assertion is therefore correct, since the angle from a to b may be marked off from c in the opposite direction, and if the lines a, b, c are parallel to one another, it follows from the existence of a line d, parallel to a, b, c for which the directed distance from d to c is equal to the distance from a to b.

1.7. Shorter Notation

Since the mappings (4) are one-to-one, points and lines may be identified, as was mentioned above, with the reflections corresponding to them by (4). Then points and lines are themselves elements of the group of motions, and the notation $\sigma_g, \ldots, \sigma_P, \ldots$ for reflections becomes superfluous: instead of σ_g we may write simply g, and instead of σ_P, simply P. The statements (ix) and (x) then read:

> $P \operatorname{I} g$ *if and only if Pg is involutory.*
>
> $g \perp h$ *if and only if gh is involutory.*

In the new notation the right side of (vii) reads $P^\alpha = Q$. Thus we have the equivalences

in the old notation		in the new notation
$P\alpha = Q$	is equivalent to	$P^\alpha = Q$
$P\sigma_a = Q$	is equivalent to	$P^a = Q$
$P\sigma_A = Q$	is equivalent to	$P^A = Q$

and correspondingly for the movement of lines instead of points. The old notation $P\alpha$ (or $g\alpha$) for the images under a motion can no longer be used, after the introduction of the new notation, since $P\alpha$ must now be taken to mean the product of the group elements P and α, which in general is different from the group element P^α.

1.8. Involutory Elements of a Group

Let us give some other formal properties, important below, of statements about involutory group elements like those on the right-hand side of (ix) and (x).

Let ρ, σ, τ, $\ldots$ be involutory elements of a group. The statement, that the product of some particular pair is also involutory, i.e.,

$$\rho\sigma \text{ is involutory,}$$

will be written in the form $\rho | \sigma$. The statement "$\rho_1 | \sigma$ and $\rho_2 | \sigma$" will be abbreviated to $\rho_1, \rho_2 | \sigma$, the statement "$\rho_1, \rho_2 | \sigma_1$ and $\rho_1, \rho_2 | \sigma_2$" to $\rho_1, \rho_2 | \sigma_1, \sigma_2$, and so forth.

This "stroke relation" is invariant under transformation: i.e., $\rho | \sigma$ implies $\rho^\alpha | \sigma^\alpha$ for every element α of the group.

For involutory elements ρ, σ the following three statements are equivalent to one another and to $\rho | \sigma$:

$$\rho\sigma = \sigma\rho \text{ and } \rho \neq \sigma; \quad \rho^\sigma = \rho \text{ and } \rho \neq \sigma; \quad \sigma^\rho = \sigma \text{ and } \rho \neq \sigma.$$

Let us also consider the following three-place relation for involutory elements ρ, σ, τ:

$$\rho\sigma\tau \text{ is involutory.}$$

This relation is *reflexive* in the sense that it always holds if ρ, σ, τ are not all distinct (e.g., if $\rho = \tau$, then $\rho\sigma\tau = \sigma^\rho$ is involutory), and it is *symmetric* in the following sense: if it holds for ρ, σ, τ, then it also holds for every permutation of ρ, σ, τ. For with $\rho\sigma\tau$ the transformed elements $(\rho\sigma\tau)^\rho = \sigma\tau\rho$ and $(\rho\sigma\tau)^\tau = \tau\rho\sigma$ are involutory and thus also their inverses (see (1)) $\tau\sigma\rho$, $\rho\tau\sigma$, $\sigma\rho\tau$.

2. The System of Axioms for Absolute Plane Geometry

2.1. *The System of Axioms*

We have seen how points, lines, and geometric relations among them can be represented in the group of motions by point reflections, line reflections, and group-theoretic relations among these reflections. Let us now exploit this situation systematically in order to construct a plane metric geometry, no longer restricted to the real Euclidean plane.

First of all we must create an axiomatic basis for such a procedure. Here we could start with a system of axioms dealing with points, lines, incidence, and so forth, and then define the mappings (4), and prove as theorems that they are one-to-one and that they satisfy the rules for translation into the new language given by the statements (vii) through (x).[7]

But instead of points, lines, and so forth, we could also from the very beginning take line reflection, motion, and group-theoretic operation (multiplication of motions) as the fundamental concepts, a procedure which, in view of our purpose of developing a geometry in the group of motions, is clearly the more direct one. So we take this second path, beginning with the following system of axioms (from [1] in the bibliography at the end of this chapter):

Basic assumption: *There is given a group $\mathfrak{G}$ generated by an invariant system $\mathfrak{S}$ of generators, in which each of the generators is involutory.*

The elements of $\mathfrak{S}$ will be denoted by lowercase Latin letters. Those involutory elements of $\mathfrak{G}$ that can be represented as the product of two elements in $\mathfrak{S}$, or in other words that can be written in the form ab with $a|b$ (i.e., the product ab is also involutory, see §1.8), will be denoted by Latin capital letters.

Axiom 1:   *For every P, Q there exists a g with P, Q|g.*
Axiom 2:   *From P, Q|g, h follows P = Q or g = h.*
Axiom 3:   *If a, b, c|P, there exists a d such that abc = d.*
Axiom 4:   *If a, b, c|g, there exists a d such that abc = d.*
Axiom D: *There exist g, h, j such that g|h but neither j|g nor j|h nor j|gh.*

For purposes of visualization, but for the time being only, let us interpret

---

[7]Cf. II4, where we proceeded in this way.

the elements of $\mathfrak{S}$ as line reflections and those of $\mathfrak{G}$ as motions, although the system of axioms does not assume that $\mathfrak{G}$ is a group of mappings (of any set). Nothing is assumed about the provenience of the elements of $\mathfrak{S}$ and $\mathfrak{G}$: in other words, $\mathfrak{G}$ is regarded as an "abstract" group.

With respect to the basic assumption let us recall that in the discussion in §1 the line reflections formed a system of involutory generators for the group of motions, and that this system of generators was an invariant complex in the group (§1.3).

Let us also recall that in §1 statement (xi) meant that a motion was a point reflection if and only if it was involutory and was representable as the product of two line reflections. The convention about the use of upper- and lowercase Latin letters is thus analogous to the shorter notation introduced in §1.7, where in place of $\sigma_a, \ldots, \sigma_P, \ldots$ we simply wrote $a, \ldots, P, \ldots$.

By means of translation rules (ix) and (x) in §1, the axioms can be interpreted in a visualizable way: Axioms 1 and 2 correspond to the statement that for every two given points there is a line joining them, which is uniquely determined if the points are distinct. Axioms 3 and 4 correspond to the three-reflections theorem (§1 (xii)), and will be cited below as the "first and second theorems of three reflections." The meaning of Axiom D is this: there exists a figure consisting of three lines, two of which are perpendicular to each other, while the third line is not perpendicular to either of them and does not pass through their intersection (Fig. 4).

The models of the system of axioms will be called *groups of motions*. Thus a group of motions is a pair $\mathfrak{G}, \mathfrak{S}$ consisting of a group $\mathfrak{G}$ and a system $\mathfrak{S}$ of generators of the group $\mathfrak{G}$ satisfying the basic assumption and the axioms. It should be emphasized that our axioms have no significance for a group $\mathfrak{G}$ alone but become significant only for $\mathfrak{G}$ together with a "distinguished" system $\mathfrak{S}$ of generators of $\mathfrak{G}$.

2.2. *The Group Plane. Motions of the Group Plane*

We now wish to give a precise form to the geometric language we shall use to describe the algebraic (group-theoretic) concepts occurring in our system of axioms. For this purpose we associate with the group of motions $\mathfrak{G}, \mathfrak{S}$ a geometric entity, the *group plane* of $\mathfrak{G}, \mathfrak{S}$, as follows.

The elements of $\mathfrak{S}$ will be called *lines* of the group plane, and those involutory group elements that can be represented as the product of two

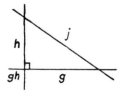

Fig. 4

elements of $\mathfrak{S}$ will be called *points* of the group plane. Two lines g and h of the group plane will be said to be *perpendicular* if $g|h$, for which we shall also write $g \perp h$. Consequently the points are those elements of the group that can written as the product of two perpendicular lines. Also, we shall say that a point P is *incident* with a line g in the group plane if $P|g$, which we may also write in the form $P \mathrel{I} g$ or $g \mathrel{I} P$. If $P \neq Q$, then by Axioms 1 and 2 the points P, Q in the group plane are joined by a unique line, which we shall denote by (P, Q).

The system of axioms allows the existence in $\mathfrak{S}$ of elements a, b, c for which $abc = 1$. The product of any two of these elements a, b, c is then equal to the third, and is thus involutory. Consequently, in the group plane the lines a, b, c are pairwise perpendicular to one another (Fig. 5); they form a *polar trilateral,* such as occurs in an elliptic plane. (Compare the sphere model of the continuous elliptic plane described in §3.) Moreover, the element ab, being the involutory product of two elements of $\mathfrak{S}$, is equal to an element C, and $C = c$; thus the same element is both a point and a line in the group plane. In general, we shall say that if $C = c$, the point C and the line c are *polar* to each other, and we shall call the point C the *pole* of the line c and the line c the *polar* of the point C.

We shall say that three lines a, b, c lie in a pencil if

(6)   *abc is a line.*

The relation (6) is symmetric, i.e., it is independent of the order in which the three lines are taken, since (6) implies that $cba = (abc)^{-1}$ is a line, the invariance of $\mathfrak{S}$ implies that $cab = (abc)^c$ is a line, and so forth (cf. §1.8). The invariance of $\mathfrak{S}$ also shows that (6) is reflexive, i.e., that it holds whenever at least two of the three lines coincide.

If $a \neq b$, the set of lines lying in the same pencil with a, b will be called the *pencil of lines* $G(ab)$, since it depends only on the product ab. This important concept will be studied in greater detail below; for example, if a, b are incident with a point P, then $G(ab)$ comprises exactly the lines through P, as follows from the first three-reflections theorem and its converse, to be proved in §4, and in general (i.e., also when a, b have no point of intersection) the reader should visualize an "ideal" point "incident" with all the

Fig. 5

lines in the pencil, and with no other lines (cf. the Klein model of hyperbolic geometry in §3 below).

The conception of pencils of lines as ideal points can be made precise from the abstract point of view and can be greatly extended; in this way we finally obtain an embedding of the group plane in a projective-metric plane, and therewith the possibility of introducing coordinates. These matters will be the subject of §6.

Now let $\alpha \in \mathfrak{G}$, and consider the transformation with α, namely, the one-to-one mapping $\gamma \to \gamma^\alpha$ of $\mathfrak{G}$ onto itself (see §1.3). Since $\mathfrak{S}$ is an invariant complex, $\mathfrak{S}$ will be mapped onto itself, and if P is a point, so that $P = gh$ with $g|h$, then $P^\alpha = g^\alpha h^\alpha$ and $g^\alpha|h^\alpha$ (see §1.8), so that P^α is also a point. Thus

$$(7) \qquad\qquad g \to g^\alpha, \quad P \to P^\alpha$$

is a one-to-one mapping of the set of lines and the set of points, each onto itself, in the group plane. Since the relation denoted by the vertical stroke is preserved under transformation (§1.8), the mapping (7) is consistent with incidence and orthogonality in the sense of §1.2. We shall call (7) the *motion of the group plane* induced by α (see §1.5, (vii) and (viii)). In particular, in the case that $\alpha = a$ is a line, we shall refer to (7) as the *reflection of the group plane* in the line a, and in case $\alpha = A$ is a point, as the reflection of the group plane in the point A.

From the invariance of $\mathfrak{S}$ it follows that every motion of the group plane takes triples of lines lying in a pencil into triples in a pencil.

If to every $\alpha \in \mathfrak{G}$ we assign the motion of the group plane induced by α, we obtain a homomorphism of $\mathfrak{G}$ onto the group of motions of the group plane. It can be shown that this homomorphism is an isomorphism (see [1], §3.7); thus we shall also, somewhat incorrectly, refer to the elements of $\mathfrak{G}$ themselves as "motions."

3.   Models of the System of Axioms. Metric Planes

In order to give some idea of the generality of this system of axioms, let us consider a number of models or, in other words, groups of motions $\mathfrak{G}, \mathfrak{S}$ in the sense of the system of axioms in §2.1. The present section omits many details, since no use is made of it in later proofs.

To obtain examples of groups of motions $\mathfrak{G}, \mathfrak{S}$, we shall in each case describe a "plane," i.e., a system $\mathfrak{E}$ of "points" and "lines," for which "incidence" of points and lines and "orthogonality" of two lines are defined and to every line g there is assigned an involutory orthogonal collineation σ_g of $\mathfrak{E}$ (cf. §1.2) leaving the points of g fixed. For $\mathfrak{S}$ we shall take this system of "line reflections" σ_g, and for $\mathfrak{G}$ the group of orthogonal collineations of $\mathfrak{E}$ generated by $\mathfrak{S}$.

For example, let us consider the continuous Euclidean plane, defined either in the usual way as a coordinate plane over the field of real numbers or as a two-dimensional Euclidean vector space. Then the discussion in §1 can be carried through in a rigorous way to show that the group $\mathfrak{G}$ generated by the system $\mathfrak{S}$ of line reflections, taken together with $\mathfrak{S}$ as the distinguished system of generators, satisfies the axioms in §2.1, so that $\mathfrak{G}$, $\mathfrak{S}$ is a group of motions. It also follows that the mapping (4) of the original plane onto the group plane consisting of $\mathfrak{G}$, $\mathfrak{S}$ is an isomorphism with respect to incidence and orthogonality (cf. §1 (ix) and (x)).

Since the same situation holds in all the following examples, we shall merely define the "planes" $\mathfrak{E}$ without repeating in every case a reference to the isomorphism of $\mathfrak{E}$ with the group plane $\mathfrak{G}$, $\mathfrak{S}$.

"Planes" of this sort, representable as an isomorphic image, with respect to incidence and orthogonality, of the group plane of a group of motions $\mathfrak{G}$, $\mathfrak{S}$, will be called *metric planes*.[8]

We first note that the above remark about the continuous Euclidean plane holds in general for every Euclidean plane in the sense of II4, §1. Consequently, the Euclidean planes discussed in that section form a subclass of the class of all metric planes; in §5 we shall characterize this subclass by additional axioms adjoined to the axiom system in §2.1.

The continuous Euclidean plane has "free mobility," i.e., any pair consisting of a point and incident line can be taken by a motion into any other such pair. A Euclidean plane without free mobility is obtained if in the continuous Euclidean plane we consider only the points with rational coordinates and the straight lines joining them; in such a plane, the lines $y = 0$ and $y = x$ cannot be carried into each other by any motion; in particular they have no angle bisector.

That there is no motion of the "rational" plane carrying these lines into each other can be proved as follows. If there were such a motion, the two lines would have an angle bisector in the "rational plane," which would be identical with one of the two angle bisectors of the two lines in the real Euclidean plane. But these latter bisectors, namely, the lines $y = (\sqrt{2} - 1)x$ and $y = -(\sqrt{2} + 1)x$, do not belong to the "rational plane," since apart from the origin they do not pass through any point with rational coordinates.

Our system of axioms contains no concepts or axioms relating to order, and there exist models (e.g., any finite model) in which it is impossible to define an order with the usual properties. Finite models are obtained as the Euclidean (coordinate) planes over finite fields, the smallest one being the plane over the prime field of characteristic 3 (see Fig. 6). This plane does not

---

[8]Metric planes can also be defined directly by a system of axioms with the fundamental concepts: point, line, incidence, and orthogonality, see [1], §2.3 in the bibliography at the end of this chapter.

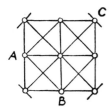

Fig. 6. A finite metric plane. Every line contains exactly three points. With respect to the inner square, standing on its vertex B, the third point on each side is the opposite vertex of the outer square; for example, the line AB also contains the point C.

have free mobility; for example, the slant lines in Fig. 6 cannot be carried by a motion into the horizontal or vertical lines.

Let us now consider metric planes with properties different from those of the ordinary Euclidean plane with respect to intersection or nonintersection of lines.

The *continuous elliptic plane* can be defined in the following way (by a model on the sphere):

Let there be given a sphere in ordinary Euclidean space. The pairs of antipodal points on the surface of the sphere (e.g., the pair consisting of the north pole and the south pole) will be called "points" and the great circles will be called "(straight) lines." By "incidence" of a "point" and a "line" we shall mean ordinary incidence of the corresponding point-pair and great circle, and "perpendicularity" of lines will be ordinary perpendicularity of the corresponding great circles. By "reflection" in the "line" g represented by the great circle K we shall then mean the one-to-one mapping of point-pairs into point-pairs and great circles into great circles induced by ordinary reflection of Euclidean space in the plane determined by K.

The group of "motions" generated by these "line reflections" consists of the mappings induced by the Euclidean rotations of the sphere.

One remarkable feature of this model is the fact that there exist polar triangles, i.e., triples of pairwise perpendicular lines (for example, the equator and two mutually perpendicular meridians). It is also noteworthy that every point reflection coincides with a line reflection; e.g., the reflection in the "point" consisting of the north and south poles (in other words, the mapping of the elliptic plane induced by a $180°$ rotation of the sphere about the north-south axis) is equal to the reflection in the equator. So we have here an example of a group of motions in whose group plane every point is equal to a line.

The Euclidean parallel axiom does not hold in this elliptic plane, since every two lines have a point of intersection, with the result that the incidence relation is the same as in a projective plane.

An incidence relation that differs from the Euclidean parallel axiom in the other direction is to be found in the *continuous hyperbolic plane,* constructed as follows (Klein model). Let there be given a circle in the projective closure of the ordinary Euclidean plane (see II3, §3.1) formed by adjunction of the line at infinity. By "points" we now mean the ordinary points in the interior of the circle (not including its circumference) and by "(straight) lines" those lines that enter the circle. Then the given circle defines a polarity by the usual pole-polar relationship for a circle.[9] Two lines are mutually "perpendicular" if each of them passes through the pole of the other (Fig. 7). The "reflection in the line g" is then the one-to-one mapping of "points" onto "points" and "lines" onto "lines" induced by the harmonic homology[10] with respect to the line g and its pole.

In this plane we see at once that through every point not on a given line there pass infinitely many lines not intersecting the given line (Fig. 8); exactly two of these lines intersect the given line on the circumference of the circle; the rest of them intersect it outside the circle.

But since points on the circumference are not "points" and tangents to it are not "lines," it follows that two lines meeting on the circumference have neither a common "point" nor a common "perpendicular" (Fig. 9), whereas two "lines" intersecting outside the circle have a common "perpendicular," namely, the polar of their point of intersection (Fig. 10). So our present model has the property that for each "line" g and each "point" P not on g there pass through P exactly two "lines" such that each of them has neither a "point" nor a "perpendicular" in common with g.

Three "lines" lie in a pencil (§2.2) if and only if they pass through the same point of the projectively closed plane. Thus in the hyperbolic plane it can happen that three lines in a pencil have neither a common point nor a common perpendicular, namely, if and only if the three lines meet at a point on the circle.

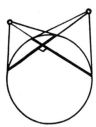

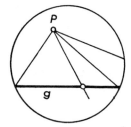

Fig. 7   Fig. 8

[9]Defined as follows: the polar of the center M is the line at infinity and for $P \neq M$, if h is the halfline from M through P, and G is the point on h for which $MP \cdot MG = 1$ (the radius of the circle is taken equal to unity), then the polar of P is the perpendicular to h at the point G.

[10]Cf. II3, §3.2.

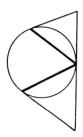

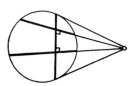

Fig. 9   Fig. 10

In §5 we shall define elliptic and hyperbolic planes by additional axioms adjoined to the system in §2.1. The continuous elliptic or hyperbolic planes defined here form only one example (from infinitely many nonisomorphic examples) of an elliptic or hyperbolic plane in the sense of §5, just as the ordinary Euclidean plane is only one example of a Euclidean plane in the sense of II4, §1.

4. Theorems in Plane Metric (Absolute) Geometry

In this section we prove a series of theorems in "absolute" geometry, i.e., theorems valid in every metric plane. We consider the metric plane to be given by definition as the group plane of a group of motions (§2.2), so that the calculus of reflections is at once at our disposal: the lines and points of the group plane are at the same time involutory elements of the group, and the geometric relations are those of the group plane. In particular, we shall make free use, side by side, of the equivalent notations $P|g$ and $P \mathrel{I} g$ for incidence, and $g|h$ and $g \perp h$ for orthogonality; and the properties of the stroke relation (§1.8) will be used without special mention.

4.1. *First Consequences of the System of Axioms*

Theorem 1 (Orthogonal intersection): *From $a, b \mathrel{I} P$ and $a \perp b$ follows $P = ab$, and conversely.*

Proof.   The notation $a \perp b$ means that ab is involutory and therefore a point. But obviously $ab \mathrel{I} a, b$. Since ab is involutory, we have $a \neq b$, so that by Axiom 2 the lines a and b can have only one common point. Thus $ab = P$. Conversely, let $P = ab$. Then the product of any pair of the elements a, b, P is equal to the remaining one, and is therefore involutory. Consequently $a \perp b, a \mathrel{I} P$, and $b \mathrel{I} P$.

Theorem 2 (Polar triangle): *If the lines a, b, c are pairwise perpendicular, then $abc = 1$, and conversely.*

Proof.   If a, b, c are pairwise permutable, then $(abc)^2 = 1$. Now if $abc \neq 1$, then abc would be involutory, i.e., ab I c, since ab is a point. Moreover, ab I b and $b \perp c$. Thus it would follow from Theorem 1 that $ab = cb$, so that $a = c$, in contradiction to $a|c$. The converse is proved in the same way as the converse statement in Theorem 1.

Theorem 3 (Existence of perpendiculars): *For every P, g there exists an l with l* I *P and l* $\perp$ *g*.

Proof.   Case 1: P I g. In this case Pg is a line: for if $P = ab$, then $a, b,$ g I P; so that by Axiom 3 there exists an l such that $abg = l$, which means that $Pg = l$. By Theorem 1 the line l has the required properties.

If P Ɨ g, then either $P \neq P^g$ or $P = g$. Thus we distinguish the two further cases:

Case 2: $P \neq P^g$ (Fig. 11). By Axiom 1 there exists an l with l I P, P^g. Since we can transform with g, we have l^g I P^g, P. By Axiom 2 it follows that $l = l^g$. Furthermore, $l \neq g$, since P I l but P Ɨ g. Therefore $l \perp g$.

Case 3: $P = g$; that is, P, g are polar. Then for arbitrary c: if c I P, then $c \perp g$ (and conversely). Now if $P = ab$, then a, b I P and therefore $a, b \perp g$. Consequently in this case there exists more than one perpendicular to g through P.

Theorem 4 (Uniqueness of the perpendiculars): *From P $\neq$ g and a,* b I *P and a, b* $\perp$ *g follows a = b*.

Proof.   Case 1: P I g. From a, g I P and $a \perp g$ it follows from Theorem 1 that $P = ag$. Similarly $P = bg$. Thus $ag = bg$, so that $a = b$.

Case 2: $P \neq P^g$. From a, b I P follows a^g, b^g I P^g. From $a, b \perp g$ follows $a^g = a, b^g = b$. Thus a, b I P, P^g, so that $a = b$, by Axiom 2.

If $P \neq g$, let (P, g) denote the (unique, by Theorem 4) perpendicular from P to g. If P I g, then $(P, g) = Pg$.

Theorem 5:   *With every line there are incident at least three distinct points.*

Proof.   Preliminary remark: If $a \not\perp b$ and if there exist three distinct points incident with a, then there also exist three incident with b. For from the points on a

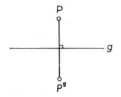

Fig. 11

draw the perpendiculars to b and consider their points of intersection with b, which must be distinct, since otherwise, by Theorem 4, two of the perpendiculars would be equal to each other and therefore, by Axiom 2, equal to a, in contradiction to $a \not\perp b$.

So if there are three points on one of the three lines g, h, j of Axiom D, there are three points on each of these lines.

Now let a, b be two arbitrary lines. Neither a nor b is perpendicular to all three lines g, h, j; for if $a \perp g, h, j$ then $a = gh$ by Theorem 2, since $g \perp h$. But then $gh | j$, contrary to hypothesis.

We now see that if there are three distinct points on a, then there are also three on b; for by the preliminary remark there must then be three distinct points on one of the three lines g, h, j, and therefore on each of these lines, and thus finally on b.

It remains to prove that there exists a line containing three distinct points. Let $l = (gh, j)$, (so that $j \neq gh$ since $j \nmid h$). On l there lie the points $gh, lj, (lj)^{gh}$ (the last because $l^{gh} = l$). Since j is incident with lj but not with gh, we see that $gh \neq lj$. Thus we also have $gh = (gh)^{gh} \neq (lj)^{gh}$. Finally, if we had $lj = (lj)^{gh}$, it would follow that $lj = l^{gh}j^{gh} = lj^{gh}$, so that $j = j^{gh}$, and thus $j = gh$ or $j | gh$, both of which are false. Thus there are at least three distinct points on the line l; this completes the proof of Theorem 5.

4.2. Supplements and Converses to the Theorems of Three Reflections

Theorem 6 (Supplement to Axiom 3): *From a, b, c I P and $abc = d$ follows d I P.*

Proof. The assertion $d|P$ means that $abcP = Pabc$ and $abc \neq P$, the first of which follows at once from the fact that a, b, c permute with P by hypothesis; and if we had $abc = P$ and therefore $ab = Pc$, then from $P|c$ it would follow that $a|b$, and then by Theorem 1 we would have $P = ab$. Setting this result in $abc = P$ shows that $Pc = P$ and thus produces the contradiction $c = 1$.

Theorem 7 (Supplement to Axiom 4): *From $a, b, c \perp g$ and $abc = d$ follows $d \perp g$.*

The proof is the same as for Theorem 6, with P replaced by g and Theorem 1 by Theorem 2.

Theorem 8 (Converse of the first three-reflections theorem): *From $a \neq b, a, b$ I P and $abc = d$ follows c I P.*

Proof (Fig. 12). Draw a perpendicular b' from P to c; then $b'c$ is a point P' and b', c I P'. By the three-reflections theorem and its supplement, abb' is a line a' through P; and $a' \neq b'$, since $a \neq b$. Also, $a'd = b'c = P'$, since $a'b'c = abc = d$, and therefore a' I P'. Altogether we have a', b' I P, P' and $a' \neq b'$; and therefore $P = P'$ because of the uniqueness of the line joining two points; thus c I P.

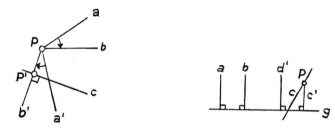

Fig. 12   Fig. 13

Theorem 9 (Converse of the second three-reflections theorem): *From* $a \neq b$, a, $b \perp g$, *and* $abc = d$ *follows* $c \perp g$.

Proof (Fig. 13). Let P be an arbitrary point on c: $c \mid P$. If $P = g$, then $c|g$, so that $c \perp g$. If $P \neq g$, draw the perpendicular c' from P to g. We now show that $c = c'$.

Since a, b, $c' \perp g$, the three-reflections theorem and its supplement show that abc' is a line d' with $d' \perp g$. Also $dc = ab = d'c'$ and therefore $cc'd' = d$. If we had $c \neq c'$, then Theorem 8 would show that d' I P and thus $P \neq g$ and $d', c'|P, g$, so that, by Theorem 4, $d' = c'$, and therefore $a = b$, contrary to hypothesis.

4.3. *Lines in a Pencil. The Theorem of Perpendiculars*

In §2.2 three lines were said, by definition, to lie *in a pencil* if they satisfy the relation (6), namely,

abc is a line.

Given two lines a, b with $a \neq b$, the set of lines c satisfying (6) was called the *pencil of lines G(ab)*. If a, b have a point V or a perpendicular v in common, then, by the three-reflections theorems and their supplements, the pencil $G(ab)$ is the set of lines through V or perpendicular to v, respectively, and we speak of a pencil of lines with the *center V* or the *axis v*. A pencil of lines with a center, i.e., a pencil consisting of all the lines through a given point, is said to be *proper,* and all other pencils are *improper*. The proper pencils are precisely those pencils that contain mutually perpendicular lines. A pencil with an axis, i.e., a pencil consisting of all the perpendiculars to a given line v, is called a *pencil of perpendiculars.*

Our system of axioms by no means precludes the existence of pencils that have no *carrier,* i.e., neither center nor axis; in other words, there exist lines a, b, c satisfying (6) for which there is no V with $V|a$, b, c and no v with $v|a$, b, c.[11]

---

[11]For example, in the Klein model for hyperbolic geometry (§3) the set of lines through any point on the circumference of the circle is a pencil without carrier.

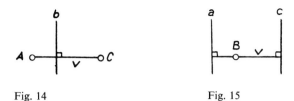

Fig. 14   Fig. 15

But the situation is different if instead of (6) we consider the analogous relations

(8)   *AbC is a line,*

(9)   *aBc is a point,*

involving both lines and points. For then we have

Theorem 10: *AbC is a line if and only if there exists a line v with $v|A, b, C$* (Fig. 14).

Theorem 11: *aBc is a point if and only if there exists a line v with $v|a, B, c$* (Fig. 15).

If $A = C$, the relation (8) always holds, and on the other hand there always exists a line v with $v|A, b, C$, namely, a perpendicular from A to b. Instead of Theorem 10 it is thus enough to prove

Theorem 10′: *AbC with $A \neq C$ is a line if and only if $b \perp (A, C)$.*

Proof of Theorem 10′. Let $(A, C) = v$ and $Av = a$, $Cv = c$. Then $AbC = (abc)^v$.

If $b \perp v$, then $a, b, c \perp v$, so that, by the three-reflections theorem and its supplement, abc is a line d with $d \perp v$, and therefore $AbC = d^v = d$.

Conversely, if AbC is a line, then $(AbC)^v = abc$ and therefore acb are also lines. Then $b \perp v$ by Theorem 9, since $a, c \perp v$ and $a \neq c$.

Moreover,

Supplement to Theorem 10: *If AbC is a line d and $A, b, C|v$, then also $d \mid v$.*

For $A \neq C$ this statement follows from the proof of Theorem 10′, and for $A = C$, we have $A, b|v$, $b^4|v^4$ and therefore $d|v$.

Proof of Theorem 11. If $v|a, B, c$ and we set $Bv = b$, then $v|a, b, c$; thus, by the three-reflections theorem and its supplement, abc is a line d with $d \perp v$. Then $aBc = abvc = abcv = dv$ is a point D with $D \mathrm{I} v$.

Conversely, if aBc is a point D, then $BcD = a$, so that by Theorem 10 and its supplement there exists a v with $v|B, c, D, a$.

From this proof we have

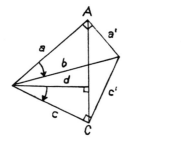

Fig. 16   Fig. 17

Supplement to Theorem 11: *If aBc is a point D and if a, B, c|v, then also D|v.*

Another formulation of Theorem 10′ is the following theorem, which Hjelmslev calls the fundamental theorem of metric geometry (Fig. 16).

Theorem 12 (Theorem of perpendiculars): *If aa′ = A, cc′ = C, A ≠ C and if abc is a line d, then:*

a′bc′ is a line if and only if d ⊥ (A, C).

The theorem of perpendiculars provides a criterion for three lines to lie in a pencil. It also shows that, given three lines a, b, c through a point, the "fourth reflection line" $d = abc$ can be constructed by drawing perpendiculars.

Proof of Theorem 12. Since $a′bc′ = a′a \cdot abc \cdot cc′ = AdC$, the assertion follows from Theorem 10′.

The theorem of perpendiculars provides the construction by means of which the following theorem is proved.

Theorem 13 (Joining a point to a pencil of lines): *Given a point P and lines a′, c′, there always exists a line incident with P and in a pencil with a′, c′* (Fig. 17).

Proof. If $a′ = c′$, take an arbitrary line through P. If $a′ ≠ c′$, draw perpendiculars a, c from P to $a′, c′$, with feet $aa′ = A, cc′ = C$. If $A = C$, then $c′$ I A and a is the desired line. If $A ≠ C$, draw a perpendicular d from P to (A, C); by the three-reflections theorem and its supplement, adc is a line b through P such that $abc = d$. By the theorem of perpendiculars, b lies in a pencil with $a′, c′$.

4.4. Metric Closure Theorems in the Triangle

With the means now at our disposal some of the well-known "triangle

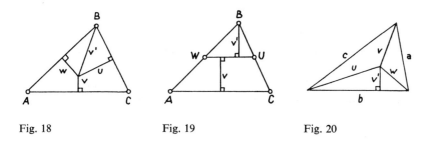

Fig. 18   Fig. 19   Fig. 20

theorems" of Euclidean geometry can be proved under our present weaker hypotheses, i.e., as theorems of absolute geometry.

Theorem 14 (Right-bisector theorem): *If $C^u = B$ and $B^w = A$, there exists a line v such that uvw is a line and $C^v = A$* (Fig. 18).

Theorem 14 states that if the right bisectors of two sides of a triangle are given, there exists a right bisector of the third side in a pencil with them.

Proof. By Theorem 13 there exists a line v' incident with B and in a pencil with u, w. Thus $uv'w$ is a line v for which $C^v = C^{uv'w} = B^{v'w} = B^w = A$.

Theorem 15: *If $C^U = B$ and $B^W = A$, then there exists a line v such that UvW is a line and $C^v = A$* (Fig. 19).

Theorem 15 states that the line joining the midpoints of two sides of a triangle has a perpendicular in common with the third side which is at the same time a right bisector of the third side.

Theorem 16 (Angle-bisector theorem): *If $c^u = b$, $b^w = a$, and S I u, w, then there exists a line v through S with $c^v = a$* (Fig. 20).

Theorem 16 states that the intersection of two angle-bisectors of a triangle lies on the bisector of the third angle.

The proofs of Theorems 15 and 16 are analogous to that of Theorem 14.

Theorem 17 (Isogonality theorem): *If a, b, c; a', b', c'; a'', b'', c'' are*

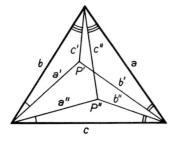

Fig. 21

lines such that $a'' = ba'c$, $b'' = cb'a$, $c'' = ac'b$, *and if* a', b', c' *are in a pencil, then* a'', b'', c'' *are also in a pencil.*

Proof. If $a'b'c'$ is a line, then $a''b''c'' = ba'c \cdot cb'a \cdot ac'b = (a'b'c')^b$ is also a line.

In particular, if a, b, c are the sides of a triangle, Theorem 17 may be stated as follows. The three fourth reflection lines (Fig. 21), with respect to the sides of a triangle, of the three lines joining a given pencil to the vertices lie in a pencil. Thus every triangle provides a one-to-one (involutory) correspondence between pencils of lines—an "isogonality" point relation.

As our last metric closure theorem for the triangle we now prove the *altitudes theorem.*

Theorem 18: *If* $abc \neq 1$ *and if*

(10) $$u \perp a, \quad v \perp b, \quad w \perp c,$$

(11) $$ubc, avc, abw \text{ are lines,}$$

then uvw *is a line.*

Theorem 18 states that the three "altitudes" in a nonpolar triangle lie in a pencil; it is not required that the sides of the triangle intersect one another.

Proof (Fig. 22). Let $au = U$, $bv = V$, $cw = W$ be the feet of the altitudes, and let p, q, r denote the lines in (11); then we have the identities

(12) $$Up = abc,$$

(13) $$Up = qV^c,$$

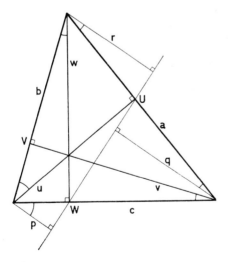

Fig. 22

(14) $$Up = rW,$$

(15) $$pqr = uvw.$$

From $abc \neq 1$ it follows by (12) that $Up \neq 1$, so that the perpendicular (U, p) is uniquely determined. By means of Theorem 11 we can show from (13) that $(U, p) \mid q$, and similarly from (14) that $(U, p) \mid r$; and then since $(U, p) \mid p, q, r$, the element pqr is a line by the second theorem of three reflections, so that (15) shows that uvw is also a line.

4.5. *The Transitivity Theorem*

From the fact that a line is always transformed into a line, we have already deduced the reflexivity and symmetry of the relation of lying "in a pencil." Another important property of this relation is its *transitivity*.

Theorem 19 (Transitivity theorem): *If $a \neq b$ and if abc and abd are lines, then acd is also a line.*

In a pencil of lines $G(ab)$ with a center V or an axis v the transitivity follows at once from the three-reflections theorems and their converses;[12] for if a, b have a point V in common, then the converse of the first three-reflections theorem shows that also c, d I V, and then, by the theorem itself, the desired assertion follows from a, c, d I V. But the great importance of Theorem 19 lies in the fact that it asserts the transitivity *in general* and thus enables us to deal effectively with pencils that have neither a center nor an axis. As we shall see, the transitivity theorem is an important tool for proofs of theorems on reflections.

The importance of the transitivity theorem for dealing with pencils of lines is shown by Theorems 20 through 23, which follow from it. (The symbol $\in$ denotes set membership.)

Theorem 20: *If $a, b, c \in G(a'b')$, then abc is a line.*

Theorem 21: *If $a \neq b$ and $a, b \in G(a'b')$ and if abc is a line, then also $c \in G(a'b')$.*

Theorems 20 and 21, which we shall call the *general theorem of three reflections and its converse,* show that the concepts of lying "in the same pencil of lines" and lying "in a pencil" coincide. The two theorems are easily proved by repeated application of the transitivity theorems.

Theorems 20 and 21 can be combined as follows.

Theorem 22: *From $a' \neq b'$ and $a', b' \in G(ab)$ follows $G(ab) = G(a'b')$.*

---

[12]Thus in metric planes satisfying Axiom V of §5.2 there is no need of any special proof of the transitivity theorem.

This latter statement shows that *any* two (distinct) lines of a pencil may be used to represent the pencil. By twofold application of Theorem 22 we have, finally:

Theorem 23: *From $a \neq b$ and $a, b \in G(a'b')$, $G(a''b'')$ follows $G(a'b') = G(a''b'')$; in other words, two distinct pencils of lines have at most one line in common.*

Also, Theorem 20 can be sharpened as follows.

Supplement to Theorem 20: *From $a, b, c \in G(a'b')$ follows $abc \in G(a'b')$.*

Proof. If $a = b$, the assertion is trivial, and for $a \neq b$ Theorem 22 shows that $G(a'b') = G(ab)$ and therefore, since abc is a line by Theorem 20, that $ab \cdot abc = ab \cdot cba = c^{ba}$ is a line, which means that $abc \in G(ab)$.

Proof of Theorem 19. First a preliminary remark:

(16) *The feet of perpendiculars to two distinct lines from a point not on both lines are distinct* (Fig. 23).

(From $a \neq b$, $P \mathrel{I} a', b', aa' = A, bb' = B$, and $A = B$ follows $P, A = B \mathrel{I} a', b'$ and $aa' = bb'$, so that $ab = a'b'$, and therefore, since $a \neq b$, we have $a' \neq b'$, so that $P = A = B$, and therefore $P \mathrel{I} a, b$.)

We now distinguish two cases:

1. *In the given metric plane there is a polar triangle.* Then, as will be proved in §5.1 without use of the transitivity theorem, every two lines have a point in common, so that by the above theorems the transitivity theorem follows at once.

2. *In the given metric plane there is no polar triangle.* Then $P \neq g$ for any point P and line g; for $P = g$ would imply, since by definition P is the product of two perpendicular lines ab, that $ab = g$, so that $abg = 1$ and therefore, by Theorem 2, there exists a polar triangle. Thus by Theorem 4 *the perpendicular from a point to a line is always uniquely determined.*

In the proof of the transitivity theorem for case 2, which we shall carry through by a sixfold application of the theorem of perpendiculars, we may assume that a, b, c, d are four distinct lines, since otherwise the assertion is trivial.

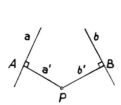

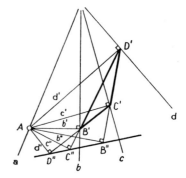

Fig. 23   Fig. 24

Choose a point A incident with a but not with b (Fig. 24); then A is not incident with either c or d. From A draw the perpendiculars b', c', d' to b, c, d, respectively. By (16) the feet B', C', D' of these perpendiculars are distinct.

The two hypotheses that bac and bad are lines each lead to a figure for the theorem of perpendiculars with A as "target." These figures show that

$$(17) \qquad b'ac' \perp (B', C'), \quad b'ad' \perp (B', D').$$

If the points B', C', D' lie on a line, then $b'ac' = b'ad'$ by (17), so that $c' = d'$. With this line $c' = d'$ the points A, C', D' are incident, and so also is B', since B', C', D' are collinear; thus $b' = c' = d'$ and the lines b, c, d have a common perpendicular through A. By the converse of the second three-reflections theorem this common perpendicular is also perpendicular to the line a, so that the desired assertion follows from the second three-reflections theorem.

Now if the lines (B', C'), (B', D'), (C', D') are distinct, let us draw perpendiculars d'', c'', b'' to them from A. By (16) the feet of these perpendiculars are distinct. Thus we obtain three more figures with A as target and with B', C', D', respectively, as center of a second pencil of lines; these figures show that

$$(18) \qquad d''b'c'' \perp (D'', C''), \quad d''c'b'' \perp (D'', B''), \quad c''d'b'' \perp (C'', B'').$$

But $b'ac' = d''$ and $b'ad' = c''$, by (17). Thus $d''c' = c''d'$, so that in (18) we have $d''c'b'' = c''d'b''$; and then $(D'', B'') = (C'', B'')$, so that B'', C'', D'' are collinear. So in (18) we also have $d''b'c'' = d''c'b''$ and therefore $b'c'' = c'b''$; substituting this result in the equation $b'ad' = c''$ we obtain $c'ad' = b''$ and thus $c'ad' \perp (C', D')$.

A sixth application of the theorem of perpendiculars, this time in the other direction, shows that cad is a line.

4.6. Counterpairing

If the lines x, x^*, y, y^* satisfy the relation $xy = y^*x^*$, we say that y, y^* are mirror images of each other with respect to x, x^* (see also §1.5 (5)).

Let it now be assumed that x, x^* are lines in a pencil $G(ab)$ and let the mapping which to every $y \in G(ab)$ assigns the line $y^* = xyx^*$ be called the *counterpairing* in the pencil $G(ab)$ determined by x, x^*. (From Theorem 20 it follows that xyx^* is a line y^* and by Theorem 21 that $y^* \in G(ab)$.) The image line y^* is also called the counterline to y with respect to x, x^*. In particular, it is clear that x^* is the counterline to x with respect to x, x^*. Moreover:

Every pair consisting of a line and its counterline u, u^* with respect to x, x^* determines the same counterpairing as x, x^*.

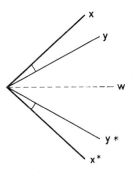

Fig. 25

Proof.   If $u^* = xux^*$ with $u, u^* \in G(ab)$, then for all $y \in G(ab)$ we have $uyu^* = u \cdot yxu \cdot x^* = u \cdot uxy \cdot x^* = xyx^*$.

If x, x^* has an angle bisector, i.e., if there exists a line $w \in G(ab)$ with $x^w = x^*$, then

$$xyx^* = xyx^w = x \cdot ywx \cdot w = x \cdot xwy \cdot w = y^w.$$

In this case the counterpairing is induced by a metric mapping of the whole plane, namely, by the reflection in w.

4.7.   Nine-Line Lemma. Counterpairing Theorem

By means of Theorems 20 and 23, proved above from the transitivity theorem, we can now prove a lemma dealing with arbitrary (not necessarily involutory) group elements with involutory products:

Nine-line lemma: *If $\alpha_1, \alpha_2, \alpha_3, \beta_1, \beta_2, \beta_3$ are motions with $\alpha_1 \neq \alpha_2$ and $\beta_1 \neq \beta_2$, and if the eight products $\alpha_i\beta_k$, $i, k = 1, 2, 3$, except for $i = k = 3$, are lines, then the ninth product $\alpha_3\beta_3$ is also a line.*

For the multiplication table of the $\alpha_i\beta_k$ the lemma states: if the eight places marked by ◯ are occupied by lines, then the place marked by * is also occupied by a line.

	$\beta_1 \neq \beta_2$		β_3
α_1	◯	◯	◯
$\neq$			
α_2	◯	◯	◯
α_3	◯	◯	*

Proof. If $\beta_3 = \beta_1$, the statement is trivial. So we assume $\beta_3 \neq \beta_1$. Then

(19) $\qquad (\alpha_1\beta_1)^{-1}(\alpha_1\beta_2) = (\alpha_2\beta_1)^{-1}(\alpha_2\beta_2) = (\alpha_3\beta_1)^{-1}(\alpha_3\beta_2),$

(20) $\qquad\qquad (\alpha_1\beta_1)^{-1}(\alpha_1\beta_3) = (\alpha_2\beta_1)^{-1}(\alpha_2\beta_3).$

By hypothesis, the parenthetical expressions are lines, so the exponents -1 can be omitted, and the group elements (19) and (20) are $\neq 1$, since $\beta_1 \neq \beta_2, \beta_3$. Thus each of the two group elements (19) and (20) determines a pencil of lines. These two pencils have in common the two lines $\alpha_1\beta_1$, $\alpha_2\beta_1$ (distinct by hypothesis) and are therefore identical by Theorem 23. The equations (19) and (20) thus show that there exists a pencil of lines G to which the eight lines of the hypothesis all belong. Since the group element (20) is also equal to $(\alpha_3\beta_1)^{-1}\alpha_3\beta_3$, we have $\alpha_3\beta_3 = (\alpha_3\beta_1)\cdot(\alpha_1\beta_1)^{-1}\cdot(\alpha_1\beta_3)$, so that $\alpha_3\beta_3$ is representable as the product of three lines from G and is therefore a line, by Theorem 20.

By the supplement to Theorem 20 this line belongs to the pencil G, so that all nine lines of the lemma belong to the same pencil.

The nine-line lemma is useful for the proof of geometric theorems, for example:

Theorem 24 (Hessenberg Counterpairing Theorem): *Let the ten lines* $a_1, a_2, a_3; b_1, b_2, b_3; c_1, c_2, c_3; g$ *satisfy the following conditions:*

1) $a_1a_2a_3$ *is a line;*
2) $a_ia_k = b_kb_i$ *for* $i, k = 1, 2, 3$ ($i \neq k$);
3) $c_ia_kc_l$ *is a line for every permutation* i, k, l *of* $1, 2, 3$;
4) $c_1b_1 \neq c_2b_2$ *(subsidiary assumption);*
5) c_1b_1g, c_2b_2g *are lines.*

Then c_3b_3g *is also a line.*

The counterpairing theorem states (cf. Fig. 26): let the vertices of a triangle be "sighted" along lines of a given pencil, and consider the counter-

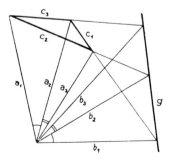

Fig. 26

lines associated with these "lines of sight" by a given counterpairing of the lines of the pencil. For each of these associated lines consider the pencil determined by it and the side of the triangle "corresponding" to it. (By the side of the triangle "corresponding" to the line associated with a given "line of sight" we mean the side opposite the vertex "sighted" by the line of sight.) Then a line that passes through two of these "intersections" also passes through the third.

Proof of Theorem 24. With the assumption 2) $a_i a_k = b_k b_i$, it is easy to verify the multiplication table

	$b_3 a_1 c_2$	$b_3 a_2 c_1$	g
$c_1 b_1$	$c_1 a_3 c_2$	$(a_3 a_1 a_2)^{c_1}$	$c_1 b_1 g$
$c_2 b_2$	$(a_3 a_2 a_1)^{c_2}$	$c_2 a_3 c_1$	$c_2 b_2 g$
$c_3 b_3$	$c_3 a_1 c_2$	$c_3 a_2 c_1$	$c_3 b_3 g$

By the assumptions 1), 3), and 5), the first eight of these products are lines. By the subsidiary assumption 4), we have $c_1 b_1 \neq c_2 b_2$, so that $b_3 a_1 c_2 \neq b_3 a_2 c_1$ by 2). Thus the desired assertion follows from the nine-line lemma.

4.8.   Theorems of Brianchon and Pappus-Pascal

Consider a hexagram, i.e., a set of six distinct lines $a_1, b_2, a_3, b_1, a_2, b_3$, such that its sides pass alternately through one (and only one) of two fixed points A, B: $a_i \mathbin{I} A$, $a_i \mathbin{I} B$, $b_i \mathbin{I} B$, $b_i \mathbin{I} A$ ($i = 1, 2, 3$). The pencils $G(a_i b_k)$ with $i \neq k$, $i, k = 1, 2, 3$, will be called the *vertices* of the hexagram, and the vertices $G(a_i b_k)$ and $G(a_k b_i)$ will be said to be *opposite* to each other. The vertices are distinct from one another and from A, B, as follows from Theorem 23 (uniqueness of a line common to two pencils). We assume that each of the three pairs of opposite vertices is joined by a unique line, called a *diagonal* of the hexagram, and (for all permutations i, k, l of 1, 2, 3) we denote the diagonal joining $G(a_i b_k)$ and $G(a_k b_i)$ by c. Then we have the

Theorem of Brianchon: *if two distinct diagonals pass through a point C, then the third diagonal also passes through C.*

If in the Brianchon figure all the lines and points are distinct and all the diagonals and points of intersection exist, the theorem of Brianchon states:

If the sides of a hexagram (Fig. 27) pass alternately through two fixed points, its diagonals that is, the lines joining pairs of opposite vertices, are copunctual.

If everywhere in the Brianchon theorem we formally replace "line" by "point" and "point" by "line" (and correspondingly, "line joining two

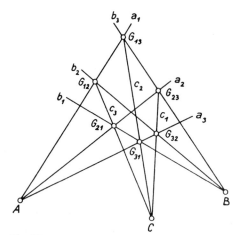

Fig. 27

points" by "point of intersection of two lines," etc.) we obtain the (dual) theorem of *Pappus-Pascal,* of basic importance for the algebraization of metric geometry (cf. §6 and 113, §1.2): *If the vertices of a hexagram lie alternately on two fixed lines,* its *diagonal points* (i.e., the intersections of pairs of pairs of opposite sides) *are collinear* (Fig. 28).

The Pappus-Pascal theorem is not only "dual" to the Brianchon theorem, but (if all the relevant points and lines are distinct) the theorems are equivalent to each other in the sense that every Pappus-Pascal figure can be regarded as a Brianchon figure and conversely, so that any closure theorem about the one figure can be reduced to a closure theorem about the other. In Fig. 28, for example, the incidence C_1 I c asserted by the Pappus-Pascal figure can be proved from the other incidences by means of the Brianchon theorem; for if we set $C_1 = C, A_1 = A, B_1 = B, g_{12} = a_1, g_{13} = a_2, a = a_3, g_{31} = b_1, g_{21} = b_2, b = b_3$, then $g_{23} = c_1, g_{32} = c_2, c = c_3$ and by the Brianchon theorem the diagonal $c_3 = c$ passes through the intersection $C = C_1$ of the other two diagonals $c_1 = g_{23}, c_2 = g_{32}$.

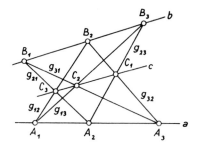

Fig. 28

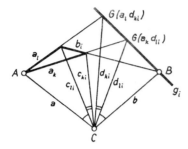

Fig. 29a

In giving proof of the Brianchon theorem (which, as we have seen, is at the same time a proof of the Pappus-Pascal theorem), we may assume that the three lines c_1, c_2, c_3 are distinct (since otherwise the assertion is trivial) and that c_2, c_3, say, intersect at a point C. Then we see that $G(C)$ is distinct from all the vertices $G(a_ib_k)$ of the hexagram, that $C \neq A$, B, and that none of the lines a_i, b_k passes through C, since all these statements follow by repeated application of Theorem 23 (uniqueness of a line common to two pencils). Thus by Theorems 13 and 23 there exists, for each vertex $G(a_ib_k)$, exactly one line $c_{ik} \in G(C)$ through $G(a_ib_k)$. Since c_3, c_2 I C, we have $c_{12} = c_{21} = c_3$, $c_{13} = c_{31} = c_2$, so that the desired assertion can be expressed in the form $c_{23} = c_{32}$ (note that $c_{23} = c_{32}$ implies $c_1 = c_{23} = c_{32}$ and thus c_1 I C).

Let a, b be the lines joining C to A, B. In the pencil $G(C)$ the lines a, b determine a counterpairing (or if $a = b$, a reflection) in which each line c_{ik} corresponds to a line d_{ik} given by $ac_{ik} = d_{ik}b$.

Now let i, k, l be a permutation of the numbers 1, 2, 3. Consider the triangle b_i, a_k, a_l (Fig. 29a), which is "sighted" along the lines a, c_{li}, c_{ki} of the pencil $G(C)$ and consider also the lines b, d_{li}, d_{ki} paired with these lines in the counterpairing in $G(C)$. Let g_i denote the line (Theorems 13 and 23) common to the two distinct pencils $G(B)$, $G(a_kd_{li})$; here $a_k \neq d_{li}$ (since a_k I C, d_{li} I C) and $G(B) \neq G(a_kd_{li})$ (since $a_k \notin G(B)$, $a_k \in G(a_kd_{li})$). Thus by the counterpairing theorem (Theorem 24) g_i belongs to the pencil $G(a_ld_{ki})$. (The subsidiary assumption $b_ib \neq a_kd_{li}$ of the counterpairing theorem is satisfied, since a_k I B.)

Thus $g_i \in G(B)$, $G(a_kd_{li})$, $G(a_ld_{ki})$, i.e.,

$$g_1 \in G(B), \quad G(a_2d_{31}), \quad G(a_3d_{21}),$$

$$g_2 \in G(B), \quad G(a_1d_{32}), \quad G(a_3d_{12}),$$

$$g_3 \in G(B), \quad G(a_2d_{13}), \quad G(a_1d_{23}).$$

Since $c_{12} = c_{21}$, $c_{13} = c_{31}$, it follows that $d_{12} = d_{21}$, $d_{13} = d_{31}$. So there exist two distinct pencils to which both g_1 and g_2 belong, and also two

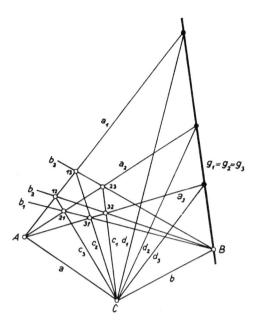

Fig. 29b

distinct pencils to which both g_1 and g_3 belong, which means that $g_1 = g_2 = g_3$. Thus $g_1 \in G(a_1 d_{23})$, $G(a_1 d_{32})$, and therefore $d_{23}, d_{32} \in G(C)$, $G(a_1 g_1)$, ($a_1 \neq g_1$, since g_1 I B, a_1 Ɨ B). From $a_1 \notin G(C)$ follows $G(C) \neq G(a_1 g_1)$, so that $d_{23} = d_{32}$ and thus $c_{23} = c_{32}$ (Fig. 29b).

This proof, due to Hessenberg, is analogous to Hilbert's proof of the affine Pappus-Pascal theorem in the Euclidean plane by means of a three-fold application of the theorem on concyclic quadrilaterals (see Hilbert [5]). Namely, Hessenberg's proof consists of introducing metric properties, by means of a special counterpairing, into a figure consisting at first entirely of incidences, and then showing, by a threefold application of the counter-pairing theorem, that this special counterpairing associates the same line through B with each of the three trilaterals b_i, a_k, a_l.

At this point we must break off our development of absolute geometry, referring the reader to Bachmann [1] for a more systematic discussion, particularly of the basic question, sketched below in §6, of embedding metric planes in projective-metric planes.

4.9. Halfrotations[13]

With respect to this embedding problem let us mention here the following concept, due to Hjelmslev:

[13]Cf. II4, §4, where halfrotations in Euclidean planes were introduced.

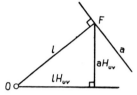

Fig. 30

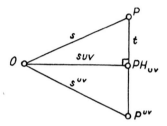

Fig. 31. $P^t = P$, $P^s = P$, and therefore $P^{PH_{uv}} = p^{t \cdot suv} = p^{suv} p = {}^{uv}$.

Let u, v be two nonperpendicular lines through a point O, so that uv is a non-involutory rotation about O (cf. §1.5). Corresponding to the rotation uv we define a *halfrotation* H_{uv} at first as a mapping of lines: for $a \mathbin{I} O$ let $aH_{uv} = auv$; and for $a \mathbin{I} O$ let l be a perpendicular from O to a, with foot F, and then let aH_{uv} be the (unique, since $u \not\perp v$) perpendicular from F to $lH_{uv} = luv$ (Fig. 30).

It can be proved (cf. Bachmann [1], §6) that the halfrotations take pencils of lines into pencils of lines or, more precisely, that they map the set of pencils of lines one-to-one onto itself, with proper pencils going into proper pencils. The importance of halfrotations lies in the fact that for every improper pencil that is not a pencil of perpendiculars for a line through O there exists a halfrotation about O taking the improper pencil into a proper pencil.

The name "halfrotation" for these mappings (which are not motions and in general preserve perpendicularity only in a Euclidean metric) is explained by the point mapping (Fig. 31): the image of the point P under the halfrotation H_{uv} is the midpoint between P and the image of P under the rotation uv.

5.   Euclidean, Hyperbolic, and Elliptic Geometry

The whole discussion up to now (in §§2 and 4) is valid for all groups of motions and is therefore part of the "absolute" geometry, i.e., a geometry which contains as special cases not only Euclidean geometry but also the two classical non-Euclidean geometries, namely, hyperbolic and elliptic. Let us now see how these three geometries fit into our present system, in which they will be defined by axioms adjoined to the system in §2.1. It must be emphasized that the Euclidean, elliptic, and hyperbolic geometries studied here are more general than the corresponding classical concepts: for example, there exist infinitely many nonisomorphic Euclidean planes, in the sense defined below, among which the classical (continuous) Euclidean plane is only a special case, and the corresponding remarks hold for the elliptic and hyperbolic planes. (Cf. also §§3 and 6 and the detailed discussion in [1].)

5.1. *Elliptic Geometry*

A principle on which we can classify the groups of motions is provided by

Axiom P: *There exists a polar trilateral* (i.e., three pairwise perpendicular lines).

A group of motions satisfying Axiom P will be called an *elliptic group of motions,* and its group plane will be called an *elliptic plane.* An example of an elliptic plane, namely the sphere model, has been given in §3. The reader is urged to visualize the following discussion by means of this model.

The fact that three lines a, b, c are pairwise perpendicular is equivalent, by Theorem 2, to $abc = 1$; that is, $ab = c$. Axiom P is thus equivalent to the requirement: there exists a point that is equal to a line. But more than that, we have:

Theorem 25:   *In an elliptic group of motions every point is equal to a line, and conversely.*

In terms of poles and polars (§2.2), Theorem 25 reads as follows: in an elliptic plane every line has a pole and every point has a polar.

In order to prove Theorem 25 we first show

(i)   *If $A|b$, then: There exists an a with $A = a$, if and only if there exists a B with $b = B$.*

Proof of (i).   Let $Ab = c$. Then $b = Ac, A = bc,$ and A, b, c are pairwise in the stroke relation.

If there exists an a with $A = a$, then $b = ac$ with $a|c$, so that ac is a point B with $b = B$.

If there exists a B with $b = B$, then $A = Bc$ with $B|c$, so that Bc is a line a with $A = a$.

Proof of Theorem 25.   In an elliptic group of motions there exists a pair A, a with $A = a$. Then if g is an arbitrary line, let G be a point on g and let h be a line through G and A. From (i) we obtain successively: h is a point, G is a line, g is a point. The proof that an arbitrary point P is equal to a line proceeds in the same way.

If $A = a$, i.e., If A and a are polar to each other, the statements $g \mid A$ (i.e., g I A) and $g \mid a$ (i.e., $g \perp a$) are equivalent; namely, all lines per-

Fig. 32   Fig. 33

pendicular to the line a intersect at the point A, and all lines through A are perpendicular to a (Fig. 32).

In an elliptic plane it is clear that the stroke relation can be interpreted geometrically in four ways. For if $A = a$ and $B = b$, then (see Fig. 33) the four statements (1) $a \perp b$; (2) A I b; (3) a I B; (4) $A \mid B$ are equivalent to one another. The fourth statement can be read as "B is a midpoint, distinct from A, of the point pair A, A." So from correct theorems of elliptic geometry we always obtain correct theorems in the same geometry if we arbitrarily replace points by lines and lines by points, in each case choosing the relevant relation from 1 through 4 ("point-line analogy").[14] For example, from the axiom "two distinct points can be joined by exactly one line," we at once obtain the following theorems: "two distinct lines[15] intersect in exactly one point," "two distinct lines have exactly one common perpendicular (the polar of their point of intersection)," and "through a given point there passes exactly one line perpendicular to a given line (nonpolar to the point)." Similar interchanges of words are often possible even in absolute geometry (cf. the two three-reflections theorems, their converses, the existence and uniqueness of perpendiculars or of lines joining points, and Theorem 1 (orthogonal intersection), Theorem 2 (polar triangle), and Theorems 10, 11, 14, 15, and 16).

Let us also note that in all nonelliptic planes the perpendicular from a point to a line is always uniquely determined, since the negation of Axiom P, namely

Axiom $\neg$P: *No polar triangle exists*

states, as we have seen, that no point coincides with a line, which by Theorem 4 implies the uniqueness of the perpendicular.

5.2.   *Euclidean Geometry*

A principle for classifying the groups of motions which, like Axiom P, deals with lines and orthogonality is as follows:

---

[14]Cf. the article by Arnold Schmidt cited in fn. 5 of this chapter.
[15]Thus the projective axioms of incidence (cf. II3, §3.1) are satisfied in elliptic planes.

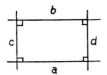

Fig. 34

Axiom R: *There exists a rectilateral* (i.e., four lines *a, b, c, d* with *a,*
b ⊥ *c, d* and *a* ≠ *b, c* ≠ *d* (Fig. 34).

This axiom states that two distinct lines can have two distinct perpendiculars in common. As we have seen, Axiom P implies the uniqueness of the
common perpendicular to two lines, which means that P implies ⌐ R (the
negation of R). Thus R implies ⌐ P.

We will call two lines *perpendicular-equal* if their pencils of perpendiculars
coincide, i.e., if every line perpendicular to one of them is also perpendicular
to the other. Since R implies ⌐ P and ⌐ P implies uniqueness of the perpendicular, it is clear that two perpendicular-equal lines cannot have a
point in common. From the second three-reflections theorem and its converse we have

(21)   *Two lines with two perpendiculars in common are perpendicular-equal.*

As we shall see, Axiom R allows us to prove the following:

Theorem of the rectilateral: *In a quadrangle with three right angles, the*
fourth angle is also a right angle: from a ⊥ *c, d and b* ⊥ *c follows b* ⊥ *d.*

This theorem means that the existence of one common perpendicular is
already enough to guarantee that two lines are perpendicular-equal.

We now have

Theorem 26:   *In a group of motions each of the following statements is*
equivalent to Axiom R:
    a)   *The theorem of the rectilateral.*
    b)   *A product of three points is always a point.*

Proof (Schütte). 1.   Statement b) follows from Axiom R. For by Axiom R
there exist lines *g, h, g', h'* with *g, h* ⊥ *g', h'* and *g* ≠ *h* and *g'* ≠ *h'*, and then for
all *x, x'*:

(22)   *From x* ⊥ *g' and x'* ⊥ *g follows x* ⊥ *x'.*

For from (21) it follows that *g', h'* are perpendicular-equal, so that *g, x* ⊥ *g', h'*,
and consequently, again by (21), *g, x* are perpendicular-equal.

Now let *A, B, C* be three arbitrary points, let *a, b, c* be the perpendiculars from
A, B, C to *g'* and let *a', b', c'* be perpendiculars from *A, B, C* to *g*. By (22) we then
have *a, b, c, g, h* ⊥ *a', b', c', g', h'*, and therefore *A = aa', B = bb', C = cc'* (by

Theorem 1) and further, by the second three-reflections theorem, abc is a line d with $d \perp g'$ and $a'b'c'$ is a line d' with $d' \perp g$. By (22) we have $d \perp d'$, so that dd' is a point D. Since a, b, c permute with a', b', c', we have $ABC = aa' \cdot bb' \cdot cc' = abc \cdot a'b'c' = dd' = D$.

2. The theorem of the rectilateral follows from statement b). For let $bc = B$, $ca = C$, $ad = D$. Then $BCD = bc \cdot ca \cdot ad = bd$, so that statement b) implies that bd is a point, i.e., that $b \perp d$.

3. The theorem of the rectilateral follows from Axiom R, as is clear from Theorem 5.

The rectilateral theorem shows that in a metric plane satisfying Axiom R the metric behaves as we would expect from Euclidean geometry. So the groups of motions satisfying Axiom R are called *metric-Euclidean groups of motions* and their planes are called *metric-Euclidean*. But it must be noted that in a metric-Euclidean plane the behavior of parallels can be quite different from the traditional Euclidean case: there exist planes with a Euclidean metric, i.e., planes satisfying R, in which more than one line can pass through a given point without intersecting a given line ("semi-Euclidean" planes). In contrast to these nonintersecting lines, however, the perpendicular-equal lines in metric-Euclidean geometry behave exactly like the parallel lines in Euclidean geometry: through a given point there passes exactly one line perpendicular-equal to a given line. The remaining nonintersecting lines that can occur in a metric-Euclidean plane must therefore have neither a point nor a perpendicular in common with the given line. If we wish to preclude such phenomena, we must add a new axiom, for example:

Axiom V: *Two lines always have either a point or a perpendicular in common.*

A metric-Euclidean group of motions or plane satisfying Axiom V is called a *Euclidean group of motions* or *plane*. If, as is customary, we call lines *parallel* when they are either equal or nonintersecting, then in a Euclidean plane the two concepts "perpendicular-equal" and "parallel" coincide.

In a metric-Euclidean plane, the product AB of two point reflections is called a *translation*. Every product AB can also be represented as a product ab with a perpendicular-equal to b, and conversely (Fig. 35). With P fixed, every translation AB can be represented in exactly one way as $AB = PQ$ (i.e., with $Q = PAB$). Since by statement b) from Theorem 26 a product of four point reflections is equal to a product of two point reflections, the

Fig. 35

translations form a group $\mathfrak{T}$, which is Abelian, since the rule $ABC = CBA$ (Theorem 26, statement b) gives $AB \cdot A'B' = ABA' \cdot B' = A' \cdot BA \cdot B' = A' \cdot BAB' = A' \cdot B'AB = A'B' \cdot AB$. Furthermore, $(AB)^\alpha = A^\alpha B^\alpha$, so that $\mathfrak{T}$ is a normal subgroup.

Let us also consider the subgroup $\mathfrak{G}_0$ of the metric-Euclidean group of motions generated by the lines through a given point O. This subgroup consists of the lines a with $a \text{ I } P$ and the "rotations" ab with $a, b \text{ I } P$, as can be seen at once from the three-reflections theorem.

The following theorem expresses the degenerate (one-dimensional) character of the Euclidean metric (cf. §6) from the point of view of absolute geometry.

Theorem 27: *If O is an arbitrary point of a metric-Euclidean group of motions $\mathfrak{G}$, then $\mathfrak{G} = \mathfrak{G}_0 \mathfrak{T}$, or more precisely: every motion $\alpha \in \mathfrak{G}$ is uniquely representable as a product $\alpha = \beta\tau$ with $\beta \in \mathfrak{G}_0$ and $\tau \in \mathfrak{T}$.*

Proof. Let the set of products $\beta\tau$ with $\beta \in \mathfrak{G}_0$, $\tau \in \mathfrak{T}$ be denoted by $\mathfrak{U}$.

(i)   *The set $\mathfrak{U}$ contains all lines.*

Proof of (i).   Let g be an arbitrary line and let g_0 be the line through O that has the same perpendiculars as g. Then $g = g_0 \cdot g_0 g$ and $g_0 \in \mathfrak{G}_0$, $g_0 g \in \mathfrak{T}$.

(ii)   *The set $\mathfrak{U}$ is multiplicatively closed.*

For we have $\beta_1 \tau_1 \cdot \beta_2 \tau_2 = \beta_1 \beta_2 \cdot \tau_1^{\beta_2} \tau_2$, where with τ_1 the normal subgroup $\mathfrak{T}$ also includes $\tau_1^{\beta_2}$.

From (i) and (ii) follows $\mathfrak{U} = \mathfrak{G}$, since the lines are involutory generators of $\mathfrak{G}$.

(iii) $\mathfrak{G}_0 \cap \mathfrak{T} = \{1\}$.

Proof of (iii).   Let $\tau \in \mathfrak{G}_0 \cap \mathfrak{T}$, and let $\tau = gh$ with g, h perpendicular-equal. Since $\tau \in \mathfrak{G}_0$, we must have $\tau = a$ with $a \text{ I } O$ or $\tau = ab$ with $a, b \text{ I } O$. The first case $gh = a$ is impossible, since we would then have $g|h$, so that gh would be a point A with $A = a$, in contradiction to the fact that Axiom $\neg$ P holds in $\mathfrak{G}$. Consequently, $gh = ab$ with $a, b \text{ I } O$. Now if we had $g \neq h$, it would follow that $a \neq b$, and from the converse and supplement to the first theorem of three reflections we would obtain $g, h \text{ I } O$ and therefore $g = h$ (since g, h have the same perpendiculars), a result which means that in fact $g = h$, so that $\tau = 1$.

From (iii) follows the uniqueness statement in Theorem 27: If $\beta_1 \tau_1 = \beta_2 \tau_2$ with $\beta_i \in \mathfrak{G}_0$, $\tau_i \in \mathfrak{T}$, it follows that $\beta_2^{-1} \beta_1 = \tau_2 \tau_1^{-1}$, so that $\beta_2^{-1} \beta_1 \in \mathfrak{G}_0 \cap \mathfrak{T}$, and thus from (iii) that $\beta_1 = \beta_2$, and also $\tau_1 = \tau_2$.

5.3.   Hyperbolic Geometry

Finally, in order to define hyperbolic geometry in our system, we first require the following axiom suggested by the Klein model in §2:

Axiom $\neg$ V: *There exist lines which have neither a point nor a perpendicular in common.*

If a, b are two such lines, we say that the pencil $G(ab)$ has no *carrier,* so that Axiom $\neg$ V asserts the existence of pencils without carrier.

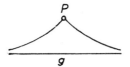

Fig. 36

From the law of transitivity and the converses of the three-reflections theorems, it follows that no two lines of a pencil without carrier can have either a point or a perpendicular in common.

In the Klein model every line belongs to exactly two pencils that have no carrier. But in an arbitrary metric plane satisfying Axiom $\neg$ V it is possible, under certain circumstances, for a line to belong to more than two such pencils (for example, in the "semi-Euclidean" planes R, $\neg$V), a possibility which must be eliminated if we wish to obtain planes with properties like the Klein model. Thus we require

Axiom H (Hyperbolic axiom):   *Given a point P and a line g there pass at most two lines through P which have neither a point nor a perpendicular in common with g* (Fig. 36).

A metric group of motions or plane satisfying Axioms $\neg$ V and H is called a *hyperbolic group of motions* or *plane*. Hyperbolic planes are non-Euclidean, i.e., they do not contain any rectilateral.

6. Extension to the Ideal Plane and Algebraization

Let us now indicate, without details of the proofs,[16] how the metric planes and groups of motions can be subsumed under the concept of a projective metric (II3, §5), for which purpose we show that every metric plane can be embedded in a projective-metric plane. The group of motions of the metric plane is then seen to be isomorphic to a subgroup of the group of motions of the projective-metric plane. If the metric plane is given as the group plane of a group of motions $\mathfrak{G}$, $\mathfrak{S}$, then on the other hand, as mentioned in §2.2, the group of motions of the metric plane is isomorphic to $\mathfrak{G}$, with the result that $\mathfrak{G}$ is now represented as a subgroup of the group of motions of a projective-metric plane.

The first step in this direction is to extend the metric plane to a projective plane by introducing ideal points and ideal lines. By an *ideal point* we here mean any pencil of lines $G(ab)$ of the metric plane. The pencils $G(P)$, which we shall call *proper ideal points,* correspond in a one-to-one way to the points P of the metric plane. By an ideal line we shall then mean a certain

[16]For a detailed discussion, see [1].

set of ideal points, namely, in the first place, a set of ideal points that have in common a line a of the metric plane. An ideal line of this sort will be called a *proper ideal line* $g(a)$. The proper ideal lines correspond one-to-one to the lines a of the metric plane. The general concept of an ideal line is then defined by means of the halfrotations H_{uv} (§4.9) about a fixed point O of the metric plane, which take a proper ideal line into a proper ideal line: $g(a)H_{uv}$ $= g(aH_{uv})$. Each of the following sets of ideal points will now be called an *ideal line:* (1) the set of pencils of perpendiculars $G(x)$ with x I O,[17] which we shall call the polar of $G(O)$; (2) each set a of ideal points that can be transformed by a halfrotation H_η about O into a proper ideal line. Here it is important to note the theorem that a halfrotation induces a one-to-one mapping of the set of ideal points onto itself. We can then show that the projective axioms of incidence,[18] and also the Fano axiom, hold for the ideal points and the ideal lines. Moreover, the theorem of Pappus-Pascal[19] for the ideal lines can easily be reduced by means of halfrotations to the same theorem for the lines of the metric plane. Thus the ideal points and the ideal lines form a projective plane (with the Fano axiom), which is called the *ideal plane* of the metric plane.

To illustrate the importance of the embedding of the metric plane into a projective plane, let us briefly discuss the three classical cases of absolute geometry, namely, the Euclidean, elliptic, and hyperbolic geometries.

A Euclidean plane is an affine plane, and its ideal plane is the projective plane arising from adjunction of the infinitely distant elements, which in this case are the parallel pencils (identical here with the pencils of perpendiculars) as infinitely distant points, and the set of infinitely distant points as the line at infinity.

An elliptic plane is itself already a projective plane and coincides with its own ideal plane, so that the embedding process is in this case superfluous.

On the other hand, a hyperbolic plane is neither an affine nor a projective plane, so that now the ideal plane represents an essential extension of the metric plane. (Consider the Klein model of a hyperbolic plane in §3, where the points of the model are the interior points of a conic section.)

In the ideal plane of a metric plane, an orthogonality for the proper ideal lines is induced by the perpendicularity in the metric plane, and the second step in the embedding into a projective-metric plane consists of the proof that this orthogonality can be represented by a projective involution without fixed points on a line "at infinity" (the singular case, see II3, §5) or by a projective polarity (the ordinary case). For a given metric plane it is clear that only one of these two cases is possible, namely, the singular case for Axiom R (§5.2) and the ordinary case for its negation $\neg$ R.

In the singular case the involution I is easy to construct, since for an

---

[17]By $G(x)$ we denote the pencil of perpendiculars with axis x (cf. §4.3).
[18]Cf. II3.
[19]*Ibid.*

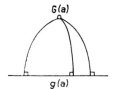

Fig. 37

arbitrary line c the set of pencils of perpendicular $G(c)$ in the metric plane forms an ideal line o (identical with the polar of the center O of the half-rotations) on which I can be defined by mapping the ideal point $G(c')$ into the ideal point $G(c)$, with $c \perp c'$. By the counterpairing theorem, this correspondence is seen to be involutory, without fixed points, and also projective.

In the ordinary case the polarity is somewhat harder to construct. We see at once that for all lines a of the metric plane the pairs $g(a)$, $G(a)$ must be pole-polar pairs (Fig. 37), which we shall call *primitive*. These pairs already provide the nucleus of a polarity, which now must be extended to arbitrary ideal points and lines, a problem to be attacked in exactly the same way as the extension of the metric plane to the projective plane, as follows. By means of the halfrotations about O we reduce the general case of pole-polar relationship to the case of primitive pole-polar pairs in the same way as the general concept of the ideal line is reduced to the proper ideal lines. Of course, we must note here that the halfrotations do not take a pole-polar pair into a pole-polar pair, i.e., perpendicularity is not invariant under halfrotations. Nevertheless, we can state a law of transformation for the primitive pole-polar pairs, as follows.

For a given H_η (where $\eta = uv$, with $u, v \perp O$) we consider the halfrotation $H_{\eta^{-1}} = H_{vu}$. It can be proved that if A, a (with A an ideal point and a an ideal line) is a primitive pole-polar pair, then $AH_{\eta^{-1}}^{-1}$, aH_η is also a primitive pole-polar pair.

In Fig. 38 this law is illustrated for polarity in the unit circle (see fn. 9 in this chapter): from the similarity of triangles OF_1F and $OA(AH_{\eta^{-1}}^{-1})$ we have $\overline{OF} \cdot \overline{OA} = \overline{OF_1} \cdot \overline{O(AH_{\eta^{-1}}^{-1})}$, which shows that if A is the pole of a, i.e., if $\overline{OF} \cdot \overline{OA} = l$, then $AH_{\eta^{-1}}^{-1}$ is the pole of aH_η.

Now a pair a, A consisting of an ideal line and an ideal point will be called a pole-polar pair if there exists a halfrotation H_η about O such that aH_η is a proper ideal line $g(b)$ and $AH_{\eta^{-1}}^{-1} = G(b)$, that is, such that aH_η and $AH_{\eta^{-1}}^{-1}$ are a primitive pole-polar pair.

Then it can be shown that the pole-polar relationship defined in this way is actually a polarity in the ideal plane, and the fact that this polarity is

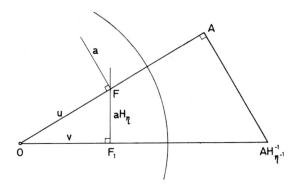

Fig. 38

projective can be shown, for example, by the counterpairing theorem, in the same way as in the singular case.

Thus we have proved that the ideal plane of a metric plane is a projective-metric plane with the property that perpendicularity in the metric plane coincides with the orthogonality in the proper part of the projective-metric plane.

The reflection in a line a in the metric plane can then be extended to a collineation in the ideal plane, namely, to a harmonic homology with $G(a)$ as center and $g(a)$ as axis, i.e., in the ordinary case, with a pole-polar pair as center and axis and, in the singular case, with an axis through the point polar to the center $G(a)$. Thus the group of motions of a metric plane is isomorphic to a subgroup of the group of motions of the projective-metric ideal plane, which gives us the

Theorem: *Every metric plane can be represented as a subplane of a projective-metric plane. The group of motions of the metric plane is isomorphic to a subgroup of the group of motions of the projective-metric plane.*

By II3, §5 there exists for every metric plane a field $\mathfrak{K}$ of characteristic $\neq 2$ and a symmetric bilinear form f over $\mathfrak{K}$ such that the metric plane can be represented as a subplane of the projective coordinate plane and the perpendicularity of lines is described by the vanishing of the form f for the corresponding triple of line coordinates. In the singular case the form f can be written

$$f(\mathfrak{u}, \mathfrak{v}) = u_1 v_1 + k u_2 v_2 \quad \text{where } -k \text{ is not a square in } \mathfrak{K}.$$

In the ordinary case the bilinear form can be written

$$f(\mathfrak{u}, \mathfrak{v}) = k_1 u_1 v_1 + k_2 u_2 v_2 + k_1 k_2 u_3 v_3 \quad \text{with} \quad k_1, k_2 \neq 0 \text{ in } \mathfrak{K}.$$

By this algebraization of the metric planes, we have shown how they are related to analytic geometry, but we have not yet given a complete survey of all metric planes. Such a survey would only be complete if conversely, in a projective-metric coordinate plane defined as above we could give an algebraic characterization of those sets of points and lines that form a metric plane (the metric subplanes). This problem has been investigated under simplifying assumptions on the given projective-metric plane or on the metric subplanes to be determined. See the inclusive survey [2].

In conclusion, let us consider how the embedding of a metric plane in its ideal plane is related to the classification (see §5) of the set of all metric planes by additional axioms. This classification is illustrated in Fig. 39.

The Euclidean planes assigned to the left-hand endpoint of the figure are precisely those planes obtainable from a singular projective-metric plane by omission of the line at infinity (and all points on it). A semi-Euclidean plane also has a singular projective-metric plane and can therefore be regarded as a subplane, not only of its ideal plane, but of the Euclidean plane arising from the ideal plane by omission of the line at infinity.

To the metric planes with Axiom $\neg$R there always corresponds an ordinary projective-metric ideal plane. The elliptic planes are precisely the elliptic[20] projective-metric planes in the sense of II3, §5. The metric planes with Axioms $\neg$R, V and $\neg$P are called half-elliptic planes. Their ideal plane is an elliptic plane and for any pole-polar pair they contain either the pole or the polar, like the Klein model for nonincident pole-polar pairs. It should be noted that an elliptic plane can also contain metric subplanes "smaller" than a half-elliptic plane in the sense that for certain pole-polar pairs they contain neither the point nor the line. An example is the non-Legendrian geometry of Dehn.[21] Such planes belong to the right-hand endpoint of the figure (Axioms $\neg$R, $\neg$V, $\neg$H).

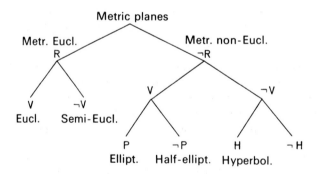

Fig. 39

[20]A projective-metric plane is called hyperbolic if it contains self-polar points, and otherwise elliptic.

[21]M. Dehn, Die Legendreschen Sätze über die Winkelsumme im Dreieck, Math. Ann. 53 (1900), 404–439. See also Hilbert [5].

The ideal plane of a hyperbolic plane is always an ordered hyperbolic projective-metric plane (the coordinate field is orderable) in which the points of the hyperbolic plane are the interior points of the fundamental conic formed by the self-polar points; and conversely, a hyperbolic plane is obtained in this way from every ordered field. Here too in the hyperbolic plane there can exist still "smaller" metric subplanes, which are then to be assigned to the right-hand end of the figure.

Finally, let us mention that there exist metric planes with a hyperbolic projective-metric ideal plane whose coordinate field is not orderable. These planes, which also belong to the class of metric planes satisfying Axioms $\neg$R. $\neg$V, $\neg$H, cannot be represented as subplanes of a Euclidean, elliptic, or hyperbolic plane.

Bibliography

[1] BACHMANN, F.: Aufbau der Geometrie aus dem Spiegelungsbegriff. Berlin, Göttingen, Heidelberg 1959.

[2] BACHMANN, F.: Modelle der ebenen absoluten Geometrie. Jber. dtsch. Math.-Ver. 66 (1964), 152–170.

[3] COXETER, H. S. M.: Non-euclidean geometry. Third ed., Toronto 1957.

[4] HESSENBERG, G.: Grundlagen der Geometrie. Berlin and Leipzig 1930. New edition, revised by J. DILLER. Berlin 1967.

[5] HILBERT, D.: The foundations of geometry. Translated from the German by E. J. Townsend. Reprint edition. The Open Court Publishing Co., La Salle, Ill. 1959.

[6] KERÉKJÁRTÓ, B.: Les fondements de la géométrie, Vol. 1: La construction élémentaire de la géométrie euclidienne. Budapest 1955.

[7] SCHUR, F.: Grundlagen der Geometrie. Leipzig 1909.

[8] WAERDEN, B. L. VAN DER: De logische grondslagen der euklidische meetkunde. Groningen 1937.

Bibliography Added in Translation

ALEKSANDROV, A. D., et al.: Mathematics, its content, methods, and meaning. Vol. 2, Chap. XVII (translated from the Russian by T. Bartha, S. H. Gould, and K. A. Hirsch). The M. I. T. Press, Cambridge, Mass. 1964.

COXETER, H. S. M.: Introduction to geometry. John Wiley and Sons, Inc., New York-London 1961.

CHAPTER 6

The Classical Euclidean and the Classical Hyperbolic Geometry

Introduction

The foundations of geometry have always been of particular interest to mathematicians. In recent times Hilbert [2] introduced a complete set of axioms for Euclidean geometry. His decisive improvement over Euclid lies in the fact that he makes no attempt to define point, line, and plane, but treats them as basic undefined elements. Of sole importance now are the mutual relationships among these basic elements, e.g., incidence, congruence, and orthogonality, which are described by a set of axioms. Every conclusion is drawn from these relationships alone.

In the present chapter we discuss a certain system[1] of axioms for classical Euclidean and classical hyperbolic geometry. By a geometry we mean a set of points and (straight) lines provided with three structures, namely, an incidence, a congruence, and an order. Each of these structures will be described axiomatically. If we wish to have a (significant) geometry, the structures cannot be completely independent; they must be "consistent" with one another, for which purpose we introduce certain "consistency axioms."[2]

---

[1]The system of axioms presented here is not exactly the same as the Hilbert system in [2], although the two are equivalent. Our system, in which properties of order and continuity are used for the first time in the present volume, comprehends exactly two geometries, namely the classical Euclidean and the classical hyperbolic geometry (over the real numbers). These two geometries are examples of the Euclidean and hyperbolic planes defined in II4 and 5.

[2]As an illustration of these concepts let us recall that a field is a set provided with two group structures, which satisfy certain consistency axioms, namely, the laws of distributivity.

In this chapter we shall confine our attention to plane geometry.[3] We shall require that any two distinct points can be joined by a straight line, that every straight line contain at least two points, and that there exist three points not on a straight line (§1). These are precisely the Hilbert axioms of *connection* for the plane.

In our formulation of the order axioms (§1) we have been guided by the intuitive fact that every line divides the plane into two sides. It is by means of this division into sides that we introduce an order. Hilbert proceeded differently but his notion of betweenness can be defined in our terms. On the basis of our axioms we can prove most of Hilbert's order axioms. For example, the statement (7) below means that exactly one of three distinct points on a straight line lies between the other two. On every line k of our geometry, a complete order can be defined by distinguishing two of the points on the line. For this order we postulate that every subset of points of k that is bounded above has a least upper bound (§2, Axiom **S**). In this way we obtain the result that our geometry is continuous.

If in addition to the axioms of §§1 and 2 we also have the parallel axiom, we obtain an affine geometry (§3). An affine geometry may be represented, provided that the theorem of Desargues is valid in it, as an analytic geometry over the real numbers (Theorem 1).

A plane geometry is said to be projective if every two lines have a point of intersection (§4). Desarguesian projective geometries can also be represented analytically (Theorem 2).

The properties of the intuitive concepts of segment and length are described by the congruence axioms (§5). We are then able to prove the well-known Archimedean axiom on line segments, namely: if from a point P on a line g we mark off a given segment sufficiently often, we will eventually get beyond any given point of the line g (Theorem 3).

The concept of angle does not occur in our system of axioms, since they enable us to construct a definition of it. Its usual properties are here introduced as theorems (cf. (21), (22), and (23)).

We show that through a given point P there exists at least one line parallel to a given line g (cf. (26)), and then it turns out that our system of axioms is satisfied by exactly two geometries (Theorem 5). One of these is the Euclidean geometry, in which the parallel line is uniquely determined, and the other is the hyperbolic geometry, in which (through any given point not on a given line) there is more than one line parallel to the given line (Theorem 4). Both of them can be represented as analytic geometries over the real numbers.

In §6 we present models for these geometries, with complete proofs, in the Euclidean case, that the axioms are valid in the model. For the hyper-

---

[3]This restriction is introduced solely for conciseness and easy visualization; the discussion for three-dimensional geometry would be completely analogous.

bolic case we carry out a great part of the argument necessary to show that the axioms are valid and point out how it can easily be completed. Finally at the end of §6, we make some mention of elliptic geometry.

In §7 we present some well-known models of great historical importance and show that they can easily be obtained by isomorphism from the models of §6.

I. Incidence and Order

Let there be given two disjoint sets $\mathfrak{P}$ and $\mathfrak{G}$ and also[4] a subset $\mathfrak{J} \subset \mathfrak{P} \times \mathfrak{G}$. The elements of $\mathfrak{P}$ will be called *points* (denoted by Latin capitals) and the elements of $\mathfrak{G}$ will be called *lines* (denoted by Latin lowercase letters). Instead of $(P, g) \in \mathfrak{J}$ we shall always write $P \mathrel{I} g$, or equivalently $g \mathrel{I} P$, and shall say that *"P is incident with g."* If $(P, g) \notin \mathfrak{J}$, we write $P \mathrel{\not{I}} g$. Then $\mathfrak{J}$ is called an *incidence relation* or *incidence structure* for $\mathfrak{P} \cup \mathfrak{G}$, provided the following axioms are satisfied:

I1: *For any two distinct points A, B, there exists exactly one line g with $A \mathrel{I} g$ and $B \mathrel{I} g$.*

I2: *For any given line g there exist at least two distinct points A and B with $A \mathrel{I} g$ and $B \mathrel{I} g$.*

I3: *There exist a line g and a point P with $P \mathrel{\not{I}} g$.* If $\mathfrak{P} \cup \mathfrak{G}$ is provided with an incidence structure $\mathfrak{J}$, then $\mathfrak{E} = (\mathfrak{P}, \mathfrak{G}, \mathfrak{J})$ is called a *plane*.

We now define what we mean by an *order structure*. With every line g in a plane $\mathfrak{E}$ we naturally associate the following intuitive idea: the line divides the points of the plane into three classes, namely, the two "sides" of g (also called halfplanes) and the points on g. Consequently we define:

A *partition into sides* is defined in the set $\mathfrak{P} \cup \mathfrak{G}$ if for every line g there exists a decomposition of set $\mathfrak{P}$ into three disjoint classes (two of these classes are called *sides* and the third will be denoted by $\bar{g}$) such that Axioms A1 and A2 are satisfied.

It is convenient to express the statement "A and B lie on the same side of g (i.e., are in the same class $\neq \bar{g}$) or are on opposite sides of g (i.e., are in the two distinct classes $\neq \bar{g}$)" by writing $(g/A, B) = 1$ and $(g/A, B) = -1$, respectively. For this function[5] we obviously have

(1) $$(g/A, B) = (g/B, A),$$

(2) $$(g/A, B)(g/B, C)(g/C, A) = 1.$$

The desired axioms can then be formulated as follows:

[4]If $\mathfrak{X}$ and $\mathfrak{Y}$ are given sets, the set $\mathfrak{X} \times \mathfrak{Y} = \{(x, y); x \in \mathfrak{X} \text{ and } y \in \mathfrak{Y}\}$ consisting of all pairs (x, y) with $x \in \mathfrak{X}$ and $y \in \mathfrak{Y}$ is called the *Cartesian product* of $\mathfrak{X}$ and $\mathfrak{Y}$.

[5]The definition of an order by means of a function, called an order function, is due to E. Sperner [5].

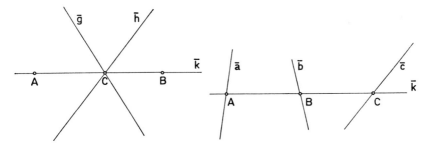

Fig. 1   Fig. 2

A1:   *If A, B, C ∈ 𝔓 and g, h, k ∈ 𝔊 with A, B, C ∈ k̄ and C ∈ ḡ, h̄ and also A, B ∉ ḡ, h̄, then (g/A, B) = (h/A, B) (cf. Fig. 1).*

A2:   *If A, B, C are three distinct points and a, b, c, k are four distinct lines with A, B, C ∈ k̄ and A ∈ ā, B ∈ b̄, C ∈ c̄, then exactly one of the values (a/B, C), (b/A, C), (c/A, B) is equal to −1 (cf. Fig. 2).*

If a division into sides is defined in 𝔓 ∪ 𝔊, we say that 𝔓 ∪ 𝔊 is provided with an *order structure* 𝔄. If 𝔓 ∪ 𝔊 is provided not only with an order structure 𝔄 but also with an incidence structure 𝔍, then 𝔈 = (𝔓, 𝔊, 𝔍, 𝔄) is called an *ordered plane* if the two structures are consistent, i.e., if the following two axioms of consistency are satisfied:

V1:   *P I g is equivalent to P ∈ ḡ.*

V2:   *If (g/A, B) = −1, then g and the line joining A and B are incident in a common point.*

If *A* and *B* are two points on opposite sides of the line *g*, in other words, if (g/A, B) = −1, then we say that the line *g* lies *between* the points *A* and *B*.

If *A*, *B*, *C* are three points on the line *k* with *C* ≠ *A*, *B*, we say that *C* lies *between A* and *B* if there exists a line *h* ≠ *k* between *A* and *B* with *h* I *C*. Axiom **A1** states that in this case every line through *C* distinct from *k* lies between *A* and *B*. Thus we may define:

(3)   *If A, B, C are three points on a line k with C ≠ A, B, then (C/A, B) = (h/A, B), where h ≠ k is an arbitrary line with h I C.*

From (1), (2), and (3) it follows at once that

(4)   *If A, B, C are three points on a line with C ≠ A, B, then (C/A, B) = (C/B, A).*

(5)   *If A, B, C, D are four points on a line with D ≠ A, B, C, then (D/A, B) (D/B, C) (D/C, A) = 1.*

From (5) we have

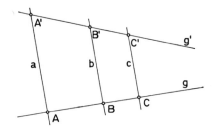

Fig. 3

(6) $(D/A, A) = 1$ *for all pairs of points* $A \neq D$.

From **A2** follows

(7)   *If A, B, C are three distinct points of a line, exactly one of the values* $(A/B, C), (B/C, A), (C/A, B)$ *is equal to* -1.

Two lines g and h are said to be *parallel* if either $g = h$ or else g and h have no point in common.

From **V2** we can draw the simple conclusion:

(8)   *Let A, B, C and A', B', C' be three points on the lines g and g', respectively, with* $A \neq B, C$ *and* $A' \neq B', C'$, *and* $g \neq g'$. *If a, b, c are three parallel lines with* $a \,\mathrm{I}\, A, A'$ *and* $b \,\mathrm{I}\, B, B'$ *and also* $c \,\mathrm{I}\, C, C'$, *then* $(A/B, C) = (A'/B', C')$ (cf. Fig. 3).

Proof.   Since $a \neq b$ and a and b are parallel, **V2** gives us the equation $(a/B, B') = 1$. Similarly, $(a/C, C') = 1$. From (3), (1), and (2) it then follows that $(A/B, C) = (a/B, C) = (a/B, C)(a/B, B')(a/C, C') = (a/C, B')(a/C, C')$ $= (a/B', C') = (A'/B', C')$.

2.   The Axiom of Continuity

The order structure of classical Euclidean and classical hyperbolic geometry satisfies a further axiom, namely, the axiom of continuity. To formulate this axiom we first show that for the points of each line on an ordered plane we can define an *order*:[6] Let k be a line and let A and E be two distinct points on k, which exist by Axiom **I2**. For the points P, Q of k we define: $P < Q$ if

(9) $(P/A, Q)(A/E, P) = -1$

or

(10) $(Q/A, P)(A/E, Q) = 1.$

[6]See III, §2 ff.

This definition provides a *total order relation,*[7] since (11) and (13) are valid:

(11) *Comparativity: Exactly one of the statements $P < Q$ or $Q < P$ or $P = Q$ is valid.*

Proof. If $P = Q$, then neither $P < Q$ nor $Q < P$, since then neither $(P/A, Q)$ nor $(Q/A, P)$ is defined.

Now let A, P, Q be three distinct points of k. From (7) we then have $(P/A, Q)(A/Q, P)(Q/P, A) = -1$, so that from (4) and (5) it also follows that $(P/A, Q)(A/P, E)(A/E, Q)(Q/A, P) = -1$, and thus

(12) $$(P/A, Q)(A/E, P) = -(Q/A, P)(A/E, Q).$$

Equation (12) states first: exactly one of the two expressions $(P/A, Q)$ $(A/E, P)$ or $(Q/A, P)(A/E, Q)$ is equal to -1, i.e., exactly one of the relations $P < Q$ or $Q < P$ holds, and second: the equations (9) and (10) are not mutually contradictory.

Finally, let $A = P \neq Q$. If $(Q/A, P)(A/E, Q) = 1$, then $P < Q$ from (10); if $(Q/A, P)(A/E, Q) = -1$, then $Q < P$ from (9).

(13) *Transitivity: If $P < Q$ and $Q < R$, then $P < R$.*

Proof. 1) $P, Q \neq A$. From $P < Q$ and $Q < R$ it follows from (10) and (9) that

$$-1 = (Q/A, P)(A/E, Q)(Q/A, R)(A/E, Q) = (Q/P, R),$$

so that $(P/Q, R) = 1$ from (7). Then from (9)

$$-1 = (P/Q, R)(P/A, Q)(A/E, P) = (P/A, R)(A/E, P),$$

i.e., $P < R$.

2) $Q = A$. Since $P < A$, i.e., $(A/E, P) = -1$ and $A < R$, i.e., $(A/E, R) = 1$, we have $-1 = (A/E, P)(A/E, R) = (A/P, R)$. Thus (7) implies the equation $(P/A, R) = 1$ so that $-1 = (P/A, R)(A/E, P)$, i.e., $P < R$.

3) $P = A$. We now have $A < Q$, i.e., $(A/E, Q) = 1$ and $Q < R$, i.e., $(Q/A, R)(A/E, Q) = -1$, so that $(Q/A, R) = -1$. From (7) we thus have $(A/Q, R) = 1$ and consequently $1 = (A/E, Q)(A/Q, R) = (A/E, R) = (R/A, A)(A/E, R)$, i.e., $A < R$.

From the definition of the order relation we at once obtain:

(14) $(Q/P, R) = -1$ *is equivalent to* $P < Q < R$ *or* $R < Q < P$.

If M is a subset of the line k, then an element $S \in k$ is called an *upper bound* of M if $X \leq S$ for all $X \in M$. If M has an upper bound, then M is said to be *bounded above.* An element T is called a *least upper bound* or *supremum*

---

[7]Compare IA, §8.3.

of M if T is an upper bound of M and if $T \leq S$ for every upper bound S of M. An order relation defined in k is said to be *continuous* if every subset of k that is bounded above has a least upper bound.

S (Axiom of continuity): *For every line k of an ordered plane $\mathfrak{E}$ the order relation defined by* (9) *and* (10) *is continuous.*

A plane E is said to be *continuous* if it is ordered and satisfies the axiom of continuity.

3. Ordered Affine Geometries

A plane $\mathfrak{E}$ is said to be *affine* if it satisfies the following axiom:

P (Parallel axiom): *For every line g and every point P there exists exactly one line, h, parallel[8] to g with h I P.*

An affine plane is said to be *Desarguesian* if the Desargues affine theorem holds in it. An affine Desarguesian plane $\mathfrak{E}$ can be represented as an affine coordinate plane over a skew field K (cf. II3, §2). We now show that the order on the plane induces an order[9] in the coordinate field K. For this purpose we mark the points $A = (0, 0)$ and $E = (1, 0)$ on the x-axis, whereupon the order of the points on the x-axis introduced in §2 provides us with an order for the elements of the field K if to every element $x \in K$ we assign the point $X = (x, 0)$ and define $a < b$ as being equivalent to $(a, 0) < (b, 0)$. Comparativity (11) and transitivity (13) are obviously satisfied for this relation. Thus it remains to prove only the following two laws of monotonicity:

(15)   *From $p < q$ and $0 < r$ follows $rp < rq$.*

(16)   *From $p < q$ follows $p + r < q + r$.*

Proof. We see from Figs. 4 and 5 that the following statement is correct:[10]

(17)   *The mappings $P \to RP$ and $P \to R + P$ of the x-axis into itself can be represented as a sequence of parallel projections[11] $P \to P' \to P'' \to RP$ and $P \to \bar{P} \to R + P$, respectively.*

By (8) and (17) we have $(P/A, Q) = (RP/A, RQ)$ and $(A/E, P) = (A/R, RP)$,

---

[8]Compare p. 178.

[9]The field K is said to be *ordered* if there is a relation $<$ defined in it such that the laws of comparativity (11), transitivity (13), and monotonicity (15) and (16), are valid.

[10]Figures 4 and 5 illustrate the addition and multiplication of points on a line as introduced in II3 for the construction of the coordinate field. Here the points A and E play, respectively, the role of the zero element and the unit element.

[11]A mapping π of a line g onto the line h with $h \neq g$ is called a *parallel projection* if for all points $P, Q \in g$ the lines joining P with $P\pi$ and Q with $Q\pi$ are parallel, if defined.

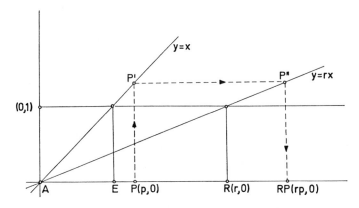

Fig. 4

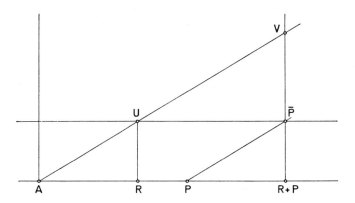

Fig. 5

so that $(P/A,Q)(A/E,P) = (RP/A,RQ)(A/R, RP)$. Now $0 < r$, i.e., by (10) we have $(A/E, R) = 1$. Consequently,

$$(P/A, Q)(A/E, P) = (RP/A, RQ)(A/R, RP)(A/E, R)$$
$$= (RP/A, RQ)(A/E, RP),$$

i.e., the statement $p < q$ is equivalent to $rp < rq$ if $0 < r$.

From (8) and (17) we have for all $S \text{ I } x$:

(18) $(P/Q, R) = (S + P/S + Q,S + R)$.

From Fig. 5 we also see that

(19) $(A/R, R + P) = (A/U, V) = (R + P/\bar{P}, V) = (R + P/P, A)$.

Now $P < Q$ is equivalent to $(P/A, Q)(A/E, P) = -1$ for $P \neq A$.[12]

---

[12]The proof for the case $P = A$ is left to the reader.

From (18) and (5) it follows that

$$-1 = (P/A, Q)(A/E, P) = (R + P/R, R + Q)(A/E, P)$$
$$= (R + P/A, R + Q)(A/E, R + P)(R + P/A, R)(A/P, R + P).$$

Now from (7), (5), (18), and (19)

$$(R + P/A, R)(A/P, R + P)$$
$$= -(R/A, R + P)(A/R, R + P)(A/R + P, P)$$
$$= -(R/A, R + P)(A/R, P) = -(R/A, R + P)(R/R + R, R + P)$$
$$= -(R/A, R + R) = (A/R, R + R)(R + R/A, R)$$
$$= (A/R, R + R)(A/R, R + R) = 1.$$

Consequently $(R + P/A, \ R + Q)(A/E, \ R + P) = -1$, i.e., $R + P < R + Q$, so that $r + p < r + q$ follows from $p < q$.

Now it is also obvious, in case $\mathfrak{E}$ is a continuous Desarguesian affine plane, that the coordinate field K also satisfies the axiom of continuity, which means that K is the field of real numbers (cf. IB1,§4.6 or [1] pp. 106 ff. in the bibliography for this chapter). We thus have the following theorem.

Theorem 1: *If $\mathfrak{E}$ is a continuous Desarguesian affine plane, then $\mathfrak{E}$ can be represented as a coordinate plane over the field of real numbers.*

4. Ordered Projective Geometries

A plane $\mathfrak{E}$ is said to be *projective* if it satisfies the following axioms:

I4:   *Two distinct lines have at least one point of intersection.*
I2′: *Every line is incident with at least three distinct points.*

We obtain examples of projective planes if we begin with a three-dimensional vector space V over field K and consider the one-dimensional linear subspaces $K\mathfrak{x}$ with $\mathfrak{x} \in V$, $\mathfrak{x} \neq 0$ as points, and the two-dimensional linear subspaces $K\mathfrak{a} + K\mathfrak{b}$ with $\mathfrak{a}, \mathfrak{b} \in V$ as lines. A point $K\mathfrak{x}$ is incident with a line $K\mathfrak{a} + K\mathfrak{b}$ if $K\mathfrak{x}$ is a subset of $K\mathfrak{a} + K\mathfrak{b}$. All these projective planes are Desarguesian. Corresponding to Theorem 1 for affine planes we have the following theorem for projective planes.

Theorem 2: *If $\mathfrak{E}$ is a continuous Desarguesian projective plane, then $\mathfrak{E}$ can be represented (as indicated above) by means of a three-dimensional vector space over the real numbers.*

We omit the proof here, since it is similar to the proof for affine planes. Instead of the function $(A/B, C)$, we now have the *separation relation:*

If A, B, C, D are four points on a line with $A, B \neq C, D$, then $[A, B/C, D] = (A/C, D)(B/C, D)$ is called a separation relation. This separation relation is invariant under perspectivities.

5. Congruence Axioms

5.1. By a congruence structure for $\mathfrak{P} \cup \mathfrak{G}$ we mean a *congruence relation* $\equiv$ for pairs of points A, B with $A \neq B$ satisfying the following axiom:

K1: *The relation of congruence is an equivalence relation* with $AB \equiv BA$.

An ordered plane is said to be *metric*[13] if there is defined in it a congruence relation consistent with the incidence and order structures in the following sense:

V3: *Let C be a point on the line g. For two points A, B with $A \neq B$ there exist on g exactly two distinct points D and D' with $AB \equiv CD \equiv CD'$ (Fig. 6).*

V4: *If C, D, D' are three distinct points on a line with $CD \equiv CD'$, then $(C/D, D') = -1$.*

V5: *If A, B, C and A', B', C' are three distinct points on the lines g and g', respectively, and if $(B/A, C) = (B'/A', C')$ and also $AB \equiv A'B'$ and $BC \equiv B'C'$ then we also have $AC \equiv A'C'$.*

V6: *If A, B, D are three noncollinear points and A', B' are two further points with $A'B' \equiv AB$, then there exist exactly two distinct points D' and D'' with $AD \equiv A'D' \equiv A'D''$ and $BD \equiv B'D' \equiv B'D''$ (Fig. 7).*

V7: *If A, B, C are three noncollinear points and D is a point of the line joining A to B, and if A', B', C', D are four points with $AB \equiv A'B'$; $BC \equiv B'C'$; $AC \equiv A'C'$ and $AD \equiv A'D'$; $BD \equiv B'D'$, then also $CD \equiv C'D'$ (Fig. 8).*

V8: *If A, B are two distinct points of a line g and D, D' are two distinct points with $AD \equiv AD'$ and $BD \equiv BD'$, then $(g/D, D') = -1$.*

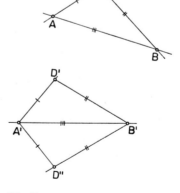

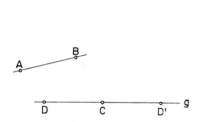

Fig. 6. Since in the plane of the paper geodesic lines are straight lines, any nontrivial diagram is necessarily false.

Fig. 7

[13]The concept of "metric plane" as defined here is more restrictive than in Chapter II5; every metric plane in our sense is also a metric plane in the sense of that chapter.

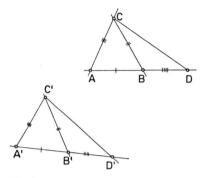

Fig. 8

5.2. We now proceed to prove several theorems for metric planes $\mathfrak{E} = (\mathfrak{B}, \mathfrak{G}, \mathfrak{J}, \mathfrak{A}, \equiv)$, i.e., planes satisfying the Axioms I1, 2, 3, A1, 2, K1, and V1 through V8 but not necessarily **P** and **S**. In §2 we have shown that an order relation $<$ can be introduced for the points of a line k. Here we begin with two fixed points A and E on k with $A \neq E$.

(20) *Let there be defined an order relation $<$ on the line k, and let P and Q be two distinct points. Then:*

a) *For every point X on k there exist exactly two points X^-, X^+ I k with $PQ \equiv XX^- \equiv XX^+$ and $X^- < X < X^+$.*

b) *If X and Y are two points on k such that $X < Y$, then $X^+ < Y^+$.*

Proof. a) By **V3** there exist on k exactly two points Y and Y', with $PQ \equiv XY \equiv XY'$, for which $(X/Y, Y') = -1$ by **V4**. Consequently it follows from (14) that either $Y < X < Y'$ or $Y' < X < Y$.

b) Let us assume that $Y^+ < X^+$. Then it follows that $Y < X^+$. By (20a) there exists a $W < X$ with $WX \equiv Y^+X^+$. Since we also have $XX^+ \equiv YY^+$ and since $W < X < X^+$ and $Y < Y^+ < X^+$ imply $(X/W, X^+) = (Y^+/Y, X^+) = -1$ (cf. (14)), we therefore obtain $WX^+ \equiv YX^+$, by Axiom **V5**. But this result is in contradiction to Axiom **V4**, since $W < X < Y < X^+$, so that $(X^+/W, Y) = 1$.

If we had $Y^+ = X^+$, it would follow by Axiom **V4** that $XX^+ \equiv YX^+$ and consequently $X = Y$, since $X, Y < X^+ = Y^+$.

Theorem 3 (Axiom of Archimedes): *Let $\mathfrak{E}$ be a continuous[14] metric plane, and let an order relation $<$ be defined on the line k. Let P and Q be two distinct points and let A be a point on the line k. Also let $A_1 = A^+$ (cf. (20)) and $A_{i+1} = A_i^+$ for all $i \geqq 1$. Then for every point B on k with $B > A$ there exists an A_k with $A_k > B$.*

Proof. Let us assume that $A_i \leqq B$ for all natural numbers i. By the

[14]That is, Axiom **S** holds.

continuity axiom **S** there exists a least upper bound C for the set M of all A_i. Then C^- is not an upper bound of M. Consequently there exists a $A_j \in M$ with $C^- < A_j$. By (20) it follows that $(C^-)^+ = C < A_j^+ = A_{j+1}$, so that C is not a least upper bound, in contradiction to our assumption.

5.3. In a metric plane we can now introduce *angles* and a *congruence relation* for them. To this end we first define a *halfline*. If A, B are two distinct points of a line g, the set of points $g_{AB} = \{X \in g; (A/B, X) = 1\}$ is called a halfline. Of course, the set of points $g_{AB}^- = \{X \in g; (A/B, X) = -1\}$ is also a halfline. Moreover, $g_{AB} = g_{AB'}$, if $B' \in g_{AB}$. The points of a line g fall into three disjoint classes $g_{AB}, \{A\}, g_{AB}^-$. A pair of halflines (g_{AB}, h_{CD}) is called an *angle* if $A = C$.[15] If A, B, C are three distinct points and g and h are the lines joining A, B and B, C, respectively, we also denote the angle (g_{BA}, h_{BC}) by $\not\prec (ABC)$.

Let $\alpha = (g_{AB}, h_{AC})$ and $\alpha' = (g'_{A'B'}, h'_{A'C'})$ be two angles. By Axioms **V3** and **V4** we may assume that $AB \equiv AC \equiv A'C' \equiv A'B'$. The two angles α and α' are said to be *congruent* $(\alpha \equiv \alpha')$ if $BC \equiv B'C'$. The concept of congruence is well defined; for if $B_1 \in g_{AB}$, $C_1 \in h_{AC}$, $B_1' \in g'_{A'B'}$, and $C_1' \in h'_{A'C'}$ with $AB_1 \equiv AC_1 \equiv A'B_1' \equiv A'C_1'$, then it follows from Axioms **V5** and **V7** that also $B_1C_1 \equiv B_1'C_1'$.

Since the relation of congruence of pairs of points is reflexive, symmetric, and transitive, it follows that this congruence of angles is also reflexive, symmetric, and transitive.

5.4. We now prove the following two congruence theorems.

(21) *If ABC and $A'B'C'$ are two triangles*[16] *with $AB \equiv A'B'$ and $AC \equiv A'C'$, then $BC \equiv B'C'$ if and only if $\alpha = (g_{AB}, h_{AC}) \equiv \alpha' = (g'_{A'B'}, h'_{A'C'})$* (Fig. 9).

Proof. We first assume $BC \equiv B'C'$. By Axiom **V3** and **V4** there exists on the halflines g_{AB} and $g'_{A'B'}$ exactly one point D and D', respectively, with $AC \equiv AD$ and $A'C' \equiv A'D'$. Then Axioms **V5** and **V7** imply that $CD \equiv C'D'$. Consequently $\alpha \equiv \alpha'$.

Conversely, if $\alpha \equiv \alpha'$, there exist exactly one $D \in g_{AB}$ and exactly one $D' \in g'_{A'B'}$ with $AC \equiv AD$, $A'C' \equiv A'D'$, and $CD \equiv C'D'$. Consequently. $CB \equiv C'B'$, again by Axioms **V5** and **V7**.

(22) *If ABC and $A'B'C'$ are two triangles with $AB \equiv A'B'$, $\alpha = \not\prec (BAC) \equiv \alpha' = \not\prec (B'A'C')$ and $\beta = \not\prec(ABC) \equiv \beta' = \not\prec(A'B'C')$, then also $AC \equiv A'C'$ and $BC \equiv B'C'$.*

---

[15]Compare also III, §12. From this concept of angle we obtain the elementary concept in a Euclidean plane.

[16]By a *triangle* we mean three noncollinear points.

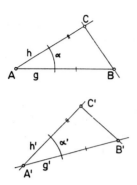

Fig. 9

Proof.   Let $D \in g'_{A'C'}$ with $A'D \equiv AC$ and $E \in h'_{B'C'}$ with $B'E \equiv BC$ (cf. Fig. 10). Then, $BC \equiv B'D$ and $AC \equiv A'E$ by (21). From $D \in g'_{A'C'}$ and $E \in h'_{B'C'}$ respectively, it follows that $(A'/C', D) = 1$ and $(B'/C', E) = 1$. Thus if k' is the line joining A' and B', then $(k'/C', D) = (A'/C', D) = (k'/C', E) = (B'/C', E) = 1$ by (3), so that $(k'/E, D) = 1$ by (2). But from $B'D \equiv B'E$ and also $A'D \equiv A'E$ and $(k'/E, D) = 1$, it follows that $E = D$ by Axiom V8, and consequently $E = D = C'$. Thus $AC \equiv A'C'$ and $BC \equiv B'C'$.

5.5.   Let g and h be two lines intersecting at a point A. Moreover, let B and C be two points, distinct from A, lying on g and h, respectively. Then the lines g and h are said to be *orthogonal* or *perpendicular* $(g \perp h)$ if $(g_{AB}, h_{AC}) \equiv (g_{\overline{AB}}, h_{AC})$. Orthogonality is well defined, since:

(23)   *Vertically opposite angles are congruent; and if two angles are congruent, their adjacent angles[17] are congruent.*

Proof (Fig. 11).   Let g and h be two lines intersecting at the point A, and let B, B' and C, C' be two distinct points on g and h, respectively, with

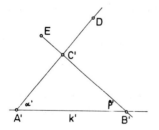

Fig. 10

[17]The angles $(g_{\overline{AB}}, h_{\overline{AC}})$ and $(g_{AB}, h_{\overline{AC}})$ are vertically opposite and adjacent to (g_{AB}, h_{AC}) respectively.

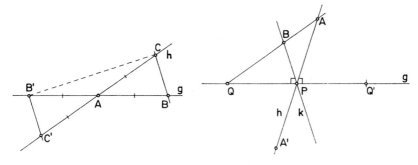

Fig. 11   Fig. 12

$AB \equiv AB' \equiv AC \equiv AC'$. Then $(A/B, B') = (A/C, C') = -1$, so that also $BB' \equiv CC'$ by Axiom **V5**. Applying Axiom **V7** to A, C', B, C and A, B, C', B', we obtain $BC \equiv B'C'$, so that $(g_{AB}, h_{AC}) = (g_{AB'}, h_{AC'})$. The second statement follows immediately from **V7** and (21).

(24)   *Let g, h, k be three lines incident with a point P. If h and k are perpendicular to g, then $h = k$.*

Proof (Fig. 12).   On g choose a point $Q \neq P$. By Axiom **V3** there exists on g a point $Q' \neq Q$ with $QP \equiv PQ'$ and on h two distinct points A and A' with $PQ \equiv PA \equiv PA'$. For these points we have $(g/A, A') = (P/A, A') = -1$ by (3) and **V4**. We now assume that $h \neq k$. By Axiom **A1** we then have $(k/A, A') = (g/A, A') = -1$. It follows that $(k/Q, A) = (k/Q, A')(k/A, A') = -(k/Q, A')$ by (2), that is, exactly one of the values $(k/Q, A)$ and $(k/Q, A')$ is -1, say, $(k/Q, A)$. From $(k/Q, A) = -1$ it follows by **V2** that k has a point B in common with the line joining Q and A. For this point we have $(g/B, A) = (Q/B, A) = 1$ by Axiom **A2**, and consequently $\measuredangle(AQQ') = \measuredangle(BQQ')$. Since $k \perp g$, we also have $(g_{PQ}, k_{PB}) \equiv (g_{\overline{PQ}}, k_{PB}) = (g_{PQ'}, k_{PB})$, and thus we obtain $BQ \equiv BQ'$ by (21). Furthermore, it follows from (21) that $\measuredangle(BQQ') \equiv \measuredangle(BQ'Q)$ and, since $h \perp g$, therefore $\measuredangle(AQQ') \equiv \measuredangle(AQ'Q)$. Consequently $\measuredangle(AQ'Q) \equiv \measuredangle(AQQ') \equiv \measuredangle(BQQ') \equiv \measuredangle(BQ'Q)$. From (22) it therefore follows that $QB \equiv QA$, in contradiction to Axiom **V4**, since also $(Q/B, A) = 1$.

(25)   *Perpendiculars: For every point P and every line g there exists exactly one line h with $h \perp g$ and $h \mathbin{I} P$.*

Proof.   We first consider the case $P \mathbin{\bar{I}} g$ (cf. Fig. 13). On the line g we choose two distinct points A and B. By Axiom **V6** there exists exactly one point $P' \neq P$ with $AP \equiv AP'$ and $BP \equiv BP'$. Thus $(g/P, P') = -1$ by Axiom **V8**. Let h be the line joining P and P'. By Axiom **V2** the lines g and h have a point of intersection S, and $SP \equiv SP'$ by **V7**, since $AP \equiv AP'$ and $BP \equiv BP'$. Consequently, the angles (g_{SA}, h_{SP}) and $(g_{SA}, h_{\overline{SP}})$ are congruent

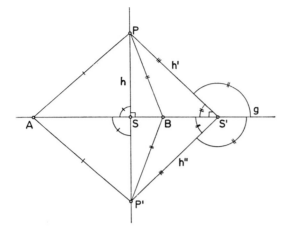

Fig. 13

by (21), so that $h \perp g$. Now let h' be another line with $h' \perp g$ and h' I P. If S' is the point of intersection of h' with g, then $PS' \equiv P'S'$ by Axiom V7. Thus, by (21) and (23), the angles $(g_{S'A}, h'_{S'P})$, $(g_{S'A}, h''_{S'P'})$, $(g^-_{S'A}, h'_{S'P})$, and $(g^-_{S'A}, h''_{S'P'})$, are congruent, where h'' is the line joining S' with P'. Consequently, with $h' \perp g$ we also have $h'' \perp g$, so that $h' = h''$ by (24), and consequently $h' = h$, since both h and h' are incident with the two distinct points P and P'.

We now consider the case P I g (cf. Fig. 14). Let Q be a point with Q I g, let k be the perpendicular from Q to g, and let S be the point of intersection of k with g. Then by V6 there exists a point R with $PR \equiv SQ$ and $SR \equiv PQ$. Consequently, by (21) and (23) the angles (g_{SP}, k_{SQ}), (g_{PS}, h_{PR}), (g^-_{SP}, k_{SQ}),

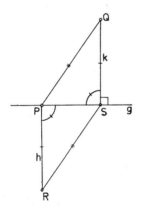

Fig. 14

and (g_{SP}^-, h_{PR}) are congruent, where h is the line joining P with R. Thus $h \perp g$.

(26)   *For every line g and every point P Ɨ g there exists at least one parallel h to g with h Ɪ p.*

Proof.   From P we draw the perpendicular k to g and at P we erect the perpendicular h to k. By (25) the lines g and h have no point of intersection.

5.6.   An affine continuous metric plane is called a *Euclidean* plane; a continuous, metric nonaffine plane is called a *hyperbolic* plane.

By (26) there exists in a hyperbolic plane at least one point P and one line g such that there is more than one line through P parallel to g. In fact, we even have the following theorem.

Theorem 4:   *In a hyperbolic plane*[18] *there exists, for every point P and every line g with P Ɨ g, more than one parallel to g through P.*

Two metric planes $\mathfrak{E} = (\mathfrak{P}, \mathfrak{G}, \mathfrak{I}, \mathfrak{A} \equiv)$ and $\mathfrak{E}' = (\mathfrak{P}', \mathfrak{G}', \mathfrak{I}', \mathfrak{A}', \equiv')$ are said to be *isomorphic* if there exists a one-to-one mapping $\varphi \colon \mathfrak{P} \cup \mathfrak{G} \to \mathfrak{P}' \cup \mathfrak{G}'$ with the following properties:

a) $\varphi(\mathfrak{P}) = \mathfrak{P}', \varphi(\mathfrak{G}) = \mathfrak{G}'$,
b) P Ɪ g is equivalent to $\varphi(P)$ Ɪ' $\varphi(g)$,
c) $(g/A, B) = (\varphi(g)/'\varphi(A), \varphi(B))$ for all $g \in \mathfrak{G}$ and all $A, B \in \mathfrak{P}$ with A, B Ɪ g,
d) $AB \equiv CD$ is equivalent to $\varphi(A)\varphi(B) \equiv' \varphi(C)\varphi(D)$.[19]

The mapping φ is then called an *isomorphism*.

In the Euclidean plane we can now prove the theorem of Desargues. Consequently, by §3 a Euclidean plane can be represented as a coordinate geometry over the real numbers, which is part of the following theorem.

Theorem 5:   *Up to isomorphism there exist exactly two continuous metric geometries, namely the Euclidean geometry, which can be represented as a coordinate geometry over the real numbers, and the hyperbolic geometry over the real numbers.*

Models for these two geometries are given in the next section.

[18]The complete proofs of Theorems 4 and 5 are not given here, since they are rather long. They will be given in a book on classical Euclidean and classical hyperbolic geometry now under preparation by H. Karzel. A proof of Theorem 4 is to be found in R. Baldus, F. Löbell: Nichteuklidische Geometrie, 4. Aufl. Berlin 1964.

[19]In the structures defined on the planes $\mathfrak{E}$ and $\mathfrak{E}'$ we thus make a distinction between the relations $\mathfrak{I}$ and $\mathfrak{I}'$ in our choice of symbols Ɪ and Ɪ', /and/', and $\equiv$ and $\equiv'$. This notation indicates that the realization of these relations by means of models may be different for the two planes.

6. Models of Euclidean, Hyperbolic, and Elliptic Planes

6.1.   We first give a model of the Euclidean plane, namely, the well-known two-dimensional coordinate geometry $\mathfrak{E}$ over the field R of real numbers.

Let $\mathfrak{P} = R \times R = \{(x_1, x_2); x_1, x_2 \in R\}$ be the set of points and let $\mathfrak{G} = \{\langle 1, m, b\rangle; m, b \in R\} \cup \{\langle 0, 1, b\rangle; b \in R\}$ be the set of lines. With a view to introducing incidence and order in a convenient way, we now define a product of lines and points. If $g = \langle 1, m, b\rangle$ or $g = \langle 0, 1, b\rangle$ and if $X = (x_1, x_2)$, let $gX = x_1 + mx_2 + b$ and $gX = x_2 + b$, respectively. Thus the product gX is a real number.

If g is a line and X is a point, then g and X are said to be incident if and only if $gX = 0$.

The incidence axioms can be verified at once. Let us first verify **I**1. If $A = (a_1, a_2)$ and $B = (b_1, b_2)$ are two distinct points, then

$$g = \left\langle 1, \frac{b_1 - a_1}{a_2 - b_2}, \frac{a_1 b_2 - a_2 b_1}{a_2 - b_2} \right\rangle \quad or \quad g = \langle 0, 1, -a_2\rangle$$

is the uniquely determined line incident with A and B, according as $a_2 \neq b_2$ or $a_2 = b_2$.

The line $g = \langle 1, m, b\rangle$ is incident with the two distinct points $A = (-b, 0)$ and $B = (-m, -b, 1)$, and the line $g = \langle 0, 1, b\rangle$, with the points $A = (0, -b)$ and $B = (1, -b)$. Thus we have also verified **I**2.

The point $(0, 0)$ is not incident with the line $\langle 0, 1, 1\rangle$, which verifies **I**3.

The coordinate geometry $\mathfrak{E}$ is affine, since we have the parallel axiom:

If $g = \langle 1, m, b\rangle$, or $g = \langle 0, 1, b\rangle$, respectively, and if $P = (p_1, p_2)$, then $h = \langle 1, m, -p_1 - mp_2\rangle$, or $h = \langle 0, 1, -p_2\rangle$, respectively, is the only line through P parallel to g.

Moreover, for every line g in $\mathfrak{E}$ we can define a partition into sides. The three mutually disjoint classes are: $\bar{g} = \{X \in \mathfrak{P}; gX = 0\}$ and the two sides $S_+ = \{X \in \mathfrak{P}; gX > 0\}$ and $S_- = \{X \in \mathfrak{P}; gX < 0\}$.

The function $(g/A, B)$ defined in §1 can now be described as follows:

$$(g/A, B) = \text{sgn}[(gA)(gB)^{-1}].^{20}$$

For $(g/A, B)$ we prove the validity of Axioms **A**1 and **A**2. If $X = (x_1, x_2)$, $Y = (y_1, y_2)$, and $v \in R$, then let $vX = (vx_1, vx_2)$ and $X + Y = (x_1 + y_1, x_2 + y_2)$. By straightforward calculation we see that

$$g(vX + (1 - v)Y) = vgX + (1 - v)gY \quad for \ all \quad g \in \mathfrak{G}, v \in R$$

$$and \quad X, Y \in \mathfrak{P}.$$

If $X = (x_1, x_2)$, $Y = (y_1, y_2)$, and $Z = (z_1, z_2)$ are three distinct points on a

[20]For a nonzero real number r we define $\text{sgn } r = 1$ if $r > 0$ and $\text{sgn } r = -1$ if $r < 0$.

line k, then there exists exactly one $v \in R$ such that $Z = vX + (1 - v)Y$. For every line $j \neq k$ incident with Z we have $jZ = j(vX + (1 - v)Y) = 0$, so that $jX(jY)^{-1} = (v - 1)v^{-1} = 1 - v^{-1}$, and consequently

$$(27) \qquad (Z/X, Y) = (j/X, Y) = \text{sgn}(1 - v^{-1}).$$

Thus we have verified **A1**.

Now let $g, h, j \neq k$ be three lines with $g \text{ I } X, h \text{ I } Y, j \text{ I } Z$. Then as before we have

$$Z = vX + (1 - v)Y,$$

$$X = (1 - v^{-1})Y + v^{-1}Z,$$

$$Y = (1 - v)^{-1}Z + (1 - (1 - v)^{-1})X,$$

$$gY(gZ)^{-1} = 1 - (1 - v^{-1})^{-1} = (1 - v)^{-1},$$

$$hZ(hX)^{-1} = 1 - (1 - v) = v,$$

$$jX(jY)^{-1} = 1 - v^{-1} = (v - 1)v^{-1}.$$

Thus we see at once that **A2** is valid.

The consistency axiom **V1** follows immediately from the definition of $\bar{g}$.

Axiom **V2** is also satisfied. For if $(j/X, Y) = -1$, so that $jX(jY)^{-1} < 0$, then j intersects the line k joining X and Y at the point $vX + (1 - v)Y$ with $v = (1 - jX(jY)^{-1})^{-1}$.

We now verify the axiom of continuity. Let A and E be two distinct points on a line k; then (9) and (10) define an order relation $<$ for k. The mapping $p \to P = pE + (1 - p)A$ of R onto the line k is one-to-one and order-preserving, i.e., $p < q$ is equivalent to $P < Q$, as we shall now show.

From $P = pE + (1 - p)A$ and $Q = qE + (1 - q)A$, it follows that $P = pq^{-1}Q + (1 - pq^{-1})A$ and $A = (1 - p)^{-1}P + p(p - 1)^{-1}E$.

By (27) we therefore have $(P/A, Q) = \text{sgn}(1 - qp^{-1})$ and also $(A/E, P) = \text{sgn } p$ and therefore $(P/A, Q)(A/E, P) = \text{sgn } p(1 - qp^{-1}) = \text{sgn}(p - q)$. Consequently $(P/A, Q)(A/E, P) = -1$ is equivalent to $p - q < 0$, so that $P < Q$ is equivalent to $p < q$.

Therefore, since the real numbers are continuous, the geometry is also continuous.

Finally, we define the congruence relation $\equiv$. If $X = (x_1, x_2)$, $Y = (y_1, y_2)$, $U = (u_1, u_2)$, and $V = (v_1, v_2)$ are four points on the plane $\mathfrak{E}$, then we take $XY \equiv UV$ to be equivalent to $(x_1 - y_1)^2 + (x_2 - y_2)^2 = (u_1 - v_1)^2 + (u_2 - v_2)^2$. The congruence axiom **K1** and the consistency axioms **V3** through **V8** are then easily verified.

The model of Euclidean geometry in Theorem 5 is now completely described.

6.2. We next turn to the Klein model of the hyperbolic plane (a visualization of this model will be given in §7). For this purpose we begin with a three-dimensional vector space $V = \{(x_1, x_2, x_3); x_i \in R\}$ over the real numbers. In V let there be defined a symmetric bilinear form which to every pair of vectors $\mathfrak{X} = (x_1, x_2, x_3)$, $\mathfrak{Y} = (y_1, y_2, y_3)$ assigns the real number $f(\mathfrak{X}, \mathfrak{Y}) = x_1 y_1 - x_2 y_2 - x_3 y_3$. The points and lines of the hyperbolic plane $\mathfrak{E}$ are then defined as one-dimensional and two-dimensional linear subspaces $R\mathfrak{x}$ with $f(\mathfrak{x}, \mathfrak{x}) > 0$ and $f(\mathfrak{x}, \mathfrak{x}) < 0$, respectively.[21] A line $g = R\mathfrak{g}$ and a point $P = R\mathfrak{p}$ are taken to be incident if and only if $f(\mathfrak{g}, \mathfrak{p}) = 0$.

In order to prove the axioms of incidence we require the following lemma.

(28)   If $\mathfrak{a} \in V$ with $f(\mathfrak{a}, \mathfrak{a}) > 0$, then $f(\mathfrak{g}, \mathfrak{g}) < 0$ for every $\mathfrak{g} \in V$ with $\mathfrak{g} \neq 0$ and $f(\mathfrak{a}, \mathfrak{g}) = 0$.

Proof.   Let $\mathfrak{a} = (a_1, a_2, a_3)$ and $\mathfrak{g} = (g_1, g_2, g_3)$. From $f(\mathfrak{a}, \mathfrak{g}) = a_1 g_1 - a_2 g_2 - a_3 g_3 = 0$ it follows that $(a_2 g_2 + a_3 g_3)^2 = (a_1 g_1)^2$. Obviously,

$$(g_2^2 + g_3^2)(a_2^2 + a_3^2) = (a_2 g_2 + a_3 g_3)^2 + (a_2 g_3 - a_3 g_2)^2$$
$$= g_1^2 a_1^2 + (a_2 g_3 - a_3 g_2)^2,$$

so that

$$(a_2^2 + a_3^2)(g_2^2 + g_3^2) \geq g_1^2 a_1^2.$$

If $g_1 \neq 0$, then $f(\mathfrak{a}, \mathfrak{a}) = a_1^2 - a_2^2 - a_3^2 > 0$ implies the inequality $a_1^2 g_1^2 > (a_2^2 + a_3^2)g_1^2$, so that

$$(a_2^2 + a_3^2)(g_2^2 + g_3^2) > (a_2^2 + a_3^2)g_1^2$$

and consequently $f(\mathfrak{g}, \mathfrak{g}) < 0$.

If $g_1 = 0$, then $f(\mathfrak{g}, \mathfrak{g}) < 0$ trivially.

If $A = R\mathfrak{a}$ and $B = R\mathfrak{b}$ are two distinct points, then $g = R\mathfrak{g}$ with $\mathfrak{g} = (a_2 b_3 - a_3 b_2, a_1 b_3 - a_3 b_1, a_2 b_1 - a_1 b_2)$ is the line joining A to B, since $f(\mathfrak{a}, \mathfrak{g}) = f(\mathfrak{b}, \mathfrak{g}) = 0$, and consequently $f(\mathfrak{g}, \mathfrak{g}) < 0$ by (28). This line is uniquely determined, since the system of equations $f(\mathfrak{a}, \mathfrak{x}) = f(\mathfrak{b}, \mathfrak{x}) = 0$ is of rank 2, because $A \neq B$.

We have thus proved that Axiom I1 is satisfied.

The simple proofs of Axioms I2 and I3 will be omitted here.

We now define for $\mathfrak{E}$ a partition into sides. If $g = R\mathfrak{g}$, then let $\bar{g} = \{X = R\mathfrak{x} \in \mathfrak{P}; f(\mathfrak{x}, \mathfrak{g}) = 0\}$. Let $X_0 = R\mathfrak{x}_0$ be a point with $X_0 \notin \bar{g}$. Then we define

$$S_+ = \{Y = R\mathfrak{y} \in \mathfrak{P}; \operatorname{sgn}[f(\mathfrak{x}_0, \mathfrak{g})f(\mathfrak{y}, \mathfrak{g})f(\mathfrak{x}_0, \mathfrak{y})] = 1\}$$

and

$$S_- = \{Y = R\mathfrak{y} \in \mathfrak{P}; \operatorname{sgn}[f(\mathfrak{x}_0, \mathfrak{g})f(\mathfrak{y}, \mathfrak{g})f(\mathfrak{x}_0, \mathfrak{y})] = -1\}.$$

[21] For every $\mathfrak{y} \in R\mathfrak{x}$ with $\mathfrak{y} \neq 0$, we have $\mathfrak{y} = r\mathfrak{x}$ with $r \in R$, $r \neq 0$. Consequently, $f(\mathfrak{y}, \mathfrak{y}) = r^2 f(\mathfrak{x}, \mathfrak{x}) > 0$ if and only if $f(\mathfrak{x}, \mathfrak{x}) > 0$.

Here S_+ and S_- are well defined, since f is bilinear.

For the verification of the order axioms we require the following lemmas.

(29) *If $A = R\mathfrak{a}$ and $B = R\mathfrak{b}$ are two points with $\mathfrak{a} = (a_1, a_2, a_3)$ and $\mathfrak{b} = (b_1, b_2, b_3)$, then* $\operatorname{sgn} f(\mathfrak{a}, \mathfrak{b}) = \operatorname{sgn}(a_1 b_1)$.

Proof. From $f(\mathfrak{a}, \mathfrak{a}) = a_1^2 - a_2^2 - a_3^2 > 0$ and $f(b, b) = b_1^2 - b_2^2 - b_3^2 > 0$ it follows that $a_1^2 b_1^2 > (a_2^2 + a_3^2)(b_2^2 + b_3^2) = (a_2 b_2 + a_3 b_3)^2 + (a_2 b_3 - a_3 b_2)^2 \geqq (a_2 b_2 + a_3 b_3)^2$, so that $|a_1 b_1| > |a_2 b_2 + a_3 b_3|$ and consequently

$$\operatorname{sgn}\{a_1 b_1 - (a_2 b_2 + a_3 b_3)\} = \operatorname{sgn}(a_1 b_1).$$

(30) *If $A = R\mathfrak{a}$, $B = R\mathfrak{b}$, and $C = R\mathfrak{c}$ are three points, then*

$$\operatorname{sgn}\{f(\mathfrak{a}, \mathfrak{b}) f(\mathfrak{b}, \mathfrak{c}) f(\mathfrak{c}, \mathfrak{a})\} = 1.$$

Proof. By (29) we have

$$\operatorname{sgn}\{f(\mathfrak{a}, \mathfrak{b}) f(\mathfrak{b}, \mathfrak{c}) f(\mathfrak{c}, \mathfrak{a})\} = \operatorname{sgn} f(\mathfrak{a}, \mathfrak{b}) \operatorname{sgn} f(\mathfrak{b}, \mathfrak{c}) \operatorname{sgn} f(\mathfrak{c}, \mathfrak{a})$$

$$= \operatorname{sgn}(a_1 b_1) \operatorname{sgn}(b_1 c_1) \operatorname{sgn}(c_1 a_1)$$

$$= \operatorname{sgn}(a_1^2 b_1^2 c_1^2) = 1.$$

The function $(g/X, Y)$ defined in §1 can now be described as follows:

(31) *If $X = R\mathfrak{x}$, $Y = R\mathfrak{y}$ are two points not incident with the line $g = R\mathfrak{g}$, then*

$$(g/X, Y) = \operatorname{sgn}\{f(\mathfrak{g}, \mathfrak{x}) f(\mathfrak{g}, \mathfrak{y}) f(\mathfrak{x}, \mathfrak{y})\}.$$

Proof. By §1 and the definition of S_+ and S_-,

$$(g/X, Y) = \operatorname{sgn}\{f(\mathfrak{x}_0, \mathfrak{g}) f(\mathfrak{x}, \mathfrak{g}) f(\mathfrak{x}_0, \mathfrak{x})\} \operatorname{sgn}\{f(\mathfrak{x}_0, \mathfrak{g}) f(\mathfrak{y}, \mathfrak{g}) f(\mathfrak{x}_0, \mathfrak{y})\}$$

$$= \operatorname{sgn}\{f(\mathfrak{x}, \mathfrak{g}) f(\mathfrak{y}, \mathfrak{g}) f(\mathfrak{x}_0, \mathfrak{x}) f(\mathfrak{x}_0, \mathfrak{y})\}.$$

But, by (30), also $\operatorname{sgn}\{f(\mathfrak{x}_0, \mathfrak{x}) f(\mathfrak{x}_0, \mathfrak{y})\} = \operatorname{sgn} f(\mathfrak{x}, \mathfrak{y})$.

We now verify Axiom **A1**.

If $A = B$, then $(g/A, B) = (h/A, B) = 1$. Now let $A = R\mathfrak{a} \neq B = R\mathfrak{b}$. Then A, B, and $\mathfrak{C} = R\mathfrak{c}$ are three distinct points of the line $k = R\mathfrak{k}$. Consequently there exist three nonzero numbers $\alpha, \beta, \gamma \in R$ with $\alpha\mathfrak{a} + \beta\mathfrak{b} + \gamma\mathfrak{c} = 0$. Since for a line $g = R\mathfrak{g}$ through $\mathfrak{C}$ with $g \neq k$ we have $f(\mathfrak{c}, \mathfrak{g}) = 0$, it follows that $0 = f(\alpha\mathfrak{a} + \beta\mathfrak{b}, \mathfrak{g}) = \alpha f(\mathfrak{a}, \mathfrak{g}) + \beta f(\mathfrak{b}, \mathfrak{g})$. Consequently, $\operatorname{sgn}\{f(\mathfrak{a}, \mathfrak{g}) f(\mathfrak{b}, \mathfrak{g})\} = -\operatorname{sgn}(\alpha\beta)$ and therefore, by (31), $(g/A, B) = \operatorname{sgn}\{f(\mathfrak{a}, \mathfrak{g}) f(\mathfrak{b}, \mathfrak{g}) f(\mathfrak{a}, \mathfrak{b})\} = -\operatorname{sgn}\{\alpha\beta f(\mathfrak{a}, \mathfrak{b})\}$. Correspondingly, $(h/A, B) = -\operatorname{sgn}\{\alpha\beta f(\mathfrak{a}, \mathfrak{b})\}$, so that $(g/A, B) = (h/A, B)$.

Axiom **A2** is verified as follows:

Let $A = R\mathfrak{a}$, $B = R\mathfrak{b}$, $C = R\mathfrak{c}$ and $a = R\bar{\mathfrak{a}}$, $b = R\bar{\mathfrak{b}}$, $c = R\bar{\mathfrak{c}}$. As in the proof of **A1**, we have $\alpha, \beta, \gamma \in R$, so that $\alpha\mathfrak{a} + \beta\mathfrak{b} + \gamma\mathfrak{c} = 0$, from which, together with the formula derived at the end of the proof of **A1**, it follows from (29) that

$$(a/B, C) = -\text{sgn } \{\beta\gamma f(\mathfrak{b}, \mathfrak{c})\} = -\text{sgn } (\beta b_1 \gamma c_1) = \text{sgn } \{\beta b_1(\alpha a_1 + \beta b_1)\},$$

$$(b/A, C) = -\text{sgn } \{\alpha\gamma f(\mathfrak{a}, \mathfrak{c})\} = -\text{sgn } (\alpha a_1 \gamma c_1) = \text{sgn } \{\alpha a_1(\alpha a_1 + \beta b_1)\},$$

$$(c/A, B) = -\text{sgn } \{\alpha\beta f(\mathfrak{a}, \mathfrak{b})\} = -\text{sgn}(\alpha a_1 \beta b_1).$$

Thus we obtain **A2** by considering the two cases sgn $\alpha a_1 = $ sgn βb_1 and sgn $\alpha a_1 = -$sgn βb_1.

Axiom **V1** follows immediately from the definitions. Let us now prove that **V2** is satisfied.

From $(g/A, B) = -1$ it follows by (31) that sgn $\{f(\mathfrak{a}, \mathfrak{g})f(\mathfrak{b}, \mathfrak{g})f(\mathfrak{a}, \mathfrak{b})\} = -1$, i.e., $f(\mathfrak{a}, \mathfrak{g})f(\mathfrak{b}, \mathfrak{g})f(\mathfrak{a}, \mathfrak{b}) < 0$. If we now set $\mathfrak{c} = f(\mathfrak{b}, \mathfrak{g})\mathfrak{a} - f(\mathfrak{a}, \mathfrak{g})\mathfrak{b}$, then

$$f(\mathfrak{c}, \mathfrak{c}) = f(\mathfrak{b}, \mathfrak{g})^2 f(\mathfrak{a}, \mathfrak{a}) + f(\mathfrak{a}, \mathfrak{g})^2 f(\mathfrak{b}, \mathfrak{b}) - 2f(\mathfrak{a}, \mathfrak{g})f(\mathfrak{b}, \mathfrak{g})f(\mathfrak{a}, \mathfrak{b}) > 0,$$

since $f(\mathfrak{a}, \mathfrak{a}), f(\mathfrak{b}, \mathfrak{b}) > 0$. But we also have $f(\mathfrak{c}, \mathfrak{g}) = 0$, so that $C = R\mathfrak{c}$ is the intersection of g with the line joining A and B.

In order to verify the axiom of continuity, we proceed in the same way as in the Euclidean plane. We show that every line k can be mapped, with preservation of the order, onto the positive real numbers, where the order relation is defined by (9) and (10), after choice of the points A and B as in §2. The proof will be omitted here.

In order to introduce the congruence relation, we first define for two points X, Y the function

$$d(X, Y) = \frac{f(\mathfrak{x}, \mathfrak{y})^2}{f(\mathfrak{x},\mathfrak{x})f(\mathfrak{y},\mathfrak{y})}.^{22}$$

Now let $X = R\mathfrak{x}$, $Y = R\mathfrak{y}$, $U = R\mathfrak{u}$, and $W = R\mathfrak{w}$ be four points in the hyperbolic plane $\mathfrak{E}$. Then $XY \equiv UW$ is equivalent to $d(X, Y) = d(U, W)$. It is obvious that this congruence is well defined.

For the proof of the axioms of consistency we require:

(32)   *For two distinct points $A = R\mathfrak{a}$ and $B = R\mathfrak{b}$*

$$d(A, B) = \frac{f(\mathfrak{a}, \mathfrak{b})^2}{f(\mathfrak{a},\mathfrak{a})f(\mathfrak{b},\mathfrak{b})} > 1.$$

Proof.   By straightforward calculation we verify that

$$f(\mathfrak{a}, \mathfrak{a})f(\mathfrak{b}, \mathfrak{b}) + f(\mathfrak{a}, \mathfrak{b})^2 = f(\mathfrak{g}, \mathfrak{g})$$

with

$$\mathfrak{g} = (a_2 b_3 - a_3 b_2, a_1 b_3 - a_3 b_1, a_2 b_1 - a_1 b_2).$$

[22] The ordinary distance function λ in the hyperbolic metric is obtained by setting $\lambda(X, Y) = |\cosh^{-1}(\sqrt{d(X, Y)})|$.

Since $f(\mathfrak{a}, \mathfrak{g}) = 0$, it follows from (28) that $f(\mathfrak{g}, \mathfrak{g}) < 0$ and therewith the assertion.

We now prove that **V3** is satisfied. Let $C = R\mathfrak{c}$, and let $P = R\mathfrak{p}$ be a point of g distinct from C. Every point of g distinct from C can be represented in the form $X = R(\lambda\mathfrak{c} + \mathfrak{p})$. We must show that the equation $d(X, C) = d(A, B)$ has exactly two solutions $D_i = R(\lambda_i\mathfrak{c} + \mathfrak{p})$. From (32) we have $d(A, B) = \kappa > 1$. Consequently the equation, quadratic in λ,

(33) $$d(X, C) = \frac{f(\mathfrak{c}, \lambda\mathfrak{c} + \mathfrak{p})^2}{f(\mathfrak{c},\mathfrak{c})f(\lambda\mathfrak{c} + \mathfrak{p}, \lambda\mathfrak{c} + \mathfrak{p})} = \kappa$$

has a positive discriminant, and therefore two real solutions λ_1 and λ_2. These solutions correspond to the two points D_1 and D_2 in g, since $f(\lambda_i\mathfrak{c} + \mathfrak{p}, \lambda_i\mathfrak{c} + \mathfrak{p}) > 0$ follows from (33). The other axioms are verified by similar calculation.

6.3.   As a third example let us mention a model of the elliptic plane $\mathfrak{E}$.[23] Let $\mathfrak{E}$ satisfy the incidence axioms of a projective plane (cf. §4). Then (26) shows that $\mathfrak{E}$ cannot be a metric plane in the sense defined here. The plane $\mathfrak{E}$ is ordered and continuous. The congruence axioms are only partly satisfied.

If V and $\overline{V}$ are two three-dimensional vector spaces over the real numbers and if $R\mathfrak{x}$ and $R\mathfrak{y}$ are one-dimensional subspaces of V and $\overline{V}$, respectively, with $\mathfrak{x} = (x_1, x_2, x_3) \in V$ and $\overline{\mathfrak{y}} = (y_1, y_2, y_3) \in \overline{V}$, let us set

$$\mathfrak{x}\overline{\mathfrak{y}} = x_1 y_1 + x_2 y_2 + x_3 y_3.$$

Then to every vector $\mathfrak{x} = (x_1, x_2, x_3) \neq 0$ in V we assign the vector $\mathfrak{x}_0$ according to the following rule:

$$\mathfrak{x}_0 = \begin{cases} (1, x_2 x_1^{-1}, x_3 x_1^{-1}), & \text{if } x_1 \neq 0, \\ (0, 1, x_3 x_2^{-1}), & \text{if } x_1 = 0, x_2 \neq 0, \\ (0, 0, 1), & \text{if } x_1 = x_2 = 0. \end{cases}$$

For the points of the elliptic plane let us take the one-dimensional subspaces of V, and for the lines, the one-dimensional subspaces of $\overline{V}$.

A point $X = R\mathfrak{x}$ with $\mathfrak{x} = (x_1, x_2, x_3)$ and the line $g = R\overline{\mathfrak{g}}$ with $\overline{\mathfrak{g}} = (g_1, g_2, g_3)$ are incident if and only if $\mathfrak{x}\overline{\mathfrak{g}} = 0$.

If $X = R\mathfrak{x}$ is a point and $g = R\mathfrak{g}$ is a line with $g \not I X$, let $S_+ = \{X = R\mathfrak{x}; \mathfrak{x}_0\overline{\mathfrak{g}}_0 > 0\}$ and $S_- = \{X = R\mathfrak{x}; \mathfrak{x}_0\overline{\mathfrak{g}}_0 < 0\}$ be the two sides of g and let $\overline{g} = \{X = R\mathfrak{x}; \mathfrak{x}_0\overline{\mathfrak{g}}_0 = 0\}$.

Definition of the congruence relation $\equiv$: Let $X = R\mathfrak{x}$, $Y = R\mathfrak{y}$, $U = R\mathfrak{u}$, and $W = R\mathfrak{w}$ be four points in the elliptic plane $\mathfrak{E}$. Then $XY \equiv UW$ is

---

[23]A general definition of elliptic geometry is to be found in [4]. In II5, §3 there is a detailed description of the well-known spherical model of elliptic geometry.

equivalent to

$$\frac{(\mathfrak{x}\mathfrak{y})^2}{(\mathfrak{x}\mathfrak{x})(\mathfrak{y}\mathfrak{y})} = \frac{(\mathfrak{u}\mathfrak{w})^2}{(\mathfrak{u}\mathfrak{u})(\mathfrak{w}\mathfrak{w})}.$$

The foregoing properties for $\mathfrak{E}$ are easily verified by calculation.

7. Construction of Models by Isomorphism

Klein, Poincaré, and other mathematicians have given visualizable models of Euclidean and hyperbolic geometry. By Theorem 5 each of them is isomorphic to the Euclidean or hyperbolic geometry in §6.

Let us now construct some of these well-known models, by applying suitable mappings to the models in §6. For this purpose we first establish that a one-to-one mapping φ of the set $\mathfrak{P}_{\mathfrak{E}}$ of points of a model $\mathfrak{E}$ onto an arbitrary set M provides us with a new model if we make the following definition: the elements of M are called points. If $g \in \mathfrak{G}_{\mathfrak{E}}$ is a line in $\mathfrak{E}$ and $\bar{g} = \{X \in \mathfrak{P}_{\mathfrak{E}};\, X \mathrel{I} g\}$, then the subset $\varphi(\bar{g}) \subset M$ is called a line. The relation $\varphi(X) \mathrel{I} \varphi(\bar{g})$ holds if and only if $X \mathrel{I} g$ holds. Let $(\varphi(\bar{g})/\varphi(A),\, \varphi(B)) = (g/A,\, B)$ (which defines the partition into sides), and finally let $\varphi(A)\varphi(B) \equiv \varphi(C)\varphi(D)$ if and only if $AB \equiv CD$, which defines the congruence in M.

Now let $\mathfrak{E}$ be the Euclidean geometry in §6, let $\mathfrak{P}_{\mathfrak{E}}$ be the set of points of $\mathfrak{E}$, let R_3 be the three-dimensional analytic geometry over the real numbers, let $R_2 = \{(x_1, x_2, 0);\, x_1, x_2 \in R\}$, let $S_2 = \{(x_1, x_2, x_3) \in R_3;\, x_1^2 + x_2^2 + x_3^2 = 1\}$ be the surface of the unit sphere with center at the origin, and let

$$S_2^* = \{(x_1, x_2, x_3) \in S_2;\, (x_1, x_2, x_3) \neq (0, 0, 1)\}$$

be the unit sphere without the "north pole" N. Let

$$\varphi_1 : \begin{cases} \mathfrak{P}_{\mathfrak{E}} \to R_2 \\ (x_1, x_2) \to (x_1, x_2, 0) \end{cases}$$

and $\varphi_2 : R_2 \to S_2^*$ be the stereographic projection $P \to p \cap S_2^*$, where p denotes the set of points on the line joining P with $N = (0, 0, 1)$. The mapping $\varphi_1 \varphi_2$ of $\mathfrak{P}_{\mathfrak{E}}$ onto S_2^* is one-to-one and provides a further model $\mathfrak{E}^*$ of Euclidean geometry on the surface of the unit sphere. If K is a circle in S_2 through N, then let $K^* = K \cap S^*$. The lines in $\mathfrak{E}^*$ are the sets K^*.

Let the hyperbolic geometry in §6 be denoted by $\mathfrak{H}$ and the set of its points by $\mathfrak{P}_{\mathfrak{H}}$. Let $D_2 = \{(x_1, x_2);\, x_1^2 + x_2^2 < 1\}$ and

$$\varphi_3 : \begin{vmatrix} \mathfrak{P}_{\mathfrak{H}} & \to & D_2 \\ R(x_1, x_2, x_3) & \to & \left(\dfrac{x_2}{x_1}, \dfrac{x_3}{x_1}\right) \end{vmatrix}.$$

It is easy to verify that this mapping is also one-to-one. The lines in the model corresponding to φ_3 are the open intervals[24] bounded by two distinct points of the circle $x_1^2 + x_2^2 = 1$.

Let $S_2^- = \{(x_1, x_2, x_3) \in S_2 ; x_3 < 0\}$ be the surface of the lower hemisphere and

$$\varphi_4 : \begin{cases} D_2 & \to S_2^- \\ (x_1, x_2) & \to (x_1, x_2, -\sqrt{1 - x_1^2 - x_2^2}) \end{cases}.$$

The mapping φ_4 is one-to-one. The mapping $\varphi_3 \varphi_4 \varphi_2^{-1} \varphi_1^{-1} : \mathfrak{P}_{\mathfrak{H}} \to D_2$ provides a well-known model of hyperbolic geometry. Here the lines are the circular arcs in D_2 that are perpendicular to the unit circle.

Finally, let us mention another model of hyperbolic geometry due to Poincaré, namely, the model corresponding to the mapping $\varphi_3 \varphi_4 \varphi_5$; here φ_5 is the stereographic projection of S_2^- into the plane $E : x_2 = -1$, from the point $(1, 0, 0)$. The points of this model fill up one side E' of a line g in E. The lines of this model are the semicircles on the side E' perpendicular to g.

Bibliography

[1] ARTIN, E.: Calculus and analytic geometry. New York 1959.
[2] HILBERT, D.: The foundations of geometry. Translated from the German by E. J. Townsend. Reprint edition. The Open Court Publishing Co., La Salle, Ill. 1959.
[3] KARZEL, H.: Metrische Geometrie. Hamburg 1963. (Lecture notes.)
[4] PODEHL, E., and REIDEMEISTER, K.: Eine Begründung der ebenen elliptischen Geometrie. Abh. math. Sem. Univ. Hamburg 10 (1934), 231–255.
[5] SPERNER, E.: Die Ordnungsfunktion einer Geometrie. Math. Ann. 121 (1949), 107–130.

Bibliography Added in Translation
COXETER, H. S. M.: Non-Euclidean geometry. 3rd ed. University of Toronto Press, Toronto 1957.
COXETER, H. S. M.: Introduction to geometry. John Wiley and Sons, Inc., New York-London 1961.
MODENOV, P. S., and PARHOMENKO, A. S.: Geometric transformations (translated from the Russian by M. B. P. Slater). 2 vols. Academic Press, New York and London 1965.
YAGLOM, I. M.: Geometric transformations (translated from the Russian by A. Shields). Random House, New York 1962.

[24]The geometry of the Klein model is discussed in detail in II5, §3.

Geometric Constructions

I. Introduction

1.   By a geometric construction we mean a problem of the following kind.

From "elements" given or already constructed (points, segments, lines, angles, circles, and so forth) other "elements" are to be derived, under these rules:

a)   Only certain well-defined instruments are to be used in each case.

b)   Each of the instruments can be used only in a preassigned way.

c)   The construction must come to an end in a finite number of steps.

The important requirement c) is always assumed, though sometimes without explicit mention. Also, we always assume that we are dealing with the classical Euclidean plane (the continuous Euclidean plane in the sense of the preceding chapters), which we shall occasionally extend to the projective plane.

2.   Besides asking how a particular problem can be solved with a preassigned instrument, it is of great interest to ask the general question: which problems can be solved with the given instrument; i.e., how can we characterize the class of problems solvable with the instrument?

Such a characterization is obtained from analytic geometry. Since segments and lines are determined by two points, and angles and circles by three points, we may consider both the given and the desired elements to be points only. Thus every problem of construction can be formulated: from given points, with the use of given instruments, to construct other points related in specified ways to the given points.

By introducing a Cartesian coordinate system on our drawing paper we can characterize both the given and the desired points by means of their coordinates. Then every construction problem leads to an analytic relation between the coordinates of the given points and those of the points desired.

Among these analytic expressions, a decisive role is played, in the examples discussed below, by extraction of square roots. In other words, we deal with expressions constructed from the coordinates of the given points by means of the four elementary operations and the extraction of square roots, where it is understood that these operations can be carried out only a finite number of times.

The present chapter deals only with constructions in the real field. Consequently, any radicand that occurs will be assumed to be nonnegative.

2. Constructions with Straightedge and Compass

2.1. *Definition and Range of Constructible Problems*

1. Definition. In a "construction with straightedge and compass" these two instruments can be used only finitely often and only in the following basic ways:

A) Through two distinct points, draw the (straight) line determined by them.

A') For two nonparallel lines, construct their point of intersection.

B) About a given point as center, draw a circle with radius equal to the distance between two given points.

The points mentioned in these rules either may have been given originally or may have been derived from the original points by allowable constructions.

These restrictions on the use of straightedge and compass are essential, since both instruments can be used in other ways.

2. Range of constructible problems. It is very easy to give an algebraic description of the range of problems, already indicated by Plato, that are solvable in this way by straightedge and compass: a point is constructible if and only if its coordinates can be expressed by means of square root signs in terms of the coordinates of the given points. (Here as always we are assuming a Cartesian coordinate system.)

The proof runs as follows. Certainly the expressions $a \pm b, a \cdot b, a : b, \sqrt{a}$ (where a and b are the coordinates of given points with $a > 0, b \neq 0$) can be constructed with straightedge and compass, and consequently also any more complicated expression involving a finite number of repetitions of these five operations.

On the other hand the determination of points of intersection of lines and circles with one another can lead only to the solution of linear and quadratic equations, with coefficients that can be expressed rationally in terms of the coordinates of given points or of points already constructed.[1]

---

[1]An elementary detailed proof is given in Breidenbach [4].

2.2. *Methods of Construction*

We distinguish two methods of construction for geometric problems: the method of geometric loci and the method of algebraic analysis.[2]

1. The method of loci. Let us assume that the solution of the problem has already been reduced to the construction of a sequence of points. Then every point P in the sequence is determined by the fact that it satisfies (at least) two conditions arising from the problem. If we retain only the first of these conditions, we obtain a set of points which (in general) make up a curve, a so-called geometric locus L_1. If we retain only the second condition, we obtain a second locus L_2. The intersections of the two curves (and only such points) can be considered as points P. If the curves L_1 and L_2 are constructible straight lines or circles, then the point P is constructible.

Examples. a)   Let there be given a circle and two points A and B in the interior. The problem is to inscribe in the circle a right-angled triangle whose sides pass through the points A and B.

Construct the circle with AB as diameter.

b)   On a circular billiard table, two balls A and B are placed on a diameter, on different sides of the midpoint M. How must one ball be struck in order to hit the other on the rebound?

Apart from the trivial solution, namely, that A is struck in the direction of the diameter BA, there is a second solution; namely, construct the fourth harmonic point N to A, B, M and draw the semicircle on MN.

2.   The method of algebraic analysis. From the given magnitudes we attempt to calculate a magnitude (e.g., a line segment) which immediately determines the desired result, and then we construct the geometric expression of the magnitude.

Examples. a)   Given a circle (with circumference U) in which a triangle XYZ is to be inscribed. Let there be given (in addition to the circle) only the midpoints of the arcs YZ, ZX, XY, which we shall call A, B, C.

Denote the arcs AB, BC, CA by c, a, b, and the arcs $AY = AZ$ by x, $BZ = BX$ by y, and $CX = CY$ by z. Then it is at once clear that $x + y + z = \dfrac{U}{2}$. But also, for example, $y + z = a$. Thus $x = \dfrac{U}{2} - a; y = \dfrac{U}{2} - b; \; z = \dfrac{U}{2} - c$.

b)   In an isosceles triangle (with vertex C) draw the three altitudes, intersecting at H. Let a be the length of the upper segment of the altitude through

---

[2]Compare, e.g., Petersen, Julius, Methoden und Theorien zur Auflösung geometrischer Konstruktionsaufgaben (Danish 1866), German translation by Fischer-Benzon, Copenhagen 1879. Cf. also B. Kerst, Methoden zur Lösung geometrischer Aufgaben, 2. Aufl. Leipzig 1925.

C, and let b be the length of the upper segment on each of the other two altitudes. Given a and b, construct a triangle.

Let H' be the reflection of H in AB and join H' to A. Let $CH' = x$, $HH' = y$. Then $\angle CAH' = 90°$ and

$$x - y = a, \quad x \cdot \frac{y}{2} = b^2,$$

so that

$$y = -\frac{a}{2} \pm \sqrt{\frac{a^2}{4} + 2b^2},$$

which is a constructible expression.

3. Discussion of the solution. Many construction problems have a solution only if the given elements satisfy certain conditions. If these conditions are included in the statement of the problem (for which purpose it is usually necessary to know the solution in advance), the solver is required to discover only the solution. But he will find the problem much more interesting and instructive if he himself must determine the conditions under which there will be no solution, one solution, or more than one solution.

The following examples, like the ones given above, are so stated that the reader must himself determine the conditions for solvability.

Exercises. 1. On the extension of a given segment AB determine a point X such that $AB \cdot AX = BX^2$,

2. Draw a transversal through a given triangle in such a way that it bisects the perimeter and the area of the triangle.

3. Construct a rectangle whose area is k^2 (where k is a given segment) and whose sides are in the ratio of 2 to 3.

4. Construct a right-angled triangle given the altitude to the hypotenuse and the sum of the legs.

5. Given the hypotenuse, construct a right-angled triangle such that one of the legs is equal to the nonadjacent segment of the hypotenuse cut off by the altitude.

6. Through the vertex A of a square $ABCD$ draw a line cutting the side BC at X and the side CD produced at Y in such a way that XY is equal to a given segment.

7. Given a line g, a segment s, and points A, B not on g, draw the circle through A, B cutting off a chord of length s on g.

8. Given their radii, draw two circles in such a position that the common outer tangents are twice as long as the common inner one.

9. In a given circle inscribe the cross formed by five equal squares.

10. Through a given point P inside a triangle draw the line bisecting the area of the triangle.

2.3. Constructions by Means of Elementary Transformations

Many beautiful problems in construction can be solved in a simple and elegant way by means of transformations. Let us consider two examples.

1. Parallel translation. a) Let there be given two intersecting straight

lines g and h and an arrow (vector) $\mathfrak{a} = \overrightarrow{AB}$. The arrow is to be inserted between the lines in such a way that A lies on g and B on h.

In a trial figure the arrow A_0B_0 suggests a parallel translation from A_0 to B_0. If we subject the line g to this translation, its image g' provides a second locus for the point B_0.

b) It is required to construct a trapezoid from the four sides a, b, c, d with $(a \parallel c)$.

Beginning with the base $a = AB$, we describe about A and B the circles with radii d and b, respectively. Then between these two circles we are required to insert the side c, given in length and direction. This side c determines the two translations, one of which is in the direction from A to B. The image of the circle about A under this translation is a second locus for C.

Exercises. 1. Given two circles (or a circle and a line) insert a given arrow (vector) between them.

2. Construct a quadrilateral, given a, b, d, α, β.

3. Given two intersecting strips (a strip is the figure consisting of two parallel lines), draw through a given point P a line on which the strips cut off equal segments.

4. Given an angle and a point P in its interior, draw a line through P such that P is the midpoint of the segment intercepted by the sides of the angle.

5. Draw three lines, one through each of three given points, in such a way as to form two strips of equal width.

6. Draw a strip of given width in such a position that its sides pass through two given points.

7. Given three points A, B, C and a segment s, draw a line g through C such that A and B are on opposite sides of g and the sum of their distances from g is equal to s.

8. In exercise 7 draw the line through C in such a way that A and B are on the same side of g and the difference between their distances from g is equal to s.

9. Let P be a point on the circumference of a given circle, and let s be a chord. Through P draw a chord bisected by s.

10. Through given points P, Q inside a circle draw parallel chords of equal length.

2. Reflection. a) Let there be given a line g and two points A and B on the same side of g. It is required to find on g a point X such that $AX + XB$ is a minimum.

Join A to the mirror image of B (in g).

b) Let there be given a line g and two points A and B on the same side of g. It is required to find a point X on g such that AX and BX make the same angle with g as with BX.

The mirror image of B in the unknown line AX can be drawn (it is the intersection of g with the circle about A with radius AB). The mirror image B', taken together with B, determines the axis AX of the reflection.

Exercises. 1. Given a point P in the interior of an angle, draw the triangle PAB of minimum perimeter, where A is on one side of the triangle and B on the other.

2. Given a triangle ABC with a fixed point Z on AB, consider the inscribed triangles with Z as fixed vertex. Draw the triangle with least perimeter.

3. Draw two tangents, one to each of two given circles, intersecting on a given line and making the same angle with the line.

4. Given a circle and an exterior point C. On the diameter through C determine a point X such that CX is equal to the tangent from X.

5. Let A, B, C be points on the line g with $AB \neq BC$, and let a be a line through A. On a determine a point X at which AB and BC subtend equal angles.

6. Given two parallel lines g, h and two points A, B between them, draw the isosceles triangle with base on g, vertex on h, and sides through A and B.

7. Given the four sides, construct a quadrilateral in which one of the diagonals bisects one of the angles.

8. Given two lines l and g and a point F on g, find on g the points X equidistant from the point F and the line l.

9. Given three lines a, a', and s, find the points A on a and A' on a' such that s is the right bisector of AA'.

10. Given a line g and two circles on the same side of g, find a point X on g from which tangents to the circles make equal angles with g.

3. Rotation. a) Let there be given a triangle ABC with a point Z on AB. In the given triangle it is required to inscribe an equilateral triangle with Z for one of its vertices.

Rotate AC about Z through $60°$. The image of AC cuts BC in a second point of the desired equilateral triangle.

b) Let there be given a circle and the points P and Q in its interior. It is required to draw two chords of equal length through P and Q perpendicular to each other.

Rotate P about the center of the circle through an angle of $90°$. The image P', together with the point Q, determines one of the two chords.

Exercises. 1. Given a segment a, a circle, and a point P in its interior, draw through P a chord of length a.

2. In a given circle inscribe an equilateral triangle, one of whose sides passes through a given point P in the interior of the circle.

3. Given a circle, a line, and an arbitrary triangle, move the triangle into such a position that two of its vertices lie on the circle and the third vertex lies on the line.

4. Draw an equilateral triangle with a vertex on each of three given parallel lines.

5. Draw an isosceles triangle with a vertex on each of three given concentric circles.

6. Given two circles and two segments s_1, s_2. With given radius draw a circle which from the given circles cuts off chords of length s_1, s_2, respectively.

7. In a given circle inscribe a chord of given length divided in a given ratio by a given secant.

8. Draw a circle, with given radius, touching two given circles.

9. Given two intersecting lines g_1, g_2 and a point P, determine points X_1 on g_1 and X_2 on g_2 such that $PX_1 = PX_2$ and $\not\prec X_1PX_2$ is equal to a given angle ϕ.

10. Draw a square such that its four sides, produced if necessary, pass through four given points A, B, C, D, respectively. Note that if we draw $BC' \perp AC$, with C' on the side of the square through D, then $BC' = AC$.

4. Rotation about a point through 180°.

a) Two circles intersect at a point S. It is required to draw a line through S on which equal chords will be cut off in each of the circles.

In addition to the trivial solution there is a second one: Rotate one of the circles through 180° about the point S. The resulting image determines with the second circle a second point for the desired line.

b) Let there be given a quadrilateral and a point M (in its interior). In the quadrilateral it is required to inscribe a parallelogram with midpoint M.

Rotate the quadrilateral through 180° about M. Each side of the quadrilateral, together with the image of its opposite side, determines a vertex of the parallelogram.

Exercises. 1. Through a given point P in the interior of a given angle, draw a line such that the segment intercepted by the sides of the angle is bisected at P.

2. Given two circles and a point P, draw through P a segment that joins some point X of the first circle (circumference) to some point Y of the second circle and is bisected at P.

3. Given three concentric circles, draw a line such that a segment intercepted by the first and second circles is equal to the neighboring segment intercepted by the second and third circles.

4. Given two opposite vertices, draw a parallelogram such that the other two vertices lie on a given circle.

5. Given a point A on a line a, a point B on a line b parallel to a, a segments s and a point Q, draw a line through Q cutting a at X and b at Y in such a way that $AX + BY = s$.

6. Through an intersection of two given circles, draw a secant such that the resulting chords are equal.

7. Through an intersection of two given circles, draw a secant such that the difference of the resulting chords is equal to a given segment.

8. Through an intersection of two given circles, draw a secant such that the sum of the resulting chords is equal to a given segment.

9. Given a strip, one of its midpoints M, an arbitrary point P, and a line g intersecting the strip. With straightedge alone, determine the image of g under rotation through 180° about M.

10. Given a strip, one of its midpoints M, and an arbitrary point P. With straightedge alone, determine the image of P under rotation through 180° about M.

5. Glide reflection.

Let there be given a line g and a vector $\mathfrak{s} \parallel g$. If we reflect a figure F in g so as to produce F' and then subject F' to the translation defined by s so as to produce the image F'', we say that F'' arises from F by a glide reflection.

a)   Let there be given a line g, two points A and B on opposite sides of g, and also a line segment s. On g it is required to determine points X and Y such that $XY = s$ and AX and BY form equal (but oppositely oriented) angles with g.

Reflect A in g to the point A', and translate A' parallel to g through the distance s to the point A''. The line BA'' provides the desired point Y on g.

b)   Given two congruent circles it is required to take K_1 into K_2 by means of a glide reflection. Here the magnitude s of the translation is given, and it is required to construct the axis of the glide reflection.

Join the midpoints M_1 and M_2 of the two circles. The axis passes through the midpoint of M_1M_2. Erect the semicircle on M_1M_2 and let it intersect the circle about M_1 with s as radius at the point N. The segment M_1N provides the direction of the desired axis.

6. Dilation. a)   In a triangle it is required to inscribe a square in such a way that one side of the square lies on a side of the triangle.

In the angle γ, say, inscribe a square with two of its sides parallel to the side c of the triangle, and then dilate this subsidiary square from C.

b)   Let there be given two lines a and b and a point O. In general, a ray through O will cut a and b in two points A and B. It is required to construct that ray for which $OB = 2 \cdot OA$.

Dilate the line a from O with magnification 2. The image A' is a second locus for B.

Exercises.   1.   Through a given point O in the interior of a given angle, draw a line cutting the sides of the angle at A, B in such a way that $OB:OA = -1$ ($-\frac{1}{2}$, $-\frac{1}{3}$, -0.7).

2.   Given a circle, an exterior point O, and a line g between O and the circle, draw a line through O meeting the line at G and the circle (for the first time) at K in such a way that $OK:OG = 2$ (2.3, 2.7).

3.   Through a given point O exterior to a given circle, draw a line meeting the circle first at A, then at B in such a way that $OB:OA = 2$ (2.2, 3).

4.   Through a point P in the interior of a circle, draw a chord divided in a given ratio at P.

5.   Through an intersection of two circles draw a line such that the resulting chords are in a given ratio.

6.   In a given semicircle inscribe a) a square with one side on the diameter, b) a rectangle with sides in the ratio 2:3, c) a rectangle whose diagonals form an angle of $60°$.

7.   Given a circle and two of its radii, draw the chord trisected by the given radii.

8.   Draw the circles (two of them) through a given point P and tangent to each of two given intersecting lines.

9. Given an angle with sides g, h and an interior point P, determine on g a point G equidistant from the line h and the point P.

10. Given a circle, a line, and a point on the line, draw the circle that is tangent to the given circle, and tangent to the given line at the given point.

7. Dilative rotation. a) Let there be given two lines a and b, a point O, and a triangle $A_0B_0C_0$. It is required to construct a triangle ABC that is similar to $A_0B_0C_0$ and such that the vertex A lies on a, the vertex B on b, and the vertex C at the point O (Fig. 1).

Construct a triangle $A'B'O$, similar to $A_0B_0C_0$, which satisfies only two of the required conditions. Then consider a sequence of similar triangles with a fixed point at O. If A' traverses the line a, then B' (by the properties of a dilative rotation) also traverses a straight line h. The angle between OB' and h is equal to the angle between OA' and a. Thus the line h can be constructed. The intersection of h with b provides the point B of the desired triangle.

b) For a quadrilateral circumscribed about a circle we are given the magnitudes of the angles and the position of the opposite vertices A and C. It is required to construct the quadrilateral (Fig. 2).

It is easy to draw a quadrilateral $AB'C'D'$ with the prescribed angles in such a position that one of its vertices lies at the prescribed point A. Let us now make a dilative rotation with A as center, $\angle C'AC = \varphi$ as angle of rotation and $AC:AC'$ as magnification factor.

Exercises. 1. Construct a triangle ABC that is similar to a given triangle $A_0B_0C_0$, has one vertex on a given circle K_a, one vertex on a given circle K_b, and one vertex at a given point C.

2. Given two parallelograms, inscribe in the first one a parallelogram similar to the second.

3. Given two circles K_a, K_b intersecting at C (and at some second point) and a triangle $A_0B_0C_0$, draw a triangle ABC similar to $A_0B_0C_0$ with A on K_a, B on K_b, and the third vertex at C.

4. In a given circle inscribe a triangle similar to a given triangle and with one vertex at a given point P on the circumference of the circle.

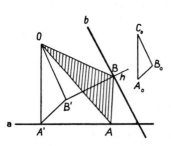

Fig. 1

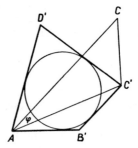

Fig. 2

5. Given a triangle ABC and a fixed point Z on the side AB, inscribe in it a triangle XYZ similar to another given triangle $A_0B_0C_0$.

6. Given points A, B on lines a, b, respectively, draw a line g intersecting a at X and b at Y in such a way that $AX{:}BY$ is equal to a given ratio $m{:}n$ and XY is equal to a given segment s.

7. Given points A, B on lines a, b, respectively, through a given point S draw a line g intersecting a at X and b at Y in such a way that $AX{:}BY$ is equal to a given ratio $m{:}n$.

8. Given two triangles ABC and DEF draw a triangle $D_0E_0F_0$, congruent to DEF, with D_0 on BC, E_0 on AC, F_0 on AB, where if necessary the sides of ABC may be produced.

9. Replace "congruent" in example 8 by "similar" and add the requirement that one side of $D_0E_0F_0$ must pass through a given point P in the interior of ABC.

10. Given two quadrilaterals, inscribe in the first one a quadrilateral similar to the second.

2.4. *Constructions by Means of Inversion*

1. We might next consider more general affine and projective mappings. But they are adequately dealt with, not only in textbooks on descriptive geometry (at least so far as they are useful for engineering drawing), but also in most of the more recent textbooks on elementary geometry, so that we shall not consider them here. On the other hand, let us give some details about inversion.[3]

2. Definition. Let there be given a circle with center O and radius r (the inversion circle). A point P in the plane of the circle determines a point P' (its image under inversion) under the two requirements: a) O, P and P' lie on the same halfline originating in O; b) $OP \cdot OP' = r^2$ (Fig. 3).

This transformation is involutory; that is, if P' is considered as the original point, then P is its image. The mapping (except for O) is one-valued. The point O itself has no image and is not the image of any other point. The interior of the circle of inversion is mapped onto the exterior, and conversely. The circle itself is pointwise fixed.

3. Theorem: *The quadrilateral consisting of two points and their images under inversion is concyclic* (Fig. 3).

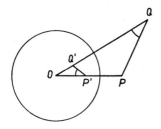

Fig. 3

[3]See also Schmidt [7].

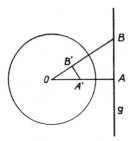

Fig. 4

For if P and Q are the two points and P' and Q' are their images, then

$$OP \cdot OP' = OQ \cdot OQ' \; (=r^2), \quad \text{so that} \quad OP : OQ = OQ' : OP'.$$

But this means that the triangles OPQ and $OQ'P'$ (note the order of the letters) are similar, from which the theorem follows.

The circle through the four points intersects the circle of inversion at right angles (theorem of secants and tangent) and is mapped into itself.

4.   The lines through O are mapped into themselves. Every line g that does not pass through O is mapped into a circle through O. The tangent to this circle at the point O is parallel to the line g (Fig. 4).

For let the perpendicular from O to g meet g at A. Let B be another point on g, and let A' and B' be the images under inversion. Then by No. 3 we have $\angle OB'A' = \angle OAB = 90°$, which holds for every point $B \neq A$ of the line. Thus the image of g is a circle through O. Since this circle is its own mirror image in OA, the tangent at O is parallel to g.

5.   The circles through O are mapped into lines. The circles that do not pass through O are mapped in circles (Fig. 5).

Let us consider the case in which the original circle lies entirely outside the circle of inversion. (The other cases may be treated similarly.) The points A and B, together with their images A' and B', lie on a halfline from O, and the point P, together with its image P', lie on another halfline from O. By No. 3 we have

$$\angle OPA = \angle OA'P' = \alpha; \quad \angle OPB = \angle OB'P' = \beta,$$

and thus $\angle APB = \beta - \alpha$ and $\angle A'P'B' = \beta - \alpha$. Thus the angles are equal in magnitude but oppositely oriented. If P traverses a circle with AB as diameter, $\angle APB$ remains constantly equal to $90°$. But $\angle A'P'B'$ has the same angle for every position of P', so that P' also traverses a circle with $A'B'$ as diameter.

6.   Construction problems (Examples). a) Let there be given a point O, a line segment k, and two lines a and b. It is required to draw lines through O cutting the given lines at A and B in such a way that $OA \cdot OB = k^2$.

Let the circle about O with radius k be taken as a circle of inversion and map the

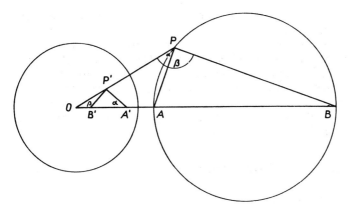

Fig. 5

line b into the circle b'. Each of the lines joining O to the intersections of a with b' is a desired line.

b)   Let there be given two circles K_1, K_2 and a point P. It is required to draw the circle K that passes through P and is tangent to the given circles.

Take P as the center of an inversion with a radius such that the circle of inversion is perpendicular to K_1 (this choice is convenient but not necessary). In this inversion K_1 is mapped into itself (see No. 3), and K_2 is mapped into a circle K_2'. The desired circle K is mapped into a line tangent to K_1 and K_2'. Thus it is only necessary to draw the common tangents of these two circles and to construct their images under the inversion.

Exercises.   1.   Given two circles with unequal radii, construct the circle of inversion under which the two given circles are transformed into each other. (Consider all possible positions for the given circles.)

2.   Two circles intersect (or touch) each other. Find an inversion taking both of them into straight lines.

3.   Draw the circle touching three given circles that have a point O in common.

4.   Given a circle with diameter AB and an exterior point P, draw the lines PA, PB intersecting the circle again at A', B'. Prove that the given circle is orthogonal to the circle through P, A', B'.

5.   Given two circles with centers A, B intersecting orthogonally at X, Y, and an arbitrary point O on the circumference of the first circle (about A). If OX, OY cut the second circle again at X', Y', prove that $X'Y'$ is a diameter of the second circle orthogonal to the diameter through O of the first circle.

6.   In a given circle let AB and XY be two mutually perpendicular chords intersecting at S. Prove that the altitude of the triangle SAX extended through S is a median of the triangle SBY.

7.   Given three (noncollinear) fixed points P, A, B and a point X variable on the segment AB. Prove that the angle of intersection of the circle through A, P, X with the circle through B, P, X is constant as X varies.

8.   Find an inversion that will transform two given circles of different radii into two circles with the same radius.

9.   Given two fixed circles with centers A, B and a variable tangent to the first circle (about A), cutting the second circle at the (variable) points X, Y. Prove that the (variable) circle through B, X, Y remains tangent to a certain fixed circle.

10.   Given a fixed circle with center O, a fixed line g, and an angle of fixed magnitude ϕ rotating about O (that is, with its vertex at O), let the sides of the rotating angle intersect g at the (variable) points X, Y. Prove that the (variable) circle through O, X, Y remains tangent to a fixed circle.

3.   Constructions with Straightedge

3.1.   *Constructions with Straightedge Alone*

1.   Possible constructions. In this section the straightedge is used exclusively for joining two given points or for constructing the intersection of two nonparallel lines.

The complete range of solvable problems is determined by the following theorem: the points and lines that can be constructed with straightedge only from given points and lines consist of those (and only those elements) that can be expressed rationally (in a given projective coordinate system) in terms of the coordinates of the given elements.[4]

2.   Examples. Thus the possibilities for construction in any given case depend on the given elements. If we are permitted to make use of them alone (and of elements constructed from them), the range of solvable problems is smaller than if we may also use arbitrary subsidiary elements (i.e., arbitrarily chosen points and lines). Here are some examples.

a)   Let there be given three points A, B, C on a line. It is required to construct a fourth point D such that A, B form a harmonic set with C, D.

If we are not allowed to use any subsidiary elements, the problem cannot be solved with the straightedge alone; for then we cannot construct any further point at all or any further line in the plane. But if subsidiary elements are allowed, then Fig. 6 provides a solution with the straightedge alone. For after choosing the two (arbitrary) lines through A, we need only draw successively the arbitrary line through C, the two (uniquely determined) lines through B, and finally the (uniquely determined) line intersecting the original line at D.

b)   Four points, no three of which are on a straight line, define a complete quadrilateral.

Without making use of subsidiary elements, we can draw (Fig. 7) the three pairs of opposite sides (a, a'; b, b'; c, c') in the quadrilateral. In this way we obtain the three diagonal points X, Y, Z. The line joining two diagonal points (e.g., YZ) intersects a definite pair of opposite sides (e.g., a and a') in two further points

---

[4]Cf. L. Bieberbach [2], p. 4, where the proof is given.

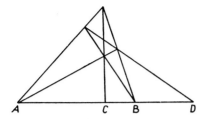

Fig. 6

(U and U'). Thus we have obtained four harmonic points on each side of the complete quadrilateral and on each of its diagonals.

c)   On a conic section, let there be given five points 1, 2, 3, 4, 5, and also a line g through the point 1. It is required to determine the intersection of g with the conic (point 6).

We apply the hexagon theorem of Pascal (cf. II10, §6.5). Let the sides 1 2 and 4 5 intersect at X and the sides 3 4 and g ($=6$ 1) at Y. The side 2 3 determines on XY the third point Z of the Pascal line, and the line from Z to 5 determines the desired point 6.

3.   Inaccessible elements.[5] Any construction that is actually carried out always takes place on a drawing surface of bounded size and with a straightedge of bounded length. Thus the problem can lead to points and lines that lie outside the page or that are too far apart to be joined by the straightedge. Both cases are included under the concept of inaccessible elements.

An inaccessible point U is given by means of two line segments that lie on

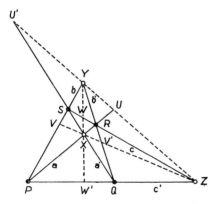

Fig. 7

[5]P. Zühlke, Konstruktionen in begrenzter Ebene, 2. Aufl. Leipzig 1930.

the page and, if produced, would pass, through U, and an inaccessible line u is determined by two inaccessible points.

Constructions with inaccessible elements mostly depend on the Desargues theorem (Fig. 8); cf. II3, §3.1: namely, if two triangles are in such a position with respect to each other that the lines joining corresponding vertices pass through a point S, then the intersections of corresponding sides lie on a line s, and conversely. The point S is the center of the figure, s is its axis, the rays through S are central rays (p, q, r), and the points on s are axial points (P, Q, R).

Let us now discuss the three fundamental problems on inaccessible elements.

4.   First fundamental problem. It is required to join an inaccessible point, determined by two lines, to an accessible point.

a)   First solution (Fig. 8). Considering the inaccessible point as the center S of a Desargues figure, take the given lines as the central rays p and q and the given point as the vertex C of one of the two triangles. Then the problem is to construct the central ray r.

Through a suitably chosen point R we draw the rays c, s, c' and thus obtain the points A, B, A', B'. By means of P and Q we find the point C' and with it the desired line r.

b)   Second solution (Fig. 9). In the first solution we could also begin with the axis s and then choose the axial points P and Q as the intersections of s with p and q. In this way the triangle ABC and the axial point R are at once determined, and the second triangle is then easy to construct.

The construction can also be based on the property of a complete quadrilateral

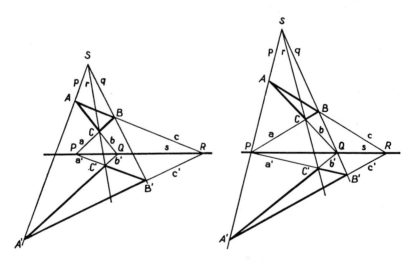

Fig. 8   Fig. 9

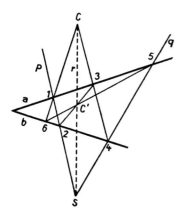

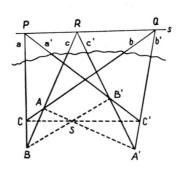

Fig. 10   Fig. 11

discussed above in No. 2b. In the quadrilateral $PQBA$, and also in $PQB'A'$, the ray r is the fourth harmonic ray for p, q and SR, so that in both quadrilaterals r passes through the third diagonal point (C and C').

c) Third solution (Fig. 10). The Pascal hexagon theorem for straight lines reads as follows (Theorem of Pappus-Pascal, cf. II3, §3.1): If the vertices of a hexagon lie alternately on two straight lines, then the intersections of the opposite sides lie on a straight line. Since this theorem deals with the construction of three points on a straight line, it is also suitable for the solution of the first fundamental problem.

If S is given as an inaccessible point by means of the lines p and q and if C is the given accessible point, let us bring p and q into intersection with the given lines a and b of the hexagon. In this way we determine the vertices 1,2 and 4,5 of the hexagon. The lines joining C with 1 and 4 determine the remaining vertices 6 and 3. The intersection of 2-3 and 5-6 determines C', and line CC' passes through S.

5. Second fundamental problem. Given an inaccessible point on an inaccessible line, a (given) point is to be joined to the inaccessible point.

The problem is reduced to the first fundamental problem by the construction of a second accessible line through the inaccessible point.

Solution. The inaccessible line is taken as the axis s of the Desargues figure (Fig. 11).

This axis s is determined by the axial points P and Q, which in their turn are given by the pairs of straight lines a, a' and b, b'. Through the third axial point R we have at the moment a single accessible line c and must draw a second accessible line c' through R. The lines a, b, c determine the Desargues triangle ABC, and the lines a' and b' determine the point C' of the second triangle. If we choose the center S arbitrarily on the line CC', the points A' and B' of the second triangle are determined at once, and with them the line $c' = A'B'$.

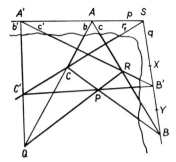

Fig. 12

Here again we could have used the Pascal hexagon between two straight lines.

6.   Third fundamental problem. Let there be given two inaccessible lines. Their intersection is to be determined by means of two accessible line segments.

Let p and q be the inaccessible lines (Fig. 8). Then the points A and A' are determined (Fig. 12) by line segments c and b, and c' and b', respectively. Similarly, let the line q be determined by inaccessible points X and Y on q. The lines determining X and Y are not drawn in the figure.

On the inaccessible line q the lines c and c' determine the inaccessible points B and B'. Moreover, the axial points $Q = (b, b')$ and $R = (c, c')$ are known. On the axis s ($= QR$) let us choose the axial point P arbitrarily and join it (second fundamental problem) to B and B'. In this way we obtain the intersections C and C', and the line $r = CC'$ is an accessible line r_1 through S. By making a second choice for P we obtain a second line r_2.

3.2.   Constructions with Auxiliary Figures Given in Advance (Steiner Constructions)

1.   Let there be given two parallel lines. With two parallel lines p and q there is also given (as their intersection) an ideal point U of the plane. Several more problems can now be solved, for example:

a)   Let the ideal point S be determined by the parallels p and q. It is required to draw through the point C the line parallel to p and q (solution by Fig. 9 or Fig. 10).

b)   It is required to bisect a segment AB on p (or on a line parallel to p). Here we need only find the fourth harmonic point to A, B and U, e.g. by Fig. 6.

c)   A segment AB on p is to be doubled, or multiplied by any integer.

d)   A segment AB (on p) is to be divided into n (e.g., $n = 3$) equal parts. Here we first add the segments $BC = CD = AB$ by problem c), obtaining q points a, b, c, d with $ab = bc = cd$, and then construct I as the intersection of Ad with Ba and project b and c from I onto AB.

e)   A segment AB (on p) is to be divided in the ratio $r: s$.

Note that parallel lines can be drawn, and line segments divided in a given (rational) ratio, only in the direction of the given parallel lines.[6]

2. Let there be given a line segment, together with its midpoint. Being given two parallel lines in advance is precisely equivalent to being given a segment AB with its midpoint M. The latter data also determine an ideal point U (as fourth harmonic point to A, B, and M).

For the proof it is only necessary to show that through a point P we can draw the line parallel to AB. To do this, we choose a point S on the line AP and join it to B and M. The line PB meets SM at a point T. The line AT determines a point Q on SB. The line PQ is the desired parallel.

More generally, we may even state that being given two parallel lines in advance is equivalent to being given a segment AB divided at a point T in a known ratio.[7]

3. Let a parallelogram be given. We now have two ideal points, and thus (indirectly) the ideal line itself. Then it is possible to draw parallel lines, and to divide segments in a prescribed rational ratio, even though they lie in arbitrary directions.

For example, if we wish to draw a line parallel to a given line g, we have a special case of the second fundamental problem (§3.1, No. 5), since the points considered there as inaccessible may now be considered as ideal points.

But we could also argue as follows. The diagonals of the given parallelogram determine its midpoint M. Either of the lines through M parallel to a pair of sides, when taken together with the two parallel sides, determines a segment and its midpoint on an arbitrary line g, so that we obtain the data of No. 2.

The fact that we can draw a line parallel to a given line with arbitrary direction can also be stated by saying that all given elements can now be subjected to an arbitrarily preassigned parallel translation.

4. Let the given element be a square. Several other elementary problems can now be solved:

a) Through a given point draw the perpendicular to a given straight line.
b) Rotate a segment AB about A through $90°$.
c) Bisect a right angle.
d) Double an angle or multiply it by any integer.

Since parallel translation is possible, it is sufficient to prove that the problem is solvable when the given elements are in special position.

a) Through the center M of a square (Fig. 13) it is required to draw the line perpendicular to XM. Here we first draw XY parallel to the side BC and YZ parallel to the diagonal BD. This latter step is equivalent to reflection in the diagonal AC. Thus $DZ = BY = XC$. Consequently, in a rotation about M through

[6]Th. Vahlen [8], p. 23 f.
[7]For the proof, see A. Adler [1], p. 79 f.

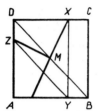

Fig. 13

$90°$ the point X goes into the point Z and ZM is perpendicular to XM with $ZM = XM$.

c) If the right angle ZMX is to be bisected, we need only to draw through M the perpendicular to ZX (problem a).

d) If the angle with vertex O and sides a and b is to be doubled, it is only necessary to draw through O the ray d perpendicular to b, and then construct the fourth harmonic ray to b, d, and a.

5.   Let a circle be given, together with its center.[8] Given a circle with its center, all constructions possible with straightedge and compass are possible with the straightedge alone. In fact it is enough to have only an arc of the circumference although the center must in every case be given. Let us prove the first of these statements.[9]

a)   We begin with the solution of two special problems:

1.   In the given circle it is required to inscribe a square.

Draw an arbitrary diameter, which will provide a line segment and its midpoint. Thus we can draw a chord parallel to the diameter. If we construct from these two lines an isosceles trapezoid, its axis of reflection (which can be constructed) provides a diameter perpendicular to the first diameter. Then we obtain the desired square by joining the endpoints of these two diameters.

2.   Through the center O of the given circle let there be drawn two rays a and b and let there be given a point A on a. It is required to determine a point B on b such that $OB = OA$.

Let the rays cut the circle at A' and B', respectively. Join A' and B' and through A draw the parallel to $A'B'$, which by the first problem is possible with straightedge alone.

The advantage here, in comparison with the case where a square was given in advance, lies in the fact that now we can rotate a line segment through an arbitrary angle.

---

[8]J. Steiner, Die geometrischen Konstruktionen, ausgeführt mittels der geraden Linie und eines festen Kreises, Berlin 1833 (Ostwalds Klassiker, Bd. 60). The theorem on the fixed circle and center is due to Poncelet, Traité des propriétés projectives des figures (1822), p. 187.

[9]For the second statement, see Bieberbach [2], p. 19 ff.

b)   Every construction with straightedge and compass can be reduced to the following fundamental problems:

1.   Draw the line joining two points.
2.   Determine the intersection of two lines.
3.   Determine the intersections of two circles.
4.   Determine the intersections of a circle with a line.

Of course the circles occurring in a given construction cannot be drawn out in full, and we consider a circle as being determined if we find, or are given, its center and a point on its circumference. This special case is easily seen to be equivalent to the more general case in which the radius is given by a line segment that does not begin at the center of the circle; for we have only to make a parallel translation of the line segment, by means of the inscribed square.

c)   The first two of these four fundamental problems are immediately solvable by the straightedge alone, and the third can be reduced to the fourth by a construction with straightedge alone.

If the two circles intersect (Fig. 14), there exists a line passing through the two points of intersection, the so-called radical axis.

The line joining the centers is divided by the point M into the segments d_1 and d_2 such that

$$r_1^2 - d_1^2 = r_2^2 - d_2^2.$$

Now if we are given two circles with centers M_1 and M_2 (Fig. 15), we draw the radius r_2 in M_1 and the radius r_1 in M_2 perpendicular to M_1M_2, as can be done with straightedge alone. Then we bisect the segment AB at C and draw the right bisector of AB, which cuts the line M_1M_2 joining the centers at the point M. Then we have $AM = BM$, so that

$$r_1^2 + d_2^2 = r_2^2 + d_1^2 \quad \text{or} \quad r_1^2 - d_1^2 = r_2^2 - d_2^2.$$

The perpendicular at M to the line joining the centers is the radical axis of the two circles and passes through their points of intersection.

It remains to solve the fourth fundamental problem. Here we are given an auxiliary circle with its center O, and a circle K (determined by its center M and the point P on its circumference, and finally the line g; and we are required to determine, with straightedge alone, the intersections of K and g (Fig. 16).

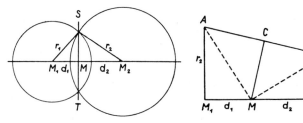

Fig. 14   Fig. 15

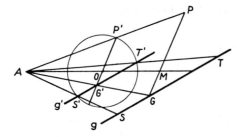

Fig. 16

The solution is provided by the dilation taking the circle about M into the circle about O. We determine the center A of the dilation by drawing through O the line parallel to MP and then joining O to M and P' to P. This dilation takes the point G (the intersection of g with PM) into the point G'. The parallel to g through G' provides the image line g', and then the intersections S' and T' provide the desired points S and T.

d)   For an arbitrary problem (with given circle and center) we can always obtain the solution with straightedge alone by considering the construction to have been carried out with straightedge and compass and then performing each step in succession with the straightedge alone. But this procedure is usually very cumbersome, so that it is natural to seek for simple solutions making immediate use of the given auxiliary circle.

For example, if a given angle is to be bisected, let us translate the angle so that its vertex S coincides with the midpoint O of the auxiliary circle. Then if A and B are the points of intersection of the sides of the angle with the auxiliary circle, we produce OA beyond O to the point C on the circumference and join C to B. The parallel CB through S bisects the given angle.

3.3.   *The Parallel Straightedge*

1.   The parallel straightedge is a straightedge with two parallel sides, the width of which we will denote by b. With this straightedge there are two fundamental operations.

Operation A.   Bring one of the two sides into incidence with two (already given) points.

Operation B.   Bring the straightedge into such a position that its two sides are incident with one each of two (already given) points.

For a given pair of points there will always be two different positions in Operation A, but in Operation B there will be none, one, or two such positions, according as the distance between the two points is smaller than, equal to, or greater than the width b. In each position we may draw a straight line along one or both of the edges so as to construct the following configurations.

By Operation A we obtain

a) the straight line determined by two given points;

b) the straight lines that are at the distance b from two given points and have the given points between them;

c) the strips of width b on whose common boundary lie two given points, and the strip of width $2b$ such that the line through the middle of it, parallel to its edges, passes through two given points.

By Operation B we obtain

d) the straight lines that pass through a given point and are at a distance b from another given point;

e) the rhombus of width b such that a pair of opposite vertices are at two given points.

The configurations determined by these two operations can also be considered as tangents to certain circles. For example, the operations determine

a′) the common tangents to two circles of zero radius;

b′) the common external tangents to two circles of radius b, with given distinct centers;

d′) the tangents from a given point to a circle of radius b with given center.

The parallel straightedge allows us to solve any problem that can be solved by straightedge and compass. In other words, the range of problems solvable by parallel straightedge is identical with the range for straightedge and compass.

2. The proof of the preceding statement will be complete if we solve the following three problems (cf. §3.2, No. 5):

a) To construct a square.

The parallel straightedge allows us to construct rhombi, whose diagonals provide us with pairs of right angles. Then each of these pairs enables us to construct a square of side b.

b) Let the lines g and h contain an angle φ at the vertex S. Let the point G lie on the line g. Then it is required to construct, with the parallel straightedge, a point H on h such that $SH = SG$. (This construction allows us to rotate a line segment through an arbitrary given angle.)

At a distance b draw the parallels to g and h in such a way that they intersect in the interior of the angle at a point T, and then join S to T. Letting A and B denote the intersections of the parallels with the given lines, join G to B and A to the intersection of GB with ST.

c) To construct the intersections of a line g with the circle with center O and radius b (width of the parallel straightedge. The circle of radius b plays the role of the Steiner auxiliary circle.)

Visualizing the circle K as already drawn, and also the tangents to it at its points of intersection with the given straight line, we see that the intersection G of the tangents with each other is the pole of the line g, so that if we can construct the point G the two tangents can be drawn with the parallel straightedge, and then the intersections of the circle K with the line g are already known, since they are the same as the intersections of the tangents with g.

But the point G can certainly be constructed; for if P is a point on g, and t_1 and t_2 are the two tangents from P to the given circle (P can always be so chosen that the two tangents exist), then the line joining P to G is the fourth harmonic ray for t_1, t_2, and g. Such a ray can be constructed with the straightedge alone (cf. §3.1, No. 2a). The same construction for a second point on g gives us a second line through G.

Remark.   The problem "through a point A, not on the given line, draw a line parallel to the given line" is very simply solved as follows:

Join a point B on a to A, and draw two adjoining strips of width b on the same side of AB. The intersections with a of the two outer parallels, taken together with the point A, form three vertices of a parallelogram whose fourth vertex can easily be constructed by means of the diagonals.

3.4.   *Straightedge and Gauge*

1.   The straightedge is used in the ordinary way to draw the line joining two given points. By a "gauge" we mean an instrument that allows us to mark off, in either direction from a given point on a given line, a fixed unit segment e.

Hilbert[10] has investigated the constructions that are possible with straightedge and gauge and has characterized them in two ways.

2.   In the first place, he has shown that the points constructible by straightedge and gauge are the same as those whose existence is guaranteed by the first four sets of axioms in the Hilbert axiom system, namely,

the axioms of connection, order, congruence, and parallelism. The axioms of continuity are not used.

These four sets of axioms assert the existence of points and lines constructible as follows:

a)   Draw the line through two given points, and determine the intersection of two given lines.

b)   Mark off, in both directions from a point P on a given line g, a segment of a given length.

c)   At a point P on a given line g construct an angle of given magnitude on either side of the line.

d)   Given a straight line g and a point P not on g, draw through P the line parallel to g.

e)   Through a given point P draw the perpendicular to a given line g.

[10]Hilbert [6], p. 115 ff.

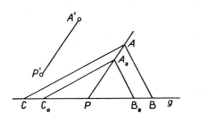

Fig. 17   Fig. 18

But these constructions can be carried out with straightedge and gauge, as we shall now show, in the order d), e), b), c).

Construction d)   The gauge allows us to mark off on a line g the unit segment e twice in succession. But then we have a line segment with its midpoint, so that parallel lines can be drawn with the straightedge alone (cf. §3.2, No. 2).

Construction e)   On a line g (Fig. 17) draw the unit segment e from O in each direction, so that $OA = OB = e$, and also construct $OX = OY = e$ in arbitrary directions. In the triangle ABC, the lines AX and BY are then altitudes (Theorem of Thales). The line CH is the third altitude and thus is perpendicular to AB. The perpendicular through P to the line g is then constructed as the parallel to CH.

Construction b)   If the segment $P'A'$ (Fig. 18) is to be marked off on g from the point P, we may translate $P'A'$ to the position PA, make $PA_0 = PB_0 = PC_0 = e$, and through A draw the parallels to A_0B_0 and A_0C_0.

Construction c)   It is sufficient to consider the case that $\measuredangle AOB = \alpha$ (Fig. 19) is to be erected on g at the point O, since otherwise we would simply make a parallel translation. (In the figure α has been drawn only on one side.) Through A draw the perpendiculars to the side h' and the line g, join B to C, and through O draw the perpendicular to BC. Since the points O, A, B, C lie on the circle (Thales), the angles OAB and OCB are equal; thus the angles COD and AOB are also equal.

3.   Hilbert's second characterization of the points constructible with straightedge and gauge is given by his theorem: only those points are con-

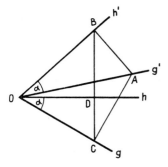

Fig. 19

structible whose coordinates can be expressed in terms of the coordinates of the given points by addition, subtraction, multiplication, and division (the four fundamental operations of arithmetic) and by taking the square root of sums of squares.

As was shown in constructions b), c), and e), the gauge enables us, in addition to to what was possible with straightedge alone, to rotate a segment about a point through a given angle. If in a rectangular coordinate system the unit segment e on the x-axis is rotated through the angle $\alpha = \tan^{-1}(b/a)$, its endpoint in the new position is the intersection of $x^2 + y^2 = e^2$ and $y = \dfrac{b}{a} \cdot x$, so that its coordinates are

$$ x = \frac{ae}{\sqrt{a^2 + b^2}}, \quad y = \frac{be}{\sqrt{a^2 + b^2}}. $$

Thus the rotation of a line segment provides, in addition to the four fundamental operations of arithmetic, precisely the extraction of the square root of a sum of squares of known magnitudes.

The significance of the preceding result is as follows.

4. Not every problem solvable with straightedge and compass is also solvable with straightedge and gauge; for example, we cannot construct a right-angled triangle with hypotenuse 1 and one side equal to $\sqrt{2} - 1$, since its other side $s = \sqrt{2\sqrt{2} - 2}$ is not constructible.

To prove this latter statement we consider the domain Ω consisting of all algebraic numbers that can be formed from the number 1 by a finite number of applications of the four rational operations and the fifth operation $\sqrt{1 + \omega^2}$, where ω is any number already constructed by the five operations. This domain of numbers contains the coordinates of all points constructible from the points $(0, 0)$ and $(1, 0)$ with straightedge and gauge, the length e of the gauge being chosen as the unit 1. The domain Ω is totally real; i.e., the conjugates of each number in the domain belong to the domain and thus are also real (cf. IB6, §8; IB7, §6). But if $s = \sqrt{2\sqrt{2} - 2}$ were constructible, its conjugate $s' = \sqrt{-2\sqrt{2} - 2}$ would also be constructible, which is not the case, since s' is not real.

5. With straightedge and gauge it is possible to construct, for example, all the regular polygons that can be constructed with straightedge and compass. Moreover, the Malfatti tangent problem is solvable, namely, to draw three circles in a triangle in such a way that each circle touches the two other circles and also two sides of the triangle. But the problem of Apollonius cannot be solved, namely, to draw the eight circles that touch three given circles.[11]

6. The construction with straightedge and compass can also be characterized axiomatically. To Hilbert's first four groups of axioms we must add the "compass-construction" axiom (the name is due to F. Schur):

[11]Bieberbach [2], p. 44.

"there exists exactly one right-angled triangle with arbitrarily given side a and hypotenuse $c > a$."

This axiom means that we can construct $\sqrt{c^2 - a^2}$, i.e., the square root of the difference of two squares. The constructibility of $\sqrt{a^2 + b^2}$ and of $\sqrt{c^2 - a^2}$ enables us to construct the solution of a quadratic equation $a_0 x^2 + a_1 x + a_2 = 0$, since the solution

$$x = -\frac{a_1}{2a_0} \pm \frac{1}{2a_0} \sqrt{a_1^2 - 4a_0 a_2}$$

can be transformed into

$$x = -\frac{a_1}{2a_0} \pm \frac{1}{2a_0} \sqrt{\{\sqrt{a_1^2 + (a_0 - a_2)^2}\}^2 - (a_0 + a_2)^2}.$$

In this way the construction of the solution is reduced to taking the square root of a sum and of a difference of squares.

3.5.   *The Mirror Straightedge*

1.   The mirror straightedge is a mirror of metal or glass that has straight edges and stands perpendicular to the plane of the drawing.[12]

Apart from the drawing of arbitrary lines, the only lines that can be constructed with the mirror straightedge are the axes of certain definite reflections (determined by given segments); in fact, precisely the following three fundamental operations can be performed:

Operation A: Joining of two points. If two points P and Q are given, we can draw the axis of the reflection in which each of the points P and Q is its own image. The axis of this reflection is the line joining the points P and Q.

Operation B: Bisection of an angle. If two lines are given, intersecting in S, we can draw the axes of two reflections in which the given lines are images of each other. For we can set the mirror so that it passes through S and then turn it until each of the two halflines in front of it is continued in a straight line by the mirror image of the other, as will happen for two positions of the mirror. The axes of the two reflections thus obtained bisect the angles between the given lines.

The following problem is a particular case (bisection of an angle of 180°): a line g is divided by a point S into two halflines and it is required to draw the axis of a reflection in which the two halflines are mirror images of each other. Thus we can draw perpendiculars with the mirror straightedge.

Operation C: Intersection of circle and line. Given the points M and P and a line g, we can find the axis, through the point M, of a reflection in which the image of P falls on g. For this purpose we turn the mirror, keeping the line g in our field of vision, until the image of P falls on g. When this axis has

---

[12]Cf. R. Proksch, Konstruktionen mit dem Spiegellineal, in: Der Mathematikunterricht, Beiträge zu seiner wissenschaftlichen und methodischen Gestaltung, edited by von Eugen Löffler, Vol. 2, 1956, No. 2, p. 20 ff.

been found, it is easy to determine the image of P (on g). The operation thus described determines the points at which the given line g intersects the circle with center M and radius MP; R. Proksch refers it as "taking bearings."

2. If only the operations A and B are allowed, the range of construction of the mirror straightedge is the same as that of the Hilbert gauge, since the gauge (together with a straightedge) is equivalent to an instrument for bisecting angles.[13]

3. If the operations A, B, and C are allowed, the mirror straightedge enables us to make exactly the same constructions as the straightedge and compass, as is easily proved from §3.2, No. 5, and §3.3.

4. Constructions with the Compass

4.1. *The Constructions of Mohr and Mascheroni*

1. The constructions of Mohr and Mascheroni are constructions with the compass alone. They were published in 1797 by Mascheroni (*Geometria del compasso*), but J. Hjelmslev has shown (1928) that they were first stated by Georg Mohr in his *Euklides Danicus,* Amsterdam, 1672.[14]

The fundamental theorem is as follows: All constructions that are possible with straightedge and compass are possible with compass alone. Here a straight line is regarded as being constructed if we have determined two of its points.

Since in a given construction two circles will in general intersect in two points, sometimes the compass must also be used to decide which of the two points of intersection is to be chosen for the remainder of the construction. For example, if through a point P we are required to draw a line parallel to the line determined by the points A and B, i.e., to find a point Q such that the line PQ is parallel to AB, the circle about P with AB as radius and the circle about B with AP as radius determine two points of intersection, but only one of them is acceptable. Here the right choice can be made by inspection.

2. Before proving the fundamental theorem let us first show: if we are given the circle with center O and radius r as inversion circle, then for every preassigned figure we can determine the inverse figure with compass alone, since we can draw the image of every point, line, and circle, as follows:

a) To construct the inverse image of a point P for which $OP = p > \dfrac{r}{2}$.

Draw the circle with center P through the point O, and let it cut the inversion circle at S and T. The circles about S and T with radius $OS = r$ determine the

---

[13]Feldblum, Über elementargeometrische Konstruktionen. Inauguraldissertation, Göttingen 1899. Cf. Adler [1], p. 143 f., Bieberbach [2], p. 49.
[14]Bieberbach [2], p. 152.

inverse image P'. (The triangles OPS and OSP' are similar, so that $OP:OS = OS:OP'$, and $OP \cdot OP' = r^2$.)

b)   To determine the inverse image of P if $OP = p = \dfrac{r}{2}$.

The image P' is at a distance $p' = 2r$ from O (since $p \cdot p = r^2$). Thus it is only necessary to multiply OP by four in order to obtain the segment OP'.

Doubling of a segment: In order to double OP draw the circle with center P through O and on this circle, beginning at the point O, mark off OP three times. (With the doubling of a segment we have obviously solved the problem of multiplying a segment by any integer.)

c)   To determine the inverse image of P if $OP = p < \dfrac{r}{2}$; i.e., to find

$$OP' = p' = \frac{r^2}{p}.$$

Construct successively: a point Q on OP with $OQ = q = n \cdot p > r/2$ (multiplication of a segment); the inverse image Q' from the equation $OQ' = q' = r^2/np$; and finally, the segment $p' = n \cdot q'$ (by multiplication).

d)   To construct the inverse image of a line through O.

Let the line g be determined by O and a point P. Since it is its own image under the inversion, nothing needs to be constructed.

e)   To construct the inverse image of a line not through O.

Let the line g be determined by the points A and B. Imagine that the perpendicular through O to the line g has been drawn and intersects g at S. If S' is the inverse image of S, the inversion will transform g into the circle with diameter OS'. For its midpoint M' we have $OM' = \frac{1}{2}OS'$. Thus for the original points M S we have $OM = 2 \cdot OS$. Multiplying these equations together gives $OM \cdot OM' = OS \cdot OS'$.

Then from the condition $OM = 2 \cdot OS$ we can construct M, since we need only reflect O in g. For this purpose we draw the circle with center A and radius AO and the circle with center B and radius BO, and then finally construct M' as the inverse image of M.

f)   To construct the inverse image of a circle K through O.

The inverse image of such a circle is a line, which will be determined by the inverse images (see c)) of two points on the circle.

g)   To draw the inverse image of a circle K (with center M) not through O.

Find the center N of the image circle in the following way: let $\bar{O}$ denote the image of O, under inversion in the circle K. Construct the inverse image N of $\bar{O}$ in the original circle of inversion. Then N is the desired center of the image circle.[15]

[15]For the proof, see A. Adler [1], p. 113, or F. Enriques [5], p. 49. In order to draw the image circle, construct by inversion one other point (arbitrary) of the circle K.

The solution of these seven fundamental problems now allows us to replace any construction with straightedge and compass by a construction with compass alone.

If this statement has been proved, the proof for the fundamental theorem in No. 1 is thereby complete. Let us prove it for a special problem.

To bisect the angle determined by its vertex S and the points A and B, one on each of its sides, the classical construction with compass and straightedge runs as follows:

1. Use the straightedge to draw the two sides.
2. With center S draw a circle K with arbitrary radius, and determine its points of intersection with the sides of the angle.
3. About each of these two points of intersection draw a circle with arbitrary radius (the same radius for both circles) and determine one of their points of intersection (T).
4. Use the straightedge to join this point T with S.

In the construction with compass alone these steps will proceed as follows:

1. No longer relevant.
2. Here we let K be the circle about S through A, so that we already have its point of intersection with one of the two sides of the angle. The intersection with the side SB is determined as follows: with an arbitrary circle (with center O not on the lines SA and SB) we construct the inverse images of the circle K and side SB, as can be done with compass alone, and then we again construct the original point of the intersection of these images, which is again possible with compass alone.
3. This step uses only the compass in any case.
4. No longer relevant.

Any construction already performed with the compass and straightedge can be dealt with in a similar way.

3. Every construction with compass and straightedge can be reduced to the following four fundamental constructions (cf. §3.2, No. 5b):

1. Draw the line through two given points.
2. Determine the intersection of two lines.
3. Determine the intersections of two circles.
4. Determine the intersections of a circle with a line.

In constructions with the compass alone the first of these problems is no longer relevant, and problem 3 can be solved at once. Thus we must ask how problems 2 and 4 can be solved with the compass alone. We shall look for a method independent of the discussion in No. 2.

a) To construct the points of intersection of a circle (of radius r) with a line not through the center O of the circle. The line is given by the points A and B.

Reflect the circle in AB. To do this, draw the circle with center A and radius AO

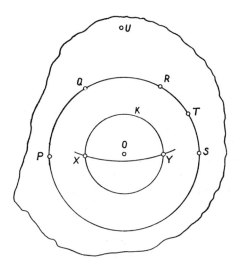

Fig. 20

and with center B and radius BO, and let O' be their second point of intersection (in all, three circles are necessary).

b) To determine[16] the points of intersection X and Y of a circle K with a line through its center O, the line being given by a point P (in addition to O). Let the radius of the circle be r, and let the distance OP be p.

Here we are given the point P and the circle K with center O. (Fig. 20). Construct the circle through P with center O, and from P mark off the radius three times (points Q, R, S) on the circumference, and also mark off the distance $ST = r$. Then $PT = \sqrt{4p^2 - r^2}$. Draw the circles about P and S with radius PT, thus obtaining the point U. Draw the circle with center U and radius $PR = p\sqrt{3}$, thus obtaining the points X and Y (in all, eight circles are necessary).

c) To determine the intersection X of two lines AB and CD, not perpendicular to each other.[17]

Reflect C in AB to C_1, then C_1 in CD to C_2, and finally C in $C_1 C_2$ to C_3. The triangles $CC_1 C_3$ and CXC_1 are similar. Thus $CC_3 : CC_1 = CC_1 : CX$ and therefore $CC_3 \cdot CX = CC_1^2$. But this equation states that X is the image of C_3 under inversion in the circle through C_1 with center C (ten circles).

d) To determine the intersection X of two lines AB and CD, perpendicular to each other.

Reflect C in AB to C', and bisect the segment CC' (see the following exercise). (Nine circles.)

[16]See Cesàro, Les problèmes de Géometrie résolus par le compas, sans la règle (Mem. de la société royale des sciences de Liège. 3 ième série Tom. 1. Brüssel 1899).
[17]See Dubois (Journ. de Math. 22. Band, 1897).

e)   To bisect the segment AB.

By means of inversion in the circle through B with center A, determine C by doubling AB to AC, and then construct the inverse image of C (seven circles).

4.   The discussion in No. 3 provides a second method of solving a problem with compass alone. Let us illustrate with the following problem, already solved with compass and straightedge in No. 2: to bisect an angle, given its vertex O and points A and B on its sides ($OA \neq OB$).

Again we draw the circle through A with center O, determine its intersection C with the second side of the angle (as in No. 3b), and then draw circles about A and C with equal but arbitrary radius (ten circles).

5.   If we wish to use the compass alone, it is usually impracticable to follow the general argument of No. 2 or No. 3. We must find methods of solution better adapted to our special instrument; i.e., we must make deeper use of the theory of the circle. Let us give some examples.

a)   There is a classical elementary construction for drawing the perpendicular at A to a segment AB.

About A draw the circle K through B and then, with the same radius, draw the following three circles: with center B (intersecting K at C), with center C (intersecting K at D), and with center D.

b)   To draw the tangent to a given circle K at a given point P on K.

Take a second point Q (arbitrary) on K. With center Q draw the circle L through P, cutting K again at R_1. Then with center P draw the circle through R_1, cutting L again at R_2. The line PR_2 is the desired tangent (two circles).

c)   To draw the circle through three given points A, B, C.

If C' is the image of C under inversion in the circle with center A and radius AB, the line BC' is the image of the desired circumcircle (ten circles).

4.2.   *Restricted Use of the Compass*

1.   Mascheroni proposed the problem of carrying out a given construction with the smallest possible number of different openings for the compass. The most stringent restriction here would be to allow only one fixed opening for all problems. The range of problems then solvable has been studied by Hjelmslev.[18] An example is the well-known doubling of a line segment with compass alone, although Adler has shown that division of the given segment into equal parts is impossible.[19]

2.   But if, in addition to the compass with fixed opening, we allow the straightedge, then we can already solve every problem solvable with straightedge and (unrestricted) compass, since Steiner has shown that,

---

[18]See Bieberbach [2], p. 37 ff.
[19]See Adler [1], p. 121.

together with the straightedge, it is enough to have one circle and its midpoint.

3.   If we imagine a construction carried out under Steiner's rule and then transform it, by means of a suitable[20] inversion, into a construction with circles only, all the circles thus used will pass through a fixed point, namely, the center of the inversion circle. So we have the following result: if we restrict the use of compass only by requiring that all circles pass through a fixed point, we have not thereby decreased the range of solvable problems, which remains the same as for straightedge and (unrestricted) compass.

5.   Constructions of Higher Degree

5.1.   *Constructions of 3rd and 4th Degree*

1.   A construction is said to be of 3rd or 4th degree if its treatment by algebraic means requires the solution of one or more irreducible equations of 3rd or 4th degree (and possibly also some linear or quadratic equations); in other words, the coordinates of the desired points are solutions of irreducible equations of 3rd or 4th degree with coefficients in the smallest field (it is to this field that the irreducibility refers; in most cases it will be the rational field) containing the coordinates of all the given points.

Since the solution of an irreducible equation of 4th degree can be reduced to that of an irreducible equation of 3rd degree[21] we need only consider constructions of 3rd degree.

To determine whether a problem is actually of 3rd degree we must decide whether the corresponding equations of 3rd degree are irreducible. This decision is made easier by the following remarks.

An equation of 3rd degree $g(x) = 0$ is reducible over the coefficient-field K if and only if $g(x) = x^3 + b_2 x^2 + b_1 x + b_0$ has a linear factor (with coefficients in K), i.e., if and only if $g(x)$ has a zero in K. So if we can prove that K does not contain any zero of $g(x)$, the given equation is certainly irreducible over K. In particular, if K is the field of rational numbers and if the coefficients b_2, b_1, b_0 of $g(x)$ are integers, then every rational zero of $g(x)$ must be an integer and must be a factor of b_0(IB5, §4.4); so in order to prove the desired irreducibility we need only show that the factors of b_0 are not zero of $g(x)$.

2.   In any case the constructions of 3rd degree can no longer be carried out with straightedge and compass. For we have the theorem: no zero of an irreducible equation of 3rd degree can be expressed by means of square-root signs. More precisely: if the coefficients of $g(x) = x^3 + b_2 x^2 + b_1 x + b_0$

---

[20]None of the straight lines that are used passes through the center of the inversion circle.

[21]Cf. Bieberbach [2], p. 72 ff.

belong to a field K and if $g(x)$ is irreducible over K, then $g(x)$ is also irreducible over every extension K' of K formed by the adjunction to K of
finitely many roots of quadratic equations: in other words, $g(x)$ has no zero
in any such extension K' (IB7, §1.5).[22] But in §2.1 we have shown that
straightedge and compass allow us to construct only those points whose
coordinates can be represented by (real) square-root signs.

3.   So for a problem of 3rd degree, we must look about for other instruments beyond the straightedge and compass. The range of possibilities with
such instruments can usually be decided very quickly and easily; for we have
the remarkable fact that all problems of 3rd (and thus of 4th) degree can be
reduced to two definite problems of 3rd degree, namely, trisection of the
angle and doubling of the cube.[23]

4.   The angle-trisection equation is obtained, for example, in the following way. Since

$$\cos 3\varepsilon = 4\cos^3\varepsilon - 3\cos\varepsilon,$$

multiplication by 2 and rearrangement gives

$$8\cos^3\varepsilon - 6\cos\varepsilon - 2\cos 3\varepsilon = 0,$$

so that by setting $\cos 3\varepsilon = a$ and $2\cos\varepsilon = x$ we obtain the trisection equation

(1) $$x^3 - 3x - 2a = 0.$$

If $x_1 = 2\cdot\cos\varepsilon$ is a solution of (1), then $x_2 = 2\cdot\cos(\varepsilon + 2\pi/3)$ and $x_3 = 2\cdot\cos(\varepsilon + 4\pi/3)$ are also solutions.

Thus for $\varphi = 3\varepsilon = 120°$ the trisection equation becomes

(2) $$x^3 - 3x + 1 = 0.$$

This equation is irreducible (over the rational field) since the factors ± 1
of the absolute term are not solutions. Moreover, it is easy to see directly that
the equation has no rational solution; for if we assume that m/n, which coprime m and n, is a solution of (2), a slight rearrangement of the equation
gives

$$m^3 = n(3mn - n^2),$$

and since m and n are coprime, we must therefore have $n = 1$; but it is easily
shown that no integer m is a solution of (2), so that our assumption is false.

This result means that since the angle $40° = 120°:3$ cannot be constructed with
straightedge and compass, it is impossible to construct the regular polygon of
nine sides.

[22]A simple proof of this theorem (if K is the rational field) is given in R. Courant,
H. Robbins, *What is Mathematics?* Oxford University Press, New York, 1941.
[23]See Bieberbach [2], p. 72 ff., or Enriques [5], p. 264 ff.

It is easy to prove the following theorem[24] about angles of the form $\varphi = 2\pi/n$: a given or constructible angle of this form can be trisected with straightedge and compass if and only if n is not divisible by 3.

5. The doubling of the cube (more generally, multiplication of a cube by an integer) leads to the equation

$$(3) \qquad\qquad x^3 - m \cdot a^3 = 0,$$

where a is the side of the given cube and x is the side of the cube desired.

Of course, the straightedge and compass can be used to multiply the cube by certain integers; for example, by $m = 8$. But it is not possible to double the cube ($m = 2$), since $x^3 - 2 = 0$ has no rational solution and is thus irreducible.[25]

6. The solution of problems of 3rd degree is possible with straightedge and compass if certain definite auxiliary curves are given in advance. The number of auxiliary curves suitable for angle trisection and for doubling the cube is almost limitless.

a) Of basic importance is the following result of Kortum and Smith: given a fixed conic section that is not a circle, all problems of 3rd degree can be solved with straightedge and compass.

For example, the fixed parabola $y = x^2$ and the variable circle $x^2 + y^2 - 2ax - 4y = 0$ lead to the trisection equation $x^3 - 3x - 2a = 0$ if we take $a = \cos 3\varepsilon$ (Descartes).

The intersection of the circle $x^2 - 2x + y^2 - y = 0$ with the hyperbola $xy = 2$ leads, after cancellation of the factor $(x + 2)$, to the equation $x^3 - 2 = 0$, and thus to the doubling of the cube of side 1.

b) London[26] has proved that all problems of 3rd degree can be solved with the straightedge alone if we are given a curve of the 3rd order, completely drawn, and also, for metric problems, a square. In particular, London recommends a fixed cissoid together with the center of its generating circle. Then the square can be constructed with straightedge alone.

The cubic parabola $y = x^3$, together with the line $y = 3x + 2a$ ($a = \cos 3\varepsilon$), leads to the trisection equation and, together with the line $y = 2$, to the equation for the doubling of the cube.

7. Instead of making use of higher curves for the solution, it is also possible to use "higher" drawing instruments; i.e., higher than the straightedge and compass. Among the many possibilities, let us describe the marked straightedge and the draftsman's triangle.

8. Parenthetically, let us also remark that there exist triangle construc-

---

[24]See Breidenbach [3], p. 21 ff.
[25]For further information, see Breidenbach [4], p. 55 ff.
[26]See Enriques [5], p. 261.

tions of higher than 2nd degree. For example, a triangle cannot be constructed with straightedge and compass alone if we are given

  a)   one side, the opposite angle, and the bisector of an adjacent angle
$(c, \gamma, \omega_\beta)$;

  b)   two sides and the radius of the incircle.[27]

5.2. *The Marked Straightedge*

1.   The marked straightedge was already used in ancient Greece for the solution of problems of the third degree. It is a straightedge on which two marks U and V have been made at the fixed distance c from each other, and it is used not only as an ordinary straightedge to join two given points but also to draw a line through a given point P in such a way that the mark U lies on the first of two given curves and the mark V on the second. In the simplest cases the curves are straight lines or circles.

In the particular case that the point P lies on the first curve and this curve is a circle of zero radius, the marked straightedge solves the following problem: to determine the points in which a circle with center P and radius c intersects a given curve.

2.   For the trisection of an angle AOB the following procedure was given by Pappus (Fig. 21): mark off the distance $OO' = r$, and through O' draw the line u parallel to OA and the line v perpendicular to OA. Then insert the segment $c = 2r$ between u and v in such a way that the straightedge passes through O.

There are two other possible positions for the marked straightedge, namely, the ones obtained by rotating through 120° and 240° the position drawn in Fig. 21. A fourth position $U_0 V_0$, unrelated to the problem, is obtained from the circle about

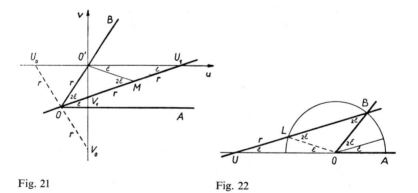

Fig. 21   Fig. 22

[27]Cf. *Praxis der Mathematik*, Vol. 1, 1959: (a) Kursawe, Unmöglichkeit einer Dreieckskonstruktion (p. 81); (b) A. Rohrberg, Ein konstruktiv nicht lösbare Dreiecksaufgabe (152).

O with radius r. The fact that this fourth position can be constructed without the marked straightedge is a clear indication that it has nothing to do with angle trisection.

Among other trisections with a marked straightedge, let us mention one due to Archimedes (Fig. 22). Draw the circle with center at the vertex O of the angle AOB and with radius r, and insert the segment r between the line OA and the circle in such a way that the straightedge passes through B. Finally, draw through O the parallel to UL.

3.   For the doubling of the cube let us describe a solution (Fig. 23) sent to us a few years ago[28] by Theophil Rossel, Leipzig. Draw the circle with center O and radius 1, through B draw the line g making an angle of 60° with the diameter AB, and insert the segment of length 1 between the circle and the line g in such a way that the straightedge runs through A. Then $BD = x = \sqrt[3]{2}$.

The chord BH makes an angle of 60° with the diameter BA, so that $BH = 1$ and $AH = \sqrt{3}$. Thus by the theorem on secants:
$$DA \cdot DF = DB \cdot DH; \text{ i.e., } DA \cdot 1 = (1 + x) \cdot x.$$
Then in the triangle AHD, by the theorem of Pythagoras,
$$DA^2 = DH^2 + AH^2; \text{ i.e., } [x(1 + x)]^2 = (1 + x)^2 + 3.$$
Simplification leads to
$$(x^3 - 2)(x + 2) = 0.$$
Here the solution (-2) corresponds to the segment BC.

A method for doubling the cube in which the constant segment c of the straightedge is inserted between straight lines is as follows:[29]

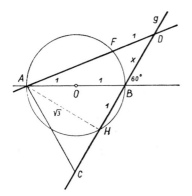

Fig. 23

[28]It has been pointed out by Hofmann that this solution is to be found already in Vieta, in his *Variorum de rebus matematicis responsorum* liber VIII, Caput V = Opera mathematica, ed. Franc. Schooten, Leiden 1646, pp. 352–353 (Tours 1593).

[29]Breidenbach, Der rechte Winkel und das Einscheibelineal. Z. math. -nat. Unterricht, Vol. 56, p. 4–13 (1925).

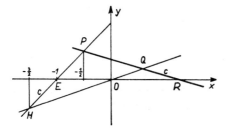

Fig. 24

In a rectangular coordinate system (Fig. 24) the point $P\left(-\frac{1}{2};\frac{1}{2}\right)$ and the auxiliary point $H(-\frac{3}{2};-\frac{1}{2})$ are marked. Then between the line HO and the x-axis the segment c of length $HE = \frac{1}{2}\sqrt{2}$ is inserted in such a position that the straightedge passes through P.

If on the x-axis we consider the variable point $R(\xi, 0)$, then the line PR has the equation

$$y - \frac{1}{2} = -\frac{\frac{1}{2}}{\frac{1}{2}+\xi}\left(x+\frac{1}{2}\right).$$

This line cuts the line HO, with equation $y = +\frac{1}{3}x$, at the point $Q\left(\frac{3\xi}{2\xi+4}\right);$ $\frac{\xi}{2\xi+4}\bigg)$. Thus ξ is determined by the fact that $QR = \frac{1}{2}\sqrt{2}$, which leads after a slight calculation to $\xi^4 + \xi^3 - 2\xi - 2 = 0$, or $(\xi^3 - 2)(\xi + 1) = 0$. The linear factor $(\xi + 1)$ corresponds to the position PE for the straightedge.

4. Construction of the regular heptagon leads to the equation $x^3 + x^2 - 2x - 1 = 0$, whose positive solution[30] we shall denote by $x_1 = 2 \cos(2\pi/7)$.

The construction (Breidenbach, Fig. 25) is quite simple: draw the unit circle about O, mark the point $P(-1;\frac{1}{2})$ and the auxiliary point $H(-1;\frac{3}{2})$, and join H to O. Between the x-axis and the line HO let the segment $\frac{3}{2}$ be inserted in such a way that the straightedge passes through P (here $QR = \frac{3}{2}$).

The proof is like No. 3, since the construction leads to $(\xi^3 + \xi^2 - 2\xi - 1)$ $(\xi + 1) = 0$, where the linear factor $(\xi + 1)$ corresponds to the position PE.

5. Construction of the regular polygon of nine sides leads to the equation $x^3 - 3x + 1 = 0$; let us denote the greater of its two positive solutions by $x_1 = 2 \cos(2\pi/9)$.

For the construction (Breidenbach, Fig. 26) we determine the point P $(\frac{1}{2};\frac{1}{2}\sqrt{3})$ on the unit circle and insert a segment of length 2 between the x-axis and the y-axis in such a way that the straightedge passes through P (here $QR = 2$).

[30]Bieberbach [2], p. 58 f.

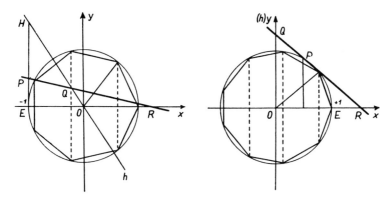

Fig. 25   Fig. 26

The proof runs as in No. 3. The construction provides $(\xi^3 - 3\xi + 1)$ $(\xi - 1) = 0$. The linear factor arises from the position PE for the straightedge.

6.   In the language of the theory of curves the insertion of the marked straightedge corresponds to the use of line-conchoids and circle-conchoids.

5.3.   The Carpenter's Square

1.   By a carpenter's square we mean the usual instrument with two sides forming a right angle at the vertex, which may be used not only to join points but also in the following way: given two points F and O and a line s, the instrument is so placed that the outer edge of one of its sides passes through F, the inner edge of the other side passes through O, and the vertex lies on the line s, and then a line is drawn along the outer edge of the second side.

The solution of the following problem is now immediate: draw the tangents common to the circle centered on O with radius b and the parabola defined by its focus F and its vertex-tangent s.

Instead of using an inner edge, we could first draw the circle, as in the following examples, and then move the carpenter's square in such a way that the outer edge of its second side is tangent to the circle.

2.   Angle Trisection (Bieberbach, Fig. 27). Let ASB be the given angle. Draw the circle $O(b)$ in the position shown in the figure and mark off $PS = SO = b$. Then place the carpenter's square as in the figure, i.e., so that the outer edge of its first side passes through P, the outer edge of its second side is tangent to the circle, and its vertex lies on SB.

3.   Doubling of the cube (Breidenbach, Fig. 28). Draw the circle with center M $(\frac{1}{2}; \frac{1}{2}\sqrt{5})$ and radius $\frac{1}{2}\sqrt{5}$. For the point F take $(-1; \frac{2}{5}\sqrt{5})$ and for the line s take the x-axis.

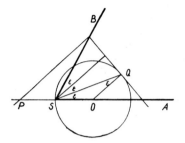

Fig. 27

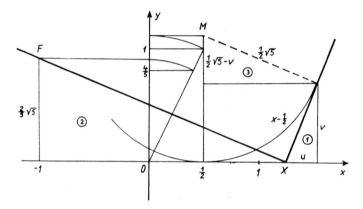

Fig. 28

Let the segment OX be denoted by x. Then from triangles 1 and 2 we obtain

$$u:v = \tfrac{2}{3}\sqrt{5}:(1 + x),$$

and from triangles 1 and 3

$$u:(x - \tfrac{1}{2}) = (\tfrac{1}{2}\sqrt{5} - v):\tfrac{1}{2}\sqrt{5}.$$

Calculating u and v and setting their values in

$$u^2 + v^2 = (x - \tfrac{1}{2})^2 \quad \text{(from triangle 1)},$$

we find

$$x^4 + x^3 - 2x - 2 = 0$$

or

$$(x^3 - 2)(x - 1) = 0.$$

4.   The regular heptagon (Breidenbach, Fig. 29). Draw the circle with center $M(-\tfrac{1}{2}; -\tfrac{1}{2})$ and radius $\tfrac{1}{2}$. For the point F take $(0; 2)$ and for the line s again take the x-axis. In the desired position the abscissa x of the point X is equal to $2\cos(2\pi/7)$, since x is a zero of the equation $x^3 + x^2 - 2x - 1 = 0$.

Let $OX = x$. Then $XB = XC = x + \tfrac{1}{2}$; $OY \cdot OF = OY \cdot 2 = x^2$, so that $OY = x^2/2$. Also, $XY^2 = x^4/4 + x^2$.

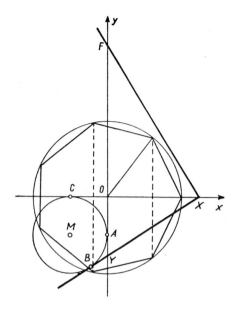

Fig. 29

But now $AY = BY$, so that $OY - OA = BX - YX$, from which we finally obtain $x^3 + x^2 - 2x - 1 = 0$.

Bibliography

[1] ADLER, A.: Theorie der geometrischen Konstruktion. Leipzig 1906.
[2] BIEBERBACH, L.: Theorie der geometrischen Konstruktionen. Basel 1952.
[3] BREIDENBACH, W.: Die Dreiteilung des Winkels, 2nd ed. Leipzig 1951.
[4] —: Das Delische Problem, 3rd ed. Stuttgart 1953.
[5] ENRIQUES, F.: Fragen der Elementargeometrie, Deutsche Ausgabe von Fleischer. II. Teil: Die geometrischen Aufgaben, ihre Lösung und Lösbarkeit. Leipzig 1907.
[6] HILBERT, D.: The foundations of geometry. Translated from the German by E. J. Townsend. Reprint edition. The Open Court Publishing Co., La Salle, Ill. 1959.
[7] SCHMIDT, H.: Die Inversion und ihre Anwendungen. Munich 1950.
[8] VAHLEN, TH.: Konstruktionen und Approximationen. Leipzig 1911.

Polygons and Polyhedra

A. Polygons

I. Introductory Remarks

1.1. *Definition*

A *polygon* (Fig. 1) is the figure formed by choosing a sequence of n arbitrary points $A_1, \ldots, A_n$, joining each point with the next by a line segment, and joining the last point with the first. The points $A_1, \ldots, A_n$ are the *vertices,* and the segments $A_1A_2, \ldots, A_nA_1$ are the *sides* of the polygon.

We speak of a *plane polygon* if all the vertices lie in one plane. Otherwise the polygon is said to be *skew.* In Part A we shall consider only plane polygons in the classical Euclidean plane.[1]

This definition, introduced by Poinsot, is more general than the one

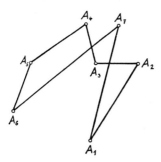

Fig. 1

[1]Cf. II2, where polygons are discussed from other points of view.

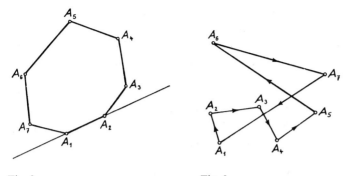

Fig. 2   Fig. 3

customary in secondary schools, where attention is confined to simple polygons. A polygon is said to be *simple* if no two vertices coincide, no vertex falls on a side, and no two sides have an inner point in common. Moreover, the emphasis is laid on convex polygons. A polygon is said to be *convex* if for each of its sides s all vertices except those by which s is bounded lie in the same open halfplane with respect to the line containing s (Fig. 2). (Cf. II15, §§1 and 2.) The intersection of all these halfplanes is called the *interior* of the polygon.

The Poinsot definition may be compared with another definition, due to Moebius, in which a polygon is defined as a system of finitely many line segments, each of which has each of its two endpoints in common with exactly one other line segment. This definition is identical with the preceding one if we also require *connectedness;* in other words, if every segment can be reached from any other segment, where passage from one segment to another is permissible only at a common endpoint.

1.2. *Angles of a Polygon*

Generalization of the concept of a polygon makes it necessary to consider the corresponding definition of an angle, which is not so obvious as for a simple polygon. Here it is convenient to introduce the notion of the *orientation* of a polygon, i.e., a cyclic ordering of the vertices in which no distinction is made between orders obtainable from one another by a cyclic permutation of all the vertices. More precisely, since by definition the vertices A_i are given in a fixed sequence, namely, $A_1, \ldots, A_n$, we define the given *orientation* and the *opposite orientation* respectively, as the class of cyclic permutations of the given sequence and the class of cyclic permutations of the A_i taken in the opposite order $A_n, \ldots, A_1$. An orientation induces a fixed orery of the two vertices bounding a side, and conversely the orientation of a given polygon is completely determined by the order of the two vertices of and one side. In a drawing the orientation is indicated by arrows (Fig. 3).

Let us first define the exterior angle at a vertex of an oriented polygon, as follows. We shall say that a side *ends* in a vertex A_i if the side joins this vertex to the vertex immediately preceding it in the orientation, and similarly the side joining A_i to the immediately following vertex is said to *issue* from A_i. Let us now consider the straight line that contains the side ending in this vertex. The order of the two vertices bounding the side determines a definite sense along the line. Let us turn the line, with preservation of its sense, around the vertex until it contains the side issuing from the vertex, the sense along the revolving line being required to agree again with the sense of traversal of the polygon. (Fig. 4a). If the angle of rotation is called positive or negative according as the rotation is clockwise or counterclockwise, then the angle of rotation β is uniquely determined by the condition

(1) $$-\pi < \beta \leq \pi,$$

where π denotes the straight angle, and this angle β is taken as the definition of the exterior angle at the vertex in question.

If we traverse the polygon in the sense of its orientation and successively determine the exterior angles at the vertices, the auxiliary line will eventually return to its original position. Thus we have proved

The sum of the exterior angles of an oriented polygon is an integral multiple of 2π.

If this sum is $2a\pi$, then a is called the *type* of the polygon. Thus the type is an integer, positive or negative. Reversal of the orientation of the polygon changes the sign of a. In particular, for a triangle $a = 1$ or $a = -1$, according to the orientation of the triangle.

The *interior angle* at a vertex is defined in a somewhat different way. We draw a straight line through the vertex along the side issuing from it (Fig. 4b) and turn the line, with preservation of the sense determined on it by the side

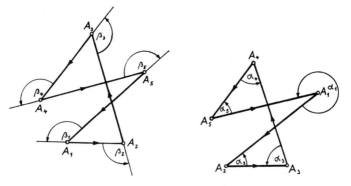

Fig. 4a   Fig. 4b

of the polygon, in the positive direction about the vertex until it coincides with the side ending in the vertex; here the sense of rotation of the rotating line is required to be the opposite of the sense of traversal of the polygon as determined by the side ending in the vertex. The angle α, uniquely determined by the condition

(2) $$0 \leqq \alpha < 2\pi,$$

is taken as the definition of the interior angle.

Since the final position of the auxiliary line in the definition of the exterior angle is the same as its initial position for the interior angle and its initial position for the exterior angle is oppositely directed to its final position for the interior angle, the sum of the exterior and the interior angles at a vertex must be equal to an odd multiple of π. From (1) and (2) it follows that

$$-\pi < \alpha + \beta < 3\pi,$$

so that

(3) $$\alpha + \beta = \pi.$$

Thus our definition has been so chosen that

The sum of the interior and exterior angles at a vertex is always equal to a straight angle, and also

The sum of the interior angles of an oriented polygon of type a is $(n - 2a)\pi$.

A general polygon is said to be *convex* if its orientation can be so chosen that every interior angle is $\leq \pi$. For polygons of type 1 this definition is identical with the elementary concept in §1.1. The polygon in Fig. 4b, is convex but not the one in Fig. 5.

The question whether a given oriented polygon is convex can be answered as follows. Under the opposite orientation the exterior angles β_i become $x_i = -\beta_i$, and therefore the corresponding interior angles $\alpha_i = \pi - \beta_i$ become $a_i = \pi - x_i = \pi + \beta_i$, where for the special case $\alpha_i = 0$, $\beta_i = \pi$ we must set $a_i = 2\pi$,

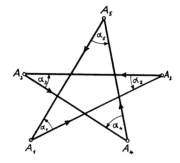

Fig. 5

$x_i = -\pi$, as an exception to our rules (1) and (2). So an oriented polygon is convex if and only if for its given orientation all the exterior angles have the same sign, or equivalently, if $0 \leq \alpha_i \leq \pi$ or $\pi \leq \alpha_i \leq 2\pi$ for all the interior angles.

2. Elementary Theory of the Regular Polygons

2.1. *Introduction*

A polygon whose sides and interior angles are equal is said to be *regular*. Here we do not exclude the case that the polygon is not simple.

The regular polygons are interesting from several points of view. On the one hand, they illustrate numerous theorems of elementary geometry, and on the other they enable us to combine geometry with algebra and the theory of numbers. We shall first consider the regular polygons from the point of view of the elementary geometry of secondary schools.

It is easy to prove that the vertices of a regular polygon lie on a circle, called the *circumcircle* of the polygon. The center of this circle is also called the *center* of the polygon and its radius is the *radius* of the polygon. Moreover, there always exists a circle, called the *incircle*, tangent to all the sides. The radii of the two circles are usually denoted by ρ and r.

In order to give a survey of the various kinds of n-gons, we divide the circumference of an arbitrary circle into n equal parts, denoting the points of division, in counterclockwise order, by

(4) $$A_1, \ldots, A_n.$$

It is convenient to introduce the symbols A_x, where x is an integer (positive, negative or zero) and to agree that A_x and A_y represent the same point if and only if $x \equiv y \pmod{n}$, that is, if $x - y$ is divisible by n.

Now let a be an integer coprime to n. Then the points

(5) $$A_0, A_a, A_{2a}, \ldots, A_{(n-1)a}$$

in this sequence are the vertices of a polygon of type a, as is easily proved.

For $a + a' = n$ the polygons of type a and of type a' are identical, except for the sense of traversal. Thus we may assume that

(6) $$0 < a < \tfrac{1}{2}n,$$

if we wish to disregard the sense.

If by $\varphi(n)$ we denote the well-known Euler totient function giving the number of positive integers $\leq n$ and prime to n (IB6, §4.2), we can say, disregarding orientation,

The number of types of regular n-gon inscribable in a circle is equal to $\tfrac{1}{2}\varphi(n)$. (See Fig. 6 for $n = 15$.)

The length of the side of a regular n-gon of type a will be denoted by $s_{n,a}$.

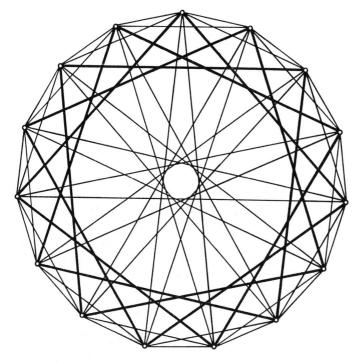

Fig. 6

For brevity, we shall often write s_n instead of $s_{n,1}$, and shall first direct our attention to polygons of type 1, since they are the only ones dealt with in elementary instruction.

2.2. *The Doubling Formula of von Ceulen*

The next three subsections deal with simple polygons. To be sure, the more general star-polygons can also be studied by the methods described here, but for a uniform treatment the algebraic method of later sections is more suitable.

One of the basic problems is to calculate the side of a regular n-gon when the radius is known. This problem cannot always be solved by the methods of elementary geometry, as becomes clear from the corresponding algebra. A further problem is to determine the side of a regular $2n$-gon, given the side of an n-gon with the same radius, and conversely, a problem which can be solved very easily, as we shall see. One of the best-known applications consists in the approximation of the number π, the ratio of the circumference to the diameter of a circle.

The most convenient solution of this latter problem depends on introduc-

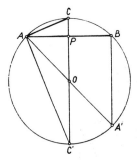

Fig. 7

ing the *complement* c_n of the side s_n, by which we mean the chord c_n subtending the arc complementary, in the semicircle, to the arc subtended by s_n. Consequently, the chords s_n and c_n are the sides of a right-angled triangle whose hypotenuse is the diameter of the circle. Since s_n and c_n are proportional to the radius r, there is no loss of generality in setting the radius equal to unity, so that

(7) $$c_n = \sqrt{4 - s_n^2}, \quad s_n = \sqrt{4 - c_n^2}.$$

Let us now consider a circle with radius 1 and center O. Let AB be the side of an inscribed n-gon and AC the side of an inscribed $2n$-gon (Fig. 7). If A' and C' are the antipodal points of A and C on the circumference, then CC' is perpendicular to AB; and if P is their point of intersection, then in the right-angled triangle CAC'

$$AC'^2 = C'P \cdot CC'.$$

Since $OP = \frac{1}{2}A'B = \frac{1}{2}c_n$, and $AC' = c_{2n}$, we may also write

$$c_{2n}^2 = (1 + \tfrac{1}{2}c_n)2$$

or

(8) $$c_{2n} = \sqrt{2 + c_n}.$$

This is the *doubling formula of von Ceulen,* to be considered as the fundamental formula in the elementary theory of polygons.

From (7) we can easily derive the usual doubling formula

(9) $$s_{2n} = \sqrt{2 - \sqrt{4 - s_n^2}}$$

but for the applications this formula is less convenient than (8). For example, if we start with $s_2 = 2$, so that $c_2 = 0$, we find at once

$$c_4 = \sqrt{2}, \; c_8 = \sqrt{2 + \sqrt{2}}, \; c_{16} = \sqrt{2 + \sqrt{2 + \sqrt{2}}}, \text{ etc.,}$$

and thus

$$s_2 = \sqrt{2}, \quad s_2 = \sqrt{2 - \sqrt{2}}, \quad s_{16} = \sqrt{2 - \sqrt{2 + \sqrt{2}}}, \quad \text{etc.,}$$

Starting with $s_6 = 1$, so that $c_6 = \sqrt{3}$, we obtain

$$c_{12} = \sqrt{2 + \sqrt{3}}, \quad c_{24} = \sqrt{2 + \sqrt{2 + \sqrt{3}}}, \quad \text{etc.}$$

and

$$s_{12} = \sqrt{2 - \sqrt{3}}, \quad s_{24} = \sqrt{2 - \sqrt{2 + \sqrt{3}}}, \quad \text{etc.}$$

Another remarkable relation can be read off from the triangle CAC' (Fig. 7); namely, from

$$AP \cdot CC' = AC' \cdot AC,$$

we have

(10) $$c_{2n} = \frac{s_n}{s_{2n}}.$$

Combining this formula with (8) we can derive an interesting homogeneous relation for s_n, s_{2n}, and s_{4n}. For if in (8) we first replace n by $2n$, writing

$$c_{4n}^2 = 2 + c_{2n},$$

and then substitute the corresponding values for c_{2n} and c_{2n} from (10), we obtain

$$\frac{s_{2n}^2}{s_{4n}^2} = 2 + \frac{s_n}{s_{2n}}$$

or

(11) $$s_{2n}^3 = s_{4n}^2 (2s_{2n} + s_n),$$

which is known as *Gregory's formula*.

2.3. The Regular Polygons of Five, Ten, and Fifteen Sides

Let us now solve the first of the two problems in §2.2, for the special cases $n = 5, 10, 15$.

We begin with $n = 10$. If AB is the side of the regular inscribed decagon, then $\angle AOB = 36°$. Thus $\angle ABO = \angle BAO = 72°$. If now at the point B we erect on AB the angle $\angle ABC = 36°$ (Fig. 8), then $\triangle AOB \sim \triangle ABC$ and

$$AO : AB = AB : AC.$$

Since $\angle COB = \angle CBO = 36°$ and $\angle CAB = \angle ACB = 72°$, therefore $CO = CB = AB$. Thus

$$AO : CO = CO : AC.$$

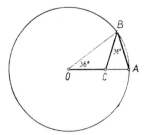

Fig. 8

So it is clear that the segment AO has been divided into two parts in such a way that the greater part CO is the mean proportional between the whole segment AO and the smaller part AC. The segment AO is then said to be divided in *golden section*, so that our result can be stated:

If the radius of a circle is divided in golden section the greater segment is equal to the side of the regular decagon inscribed in the circle.

For the side s_{10} we therefore have

$$1 : s_{10} = s_{10} : (1 - s_{10}),$$

so that s_{10} is s root of the equation

(12) $$s^2 + s - 1 = 0.$$

The positive root of this equation is

(13) $$s_{10} = \tfrac{1}{2}(\sqrt{5} - 1).$$

Thus we find at once

(14) $$c_{10} = \sqrt{4 - s_{10}^2} = \tfrac{1}{2}\sqrt{10 + 2\sqrt{5}}.$$

From (10) we now have

(15) $$s_5 = c_{10} s_{10} = \tfrac{1}{2}\sqrt{10 - 2\sqrt{5}}.$$

So we have solved the first problem for $n = 10$ and $n = 5$. Furthermore, we can calculate c_5 as

(16) $$c_5 = \tfrac{1}{2}(\sqrt{5} + 1).$$

Thus we have shown that the roots of equation (12) are given by s_{10} and $-c_5$.

From (13) and (16) it follows that

$$s_{10}^2 + c_5^2 = 3$$

or

(17) $$s_{10}^2 + 1 = s_5^2.$$

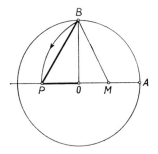

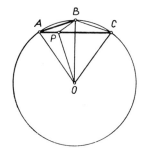

Fig. 9   Fig. 10

If now through the center O of a circle of radius 1 we draw two perpendicular radii OA and OB (Fig. 9) and let M be the midpoint of OA, then obviously $MB = \frac{1}{2}\sqrt{5}$ and $OP = s_{10}$, where P lies on the extension of AO with $MP = MB$. From (17) we then have $PB = s_5$ and have thereby provided a proof for one of the classical constructions for the 5-gon and the 10-gon.

Let us also give a geometric proof of (17). In the circle about O let AB be the side of the 10-gon and AC the side of the 5-gon (Fig. 10), where B and C lie on the same side of the line OA. Let P be the intersection of AC with the bisector of the angle AOB. Then $\triangle APB \sim \triangle ABC$, since $PA = PB$, and we have

$$AP : AB = AB : AC.$$

Further, $\sphericalangle CAB = \frac{1}{2} \cdot 36° = 18°$, so that $\sphericalangle OAC = 72° - 18° = 54°$, and $\sphericalangle COP = 36° + 18° = 54°$. Thus we also have $\triangle CPO \sim \triangle COA$ and

$$CP : CO = CO : CA.$$

As a result

$$AB^2 + CO^2 = AP \cdot AC + CP \cdot AC = AC^2,$$

which is the desired relation.

Finally, let us calculate the side of the regular 15-gon. In the circle with center O let AB be a chord of length equal to the radius, and let AC be the side of the 10-gon (Fig. 11). Here B and C lie on the same side of the line AC', where C' is the point antipodal to C. The arc AB is a sixth of the circumference of the circle and the arc AC is a tenth. Since $1/6 - 1/10 = 1/15$, the chord BC is the side of the 15-gon. Thus we have already given the construction of this polygon.

Let P be the foot of the perpendicular from A to BC. Since $\sphericalangle CBA = \sphericalangle CC'A$, we have $\triangle PBA \sim \triangle AC'C$. From $AB = \frac{1}{2}CC'$ it thus follows that

$$BP = \frac{1}{2}AC' = \frac{1}{2}c_{10}.$$

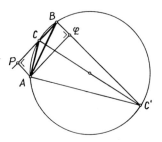

Fig. 11

Since the arc AB is a sixth of the circumference of the circle and $\measuredangle PCA$ is equal to the angle subtended by this chord at the circumference, we have $\measuredangle PCA = 30°$, from which follows

$$CP = \tfrac{1}{2}AC\sqrt{3} = \tfrac{1}{2}s_{10}\sqrt{3}.$$

Thus

(18) $$s_{15} = \tfrac{1}{2}(c_{10} - s_{10}\sqrt{3})$$

or, from (13) and (14),

(19) $$s_{15} = \tfrac{1}{4}(\sqrt{10 + 2\sqrt{5}} - \sqrt{15} + \sqrt{3}).$$

It is also easy to determine c_{15} from the figure. For let Q be the foot of the perpendicular from A to BC'. Then $BQ = PA = \tfrac{1}{2}s_{10}$ also, since $\measuredangle AC'B = 30°$,

$$C'Q = \tfrac{1}{2}AC'\sqrt{3} = \tfrac{1}{2}c_{10}\sqrt{3}.$$

For $BC = c_{15}$ we find

(20) $$c_{15} = \tfrac{1}{2}(c_{10}\sqrt{3} + s_{10})$$

or with the numerical values

(21) $$c_{15} = \tfrac{1}{4}(\sqrt{30 + 6\sqrt{5}} + \sqrt{5} - 1).$$

2.4.   Measurement of the Circle

The tangents to a circle parallel to an inscribed regular n-gon form a regular n-gon circumscribed about the circle. So if we begin, for example, with the side of the inscribed regular 6-gon and successively calculate the sides of the 12-gon, 24-gon, and so forth, and similarly for the circumscribed polygons, we obtain upper and lower bounds (after multiplication by 12, 24, etc.) for the circumference of the circle. The length of the circumference is the common limit approached by the circumferences of the two sequences of polygons as the number of sides increases. But this procedure, due to

Archimedes, is extremely laborious, since the 96-gon provides only the first two decimal places for the number π.

The following, more convenient method was developed by Huyghens and Snell for calculating these sequences

(22)
$$u_n, u_{2n}, \ldots,$$

and

(23)
$$U_n, U_{2n}, \ldots,$$

where u_n and U_n are the perimeters, for arbitrary n, of the inscribed and circumscribed n-gons.

Let AB be the side of a regular n-gon (Fig. 12) inscribed in a circle with center O, and let C be the midpoint of the arc corresponding to AB. Let the tangent at C cut OA and OB in D and E. Then DE is the side of the circumscribed n-gon. Now if P is the foot of the perpendicular from O to AB, then

$$DE : AB = OC : OP = OC : \tfrac{1}{2}A'B,$$

where A' is the point antipodal to A. Let the side of the circumscribed n-gon be denoted by S_n. Then

$$S_n : s_n = 1 : \tfrac{1}{2}c_n$$

or

(24)
$$c_n = 2\frac{s_n}{S_n}, \quad S_n = 2\frac{s_n}{c_n},$$

Since $c_n \to 2$ as $n \to \infty$, we have

$$\frac{u_n}{U_n} = \frac{s_n}{S_n} \to 1,$$

which means that the sequences (22) and (23) approach the same limit.

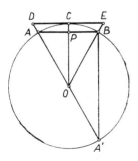

Fig. 12

From (10) and (24) we easily obtain

$$2\frac{s_{2n}}{S_{2n}} = c_{2n} = \frac{s_n}{s_{2n}}$$

or

$$2s_{2n}^2 = s_n S_{2n},$$

which is conveniently written in the form

(25) $$\qquad\qquad u_{2n}^2 = u_n U_{2n}.$$

In words:

The perimeter of the regular inscribed 2n-gon is the geometric mean between the perimeters of the inscribed regular n-gon and the circumscribed regular 2n-gon.

Formula (8) can also be written as

$$c_{2n} \cdot c_{2n} = 2 + c_n.$$

From (10) and (24) we see that

$$2\frac{s_{2n}}{S_{2n}}\frac{s_n}{s_{2n}} = 2 + 2\frac{s_n}{S_n}$$

or

$$\frac{1}{S_{2n}} = \frac{1}{s_n} + \frac{1}{S_n},$$

which again is conveniently written as

(26) $$\qquad\qquad \frac{2}{U_{2n}} = \frac{1}{u_n} + \frac{1}{U_n}.$$

In words:

The perimeter of the regular circumscribed 2n-gon is the harmonic mean between the perimeters of the regular circumscribed n-gon and regular inscribed n-gon.

Starting from $u_6 = 6$, $U_6 = 6.9282032$ it is quite easy to calculate the terms of the two numerical sequences (22) and (23); eight steps already give five decimal places for π.

Up to now we have approximated the circumference of a given circle with inscribed and circumscribed polygons of an increasing number of sides; but we can also derive interesting results about the number π by a different course, beginning with any regular polygon and deriving another from it, with twice as many sides but the same perimeter, so that as the procedure is repeated the circumferences of the inscribed and circumscribed circles will approach the perimeter of the original polygon. Here the following is of fundamental importance.

If r_n and ϱ_n are the radii of the inscribed and circumscribed circles for a regular n-gon and r_{2n} and ϱ_{2n} are the corresponding radii for the 2n-gon of the same perimeter, then

(27) $$\varrho_{2n} = \tfrac{1}{2}(\varrho_n + r_n)$$

and

(28) $$r_{2n} = \sqrt{r_n \varrho_{2n}}.$$

These are the so-called *formulas of J. C. Schwab*. Let AB be a side of a regular n-gon, and let C be the midpoint of the arc corresponding to AB on the circumscribed circle with center O. If A^* and B^* are the midpoints of CA and CB (Fig. 13) then obviously $A^*B^* = \tfrac{1}{2}AB$ and $\measuredangle A^*OB^* = \tfrac{1}{2}\measuredangle AOB$. Thus A^*B^* is the side of a regular $2n$-gon which has the same perimeter as the original n-gon. Then if P is the foot of the perpendicular from O to AB and P^* is the corresponding point on A^*B^*, it is clear that P^* is the midpoint of PC, so that

$$2OP^* = OP + OC$$

or

$$2\varrho_{2n} = \varrho_n + r_n.$$

Thus we have already proved formula (27).

In the right-angled triangle OA^*C we have

$$OA^{*2} = OP^* \cdot OC,$$

or

$$r_{2n}^2 = \varrho_{2n} r_n,$$

which is equivalent to formula (28).

An attractive application of the theorem just proved is as follows. Let us

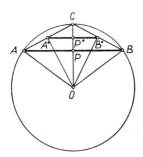

Fig. 13

start from a square with perimeter equal to 2 and form successively the numbers $\varrho_4, r_4, \varrho_8, r_8, \ldots$. It is clear that the two sequences $2\pi\varrho_4, 2\pi\varrho_8, \ldots$ and $2\pi r_4, 2\pi r_8, \ldots$ approach the same value 2 and that the numbers of the first sequence are all smaller, and those of the second all greater, than this limiting value. Thus we have proved the *theorem of Schwab*;

The number $1/\pi$ is the limiting value of the sequence

$$c_0, c_1, c_2, \ldots$$

obtained by setting $c_0 = 0$, $c_1 = \frac{1}{2}$ and alternately taking the arithmetic and harmonic means of the two preceding numbers. The numbers are alternately smaller and greater than $1/\pi$.

The resources of higher analysis allow us to construct much better methods of approximating π, but it is interesting that even the simplest elementary geometry already provides a useful approximation.

3. The Algebraic Theory of Regular Polygons

3.1. *Cyclotomy*[2]

It is well known that the complex number

(29) $$z = x + iy$$

can be represented by a point with rectangular coordinates $\langle x, y \rangle$ in a plane. In this representation the n solutions of the binomial equation

(30) $$z^n - 1 = 0,$$

namely, the so-called nth roots of unity, lie on the circle with radius 1 and center at the origin. Often it is convenient to write z in the trigonometric form

(31) $$z = e^{i\theta} = \cos \theta + i \sin \theta.$$

Then the number z satisfies equation (30) if and only if

(32) $$n\theta \equiv 0 \pmod{2\pi}.$$

A complete system of solutions is given by

(33) $$\theta = \frac{2a\pi}{n}, \quad a = 0, 1, \ldots, n - 1.$$

Consequently,

The points representing the n-th roots of unity are the vertices of a simple regular n-gon.

[2]Cf. IB7, §5.

Since $z = 1$ is always a solution of (30), we can divide the left side by $z - 1$. The remaining roots satisfy the equation

$$(34) \qquad z^{n-1} + z^{n-2} + \cdots + z + 1 = 0,$$

which will be the starting point for our further discussion.

As an application of this equation, which will also show that sometimes there is no need to solve it explicitly (e.g., if we wish to verify the correctness of a construction found in some other way), let us consider von Staudt's construction of the regular pentagon.

At the points $z = 1$ and $z = i$ draw the tangents to the unit circle, and let A be their point of intersection (Fig. 14). At the point $z = -i$ draw a second horizontal tangent and mark off on it the radius four times to the left from the point of contact, letting B be the point thus obtained. Let the line AB intersect the circle at P and Q, join P, Q to the point $z = -i$, and let P_1, Q_1 be the intersections of these lines with the horizontal axis. Then the vertices of a regular pentagon are given by the point $z = 1$ and the intersections with the circle of the lines through P_1 and Q_1 perpendicular to the horizontal axis.

In order to prove the correctness of this construction we first note that A represents the complex number $1 + i$ and B the complex number $-4 - i$. Thus every point on the line AB is given by

$$(35) \qquad z = \lambda(1 + i) - (1 - \lambda)(4 + i) = (5\lambda - 4) + (2\lambda - 1)i,$$

where λ is a real parameter. For the intersections of the line AB with the circle we have

$$(5\lambda - 4)^2 + (2\lambda - 1)^2 = 1,$$

or

$$(36) \qquad 29\lambda^2 - 44\lambda + 16 = 0.$$

Any point on the line joining the point (35) with the point $z = -i$ represents the complex number

$$(5\lambda - 4)\mu + (2\lambda - 1)\mu i - (1 - \mu)i = (5\lambda - 4)\mu + (2\lambda\mu - 1)i.$$

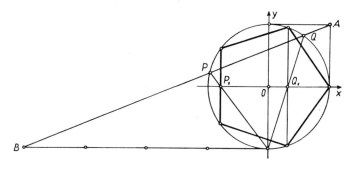

Fig. 14

Thus such a point lies on the real axis if $2\lambda\mu = 1$, when it is given by

$$x = \frac{5\lambda - 4}{2\lambda}.$$

Substituting

$$\lambda = \frac{4}{5 - 2x}$$

into the left side of equation (36), we find for x the equation

(37) $4x^2 + 2x - 1 = 0.$

But for every point on the circumference of the unit circle we have

$$z\bar{z} = 1,$$

(where $\bar{z}$ denotes the complex number conjugate to z) and therefore

$$2x = z + \bar{z} = z + z^{-1}.$$

Consequently the lines P_1 and Q_1 perpendicular to the real axis cut the circle at points representing the solutions of the equation

$$(z + z^{-1})^2 + (z + z^{-1}) - 1 = 0$$

or

(38) $z^4 + z^3 + z^2 + z + 1 = 0.$

But this is equation (34) for $n = 5$.

3.2. Primitive nth Roots of Unity

The roots of equation (30) are the numbers

(39) $\varepsilon_{n,a} = e^{2a\pi i/n}, \quad a = 0, 1, \ldots, n - 1.$

Of course,

$$\varepsilon_{n,a}^n = 1,$$

but it is also possible that

$$\varepsilon_{n,a}^m = 1$$

with $0 < m < n$. The smallest positive integer m for which equation (40) holds is called the *period* of the root $\varepsilon_{n,a}$. Let us determine this period. Since

(40) $\varepsilon_{n,a}^m = e^{2ma\pi i/n},$

equation (40) will hold if and only if ma/n is an integer. Let g be the greatest common divisor of a and n, so that $a = a_1 g, n = n_1 g$, with coprime a_1 and n_1. Then it is clear that ma_1/n_1 will be an integer if and only if m is a multiple of $n_1 = n/g$. In fact, $m = n/g$, since m was chosen as the smallest number of this sort. Thus we have the desired result:

The period of every nth root of unity is a divisor of n. The period of the root $\varepsilon_{n,a}$ is found by dividing n by the greatest common divisor of a and n.

The period is equal to n if a and n are coprime, since g is then equal to 1.

In this case the n-th root of unity is said to be *primitive;* in particular $\varepsilon_{n,1}$ is always primitive. *It is clear that the roots with period m are primitive roots of the equation $z^m - 1 = 0$.*

The number of primitive n-th roots of unity is equal to $\varphi(n)$, where again $\varphi(n)$ is the Euler totient function (see §2.1 and IB6, §4.2).

From this theorem it is clear that the number of n-th roots of unity is equal to the sum $\sum \varphi(d)$, where d runs through all positive divisors of n (including 1 and n). Thus we have the well-known *formula of Gauss:*

$$(41) \qquad\qquad \sum \varphi(d) = n$$

with $0 < d$, where d is a divisor of n.

3.3. *Equation for the Primitive nth Roots of Unity*

The primitive n-th roots of unity satisfy an equation

$$(42) \qquad\qquad \Phi_n(z) = 0,$$

where $\Phi_n(z)$ is a polynomial of degree $\varphi(n)$ with highest coefficient equal to unity. From the preceding results we find

$$(43) \qquad\qquad z^n - 1 = \Phi_{d_1}(z) \ldots \Phi_{d_t}(z),$$

where $d_1, \ldots, d_t$ are the (positive) divisors of n, arranged in order of magnitude. In particular, $d_1 = 1$, $d_t = n$. From (43) it follows that

$$(44) \qquad\qquad \Phi_n(z) = \frac{z^n - 1}{\Phi_1(z) \cdots \Phi_{d_{t-1}}(z)},$$

and thus we can easily calculate $\Phi_n(z)$ if all the $\Phi_d(z)$, where $d < n$ is a divisor of n, are already known. If n is a prime p, then only $\Phi_1(z) = z - 1$ occurs in the denominator, and we have

$$(45) \qquad\qquad \Phi_p(z) = z^{p-1} + \cdots + z + 1.$$

It is easy to show that *all the polynomials must have integral coefficients,* since this assertion is correct with $n = 1$, and if we assume its correctness for all $m < n$, then from (44) it is also correct for $m = n$, since the highest coefficients of the polynomials in the denominator are equal to unity.

The coefficients in $\Phi_n(z)$ are the same from right to left as from left to right.

For with ε the number ε^{-1} is also a primitive root.

3.4. *Practical Calculation of the Polynomials $\Phi_n(z)$*

The formula (44) describes a recursive procedure by means of which, starting from $\Phi_1(z) = z - 1$, we can successively determine the polynomials $\Phi_2(z)$, $\Phi_3(z)$, ... But for large values of n this procedure is quite laborious, and does not give us a clear picture of the properties of the polynomials $\Phi_n(z)$. However, we can prove certain general theorems which will shorten the computation a great deal. We first prove:

If p is a prime divisor of n, then

$$(46) \qquad\qquad \Phi_{np}(z) = \Phi_n(z^p).$$

The roots of $\Phi_{np}(z) = 0$ are

$$(47) \qquad\qquad \varepsilon_{np,a} = e^{2a\pi i/np},$$

where a and np are coprime. But then a and n are also coprime, and the numbers $\varepsilon^p_{np,a}$ are precisely the $\varepsilon_{n,a}$.

If p is a prime that does not divide n, then

$$(48) \qquad\qquad \Phi_{np}(z) = \frac{\Phi_n(z^p)}{\Phi_n(z)}.$$

In this case the numbers (47) are characterized by the conditions that n and a are coprime and a is not divisible by p. If we omit the second condition, we obtain the roots of $\Phi_n(z) = 0$ as well, which means that the zeros of $\Phi_{np}(z)$ are given by the numbers $\varepsilon_{np,a}$, together with the $\varepsilon_{n,a}$.

The case $p = 2$ deserves further attention.

If the number n is even, then $\Phi_{2n}(z)$ is formed from $\Phi_n(z)$ by replacing z by z^2.

This theorem is only a special case of the first theorem above.

If the number $n > 1$ is odd, then $\Phi_{2n}(z)$ is obtained from $\Phi_n(z)$ by replacing z by $-z$:

$$(49) \qquad\qquad \Phi_{2n}(z) = \Phi_n(-z), \quad n > 1 \text{ and odd.}$$

For if $d_1, \ldots, d_t$ are the divisors of n, then from (45) and (44) we have

$$\Phi_{2n}(z) = \frac{z^{2n} - 1}{\Phi_{d_1}(z)\cdots\Phi_{d_t}(z)\Phi_{2d_1}(z)\cdots\Phi_{2d_{t-1}}(z)} = \frac{z^n + 1}{\Phi_{2d_1}(z)\cdots\Phi_{2d_{t-1}}(z)}.$$

If we assume the result for $1 < m < n$ and note that $\Phi_{2d_1}(z) = \Phi_2(z) = z + 1$, we can write the last term in the above equation as

$$\frac{-(-z)^n + 1}{\{-(-z) + 1\}\Phi_{2d_2}(z)\cdots\Phi_{2d_{t-1}}(z)} = \frac{(-z)^n - 1}{(-z - 1)\Phi_{d_1}(-z)\cdots\Phi_{d_{t-1}}(-z)}$$

$$= \Phi_n(-z).$$

From this theorem and the second theorem we also have the identity:

$$(50) \qquad\qquad \Phi_n(z^2) = \Phi_n(z)\Phi_n(-z), \quad n < 1 \text{ and odd.}$$

Repeated application of the first theorem leads to the result:

If m is product of the distinct prime factors of n, then $\Phi_n(z)$ is obtained from $\Phi_m(z)$ by multiplying all exponents with n/m:

$$(51) \qquad\qquad \Phi_n(z) = \Phi_m(z^{n/m}).$$

For example,

$$\Phi_6(z) = \frac{\Phi_2(z^3)}{\Phi_2(z)} = \frac{z^3 + 1}{z + 1} = z^2 - z + 1,$$

from which

$$\Phi_{24}(z) = \Phi_6(z^4) = z^8 - z^4 + 1.$$

In conclusion, let us interpret these theorems from the point of view of the theory of numbers. From (49) it follows at once by comparison of the degrees that

(52) $$\varphi(2n) = \varphi(n), \quad n \text{ odd.}$$

Of course, the conclusion follows only under the assumption $n > 1$, but the equation is also correct for $n = 1$.

Now let the number n be decomposed into its prime factors

(53) $$n = p_1^{\alpha_1} \cdots p_r^{\alpha_r}.$$

From (51) it follows by comparison of degrees that

$$\varphi(n) = \frac{n}{m} \varphi(m) = p_1^{\alpha_1 - 1} \cdots p_r^{\alpha_r - 1} \varphi(p_1 \cdots p_r).$$

From (48) we find

$$\varphi(p_1 \cdots p_r) = p_r \varphi(p_1 \cdots p_{r-1}) - \varphi(p_1 \cdots p_{r-1}) = (p_r - 1)\varphi(p_1 \cdots p_{r-1}).$$

Preceding this way we finally obtain

(54) $$\varphi(n) = p_1^{\alpha_1 - 1}(p_1 - 1) \cdots p_r^{\alpha_r - 1}(p_r - 1),$$

which is the well-known explicit representation of the function $\varphi(n)$ (Cf. IB6, §5, Theorem 7).

3.5. Equation for the Length of the Side of a Regular Polygon

The d powers of a primitive root $\varepsilon_{n,a}, \varepsilon_{n,a}^2, \ldots, \varepsilon_{n,a}^n$ are represented by n distinct points on the unit circle. If we start from such a point and join it to the next, in counterclockwise order around the circle, we obtain a regular polygon of type a. This polygon is already determined by any one of its vertices, since the sides in each case subtend a of the n arcs into which the circumference is divided.

Then it is clear that the following assertion holds:

Each of the solutions of the equation

(55) $$\Phi_n(z) = 0$$

determines a regular n-gon. The type is in each case given by a number co-prime to n.

The roots $\varepsilon_{n,a}$ and $\varepsilon_{n,a'}$ with $a + a' = n$ are reciprocal to each other. For

$$\varepsilon_{n,a'} = e^{2a'\pi i/n} = e^{2(n-a)\pi i/n} = e^{-2a\pi i/n} = \varepsilon_{n,a}^{-1}.$$

Thus $\varepsilon_{n,a}$ and $\varepsilon_{n,a'}$ determine the same polygons, but with opposite sense. Disregarding the sense, we see (cf. §2.1) that the number of distinct types of regular inscribed n-gons is equal to $\frac{1}{2}\varphi(n)$.

Whether a regular n-gon can be constructed with ruler and compass depends on whether equation (45) can be solved in square roots. This purely algebraic question will not be discussed here, the final result (cf. IB7, §10.3) being as follows (Gauss):

A regular n-gon is constructible if and only if the number n can be repre-

sented in the form $n = ab$, *with* $a = 2^\alpha$, $\alpha \geqq 0$, *and* $b = 1$ *or* $= p_1 \cdots p_r$, *where* $p_1, \ldots, p_r$ *are old primes of the form* $2^\beta + 1$ (IB6, §2.12).

In general, the sides of an n-gon cannot be expressed in terms of square roots but, as we shall see, their lengths are algebraic irrationalities, so that *the natural representation of the side of an n-gon is simply the equation of which it is a root.* We shall see that this equation includes all types for a given n.

For $n > 2$ the number $\phi(n)$ is even. From the last theorem of §3.3, we see that division of $\Phi_n(z)$ by $z^{\frac{1}{2}\phi(n)}$ leads to a linear combination of expressions of the form

$$z^k + z^{-k}$$

with integral coefficients. But these expressions can be represented (trivially for $z = 1$) as entire rational functions in $z + z^{-1}$ by making use of the relation

$$(z^k + z^{-k})(z + z^{-1}) = (z^{k+1} + z^{-(k+1)}) + (z^{k-1} + z^{-(k-1)}).$$

Thus the equation (45) can be replaced by another equation whose roots are the numbers

$$(56) \qquad\qquad \varepsilon_{n,a} + \varepsilon_{n,a}^{-1} = 2\cos\frac{2a\pi}{n}.$$

The side of a regular n-gon of type a will be calculated from

$$(57) \qquad\qquad s_{n,a} = 2\sin\frac{a\pi}{n},$$

and thus, since

$$(58) \qquad\qquad 2\cos\frac{2a\pi}{n} = 2 - 4\sin^2\frac{a\pi}{n},$$

we have

The sides of a regular n-gon are the roots of an equation

$$(59) \qquad\qquad \Psi_n(s) = 0,$$

where the polynomial $\Psi_n(s)$ *can be obtained from* $\Phi_n(z)$ *by the transformation*

$$(60) \qquad\qquad z + z^{-1} = 2 - s^2.$$

The degree of the polynomial $\Psi_n(s)$ is again $\varphi(n)$. However, the roots are numerically equal in pairs but of opposite sign, so that it is only necessary to consider the $\frac{1}{2}\varphi(n)$ positive roots.

For $n = 6$ we start from

$$\Phi_6(z) = z^2 - z + 1,$$

and find the polynomial

$$\Psi_6(z) = -s^2 + 1.$$

The only positive zero of this polynomial is $s = 1$, as was to be expected.
For $n = 5$,

$$\Phi_5(z) = z^4 + z^3 + z^2 + z + 1,$$

from which we obtain

$$\Psi_5(s) = s_5(s) = s^4 - 5s^2 + 5.$$

The positive zeros are

$$s_{5,1} = \tfrac{1}{2}\sqrt{10 - 2\sqrt{5}}, \quad s_{5,3} = \tfrac{1}{2}\sqrt{10 + 2\sqrt{5}}.$$

In general, it is extremely laborious to calculate a solution of (58) directly, even when *a priori* we can expect an answer in square roots. *Thus we must regard the equation as the final expression of the length of the sides.* For numerical purposes it is most convenient to calculate directly from (57); a formula will sometimes allow us to obtain the solutions in algebraic form. Let us illustrate by the example of the regular 15-gon. To begin with, we have

$$\Phi_{15}(z) = z^8 - z^7 + z^5 - z^4 + z^3 - z + 1.$$

Then by the transformation (60) we find

(61) $$\Psi_{15}(z) = s^8 - 7s^6 + 14s^4 - 8s^2 + 1.$$

The positive zeros of this polynomial are

$$s_{15,1} = 2\sin\frac{\pi}{15}, \quad s_{15,2} = 2\sin\frac{2\pi}{15}, \quad s_{15,4} = 2\sin\frac{4\pi}{15}, \quad s_{15,7} = 2\sin\frac{7\pi}{15}.$$

Noting that

$$\frac{1}{15} = \frac{1}{6} - \frac{1}{10}, \quad \frac{2}{15} = \frac{3}{10} - \frac{1}{6}, \quad \frac{4}{15} = \frac{1}{10} + \frac{1}{6}, \quad \frac{7}{15} = \frac{3}{10} + \frac{1}{6}.$$

we thus obtain

$$\sin\frac{\pi}{15} \qquad\qquad\qquad = \sin\frac{\pi}{6}\cos\frac{\pi}{10} - \cos\frac{\pi}{6}\sin\frac{\pi}{10},$$

$$\sin\frac{2\pi}{15} = -\sin\frac{\pi}{6}\cos\frac{3\pi}{10} + \cos\frac{\pi}{6}\sin\frac{3\pi}{10} = -\sin\frac{\pi}{6}\sin\frac{\pi}{5} + \cos\frac{\pi}{6}\cos\frac{\pi}{5},$$

$$\sin\frac{4\pi}{15} = \sin\frac{\pi}{10}\cos\frac{\pi}{6} + \cos\frac{\pi}{10}\sin\frac{\pi}{6} = \sin\frac{\pi}{6}\cos\frac{\pi}{10} + \cos\frac{\pi}{6}\sin\frac{\pi}{10},$$

$$\sin\frac{7\pi}{15} = \sin\frac{3\pi}{10}\cos\frac{\pi}{6} + \cos\frac{3\pi}{10}\sin\frac{\pi}{6} = \sin\frac{\pi}{6}\sin\frac{\pi}{5} + \cos\frac{\pi}{6}\cos\frac{\pi}{5}$$

or

$$s_{15,1} = \tfrac{1}{2}(c_{10} - c_6 s_{10}),$$

$$s_{15,2} = \tfrac{1}{2}(-s_5 + c_6 c_5),$$

$$s_{15,4} = \tfrac{1}{2}(c_{10} + c_6 s_{10}),$$

$$s_{15,7} = \tfrac{1}{2}(s_5 + c_6 c_5).$$

From the numerical values (13), (14), (15), and (16), we can write

$$s_{15,1} = \tfrac{1}{4}(\sqrt{10 + 2\sqrt{5}} - \sqrt{15} + \sqrt{3}),$$

$$s_{15,2} = \tfrac{1}{4}(-\sqrt{10 - 2\sqrt{5}} + \sqrt{15} + \sqrt{3}),$$

$$s_{15,4} = \tfrac{1}{4}(\sqrt{10 + 2\sqrt{5}} + \sqrt{15} - \sqrt{3}),$$

$$s_{15,7} = \tfrac{1}{4}(\sqrt{10 - 2\sqrt{5}} + \sqrt{15} + \sqrt{3}),$$

which gives us all the positive roots of the equation $\Psi_{15}(s) = 0$.

The complements $c_{n,a}$, being related to $s_{n,a}$ by

$$c_{n,a}^2 = 4 - s_{n,a}^2,$$

satisfy an equation obtained from $\Phi_n(z)$ by the substitution

(62) $z + z^{-1} = s^2 - 2,$

and like the sides they are characterized by this equation.

B. Polyhedra

4. Morphology of Polyhedra

4.1. *Definitions*

By a polyhedron we mean a system of plane polygons with the property that every side of each of these polygons is also the side of exactly one other polygon in the system. In this definition the polygons are understood in the sense of Moebius (§1.1). We shall not at first require the polygons to be connected.[3]

The polygons are called the *faces* of the polyhedron, their sides are its

---

[3]For a more general (abstract) definition of "polyhedron," see IB9, §3.

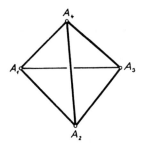

Fig. 15

edges, and their vertices are its *vertices*. As a result of the definition, every edge has exactly two vertices, and every edge belongs to exactly two faces.

If the vertices of a polyhedron, in some definite order, are denoted by $A_1, \ldots, A_e$, its *schema* is obtained by grouping together the vertices of each of the faces. A simple example being the tetrahedron in Fig. 15, with the schema

(63) $A_1A_2A_3, \quad A_1A_2A_4, \quad A_1A_3A_4, \quad A_2A_3A_4.$

Thus in a schema the vertices are assembled in certain subsets. Polyhedra corresponding to the same schema are called *isomorphic* and are considered to be polyhedra of the same type (cf. also II16, Introduction). That part of the theory of polyhedra in which no distinction is made between isomorphic polyhedra is called the *morphology* of polyhedra.

If the faces are connected polygons, we can represent the schema, in an easily visualized way, by means of a system of polygons lying in a plane, where each of the polygons corresponds to one of the subsystems of the schema. It follows that vertices of distinct polygons may be denoted by the same symbol. The act of combining vertices which have the same notation, and which therefore correspond to the same vertex of the polyhedron, is called *identification*. In the same way we can identify the sides of the polygons by grouping together the sides which correspond to the same edge of the polyhedron. The way in which sides are to be identified is immediately clear from the notation for the vertices.

If a system of f connected polygons is given in the plane and if identification of certain vertices leads to the schema of a polyhedron, the schema is said to be a *polyhedral net*. In making a diagram it is customary to have as many vertices come together as possible. This can be done in various ways, as shown in Fig. 16, which represents two different nets for a tetrahedron.

4.2.   *Orientable and Nonorientable Polyhedra*

From now on let us consider polyhedra whose faces are connected polygons, oriented by a definite sequence of their vertices (§1.2). We say that the faces are *coherently oriented* if the following situation holds: consider a side common to two adjacent faces; the orientation of each of these faces will determine an order for the two endpoints of the side; if these two orders are opposite to each other, the faces are said to be coherently oriented. For example, the orientation of the faces of the tetrahedron in Fig. 15 is coherent if it is determined by the sequences

(64) $A_1A_2A_3,\quad A_4A_2A_1,\quad A_4A_1A_3,\quad A_4A_3A_2.$

A polyhedron whose faces can be coherently oriented is said to be *orientable* (cf. II16 §1), or to obey the *Moebius edge rule*.

All the polyhedra occurring in elementary geometry are orientable, but it is not difficult to construct nonorientable examples. Let us make a somewhat closer study of two of them, most easily described by constructing a net.

In the first example (Fig. 17) isosceles triangles are drawn outwards from the sides of a regular pentagon and the free vertex of each of the triangles is identified with the vertex of the pentagon opposite the common side. If we also consider the five triangles determined by the sides of the pentagon and its midpoint, we obtain a net for a polyhedron whose nonorientability is easily proved. Let us also show how we can construct a model of it in space.

The most convenient way is to begin by ignoring the triangles lying inside

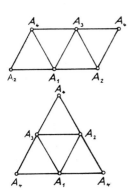

Fig. 16

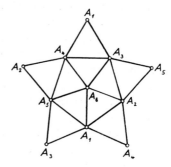

Fig. 17

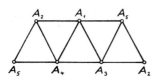

Fig. 18

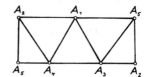

Fig. 19

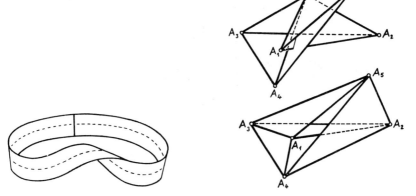

Fig. 20   Fig. 21

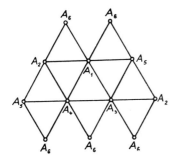

Fig. 22

the pentagon. Thus we first consider a strip of triangles as drawn in Fig. 18. By means of a continuous deformation we can transform the strips into a rectangle in which two of the opposite sides are identified (Fig. 19). A visualizable model is obtained by fastening the ends of a strip of paper together to form a closed band after first turning the strip through 180° (Fig. 20). Such a configuration is called a *Moebius strip*.

The line segments in Fig. 18 that belong to only one triangle form a polygon $A_1A_2A_3A_4A_5$ representable (see Fig. 21) as a skew polygon in space in the manner shown in Fig. 21. By the addition of a sixth vertex A_6 the strip of triangles can be extended to form to a polyhedron with ten faces; in other words, the triangles $A_6A_1A_2, \ldots, A_6A_5A_1$ must be added. The polyhedron thus obtained is called the *nonorientable decahedron of Moebius*.

In Fig. 22 the net for this decahedron is represented in a somewhat different form, consisting of half of the net of twenty triangles in Fig. 23, which represents an orientable polyhedron that can be realized in space as a dodecahedron (Fig. 24).

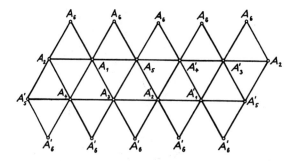

Fig. 23

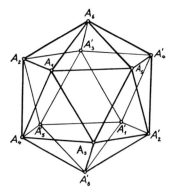

Fig. 24

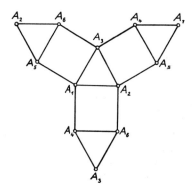

Fig. 25

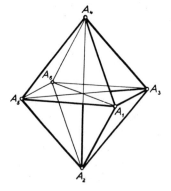

Fig. 26

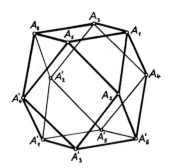

Fig. 27

Now there exists a remarkable relationship between the decahedron and the icosahedron, as follows: the vertices, edges, and faces of the icosahedron can be combined in pairs in such a way that to each pair there corresponds a vertex, edge, or face, respectively, of the decahedron. In space it is precisely the opposite elements of the polyhedron in Fig. 24 that are combined in pairs. The icosahedron is called a *two-sheeted covering* of the decahedron. The notation has been so chosen that A_i and A_i' are opposite to each other and correspond to the same vertex of the decahedron.

Our second example of a nonorientable polyhedron, characterized by the net in Fig. 25, has only seven faces. It can be realized in space as a regular octahedron in which four suitably chosen triangles have been removed (Fig. 26) and three diagonal faces have been adjoined. The two-sheeted covering corresponding to this *nonorientable heptahedron* is the semiregular polyhedron of fourteen sides in Fig. 27. The above relationship between the heptahedron and its covering is easily verified in the space models.

5. The Euler Formula for Polyhedra and the Classification of Certain Types of Polyhedra

5.1. *Convex Polyhedra*

The concept of a polyhedron in the preceding section is still too general. In order to make more precise statements and to gain a clearer picture of the various forms of polyhedra, we must confine our attention to a subset of the general polyhedra. In elementary geometry it is customary to consider only convex polyhedra, which are relatively easy to study.

A *convex* polyhedron is a system of finitely many convex polygons (cf. §1.1) with the property that each side of every polygon is also a side of a second polygon and for every polygon in the system the vertices of the polyhedron that do not lie on the given polygon are all in the same halfspace determined by the plane of the polygon (cf. II15, §§1 and 2).

The plane of a polygon bounds two halfspaces, one of which contains the vertices of the polyhedron that do not lie in the given plane. The intersection of all these halfspaces forms the *interior* of the polyhedron.

Let us now project a convex polyhedron orthogonally onto a plane ε such that ε is not perpendicular to any of the faces and no two vertices of the polyhedron have the same projection. Since the polyhedron has only finitely many vertices, edges, and faces, it is clear that ε can always be chosen to satisfy these conditions, in which case we shall say that ε is in general position.

The projections of the vertices of the polyhedron form a finite system of points. Let us consider the convex hull of this system, by which we mean the smallest convex polygon containing the points in its interior or on its boundary. It is easy to see that the points not lying in its interior are vertices of the hull, since otherwise ε would be perpendicular to a face of the polyhedron. Thus the sides of the hull are projections of certain edges, which taken together form a (generally skew) polygon. The convex hull is called the *outline* of the polyhedron on the plane ε. Summing up, we can say

If a convex polyhedron is orthogonally projected onto a plane in general position, the vertices of the polyhedron are projected into the interior or onto the boundary of a convex polygon.

Without proof we also mention the following theorem of Cauchy:

Two isomorphic convex polyhedra are congruent if their corresponding faces are congruent.

Let us now make a further study of convex polyhedra, which will illustrate the essential features of the general theory, even though many of our results could be obtained under less restrictive conditions.

5.2. The Euler Formula for Polyhedra

For convex polyhedra there is a simple relationship connecting the number e of vertices, k of edges, and f of faces, expressed by the well-known Euler formula

(65)
$$e - k + f = 2,$$

which is most easily proved for convex polyhedra, although it holds more generally.

To prove this formula, let us project the polyhedron orthogonally onto a plane ε in general position (cf. §5.1). The sum ω of the interior angles of the faces of the polyhedron is equal to the sum of their projections; for although the individual angles are changed in the mapping, every n-gon is projected again into a convex n-gon, the sum of whose angles remains the same, namely, $(n - 2)\pi$. If the polygon forming the outline is an m-gon, its vertices are the projections of m vertices of the polyhedron, while the remaining $e - m$ vertices are projected into the interior of the outline. The sum of the projections of the angles at each of the $e - m$ vertices is 2π, but for each of the m vertices projected into the outline, this sum is equal to twice the angle in the outline since any polyhedral edge that is projected as a side of the outline is a side in two faces of the polyhedron, so that in the projection the angle in the outline is covered twice by the plane angles of the polyhedron.

For the sum ω we thus obtain

(66) $$\omega = (e - m)2\pi + 2(m - 2)\pi = 2(e - 2)\pi.$$

The sum of the angles of an n-gon of the polyhedron is $(n - 2)\pi$, so that if the faces have $n_1, \ldots, n_f$ sides, respectively, then

$$\omega = (n_1 + \cdots + n_f - 2f)\pi,$$

where the sum $n_1 + \cdots + n_f$ is obviously twice the number of edges. Thus

(67) $$\omega = 2(k - f)\pi.$$

From (66) and (67) we at once obtain (65).

5.3. *Some Necessary Conditions for the Existence of a Convex Polyhedron*

Since the numbers e, k, and f are interrelated, they cannot be arbitrarily prescribed. Let us deduce some necessary conditions.

Since each vertex is the endpoint of at least three edges and each face contains at least three edges, while on the other hand each edge has exactly two vertices and belongs to exactly two faces, the following inequalities are obvious:

(68) $$2k \geq 3e, \quad 2k \geq 3f.$$

By the Euler formula (65) we can now eliminate any one of the numbers $e, f,$ or k. Multiplying both sides of equation (65) by 3, we see that

(69) $$3e \geq k + 6, \quad 3f \geq k + 6,$$

and multiplying both sides by 2,

(70) $$2e \geq f + 4, \quad 2f \geq e + 4.$$

From (68) and (69) it follows that

$$6e \geq 2k + 12 \geq 3e + 12, \quad 6f \geq 2k + 12 \geq 3f + 12,$$

or

(71) $$e \geq 4, \quad f \geq 4, \quad k \geq 6.$$

The smallest possible numbers occur in the tetrahedron.

A further conclusion is:

There does not exist any convex polyhedron with exactly 7 edges. For if $k = 7$, it would follow from (69) that $3e \geq 13$, and from (68) that $3e \leq 14$. But no integer e satisfies both these conditions.

The following two theorems are also important:

There is no convex polyhedron in which each face has more than five sides.

For otherwise we would have $2k \geq 6f$, which implies the contradiction $k \geq 3f \geq k + 6$. The argument is the same for the theorem:

There is no convex polyhedron in which more than five edges issue from each vertex.

In the preceding theorems and formulas the numbers e and f have played the same role and can therefore be interchanged without affecting the validity of the theorems. In the theory of convex polyhedra we recognize a *duality*, based on the following fact:

For every convex polyhedron with e vertices, f faces, and k edges, there exists a dual polyhedron with f vertices, e faces, and k edges.

A polyhedron of this sort can be derived from a given polyhedron by taking its polar reciprocal with respect to a sphere.[4]

5.4. Convex Polyhedra with Triangular Faces

Polyhedra with certain regularities are naturally the most interesting. Let us first consider convex polyhedra in which every face is a triangle. Without further restriction it is not easy to enumerate the possible types, since many different forms can be obtained by a suitable combination of triangles. Thus we make the restrictive assumption: the faces of the polyhedron are equilateral triangles.

In this case there are only finitely many polyhedra (with sides of given length), three of which are regular (§5.5). The corresponding question for faces of four or five sides leads only to regular polyhedra.

It is clear that not more than five faces can come together at a vertex. For if we denote the number of vertices with three edges by e_3, of vertices with four edges by e_4, and of vertices with five edges by e_5, then twice the number of edges can be written in the form

$$(72) \qquad\qquad 2k = 3e_3 + 4e_4 + 5e_5,$$

and also, of course,

$$(73) \qquad\qquad 2k = 3f.$$

Writing the Euler theorem in the form

$$6e_3 + 6e_4 + 6e_5 + 6f = 6k + 12,$$

---

[4]The polar reciprocal is obtained by replacing each vertex of the given polyhedron by its polar plane and each face by its pole. It can be proved that the result is again a convex polyhedron.

we see from (73) and (72) that

$$6e_3 + 6e_4 + 6e_5 = 2k + 12 = 3e_3 + 4e_4 + 5e_5 + 12$$

or

(74) $$3e_3 + 2e_4 + e_5 = 12.$$

This Diophantine equation has 19 nonnegative solutions. But not every solution corresponds to a convex polyhedron with triangular faces. A more detailed investigation, which we shall omit, shows that precisely the following eight polyhedra are possible:

	e_2	e_4	e_5	e	f	k
I	4	0	0	4	4	6
II	2	3	0	5	6	9
III	0	6	0	6	8	12
IV	0	5	2	7	10	15
V	0	4	4	8	12	18
VI	0	3	6	9	14	21
VII	0	2	8	10	16	24
VIII	0	0	12	12	20	30

The nets for these polyhedra are given in Fig. 28, and the polyhedra themselves are sketched in Fig. 29. The polyhedra I, III, and VIII are regular. The fact that in addition to these regular polyhedra there also exist five other convex polyhedra II, IV through VII with equilateral triangular faces was first noted by O. Rausenberger.[5] The result can be summed up as follows:

In addition to the three regular polyhedra, there exist five other types of convex polyhedra with equilateral triangular faces.

From the Cauchy theorem (§5.1) it follows that this enumeration is complete.

5.5. *Polyhedra with n Vertices on Each Face and m Edges at Each Vertex.*

It is easy to determine all polyhedra with the same number n of sides on

[5]Konvexe pseudoreguläre Polyeder. Zeitschr. für math. u. naturwiss. Unterricht (1915) S. 135–142. Cf. also Freudenthal and van der Waerden, Over een bewering van Euclides. Simon Stevin 25 (1946/47), p. 115–121.

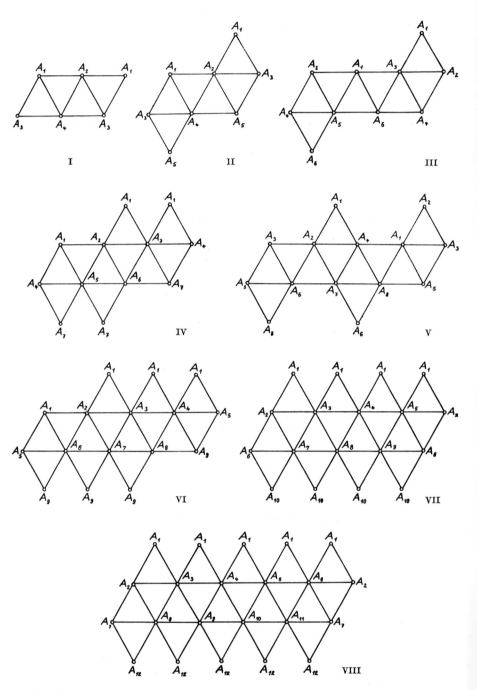

Fig. 28

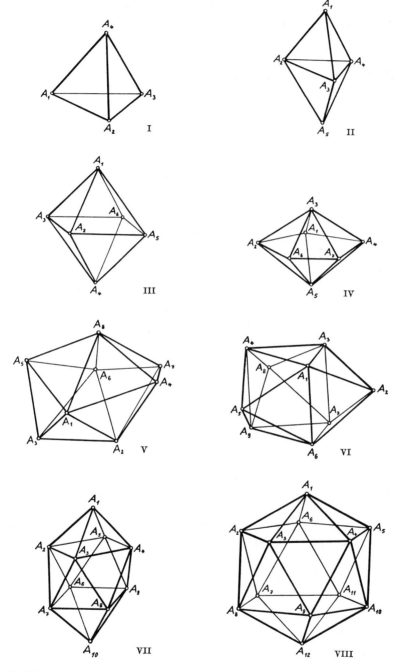

Fig. 29

each face and the same number m of edges at each vertex. Since twice the number of edges is given by

(75) $$2k = nf = me,$$

it follows from the Euler theorem that

$$\frac{2k}{m} - k + \frac{2k}{n} = 2$$

or

(76) $$\frac{1}{m} + \frac{1}{n} = \frac{1}{2} + \frac{1}{k}.$$

This equation and the conditions $m \geq 3, n \geq 3, k \geq 6$, i.e.,

$$\frac{1}{2} < \frac{1}{m} + \frac{1}{n} \leq \frac{2}{3},$$

determine the numbers m, n, and k, and the corresponding numbers e and f can be calculated from (75). The following list contains all possible cases:

	m	n	e	f	k
I	3	3	4	4	6
II	3	4	8	6	12
III	4	3	6	8	12
IV	3	5	20	12	30
V	5	3	12	20	30

The polyhedra I, III, V have already occurred in §5.4. Note that I is selfdual and II, III and IV, V form dual pairs.

Thus there are at most five types of convex polyhedra with the same number of sides on each face and the same number of edges at each vertex.

These polyhedra can all be realized, and in fact even as *regular* polyhedra, i.e.,with regular polygons as faces and with the same number of faces at each vertex. The regular dodecahedron is sketched in Fig. 30, but we omit the proof of its existence.

5.6. *Polyhedra with Uniform Vertices*

The vertices of a polyhedron are said to be uniform if

i) the number of edges is the same at every vertex;

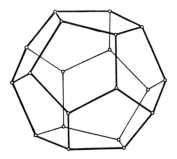

Fig. 30

ii) for every positive integer n the number of n-faces (i.e., faces with n-sides) is the same for every vertex.

At a given vertex let the number of a-vertices be m_a, the number of b-vertices be m_b, and so forth. Twice the number of edges is obviously equal to

(77) $$2k = (m_a + m_b + \cdots)e = me,$$

where

(78) $$m = m_a + m_b + \cdots$$

is the number of edges at each vertex. Now let f_a be the number of a-vertices, f_b the number of b-vertices, and so forth in the whole polyhedron. Then

(79) $$f_a = \frac{m_a e}{a}, \quad f_b = \frac{m_b e}{b}, \ldots,$$

and

(80) $$f = f_a + f_b + \cdots$$

is the number of faces of the polyhedron.

Application of the Euler theorem gives

(81) $$\left[2 - \left\{ m_a\left(1 - \frac{2}{a}\right) + m_b\left(1 - \frac{2}{b}\right) + \cdots \right\} \right]e = 4.$$

Thus

(82) $$m_a\left(1 - \frac{2}{a}\right) + m_b\left(1 - \frac{2}{b}\right) + \cdots < 2.$$

If the numbers $a, b, \ldots$ are arranged in order of magnitude, we have $a \geqq 3, b \geqq 4, \ldots$. But then not more than three summands can actually occur on the left side of the inequality (82). For with four summands, even as small as possible, we would already have

$$\frac{1}{3} + \frac{2}{4} + \frac{3}{5} + \frac{4}{6} > 2.$$

Thus we have proved that

Not more than three distinct types of n-vertices can occur at a vertex.

Let us now write equation (81) in the form

(83) $$\left\{(2 - m) + 2\left(\frac{m_a}{a} + \frac{m_b}{b} + \frac{m_c}{c}\right)\right\}e = 4.$$

The second-to-last theorem in §5.3 shows that $m \leq 5$, which means that m is one of the numbers 3, 4, or 5. Not more than one of the numbers m_a, m_b, m_c can be equal to zero, since otherwise we would have the case discussed above. Let us now examine (83) more closely, on the basis of the following theorem:

If a is odd, then at least one of the numbers $m_a - 1$, m_b, m_c is greater than unity.

For let us consider the sides of an a-gon arranged in cyclic order, beginning with a certain vertex, and let us first take $m_a = 1$. If the first side belongs to a p-gon, where p is equal to b or c, then the third side, the fifth, and so forth, also belong to a p-gon and thus in particular the last side, since a is odd. For if the first side belongs to a p-gon, the first vertex is a vertex of a p-gon, and consequently also the third vertex. Now if the third side did not belong to a p-gon, it is obvious that the second side would belong to one, so that two p-gons would meet at the second vertex and consequently at every vertex, which means that the third side does in fact belong to a p-gon. Thus at least two p-gons meet at each vertex, which implies the assertion.

The case $m_a = 2$ is somewhat more difficult. We show that the assumptions a is odd, $m_a = 2$, $m_b = m_c = 1$ lead to a contradiction. From (83) it follows that $2/a + 1/b + 1/c > 1$. If we again take a, b, c to be arranged in order of magnitude, the only possibility is that $a = 3$. But then the following, easily verifed remark is decisive: every triangle is adjacent to exactly one 3-gon, one b-gon, and one c-gon. But now consider the wreath, let us call it, of $2b$-gons containing a vertex or a side of a given b-gon. A side of a b-gon necessarily belongs to a 3-gon. Thus one endpoint of this side belongs to a second adjacent triangle, and the other endpoint to a c-gon, as follows from the above remark. But in every sequence two triangles adjacent to each other are followed by exactly one c-gon, a fact which in the last triangle produces a contradiction, since the number of polygons in the wreath is even.

1. Let $m = 3$. Then there are two subcases:

a) $m_a = 2, m_b = 1, m_c = 0$. From (83) we must then have

$$(84) \qquad \left(\frac{4}{a} + \frac{2}{b} - 1\right)e = 4,$$

so that

$$\frac{2}{b} > 1 - \frac{4}{a}.$$

If a were odd, it would follow, since m_b is equal to 1 and $m_c = 0$, that $m_a - 1 \geq 2$, which would imply the contradiction $m_a \geq 3$. Consequently a is even, and $a \geq 4$.

If $a = 4$, then b can take any value except 4, since a and b are unequal.

For $a = 6$ we have $b < 6$, and therefore $b = 3, 4$, or 5.
For $a = 8$ we have $b < 4$, so that $b = 3$.
For $a = 10$ we still have $b < 4$, so that $b = 3$.
For $a \geq 12$ we would get the result $b < 3$, so that necessarily $a < 12$.

The corresponding numbers e are easily determined from (84), and the numbers k and f from (77), (79), and (80). It turns out that an integer value can be found for e.

b) $m_a = 1, m_b = 1, m_c = 1$. Then

$$(85) \qquad \left(\frac{2}{a} + \frac{2}{b} + \frac{2}{c} - 1\right)e = 4,$$

so that

$$\frac{2}{a} + \frac{2}{b} + \frac{2}{c} > 1.$$

None of the numbers a, b, c is odd. If they are arranged in order of magnitude, we have $a \geq 4, b \geq 6, c \geq 8$, and then it is easy to verify that the only possibilities are $a = 4, b = 6, c = 8$ and $a = 4, b = 6, c = 10$, and integer values can be found for e.

2. $m = 4$. Here we must distinguish three subcases:
a) $m_a = 2, m_b = 2, m_c = 0$. Then

$$(86) \qquad \left(\frac{4}{a} + \frac{4}{b} - 2\right)e = 4,$$

so that

$$\frac{2}{a} + \frac{2}{b} > 1.$$

The only possibilities are $a = 3, b = 4; a = 3, b = 5$.

b) $m_a = 3$, $m_b = 1$, $m_c = 0$. Then

(87)
$$\left(\frac{6}{a} + \frac{2}{b} - 2\right)e = 4,$$

so that

$$\frac{3}{a} + \frac{1}{b} > 1.$$

For $a = 3$ the value of b is indeterminate but > 3. For $a = 4$ we have $b = 3$.

c) $m_a = 2$, $m_b = 1$, $m_c = 1$. Then

(88)
$$\left(\frac{4}{a} + \frac{2}{b} + \frac{2}{c} - 2\right)e = 4,$$

so that

$$\frac{2}{a} + \frac{1}{b} + \frac{1}{c} > 1.$$

The only available solution is $a = 4$, $b = 3$, $c = 5$, since a cannot be odd and is thus $\geqq 4$.

3. $m = 5$. Assuming $a < b$, we have from (83)

$$2\left(\frac{m_a}{3} + \frac{m_b}{4} + \frac{m_c}{4}\right) > m - 2 = 3$$

or

$$2\frac{m_a}{3} > 3 - \frac{1}{2}(5 - m_a) = \frac{1}{2} + \frac{1}{2}m_a,$$

which means that

$$\frac{1}{6}m_a > \frac{1}{2}.$$

Thus $m_a > 3$, which leads to $m_a = 4$, $m_b = 1$, $m_c = 0$.

Thus

(89)
$$\left(\frac{8}{a} + \frac{2}{b} - 3\right)e = 4,$$

so that

$$\frac{8}{a} + \frac{2}{b} > 3.$$

The only solutions are $a = 3$, $b = 4$ and $a = 3$, $b = 5$. For these numbers an integer value can be found for e.

The results are displayed in the following table, where we have distinguished between polyhedra in which only two types of faces meet at a vertex and polyhedra in which three types of faces meet:

	m	m_a	m_b	a	b	f_a	f_b	e	f	k
I	3	2	1	4	$n \neq 4$	n	2	$2n$	$n+2$	$3n$
II	3	2	1	6	3	4	4	12	8	18
III	3	2	1	6	4	8	6	24	14	36
IV	3	2	1	6	5	20	12	60	32	90
V	3	2	1	8	3	6	8	24	14	36
VI	3	2	1	10	3	12	20	60	32	90
VII	4	2	2	3	4	8	6	12	14	24
VIII	4	2	2	3	5	20	12	30	32	60
IX	4	3	1	3	$n \neq 3$	$2n$	2	$2n$	$2n+2$	$4n$
X	4	3	1	4	3	18	8	24	26	48
XI	5	4	1	3	4	32	6	24	38	60
XII	5	4	1	3	5	80	12	60	92	150

	m	m_a	m_b	m_c	a	b	c	f_a	f_b	f_c	e	f	k
XIII	3	1	1	1	4	6	8	12	8	6	48	26	72
XIV	3	1	1	1	4	6	10	30	20	12	120	62	180
XV	4	2	1	1	4	3	5	30	20	12	60	62	120

All these cases can be realized as convex polyhedra in space.

5.7. *Semiregular Polyhedra*

We have already remarked that the polyhedra enumerated in §5.6 exist in space as convex polyhedra, and we can even demand that all the faces be regular, in which case the polyhedra are said to be *semiregular* (or Archimedean, since they were known to Archimedes). We shall not give a complete proof of their existence, which in some cases is impossible by elementary means, but shall content ourselves with a brief description (see Fig. 31 in each case).

I. *Regular prism.* The polyhedron is an *n*-sided regular prism, whose vertical faces are squares. The excluded case $n = 4$ corresponds to the cube.

II. *Truncated tetrahedron*. It is possible to choose two points on each of the edges of a regular tetrahedron in such a way that the (triangular) faces are, so to speak, truncated into regular hexagons, and we obtain a convex polyhedron with triangles and hexagons for faces.

III. *Truncated octahedron*. The same truncation is applied to an octahedron.

IV. *Truncated icosahedron*. The truncation is applied to an icosahedron.

V. *Truncated cube*. The truncation is applied to a cube.

VI. *Truncated dodecahedron*. The truncation is applied to a dodecahedron.

VII. *Cubo-octahedron*. The vertices consist of the midpoints of the edges of a cube.

VIII. *Icosi-dodecahedron*. The vertices are the midpoints of the edges of a dodecahedron.

Up to now the existence of the various polyhedra has been a rather trivial question. But in the following examples the proof of existence involves certain difficulties which we shall avoid by plausible arguments from continuity and visualization.

IX. *Anti-prism*. Consider a regular n-gon with a congruent n-gon in a parallel plane above it (the line joining the two centers is to be perpendicular to the two planes), the second n-gon being rotated through half the angle at its center. By suitable choice of the distance between the planes we can arrange that each side of the original polygon forms an equilateral triangle with the corresponding vertex of the upper polygon; in other words, the vertical faces are equilateral triangles and the polyhedron is a special kind of prismoid. The excluded case $n = 3$ corresponds to the octahedron.

X. *Rhombi-cubo-octahedron*. The midpoints of the edges of a cubo-octahedron are the vertices of a convex polyhedron consisting of six squares, eight equilateral triangles, and twelve rectangles. By moving the squares outward a short distance, we can transform the rectangles into squares and this way obtain the desired polyhedron.

XI. *Obtuse cube*. Move each of the faces of a cube out a little along the perpendicular to its plane, and then turn it a little in its plane (all the faces are to be turned in the same sense as viewed from the midpoint of the cube). For a suitable choice of the outward motion and of the rotation, the squares can be joined to one another by equilateral triangles in such a way that four triangles meet at each vertex.

XII. *Obtuse dodecahedron*. The same procedure is applied to the dodecahedron.

XIII. *Truncated cubo-octahedron*. By truncation of the vertices of a cubo-octahedron, we can obtain a polyhedron consisting of six regular 8-gons, eight nonregular 6-gons, and twelve rectangles. By moving the 8-gons slightly outwards, we can arrange that the rectangles become squares. Then the 6-gons also become regular.

XIV. *Truncated icosi-dodecahedron.* The same procedure is applied to an icosi-dodecahedron.

XV. *Rhombi-icosi-dodecahedron.* The midpoints of the edges of an icosi-dodecahedron form the vertices of a convex polyhedron consisting of twelve regular 5-gons, twenty equilateral triangles, and thirty rectangles. If we move the pentagons slightly outwards, the rectangles are transformed into squares.

Cardboard models of these figures are shown in Fig. 31.

6. Symmetric Polyhedra

6.1. *The Concept of Symmetry*

The need for a more exact study of the various kinds of symmetry possible in polyhedra first arose in crystallography, and the question was given a great deal of attention by Moebius, who set up the following definition: a figure is said to be *symmetric* if it is congruent to itself in more than one way.

Let us first consider the concept of congruence from a very general point of view. Let us imagine a system of finitely many operations in space taking points into points. The system is to include the operation that leaves every point fixed, the operation inverse to each given operation, and the operation resulting from successive performance of any two operations. In other words, the operations are to form a *finite group.*[6]

Two figures are said to be *congruent* with respect to this group if they can be taken into each other by an operation of the group.

An operation of a given group that carries a given figure (i.e., a collection of points) into coincidence with itself (with permutation, in general, of the points) is called a *superposition* of the figure; the superpositions of a given figure form a subgroup of the group of operations.

The simplest operations are the rigid motions, which we may extend by including reflections in a plane. In the first case we obtain congruence in the narrower sense; in the second, mirror images as well.

If S is the set consisting of all the points into which a given point P is carried by the operations of a group, then S is carried into itself by every operation of the group and is therefore symmetric with respect to the group.

If a finite group consists only of motions and reflections, such a system of points is finite and its center of gravity O remains invariant under all the

---

[6]For the concepts of group theory see IB2.

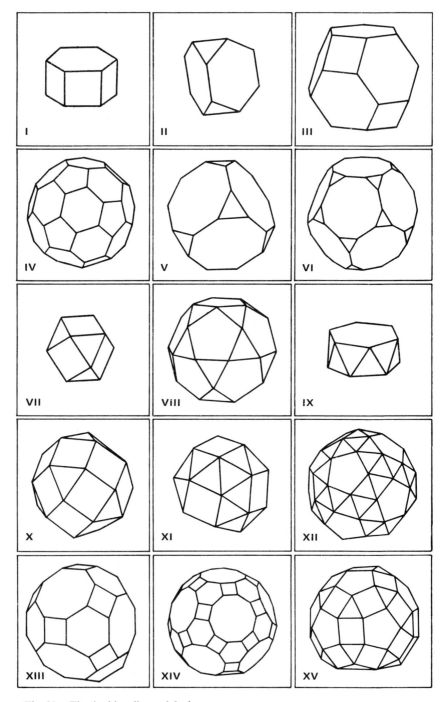

Fig. 31. The Archimedian polyhedra

operations. The motions in the group are rotations about O, and every reflection consists of reflection in a plane through O followed by rotation about an axis perpendicular to the plane through O.

6.2. The Polyhedral Groups

The finite groups of motions in space are very closely related to the regular polyhedra. So let us first study the superpositions of polyhedra.

1.   We begin with a regular tetrahedron, with vertices A_1, A_2, A_3, and A_4. The tetrahedron can be made to coincide with itself by a rotation if we bring the vertex A_1 into the position of any one of the four vertices and then rotate the polyhedron around the altitude through A_1 by an angle of 0, $(2/3)\pi$, or $(4/3)\pi$. Here we have $4 \cdot 3 = 12$ possibilities, including the identity.

The structure of the *tetrahedral group* thus defined is easy to describe. Every motion that takes the tetrahedron into coincidence with itself is equivalent either to a rotation about an altitude (which generates a three-term cycle, namely, the cyclic permutations of the vertices not on the altitude), or to a rotation (through an angle π) about a bimedian (i.e., a line joining the midpoints of two opposite edges); for the edges $A_1 A_2$ and $A_3 A_4$, for example, this rotation generates the double transposition (12) (34). Thus the group is isomorphic to the group (of order 12) of even permutations of four objects. In other words:

The tetrahedral group is isomorphic to the alternating group of four elements (cf. IB2, §15.3).

The double transpositions contained in this group, taken together with the identity, form an invariant subgroup (a normal divisor with index 3), the so-called *Klein four-group* (IB2, §15.3.3).

If we let a, b, c denote the bimedians of the tetrahedron and also the corresponding operations of the group of rotations (so that a denotes rotation about the line a, and so forth), it is geometrically evident that

$$a^2 = b^2 = c^2 = 1, \quad ab = ba = c, \quad bc = cb = a, \quad ca = ac = b,$$

where 1 denotes the unit element of the group (the identity).

2.   Let us now consider a cube with the vertices $A_1, \ldots, A_8$. The cube may be brought into coincidence with itself by moving the vertex A_1 into the position of each of the eight vertices and then rotating the cube about the (spacial) diagonal through A_1 by an angle of 0, $(2/3)\pi$, $(4/3)\pi$. Altogether, this produces $8 \cdot 3 = 24$ operations.

The midpoints of the six faces of the cube are the vertices of a regular octahedron. A rotation taking the cube into itself also takes the octahedron into itself, and the various positions of the octahedron correspond in a one-to-one way with the positions of the cube. Thus the superpositions of the cube are the same as those of the octahedron. (There is a corresponding relationship, as we shall see, between the dodecahedron and the icosahedron.) The groups are named after the polyhedra with triangular faces, so that the group now in question is called the *octahedral group*.

Here, too, the structure of the group is easily described. The cube has 4 (spatial) diagonals and in none of the rotations (except the identity) does each of these diagonals go into itself. Corresponding to the 24 operations we thus obtain 24 permutations of the 4 diagonals. Since there are exactly 4! = 24 permutations of 4 elements, we thus obtain all the operations. Consequently

The octahedral group is isomorphic to the symmetric group of permutations of four elements.

The octahedral group is obviously the same as the tetrahedral group extended by the operation of "turning the figure inside out" (i.e., subjecting the vertices of the tetrahedron to an odd permutation; compare the operation of turning a left glove inside out so as to obtain a right one). This fact can easily be visualized by noting that in a cube we can inscribe two tetrahedra that will be interchanged if the whole figure is turned inside out (for example, from top to bottom; compare I and III in Fig. 29).

3. We now consider the dodecahedron, with vertices $A_1, \ldots, A_{20}$. The dodecahedron can be brought into coincidence with itself if we move the vertex A_1 into the position of each of the twenty vertices and then turn the figure through an angle of 0, $(2/3)\pi$, $(4/3)\pi$ about the (spatial) diagonal joining A_1 to the midpoint of the dodecahedron. Altogether there are $20 \cdot 3 = 60$ possibilities.

The midpoints of the twelve faces of the dodecahedron are the vertices of a regular icosahedron, so that, just as for the cube and octahedron, as we saw above, the group of motions for the dodecahedron is the same as for the icosahedron. The group thus obtained is called the *icosahedral group*.

Eight suitably chosen vertices of the dodecahedron are the vertices of a cube (Fig. 32). The twelve edges of the cube are diagonals in the twelve faces of the dodecahedron and the whole cube is already determined by one such diagonal. Since a face of the dodecahedron has five diagonals, there are in all five such cubes.

The rotations of the dodecahedron permute the five inscribed cubes among themselves, and none of them (except the identity) leaves every cube unchanged. A superposition of the dodecahedron is a rotation about a (spatial) diagonal of the dodecahedron, or about a diagonal of the corresponding icosahedron, or else about the line joining the midpoints of two opposite edges. Under the first kind of rotation two of the cubes are transformed into each other, while the three others are cyclically permuted; under the second the five cubes are cyclically permuted, and under the third

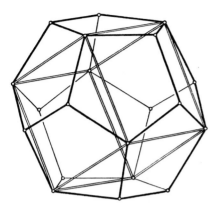

Fig. 32

one cube is transformed into itself while the four others are exchanged pair-wise in a double transposition. The entire group of permutations of the five cubes thus contains only even permutations. Consequently, since the group is of order 60,

The icosahedral group is isomorphic to the alternating group of five elements.

4.   But the groups enumerated above do not yet exhaust all the finite groups of motions in space. We can obtain another if we start from any regular n-gon $A_1, \ldots, A_n$. The rotations about its midpoint form a cyclic group of order n. But it can also be turned over about the line joining its midpoint to one of its vertices, and then again rotated. Instead of a line through a vertex we can also choose a line joining the midpoint of the polygon to the midpoint of a side. Only when n is even are these latter lines distinct from the lines through the vertices. In this way we obtain $2n$ operations, forming a group, the so-called *dihedral group*.

In order to understand the choice of name, let us imagine that the surface of a polygon is covered with two membranes, fastened together around the edge. In this way we obtain a dihedral (i.e., two-faced) polyhedron which is, so to speak, infinitely thin and may be regarded as a limiting case of a regular polyhedron. Such a polyhedron is called a *dihedron*.

The dihedral group can also be regarded as the superposition group of a convex polyhedron, which is no longer required to be regular (except for $n = 3, 4$), namely, a *double pyramid,* i.e., the figure consisting of two regular pyramids with a common base. We note that the dihedral group of order 4 is isomorphic to the Klein four-group.

In the theory of groups it is proved that, up to isomorphism, the groups described above are the only finite rotation groups in space.

Let us give a brief sketch of the proof. Every rotation of space is uniquely determined by its effect on the points of the unit sphere about O. A rotation about O that is not the identity leaves fixed exactly two antipodal points on the sphere. These fixed points are called the *poles* of the rotation.

Now let $\mathfrak{G}$ be a finite group of rotations about O. Let the order of $\mathfrak{G}$ be $N > 1$. Let two points be called conjugate if they are taken into each other by a rotation in $\mathfrak{G}$. The finitely many poles of the rotations in $\mathfrak{G}$ fall into r classes of conjugate poles.

All the rotations of $\mathfrak{G}$ that have the same pole P form a subgroup $\mathfrak{g}$, if we include the identity. If the order of $\mathfrak{g}$ is n, we say that P is an *n-fold pole*. The number of poles conjugate to P is $p = N/n$, as is shown in group theory.

Let τ be a rotation in $\mathfrak{G}$, and let the pole conjugate to P in this rotation be denoted by τP. A rotation ρ in $\mathfrak{G}$ leaves τP fixed if and only if $\rho \tau P = \tau P$, so that $\tau^{-1}\rho\tau P = P$; in other words, if $\sigma = \tau^{-1}\rho\tau$ belongs to $\mathfrak{g}$. Then $\rho = \tau\sigma\tau^{-1}$, where τ is fixed and for σ we may choose an arbitrary element from the group **a**. The number of rotations leaving τP fixed is thus also equal to n. So in $p(n-1)$ cases a nonidentical rotation in $\mathfrak{G}$ will leave fixed a pole conjugate to P. But the total number of distinct nonidentical rotations in $\mathfrak{G}$ is $N-1$, each of them associated with two poles. So if we index the classes of conjugate poles by $1, \ldots, r$, we have

$$2N - 2 = p_1(n_1 - 1) + \cdots + p_r(n_r - 1)$$

or

(90)
$$2\left(1 - \frac{1}{N}\right) = \left(1 - \frac{1}{n_1}\right) + \cdots + \left(1 - \frac{1}{n_r}\right).$$

But $N \geqq 2, n_1 \geqq 2, \ldots, n_r \geqq 2$, so that

(91)
$$1 \leqq \left(1 - \frac{1}{n_1}\right) + \cdots + \left(1 - \frac{1}{n_r}\right) < 2.$$

If $r = 1$, we obtain a contradiction. Thus $r \geqq 2$. From $r \geqq 4$ it would follow that

$$\left(1 - \frac{1}{n_r}\right) + \cdots + \left(1 - \frac{1}{n_r}\right) \geqq \frac{1}{2} + \frac{1}{2} + \frac{1}{2} + \frac{1}{2} = 2.$$

So we must have $r \leqq 3$, which means that only the values $r = 2$ and $r = 3$ are possible.

a) Let us first take $r = 2$. From $2 \leqq n_1 \leqq N$ it follows from

$$\frac{1}{n_1} + \frac{1}{n_2} = \frac{2}{N}.$$

that $n_1 = n_2 = N$.

b) Now take $r = 3$. Then

$$\frac{1}{n_1} + \frac{1}{n_2} + \frac{1}{n_3} = 1 + \frac{2}{N}.$$

At least one of the numbers n_1, n_2, n_3 is equal to 2, since otherwise every summand on the left side would be $\leqq 1/3$, whereas the right hand side is > 1. So let $n_1 = 2$. Then

$$\frac{1}{n_2} + \frac{1}{n_3} = \frac{1}{2} + \frac{2}{N}.$$

If also $n_2 = 2$, then $n_3 = \frac{1}{2}N$, so that N is an even number.

Let us now assume $n_2 > 2, n_3 > 2$. Then at least one of these numbers must be equal to 3, since the assumption that $n_2 \geqq 4, n_3 \geqq 4$, leads to a contradiction. Let us take $n_2 = 3$. Then

$$\frac{1}{n_3} = \frac{1}{6} + \frac{2}{N},$$

so that $n_3 < 6$. To $n_3 = 3$ corresponds $N = 12$, to $n_3 = 4$ corresponds $N = 24$, and to $n_3 = 5$ corresponds $N = 60$, which exhausts all possible cases. Thus we have proved

There exist only five types of finite groups of rotations.

A more detailed discussion would show that the first group is cyclic. The others are isomorphic to the polyhedral groups described above. It is clear that in the last three groups the vertices of the polyhedra are projected from the midpoint of the sphere into the n_3-fold poles on the spherical surface. In this projection the midpoints are taken into the n_1-fold poles and the midpoints of the faces into the n_2-fold poles. In the case of a cyclic group, there are only two N-fold poles, which are not conjugate to each other, and they may be taken as the vertices of a double pyramid. But these poles are conjugate with respect to the dihedral group, and in this case there are two classes of 2-fold poles as well. One class consists of the vertices, and the other of the midpoints of the sides.

6.3. *The Polyhedra Symmetric with Respect to the Polyhedral Groups*

Let there now be given a polyhedral group. We ask about the properties of systems of conjugate points. Except for certain exceptional cases we will find that they are identical with the systems of vertices of the polyhedra enumerated in §5.6., a result which solves the existence problems for those polyhedra.

If P is not a pole, the number of points conjugate to P is equal to N, i.e., the order of the group. This number is reduced to N/n if P is an n-fold pole. In any case we shall take into account only points on the surface of a unit sphere.

Let us first investigate the cyclic group. If P is not a pole, the result is a regular polygon. If P is a pole, then P is conjugate only to itself.

The dihedral group is of greater interest. If P is an n-fold pole (where $2n$ is the order of the group), then the figure consisting of the points conjugate to P is a pair of points. If P is a 2-fold pole, we obtain a regular n-gon.

But if P is not a pole and if P lies on the (shortest) arc joining two adjacent 2-fold poles, we obtain a $2n$-gon. If P lies on the arc joining an n-fold pole and a 2-fold pole, we obtain a polyhedron of type I, and otherwise a polyhedron of type IX. Limiting cases occur for $n = 2$, when we obtain a rectangle or a tetrahedron with congruent faces (sphenoid).

Let us now turn to the last three groups. It is clear that the n_3-fold poles are the vertices of a tetrahedron, octahedron, or icosahedron, while the n_2-fold poles are the vertices of the dual polyhedra, i.e., the tetrahedron, the cube, and the dodecahedron. In the tetrahedral group the 2-fold poles are the vertices of an octahedron, in the octahedral group they form a cubo-octahedron or a polyhedron of type VII, and in the icosahedral group, an icosi-dodecahedron, i.e., a polyhedron of type VIII.

Now let us study successively the systems of conjugate non-poles in the following cases:

1. Tetrahedral group. Let P lie on the arc joining a 2-fold and a 3-fold pole. It is easy to see that we obtain a polyhedron of type II. If P lies on the arc joining two adjacent 3-fold poles (in distinct classes), we obtain a polyhedron of type VII, and otherwise a polyhedron with twenty triangular faces of the type of the icosahedron.

2. Octahedral group. If P lies on the arc joining a 2-fold pole and an adjacent 3-fold pole, we obtain a polyhedron of type V, and if P lies on the arc joining a 2-fold pole and an adjacent 4-fold pole, a polyhedron of type III. But if we take P on the arc joining a 3-fold pole and an adjacent 4-fold pole, we obtain a polyhedron of type X, and otherwise a polyhedron of type XI.

3. Icosahedral group. If P lies on the arc joining a 2-fold and an adjacent 3-fold pole, we obtain a polyhedron of type VI, and if P lies on the arc joining a 2-fold pole and an adjacent 5-fold pole, a polyhedron of type IV. But if P is on the arc joining a 3-fold pole and an adjacent 5-fold pole, we obtain a polyhedron of type XV, and otherwise a polyhedron of type XII.

Thus we have obtained the polyhedra in the table in §5.6 with the exception of the types XIII and XIV. Since the number of vertices here is 48 and 120, respectively, it is natural to suppose that we will obtain these polyhedra if, by including reflections, we extend the octahedral and icosahedral groups to complete groups of permutations. This is in fact the case, since it is easy to verify that we obtain the desired polyhedra by taking P in general position, i.e., not on an arc joining two adjacent poles. The situation does not arise in the tetrahedral group, since the octahedral group is already the completed tetrahedral group.

This last result is in agreement with the fact that in a cube we can inscribe two tetrahedra which are taken into each other by a reflection. Correspondingly, in a truncated cubo-octahedron we can inscribe two obtuse cubes, and in a truncated icosi-dodecahedron two obtuse dodecahedra.

Exercises.   1.   Using formula (24), prove

$$S_5 = 2\sqrt{5 - 2\sqrt{5}}, \quad S_{10} = \tfrac{2}{3}\sqrt{25 - 10\sqrt{5}}.$$

2. Prove that four regular 30-gons can be inscribed in a circle, with sides determined by

$$s_{30,1} = s_5 c_6 - c_5 s_6, \quad s_{30,7} = c_6 c_{10} - s_6 s_{10},$$
$$s_{30,11} = s_5 c_6 + c_5 s_6, \quad s_{30,13} = c_6 c_{10} + s_6 s_{10}.$$

Thus, taking unity for the radius of the circle, show that

$$s_{30,1} = \tfrac{1}{4}(\sqrt{30 - 6\sqrt{5}} - \sqrt{5} - 1), \quad s_{30,0} = \tfrac{1}{4}(\sqrt{30 + 6\sqrt{5}} - \sqrt{5} + 1),$$
$$s_{30,11} = \tfrac{1}{4}(\sqrt{30 - 6\sqrt{5}} + \sqrt{5} + 1), \quad s_{30,13} = \tfrac{1}{4}(\sqrt{30 + 6\sqrt{5}} + \sqrt{5} - 1).$$

3. Express equation (9) in the form

$$s_{2n} = \sqrt{1 + \tfrac{1}{2}s_n} - \sqrt{1 - \tfrac{1}{2}s_n}.$$

4. Use formula (10) to prove

(a) $s_n c_{2n} = s_{2n}(c + 2),$

and thus, by (24)

(b) $S_n c_n = S_{2n}(c_n + 2).$

5. Prove

$$c_n = \frac{4}{\sqrt{4 + S_n^2}},$$

and thus, by (b) of exercise 4,

$$S_{2n} = \frac{2S_n}{2 + \sqrt{4 + S_n^2}}$$

6. Prove

$$S_{2n} = \frac{2s_n}{c_n + 2} = \frac{2s_n}{2 + \sqrt{4 - s_n^2}}.$$

7. From (10) and (24) show that

$$S_n = \frac{2s_{2n}c_{2n}}{c_n}$$

and thus that

$$S_n = \frac{2s_{2n}\sqrt{4 - s_{2n}^2}}{2 - s_{2n}^2}.$$

8. Prove

$$S_{2n}s_n = 2s_{2n}^2.$$

9. Letting the perimeters of the inscribed and circumscribed regular n-gons be denoted by u_n and U_n, respectively, show that Gregory's formula (11) can be written

$$u_{4n}^2(u_n + u_{2n}) = 2u_{2n}^3,$$

and prove that

$$U_{2n}(u_n + U_n) = 2u_nU_n$$

and

$$U_{4n} = \frac{2u_{2n}^2}{u_n + u_{2n}}.$$

10. Using Gregory's formula, show that

$$4u_{2n} - u_n < 4u_{4n} - u_{2n}.$$

11. Use the inequality in exercise 10 to prove the Huyghens inequality

$$2\pi > \tfrac{1}{3}(4u_{2n} - u_n) = u_{2n} + \tfrac{1}{3}(u_{2n} - u_n).$$

12. Use Gregory's formula to prove that

$$\frac{u_n^2 + 2u_{2n}^2}{u_n} > \frac{u_{2n}^2 + 2u_{4n}^2}{u_{2n}}.$$

13. Use the inequality in exercise 12 to prove the following inequality (also due to Huyghens):

$$2\pi < \tfrac{1}{3}(u_n + 2U_{2n}) = U_{2n} - \tfrac{1}{3}(U_{2n} - u_n).$$

14. Let C be the midpoint of a circular arc AB (less than a semicircle), and let D be the midpoint of the arc AC. Prove that the area of the triangle ACB is less

than eight times as great as that of the triangle ADC, and thus show that the area of the circular segment bounded by the arc AB (and the chord AB) is more than $\frac{4}{3}$ as great as the area of the triangle ACB. Letting a_n denote the area of the regular n-gon inscribed in a circle of unit radius, use the above results to prove

$$\pi > a_{2n} + \tfrac{1}{3}(a_{2n} - a_n).$$

Show that this last inequality is a corollary of the Huyghens inequality in exercise 11.

15. Let C be the midpoint of the circular arc AB (less than a semicircle), and let F be the point of intersection of the tangents to the circle at A, B. Then let D be the intersection of the tangents at A, C, and let E be the intersection of the tangents at B, C. Prove that the area of the triangle ACB is less than twice the area of the triangle DFE, and then show that the area of the circular segment is less than 2/3 as great as that of the triangle AFB. Letting A_n denote the area of the regular n-gon circumscribed about a circle of unit radius, use the above results to prove

$$\pi < A_n - \tfrac{1}{3}(A_n - a_n).$$

Show that this last inequality is a corollary of the Huyghens inequality in exercise 13.

16. Show that the difference $A_n - a_n$ (as in the foregoing exercises) is equal to the area of a regular n-gon inscribed in a circle with radius equal to the side of the given circumscribed n-gon, or equal to the area of a regular n-gon circumscribed about a circle with radius equal to the side of the given circumscribed n-gon, or equal to the area of a regular n-gon circumscribed about a circle with radius equal to the side of the given inscribed n-gon.

17. Letting r_n and ρ_n denote, respectively, the radii of the circumscribed and inscribed circle of a regular n-gon, and r_{2n}, ρ_{2n} the corresponding radii for the $2n$-gon of the same area, show that

$$\rho_{2n} = \sqrt{\tfrac{1}{2}\rho_n(\rho_n + r_n)},$$
$$r_{2n} = \sqrt{r_n \rho_n}.$$

Beginning with a square of area 2 construct a sequence (compare the theorem of Schwab) converging to the limit $\sqrt{2/\pi}$.

18. For odd n prove that

$$\Phi_n(z)\,\Phi_n(-z) = \Phi_n(z^2).$$

19. Calculate the polynomial $\Phi_{30}(z)$ and from it deduce the polynomial $\Psi_{30}(z)$.

20. From (46) and (48) prove

$$\phi(pn) = p\phi(n), \text{ if } n \text{ is divisible by } p,$$
$$= (p - 1)\phi(n) \text{ if } n \text{ is not divisible by } p.$$

21. Prove that in a convex polyhedron the sum of the number of faces with three vertices (triangles) and of vertices with three edges is greater than 8.

22. A convex polyhedron, all of whose faces have n vertices and all of whose vertices have m edges, admits only the following possibilities:

$$n = 3, m = 3, 4, 5; n = 4, m = 3; n = 5, m = 3.$$

Determine the number of vertices, edges and faces in each case.

23. A convex polyhedron in which every vertex has at least five edges must have at least twenty triangular faces.

24. A polyhedron all of whose faces are triangular must have at least one vertex with more than four edges.

25. The radius of the sphere circumscribed about the dodecahedron of unit side is given by

$$R = \tfrac{1}{4}(\sqrt{15} + \sqrt{3}),$$

and the corresponding radius for the icosahedron of unit side is

$$R = \tfrac{1}{4}\sqrt{10 + 2\sqrt{5}}.$$

26. The radius of the sphere inscribed in the dodecahedron of unit side is

$$\rho = \tfrac{1}{20}\sqrt{250 + 110\sqrt{5}},$$

and the corresponding radius for the icosahedron is

$$\rho = \tfrac{1}{12}(3\sqrt{3} + \sqrt{15}).$$

27. The radius of the sphere tangent to the edges of a dodecahedron of unit side is

$$r = \tfrac{1}{4}(3 + \sqrt{5}),$$

and the corresponding radius for the icosahedron is

$$r = \tfrac{1}{4}(1 + \sqrt{5}).$$

28. Show that the icosahedron group has the following subgroups:

1) Rotations about an axis through two opposite vertices (determine the order of these subgroups and how many of them there are);

2) Rotations about an axis through the midpoints of two opposite faces (again determine the order and the number);

3) Rotations about an axis through the midpoints of two opposite edges (order and number);

4) The above axes also determine dihedral groups (find the order and number in each case);

5) The rotations taking an inscribed octahedron (in the icosahedron) into itself form a subgroup. Prove that this subgroup is isomorphic to the tetrahedron group and is thus of order 12. In this way we have found 57 proper subgroups, which can be proved to exhaust all such subgroups.

29. Solve the same problem for the octahedron group.

30. Show that the 60 elements of the icosahedron group can be divided into (distinct) classes of conjugate elements (2 elements a and b of a group are conjugate if there exists an element c with $b = cac^{-1}$) as follows:

a) the rotations through $2\pi/5$ about an axis through a vertex (12 elements);

b) the same rotations through $4\pi/5$ (12 elements);

c) the 20 elements of order 3;

d) the 15 elements of order 2;

e) the identity element.

Prove that the icosahedron group has no proper normal subgroup, that is a group containing all the conjugates of each of its elements.

31. Solve the same problem for the octahedron group.

Bibliography

[1] BRÜCKNER, M.: Vielecke und Vielflache. Leipzig 1900. (Standardwerk mit vollständiger Aufzählung und vielen Bildern.)

[2] COXETER, H. S. M.: Regular Polytopes. London 1948.

[3] ROUSE BALL, W. W., COXETER, H. S. M.: Mathematical Recreations and Essays. New York 1947.

[4] STEINITZ, E., und RADEMACHER, H.: Vorlesungen über die Theorie der Polyeder. Berlin 1934.

Bibliography Added in Translation

COXETER, H. S. M.: Introduction to geometry. John Wiley and Sons, Inc., New York-London 1961.

LYUSTERNIK, L. A.: Convex figures and polyhedra (translated from the Russian by T. J. Smith). Dover Publications, New York 1963.

Analytic Treatment of Geometry

Affine and Euclidean Geometry

Of central importance in modern *analytic geometry* is the concept of a *vector space*. However, it is not the vector space itself that serves directly as a space for the geometry, since in it the zero vector plays a special role, whereas in geometry all points are usually of equal importance. In order to arrive at a *homogeneous* space, we must introduce not only vectors but also points, the connection between them being determined by a few simple, easily visualized axioms. The procedure followed here was first described by Hermann Weyl in his book *Space, Time and Matter*, 1918 [17].

The concept of invariance under coordinate transformations, which was indispensable in the older literature, is no longer needed, since from the very beginning we do not distinguish any particular basis in the underlying vector space; we calculate, not with the various possible coordinates for the vectors, but with the vectors themselves. What is important is not the coordinate transformations, but the automorphisms, i.e., the structure-preserving, one-to-one mappings that belong to the geometry in question. Here too, as with vectors, we do not calculate with representations for a given basis, but with the mappings themselves, so that we have no need for matrices.

Among the many possible geometries we shall discuss the two simplest ones, which are also the most important, in view of their connection with other parts of mathematics; namely, *affine and Euclidean geometry of n dimensions*. The assumption of finite dimensionality, though unnecessary for many of our results, will be made from the very beginning, since it is important in the study of automorphisms and is helpful for visualization (with $n = 2$ or 3).

The domain of scalars for the underlying vector space will be a field K, which the reader can usually take to be the field of real numbers. At first, we require only the properties of a field, including commutativity, but order properties will be needed later. The property of the real field that every subset bounded from above has a least upper bound (completeness) will be needed in its full extent only in the measurement of angles; for the construction of Euclidean geometry a weaker assumption would be sufficient; however, we shall not be interested in these intermediate possibilities.

Before proceeding to our actual task, let us assemble the most important concepts of linear algebra in the form in which we shall use them. For a detailed discussion the reader may consult Chapter IB3. Here we give only the fundamental concepts, postponing the more complicated constructions to the later sections where they will naturally occur. The *exterior algebra* of Grassmann, which has proved to be very useful in analytic geometry, must be omitted here for lack of space.

I. Basic Concepts of Linear Algebra

1.1. *Vector Spaces*

A *vector space* $\mathfrak{B}$ over a (commutative) field K is an additively written Abelian group, together with an "S-multiplication" which to every pair consisting of an element a from K and an element $\mathfrak{x}$ from $\mathfrak{B}$ assigns an element $a\mathfrak{x}$ from $\mathfrak{B}$, in such a way that the following axioms are satisfied:

1. $a(\mathfrak{x} + \mathfrak{y}) = a\mathfrak{x} + a\mathfrak{y}$ for all $a \in K$ and all $\mathfrak{x}, y \in \mathfrak{B}$,
2. $(a + b)\mathfrak{x} = a\mathfrak{x} + b\mathfrak{x}$ for all $a, b \in K$ and all $\mathfrak{x} \in \mathfrak{B}$,
3. $(ab)\mathfrak{x} \quad = a(b\mathfrak{x})$ for all $a, b \in K$ and all $\mathfrak{x} \in \mathfrak{B}$,
4. $\qquad 1\mathfrak{x} = \mathfrak{x}$ for all $\mathfrak{x} \in \mathfrak{B}$.

The elements of $\mathfrak{B}$ are usually called *vectors* and are denoted by lower-case Gothic letters, as in the axioms above. In particular, the neutral element o of the Abelian group is called the *zero vector*. The elements of K are called *scalars* and are written with lower-case Latin letters.

A *vector subspace* is naturally defined as follows.

A (nonempty) subset $\mathfrak{U} \subset \mathfrak{B}$ *is called a vector subspace if with the above operations it is itself a vector space.*

Here it is necessary and sufficient that if $\mathfrak{x}, \mathfrak{y} \in \mathfrak{U}$ and $a \in K$, then $\mathfrak{x} + \mathfrak{y} \in \mathfrak{U}$ and $a\mathfrak{x} \in \mathfrak{U}$, which implies that the *intersection* $U_1 \cap U_2$ of two vector subspaces U_1, U_2 is again a subspace but does not imply that the set-theoretic union $U_1 \cup U_2$ is necessarily a subspace. Instead, the *sum* $U_1 + U_2$ of two subspaces U_1, U_2 is defined as the intersection of all subspaces containing both U_1 and U_2; in other words, $U_1 + U_2 = \{x \mid x = u_1 + u_2, u_1 \in U_1, u_2 \in U_2\}$. If also $U_1 \cap U_2 = \{o\}$, the sum is said to be *direct*, and then we write $U_1 \oplus U_2$.

Let us give two examples of vector spaces.

1)   The set $K \times K \times \cdots \times K = K^n$ of all n-tuples of elements of K with the following operations; if $\mathfrak{x} = (x_1, \ldots, x_n)$, $\mathfrak{y} = (y_1, \ldots, y_n)$ and $a \in K$, then

$$\mathfrak{x} + \mathfrak{y} = (x_1 + y_1, \ldots, x_n + y_n),$$

$$a\mathfrak{x} = (ax_1, ax_2, \ldots, ax_n).$$

The set of all n-tuples with vanishing first k components is a subspace of this vector subspace.

2)   Every n-tuple of K^n in example 1) is a mapping of the set $\{1, 2, \ldots, n\}$ into the field K. Let us now replace the set $\{1, 2, \ldots, n\}$ by an arbitrary nonempty set D and consider the set $\mathfrak{F}$ of all mappings of D into the field K, the mappings being denoted by f, g, etc., and the images by $f(x)$, $g(x)$, etc. If addition of two elements f and g in $\mathfrak{F}$ is defined by

$$(f + g)(x) = f(x) + g(x) \quad \text{for every} \quad x \quad \text{in} \quad D,$$

and S-multiplication of an element f in $\mathfrak{F}$ with a scalar a in K is defined by

$$(af)(x) = a(f(x)) \quad \text{for every} \quad x \quad \text{in} \quad D,$$

it is easy to show that with these operations $\mathfrak{F}$ is a vector space over K. If the set D is infinite, the vector spaces thus obtained are actual generalizations of example 1), and an interesting subspace $\mathfrak{E} \subset \mathfrak{F}$ consists of the set of those mappings in $\mathfrak{F}$ that take at most finitely many elements x into a nonzero value.

The special case $D = K$, namely, the set of all mappings of K into itself, is particularly important. Subspaces of $\mathfrak{F}$ are then given, for example, by the set $\mathfrak{P}$ of all entire rational functions, or the set $\mathfrak{P}_m$ of all entire rational functions of degree $\leq m$, where $\mathfrak{P}_m \subset \mathfrak{P} \subset \mathbf{A}$.

1.2.   Linear Independence: Dimension

If D is infinite, the basic distinction between examples 1) and 2) is that they are of different "dimension." In order to define this concept, we begin with *linear dependence*:

The vectors $\mathfrak{x}_1, \mathfrak{x}_2, \ldots, \mathfrak{x}_k$ *are said to be linearly dependent if there exist scalars* $a_1, a_2, \ldots, a_k$, *not all equal to zero, such that*

$$a_1\mathfrak{x}_1 + a_2\mathfrak{x}_2 + \cdots + a_k\mathfrak{x}_k = \mathfrak{o}.$$

Vectors $\mathfrak{x}_1, \ldots, \mathfrak{x}_k$ that are not *linearly* dependent are said to be *linearly independent,* which will be the case if and only if the equation

$$a_1\mathfrak{x}_1 + a_2\mathfrak{x}_2 + \cdots + a_k\mathfrak{x}_k = \mathfrak{o}$$

holds only for $a_1 = a_2 = \cdots = a_k = 0$; the left side of this equation is often called a *linear combination* of the given vectors.

The maximal number of linearly independent vectors in $\mathfrak{V}$ is called the dimension of $\mathfrak{V}$. If this number is infinite, i.e., if for every natural number N there exists a set of N linearly independent vectors, then $\mathfrak{V}$ is said to be infinite-dimensional. For lack of space, we confine our attention here to finite-dimensional spaces, although many of our results are equally valid for the infinite-dimensional case. In other words, we assume that there is a natural number n such that *there exist n linearly independent vectors in V but every set of $(n + 1)$ vectors is linearly dependent.* Then $\mathfrak{V}$ is said to be n-dimensional and will be denoted by $\mathfrak{P}_n$. In what follows, the reader should think chiefly of the easily visualized cases $n = 2$ and $n = 3$.

In example 1) above, the vector space K^n is n-dimensional; but in example 2), if D is infinite, then $\mathfrak{F}$ and all the other subspaces are infinite-dimensional, except $\mathfrak{P}_m$, which is $(m + 1)$-dimensional.

1.3. *Basis*

If in an n-dimensional vector space $\mathfrak{V}_n$ the n vectors $a_1, \ldots, a_n$ are linearly independent, then an arbitrary vector $\mathfrak{x}$ in $\mathfrak{V}_n$ is a linear combination of these n vectors, since the $n + 1$ vectors $\mathfrak{x}, a_1, \ldots, a_n$ are linearly dependent and the scalar factor for $\mathfrak{x}$ cannot be zero, since otherwise the $a_1, \ldots, a_n$ would be linearly dependent. So the linear dependence of $\mathfrak{x}$ can be written in the form

$$\mathfrak{x} = x_1 a_1 + \cdots + x_n a_n,$$

where the scalars $x_1, \ldots, x_n$ are uniquely determined by the vector $\mathfrak{x}$, since a second representation $x = y_1 a_1 + \cdots + y_n a_n$ would imply

$$(x_1 - y_1)a_1 + \cdots + (x_n - y_n)a_n = \mathfrak{o},$$

with at least one nonvanishing difference $x_i - y_i$; i.e., $a_1, \ldots, a_n$ would be linearly dependent. More concisely, *any set of n linearly independent vectors $a_1, \ldots, a_n$ of an n-dimensional vector space $\mathfrak{V}_n$ forms a basis of $\mathfrak{V}_n$,* i.e., every vector in $\mathfrak{V}_n$ is uniquely representable as a linear combination of the $a_1, \ldots, a_n$.

The concept of a basis can be defined independently of the dimension of the vector space $\mathfrak{V}$, as follows. If $\mathfrak{B}$ is a subset of $\mathfrak{V}$ such that every $\mathfrak{x}$ in $\mathfrak{V}$ can be represented as a linear combination of vectors from $\mathfrak{B}$, then $\mathfrak{B}$ is called a system of generators for $\mathfrak{V}$, and if $\mathfrak{B}$ has the further property that every finite subset of $\mathfrak{B}$ is linearly independent, then $\mathfrak{B}$ is said to be *linearly independent*; and then a *basis* of $\mathfrak{V}$ is defined as a linearly independent system of generators of $\mathfrak{V}$. The representation of a vector in $\mathfrak{V}$ as a linear combination of elements of basis $\mathfrak{B}$ is uniquely determined.

It can be shown, e.g., from the Zorn lemma (cf. IB11) that every vector space has a basis.

2. Affine Point Spaces

2.1. *Definition and Immediate Consequences*

From now on we consider an n-dimensional vector space $\mathfrak{V}_n$ over a commutative field K. By an *n-dimensional affine point space*[1] $\mathfrak{A}_n$ *with respect to $\mathfrak{V}_n$* we mean a set of elements, called *points* and denoted by capital italic letters, with the following property:

To each ordered pair of points (P, Q) there corresponds a unique vector $\overrightarrow{PQ}$ satisfying the two axioms

Axiom 1:   *For every $P \in \mathfrak{A}_n$ and $\mathfrak{v} \in \mathfrak{V}_n$ there exists exactly one Q with* $\overrightarrow{PQ} = \mathfrak{v}$.

Axiom 2. $\overrightarrow{PQ} + \overrightarrow{QR} = \overrightarrow{PR}$ *for all $P, Q, R \in \mathfrak{A}_n$.*

The first of these axioms can be visualized as follows: The point Q is obtained by "attaching" the vector $\mathfrak{v}$ to the point P. In other words, every vector $\mathfrak{v}$ can be regarded as a mapping of $\mathfrak{A}_n$ onto itself.

Before drawing any conclusions from this system of axioms, let us first show that it is consistent. To this end we construct the following model. As the set $\mathfrak{A}_n$ we take the elements of the vector space $\mathfrak{V}_n$ itself, and to every pair of points $(\mathfrak{p}, \mathfrak{q})$ we assign the vector $\mathfrak{q} - \mathfrak{p}$. Axioms 1 and 2 are then satisfied, since 1. is implied by $\mathfrak{q} = \mathfrak{p} + \mathfrak{v}$, and 2. by the equation $(\mathfrak{q} - \mathfrak{p}) + (\mathfrak{r} - \mathfrak{q}) = \mathfrak{r} - \mathfrak{p}$, valid for all $\mathfrak{p}, \mathfrak{q}, \mathfrak{r}$ in $\mathfrak{V}_n$. In the following discussion, however, we will not use this model, but will make a clear distinction between the affine point space $\mathfrak{A}_n$ and the corresponding vector space $\mathfrak{V}_n$.

The conclusions that can be drawn from these axioms for the n-dimensional affine point space (including the axioms for the underlying vector space) form the content of n-dimensional *affine geometry* over the field K. As long as we introduce no special assumptions about the field or the dimension of the space, our results are valid for every affine space.

We begin with the proof of some simple, immediate consequences of the axioms.

a)   The pair (P, P) corresponds to the zero vector, since for $Q = P$ Axiom 2 implies that

$$\overrightarrow{PP} + \overrightarrow{PR} = \overrightarrow{PR}, \quad \text{i.e.,} \quad \overrightarrow{PP} = \mathfrak{v}.$$

Conversely, $\overrightarrow{PQ} = \mathfrak{v}$ implies $P = Q$, since by Axiom 1 there exists exactly one Q with $PQ = \mathfrak{v}$, and $Q = P$ satisfies this equation.

---

[1] In the next section we shall see how far the present concept coincides, for $n = 2$, with the affine plane defined in II3, §1.

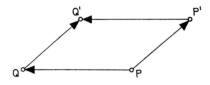

Fig. 1

b)   Also, $\overrightarrow{PQ} = -\overrightarrow{QP}$, for all P and Q, since for $P = R$ Axiom 2 and $\overrightarrow{PP} = o$ imply that

$$\overrightarrow{PQ} + \overrightarrow{QP} = \overrightarrow{PP} = o.$$

c)   Finally, for four points P, Q, P', Q', the equation $\overrightarrow{PQ} = \overrightarrow{P'Q'}$ holds if and only if $\overrightarrow{PP'} = \overrightarrow{QQ'}$. For by Axiom 2 the vector $\overrightarrow{PQ'}$ can be represented in the two different ways (Fig. 1)

$$\overrightarrow{PQ'} = \overrightarrow{PQ} + \overrightarrow{QQ'} = \overrightarrow{PP'} + \overrightarrow{P'Q'}.$$

Then if $\overrightarrow{PP'} = \overrightarrow{QQ'}$, it follows that $\overrightarrow{PQ} = \overrightarrow{P'Q'}$; and conversely, if $\overrightarrow{PQ} = \overrightarrow{P'Q'}$, then $\overrightarrow{PP'} = \overrightarrow{QQ'}$.

Two pairs of points (P, Q) and $(P'Q')$ for which $\overrightarrow{PQ} = \overrightarrow{P'Q'}$ are said to be *parallel-equal* (or parallel-congruent, translation-congruent, etc.) and the (ordered) set of four points (P, Q, Q', P') is called a *parallelogram,* a name that will be justified by our defining what we mean by a line and by the parallelism of two lines.

Parallel-equality of pairs of points is an equivalence relation, as follows at once from the equivalence of equal vectors.

2.2.   Subspaces

A *line* is the set of points obtained by "attaching" the vectors of a one-dimensional subspace $\mathfrak{U}$ of $\mathfrak{V}_n$ to a fixed point. For example, if A is the point and if $\mathfrak{a} \neq o$ is any vector in $\mathfrak{U}$, then the line consists of all points X for which $\overrightarrow{AX} = x\mathfrak{a}$ with $x \in K$. More generally, by a k-dimensional affine subspace $\mathfrak{B}$ of $\mathfrak{A}_n$ we mean the set of points generated by attaching all the vectors of a k-dimensional vector subspace $\mathfrak{U}$ of $\mathfrak{V}_n$ to a fixed point A. In symbols:

$$\mathfrak{B} = \{X \mid \overrightarrow{AX} \in \mathfrak{U} \subset \mathfrak{V}_n\}.$$

Equivalently, if B is any other point of $\mathfrak{B}$, i.e., if $\overrightarrow{AB} \in \mathfrak{U}$, then also

$$\mathfrak{B} = \{X \mid \overrightarrow{BX} \in \mathfrak{U} \subset \mathfrak{B}_n\},$$

since $\overrightarrow{AX} \in \mathfrak{U}$ holds only if $\overrightarrow{BX} = \overrightarrow{AX} - \overrightarrow{AB} \in \mathfrak{U}$. It is easy to show that affine subspaces, together with the mapping of pairs of points onto vectors induced by the original mapping, are themselves affine point spaces with respect to their defining vector subspaces. For $k = 1$ the subspaces, as mentioned above, are called *lines*,[2] for $k = 2$ they are *planes*, and for $k = n - 1$ they are *hyperplanes*.

If U consists only of the zero vector, i.e., if $k = 0$, the corresponding affine subspaces consist of a single point $\mathfrak{B} = \{A\}$, in which case the subspace will also be denoted by A.

For two affine subspaces $\mathfrak{B}$ and $\mathfrak{C}$ in $\mathfrak{A}_n$ the intersection $\mathfrak{B} \cap \mathfrak{C}$ (if nonempty) is again an affine subspace, and similarly for the intersection of arbitrarily many affine subspaces. On the other hand, the set-theoretic union of two affine subspaces is not, in general, an affine subspace. But if we define the *sum* $\mathfrak{B} \vee \mathfrak{C}$ as the intersection of all affine subspaces containing both $\mathfrak{B}$ and $\mathfrak{C}$, then $\mathfrak{B} \vee \mathfrak{C}$ is itself a subspace, which is said to be *spanned* by $\mathfrak{B}$ and $\mathfrak{C}$. For example, two intersecting lines span an affine plane, and two skew lines span a three-dimensional affine space.

2.3. *Parallelism*

Definition. *Two affine subspaces are said to be parallel if one of their generating vector spaces is contained in the other.*

In particular, two lines are parallel if their generating one-dimensional vector spaces coincide. We now show that *two parallel lines either have no point in common or else they coincide*. For if from a common point A we draw the vectors of the generating vector space, which by assumption is the same for the two lines, we obtain the same set of points. Similarly, any two parallel subspaces either have no point in common or else one of them is contained in the other.

Parallelism of lines, or of any two subspaces of equal dimension, is an equivalence relation, but for subspaces of unequal dimension transitivity may not hold; for example, two lines parallel to the same plane may not be parallel to each other (Fig. 2).

A parallelogram (see §2.1) is an ordered set of four points (P, Q, P', Q') with $\overrightarrow{PQ} = \overrightarrow{P'Q'}$. If P and Q do not coincide, it follows at once that the line joining P and Q is parallel to the line joining P' and Q'. Also, if $P \neq P'$, the

---

[2] Lines are traditionally denoted by lower-case italic letters g, h, ...

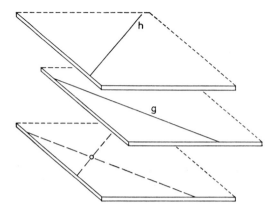

Fig. 2

line joining P and P' is parallel to the line joining Q and Q'. The possibility that all four points are collinear or that some or all of them coincide is not excluded, in which case the parallelogram is said to be degenerate.

If we call $P \vee Q, Q \vee Q', Q' \vee P', P' \vee P$ the *sides* of the parallelogram (P, Q, Q', P'), the preceding results states that the opposite sides of a parallelogram are parallel, a property which is often taken as a definition for the parallelogram but which coincides with our definition only if the four points are not all collinear (see Fig. 3).

2.4. *Ratio of Collinear Vectors or Points*

If A, B, C are three points given in that order on a line, the vectors $\overrightarrow{AB}$ and $\overrightarrow{AC}$ belong to the same one-dimensional vector space, and if $A \neq B$, i.e., $\overrightarrow{AB} \neq \mathfrak{o}$, there exists exactly one scalar c such that

$$\overrightarrow{AC} = c \cdot \overrightarrow{AB}.$$

Then c is called the *affine coordinate* of the point C on the given line, with respect to the *origin* A and the *unit point* B. The scalar c is also called the

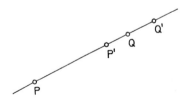

Fig. 3

ratio of the three points (the ratio in which C divides AB), for which we write

$$c = (A, B; C).$$

If A and B are fixed, this relation provides a one-to-one correspondence between the points of the line $A \vee B$ and the scalars in the field K; in particular, the point A corresponds to the scalar 0 and the point B to the scalar 1.

Let us emphasize that the "ratio in which C divides $\overrightarrow{AB}$" is often given a different definition. For example, in II16, §2 it was taken to be the scalar u defined by

$$\overrightarrow{AC} = u \cdot \overrightarrow{BC}$$

for which, noting that $\overrightarrow{BC} = \overrightarrow{AC} - \overrightarrow{AB}$, we see at once that

$$u = \frac{c}{c - 1}, \quad c \neq 1.$$

Again, the scalar $-u$ is often preferred, and many other definitions could be suggested. Permuting the three points gives, in addition to c, the scalars

$$\frac{1}{1 - c}, \quad 1 - \frac{1}{c}, \ 1 - c, \quad \frac{c}{c - 1}, \quad \frac{1}{c}.$$

2.5. *Position Vectors*

Our purpose in defining an affine point space in terms of a vector space is to use linear algebra in the construction of affine geometry. To this end, we represent the points and subspaces of the affine space by vectors, as will be possible if we distinguish a point, call it O (for *origin*) in $\mathfrak{A}_n$, whereupon the points X of $\mathfrak{A}_n$ correspond one-to-one to the vectors $\mathfrak{x} = \overrightarrow{OX}$, the vector $\mathfrak{x}$ being called the position vector of the point X with respect to O. If another point O' is chosen instead of O, the position vectors $\mathfrak{x}$ and $\mathfrak{x}'$ for the point X with respect to O and O' are related to each other by the equation

$$\mathfrak{x} = \mathfrak{x}' + \mathfrak{a}, \quad \text{with} \quad \mathfrak{a} = \overrightarrow{OO'},$$

as follows at once from Axiom 2 (see Fig. 4). The (arbitrary) choice of origin O is usually determined by the desire to simplify the resulting formulas as much as possible.

Now if $\mathfrak{B} \subset \mathfrak{A}_n$ is an arbitrary subspace with generating vector space U, and if P is an arbitrary point in $\mathfrak{B}$ with position vector $\mathfrak{p} = \overrightarrow{OP}$, then the variable vector

$$\mathfrak{x} = \mathfrak{p} + \mathfrak{u}, \quad \text{with} \quad \mathfrak{u} \in \mathfrak{U}$$

traverses the totality of position vectors of points of $\mathfrak{B}$ (see Fig. 5, where $\mathfrak{B}$ is a plane).

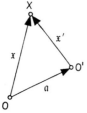

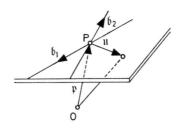

Fig. 4   Fig. 5

If the vector space itself (see the model on p. 297) is used as the affine point space, the subspace $\mathfrak{B}$ is the coset (cf. IB2, §3.5) of $\mathfrak{U}$ generated by $\mathfrak{p}$. If $\{b_1, \ldots, b_k\}$ is a basis for $\mathfrak{U}$, we can also write

$$\mathfrak{x} = \mathfrak{p} + x_1 b_1 + \cdots + x_k b_k, \quad x_1, \ldots, \quad x_k \in K$$

and speak of a parametric representation of $\mathfrak{B}$. In particular, for $k = 1$, we obtain the parametric representation of a *line*

$$\mathfrak{x} = \mathfrak{p} + x\mathfrak{b}, \quad x \in K$$

where $\mathfrak{b} \neq \mathfrak{o}$ is a basis vector. Here $\mathfrak{b}$ is also called a *direction vector* for the line, and the equation is said to be in *point-direction* form. Two lines are parallel if and only if their direction vectors are linearly dependent, as follows at once from the definition in §2.3.

Similarly, for a plane spanned by two direction vectors b_1 and b_2 intersecting at the point P, the parametric representation is (see Fig. 5)

$$\mathfrak{x} = \mathfrak{p} + x_1 b_1 + x_2 b_2 \quad x_1, \quad x_2 \in K.$$

We now prove the following theorem.

Two nonparallel lines in the same plane have exactly one point in common.

For if $\mathfrak{x} = \mathfrak{p} + x\mathfrak{b}$, and $\mathfrak{y} = \mathfrak{q} + y\mathfrak{c}$ are the parametric representations of the two lines, we seek scalars x_0, y_0 for which the vectors $\mathfrak{x}$ and $\mathfrak{y}$ coincide:

$$\mathfrak{p} + x_0 \mathfrak{b} = \mathfrak{q} + y_0 \mathfrak{c}; \quad \text{i.e.,} \quad \mathfrak{q} - \mathfrak{p} = x_0 \mathfrak{b} - y_0 \mathfrak{c}.$$

Since the vectors b and c are linearly independent (i.e., the lines are not parallel), they form a basis for the $\mathfrak{B}_2$ underlying the affine plane. Then the vector $\mathfrak{q} - \mathfrak{p}$ is uniquely representable as a linear combination of b and c, which means that the scalars x_0 and y_0 are uniquely determined. Thus there is exactly one point of intersection

$$\mathfrak{x}_0 = \mathfrak{y}_0 = \mathfrak{p} + x_0 \mathfrak{b} = \mathfrak{q} + y_0 \mathfrak{c}.$$

3.   Relationship to Synthetic Geometry. Further Theorems

3.1.   *Pappus Affine Planes*

We now show[3] that for $n = 2$ the affine point space introduced above provides a model for the system of axioms A1, A2, A3, (P) in II3, so that, wherever it seems advantageous, we can then make use of the methods and results of synthetic geometry. For convenience, we repeat the (four) axioms for a Pappus plane.

A1:   For two distinct points P, Q there exists exactly one line incident with P and Q.

A2:   For every line g and every point P there exists exactly one line incident with P and parallel to g.

A3:   There exist three points not incident with the same line.

(P)   If, in a hexagon $P_1Q_2P_3Q_1P_2Q_3$ whose vertices lie alternately on two lines g and h (with no vertex on both of them), two pairs of opposite sides are parallel, then the third pair of opposite sides is also parallel. In our model the "incidence" of P and g means that $P \in g$.

Now Axiom A1 is satisfied in our analytic affine geometry. For in the first place, there is *at most* one line incident with P and Q, since we have just proved (in the preceding section) that if two distinct lines are parallel, they have no point in common, and if they are nonparallel, they have only one; and secondly, there is *at least* one such line, since in the parametric equation

$$\mathfrak{x} = \mathfrak{p} + x(\mathfrak{q} - \mathfrak{p})   x \in K$$

the values $x = 0$ and $x = 1$ correspond to the position vectors of P and Q, respectively (two-point form for the equation of a line, Fig. 6).

But A2 also holds, since the line h through P parallel to g has the parametric representation

$$\mathfrak{x} = \mathfrak{p} + x\mathfrak{b}   x \in K$$

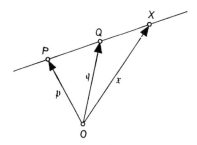

Fig. 6.   Two-point form

[3] A proof from a somewhat different point of view is given in II3, §2.1.

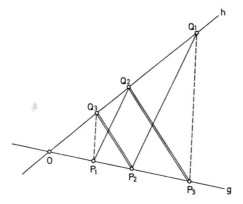

Fig. 7. Pappus-Pascal theorem

with $\mathfrak{b}$ as direction vector for g, and $\mathfrak{p}$ as position vector for P; and h is unique, since it must have the same generating vector space as g.

The proof of A3 is as follows. In $\mathfrak{B}_2$ there exist two linearly independent vectors $\mathfrak{p}$ and $\mathfrak{q}$, and therefore, if O is any point with $\overrightarrow{OP} = \mathfrak{p}$ and $\overrightarrow{OQ} = \mathfrak{q}$, the three points O, P, Q are not collinear.

Finally, we prove (P) by dividing it into two cases:

1) g and h are not parallel.

2) g and h are parallel.

In the first case (Fig. 7) we take as origin the uniquely determined intersection O of g and h and write $\mathfrak{a}$, $\mathfrak{b}$ for the direction vectors for g and h. Then the points P_i and Q_i have the position vectors

$$\mathfrak{p}_i = p_i \mathfrak{a}, \ \mathfrak{q}_i = q_i \mathfrak{b}, \quad (i = 1, 2, 3)$$

where p_i and q_i are suitable scalars, and the fact that the line $P_1 \vee Q_2$ is parallel to the line $Q_1 \vee P_2$ means that the vectors $\overrightarrow{P_1 Q_2} = q_2 \mathfrak{b} - p_1 \mathfrak{a}$ and $\overrightarrow{Q_1 P_2} = p_2 \mathfrak{a} - q_1 \mathfrak{b}$ are linearly dependent, which in turn, since $\mathfrak{a}$ and $\mathfrak{b}$ are linearly independent and (by assumption) neither p_i nor q_i vanishes, means that

$$p_1 q_1 = p_2 q_2.$$

Similarly, from the parallelism of $Q_2 \vee P_3$ and $P_2 \vee Q_3$ we have

$$p_2 q_2 = p_3 q_3,$$

so that

$$p_1 q_1 = p_3 q_3,$$

which implies, as desired, that the lines $P_1 \vee Q_3$ and $Q_1 \vee P_3$ are parallel.

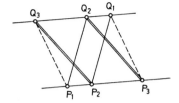

Fig. 8. Pappus-Pascal theorem $(g \parallel h)$

In the second case (Fig. 8) the four noncollinear points P_1, P_2, Q_1, Q_2 form a parallelogram and therefore

$$\overrightarrow{P_1 P_2} = \overrightarrow{Q_1 Q_2}.$$

Similarly, P_2, P_3, Q_2, Q_3 is a parallelogram, so that

$$\overrightarrow{P_2 P_3} = \overrightarrow{Q_3 Q_2},$$

and by addition

$$\overrightarrow{P_1 P_3} = \overrightarrow{P_1 P_2} + \overrightarrow{P_2 P_3} = \overrightarrow{Q_2 Q_3} + \overrightarrow{Q_3 Q_2} = \overrightarrow{Q_3 Q_1}.$$

Thus P_1, P_3, Q_1, Q_3 form a parallelogram, which means, as desired, that the lines $P_1 \vee Q_3$ and $Q_1 \vee P_3$ are parallel.

In Case 1 of the above proof for Axiom (P) we have made tacit use of the commutativity of the field K. By examining the proof, say for left vector spaces over a skew field, the interested reader will find that in this case (P) is not valid in general.

Thus an affine point space of dimension 2 (over a commutative field K) is always a Pappus affine plane in the sense of II3, and since the converse was proved in that section, the two concepts are in fact equivalent. Thus if we wish to avoid algebraic concepts in our construction of geometry, we can obtain our present two-dimensional affine point space by starting from the axioms for a Pappus affine plane. But it must be pointed out that this approach demands a considerable amount of work, since for every Pappus affine plane the corresponding commutative field K must be constructed from the beginning.

As follows from the above proof, every commutative field K can in fact be constructed in this way. In particular, we may choose for K the prime field of characteristic p, or more generally the Galois fields $GF(p^m)$, whereupon we obtain examples, and in fact all possible examples, of Pappus affine planes (and n-dimensional affine point spaces) with only finitely many points and lines. The two simplest finite affine planes contain four and nine points respectively, with six and twelve lines, as the reader may easily verify. These

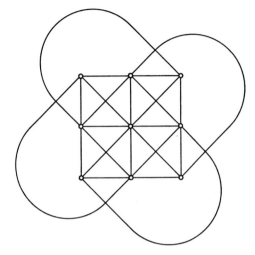

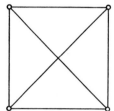

Fig. 9.   Affine plane with 4 points   Fig. 10.   Affine plane with 9 points

two planes are illustrated in Figs. 9 and 10, where the lines are indicated by line segments or curves, and only the circled intersections are actually points of the geometry. In the four-point geometry every line contains two points, and in the nine-point, three.

The above proof (that every affine point space is a model for the axioms of synthetic affine geometry) has been given only for $n = 2$, in view of the fact that in II3 there is no discussion of n-dimensional spaces with $n > 3$. So let us mention here that H. Lenz ("Ein kurzer Weg zur analytischen Geometrie," Mathematisch-Physikalische Semesterberichte, Bd. VI (1958), pp. 57–67) gives a system of axioms for affine spaces of arbitrary (including infinite) dimension and proves that every model for his system is in fact an affine point space over some (not necessarily commutative) field K. If we wish to obtain precisely the same concept of an affine point space as in the present chapter, we must add axioms ensuring that the space is finite-dimensional and that K is commutative.

3.2.   A Criterion for Collinearity

The following criterion is often useful for deciding whether or not three points are collinear.

Three points P, Q, R of an affine point space (of arbitrary dimension) are collinear if and only if there exist scalars u, v, w (not all zero) such that $u + v + w = 0$ and $u\mathfrak{p} + v\mathfrak{q} + w\mathfrak{r} = \mathfrak{o}$, where $\mathfrak{p}$, $\mathfrak{q}$, $\mathfrak{r}$ are the position vectors of P, Q, R with respect to an arbitrary point O.

For if P, Q, R are collinear, then $\mathfrak{q} - \mathfrak{p}$ and $\mathfrak{r} - \mathfrak{p}$ are linearly dependent, so that with suitable scalars v and w, not both zero, we have

$$v(\mathfrak{q} - \mathfrak{p}) + w(\mathfrak{r} - \mathfrak{p}) = \mathfrak{o}$$

or $\qquad\qquad -(v + w)\mathfrak{p} + v\mathfrak{q} + w\mathfrak{r} = \mathfrak{o}.$

Setting $u = -(v + w)$ gives the first half of the criterion.

Conversely, given the three scalars u, v, w, let us assume $u \neq 0$. Then v and w cannot both be zero, and

$$-(v + w)\mathfrak{p} + v\mathfrak{q} + w\mathfrak{r} = v(\mathfrak{q} - \mathfrak{p}) + w(\mathfrak{r} - \mathfrak{p}) = \mathfrak{o}$$

implies that the vectors PQ and PR are linearly dependent.

3.3. The Two-Line Theorem and Desargues Theorems

The two-line theorem, which is essentially contained in the axiom of distributivity (§1.1) may be stated as follows (Fig. 11).

If the two lines g and g' intersect at the point O, and if h and k intersect g and g' at A, B and A', B' respectively, then the lines h and k are parallel if and only if they divide the segments OA and OA' in the same ratio, i.e., $(O, A; B) = (O, A'; B')$.

For the proof we set $\overrightarrow{OA} = \mathfrak{a}$ and $\overrightarrow{OA'} = \mathfrak{a}'$, where $\mathfrak{a}$ and $\mathfrak{a}'$ must be linearly independent, since g and g' are not parallel. Setting $(O, A; B) = b$ and $(A, A'; B') = b'$ gives

$$\overrightarrow{OB} = b\mathfrak{a} \quad \text{and} \quad \overrightarrow{OB'} = b'\mathfrak{a}'.$$

Since $\mathfrak{a} - \mathfrak{a}'$ is a direction vector for h, and $b\mathfrak{a} - b'\mathfrak{a}'$ for k, the lines h and k are parallel if and only if these two vectors are linearly dependent, i.e., if there exist two scalars x, y, not both zero, such that

$$x(\mathfrak{a} - \mathfrak{a}') + y(b\mathfrak{a} - b'\mathfrak{a}) = \mathfrak{o},$$

which by the linear independence of a and a' implies the scalar equations

$$x + yb = 0$$

$$x + yb' = 0.$$

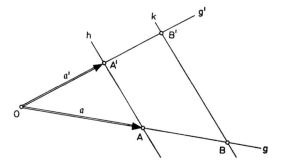

Fig. 11.   Two-line theorem

But this latter system has a nontrivial solution if and only if $b = b'$, which completes the proof of the theorem.

From this theorem we at once obtain a theorem of great importance in synthetic affine geometry (cf. II3, §1) namely the Desargues dilation theorem (Desargues theorem for triangles with parallel sides). Recalling that two triangles ABC and $A'B'C'$ are said to be in *point perspective* when the three lines $A \vee A'$, $B \vee B'$ and $C \vee C'$ are distinct and pass through one point, the so-called center of perspectivity, we may state the theorem as follows.

If ABC and $A'B'C'$ are two triangles in point perspective (Fig. 12), *with $A \vee B$ parallel to $A' \vee B'$ and $B \vee C$ parallel to $B' \vee C'$, then $A \vee C$ is also parallel to $A' \vee C'$.*

Proof. By the two-line theorem

$$(O, A; A') = (O, B; B') \quad \text{and} \quad (O, B; B') = (O, C; C'),$$

so that

$$(O, A; A') = (O, C; C'),$$

and therefore $A \vee C$ is parallel to $A' \vee C'$. The reader should consider all possible converses.

Omitting the assumption that corresponding sides of the triangles are parallel gives the so-called "greater" or "projective" Desargues theorem (Fig. 13).

If two triangles ABC and $A'B'C'$ are in point perspective, then the three points of intersection of corresponding sides, provided they exist, are collinear.

If the center of perspectivity is taken as the origin O, the criterion for collinearity gives the following equations, with suitable scalars a, a', b, b', c, c',

$$a\mathfrak{a} + a'\mathfrak{a}' = \mathfrak{o}, a + a' = 1,$$
$$b\mathfrak{b} + b'\mathfrak{b}' = \mathfrak{o}, b + b' = 1,$$
$$c\mathfrak{c} + c'\mathfrak{c}' = \mathfrak{o}, c + c' = 1$$

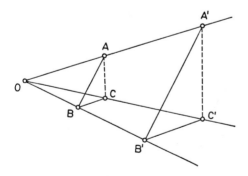

Fig. 12.   Desargues dilation theorem

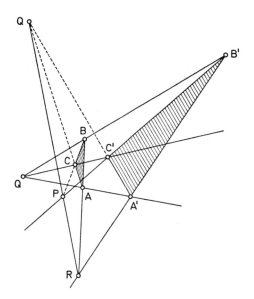

Fig. 13. "Greater" Desargues

for the position vectors of A, A', B, B', C, C'. To determine the position vector $\mathfrak{p}$ for the intersection P of the sides, say BC and $B'C'$, we take the difference of the last two equations

$$b\mathfrak{b} - c\mathfrak{c} = c'\mathfrak{c}' - b'\mathfrak{b}'.$$

Since $\mathfrak{p}$ must be a linear combination of $\mathfrak{b}$ and $\mathfrak{c}$, and also of $\mathfrak{b}'$ and $\mathfrak{c}'$, with coefficient sum equal to 1, we multiply this equation by $(b - c)^{-1} = (c' - b')^{-1}$ (note that $b - c \neq 0$ by our assumption that the point of intersection exists):

$$(b - c)^{-1}b\mathfrak{b} - (b - c)^{-1}c\mathfrak{c} = (c' - b')^{-1}c'\mathfrak{c}' - (c' - b')^{-1}b'\mathfrak{b}',$$

thereby obtaining a vector which must be the position vector $\mathfrak{p}$, and the vectors $\mathfrak{q}$ and $\mathfrak{r}$ are obtained by cyclic permutation. Thus

$$\mathfrak{p} = (b - c)^{-1}b\mathfrak{b} - (b - c)^{-1}c\mathfrak{c},$$

$$\mathfrak{q} = (c - a)^{-1}c\mathfrak{c} - (c - a)^{-1}a\mathfrak{a},$$

$$\mathfrak{r} = (a - b)^{-1}a\mathfrak{a} - (a - b)^{-1}b\mathfrak{b}$$

are the position vectors of P, Q, R. But these three points satisfy the condition for collinearity of §3.2, since the scalars $u = b - c$, $v = c - a$, $w = a - b$ are such that $u\mathfrak{p} + v\mathfrak{q} + w\mathfrak{r} = \mathfrak{o}$ and $u + v + w = 0$, which completes the proof.

Let us state a converse of this theorem of Desargues, leaving the proof to the reader.

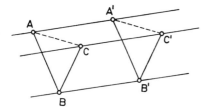

Fig. 14.   Translation Desargues

*If in the triangles A B C and A′ B′ C′ the intersections of corresponding sides
(A ∨ B) ∩ (A′ ∨ B′) . . . are collinear, then the lines joining corresponding
vertices are either parallel or concurrent (i.e., pass through one point).*

Also of importance for synthetic affine geometry (i.e., for translation
planes) is the so-called "little" Desargues theorem (cf. II3), which we shall
call the "translation Desargues" (Fig. 14).

*If in the triangles A B C and A′B′C′ the lines A ∨ A′, B ∨ B′, C ∨ C′ are
parallel and distinct, and if the sides A ∨ B and B ∨ C are parallel respectively to the sides A′ ∨ B′ and B′ ∨ C′, then A ∨ C is also parallel to
A′ ∨ C′.*

The theorem is an immediate consequence of the fact that parallel-
equality of point pairs is transitive (cf. §2.1).

Note that in the proofs of the three Desargues theorems we have made no use of
the two-dimensionality of the space (which also played no role in the two-line
theorem) nor of commutativity of multiplication in the field of scalars. Thus the
fields resulting from algebraization (cf. II3) of Desargues affine planes (in contrast
to Pappus affine planes) are not necessarily commutative.

4.   Affine Mappings

4.1.   *The Affine Group*

A geometric concept of fundamental importance is the "equivalence" of
figures (cf. II1, §6 and II12 and 13) in a given geometry, two figures being
said to be equivalent if the group of automorphisms of the geometry in
question contains an automorphism taking one figure into the other.

It is clear, from the properties of a group, that this concept of equivalence
has the general properties of an equivalence relation, namely reflexivity,
symmetry, and transitivity.

So we now have the problem of determining the automorphisms of an
affine point space $\mathfrak{A}_n$, where by an automorphism of any set with structure
we mean a one-to-one mapping of the set onto itself under which the structure relations are preserved. In our case the structure of $\mathfrak{A}_n$ is determined by
the two axioms for an affine point space, together with the structure of the

vector space $\mathfrak{B}_n$. The automorphisms of $\mathfrak{B}_n$ are already known, namely, the one-to-one *linear* mappings of $\mathfrak{B}_n$ onto itself, i.e., the mappings A such that

1) $A(\mathfrak{x} + \mathfrak{y}) = A\mathfrak{x} + A\mathfrak{y}$ for all $\mathfrak{x}, \mathfrak{y} \in \mathfrak{B}_n$,

2) $A(c\mathfrak{x}) = cA\mathfrak{x}$ for all $c \in K$ and all $\mathfrak{x} \in \mathfrak{B}_n$.

Also, a one-to-one mapping α of $\mathfrak{A}_n$ onto itself is an automorphism of the affine point space $\mathfrak{A}_n$ if it corresponds to an automorphism of the vector space $\mathfrak{B}_n$, i.e., if there exists a one-to-one linear mapping A of $\mathfrak{B}_n$ onto itself, such that

$$\overrightarrow{\alpha X \alpha Y} = A(\overrightarrow{XY}) \quad \text{for all } X, Y \in \mathfrak{A}_n.$$

Then A is uniquely determined.

In particular, parallel-equal pairs of points are taken by an automorphism of $\mathfrak{A}_n$ into pairs of points with the same property; the images of parallelograms are again parallelograms.

The automorphisms of an affine point space $\mathfrak{A}_n$ are called *regular affine mappings,* or *affinities.* The following properties are an immediate result of the definition.

a)   If three points P, Q, R are collinear, their images αP, αQ, αR are also collinear; in other words, affinities are line-preserving. More generally, the image of a k-dimensional subspace of $\mathfrak{A}_n$ is again a k-dimensional subspace.

b)   If $\mathfrak{B}$ and $\mathfrak{C}$ are parallel subspaces of $\mathfrak{A}_n$, then so also are their images $\alpha\mathfrak{B}$ and $\alpha\mathfrak{C}$; in particular, the images of parallel lines are parallel.

c)   The ratio of three collinear points P, Q, R remains unchanged, i.e., $(\alpha P, \alpha Q; \alpha R) = (P, Q; R) = c$.

Conversely, these properties could have been taken as the definition of an affinity; in fact, it is enough to require that lines go into lines and that ratios of collinear points remain unchanged.

The set of all affinities of the space $\mathfrak{A}_n$ forms a group under composition (i.e., successive application) of transformations.

Proof. If α and β are affinities, and A and are the corresponding linear mappings, then

$$\overrightarrow{\alpha \circ \beta X \alpha \circ \beta Y} = A(\overrightarrow{\beta X \beta Y}) = A(B(\overrightarrow{XY})) = A \circ B(\overrightarrow{XY})$$

for all X, $Y \in A_n$. Thus $\alpha \circ \beta$ is also an affinity and the corresponding linear mapping is $A \circ B$.

Moreover, since α and A are one-to-one, it follows from $\overrightarrow{\alpha X \alpha Y} = A(\overrightarrow{XY})$ that

$$A^{-1}(\overrightarrow{X'Y'}) = \overrightarrow{\alpha^{-1}X'\alpha^{-1}Y'} \quad \text{for all} \quad X', Y' \in \mathfrak{A}_n,$$

so that α^{-1} is also an affinity and its corresponding linear mapping is A^{-1}.

The automorphism group of $\mathfrak{A}_n$ is also called the *affine group* $\mathfrak{A}_n$ of $\mathfrak{A}_n$.

In our construction the affine point space is defined by the axioms on p. 297, and the affine group is derived from it. Conversely, the affine group could be used to define the affine geometry, as was done by F. Klein in his "Erlanger Program" (cf. II12 and 13).

4.2. *Algebraic Description of Affinities*

If after arbitrary choice of an origin O in $\mathfrak{A}_n$ we describe the points of $\mathfrak{A}_n$ by their position vectors with respect to O, i.e., $\mathfrak{x} = \overrightarrow{OX}$ corresponds to an arbitrary point X and $\mathfrak{a} = \overrightarrow{O\alpha O}$ to the image αO of the origin, then for the position vector $\mathfrak{y}$ of the image $Y = \alpha X$ of X we have the equation $\overrightarrow{OY} = \overrightarrow{O\alpha O} + \overrightarrow{\alpha O \alpha X}$ or

$$(1) \qquad\qquad \mathfrak{y} = A\mathfrak{x} + \mathfrak{a}$$

which is the desired algebraic representation of the affinity α. Here A is the automorphism of $\mathfrak{B}_n$ corresponding to α.

Since the choice of O is arbitrary, we must ask how this representation is changed by a different choice of origin O^* (Fig. 15). Denoting $\overrightarrow{OO^*}$ by $\mathfrak{c}$ and the new position vectors of X and Y by $\mathfrak{x}^*$ and $\mathfrak{y}^*$, we have $\mathfrak{x} = \mathfrak{c} + \mathfrak{x}^*$ and $\mathfrak{y} = \mathfrak{c} + \mathfrak{y}^*$, or

$$\mathfrak{y}^* + \mathfrak{c} = A\mathfrak{y}^* + A\mathfrak{c} + \mathfrak{a},$$

$$\mathfrak{y}^* = A\mathfrak{x}^* + \mathfrak{a}^* \qquad \text{with } \mathfrak{a}^* = \mathfrak{a} + A\mathfrak{c} - \mathfrak{c},$$

which shows that in (1) the arbitrary choice of origin affects only the additive vector a and not the linear mapping A (as is clear, in any case, from the definition of affine mappings).

From the coordinate-free representation (1) of the affine mapping α we obtain the usual coordinate representation (cf. II10, §2) by choosing a basis $\{\mathfrak{e}_1, \ldots, \mathfrak{e}_n\}$ in $\mathfrak{B}_n$. Setting

$$\mathfrak{y} = \sum_{j=1}^{n} y_j \mathfrak{e}_j, \ \mathfrak{x} = \sum_{k=1}^{n} x_k \mathfrak{e}_k, \ \mathfrak{a} = \sum_{j=1}^{n} a_j \mathfrak{e}_j \quad \text{and} \quad A\mathfrak{e}_k = \sum_{j=1}^{n} a_{jk} \mathfrak{e}_j$$

in equation (1) gives

$$\sum_{j=1}^{n} y_j \mathfrak{e}_j = \sum_{j=1}^{n} \left(\sum_{k=1}^{n} a_{jk} x_k \right) \mathfrak{e}_j + \sum_{j=1}^{n} a_j \mathfrak{e}_j,$$

from which, since the vectors $\mathfrak{e}_1, \ldots, \mathfrak{e}_n$ are linearly independent, we have

$$(2) \qquad\qquad y_j = \sum_{k=1}^{n} a_{jk} x_k + a_j, \quad (j = 1, \ldots, n)$$

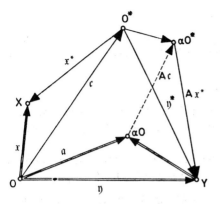

Fig. 15

so that the coordinate representation depends not only on the choice of origin but also on the choice of basis $e_1, \ldots, e_n$.

The set $\{O, e_1, \ldots, e_n\}$ is called a *coordinate system* for the affine point space $\mathfrak{A}_n$. The points obtained by "attaching" the vectors $e_1, \ldots, e_n$ to O will be denoted by $E_1, \ldots, E_n$. The ordered $(n + 1)$-tuple $(O, E_1, \ldots, E_n)$ is then called a *coordinate simplex* (cf. Fig. 16). For two given coordinate simplexes, there exists exactly one affinity taking the first simplex into the second.

The representations (1) and (2) of the affinity α do not immediately indicate its geometric properties. So the purpose of the following sections is to classify the affinities from a geometric point of view and then, for each type of affinity, to introduce a coordinate system $\{O, e_1, \ldots, e_n\}$ such that from the algebraic representation of α we can read off its geometric properties as easily as possible.

We first classify the affinities on the basis of fixed points, i.e., points remaining unchanged.

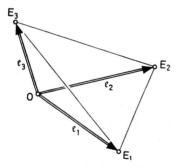

Fig. 16. Coordinate simplex

4.3. *Affine Mappings with Fixed Point*

If the mapping α has fixed points, it is convenient to choose one of them for the origin O. Then in the representation (1) we have $\mathfrak{a} = \mathfrak{o}$, so that the mapping is given by

$$\mathfrak{y} = A\mathfrak{x}.$$

Consequently, classification of the affinities of $\mathfrak{A}_n$ with at least one fixed point is equivalent to classifying the automorphisms of $\mathfrak{B}_n$, a problem which is solved in linear algebra.

Let us give the chief results (i.e., the theory of elementary divisors) for a single linear mapping with real or complex scalar field K (cf. [5], [8]).

Theorem (of Jordan): *Let A be a linear mapping of the real vector space $\mathfrak{B}_n$ into itself. Then $\mathfrak{B}_n$ can be represented as the direct sum*

$$\mathfrak{B}_n = \mathfrak{U}_1 \oplus \mathfrak{U}_2 \oplus \cdots \oplus \mathfrak{U}_r$$

of subspaces of the following kind.

1) *Each of the subspaces $\mathfrak{U}$ is invariant under A, i.e., $A\mathfrak{U} \subset \mathfrak{U}$, and is irreducible, i.e., cannot be represented as a direct sum of invariant subspaces.*

2) *For each of these subspaces, either there exist two real numbers a, b with $b \neq 0$ and a basis $\{\mathfrak{e}_1, \ldots, \mathfrak{e}_{2k}\}$ such that*

$$
\begin{aligned}
A\mathfrak{e}_1 &= a\mathfrak{e}_1 + b\mathfrak{e}_2 \\
A\mathfrak{e}_2 &= -b\mathfrak{e}_1 + a\mathfrak{e}_2 \\
A\mathfrak{e}_3 &= \mathfrak{e}_2 + a\mathfrak{e}_3 + b\mathfrak{e}_4 \\
(*) \quad A\mathfrak{e}_4 &= -b\mathfrak{e}_3 + a\mathfrak{e}_4 \\
&\vdots \\
A\mathfrak{e}_{2k-1} &= \mathfrak{e}_{2k-2} + a\mathfrak{e}_{2k-1} + b\mathfrak{e}_{2k} \\
A\mathfrak{e}_{2k} &= -b\mathfrak{e}_{2k-1} + a\mathfrak{e}_{2k},
\end{aligned}
$$

or else there exists a real number c and a basis $\{\mathfrak{e}_1, \ldots, \mathfrak{e}_k\}$ such that

$$
\begin{aligned}
A\mathfrak{e}_1 &= c\mathfrak{e}_1 \\
A\mathfrak{e}_2 &= \mathfrak{e}_1 + c\mathfrak{e}_2 \\
(**) \quad A\mathfrak{e}_3 &= \mathfrak{e}_2 + c\mathfrak{e}_3 \\
&\vdots \\
A\mathfrak{e}_k &= \mathfrak{e}_{k-1} + c\mathfrak{e}_k.
\end{aligned}
$$

For a vector space $\mathfrak{B}_n$ over the field of complex numbers, the corresponding statement is simpler; 1) reads as before and 2) is replaced by

2') For each of the subspaces $\mathfrak{U}$ there exist a complex number c and a basis $\{\mathfrak{e}_1, \ldots, \mathfrak{e}_k\}$ such that (**) holds.

If as a basis for the whole $\mathfrak{A}_n$ we take the union of the bases for $\mathfrak{U}_1, \ldots, \mathfrak{U}_r$, as listed in the theorem, the resulting coordinate representation of A is called the

Jordan normal form of A. In the present case we have $c \neq 0$, since we are discussing only automorphisms of $\mathfrak{A}_n$, but the theorem holds for noninvertible mappings as well.

The various numbers (a_j, b_j), or c_j, are not necessarily distinct, so that the bases for the various subspaces $\mathfrak{U}_j$ are by no means uniquely determined by A, since for a fixed pair (a, b), or a fixed c, it is only the direct sum of the corresponding subspaces that is determined by A alone, for which reason this sum of spaces is called the (generalized) *eigenspace* of A belonging to (a, b), or c. Since the dimensions of the various eigenspaces are thus determined by A, we see the automorphisms of $\mathfrak{A}_n$ can be classified by a system of scalars and natural numbers, i.e., the scalars (a_j, b_j) or c_j, and the dimensions of the various eigenspaces.

Examples. For the real vector space $\mathfrak{A}_1$ there is only one class of automorphisms with fixed point, namely the dilations.

$$\mathfrak{y} = cx \qquad (c \neq 0).$$

In the real $\mathfrak{A}_2$ there are three such classes, which can be further subdivided according to the scalars involved, since the above Jordan theorem admits the following possibilities.

1) $A\mathfrak{e}_1 = a\mathfrak{e}_1 + b\mathfrak{e}_2,$
    $A\mathfrak{e}_2 = -b\mathfrak{e}_1 + a\mathfrak{e}_2.$ $\qquad (b \neq 0)$

Here O is the only fixed point, and there is no *fixed line*, i.e., no line that is mapped onto itself by α.

2) $A\mathfrak{e}_1 = c\mathfrak{e}_1,$
    $A\mathfrak{e}_2 = \mathfrak{e}_1 + c\mathfrak{e}_2.$ $\qquad (c \neq 0)$

For $c \neq 1$ there is exactly one fixed line, i.e., fixed as a whole, namely the line through O with direction vector $\mathfrak{e}_1$, but for the special case $c = 1$ this line is even pointwise fixed, i.e., each of its points is fixed and all other lines parallel to the pointwise fixed line are themselves fixed lines (though not pointwise fixed). Such a mapping is called a *shear* (Fig. 17).

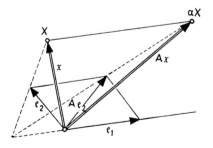

Fig. 17.   Shear

3) $A\mathfrak{e} = c_1\mathfrak{e}_1,$
   $A\mathfrak{e}_2 = c_2\mathfrak{e}_2.$ $(c_1c_2 \neq 0)$

Here there are at least two fixed lines through O, with direction vectors e_1 and e_2 (*Euler affinity*).

Special case a). If $c_1 = 1$, the line with direction vector $\mathfrak{e}_1$ is pointwise fixed, and e_2 is the direction vector of any line joining a nonfixed point to its image (*axial affinity*) (Fig. 18). In particular, for $c_2 = -1$ the result is an *affine reflection* about the pointwise fixed line in the direction $\mathfrak{e}_2$ (Fig. 18a).

Special case b). If $c_1 = c_2 = c$, then every line through O is a fiixed line (*central dilation*). For $c \neq 1$, the only fixed point is O, and for $c = -1$ we have the *point reflection* in O.

The two-dimensional case can also be discussed independently of the Jordan theorem.

In the real $\mathfrak{A}_3$ there are four classes of automorphisms with fixed point, and the bases mentioned in the Jordan theorem again provide us with suitable coordinate systems.

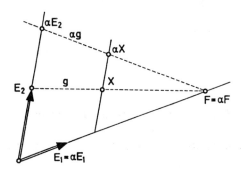

Fig. 18.   Axial affinity

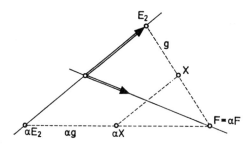

Fig. 18a.   Affine reflection

1) $\quad A\mathfrak{e}_1 = \quad a\mathfrak{e}_1 + b\mathfrak{e}_2$

$\quad A\mathfrak{e}_2 = -b\mathfrak{e}_1 + a\mathfrak{e}_2 \qquad\qquad (bc \neq 0)$

$\quad A\mathfrak{e}_3 = \qquad\qquad c\mathfrak{e}_3$

2) $\quad A\mathfrak{e}_1 = \quad c_1\mathfrak{e}_1$

$\quad A\mathfrak{e}_2 = \qquad\quad c_2\mathfrak{e}_2 \qquad\qquad (c_1c_2c_3 \neq 0)$

$\quad A\mathfrak{e}_3 = \qquad\qquad c_3\mathfrak{e}_3$

3) $\quad A\mathfrak{e}_1 = \quad c_1\mathfrak{e}_1$

$\quad A\mathfrak{e}_2 = \quad \mathfrak{e}_1 + c_1\mathfrak{e}_2 \qquad\qquad (c_1c_3 \neq 0)$

$\quad A\mathfrak{e}_3 = \qquad\qquad c_3\mathfrak{e}_3$

4) $\quad A\mathfrak{e}_1 = \quad c\mathfrak{e}_1$

$\quad A\mathfrak{e}_2 = \quad \mathfrak{e}_1 + \quad c\mathfrak{e}_2 \qquad\qquad (c \neq 0)$

$\quad A\mathfrak{e}_3 = \qquad\quad \mathfrak{e}_2 + c\mathfrak{e}_3.$

The geometric interpretation of these four types of mappings can easily be carried out by the reader, in analogy with our two-dimensional results.

For n-dimensional space we mention only the particularly important *Euler affinity*

$$A\mathfrak{e}_k = c_k\mathfrak{e}_k \qquad (k = 1, \ldots, n).$$

Through the fixed point O there are n distinct fixed lines, spanning the entire space $\mathfrak{A}_n$, on each side of which the affinity acts as a dilation with magnification factor c_k. For a linear mapping A to be an Euler affinity it is sufficient, but not necessary, that all these eigenvalues c_k be distinct. If all the c_k have the same value c, the affinity is called a *central dilation* about O with magnification factor c, and the mapping is given by $\mathfrak{y} = c\mathfrak{x}$. For $c = -1$ the result is a *point reflection* in O, where we note that if a figure is taken into itself by the point reflection in O, then O, which may or may not belong to the figure, is called its *center*. The set of all affinities with the same fixed point O is a subgroup $\mathbf{A}_0$ of the affine group $\mathbf{A}$, and the set $\sum_0$ of all dilations with the same fixed point O is a normal subgroup of $\mathbf{A}$ isomorphic to the multiplicative group of the field of scalars.

4.4. Affine Mappings without Fixed Point

The general representation of an affine mapping α

$$\mathfrak{y} = A\mathfrak{x} + \mathfrak{a}$$

shows that for α to have no fixed point it is necessary and sufficient that the equation $x = Ax + a$ or

$$(I - A)\mathfrak{x} = \mathfrak{a}$$

should have no solution $\mathfrak{x}$. Thus the vector $\mathfrak{a}$ cannot be an image for the linear mapping $I - A$, which means that $I - A$ cannot have an inverse; or in other words, that A has the eigenvalue 1.

The Jordan theorem provides us with an over-all view of such mappings. For let $\mathfrak{U}_1, \ldots, \mathfrak{U}_p$ be the generalized eigenspaces belonging to the eigenvalue 1 and let

$$\mathfrak{U}' = \mathfrak{U}_1 \oplus \cdots \oplus \mathfrak{U}_p$$

be their direct sum. Then the other invariant subspaces $\mathfrak{U}_{p+1}, \ldots, \mathfrak{U}_r$ correspond to pairs of numbers (a, b) with $b \neq 0$, or to numbers $c \neq x$; and, like $\mathfrak{U}'$, their direct sum

$$\mathfrak{U}'' = \mathfrak{U}_{p+1} \oplus \cdots \oplus \mathfrak{U}_r$$

is invariant under A and uniquely determined by A.

Thus $\mathfrak{B}_n$ admits the following representation as a direct sum of invariant subspaces

$$\mathfrak{B}_n = \mathfrak{U}' \oplus \mathfrak{U}''.$$

The restriction of $I - A$ to $\mathfrak{U}'$, which we shall call $(I - A)'$ is *not* invertible; on the other hand, its restriction to $\mathfrak{U}''$, call it $(I - A)''$, does have an inverse.

We now wish, by a suitable choice of origin O, to make the vector $\mathfrak{a}$ as simple as possible. Choice of another origin O^* (see §4(2)) gives

$$\mathfrak{a}^* = \mathfrak{a} - (I - A)\mathfrak{c},$$

where $\mathfrak{c} = \overrightarrow{OO^*}$, so that for the components in $\mathfrak{U}'$ and $\mathfrak{U}''$, we have

$$\mathfrak{a}^{*\prime} = \mathfrak{a}' - (I - A)'\mathfrak{c}',$$

$$\mathfrak{a}^{*\prime\prime} = \mathfrak{a}'' - (I - A)''\mathfrak{c}''.$$

Since $(I - A)''$ is invertible, we can arrange that $\mathfrak{a}^{*\prime\prime} = \mathfrak{o}$, and may thus assume from the beginning that $\mathfrak{a}$ is in the subspace $\mathfrak{U}'$.

We have seen above that $\mathfrak{a}$ cannot be an image vector for the transformation $I - A$, and therefore not for $(I - A)'$. But the set of images for $I - A$ is determined by the Jordan theorem. For, if $c = x$, we have (by p.)

$$(I - A)\mathfrak{e}_1 = \mathfrak{o},$$

$$(I - A)\mathfrak{e}_j = -\mathfrak{e}_{j-1} \qquad (j = 2, \ldots, k),$$

which means that the set of images under $I - A$ is spanned by $\mathfrak{e}_1, \ldots, \mathfrak{e}_{k-1}$.

Since $\mathfrak{U}'$ is the direct sum of $\mathfrak{U}_1, \ldots, \mathfrak{U}_p$, this result can be applied to each of these subspaces; in other words $\mathfrak{a}$ cannot lie in the direct sum

$$(I - A)\mathfrak{U}_1 \oplus \cdots \oplus (I - A)\mathfrak{U}_p.$$

Consequently, in the representation of a in terms of basis vectors

$$\mathfrak{a} = (a_{11}\mathfrak{e}_{11} + \cdots + a_{1k_1}\mathfrak{e}_{1k_1}) + \cdots + (a_{p1}\mathfrak{e}_{p1} + \cdots + a_{pk_p}\mathfrak{e}_{pk_p})$$

the scalars $a_{1k_1}, \ldots, a_{pk_p}$ (the final coefficients in each of the parentheses) *cannot* all vanish but, by a suitable choice of origin, all the other coefficients can be made to vanish.

Examples. In the real space $\mathfrak{A}_1$ there is only one type of affinities without fixed point, namely the translations $\mathfrak{y} = \mathfrak{x} + \mathfrak{a}$, with $\mathfrak{a} \neq 0$.

In the real $\mathfrak{A}_2$ we must first distinguish the two cases, dim $\mathfrak{U}' = 2$ and dim $\mathfrak{U}' = 1$. If dim $\mathfrak{U}' = 2$, there are two possibilities for the linear mapping A. In the first place, we may have

$$\begin{aligned} A\mathfrak{e}_1 &= \mathfrak{e}_1 \\ A\mathfrak{e}_2 &= \mathfrak{e}_2 \end{aligned} \quad \text{i.e. } A = I$$

or

$$\begin{aligned} A\mathfrak{e}_1 &= \mathfrak{e}_1 \\ A\mathfrak{e}_2 &= \mathfrak{e}_1 + \mathfrak{e}_2, \end{aligned}$$

with the corresponding affine transformations

1) $\mathfrak{y} = \mathfrak{x} + \mathfrak{a}$ with $\mathfrak{a} \neq \mathfrak{o}$.

This type is called a *translation* or *shift* (Fig. 19), where it is to be noted that, although there are no fixed points, all lines in the direction of $\mathfrak{a}$ are fiixed lines, and any pair of points (P, Q) is parallel-equal to its image pair $(\alpha P, \alpha Q)$.

Secondly, we may have

2) $\mathfrak{y} = A\mathfrak{x} + \mathfrak{a}$

where $A\mathfrak{e}_1 = \mathfrak{e}_1, A\mathfrak{e}_2 = \mathfrak{e}_1 + \mathfrak{e}_2,$

and $\mathfrak{a} = a_2\mathfrak{e}_2, a_2 \neq 0.$

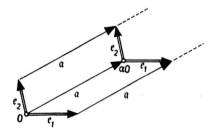

Fig. 19.   Translation

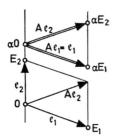

Fig. 20

Here the corresponding affine mapping α may be regarded as a shear $x \to Ax$, followed by a translation $\mathfrak{x} \mapsto \mathfrak{x} + a$ (Fig. 20).

If dim $\mathfrak{U}' = 1$, there is only one possibility for the linear mapping A

$$\begin{matrix} A\mathfrak{e}_1 = \mathfrak{e}_1 \\ A\mathfrak{e}_2 = c\mathfrak{e}_2 \end{matrix} \quad \text{with } c \neq 1,$$

with the corresponding affine mapping

3) $\mathfrak{y} = A\mathfrak{x} + \mathfrak{a}$

where $A\mathfrak{e}_1 = \mathfrak{e}_1, \quad A\mathfrak{e}_2 = c\mathfrak{e}_2,$

and $c \neq 1, \quad \text{and } \mathfrak{a} = a_1\mathfrak{e}_1 \neq \mathfrak{o}.$

An affinity of this type may be regarded as an axial affinity followed by a translation along the axis of the axial affinity. The axis is a fixed line, and there are no other fixed lines (Fig. 21). In the special case $c = -1$, the transformation α is also called an (affine) *glide-reflection* (Fig. 22).

In the real A_3, there are six types of affine transformations without fixed point. In 1), 2), and 3) below, dim $\mathfrak{U}' = 3$; in 4) and 5), dim $\mathfrak{U}' = 2$; and in 6), dim $\mathfrak{U}' = 1$.

1) $\mathfrak{y} = \mathfrak{x} + \mathfrak{a}$ with $\mathfrak{a} \neq \mathfrak{o}.$

Here $A = I$, so that the affinity is a translation.

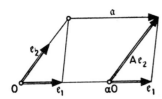

Fig. 21

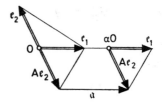

Fig. 22.   Affine glide reflection

2) $\eta = A\mathfrak{x} + \mathfrak{a}$

  $A\mathfrak{e}_1 = \mathfrak{e}_1$

with $A\mathfrak{e}_2 = \mathfrak{e}_1 + \mathfrak{e}_2$ and $\mathfrak{a} = a_2\mathfrak{e}_2 + a_3\mathfrak{e}_3 \neq \mathfrak{o}$

  $A\mathfrak{e}_3 = \qquad \mathfrak{e}_3$

3) $\eta = A\mathfrak{x} + \mathfrak{a}$

  $A\mathfrak{e}_1 = \mathfrak{e}_1$

with $A\mathfrak{e}_2 = \mathfrak{e}_1 + \mathfrak{e}_2$ and $\mathfrak{a} = a_3\mathfrak{e}_3 \neq \mathfrak{o}$

  $A\mathfrak{e}_3 = \qquad \mathfrak{e}_2 + \mathfrak{e}_3$

4) $\eta = A\mathfrak{x} + \mathfrak{a}$

  $A\mathfrak{e}_1 = \mathfrak{e}_1$

with $A\mathfrak{e}_2 = \qquad \mathfrak{e}_2 \quad (c \neq 1)$ and $\mathfrak{a} = a_1\mathfrak{e}_1 + a_2\mathfrak{e}_2 \neq \mathfrak{o}$

  $A\mathfrak{e}_3 = \qquad c\mathfrak{e}_3$

5) $\eta = A\mathfrak{x} + \mathfrak{a}$

  $A\mathfrak{e}_1 = \mathfrak{e}_1$

with $A\mathfrak{e}_2 = \mathfrak{e}_1 + \mathfrak{e}_2 \quad (c \neq 1)$ and $\mathfrak{a} = a_2\mathfrak{e}_2 \neq \mathfrak{o}$

  $A\mathfrak{e}_3 = \qquad c\mathfrak{e}_3$

6) $\eta = A\mathfrak{x} + \mathfrak{a}$

  $A\mathfrak{e}_1 = \mathfrak{e}_1$

with $A\mathfrak{e}_2 = \qquad a\mathfrak{e}_2 + b\mathfrak{e}_3$ and $\mathfrak{a} = a_1\mathfrak{e}_1 \neq \mathfrak{o}.$

  $A\mathfrak{e}_3 = \qquad c\mathfrak{e}_2 + d\mathfrak{e}_3$

The restriction to U'' cannot have the eigenvalue 1, i.e., $(a - 1)(d - 1) - bc \neq 0$.

The reader, guided by our two-dimensional results, should interpret these six types geometrically for himself, giving special values to the scalars.

For the n-dimensional space, let us mention only the *translations,* as being the most important special case of affinities without fixed point. As in two two and three dimensions, their algebraic representation is

$$\eta = \mathfrak{x} + \mathfrak{a},$$

with no change in $\mathfrak{a}$ for a change of origin. Since $\overrightarrow{X\mathfrak{a}X} = \mathfrak{a}$ for arbitrary X, the translations correspond one-to-one to the vectors $\mathfrak{a} \neq \mathfrak{o}$, which are therefore called *translation vectors.* For convenience, the identical transformation of $\mathfrak{A}_n$ is also called a translation, with translation vector $\mathfrak{o}$.

Then the set **T** of all translations forms a subgroup of the affine group **A**. For if α is the translation $\mathfrak{x} \leftrightarrow \mathfrak{x} + \mathfrak{a}$ and β is the translation $\mathfrak{x} \leftrightarrow \mathfrak{x} + \mathfrak{b}$, then the combined transformation $\alpha \circ \beta$ is given by $\mathfrak{x} \leftrightarrow \mathfrak{x} + \mathfrak{a} + \mathfrak{b}$, which is also a translation; and it is equally easy to see that $\mathfrak{x} \leftrightarrow \mathfrak{x} - \mathfrak{a}$ is the inverse of $x \mapsto x + \mathfrak{a}$. The above proof also shows that the group **T** is isomorphic to the additive group of the vector space $\mathfrak{B}_n$, and is therefore Abelian. Furthermore, **T** is a *normal subgroup* of **A**. For is α is the translation $x \mapsto x + \mathfrak{a}$ and β is the affinity $x \mapsto Bx + \mathfrak{b}$, then the affinity $\beta^{-1} \circ \alpha \circ \beta$ is given by $x \mapsto x + B^{-1}a$, which is again a translation.

The factor group **A**/**T** is isomorphic to the group $\mathbf{A}_n^0$ of affinities with fixed point O, where O is arbitrary.

4.5. *Generalizations*

In the preceding sections we have dealt with automorphisms, i.e., one-to-one affine mappings of $\mathfrak{A}_n$ *onto* itself. But now, discarding the assumption of invertibility, we consider more general mappings α of $\mathfrak{A}_n$ into itself, corresponding to linear mappings A of $\mathfrak{B}_n$ that are not necessarily invertible. Here too the "affine" mapping α is represented algebraically by

$$\mathfrak{y} = A x + \mathfrak{a},$$

which, if α is not invertible, is called a *singular affinity*. The reader should ask himself what changes must now be made in the properties a), b), and c) on p. 311.

If $\mathfrak{a}$ is singular, the corresponding linear mapping A is not one-to-one. Thus there exist at least two vectors $x_1 \neq x_2$ with the same image $Ax_1 = Ax_2$, which means that their difference $x_1 - x_2$ has the image $A(x_1 - x_2) = \mathfrak{o}$. The set of all vectors x that are mapped onto the zero vector form a subspace, the *kernel* $\mathfrak{R}$ of the mapping A, where $\mathfrak{R} = \{x \mid Ax = \mathfrak{o}\}$. Two vectors x_1 and x_2 have the same image *if and only if* their difference lies in the kernel of A.

Important examples of singular affinities are the parallel projections, defined as follows. Let $\mathfrak{B} \subset \mathfrak{A}_n$ be a subspace, generated by $\mathfrak{U} \subset \mathfrak{B}_n$, and let $\mathfrak{P}$ be a subspace of $\mathfrak{B}_n$ such that $\mathfrak{B}_n$ is the direct sum of $\mathfrak{U}$ and $\mathfrak{P}$. If now, through an arbitrary point $X \in \mathfrak{A}_n$, we draw the subspace generated by $\mathfrak{P}$, this subspace intersects B in exactly one point Y, since the position vectors $\mathfrak{y}$ for any points of intersection Y must satisfy

$$\mathfrak{y} = \mathfrak{b} + \mathfrak{u} = x + \mathfrak{p}, \quad \mathfrak{u} \in \mathfrak{U}, \mathfrak{p} \in \mathfrak{P}$$

where $\mathfrak{b}$ and x are given and $\mathfrak{u}$ and $\mathfrak{p}$ are to be determined. Since $\mathfrak{B}_n = \mathfrak{U} \oplus \mathfrak{P}$, there is exactly one element $\mathfrak{u} \in \mathfrak{U}$ and exactly one element $\mathfrak{p} \in \mathfrak{P}$ such that $\mathfrak{u} - \mathfrak{p} = x - \mathfrak{b}$, and therefore exactly one point of intersection Y with $\mathfrak{y} = \mathfrak{b} + \mathfrak{u} = x + \mathfrak{p}$. The mapping $X \to Y$ is called the *parallel projection* of A_n in the *direction* $\mathfrak{P}$ onto the *image space* $\mathfrak{B}$.

Example 1. Let $\mathfrak{B}$ be a plane in $\mathfrak{A}_3$, and let $\mathfrak{P}$ be a one-dimensional direction (Fig. 23).

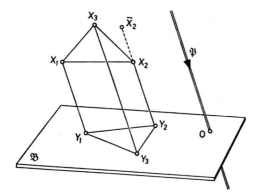

Fig. 23.   Affine projection onto a plane

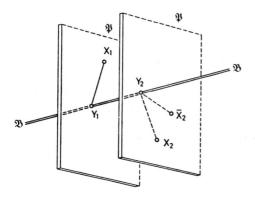

Fig. 24.   Parallel projection onto a line

Example 2.   Let $\mathfrak{B}$ be a line in $\mathfrak{A}_3$, and let $\mathfrak{P}$ be a two-dimensional direction (Fig. 24).

It remains to show that the parallel projections are singular affinities. If the origin O is in the image space $\mathfrak{B}$, we have $\mathfrak{b} = \mathfrak{o}$, so that $\mathfrak{x}$ corresponds to a unique vector u, with

$$u = A\mathfrak{x},$$

where A is the corresponding projection.

Then A is linear, since $\mathfrak{x}_1 = \mathfrak{u}_1 - \mathfrak{p}_1$ and $\mathfrak{x}_2 = \mathfrak{u}_2 - \mathfrak{p}_2$ imply $\mathfrak{x}_1 + \mathfrak{x}_2 = (\mathfrak{u}_1 + \mathfrak{u}_2) - (\mathfrak{p}_1 + \mathfrak{p}_2)$, with $\mathfrak{u}_1 + \mathfrak{u}_2 \in \mathfrak{U}$ and $\mathfrak{p}_1 + \mathfrak{p}_2 \in \mathfrak{P}$, so that

$$A(\mathfrak{x}_1 + \mathfrak{x}_2) = \mathfrak{u}_1 + \mathfrak{u}_2 = A\mathfrak{x}_1 + A\mathfrak{x}_2,$$

since $\mathfrak{B}_n$ is the direct sum of $\mathfrak{U}$ and $\mathfrak{P}$; and similarly, A is linear homogeneous.

Moreover, A is not invertible, since any pair of points X and $\overline{X}$ are mapped onto the same point Y if the vector joining them lies in $\mathfrak{P}$.

Again, A is *idempotent*, i.e., $A \circ A = A$, since

$$A \circ A(\mathfrak{x}) = A(\mathfrak{u}) = \mathfrak{u} = A\mathfrak{x} \quad \text{for every} \quad \mathfrak{x} \in \mathfrak{B}_n.$$

Thus parallel projection can be represented in the form

$$\mathfrak{y} = A\mathfrak{x} \quad \text{with} \quad A \circ A = A.$$

Conversely, we now prove that if A is idempotent, then every corresponding affine mapping α is a parallel projection. For if A satisfies $A \circ A = A$, then for the image space $\mathfrak{U}$ of A and the subspace $\mathfrak{P} = \{\mathfrak{x} \mid A\mathfrak{x} = \mathfrak{o}\}$, i.e., for the *kernel* of A, we have

$$\mathfrak{B}_n = \mathfrak{U} \oplus \mathfrak{P}$$

and since, as is shown by the decomposition

$$\mathfrak{x} = A\mathfrak{x} + (\mathfrak{x} - A\mathfrak{x})$$

every vector $\mathfrak{x}$ can be represented in the form $\mathfrak{u} + \mathfrak{p}$ with $\mathfrak{u} \in \mathfrak{U}$ and $\mathfrak{p} \in \mathfrak{P}$, it therefore follows, for a vector $\mathfrak{x}$ in $\mathfrak{U} \cap \mathfrak{P}$, that $\mathfrak{x} = A\mathfrak{z}$ for some $\mathfrak{z} \in \mathfrak{B}_n$ and $A\mathfrak{x} = \mathfrak{o}$, which means that $\mathfrak{x} = A\mathfrak{z} = (A \circ A)(\mathfrak{z}) = A\mathfrak{x} = \mathfrak{o}$. Thus $\mathfrak{B}_n$ is the direct sum of $\mathfrak{U}$ and $\mathfrak{P}$, and A is the parallel projection onto $\mathfrak{U}$ in the direction P.

With a suitable choice of basis, the mapping for parallel projection can be written

$$A\mathfrak{e}_j = \mathfrak{e}_j \quad \text{for } j = 1, \ldots, r \quad \text{and} \quad A\mathfrak{e}_j = \mathfrak{o} \quad \text{for } j = r + 1, \ldots, n.$$

A further generalization of affine mappings is obtained by discarding the restriction that an affine subspace must be mapped into itself. Let us now agree that a mapping α of a given affine space $\mathfrak{A}_n$ into another affine space $\mathfrak{A}'_k$ is also to be called an affinity if it corresponds to a linear mapping A of $\mathfrak{B}_n$ into $\mathfrak{B}'_n$, i.e.,

$$\overrightarrow{\alpha X \alpha Y} = A(\overrightarrow{XY}) \quad \text{for all} \quad X, Y \in \mathfrak{A}_n.$$

If the image αO of the (arbitrary) origin O in $\mathfrak{A}_n$ is taken as the origin O' in A'_k, the mapping becomes

$$\mathfrak{y} = A\mathfrak{x}.$$

Since the bases for $\mathfrak{B}_n$ and $\mathfrak{B}'_k$ can now be chosen independently, the classification problem becomes very easy. For if r is the *rank* of A, i.e., the dimension of the image space of A, there exist r linearly independent vectors $\mathfrak{e}_1, \ldots, \mathfrak{e}_r$ in $\mathfrak{B}_n$ whose images $\mathfrak{e}'_1 = A\mathfrak{e}_1, \ldots, \mathfrak{e}'_r = A\mathfrak{e}_r$ are also linearly independent. Thus we can choose $n - r$ other vectors $\mathfrak{e}_{r+1}, \ldots, \mathfrak{e}_n$ in $\mathfrak{B}_n$

such that $\{c_{r+1}, \ldots, c_n\}$ is a basis for the kernel of A and $\{c_1, \ldots, c_n\}$ is a basis for $\mathfrak{V}_n$. Since as basis vectors for $\mathfrak{V}_k$ we can first take the vectors $c_1, \ldots, c'_r$ and then the arbitrary, linearly independent vectors $c'_{r+1}, \ldots, c'_k$, it is clear that the only characteristic feature of the mapping α is its rank r.

Affinities of this kind do not form a group, since in general they have no inverses and cannot be multiplied by one another.

5. Quadrics in Affine Geometry

5.1. *Linear and Bilinear Forms: Quadrics*

For the definition of *quadrics* (for $n = 2$, also called *conic sections*) we require the concepts of *linear form* and *bilinear form*, defined as follows (cf. IB3, §2.6).

A *linear form* l is a mapping of V_n into K such that

1) $l(x_1 + x_2) = l(x_1) + l(x_2)$ $\quad$ for all $x_1, x_2 \in \mathfrak{V}_n$,

2) $\qquad l(sx) = sl(x)$ $\qquad\qquad$ for all $x \in \mathfrak{V}_n$ and $s \in K$.

For a given (nonzero) linear form l, a given scalar a and an arbitrary origin O (cf. §2.2), the solutions of

$$l(x) = a$$

determine a *hyperplane*

$$H = \{X \mid x = \overrightarrow{OX}, l(x) = a\}$$

and every hyperplane can be described in this way. If (with fixed l) a runs through the entire field of scalars, we obtain all the hyperplanes parallel to H.

A *bilinear form* f is a mapping of $\mathfrak{V}_n \times \mathfrak{V}_n$ into K which, if either argument is fixed, becomes a linear form in the other argument. In particular, f is a *symmetric bilinear* form if $f(x, \mathfrak{y}) = f(\mathfrak{y}, x)$ for all x and $\mathfrak{y}$. Thus a symmetric bilinear form is characterized by the following properties

1) $\qquad f(x, \mathfrak{y}) = f(\mathfrak{y}, x)$ $\qquad\qquad$ for all $x, \mathfrak{y} \in \mathfrak{V}_n$,

2) $\quad f(x_1 + x_2 y) = f(x_1, y) + f(x_2 \mathfrak{y})$ $\quad$ for all $x_1, x_2, \mathfrak{y} \in \mathfrak{V}_n$,

3) $\qquad f(sx, \mathfrak{y}) = sf(x, \mathfrak{y})$ $\qquad\qquad$ for all $x, \mathfrak{y} \in \mathfrak{V}_n$ and all $s \in K$.

The mapping F of $\mathfrak{V}_n$ into K defined by $F(x) = f(x, x)$ is a *quadratic form*. Conversely, the quadratic form F determines the symmetric bilinear form f:

$$f(x, \mathfrak{y}) = \tfrac{1}{2}(F(x + \mathfrak{y}) - F(x) - F(\mathfrak{y})),$$

where it is assumed that K is not of characteristic 2 and that f does not vanish for all $x, \mathfrak{y}$.

But it may happen that for some $\mathfrak{u} \neq \mathfrak{o}$ the linear form $\mathfrak{x} \mapsto f(\mathfrak{u}, \mathfrak{x})$ is the zero form. The set of all these $\mathfrak{u}$ (including the zero vector) forms a subspace $\mathfrak{U}_f$, called the *null space* of f. By symmetry, $\mathfrak{x} \mapsto f(\mathfrak{x}, \mathfrak{u})$ is also the zero form for all $\mathfrak{u} \in \mathfrak{U}_f$, and in particular, $f(\mathfrak{u}, \mathfrak{u}) = 0$ for all $\mathfrak{u} \in \mathfrak{U}_f$. If $\mathfrak{U}_f$ is of dimension d, the natural number $r = n - d$ is called the *rank* of f. Bilinear forms with $d = 0$, and therefore $r = n$, are said to be *nondegenerate*. If f is nondegenerate, there exists, for every linear form l, exactly one vector $\mathfrak{a}$ such that $l(\mathfrak{x}) = f(\mathfrak{a}, \mathfrak{x})$. For degenerate f there exists (at least) one r-dimensional subspace $\mathfrak{W}$ such that the restriction of f to $\mathfrak{W}$ is nondegenerate and $\mathfrak{B}_n = \mathfrak{W} \oplus \mathfrak{U}_f$ (cf., for example, [5]).

An example of a symmetric bilinear form f of rank r in the vector space of n-tuples in K is given by

$$f(\mathfrak{x}, \mathfrak{y}) = \sum_{j=1}^{r} x_j y_j.$$

The null space $\mathbf{A}_f$ consists of the vectors with vanishing first r components.

For a given symmetric bilinear form f, linear form l and scalar c, the set of solutions $x = \overrightarrow{OX}$ (with arbitrary origin O in $\mathfrak{A}_n$) of the equation

(1) $f(\mathfrak{x}, \mathfrak{x}) = 2l(\mathfrak{x}) + c = 0$

determines a subset $\mathfrak{Q}$ of $\mathfrak{A}_n$

$$\mathfrak{Q} = \{X \mid \mathfrak{x} = \overrightarrow{OX}, f(\mathfrak{x}, \mathfrak{x}) + 2l(\mathfrak{x}) + c = 0\}$$

which, provided it is not empty, is called a *quadric*; let us note that it is determined by f, l, c, and O.

The requirement that $\mathfrak{Q}$ is to be independent of the choice of origin O determines how f, l, c will change for a new choice of origin O'. Substituting $\mathfrak{x} = \mathfrak{x}' + \mathfrak{a}$ (cf. p. 11) in (1) gives the following condition on $\mathfrak{x}'$

$$f(\mathfrak{x}' + \mathfrak{a}, \mathfrak{x}' + \mathfrak{a}) + 2l(\mathfrak{x}' + \mathfrak{a}) + c = 0$$

or

(1') $f(\mathfrak{x}', \mathfrak{x}') + 2(f(\mathfrak{a}, \mathfrak{x}') + l(\mathfrak{x}')) + (f(\mathfrak{a}, \mathfrak{a}) + 2l(\mathfrak{a}) + c) = 0,$

which is an equation of the same type as (1). Thus the bilinear form f remains unchanged, but the linear form l and the scalar c become

(2) $l' = l + f(\mathfrak{a}),$

(3) $c' = f(\mathfrak{a}, \mathfrak{a}) + 2l(\mathfrak{a}) + c.$

Then equation (1) and the equation

$$f(\mathfrak{x}', \mathfrak{x}') + 2l'(\mathfrak{x}') + c' = 0$$

describe the same set of points.

After choice of origin, every equation of the type (1) uniquely determines the corresponding *set of points* $\mathfrak{Q}$, but various equations can determine the same set $\mathfrak{Q}$, e.g., if two equations differ only by a nonzero factor, and this condition is by no means necessary for the two equations to define the same quadric. For example, if $K = \mathbb{R}$, the equations $f(x, x) = 0$ and $\bar{f}(x, x) = 0$ with *positive definite* (cf. IB3, §3.2) forms f and $\bar{f}$ define the same set (consisting of a single point), even if f and $\bar{f}$ are not proportional. However, if the quadric $\mathfrak{Q}$ is not contained in a hyperplane, then any two of its representations differ only by a nonzero factor (cf. §§5.2 and 5.3). These *proper* quadrics, i.e., the ones that "span" the whole space $\mathfrak{A}_n$, will be the chief subject for discussion in the following sections.

In order to classify quadrics into types, we begin with the two possibilities for an arbitrary quadric $\mathfrak{Q}$, namely

a) $\mathfrak{Q}$ has a representation in which the linear form l is the zero form.

b) $\mathfrak{Q}$ has no such representation.

In the first case $\mathfrak{Q}$ has a center (cf. §4.3), but not in the second, as we shall see.

5.2. Quadrics with Center

Let us consider the first case, namely that $\mathfrak{Q}$ can be described by

$$(4) \qquad\qquad f(x, x) + c = 0,$$

where the orgin O must be a center of $\mathfrak{Q}$, since if x satisfies equation (4), so does $-x$.

Since multiplication of (4) by a nonzero scalar does not change its set of solutions, we may assume that the defining equation for $\mathfrak{Q}$ is either

$$(4') \qquad\qquad f(x, x) = 1$$

or

$$(4'') \qquad\qquad f(x, x) = 0,$$

where in (4″) the bilinear form f can be determined only up to a nonzero factor.

Let us now examine the possible representations of a *proper* quadric that can be described by an equation of the type (4′). By hypothesis, there exist linearly independent vectors $b_1, \ldots, b_n$ for which $f(b_j, b_j) = 1$ and thus also $f(-b_j, -b_j) = 1$. Now if

$$\bar{f}(x, x) + 2\bar{l}(x) + \bar{c} = 0$$

is to have the same set of solutions as $f(x, x) = 1$, we see that $\bar{l}$ be the zero form; for if we substitute b_j in this equation, and then $-b_j$, and subtract, we get $\bar{l}(b_j) = 0$ for $j = 1, \ldots, n$. Since $f(b_j, b_j) = 1$, it follows that $\bar{f}(b_j, b_j) = \bar{c}f(b_j, b_j)$. If we now determine the intersection of the quadric with the plane (through O) spanned by b_i and b_k, we see, as will be left unproved here, that also $\bar{f}(b_i, b_k) = -\bar{c}f(b_i, b_k)$ for $i \neq k$, which means, as desired, that $\bar{f} = -\bar{c}f$.

If $f(x, x) = 1$ has any solution at all, there are n linearly independent vectors among its solutions, so that a quadric of type $(4')$ is necessarily proper.

Similarly, for the *proper* quadrics of type $(4'')$ we can show that in any representation (with fixed origin O) the linear form and the constant must both be missing, so that the bilinear forms are identical with each other up to a nonzero scalar factor, as can be seen by considering the intersection of the quadric with the three-dimensional subspace through O spanned by any three of the linearly independent vectors $b_1, \ldots, b_n$.

We thus have the desired result that any two equations of a *proper* quadric with center can differ from each other only by a nonzero factor, which means that the null space $\mathfrak{U}_f$ and the rank r of the bilinear form f are associated with the quadric (and not only with its representation). To see the geometirc significance of $\mathfrak{U}_f$, we consider the subspace through the origin generated by $\mathfrak{U}_f$, say $\mathfrak{M} \subset \mathfrak{A}_n$. Every point in $\mathfrak{M}$ is a center of the quadric, since a new choice of origin $O' \in \mathfrak{M}$ shows, by (2) and (3) that $l' = l = 0$, $c' = c$, and therefore, with O' as origin, $f(x', x') + c = 0$ also describes the quadric, which means that O' is a center. Furthermore, the subspace generated by $\mathfrak{U}_f$ is completely contained in the quadric. Consequently, quadrics with $r < n$ are also called *cylinders*.

Quadrics of type $(4')$ do not contain the origin O, and therefore any line through O has either no point or two points in common with the quadric.

Any central quadric of type $(4'')$ contains the origin O and, if X is a point on the quadric, also contains the whole line $O \vee X$ joining O to X; for if $f(x, x) = 0$, then $f(tx, tx) = 0$ for every $t \in K$. So these quadrics are called *hypercones*.

For the bilinear form f let us now determine a *conjugate* basis of $\mathfrak{B}_n$, i.e., a basis $\{e_1, \ldots, e_n\}$ for which

$$f(e_i, e_k) = 0$$

if $i \neq k$. The existence of such bases is easily shown by successive construction of $e_1, \ldots, e_n$.

If r is the rank of f, then exactly r of the scalars $a_i = f(e_i, e_i)$ are different from zero. Let us index the basis vectors in such a way that

$$a_1 \neq 0, \ldots, a_r \neq 0 \quad \text{and} \quad a_{r+1} = \cdots = a_n = 0,$$

where we note that if the vectors $e_1, \ldots, e_r$ are replaced by any convenient nonzero multiples $b_1 e_1, \ldots, b_r e_r$, the scalars $a_1, \ldots, a_r$ will in general be changed but the conjugacy will not be disturbed. Thus for $K = \mathbb{R}$, we can construct a conjugate basis $\{e_1, \ldots, e_n\}$ such that

$$f(e_i, e_i) = 1 \quad \text{for } i = 1, \ldots, p,$$
$$f(e_i, e_i) = -1 \quad \text{for } i = p + 1, \ldots, r,$$
$$f(e_i, e_i) = 0 \quad \text{for } i = r + 1, \ldots, n.$$

By the *Sylvester* law of inertia (cf. IB3, §3.2) for $K = \mathbb{R}$, the rank r, and the natural number p (called the *index* of f) are uniquely associated with the bilinear form f.

In the coordinate system $\{O, e_1, \ldots, e_n\}$ the equations (4') and (4'') for $x = x_1 e_1 + \cdots + x_n e_n$ become

(4') $$x_1^2 + \cdots + X_p^2 - X_{p+1}^2 - \cdots - X_r^2 = 1$$

(4'') $$X_1^2 + \cdots + X_p^2 - X_{p+1}^2 - \cdots - X_r^2 = 0.$$

So for every central quadric of type (4') in the real affine point space A_n the two natural numbers r and p are uniquely determined, but for the hypercones (4'') we must note that the defining equations cannot be normalized in this way. If we pass from f to $-f$, the index p is replaced by $r - p$, so that with every hypercone we can uniquely associate the two natural numbers r and $\max \{p, r - p\}$.

For $K = \mathbb{C}$ we may assume that all the scalars $a_1, \ldots, a_r$ are equal to 1. The central quadrics of both types are distinguished only by their rank.

Examples. In the real affine plane there are four types of proper central conics

1) $x_1^2 + x_2^2 = 1$ ellipse
2) $x_1^2 - x_2^2 = 1$ hyperbola
3) $x_1^2 = 1$ pair of parallel lines
4) $x_1^2 - x_2^2 = 0$ pair of intersecting lines

but in the complex affine plane there are only three, with the equations

$$x_1^2 + x_2^2 = 1, \quad x_1^2 = 1, \quad x_1^2 + x_2^2 = 0.$$

In the real $\mathfrak{A}_3$ there are the following types of proper central quadrics

1) $x_1^2 + x_2^2 + x_3^2 = 1$ ellipsoid, Fig. 25
2) $x_1^2 + x_2^2 - x_3^2 = 1$ one-sheeted hyperboloid, Fig. 26
3) $x_1^2 - x_2^2 - x_3^2 = 1$ two-sheeted hyperboloid, Fig. 27
4) $x_1^2 + x_2^2 = 1$ elliptic cylinder
5) $x_1^2 - x_2^2 = 1$ hyperbolic cylinder
6) $x_1^2 = 1$ pair of parallel planes
7) $x_1^2 + x_2^2 - x_3^2 = 0$ cone
8) $x_1^2 - x_2^2 = 0$ pair of intersecting planes.

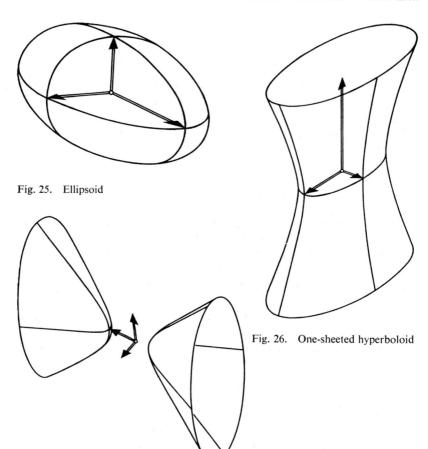

Fig. 25. Ellipsoid

Fig. 26. One-sheeted hyperboloid

Fig. 27. Two-sheeted hyperboloid

5.3. *Quadrics without Center*

Continuing our discussion of the possible cases in §5.1 we now assume that $\mathfrak{Q}$ has no representation without linear part. Then the bilinear form f must be degenerate, since otherwise, for suitable choice of the vector $\mathfrak{a}$, the linear form $l' = l + f(\mathfrak{a}, \ )$ would be the zero form, so that by (2) a representation of $\mathfrak{Q}$ without linear part would still be possible.

Since l is not the zero form, the equation $l(x) = 0$ determines in $\mathfrak{V}_n$ an $(n - 1)$-dimensional vector subspace $\mathfrak{U}$. To the bilinear form f there corresponds a degeneration space $\mathfrak{U}_f$ that is at least one-dimensional and cannot be completely contained in $\mathfrak{U}$, since the following argument shows that $\mathfrak{U}_f \subset \mathfrak{U}$ leads to a contradiction. By §5.1, every vector x can be decomposed as follows:

$$\mathfrak{x} = \mathfrak{y} + \mathfrak{u} \quad \text{with} \quad \mathfrak{y} \in \mathfrak{W} \quad \text{and} \quad \mathfrak{u} \in \mathfrak{U}_f.$$

Then equation (1) becomes

$$f(\mathfrak{y}, \mathfrak{y}) + 2l(\mathfrak{y}) + 2l(\mathfrak{u}) + c = 0,$$

since $f(\mathfrak{u}, y) = f(\mathfrak{u}, \mathfrak{u}) = 0$; and therefore $\mathfrak{U}_f \subset \mathfrak{U}$ would imply $l(\mathfrak{u}) = 0$.

Since f is not degenerate on $\mathfrak{W}$, the linear part of the equation $f(\mathfrak{y}, \mathfrak{y}) + 2l(\mathfrak{y}) + c = 0$ for $\mathfrak{y} \in \mathfrak{W}$ could be made equal to zero, and the quadric would after all have an equation without linear part.

So there exists a vector $\mathfrak{e}_n \in \mathfrak{U}_f$ with $l(\mathfrak{e}_n) \neq 0$. Furthermore, there esists for f a conjugate basis $\{\mathfrak{e}_1, \ldots, \mathfrak{e}_{n-1}\}$ in $\mathfrak{U}$. Since $\mathfrak{e}_n \in \mathfrak{U}_f$, it follows that $\{\mathfrak{e}_1, \ldots, \mathfrak{e}_n\}$ is a conjugate basis of $\mathfrak{V}_n$. Finally, if we choose the origin O on the quadric $\mathfrak{Q}$, set $f(\mathfrak{e}_i, \mathfrak{e}_i) = a_i$ as in the preceding section, and normalize the $\mathfrak{e}_n$ so that $l(\mathfrak{e}_n) = -\frac{1}{2}$, the equation (1) for the position vector $\mathfrak{x} = x_1 \mathfrak{e}_1 + \cdots + x_n \mathfrak{e}_n$ becomes

$$(5) \qquad x_n = \sum_{i=1}^{r} a_i x_i^2 \quad (1 \leqq r = n \leqq 1).$$

From this representation it follows that all the quadrics described here are proper. Moreover, the coordinate system we have just constructed shows that every other representation differs from (5) only by a nonzero factor. But in contrast to the central quadrics of type (4'), these quadrics do not have a uniquely determined equation.

To every quadric of the present type we therefore associate a natural number, the rank r of the bilinear form f. If this rank is maximal, i.e., if $r = n - 1$, we speak of *paraboloids*, otherwise, of *parabolic cylinders*.

For $K = \mathbb{R}$, we can further normalize the conjugate basis in such a way that

$$\begin{aligned} f(\mathfrak{e}_i, \mathfrak{e}_i) &= 1 && \text{for } i = 1, \ldots, p, \\ f(\mathfrak{e}_i, \mathfrak{e}_i) &= -1 && \text{for } i = p+1, \ldots, r \quad \text{with } 1 \leqq r \leqq n-1, \\ f(\mathfrak{e}_i, \mathfrak{e}_i) &= 0 && \text{for } i = r+1, \ldots, n. \end{aligned}$$

Then (5) becomes

$$x_n = x_1^2 + \cdots + x_p^2 - x_{p+1}^2 - \cdots - x_r^2 \quad (1 \leqq r \leqq n-1).$$

In this case we can associate with the quadric not only its rank r but also the natural number $\max\{p, r - p\}$.

Example. In the real affine plane there is only one type of noncentral conic, the parabola. Every parabola can be described, in a suitable coordinate system, by the equation

$$x_2 = x_1^2.$$

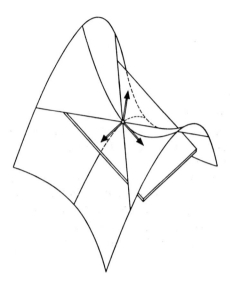

Fig. 28.   Hyperbolic hyperboloid

In the real affine three-dimensional space the number of types increases to three

1) $x_3 = x_1^2 + x_2^2$ elliptic paraboloid
2) $x_3 = x_1^2 - x_2^2$ hyperbolic paraboloid, Fig. 28
3) $x_3 = x_1^2$ parabolic cylinder.

The name *conic section* for quadrics in the affine plane requires some explanation. From the normal form for the equation of a cone in $\mathfrak{A}_3$ (see §5.2) it can be shown that the ellipse, hyperbola, parabola, and pair of intersecting lines are the intersection of a cone with a plane. The details are left to the reader.

We must still show that the quadrics discussed in this section are characterized by the fact that they have no center. If a quadric $\mathfrak{Q}$ has no center, the linear form in its defining equation cannot be reduced to zero by any choice of origin, since otherwise the new origin would be a center. Conversely, if the linear form cannot be reduced to zero, $\mathfrak{Q}$ cannot have a center, since a line with e_n as direction vector has exactly one point in common with $\mathfrak{Q}$, so that only points of $\mathfrak{Q}$ itself could be centers, and the origin (arbitrary on $\mathfrak{Q}$) could be so chosen as to be a center, which is obviously impossible (since the linear form is not zero).

5.4.   Tangents and Tangential Hyperplanes

A line has either no point, one point, or two points in common with a quadric, or else it is entirely contained in the quadric. For if the quadric $\mathfrak{Q}$ is represented by

$$f(x, x) + 2l(x) + c = 0$$

and the line by

$$x = p + xb, \quad (x \in K)$$

the scalar corresponding to a point of intersection must satisfy the condition

(6) $x^2 f(b, b) + 2x(f(p, b) + l(b)) + (f(p, p) + 2l(p) + c) = 0,$

which proves the assertion. The line will be contained in $\mathfrak{Q}$ if and only if

$$f(b, b) = 0, f(p, b) + l(b) = 0 \text{ and } f(p, p) + 2l(p) + c = 0;$$

there will be exactly two points of intersection if $f(b, b) \neq 0$, and exactly one point of intersection if $f(b, b) \neq 0$ and the quadratic equation (6) has two equal solutions, or else $f(b, b) = 0$ and $f(p, b) + l(b) \neq 0$. In the first case, the line is called a *tangent* to the quadric, and the single point of intersection is the *point of contact* of the tangent. At first sight, this definition seems to depend on the choice of representation of the quadric $\mathfrak{Q}$, but for *proper* quadrics, as we have seen, any two representations differ only by a nonzero factor, so that a tangent for one representation is also a tangent for any other.

The lines lying entirely on the quadric are also called tangents.

Let us now assume that the point P with position vector p lies on the proper quadric $\mathfrak{Q}$ and that $f(b, b) \neq 0$. Then $x = 0$ is the only solution of (6) if and only if

(7) $f(p, b) + l(b) = 0.$

Thus the lines through P with direction vectors b satisfying (7) are tangents to the quadric, all of them lying in a hyperplane, and the position vector x of a point X on any of these tangents must, by (7), satisfy the equation

$$f(p, x - p) + l(x - p) = 0.$$

Noting that p must satisfy the equation of the quadric, we obtain the equivalent condition

(8) $f(p, x) + l(x) + l(p) + c = 0.$

If the linear form $x \mapsto f(p, x) + l(x)$ is not the zero form, equation (8) represents a hyperplane, called the *tangential plane* at the point of contact P, which is then called a *regular point* of the quadric. Any tangent through P (cf. Fig. 28) lying entirely on $\mathfrak{Q}$ is contained in this hyperplane.

The definition of a regular point, which appears at first sight to depend on the choice of representation, is in fact independent of this choice, so that we may legitimately speak of a regular, or nonregular, point of the quadric.

Nonregular points occur only for central quadrics of the type (4″), namely the hypercones; for if the linear form $x \mapsto f(\mathfrak{p}, x) + l(x)$ is the zero form, then by the classification on p. 327 the origin O can be so chosen that $\mathfrak{Q}$ has the equation

$$f(x, x) + c = 0.$$

Then $f(\mathfrak{p}, x) = 0$ for all $x \in \mathfrak{B}_n$ and so in particular for $f(\mathfrak{p}, \mathfrak{p}) = 0$. Since $\mathfrak{p}$ is the position vector of a point on $\mathfrak{Q}$, we must have $c = 0$.

The nonregular points of the hypercone with equation $f(x, x) = 0$ are characterized by the two conditions $f(\mathfrak{p}, \mathfrak{p}) = 0$ and $f(\mathfrak{p}, x) = 0$ for all $x \in \mathfrak{B}_n$. It follows that the defining equation can also be written in the form $f(x - p, x - p) = 0$, which means that every nonregular point is a center of the hypercone. Conversely, every center of the hypercone is nonregular. These *singular* points are also called *vertices* of the hypercone.

5.5. *Affine Equivalence of Quadrics*

If we transform a quadric by a regular affinity, we again obtain a quadric; for we may assume, for simplicity, that the relation between x and its image $\mathfrak{y}$ is already written in the form $x = A\mathfrak{y} + \mathfrak{a}$, whereupon (1) becomes

$$f(A\mathfrak{y}, A\mathfrak{y}) + 2(f(\mathfrak{a}, A\mathfrak{y}) + l(A\mathfrak{y})) + (f(\mathfrak{a}, \mathfrak{a}) + 2l(\mathfrak{a}) + c) = 0,$$

which is again an equation of the form (1).

Thus the set of all quadrics falls into classes of affine equivalent quadrics. For the particular case $K = \mathbb{R}$, we assert that this division into classes by affine equivalence is identical with the classification by means of "normalized representations" in §§2 and 3 above.

For the proof we first note that any two quadrics belonging in the same class by §§2 and 3 must also be affinely equivalent. For in the "suitably chosen" coordinate systems of those sections, the equations of the two quadrics are identical, so that the regular affinity taking one coordinate system into the other will also map the one quadric onto the other.

Conversely, if two quadrics are affinely equivalent, then either they both have a center, or neither does, and in the first case, they must both be hypercones, or else they must both have an equation of type (4′); and finally, the corresponding numbers r and p, or r and max $\{p, r - p\}$, must also agree, since the corresponding bilinear forms $f(x, \mathfrak{y})$ and $f(Ax, A\mathfrak{y})$ must have the same numbers r and p (cf. [5]).

6. Affine Subspaces over Ordered Fields

6.1. *Determinants*

We first recall three concepts of linear algebra, all of which have the name *determinant*. The fact that the field of scalars is ordered will not be used here.

A *determinant function* D over the vector space $\mathfrak{B}_n$ is a nontrivial alternating n-fold multilinear form (cf. IB3, §3.4). The arguments of the function D are therefore n-tuples of vectors $(\mathfrak{x}_1, \ldots, \mathfrak{x}_n)$ and its values are scalars, for which we have the following rules:

1. $D(\ldots, \mathfrak{x}_i' + \mathfrak{x}_i'', \ldots) = D(\ldots, \mathfrak{x}_i', \ldots) + D(\ldots, \mathfrak{x}_i'', \ldots)$
for $i = 1, \ldots, n$,

2. $D(\ldots, c\mathfrak{x}_i, \ldots) = cD(\ldots, \mathfrak{x}_i, \ldots)$ for $i = 1, \ldots, n$,

3. $D(\ldots, \mathfrak{x}_i, \ldots, \mathfrak{x}_k, \ldots) = -D(\ldots, \mathfrak{x}_k, \ldots, \mathfrak{x}_i, \ldots)$ for $i \neq k$.

4. D is not the zero mapping.

Determinant functions of this sort exist for every vector space $\mathfrak{B}_n$ with a commutative field of scalars, and any two of them differ only by a nonzero factor.

By means of an arbitrary determinant function D we can test whether the n vectors $\mathfrak{x}_1, \ldots, \mathfrak{x}_n$ are linearly dependent, which will be the case if and only if

$$D(\mathfrak{x}_1, \ldots, \mathfrak{x}_n) = 0,$$

from which we also obtain a criterion that $n + 1$ points $X_0, X_1, \ldots, X_n$ of $\mathfrak{A}_n$ lie in a hyperplane. For if $\mathfrak{x}_0, \mathfrak{x}_1, \ldots, \mathfrak{x}_n$ are the position vectors of these points (with arbitrary origin O), the equation

$$D(\mathfrak{x}_1 - \mathfrak{x}_0, \ldots, \mathfrak{x}_n - \mathfrak{x}_0) = 0$$

(which can also be put in symmetric form) is satisfied if and only if there is a hyperplane containing $X_0, X_1, \ldots, X_n$.

Thus for $n = 2$ we obtain the following *condition for collinearity*.
Three points X_0, X_1, X_2 of $\mathfrak{A}_2$ lie on a line if and only if

$$D(\mathfrak{x}_0, \mathfrak{x}_1) + D(\mathfrak{x}_1, \mathfrak{x}_2) + D(\mathfrak{x}_2, \mathfrak{x}_0) = 0.$$

Applications. We first prove a theorem (cf. Gauss, Werke IV, pp. 385–392) about the complete quadrilateral in affine planes with a scalar field of characteristic not equal to 2.
In a complete quadrilateral, the midpoints of the diagonals are collinear.
For by hypothesis, in the notation of Fig. 29,

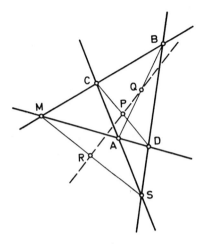

Fig. 29.   Complete quatrilateral

$$D(\mathfrak{s}, \mathfrak{c}) + D(\mathfrak{c}, \mathfrak{a}) + D(\mathfrak{a}, \mathfrak{s}) = 0,$$

$$D(\mathfrak{s}, \mathfrak{d}) + D(\mathfrak{d}, \mathfrak{b}) + D(\mathfrak{b}, \mathfrak{s}) = 0,$$

$$D(\mathfrak{m}, \mathfrak{d}) + D(\mathfrak{d}, \mathfrak{a}) + D(\mathfrak{a}, \mathfrak{m}) = 0,$$

$$D(\mathfrak{m}, \mathfrak{c}) + D(\mathfrak{c}, \mathfrak{b}) + D(\mathfrak{b}, \mathfrak{m}) = 0,$$

so that, since the position vectors of the midpoints P, Q, R of the diagonals have the form

$$\mathfrak{p} = \tfrac{1}{2}(\mathfrak{c} + \mathfrak{d}),$$

$$\mathfrak{q} = \tfrac{1}{2}(\mathfrak{a} + \mathfrak{b}),$$

$$\mathfrak{r} = \tfrac{1}{2}(\mathfrak{m} + \mathfrak{s}),$$

we obtain the condition for collinearity of P, Q, R

$$D(\mathfrak{c} + \mathfrak{d}, \mathfrak{a} + \mathfrak{b}) + D(\mathfrak{a} + \mathfrak{b}, \mathfrak{m} + \mathfrak{s}) + D(\mathfrak{m} + \mathfrak{s}, \mathfrak{c} + \mathfrak{d}) = 0.$$

Carrying out the calculation leads to the sum of the twelve summands in the four equations above, which are valid by hypothesis. Thus the theorem is proved.

As a second application we prove the following theorem (Menelaus).

Let X, Y, Z be points (distinct from the vertices) on the sides $B \vee C$, $C \vee A$, $A \vee B$ of the triangle ABC, and let x, y, z denote the ratios $(X, B; C)$, $(Y, C; A)$, $(Z, A; B)$ in which the sides are divided by X, Y, Z. Then the three points X, Y, Z are collinear if and only if $xyz = 1$ (Fig. 30).

By our definition of these ratios (p. 300), the position vectors, with arbitrary origin, of the points X, Y, Z (note that the scalars x, y, z are all different from 1) are given by

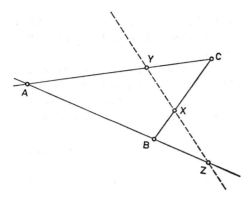

Fig. 30. Theorem of Menelaus

$$\mathfrak{x} = -\frac{x}{1-x}b + \frac{1}{1-x}c,$$

$$\mathfrak{y} = -\frac{y}{1-y}c + \frac{1}{1-y}a,$$

$$\mathfrak{z} = -\frac{z}{1-z}a + \frac{1}{1-z}b$$

which means that

$$D(\mathfrak{x}, \mathfrak{y}) + D(\mathfrak{y}, \mathfrak{z}) + D(\mathfrak{z}, \mathfrak{x})$$
$$= \frac{1 - xyz}{(1-x)(1-y)(1-z)}(D(a, b) + D(b, c) + D(c, a)).$$

Since A, B, C are not collinear, we have $D(a, b) + D(b, c) + D(c, a) \neq 0$, so that X, Y, Z are collinear if and only if $1 - xyz = 0$.[4]

The determinant functions (whose arguments are n-tuples of vectors) shuld be clearly distinguished from the *determinant of a linear mapping of a vector space into itself.* The latter determinant will be here defined as follows: If in $D(\mathfrak{x}_1, \ldots, \mathfrak{x}_n)$ we replace the arguments by their images $A\mathfrak{x}_1, \ldots, A\mathfrak{x}_n$ under the mapping A, we again obtain a determinant function in the arguments $\mathfrak{x}_1, \ldots, \mathfrak{x}_n$. So there must be a scalar, which we shall denote by det A, such that

$$D(A\mathfrak{x}_1, \ldots, A\mathfrak{x}_n) = \det A \cdot D(\mathfrak{x}_1, \ldots, \mathfrak{x}_n)$$

for all $\mathfrak{x}_1, \ldots, \mathfrak{x}_n \in \mathfrak{V}_n$. The determinant det A of the linear mapping of A into itself is independent of the determinant function used to define it, and from the definition we at once have the *multiplication theorem*

$$\det (A \circ B) = \det A \cdot \det B.$$

[4]If K is ordered (cf. p. 339), this result can be deduced very simply from the "Axiom of Pasch."

Since it follows at once that $\det I = 1$; $\det A^{-1} = (\det A)^{-1}$, the mapping det of the group $\mathbf{A}_n^0$ of affinities with fixed point O into the multiplicative group of the scalar field is a *homomorphism*.

The affinities with $\det A = 1$ form a subgroup Γ_n of the affine group $\mathbf{A}_n$, the so-called *equiaffine* group, and the corresponding equiaffine geometry has the additional scalar invariant $D(\mathbf{x}_1, \ldots, \mathbf{x}_n)$, which does not change in value under any equiaffine mapping. In the next section this invariant will be interpreted as a volume.

Finally, let us give a definition for the *determinant of an $n \times n$ matrix* (a_{ik}) which will be needed to formulate the *Lagrange identity*.

After choice of a basis $\mathfrak{e}_1, \ldots, \mathfrak{e}_n$ for $\mathfrak{B}_n$, the matrices (a_{ik}) correspond one-to-one to the linear mappings A with

$$A\mathfrak{e}_i = \sum_{k=1}^{n} a_{ki}\mathfrak{e}_k \quad (i = 1, \ldots, n).$$

We now define

$$\det (a_{ik}) = \det A$$

and must show that this definition does not depend on the choice of basis. Another basis $\mathfrak{e}'_1, \ldots, \mathfrak{e}'_n$ corresponds to the linear mapping A' with

$$A'\mathfrak{e}'_i = \sum_{k=1}^{n} a_{ki}\mathfrak{e}'_k \quad (i = 1, \ldots, n)$$

and on the other hand there exists exactly one automorphism B of $\mathfrak{B}_n$ taking the basis $\mathfrak{e}_1, \ldots, \mathfrak{e}_n$ into the basis $\mathfrak{e}'_1, \ldots, \mathfrak{e}'_n$

$$\mathfrak{e}'_i = B\mathfrak{e}_i \quad (i = 1, \ldots, n).$$

Thus

$$(A' \circ B)\mathfrak{e}_i = \sum_{k=1}^{n} a_{ki}B\mathfrak{e}_k = B(\sum_{k=1}^{n} a_{ki}\mathfrak{e}_k) = (B \circ A)\mathfrak{e}_i$$

for $i = 1, \ldots, n$; i.e.,

$$A' = B \circ A \circ B^{-1}$$

and from the multiplication theorem we also have

$$\det A' = \det B \cdot \det A \cdot \det B^{-1} = \det A$$

So from our definition we obtain the explicit representation (which could have been used as a definition)

$$\det (a_{ik}) = \sum_{\sigma \in S_n} \varepsilon(\sigma)a_{1\sigma(1)} \cdots a_{n\sigma(n)}$$

where we must sum over all permutations σ of the numbers $1, \ldots, n$ and take $\varepsilon(\sigma)$ equal to $+1$ or -1 according as the permutation σ is even or odd. We also write

$$\det (a_{ik}) = \begin{vmatrix} a_{11} & \cdots & a_{1n} \\ \vdots & & \vdots \\ a_{n1} & \cdots & a_{nn} \end{vmatrix}.$$

6.2. *Orientation of Affine Spaces*

With the concepts introduced up to now it is not yet possible, for example, to give any precise meaning to the following simple statements of intuitive geometry (cf. II1, Nos. 2–8).

A line can be traversed in two senses.

In the plane there are two senses of rotation, each of which can be characterized by a direction of passage around a triangle.

In space there are two senses of screw motion, each of which can be characterized by an ordered set of three lines.

The situation can be given a precise description if we assume that the scalar field K is *ordered* (cf. IB1, §3), i.e., there is a relation $<$ in K satisfying the following *axioms of order.*

1.   *For $a, b \in K$ exactly one of $a = b, a < b, b < a$ is valid.*
2.   *$a < b$ and $b < c$ implies $a < c$ for all $a, b, c \in K$.*
3.   *$a < b$ implies $a + c < b + c$ for all $a, b, c \in K$.*
4.   *$a < b$ implies $ac < bc$ for all $a, b \in K$ and all $c < 0$ in K.*

The elements of K for which $a > 0$ are called positive, and those for which $a < 0$ are negative.

If the domain of scalars is an ordered field, as we shall assume from now on, finite and complex affine point-spaces are excluded, since their scalar fields cannot be ordered in such a way that properties 3 and 4 also hold.

The two ordered fields of greatest importance to us are the field $\mathbb{R}$ of real numbers and the field $\mathbb{Q}$ of rational numbers.

For one dimension the orientation problem is easily solved as follows. The parametric representation

$$\mathfrak{x} = \mathfrak{p} + x\mathfrak{b} \quad (x \in K)$$

puts the points X on a given line into one-to-one correspondence with the scalars x. We then say that X_1 lies *before* X_2 if $x_1 < x_2$, where it must be noted that this correspondence is based on an arbitrarily chosen coordinate system $\{P, b\}$. If x and x' are the scalars corresponding to the same point X in the two coordinate systems $\{P, b\}$ and $\{P', b'\}$, then

$$x' = cx + d \quad (c \neq 0)$$

with suitable scalars c and d; and now, if $c > 0$, then $x'_1 < x'_2$ if and only if

$x_1 < x_2$, but if $c < 0$, then $x_1' < x_2'$ if and only if $x_1 > x_2$. Thus the relation "smaller than" in K can be transferred to the line in two different ways, so that we may speak of the two *orientations* or *senses* (of traversal) on the line. If one of the two senses is distinguished, we say that the line has been *oriented.*

The two possibilities for *orienting* a line are reflected in the corresponding affine group since an affinity either preserves the orientation or changes it, according as the magnification factor c is positive or negative. The set $\mathbf{A}_1^+$ of affinities preserving orientation is a normal subgroup of index 2 in the affine group $\mathbf{A}_1$.

To define orientation for affine point space A_n with $n \geqq 2$, we start from the latter point of view; namely, from the fact that, just as in the one-dimensional case, the set $\mathbf{A}_n^+$ of affinities with *positive determinant* is a normal subgroup of index 2 in the affine group $\mathbf{A}_n$.

We now say that two coordinate simplexes $(O, E_1, \ldots, E_n)$ and $(O', E_1', \ldots, E_n')$ have the *same orientation* if the group $\mathbf{A}_n^+$ contains an affine mapping taking the first simplex into the second, where the reader should verify that we thereby set up an *equivalence relation* in the coordinate simplexes. Since the translations belong to A_+ we can test whether two coordinate simplexes have the same orientation by examining the corresponding *ordered* bases. By definition, the coordinate complexes corresponding to $(e_1, \ldots, e_n)$ and $(e_1', \ldots, e_n')$ will have the same orientation if

$$\frac{D(e_1', \ldots, e_n')}{D(e_1, \ldots, e_n)} > 0,$$

where D is an arbitrary determinant function. If neither of two given bases $(e_n', \ldots, e_n')$ and $(e_1'', \ldots, e_n'')$ has the same orientation as the basis $(e_1, \ldots, e_n)$, they will have the same orientation as each other. Thus there are exactly two equivalence classes of ordered bases. If we select one of the two classes as *positively oriented* (the choice between them is arbitrary), then $\mathfrak{B}_n$ and with it $\mathfrak{A}_n$ are said to be *oriented.* The group $\mathbf{A}_n^+$ then contains preserve *orientation.*

This arbitrary choice between the two equivalence classes can be made, for example, by prescribing a determinant function D and then saying that an ordered base is positively oriented if it gives a positive value for D. In other words, a vector space is oriented by the choice of a determinant function.

Let us now consider the geometric significance of the scalar $D(b_1, \ldots, b_n)$. If $b_1, \ldots, b_n$ are linearly independent and if P is an arbitrary point, the set of all points X whose position vectors are given by

$$\mathfrak{x} = \mathfrak{p} + x_1 b_1 + \cdots + x_n b_n \quad \text{with} \quad 0 \leqq x_1 \leqq 1, \ldots 0 \leqq x_n \leqq 1$$

is called a *parallelotope* (or *parallelepiped*, Fig. 31). If the order of the vectors

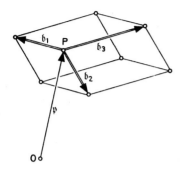

Fig. 31. Parallelotope

is important, we speak of an *oriented* parallelotope, which is therefore determined by P and the ordered basis $(b_1, \ldots, b_n)$, whose vectors b_i are called *edges*. To the oriented parallelotope spanned by $b_1, \ldots, b_n$ we now wish to assign an oriented (i.e., signed) *volume* $V(b_1, \ldots, b_n)$, for which we shall require that it shall not be changed by a shear and that dilation of an edge multiplies the volume by the same factor. Moreover, a unit of volume must be determined, i.e., the volume 1 must be assigned to some definite parallelotope. In other words the function V must be such that

1. $V(\ldots, x_i, \ldots, x_k, \ldots) = V(\ldots, x_i + x_k, \ldots, x_k, \ldots)$

for $i \neq k$,

2. $V(\ldots, cx_i, \ldots) = cV(\ldots, x_i, \ldots)$ for $i = 1, \ldots, n$,

3. $V(e_1, \ldots, e_n) = 1$ for a prescribed basis $(e_1, \ldots, e_n)$.

It is convenient, in order to avoid special cases, to regard V as a function of an arbitrary (perhaps linearly dependent) n-tuple of vectors, whereupon it can be proved that V is additive in each argument and that interchange of two arguments changes the sign of V; in other words, V is a *determinant* function (uniquely determined by property 3). For the proof see, e.g., [15], pp. 100 ff.

Thus the scalar $D(b_1, \ldots, b_n)$ can be interpreted as the oriented volume of the parallelotope with edges $b_1, \ldots, b_n$, and in fact its absolute value $|D(b_1, \ldots, b_n)|$ coincides with the *Jordan content* of a parallelotope (cf. III3, §1). Definition of the volumes of other polyhedra will not be discussed here.

The image of a parallelotope under a regular affinity α is again a parallelotope, whose oriented volume is obtained from the oriented volume of its preimage by multiplication with the determinant det A of the corresponding linear mapping A. Thus it is not the volume of a parallelotope that is invariant under the affine group, but the quotient of the volumes of two parallelotopes.

The *equiaffine group* Γ_n (cf. §6.1) is the largest subgroup of $\mathbf{A}_n$ that preserves the oriented volume of parallelotopes. It is characterized by the property that $\det A = 1$ and is therefore contained in $\mathbf{A}_n^+$. Of interest also is the *extended equiaffine group* Γ_n^*, which leaves invariant the *absolute value* of the oriented volume, i.e., the Jordan content of the parallelotope. The group Γ_n is a normal subgroup of Γ_n^* of index 2, and Γ_n^* is a normal subgroup of $\mathbf{A}_n$.

Example. In the real $\mathfrak{A}_2$ we distinguished six types of mappings (cf. p. 315 and p. 319). For the mappings with fixed point the corresponding determinants are

1) $\det A = a^2 + b^2 > 0,$ 2) $\det A = c^2 > 0,$ 3) $\det A = c_1 c_2,$

and for the mappings without fixed point

   1) $\det A = 1$ 2) $\det A = 1$ 3) $\det A = c \neq 1.$

Thus the four types under 1) and 2) preserve orientation in every case, whereas the two types under 3) preserve orientation if and only if $c_1 c_2 > 0$ or $c > 0$, respectively. In particular, translations, shears, and point reflections belong to the equiaffine group Γ_2. The affine reflections ($c_1 = 1$, $c_2 = -1$) do not preserve orientation, since $\det A = -1$, but belong to the group Γ_n^*.

In the affine plane an ordered basis $(\mathfrak{e}_1, \mathfrak{e}_2)$, to be called positively oriented, can be chosen arbitrarily, as was shown above, so that in the three-dimensional space of every day experience the choice can be determined only by physical properties.

In order to orient the "plane of the paper," for example, we may note that a line with the direction vector $\mathfrak{e}_1$ divides the plane into a "left" and a "right" halfplane (cf. the following section) and then say that $(\mathfrak{e}_1, \mathfrak{e}_2)$ is positively oriented if $\mathfrak{e}_2$ points into the left halfplane; and the familiar conventions for orienting three-dimensional space are likewise of a physical nature (cf. III, No. 6).

6.3.   Convex Sets

The order in the field of scalars can also be used to define concepts like line segment, halfspace, and convex set and to give a precise meaning to statements like

A line divides the plane into convex parts
A plane divides space into two halfspaces
A point lies in the interior of an ellipse.

We first define a (line) segment. If P and Q are two points of $\mathfrak{A}_n$ (over an ordered field K), the set of all points X with position vectors of the form

$$\mathfrak{x} = \mathfrak{p} + x(\mathfrak{q} - \mathfrak{p}) \quad \text{with} \quad 0 \leqq x \leqq 1$$

is called the *segment* $\overline{PQ}$ (the "segment joining" P and Q). (Note that the symbol $\overline{PQ}$ denotes a set of points and not, as is often the case, a scalar associated with it.) We can now give the definition of *convexity* of a subset of $\mathfrak{A}_n$.

   The subset $\mathfrak{K} \subset \mathfrak{A}_n$ *is said to be convex if for all points* $P, Q \in \mathfrak{K}$ *the segment* $\overline{PQ}$ *also belongs to K.*

All subspaces of A_n are convex, e.g., the lines and the planes, and it is immediately clear that the intersection of arbitrarily many convex sets is itself convex.

Halfspaces, defined as follows, are another important example of convex sets. For given origin O, the equation $l(\mathfrak{x}) = a$ determines a hyperplane $\mathfrak{H}$, which can also be represented in the form (independent of the choice of origin)

$$l(\mathfrak{x} - \mathfrak{x}_0) = 0,$$

where $\mathfrak{x}_0$ is the position vector of an arbitrary point in the hyperplane. The set of all points in $\mathfrak{A}_n$ can then be divided into three classes, according to whether $l(\mathfrak{x} - \mathfrak{x}_0) = 0$ or $l(\mathfrak{x} - \mathfrak{x}_0) > 0$ or $l(\mathfrak{x} - \mathfrak{x}_0) < 0$. The two latter sets are called *open* halfspaces, so that a hyperplane may be said to divide $\mathfrak{A}_n$ into two open open halfspaces. The sets defined by $l(\mathfrak{x} - \mathfrak{x}_0) \geq 0$ or $l(\mathfrak{x} - \mathfrak{x}_0) \leq 0$ are called *closed* halfspaces. In what follows, the term "halfspace" is an abbreviation for "closed halfspace."

   Note that

$$cl(\mathfrak{x} - \mathfrak{x}_0) = 0$$

describes the same hyperplane for every scalar $c \neq 0$, so that for negative c the inequalities determining the halfspaces are interchanged. Thus we cannot distinguish a "positive" halfspace (unless $\mathfrak{H}$ and $\mathfrak{A}_n$ are already *oriented*).

Halfspaces are convex sets, since $l(\mathfrak{p} - \mathfrak{x}_0) \geq 0$ and $l(\mathfrak{q} - \mathfrak{x}_0) \geq 0$ imply

$$l((1 - x)\mathfrak{p} + x\mathfrak{q} - \mathfrak{x}_0) = (1 - x)l(\mathfrak{p} - \mathfrak{x}_0) + xl(\mathfrak{q} - \mathfrak{x}_0) \geq 0$$

$$\text{for all } x \in K \quad \text{with } 0 \leq x \leq 1.$$

A convex *polyhedron* can now be defined as the intersection of a finite number of halfspaces. For example, a convex quadrangle in the plane is the intersection of four halfplanes. If two of the lines "bounding" the halfplanes are parallel, the convex quadrangle is called a *trapezoid*; if the other two bounding lines are also parallel, the vertices of the quadrangle form a parallelogram in the sense of our definition on p. 298, and for brevity we shall also call this convex quadrangle a *parallelogram*. The n-dimensional generalization of the parallelogram, namely the parallelotope, is a convex polyhedron representable as the intersection of $2n$ halfspaces.

As a final example of a convex set, let us mention the *simplex*, namely the n-dimensional generalization of a triangle. Let $A_0, A_1, \ldots, A_n$ be points in

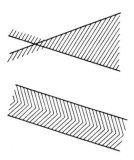

Fig. 32.   Angle and strip as intersection of two halfplanes

$\mathfrak{A}_n$, not all in the same hyperplane. Then the set of all points X with position vectors (for an arbitrary origin) given by

$$\mathfrak{x} = \mathfrak{a}_0 + x_1(\mathfrak{a}_1 - \mathfrak{a}_0) + \cdots + x_n(\mathfrak{a}_n - \mathfrak{a}_0)$$

with $x_1 \geq 0, \ldots, x_n \geq 0, x_1 + \cdots + x_n \leq 1$, is called a *simplex* with the vertices $A_0, A_1, \ldots, A_n$. A simplex is a convex polyhedron representable as the intersection of $n + 1$ halfspaces (Fig. 33).

This definition can also be put in the following form, symmetric in the vertices,

(1) $$\mathfrak{x} = x_0\mathfrak{a}_0 + x_1\mathfrak{a}_1 + \cdots + x_n\mathfrak{a}_n$$
with $x_0 + x_1 + \cdots + x_n = 1$ and $x_0 \geq 0, x_1 \geq 0, \ldots, x_n \geq 0$,

which admits a mechanical interpretation. If the masses $x_0, x_1, \ldots, x_n$ (with total mass 1) are assigned to the points $A_0, A_1, \ldots, A_n$ respectively, then X is the center of gravity of the point system and, in the terminology of Möbius, the scalars $x_0, x_1, \ldots, x_2$ are the *barycentric coordinates* of X with respect to $A_0, A_1, \ldots, A_n$. It is easy to show that these coordinates are independent of the choice of origin.

If the condition $x_0 \geq 0, x_1 \geq 0, \ldots, x_n \geq 0$ is omitted, then (1) represents all the points of $\mathfrak{A}_n$; in other words, the $(n + 1)$-tuples $(x_0, x_1, \ldots, x_n)$ with $x_0 + x_1 + \cdots + x_n = 1$ are in one-to-one correspondence with the points of $\mathfrak{A}_n$; and here too we speak of barycentric coordinates. Since the order in K no longer plays a role, these coordinates can be defined in any affine space.

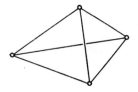

Fig. 33.   The tetrahedron as a simplex in $\mathfrak{A}_3$

Similarly, the parametric representation of subspaces (p. 302) can be written in symmetric form. If $\mathfrak{a}_1 - \mathfrak{a}_0, \ldots, \mathfrak{a}_k - \mathfrak{a}_0$ are linearly independent, the equation

$$\mathfrak{o} = x_0\mathfrak{a}_0 + x_1\mathfrak{a}_1 + \cdots + x_k\mathfrak{a}_k \quad \text{with} \quad x_0 + x_1 + \cdots + x_k = 1$$

represents a k-dimensional subspace of $\mathfrak{A}_n$. For example, the equation of the line through the points A_0, A_1 becomes

$$\mathfrak{o} = x_0\mathfrak{a}_0 + x_1\mathfrak{a}_1 \quad \text{with} \quad x_0 + x_1 = 1$$

(symmetric form of the "two-point" equation of a line).

By weakening the normalization condition to $x_0 + \cdots + x_n \neq 0$, we obtain the so-called *homogeneous coordinates* of X. All proportional $(n + 1)$-tuples $(cx_0, \ldots, cx_n)$, $c \neq 0$, and only these correspond to the same point X, the $(n + 1)$-tuples $(x_0, \ldots, x_n)$ with $x_0 + \cdots + x_n = 0$ do not correspond to any point in $\mathfrak{A}_n$ and therefore enable us to introduce the "improper" points of *projective geometry* (cf. the following chapter).

Examples of convex sets other than polyhedra are provided by the quadrics. If f is a *positive semidefinite* bilinear form (i.e., if the rank r and the index p are identical), the set of points defined by

$$f(\mathfrak{x}, \mathfrak{x}) \leqq 1$$

is convex, and is called the *interior* of the quadric $\mathfrak{Q}$ with equation $f(\mathfrak{x}, \mathfrak{x}) = 1$. The points interior to $\mathfrak{Q}$ are characterized geometrically by the fact that no tangent to $\mathfrak{Q}$ passes through them, a fact that is used to define the "interior" of other quadrics; but then the interior is not necessarily convex.

Euclidean Geometry

7. Euclidean Vector and Point Spaces

7.1. *Euclidean Vector Space. Scalar Product*

In affine geometry such concepts as *length, angle, magnitude,* and *volume* are not defined; one may speak only of "ratio of lengths" on a line and "ratio of volumes" in $\mathfrak{A}_n$. If these familiar concepts from elementary Euclidean geometry are to be introduced into our deductive theory, the system of axioms for affine geometry must be suitably enlarged. A formally simple approach to Euclidean geometry is based on the concept of a *Euclidean vector space,* i.e., a vector space $\mathfrak{V}_n$ over the field $\mathbb{R}$ of real numbers in which a *positive definite symmetric bilinear form* f (cf. p. 325) is distinguished, with the result that to every vector pair $(\mathfrak{x}, \mathfrak{y})$ there corresponds a real number $f(\mathfrak{x}, \mathfrak{y})$, denoted for brevity by $\mathfrak{x} \cdot \mathfrak{y}$, such that

1. $\mathfrak{x} \cdot \mathfrak{y} = \mathfrak{y} \cdot \mathfrak{x}$ for all $\mathfrak{x}, \mathfrak{y} \in \mathfrak{V}_n$,

2. $(\mathfrak{x}_1 + \mathfrak{x}_2) \cdot \mathfrak{y} = \mathfrak{x}_1 \cdot \mathfrak{y} + \mathfrak{x}_2 \mathfrak{y}$ for all $\mathfrak{x}_1, \mathfrak{x}_2, \mathfrak{y} \in \mathfrak{V}_n$,

3. $(c\mathfrak{x}) \cdot \mathfrak{y} = c(\mathfrak{x} \cdot \mathfrak{y})$ for all $\mathfrak{x}, \mathfrak{y} \in \mathfrak{V}_n$ and all $c \in \mathbb{R}$,

4. $\mathfrak{x} \cdot \mathfrak{x} > 0$ for all $\mathfrak{x} \neq \mathfrak{o}$ in V_n.

The real number $\mathfrak{x} \cdot \mathfrak{y}$ is called the *scalar product* (also the *inner product*) of the two vectors $\mathfrak{x}, \mathfrak{y} \in \mathfrak{V}_n$ and the *length* (or *magnitude*) $|\mathfrak{x}|$ of $\mathfrak{x}$ is defined by

$$(1) \qquad\qquad |\mathfrak{x}| = \sqrt{\mathfrak{x} \cdot \mathfrak{x}},$$

so that, by 4, the length of any nonzero vector is positive.

Vectors of length 1 are called *unit vectors*; for $\mathfrak{x} \neq \mathfrak{o}$ there is exactly one multiple $\mathfrak{x}^0$ of x by a positive scalar, namely

$$(2) \qquad\qquad \mathfrak{x}^0 = \frac{1}{|\mathfrak{x}|}\mathfrak{x},$$

that of unit length.

From property 4 (positive definiteness) of the scalar product we deduce the *Cauchy-Schwarz inequality*

$$(3) \qquad\qquad -|\mathfrak{x}| \cdot |\mathfrak{y}| \leq \mathfrak{x} \cdot \mathfrak{y} \leq |\mathfrak{x}| \cdot |\mathfrak{y}|.$$

For if x and y are both nonzero, then for the corresponding unit vectors $\mathfrak{x}^0, \mathfrak{y}^0$ we have

$$(\mathfrak{x}^0 - \mathfrak{y}^0) \cdot (\mathfrak{x}^0 - \mathfrak{y}^0) = 2 - 2\mathfrak{x}^0 \cdot \mathfrak{y}^0 \geq 0,$$

$$(\mathfrak{x}^0 + \mathfrak{y}^0) \cdot (\mathfrak{x}^0 + \mathfrak{y}^0) = 2 + 2\mathfrak{x}^0 \cdot \mathfrak{y}^0 \geq 0,$$

which taken together give assertion (3). If at least one of $\mathfrak{x}$ and $\mathfrak{y}$ is the zero vector, then (3) also holds, which completes the proof.

At the same time we see that equality cannot occur in (3) if the vectors $\mathfrak{x}$ and $\mathfrak{y}$ are linearly independent.

A direct consequence of (3) is the *triangle inequality*

$$(4) \qquad\qquad |\mathfrak{x} + \mathfrak{y}| \leq |\mathfrak{x}| + |\mathfrak{y}|,$$

since (4) is equivalent to

$$|\mathfrak{x} + \mathfrak{y}|^2 \leq |\mathfrak{x}|^2 + 2|\mathfrak{x}| \cdot |\mathfrak{y}| + |\mathfrak{y}|^2,$$

which follows from (1) and the right half of (3).

To arbitrary nonzero vectors $\mathfrak{x}, \mathfrak{y}$ we now wish to assign a real number $\varphi(\mathfrak{x}, \mathfrak{y})$, as the measure of the *angle* between them, in such a way that $\varphi(\mathfrak{x}, \mathfrak{y})$ does not depend on the length of the vectors $\mathfrak{x}, \mathfrak{y}$. In other words, φ is to be a function of the corresponding unit vectors $\mathfrak{x}^0, \mathfrak{y}^0$.

It would be natural to assign the real number $x^0 \cdot y^0$ to the vector pair $\mathfrak{x}, \mathfrak{y}$. But our angle measure φ will be required (in the two-dimensional case) to be in a certain sense additive. In order to define φ in such a way that it has this property (for the proof see §8.3) we make use of the familiar arc cos function from analysis

(5) $$\varphi(\mathfrak{x}, \mathfrak{y}) = \text{arc cos } (\mathfrak{x}^0, \mathfrak{y}^0) = \text{arc cos } \frac{\mathfrak{x} \cdot \mathfrak{y}}{|\mathfrak{x}| \cdot |\mathfrak{y}|},$$

which is meaningful here, since $-1 \leq \mathfrak{x}^0 \cdot \mathfrak{y}^0 \leq 1$, by the Cauchy-Schwarz inequality, so that the range of values of φ is the interval $[0, \pi]$ and (5) can also be written in the familiar form

(6) $$\cos \varphi(\mathfrak{x}, \mathfrak{y}) = \frac{\mathfrak{x} \cdot \mathfrak{y}}{|\mathfrak{x}| \cdot |\mathfrak{y}|} \quad \text{with } 0 \leq \varphi(\mathfrak{x}, \mathfrak{y}) \leq \pi.$$

From (5) and (6) we see at once that for all $\mathfrak{x} \neq \mathfrak{o}$ and all $\mathfrak{y} \neq \mathfrak{o}$,

$$\varphi(\mathfrak{x}, \mathfrak{y}) = \varphi(\mathfrak{y}, \mathfrak{x}), \; \varphi(-\mathfrak{x}, -\mathfrak{y}) = \varphi(\mathfrak{x}, \mathfrak{y}),$$
$$\varphi(-\mathfrak{x}, \mathfrak{y}) = \varphi(\mathfrak{x}, -\mathfrak{y}) = \pi - \varphi(\mathfrak{x}, \mathfrak{y}).$$

7.2.   Orthogonality

The vectors $\mathfrak{x}$, $\mathfrak{y}$ are said to be *orthogonal* to each other, written $\mathfrak{x} \perp \mathfrak{y}$, if

$$\mathfrak{x} \cdot \mathfrak{y} = 0$$

i.e., if $\varphi(\mathfrak{x}, \mathfrak{y}) = \pi/2$ or at least one of the two vectors $\mathfrak{x}$, $\mathfrak{y}$ is zero.

Since the scalar product is positive definite, a nonzero vector cannot be orthogonal to itself, and since it is symmetric, the relation of orthogonality is also symmetric.

The vectors of a set $\mathfrak{a}_1, \ldots, \mathfrak{a}_k$ (not including the zero vector) are linearly independent if they are pairwise orthogonal, since the equation

$$x_1 \mathfrak{a}_1 + \cdots + x_k \mathfrak{a}_k = \mathfrak{o}$$

implies, if we take the scalar product with $\mathfrak{a}_j$, that

$$x_j(\mathfrak{a}_j \cdot \mathfrak{a}_j) = 0,$$

so that $x_j = 0$ for $j = 1, \ldots, k$.

A basis $\{\mathfrak{b}_1, \ldots, \mathfrak{b}_n\}$ of a Euclidean $\mathfrak{B}_n$ is called an *orthogonal basis* if its vectors $\mathfrak{b}_j$ are pairwise orthogonal. Given a (not necessarily orthogonal) basis $\{\mathfrak{a}_1, \ldots, \mathfrak{a}_n\}$, we can always construct an orthogonal basis $\{\mathfrak{b}_1, \ldots, \mathfrak{b}_n\}$ such that, for $k = 1, 2, \ldots, n$, the vectors $\mathfrak{a}_1, \ldots, \mathfrak{a}_k$ span the same subspace as $\mathfrak{b}_1, \ldots, \mathfrak{b}_k$.

The proof follows from the *orthogonalization process* of E. Schmidt. As a first step we take $\mathfrak{b}_1 = \mathfrak{a}_1$, and then, assuming that satisfactory $\mathfrak{b}_1, \ldots, \mathfrak{b}_j$ have already been found, for $\mathfrak{b}_{j+1}$ we set

$$\mathfrak{b}_{j+1} = x_1 \mathfrak{b}_1 + \cdots + x_j \mathfrak{b}_j + \mathfrak{a}_{j+1}$$

with scalars $x_1, \ldots, x_j$ to be determined from $b_1 \cdot b_{j+1} = \cdots = b_j \cdot b_{j+1} = 0$, i.e., from the equations

$$x_1(b_1 \cdot b_1) + (b_1 \cdot a_{j+1}) = 0$$
$$\vdots \qquad\qquad \vdots$$
$$x_j(b_j \cdot b_j) + (b_j \cdot a_{j+1}) = 0,$$

which must have a unique solution $x_1, \ldots, x_j$, since by hypothesis the $b_1, \ldots, b_j$ are all nonzero (b_{j+1} cannot be zero, since the $a_1, \ldots, a_{j+1}$ would then be linearly dependent), so that we finally obtain n pairwise orthogonal vectors with the desired property.

Let us also remark that

$$b_k \cdot b_k \leqq a_k \cdot a_k \quad \text{for } k = 1, \ldots, n,$$

with equality only if a_k is already orthogonal to $b_1, \ldots, b_k$, as follows at once (since the $b_1, \ldots, b_n$ are pairwise orthogonal) from the equation

$$b_{j+1} \cdot b_{j+1} + (x_1^2 b_1 \cdot b_1 + \cdots + x_j^2 b_j \cdot b_j) = a_{j+1} \cdot a_{j+1}.$$

The basis vectors of an orthogonal basis $\{b_1, \ldots, b_n\}$ are of arbitrary (nonzero) length, and their pairwise orthogonality is not disturbed if all the b_i are replaced by the vectors $b_i/|b_i|$ of unit length. A basis $\{b_1, \ldots, b_n\}$ with

$$|b_i| = 1 \quad \text{for all } i \quad \text{and } b_i \cdot b_k = 0 \quad \text{for } i \neq k$$

is called an *orthonormal basis*.

The concept of orthogonality can be extended to subspaces $\mathfrak{U}'$ and $\mathfrak{U}''$ in $\mathfrak{B}_n$ by saying that $\mathfrak{U}'$ and $\mathfrak{U}''$ are orthogonal to each other if every vector in $\mathfrak{U}'$ is orthogonal to every vector in $\mathfrak{U}''$.

For an arbitrary subspace $\mathfrak{U} \subset \mathfrak{B}_n$ the set of vectors

$$\mathfrak{U}_\perp = \{\mathfrak{x} \mid \mathfrak{u} \cdot \mathfrak{x} = 0 \text{ for all } \mathfrak{u} \in \mathfrak{U}\}$$

is a vector subspace, orthogonal to $\mathfrak{U}$, which is called the *orthogonal complement* of $\mathfrak{U}$.

If k is the dimension of $\mathfrak{U}$, then $\mathfrak{U}_\perp$ has the dimension $n - k$; moreover, since the scalar product is positive definite, the spaces $\mathfrak{U}$ and $\mathfrak{U}_\perp$ can have only the zero vector in common. Thus $\mathfrak{B}_n$ is the direct sum of $\mathfrak{U}$ and $\mathfrak{U}_\perp$:

$$\mathfrak{B}_n = \mathfrak{U} \oplus \mathfrak{U}_\perp.$$

For one-dimensional $\mathfrak{U}$, it follows that $\mathfrak{U}_\perp$ is of dimension $n - 1$, and consists of the vectors x with

$$\mathfrak{u} \cdot \mathfrak{x} = 0,$$

where u is a nonzero vector in $\mathfrak{U}$.

The mapping which to the vector x assigns the scalar $u \cdot x$ is a linear form (cf. p. 325). So every vector u in a Euclidean vector space corresponds to a linear form, and distinct vectors correspond to distinct linear forms. Also, all linear forms are obtained in this way, since if l is an arbitrary linear form, then

$$u = l(b_1)b_1 + \cdots + l(b_n)b_n,$$

where $\{b_1, \ldots, b_n\}$ is an orthonormal basis of $\mathfrak{V}_n$, satisfies the equation

$$l(x) = u \cdot x$$

for all x in $\mathfrak{V}_n$.

A Euclidean vector space $\mathfrak{V}_n$ can thus be identified in a natural way with its *dual space*, i.e., the vector space of its linear forms.

The concept of a Euclidean vector space admittedly allows us to construct Euclidean geometry in a rapid and simple way, but it is often felt to have the disadvantage that its axioms are not readily visualizable. Of the numerous attempts that have been made to bring the axioms closer to our intuitive ideas, let us here give a few examples.

1. *Congruence* and *orthogonality* are taken as basic concepts, intuitively natural axioms are set up, and the existence of a symmetric, positive definite, bilinear form is deduced (cf. [14]).

2. It is taken for granted from the beginning that $\mathfrak{V}_n$ is a *normed vector space*, i.e., that there exists a mapping $x \to \|x\|$ of $\mathfrak{V}_n$ into $\mathbb{R}$ such that

(1) $x \neq o$ implies $\|x\| > 0$
(2) $c \in \mathbb{R}$ and $x \in \mathfrak{V}_n$ imply $\|cx\| = |c| \cdot \|x\|$
(3) $x, \mathfrak{y} \in \mathfrak{V}_n$ implies $\|x + \mathfrak{y}\| \leq \|x\| + \|\mathfrak{y}\|$.

A vector space of this sort is called a (finite-dimensional) *Banach space* (cf. III11, §2) or alternatively a *Minkowski space* (not to be confused with the Minkowski space of special relativity).

The *unit sphere* of a normed vector space $\mathfrak{V}_n$ is defined by $\|x\| \leq 1$. It is a compact convex set with interior points and with zero as its center, and can be used, conversely, to determine the norm.

The Euclidean vector spaces are then distinguished among the normed vector spaces by means of suitable additional conditions; e.g.,

a)   The norm $\| \ \|$ satisfies the *parallelogram identity*

$$\|x + \mathfrak{y}\|^2 + \|x - \mathfrak{y}\|^2 = 2\|x\|^2 + 2\|\mathfrak{y}\|^2.$$

b)   There are *sufficiently many* linear mappings taking the *unit sphere* into itself, i.e., for every vector pair $x, \mathfrak{y}$ with $\|x\| = \|\mathfrak{y}\| = 1$ there exists a linear mapping A, taking the unit sphere into itself, such that $x = A\mathfrak{y}$ (cf. [10]).

The unit spheres in Euclidean vector spaces are central quadrics of the type $p = r = n$ and $c \neq 0$, and are therefore ellipsoids (cf. p. 329).

3. It is also possible to avoid "metric" concepts altogether in the axioms, giving precedence instead to the group concept. Here the intuitive starting point

is the "free mobility of rigid bodies" in everyday space (Helmholtz). To define rigid bodies mathematically (without use of a metric) we have recourse to generalized line elements, i.e., direction elements of rank 1 to ($n - 1$). If now for $n \geq 3$, we are given a group of linear mappings of $\mathfrak{V}_n$ onto itself that is simply transitive on the set of these generalized line elements, there exists a positive definite symmetric bilinear form such that the given group is precisely the orthogonal group (for this concept see §10) corresponding to the bilinear form. A proof of this theorem, i.e., a solution of the "Helmholtz space problem" for real vector spaces, is to be found in [18] and [19].

In all these cases we finally arrive at the "Euclidean vector space" taken here as the starting point.

The reader should compare the above discussion with II10, §5, and with the "synthetic systems of axioms in II4 and II6.

7.3. *Euclidean Point Spaces*

An affine point space corresponding to a Euclidean vector space is called a *Euclidean point space*, and will here be denoted by $\mathfrak{E}_n$ instead of $\mathfrak{A}_n$. All the definitions and theorems (i.e., consequences of the axioms) of affine geometry remain valid in Euclidean geometry, but now there are new concepts and theorems that cannot be formulated in affine geometry.

If distance is defined by

$$d(X, Y) = |\overrightarrow{XY}|,$$

a Euclidean space becomes a *metric space*. The properties (cf. II16, §9)

1. $d(X, Y) = 0$ if and only if $X = Y$
2. $d(X, Y) = d(Y, X)$ for all $X, Y \in \mathfrak{E}_n$
3. $d(X, Y) \leq d(X, Z) + d(Z, Y)$ for all $X, Y, Z \in \mathfrak{E}_n$

follow directly from the definition of d and the triangle inequality (4) (see p. 346). This metric is invariant under translation, i.e., $d(\tau X, \tau Y) = d(X, Y)$ for all $X, Y \in \mathfrak{E}_n$ and all τ (cf. p. 321).

Every subspace $\mathfrak{U}$ of a Euclidean vector space is itself a Euclidean vector space, if the scalar product is restricted to $\mathfrak{U}$. Thus the affine subspaces of $\mathfrak{E}_n$ are also Euclidean point spaces. By p. 325 the hyperplanes can be represented by linear forms

$$l(\mathfrak{x}) = a.$$

Since the linear forms l in a Euclidean space are in one-to-one correspondence with the vectors $\mathfrak{u}$, the equation of a hyperplane (which is defined only up to a scalar factor) can be written in the so-called *Hesse normal form*

(1) $$\mathfrak{u} \cdot \mathfrak{x} = a \quad \text{with } |\mathfrak{u}| = 1,$$

where the geometric interpretation of $\mathfrak{u}$ and a is clear; namely, the line

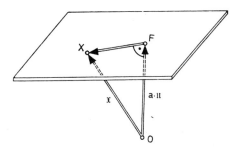

Fig. 34. The Hesse normal form for a plane

through O in the direction $\mathfrak{u}$, which cuts the hyperplane in exactly one point F, is perpendicular to every line (through F) in the hyperplane, and therefore the (unit) vector $\mathfrak{u}$ is called the *normal* to the hyperplane. Also, $|\overrightarrow{OF}| = |a|$, i.e., $|a|$ is the *distance* from the origin O to the hyperplane (Fig. 34).

The unit vector $-\mathfrak{u}$, is also normal to the same hyperplane, and the choice of one of these two vectors in (1) distinguishes the halfspace (cf. p. 343) into which $\mathfrak{u}$ *points*; i.e., for the position vectors x of points in this halfspace we have $\mathfrak{u} \cdot x \geqq a$. If the origin O lies in the distinguished halfspace, then a is negative.

Choice of another origin P changes (1); for if we set $x = x' + \mathfrak{p}$, with $\overrightarrow{OP} = \mathfrak{p}$, we obtain

(1′) $$\mathfrak{u} \cdot x' = a - \mathfrak{u} \cdot \mathfrak{p} = a',$$

which shows that a' is the *directed* distance from P to the hyperplane (1).

The equation for the hyperplane can be written in a form independent of the choice of origin, namely

$$\mathfrak{u} \cdot (x - x_0) = 0,$$

where x_0 is the position vector of an arbitrary point X_0 on the hyperplane. Similarly, any k-dimensional subspace $\mathfrak{B}$ of $\mathfrak{E}_n$ can be described by the $n - k$ equations

(2) $$\mathfrak{u}_j \cdot (x - x_0) = 0, \quad j = 1, 2,, \ldots, n - k,$$

with linearly independent $\mathfrak{u}_j$. The subspace $\mathfrak{U}$ spanned by the $\mathfrak{u}_1$, $\mathfrak{u}_2, \ldots, \mathfrak{u}_{n-k}$ is said to be *normal* to $\mathfrak{B}$.

Along with the subspaces of affine geometry, there exists in Euclidean space, as in any metric space, another important class of figures, the *spheres* or *hyperspheres* (for $n = 2$ they are called *circles*), distinguished by the simplicity of their defining property. For a given point $M \in \mathfrak{E}_n$ and given number $r \geqq 0$, the set

$$\{X \mid d(M, X) = r\}$$

is called a sphere with *center* M and *radius* r. With arbitrary origin O the equation of this sphere is

(3) $(x - m) \cdot (x - m) = r^2.$

Thus an equation of the form

(4) $x \cdot x + 2a \cdot x + c = 0$

represents a sphere if and only if $a \cdot a \geqq c$, since (4) can be transformed into

$$(x + a) \cdot (x + a) = a \cdot a - c.$$

Finally, we now state what we mean by an *angle*. If $\bar{g}$ and $\bar{h}$ are halfline: issuing from the same point P, we call the unordered pair $\{\bar{g}, \bar{h}\}$ an *angle* with *vertex* P. If the parametric equations of $\bar{g}$ and $\bar{h}$ are

$$x = p + xa \quad \text{with } x \geqq 0 \quad \text{and.} \quad \mathfrak{y} = p + yb \quad \text{with } y \geqq 0$$

then, by definition, the number $\varphi(a, b)$ measuring the angle between the vectors a, b also measures the angle $\{\bar{g}, \bar{h}_1\}$ i.e., $\varphi(\bar{g}, \bar{h}) = \varphi(a, b)$, and the angle between two lines *oriented* by their direction vectors (in which it is clearly unnecessary that they have a point of intersection) is defined in the same way. If the lines are orthogonal, it is not even necessary that they be oriented.

Similarly, for two hyperplanes, oriented by the choice of their normal unit vectors u_1, u_2 (Fig. 35), the angle between them is equal, by definition, to the angle between u_1, u_2, and again the orientation is unnecessary if the hyperplanes are orthogonal.

7.4.   Orientation. Volume

An affine point space can be oriented by choosing a determinant function D in the corresponding vector space and agreeing that the ordered bases for

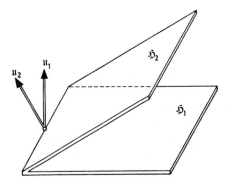

Fig. 35.   Angle between two oriented planes

which D is positive are positively oriented, which means that determinant functions differing only by a positive factor define the same orientation (cf. p. 342).

If the vector space is Euclidean, then among the determinant functions defining the same orientation there will be exactly one, to be denoted by D, which for an arbitrarily preassigned *positively oriented, orthonormalized* basis (and thus for all such bases) takes the value 1. Thus

$$D(\mathfrak{b}_1, \ldots, \mathfrak{b}_n) = 1,$$

for all ordered orthonormal bases $(\mathfrak{b}_1, \ldots, \mathfrak{b}_n)$ that are positively oriented.

For this distinguished determinant function D we have the Lagrange identity

$$(1) \qquad D(\mathfrak{x}_1, \ldots, \mathfrak{x}_n) \cdot D(\mathfrak{y}_1, \ldots, \mathfrak{y}_n) = \begin{vmatrix} \mathfrak{x}_1 \cdot \mathfrak{y}_1 & \cdots & \mathfrak{x}_1 \cdot \mathfrak{y}_n \\ \vdots & & \vdots \\ \mathfrak{x}_n \cdot \mathfrak{y}_1 & \cdots & \mathfrak{x}_n \cdot \mathfrak{y}_n \end{vmatrix},$$

which is proved as follows. For fixed $\mathfrak{y}_1, \ldots, \mathfrak{y}_n$ each side of the equation is a determinant function with arguments $\mathfrak{x}_1, \ldots, \mathfrak{x}_n$. Thus the two sides can differ only by a factor, which up to now may depend on the $\mathfrak{y}_1, \ldots, \mathfrak{y}_n$. But the same argument for fixed $\mathfrak{x}_1, \ldots, \mathfrak{x}_n$ shows that this factor must be a constant, and substitution of a positively oriented orthonormal basis $\mathfrak{b}_1, \ldots, \mathfrak{b}_n$, for $\mathfrak{x}_1, \ldots, \mathfrak{x}_n$ and $\mathfrak{y}_1, \ldots, \mathfrak{y}_n$ shows that it must be equal to 1.

The fact that there is exactly one distinguished determinant function in an oriented Euclidean vector space means that the concept of oriented volume of an oriented parallelotope is meaningful in Euclidean geometry (in distinction to affine geometry, since it was only in equiaffine geometry that volume could be defined, cf. p. 342); so we shall now write V instead of D. By the Lagrange identity, the square of the volume of a parallelepiped is given by

$$(2) \qquad V(\mathfrak{x}_1, \ldots, \mathfrak{x}_n)^2 = \begin{vmatrix} \mathfrak{x}_1 \cdot \mathfrak{x}_1 & \cdots & \mathfrak{x}_1 \cdot \mathfrak{x}_n \\ \vdots & & \vdots \\ \mathfrak{x}_n \cdot \mathfrak{x}_1 & \cdots & \mathfrak{x}_n \cdot \mathfrak{x}_n \end{vmatrix},$$

which remains valid under a change of orientation, so that the magnitude of the *volume*, as defined by (2), is independent of orientation. Equation (2) shows at once that zero is a lower bound for its right-hand side (called the *Gram determinant* of the n-tuple of vectors $\mathfrak{x}_1, \ldots, \mathfrak{x}_n$)

$$(3) \qquad \begin{vmatrix} \mathfrak{x}_1 \cdot \mathfrak{x}_1 & \cdots & \mathfrak{x}_1 \cdot \mathfrak{x}_n \\ \vdots & & \vdots \\ \mathfrak{x}_n \cdot \mathfrak{x}_1 & \cdots & \mathfrak{x}_n \cdot \mathfrak{x}_n \end{vmatrix} \geqq 0,$$

and an upper bound can also be obtained from (2). For if the $\mathfrak{x}_1, \ldots, \mathfrak{x}_n$ are linearly independent and if, by the Schmidt orthogonalization process (cf. p. 348), we construct the corresponding orthogonal basis $\mathfrak{z}_1, \ldots, \mathfrak{z}_n$, then $V(\mathfrak{z}_1, \ldots, \mathfrak{z}_n) = V(\mathfrak{x}_1, \ldots, \mathfrak{x}_n)$ and, from (2),

$$(V(\mathfrak{x}_1, \ldots, \mathfrak{x}_n))^2 = \begin{vmatrix} \mathfrak{z}_1 \cdot \mathfrak{z}_1 & \cdots & 0 \\ \vdots & & \vdots \\ 0 & \cdots & \mathfrak{z}_n \cdot \mathfrak{z}_n \end{vmatrix} = (\mathfrak{z}_1 \mathfrak{z} \cdot_1) \cdots (\mathfrak{z}_n \cdot \mathfrak{z}_n).$$

But by p. 347, $\mathfrak{z}_j \cdot \mathfrak{z}_j \leq \mathfrak{x}_j \cdot \mathfrak{x}_j$ so that for the determinant of linearly independent vectors $x_1, \ldots, x_n$ we have proved the *Hadamard inequality for determinants*

(4)
$$\begin{vmatrix} \mathfrak{x}_1 \cdot \mathfrak{x}_1 & \cdots & \mathfrak{x}_1 \cdot \mathfrak{x}_n \\ \vdots & & \vdots \\ \mathfrak{x}_n \cdot \mathfrak{x}_1 & \cdots & \mathfrak{x}_n \cdot \mathfrak{x}_n \end{vmatrix} \leq (\mathfrak{x}_1 \cdot \mathfrak{x}_1) \cdots (\mathfrak{x}_n \cdot \mathfrak{x}_n)$$

and if the $\mathfrak{x}_1, \ldots, \mathfrak{x}_n$ are linearly dependent, the Gram determinant vanishes, so that (4) is trivially satisfied.

Since for linearly independent vectors $\mathfrak{x}_1, \ldots, \mathfrak{x}_n$ the equality in (4) holds if and only if (see p. 348) they are pairwise orthogonal, we have the result that among all parallelotopes with edges of given length, the *rectangular parallelotope* (i.e., with pairwise orthogonal edges) has the greatest volume.

The above arguments can be applied to a k-dimensional parallelotope. By p. 350, the k-dimensional subspace $\mathfrak{U}$ spanned by its edges is again a Euclidean vector space, but the orientation of the whole space $\mathfrak{B}_n$ does not induce an orientation in $\mathfrak{U}$, which is frequently oriented in the following way. The orthogonal complement $\mathfrak{U}_\perp$ is oriented by choice of an orthonormal basis $\mathfrak{u}_1, \ldots, \mathfrak{u}_{n-k}$ and then the distinguished determinant function $D_\mathfrak{u}$ for $\mathfrak{U}$ is determined from the D for $\mathfrak{B}_n$ by setting

(5) $$D_\mathfrak{u}(\mathfrak{x}_1, \ldots, \mathfrak{x}_k) = D(\mathfrak{u}_1, \ldots, \mathfrak{u}_{n-k}, \mathfrak{x}_1, \ldots, \mathfrak{x}_k).$$

For example, the "oriented" hyperplanes in the preceding section can be oriented in this way.

If we now denote the "k-dimensional" volume by V_k, it follows from (5) and simple rules for determinants that

(6)
$$(V_k(\mathfrak{x}_1, \ldots, \mathfrak{x}_k))^2 = \begin{vmatrix} x_1 \cdot x_1 & \cdots & x_1 \cdot x_k \\ \vdots & & \vdots \\ x_k \cdot x_1 & \cdots & x_k \cdot x_k \end{vmatrix}$$

and again we have the upper and lower bounds

(7)
$$0 \leq \begin{vmatrix} x_1 \cdot x_1 & \cdots & x_1 \cdot x_k \\ \vdots & & \vdots \\ x_k \cdot x_1 & \cdots & x_k \cdot x_k \end{vmatrix} \leq (x_1 \cdot x_1) \cdots (x_k \cdot x_k),$$

which is a generalization of the Cauchy-Schwarz inequality ($k = 2$).

8. Topics in the Euclidean Geometry of the Plane and of Three-Dimensional Space

8.1. *Some Theorems in Plane Geometry*

The following examples will not, to begin with, involve orientation of the plane, since the proofs will depend only on the scalar product, without reference to determinant functions.

1) For every line g and every point P there exists exactly one line through P perpendicular to g, and this line (called the *perpendicular* from P to g) has exactly one point L (called the *foot* of the perpendicular) in common with g.

Proof. If $\mathfrak{U}$ is the one-dimensional subspace of $\mathfrak{V}_2$ corresponding to g, the orthogonal complement $\mathfrak{U}_\perp$ is also one-dimensional, and the line through P generated by $\mathfrak{U}_\perp$ is the only line that satisfies the given conditions. In particular, this line has exactly one point of intersection with g, since it is not parallel to g (cf. p. 302).

The point L on g is also characterized by the fact that it is at a minimum distance from P, as is proved by setting $\overrightarrow{LP} = \mathfrak{p}$ and $\overrightarrow{LX} = \mathfrak{x}$, whereupon, since $X \in g$, we have $\mathfrak{p} \cdot \mathfrak{x} = 0$ and thus

$$d(P, X) = \sqrt{\mathfrak{p} \cdot \mathfrak{p} + \mathfrak{x} \cdot \mathfrak{x}} \geq |\mathfrak{p}| = d(P, L)$$

with equality if and only if $\mathfrak{x} = \mathfrak{o}$, so that $X = L$.

2) The set of all points X equidistant from two given points A, B is the line through the midpoint of AB perpendicular to $A \vee B$ ("perpendicular bisector").

For

$$|\mathfrak{x} - \mathfrak{a}| = |\mathfrak{x} - \mathfrak{b}|$$

implies

$$\mathfrak{x} \cdot \mathfrak{x} - 2\mathfrak{a} \cdot \mathfrak{x} + \mathfrak{a} \cdot \mathfrak{a} = \mathfrak{x} \cdot \mathfrak{x} - 2\mathfrak{b} \cdot \mathfrak{x} + \mathfrak{b} \cdot \mathfrak{b}$$

or

$$(2\mathfrak{a} - 2\mathfrak{b}) \cdot \mathfrak{x} = \mathfrak{a} \cdot \mathfrak{a} - \mathfrak{b} \cdot \mathfrak{b}$$

which is the equation of a line perpendicular to $A \vee B$ through the point with position vector $\mathfrak{m} = (1/2)(\mathfrak{a} + \mathfrak{b})$, i.e., through the midpoint of $\overline{AB}$.

3) The "altitudes" of a triangle ABC intersect at a point H.

Since any two altitudes, say the perpendiculars through A to $B \vee C$ and through B to $C \vee A$, are not parallel and therefore intersect in exactly one point H, we may choose as origin for the position vectors of A, B and C

$$\mathfrak{a} = \overrightarrow{HA}, \quad \mathfrak{b} = \overrightarrow{HB}, \quad \mathfrak{c} = \overrightarrow{HC}.$$

By hypothesis $\mathfrak{a} \cdot (\mathfrak{b} - \mathfrak{c}) = 0$ and $\mathfrak{b} \cdot (\mathfrak{c} - \mathfrak{a}) = 0$, so that $\mathfrak{a} \cdot \mathfrak{b} = \mathfrak{b} \cdot \mathfrak{c} = \mathfrak{c} \cdot \mathfrak{a}$. But then

$$\mathfrak{c} \cdot (\mathfrak{a} - \mathfrak{b}) = 0,$$

which implies the desired assertion, since for $\mathfrak{c} = \mathfrak{o}$ (i.e., $C = H$) there is nothing to prove and for $\mathfrak{c} \neq \mathfrak{o}$ it implies that $C \vee H$ is perpendicular to $A \vee B$.

4) The "perpendicular bisectors" of a triangle ABC intersect at a point M.

The proof follows at once from 2), and at the same time implies the existence and uniqueness of the "circumcircle" of ABC.

Let us examine more closely the connection between 3) and 4). For this purpose we need the *center of gravity* S of ABC, i.e., the point which, for arbitrary choice of origin O, has the position vector

$$\mathfrak{s} = \tfrac{1}{3}(\mathfrak{a} + \mathfrak{b} + \mathfrak{c}).$$

Then let σ be the dilation with fixed point S and magnification factor (-2), i.e., such that if $\mathfrak{y} = \sigma(\mathfrak{x})$ then $\mathfrak{y} - \mathfrak{s} = -2(\mathfrak{x} - \mathfrak{s})$ or $\mathfrak{y} = -2\mathfrak{x} + 3\mathfrak{s}$. If we now take the point M as origin, the position vector of its image $\sigma(M)$ is $3\mathfrak{s} = \mathfrak{a} + \mathfrak{b} + \mathfrak{c}$. But $\sigma(M)$ is the point H of intersection of the altitudes, in view of the equations

$$(\mathfrak{a} + \mathfrak{b} + \mathfrak{c} - \mathfrak{a}) \cdot (\mathfrak{b} - \mathfrak{c}) = \mathfrak{b} \cdot \mathfrak{b} - \mathfrak{c} \cdot \mathfrak{c} = 0,$$

$$(\mathfrak{a} + \mathfrak{b} + \mathfrak{c} - \mathfrak{b}) \cdot (\mathfrak{c} - \mathfrak{a}) = \mathfrak{c} \cdot \mathfrak{c} - \mathfrak{a} \cdot \mathfrak{a} = 0,$$

$$(\mathfrak{a} + \mathfrak{b} + \mathfrak{c} - \mathfrak{c}) \cdot (\mathfrak{a} - \mathfrak{b}) = \mathfrak{a} \cdot \mathfrak{a} - \mathfrak{b} \cdot \mathfrak{b} = 0,$$

which follow from the fact that $\mathfrak{a} \cdot \mathfrak{a} = \mathfrak{b} \cdot \mathfrak{b} = \mathfrak{c} \cdot \mathfrak{c}$ (with M as origin). Thus M, S and H lie on a line (the Euler line) with $(M, S; H) = 3$.

The point $\sigma(M)$ is the intersection of the perpendicular bisectors in the image triangle $\sigma(A)$ $\sigma(B)$ $\sigma(C)$, since ratios of (collinear) line segments, and also orthogonality, are preserved under dilations (Fig. 36).

5) The set of all points equidistant (but with oppositely oriented distances, cf. p. 350) from two given oriented intersecting lines g, h is a line w through the intersection O of g and h. If w is also oriented, then $\varphi(g, w) = \varphi(w, h)$, so that w is the *angle bisector* of g and h.

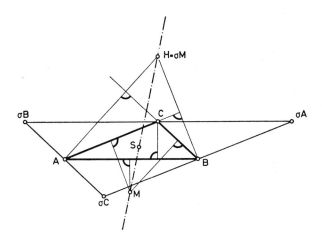

Fig. 36.   The Euler line

The proof follows at once from the equations of g, h in Hesse normal form, say $\mathfrak{n} \cdot \mathfrak{x} = 0$ and $\mathfrak{m} \cdot \mathfrak{x} = 0$. For by p. 351 the "oriented" distances of a point P from g and h are $\mathfrak{n} \cdot \mathfrak{p}$ and $\mathfrak{m} \cdot \mathfrak{p}$, respectively. Thus the desired set w is described by $\mathfrak{n} \cdot \mathfrak{p} = -\mathfrak{m} \cdot \mathfrak{p}$ or $(\mathfrak{n} + \mathfrak{m}) \cdot \mathfrak{p} = 0$, so that w is a line through the vertex O. If we take

$$\frac{1}{|\mathfrak{n} + \mathfrak{m}|}(\mathfrak{n} + \mathfrak{m}) \cdot \mathfrak{r} = 0$$

as its Hesse normal form, then w is oriented and for $\varphi(g, w)$ and $\varphi(w, g)$ we have

$$\cos \varphi(g, w) = \frac{\mathfrak{n} \cdot \mathfrak{n} + \mathfrak{n} \cdot \mathfrak{m}}{|\mathfrak{n} + \mathfrak{m}|} \quad \text{and} \quad \cos \varphi(w, h) = \frac{\mathfrak{n} \cdot \mathfrak{m} + \mathfrak{m} \cdot \mathfrak{m}}{|\mathfrak{n} + \mathfrak{m}|}.$$

Since $\mathfrak{m}$ and $\mathfrak{n}$ are unit vectors, the assertion follows.

8.2.   *Examples for the Geometry of Space*

Here too we first discuss theorems in which the orientation plays no role. Examples 1) and 2) above can be transferred immediately to space (even to n-dimensional space). Thus through every point there is exactly one perpendicular to a given plane, and the foot of the perpendicular is at a minimum distance from the plane. The set of all points equidistant from two given points A, B is a plane orthogonal to $A \vee B$ through the midpoint of $\overline{AB}$. The proof is the same for every dimension.

Let us give an example in which three-dimensionality is essential. The perpendicular from a vertex of a tetrahedron (simplexes in $\mathfrak{A}_3$ are also called tetrahedra, cf. p. 344) to the plane spanned by its other three vertices is called

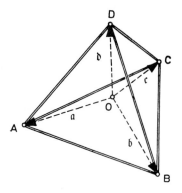

Fig. 37. Orthocentric tetrahedron

an "altitude," and in general the four altitudes are not concurrent. But we have the following theorem (Fig. 37).

6) If three of the altitudes of a tetrahedron intersect in a point, then the fourth also passes through the point. This "intersection of the altitudes" exists if and only if the tetrahedron has two pairs of mutually orthogonal "opposite edges" (orthotetrahedron).

To prove the first assertion we take for origin the common point O of the three intersecting altitudes (say the altitudes from A, B, C). The position vectors $\mathfrak{a}$, $\mathfrak{b}$, $\mathfrak{c}$ then satisfy the conditions

$$\mathfrak{a} \cdot (\mathfrak{b} - \mathfrak{c}) = 0 \quad \text{and} \quad \mathfrak{a} \cdot (\mathfrak{c} - \mathfrak{d}) = 0,$$

$$\mathfrak{b} \cdot (\mathfrak{a} - \mathfrak{c}) = 0 \quad \text{and} \quad \mathfrak{b} \cdot (\mathfrak{c} - \mathfrak{d}) = 0,$$

$$\mathfrak{c} \cdot (\mathfrak{a} - \mathfrak{b}) = 0 \quad \text{and} \quad \mathfrak{c} \cdot (\mathfrak{b} - \mathfrak{d}) = 0$$

from which it follows that the following six scalar products are all equal

(1) $$\mathfrak{a} \cdot \mathfrak{b} = \mathfrak{a} \cdot \mathfrak{c} = \mathfrak{a} \cdot \mathfrak{d} = \mathfrak{b} \cdot \mathfrak{c} = \mathfrak{b} \cdot \mathfrak{d} = \mathfrak{c} \cdot \mathfrak{d}.$$

But then

$$\mathfrak{d} \cdot (\mathfrak{a} - \mathfrak{b}) = 0 \quad \text{and} \quad \mathfrak{d} \cdot (\mathfrak{b} - \mathfrak{c}) = 0,$$

which for $\mathfrak{d} \neq \mathfrak{o}$ means that $O \vee D$ is an altitude of the tetrahedron. (For $\mathfrak{d} = \mathfrak{o}$, i.e., $O = D$, the assertion is trivially valid.)

It is also easy to see that any two opposite edges are orthogonal to each other

$$(\mathfrak{a} - \mathfrak{b}) \cdot (\mathfrak{c} - \mathfrak{d}) = \mathfrak{a} \cdot \mathfrak{c} - \mathfrak{b} \cdot \mathfrak{c} - \mathfrak{a} \cdot \mathfrak{d} + \mathfrak{b} \cdot \mathfrak{d} = 0,$$

(2) $$(\mathfrak{b} - \mathfrak{c}) \cdot (\mathfrak{a} - \mathfrak{d}) = \mathfrak{a} \cdot \mathfrak{b} - \mathfrak{a} \cdot \mathfrak{c} - \mathfrak{b} \cdot \mathfrak{d} + \mathfrak{c} \cdot \mathfrak{d} = 0,$$

$$(\mathfrak{c} - \mathfrak{a}) \cdot (\mathfrak{b} - \mathfrak{d}) = \mathfrak{b} \cdot \mathfrak{c} - \mathfrak{a} \cdot \mathfrak{c} - \mathfrak{c} \cdot \mathfrak{d} + \mathfrak{a} \cdot \mathfrak{d} = 0.$$

Conversely, if the tetrahedron has two pairs of mutually orthogonal opposite edges, the third pair has the same property; the third equation in (2) results from addition of the first two (with an arbitrary choice of origin).

In order to show that the intersection of the altitudes exists, we consider the three planes orthogonal respectively to $A \vee D, B \vee D, C \vee D$ (which in each case must contain the lines $B \vee C, C \vee A, A \vee B$, orthogonal respectively to $A \vee D, B \vee D, C \vee D$) with the equations

$$(\mathfrak{a} - \mathfrak{d})\cdot(\mathfrak{x} - \mathfrak{b}) = 0, \quad (\mathfrak{b} - \mathfrak{d})\cdot(\mathfrak{x} - \mathfrak{c}) = 0, \quad (\mathfrak{c} - \mathfrak{d})\cdot(\mathfrak{x} - \mathfrak{a}) = 0.$$

Since $\mathfrak{a} - \mathfrak{d}, \mathfrak{b} - \mathfrak{d}, \mathfrak{c} - \mathfrak{d}$ are linearly independent, these three planes must have exactly one point in common. Taking this point for origin, we have

$$\mathfrak{a}\cdot\mathfrak{b} = \mathfrak{b}\cdot\mathfrak{d}, \quad \mathfrak{b}\cdot\mathfrak{c} = \mathfrak{c}\cdot\mathfrak{d}, \quad \mathfrak{a}\cdot\mathfrak{c} = \mathfrak{a}\cdot\mathfrak{d}$$

and also, from the second, third, and first equation of our assumption (2),

$$\mathfrak{a}\cdot\mathfrak{c} = \mathfrak{c}\cdot\mathfrak{d}, \quad \mathfrak{a}\cdot\mathfrak{b} = \mathfrak{a}\cdot\mathfrak{d}, \quad \mathfrak{b}\cdot\mathfrak{c} = \mathfrak{b}\cdot\mathfrak{d}.$$

Thus equations (1) are valid, and the origin O is the point of intersection of the altitudes.

7) Intersection of a line and a sphere. Intersection of two spheres.

For the equation of the sphere we take

(3) $$\mathfrak{x}\cdot\mathfrak{x} = \mathfrak{r}^2$$

and for the line its parametric representation

(4) $$\mathfrak{x} = \mathfrak{p} + s\mathfrak{a} \quad \text{with} \quad |\mathfrak{a}| = 1.$$

The values of the parameter corresponding to points of intersection of the sphere and the line are determined by the equation

(5) $$s^2 + 2(\mathfrak{p}\cdot\mathfrak{a})s + ((\mathfrak{p}\cdot\mathfrak{p}) - r^2) = 0.$$

For $(\mathfrak{p}\cdot\mathfrak{a})^2 - (\mathfrak{p}\cdot\mathfrak{p}) + r^2 < 0$ there are two real solutions s_1 and s_2, and for $|\mathfrak{a}| = 1$ we have the following geometric interpretation of s_1 and s_2

$$d(P, S_1) = s_1, \quad d(P, S_2) = s_2,$$

where S_1 and S_2 are the points of intersection of the sphere and the line (Fig. 38). The product

(6) $$s_1 s_2 = \mathfrak{p}\cdot\mathfrak{p} - r^2$$

which by (5) is independent of $\mathfrak{a}$, is called the "power" of the point P with respect to the sphere (3).

If $(\mathfrak{p}\cdot\mathfrak{a})^2 - (\mathfrak{p}\cdot\mathfrak{p}) + r^2 = 0$, the line (4) is tangent to the sphere (3), and $s_1 = s_2$ is equal to the distance t from P to the point of contact of the tangent. Then (6) implies $s_1\cdot s_2 = t^2$ (secant-tangent theorem).

Tangents from P to the sphere (3) exist if and only if the power of P is

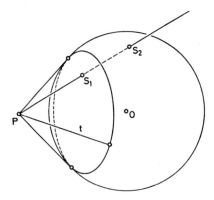

Fig. 38.   Secant theorem for the sphere

nonnegative. The set of all corresponding points of contact lie on a plane

$$\mathfrak{p} \cdot \mathfrak{x} = r^2.$$

For if (4) represents a tangent, then $s = -(\mathfrak{p} \cdot \mathfrak{a})$ is the unique solution of (5) and $\mathfrak{x} = \mathfrak{p} - (\mathfrak{p} \cdot \mathfrak{a}) \mathfrak{a}$ is the position vector of the point of contact of the tangent. But then $\mathfrak{p} \cdot \mathfrak{x} = \mathfrak{p} \cdot \mathfrak{p} - (\mathfrak{p} \cdot \mathfrak{a})^2 = r^2$, which is the desired assertion.

Finally we note that the intersection of any two spheres lies on a plane, since from $(\mathfrak{x} - \mathfrak{m}_1) \cdot (\mathfrak{x} - \mathfrak{m}_1) = r_1^2$ and $(\mathfrak{x} - \mathfrak{m}_2) \cdot (\mathfrak{x} - \mathfrak{m}_2) = r_2^2$ it follows by subtraction that

$$2(\mathfrak{m}_2 - \mathfrak{m}_1) \cdot \mathfrak{x} = (\mathfrak{m}_2 \cdot \mathfrak{m}_2 - r_2^2) - (\mathfrak{m}_1 \cdot \mathfrak{m}_1 - r_1^2)$$

which is the equation of a plane. All points on this plane have the same power with respect to the two spheres.

The results of example 7) apply without essential change to any dimension.

8.3.   *Alternating Vector Products*

In the following discussion, which involves orientation, the determinant function will again be denoted by D.

In the two-dimensional case the mapping

$$(\mathfrak{x}, \mathfrak{y}) \mapsto D(\mathfrak{x}, \mathfrak{y})$$

is bilinear and alternating, so that $D(\mathfrak{x}, \mathfrak{y})$, usually written $[\mathfrak{x}, \mathfrak{y}]$, is called the *alternating product* (or *outer product*) of the two vectors $\mathfrak{x}$ and $\mathfrak{y}$; by §7.4 p. 353, it gives the oriented area of the oriented parallelogram with edges $\mathfrak{x}, \mathfrak{y}$, and from the Lagrange identity we have

$$[\mathfrak{x}, \mathfrak{y}]^2 = (\mathfrak{x} \cdot \mathfrak{x})(\mathfrak{y} \cdot \mathfrak{y}) - (\mathfrak{x} \cdot \mathfrak{y})^2.$$

Thus for the unit vectors $\mathfrak{x}^0$, $\mathfrak{y}^0$

$$(\mathfrak{x}^0 \cdot \mathfrak{y}^0)^2 + [\mathfrak{x}^0, \mathfrak{y}^0]^2 = 1,$$

and from $\cos \varphi(\mathfrak{x}^0, \mathfrak{y}^0) = \mathfrak{x}^0 \cdot \mathfrak{y}^0$ it follows that

$$|[\mathfrak{x}^0, \mathfrak{y}^0]| = \sin \varphi(\mathfrak{x}^0, \mathfrak{y}^0).$$

So in the oriented plane it is natural to refine the concept of angle measure introduced in §8.1. To the ordered pair $(\mathfrak{x}^0, \mathfrak{y}^0)$ we now assign, by means of the equations

$$\cos \omega(\mathfrak{x}^0, \mathfrak{y}^0) = \mathfrak{x}^0 \cdot \mathfrak{y}^0,$$

$$\sin \omega(\mathfrak{x}^0, \mathfrak{y}^0) = [\mathfrak{x}^0, \mathfrak{y}^0],$$

a unique real number $\omega(\mathfrak{x}^0, \mathfrak{y}^0)$ with $0 \leq \omega < 2\pi$, called the *measure of the oriented angle* between the vectors of the ordered pair $(\mathfrak{x}^0, \mathfrak{y}^0)$.

Here φ can be calculated from ω, since

$$\varphi(\mathfrak{x}^0, \mathfrak{y}^0) = \min \{\omega(\mathfrak{x}^0, \mathfrak{y}^0), 2\pi - \omega(\mathfrak{x}^0, \mathfrak{y}^0)\}$$

but not conversely, since e.g. the vectors $(\mathfrak{x}^0, \mathfrak{y}^0)$ and $(\mathfrak{y}^0, \mathfrak{x}^0)$ determine φ uniquely but admit two values, in general, for ω.

Like φ, the function ω can be defined for arbitrary vectors $\mathfrak{x}$, $\mathfrak{y}$ not both zero. Then we have

$$\cos \omega(\mathfrak{x}, \mathfrak{y}) = \frac{\mathfrak{x} \cdot \mathfrak{y}}{|\mathfrak{x}| \cdot |\mathfrak{y}|},$$

$$\sin \omega(\mathfrak{x}, \mathfrak{y}) = \frac{[\mathfrak{x}, \mathfrak{y}]}{|\mathfrak{x}| \cdot |\mathfrak{y}|}.$$

We now show that the oriented angle measure is additive, by which we mean that for any three nonzero vectors $\mathfrak{x}$, $\mathfrak{y}$, $\mathfrak{z}$ we have

$$\omega(\mathfrak{x}, \mathfrak{y}) + \omega(\mathfrak{y}, \mathfrak{z}) = \omega(\mathfrak{x}, \mathfrak{z}) \bmod 2\pi,$$

where for $\omega(\mathfrak{x}, \mathfrak{y}) + \omega(\mathfrak{y}, \mathfrak{z}) < 2\pi$ the congruence may be replaced by equality.

For the proof we may assume that $\mathfrak{x}$, $\mathfrak{y}$, $\mathfrak{z}$ are *unit vectors*, so that

$$\mathfrak{x} \cdot \mathfrak{z} = (\mathfrak{x} \cdot \mathfrak{y})(\mathfrak{y} \cdot \mathfrak{z}) - [\mathfrak{x}, \mathfrak{y}][\mathfrak{y}, \mathfrak{z}],$$

$$[\mathfrak{x}, \mathfrak{z}] = [\mathfrak{x}, \mathfrak{y}](\mathfrak{y} \cdot \mathfrak{z}) + (\mathfrak{x} \cdot \mathfrak{y})[\mathfrak{y}, \mathfrak{z}],$$

Since the first equation follows immediately from the Lagrange identity applied to the determinant product $[\mathfrak{x}, \mathfrak{y}]\,[\mathfrak{y}, \mathfrak{z}]$, and in the second, if for linearly independent $\mathfrak{x}$, $\mathfrak{y}$ we set $\mathfrak{z} = u\mathfrak{x} + v\mathfrak{y}$, the left side gives $v[\mathfrak{x}, \mathfrak{y}]$, and the right side $v(\mathfrak{y} \cdot \mathfrak{y})[\mathfrak{x}, \mathfrak{y}]$, which is again equal to $v[\mathfrak{x}, \mathfrak{y}]$, since $\mathfrak{y}$ is a unit vector; for linearly dependent $\mathfrak{x}$, $\mathfrak{y}$ we may set $\mathfrak{y} = a\mathfrak{x}$, thereby obtaining $a^2(\mathfrak{x} \cdot \mathfrak{x})[\mathfrak{x}, \mathfrak{z}]$

on the right side, which at once implies the second of the above equations, since $\mathfrak{x}$ and $\mathfrak{y}$ are unit vectors.

If for brevity we set

$$\omega(\mathfrak{x}, \mathfrak{y}) = \alpha, \quad \omega(\mathfrak{y}, \mathfrak{z}) = \beta, \quad \omega(\mathfrak{x}, \mathfrak{z}) = \gamma,$$

the two equations become

$$\cos \gamma = \cos \alpha \cos \beta - \sin \alpha \sin \beta,$$

$$\sin \gamma = \sin \alpha \cos \beta + \cos \alpha \sin \beta,$$

i.e., $\cos (\alpha + \beta) = \cos \gamma$ and $\sin (\alpha + \beta) = \sin \gamma$, which implies the desired additivity

$$\alpha + \beta \equiv \gamma \bmod 2\pi.$$

Thus we have shown that definition (5) on p. 347 leads, when extended in the manner just described, to an additive angle measure in $\mathfrak{B}_2$. In §10 we shall see that, if continuity is assumed, this angle measure is uniquely determined up to a normalization factor.

As an application of the alternating product let us prove the theorem on the *angle at the circumference* of a circle, namely

The set of all points X such that for fixed distinct points A, B the oriented lines A $\vee$ X and B $\vee$ X enclose a fixed oriented angle γ is a circle through A and B.

For if the midpoint of $\overline{AB}$ is taken as origin, the condition on X becomes

$$(\mathfrak{x} + \mathfrak{a}) \cdot (\mathfrak{x} - \mathfrak{a}) = |\mathfrak{x} + \mathfrak{a}|\, |\mathfrak{x} - \mathfrak{a}|\, \cos \gamma,$$

$$[\mathfrak{x} + \mathfrak{a}, \mathfrak{x} - \mathfrak{a}] = |\mathfrak{x} + \mathfrak{a}|\, |\mathfrak{x} - \mathfrak{a}|\, \sin \gamma,$$

which means that

$$\frac{\mathfrak{x} \cdot \mathfrak{x} - \mathfrak{a} \cdot \mathfrak{a}}{2[\mathfrak{a}, \mathfrak{x}]} = \cot \gamma$$

or

$$\mathfrak{x} \cdot \mathfrak{x} - 2 \cot \gamma [\mathfrak{a}, \mathfrak{x}] - \mathfrak{a} \cdot \mathfrak{a} = 0.$$

Now there is exactly one vector $\mathfrak{m}$ with $\cot \gamma [a, y] = \mathfrak{m} \cdot \mathfrak{x}$, for all $\mathfrak{x} \in \mathfrak{B}_2$, so that the condition can be written in the form

(1) $$(\mathfrak{x} - \mathfrak{m}) \cdot (\mathfrak{x} - \mathfrak{m}) = \mathfrak{m} \cdot \mathfrak{m} + \mathfrak{a} \cdot \mathfrak{a}.$$

Thus the desired set is a subset of the circle (1), and in fact, as the reader may easily prove, all points of the circle (except A and B) satisfy the given condition. Since $\mathfrak{m} \cdot \mathfrak{a} = 0$, the length of $\mathfrak{m}$ is given by

$$\cot \gamma \cdot |\mathfrak{a}| \cdot |\mathfrak{m}| = |\mathfrak{m}|^2,$$

and therefore the radius r of the circle by

(2) $$r^2 = |a|^2(1 + \cot^2 \gamma) = \frac{|a|^2}{\sin^2 \gamma}$$

or

(3) $$2r|\sin \gamma| = 2|a| = d(A, B).$$

Setting $\gamma = \pi/2$ or $\gamma = 3\pi/2$ given the theorem of Thales, that the midpoint of $\overline{AB}$ is then the center of the circle (1).

In the three-dimensional case we can define an alternating vector product as follows. Let D be the determinant function establishing the orientation of $\mathfrak{B}_3$. Then for fixed $a, b \in \mathfrak{B}_3$ the mapping

$$x \mapsto D(a, b, x)$$

of $\mathfrak{B}_3$ into $\mathbb{R}$ is a linear form, so that by §7.2 there is exactly one vector $c \in \mathfrak{B}_3$ such that

$$D(a, b, x) = c \cdot x$$

for all $x \in \mathfrak{B}_3$. This vector c is called the *vector product* (also *cross product* or *outer product*) of the two vectors a, b and is written

$$c = a \times b.$$

The following rules are an immediate consequence of the definition.

1) $$a \times b = -(b \times a),$$
2) $$(a_1 + a_2) \times b = a_1 \times b + a_2 \times b,$$
3) $$(ca) \times b = c(a \times b),$$
4) $$(a \times b) \cdot a = (a \times b) \cdot b = 0.$$

Thus the vector product provides an alternating bilinear mapping of $\mathfrak{B}_3 \times \mathfrak{B}_3$ into $\mathfrak{B}_3$ in which, by rule 4), the product is orthogonal to each of its factors.

From the defining equation $D(a, b, x) = (a \times b) \cdot x$ it follows that $a \times b = o$ if and only if a and b are linearly dependent.

In order to calculate the length of the vector $a \times b$ we note that by the Lagrange identity

$$D(a, b, x)D(a, b, x) = \begin{vmatrix} a \cdot a & a \cdot b & a \cdot x \\ a \cdot b & b \cdot b & b \cdot x \\ a \cdot x & b \cdot x & x \cdot x \end{vmatrix}$$

for all $a, b, x \in \mathfrak{B}_3$, so that, setting $x = a \times b$, we obtain

$$((a \times b) \cdot (a \times b))^2 = ((a \times b) \cdot (a \times b)) \begin{vmatrix} a \cdot a & a \cdot b \\ a \cdot b & b \cdot b \end{vmatrix}.$$

Thus for $a \times b \neq o$

5) $\qquad (a \times b) \cdot (a \times b) = (a \cdot a)(b \cdot b) - (a \cdot b)^2$

and for $a \times b = o$, so that a and b are linearly dependent, rule 5) still holds (since both sides are then equal to zero). Thus the length of the vector $|a \times b|$ is equal to the area of a parallelogram with edges a and b (p. 354).

From (5) it follows at once that $|a \times b| = |a|\,|b|\,|\sin \varphi(a, b)|$, and if in 5) we replace a by $a + c$, we obtain

$$(a \times b) \cdot (a \times b) + 2(a \times b) \cdot (c \times b) + (c \times b) \cdot (c \times b)$$

$$= ((a \cdot a)(b \cdot b) - (a \cdot b)^2) + 2((a \cdot c)(b \cdot b) - (a \cdot b)(b \cdot c))$$

$$+ ((c \cdot c)(b \cdot b) - (b \cdot c)^2),$$

and thus the following generalization of 5)

5') $\qquad (a \times b) \cdot (c \times b) = (a \cdot c)(b \cdot b) - (a \cdot b)(b \cdot c).$

Expansion of the left side now gives

$$-((a \times b) \times b) \cdot c = ((b \cdot b)a - (a \cdot b)b) \cdot c$$

for all $c \in \mathfrak{B}_3$, which means that

$$(a \times b) \times b = -(b \cdot b)a + (a \cdot b)b,$$

and, by interchange of a and b,

$$(a \times b) \times a = -(b \times a) \times a = -(a \cdot b)a + (a \cdot a)b.$$

These two equations are special cases of the *expansion rule*

6) $\qquad (a \times b) \times c = -(b \cdot c)a + (a \cdot c)b,$

which can easily be proved in general on the basis of the two special cases. For if the vectors a, b are linearly independent, then so are $a, b, a \times b$, so that any $c \in \mathfrak{B}_3$ can be written in the form

$$c = xa + yb + z(a \times b)$$

from which 6) follows by direct calculation.

From 6) we see that the vector product is not associative; but we have the *Jacobi identity*

$$(a \times b) \times c + (b \times c) \times a + (c \times a) \times b = o.$$

Finally the expansion rule 6) leads to the following generalization of 5) and 5')

7) $\qquad (a \times b) \cdot (c \times \mathfrak{d}) = (a \cdot c)(b \cdot \mathfrak{d}) - (a \cdot \mathfrak{d})(b \cdot c).$

For the proof it is only necessary to multiply 6) by the scalar $\mathfrak{d}$.

Applications of the vector product

1) *Plücker equation of a line* $\mathfrak{a} \times \mathfrak{x} = \mathfrak{b}$, with $|\mathfrak{a}| = 1$ and $\mathfrak{a} \cdot \mathfrak{b} = 0$. The parametric representation of a line $\mathfrak{x} = \mathfrak{p} + x\mathfrak{a}$, with $|\mathfrak{a}| = 1$, implies for the vector product $\mathfrak{a} \times \mathfrak{x}$ that $\mathfrak{a} \times \mathfrak{x} = \mathfrak{a} \times \mathfrak{p}$ for all points on the line, and setting $\mathfrak{b} = \mathfrak{a} \times \mathfrak{p}$ shows that $\mathfrak{a} \cdot \mathfrak{b} = 0$. Conversely, for every $\mathfrak{b}$ with $\mathfrak{a} \cdot \mathfrak{b} = 0$, there is at least one vector $\mathfrak{p}$ with $\mathfrak{b} = \mathfrak{a} \times \mathfrak{p}$; for example, $\mathfrak{p} = \mathfrak{b} \times \mathfrak{a}$, as is seen from the expansion rule. The equation $\mathfrak{a} \times \mathfrak{x} = \mathfrak{b}$ with $|\mathfrak{a}| = 1$ and $\mathfrak{a} \cdot \mathfrak{b} = 0$ can then be written in the form $\mathfrak{a} \times \mathfrak{x} = \mathfrak{a} \times \mathfrak{p}$ or $\mathfrak{a} \times (\mathfrak{x} - \mathfrak{p}) = \mathfrak{o}$. Thus $\mathfrak{x} - \mathfrak{p}$ must be a multiple of $\mathfrak{a}$, which means that all points satisfying $\mathfrak{a} \times \mathfrak{x} = \mathfrak{b}$ lie on the line in the direction $\mathfrak{a}$.

2) *Line of intersection of two planes.* If the two planes are given in the Hesse normal form

$$\mathfrak{u}_1 \cdot \mathfrak{x} = a_1 \quad \text{and} \quad \mathfrak{u}_2 \cdot \mathfrak{x} = a_2,$$

then every x satisfying both these equations must also satisfy

$$-(\mathfrak{u}_2 \cdot \mathfrak{x})\mathfrak{u}_1 + (\mathfrak{u}_1 \cdot \mathfrak{x})\mathfrak{u}_2 = a_1\mathfrak{u}_2 - a_2\mathfrak{u}_1$$

or

$$(\mathfrak{u}_1 \times \mathfrak{u}_2) \times \mathfrak{x} = a_1\mathfrak{u}_2 - a_2\mathfrak{u}_1,$$

which for linearly independent vectors $\mathfrak{u}_1, \mathfrak{u}_2$ is the equation of a line in Plücker form (after multiplication by $1/|\mathfrak{u}_1 \times \mathfrak{u}_2|$).

For the direction vector $\mathfrak{a}$ and the "moment vector" $\mathfrak{b}$ (with respect to O) we have

$$\mathfrak{a} = \frac{\mathfrak{u}_1 \times \mathfrak{u}_2}{|\mathfrak{u}_1 \times \mathfrak{u}_2|} \quad \text{and} \quad \mathfrak{b} = \frac{a_1\mathfrak{u}_2 - a_2\mathfrak{u}_1}{|\mathfrak{u}_1 \times \mathfrak{u}_2|}$$

from which it is obvious that $\mathfrak{a} \cdot \mathfrak{b} = 0$.

3) *Common perpendicular to two skew lines.* If $\mathfrak{x} = \mathfrak{p} + x\mathfrak{a}$ and $\mathfrak{y} = \mathfrak{q} + y\mathfrak{b}$ are the parametric representations of the two lines, we wish to determine two scalars x, y such that $\mathfrak{x} - \mathfrak{y}$ is orthogonal to $\mathfrak{a}$ and to $\mathfrak{b}$

$$(\mathfrak{x} - \mathfrak{y}) \cdot \mathfrak{a} = x(\mathfrak{a} \cdot \mathfrak{a}) - y(\mathfrak{a} \cdot \mathfrak{b}) + (\mathfrak{p} - \mathfrak{q}) \cdot \mathfrak{a} = 0,$$

$$(\mathfrak{x} - \mathfrak{y}) \cdot \mathfrak{b} = x(\mathfrak{a} \cdot \mathfrak{b}) - y(\mathfrak{b} \cdot \mathfrak{b}) + (\mathfrak{p} - \mathfrak{q}) \cdot \mathfrak{b} = 0.$$

If $(\mathfrak{a} \cdot \mathfrak{a})(\mathfrak{b} \cdot \mathfrak{b}) - (\mathfrak{a} \cdot \mathfrak{b})^2 \neq 0$, i.e., $\mathfrak{a} \times \mathfrak{b} \neq \mathfrak{o}$, this system of linear equations has a unique solution, and the line joining the corresponding points, i.e., the common perpendicular to the given lines, lies along the vector $\mathfrak{a} \times \mathfrak{b}$ (Fig. 39).

Thus if $\mathfrak{p}_0$ and $\mathfrak{q}_0$ are the position vectors of the points of intersection (of the lines with this common perpendicular) then

$$\mathfrak{p}_0 - \mathfrak{q}_0 = d\frac{\mathfrak{a} \times \mathfrak{b}}{|\mathfrak{a} \times \mathfrak{b}|},$$

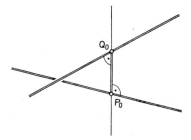

Fig. 39. Common perpendicular to two lines

so that the distance between the lines is

$$|\mathfrak{p}_0 - \mathfrak{q}_0| = |d| = \frac{|D(\mathfrak{a}, \mathfrak{b}, \mathfrak{p}_0 - \mathfrak{q}_0)|}{|\mathfrak{a} \times \mathfrak{b}|} = \frac{|D(\mathfrak{a}, \mathfrak{b}, \mathfrak{x} - \mathfrak{y})|}{|\mathfrak{a} \times \mathfrak{b}|}$$

(which holds for arbitrary $\mathfrak{x} = \mathfrak{p}_0 + x\mathfrak{a}$ and $\mathfrak{y} = \mathfrak{q}_0 + y\mathfrak{b}$).

9. Trigonometry

9.1. *Triangles in the Euclidean Plane*

As the simplest "elements" of a triangle ABC in the Euclidean plane it is traditional to consider the lengths of the sides and the magnitudes of the so-called "interior angles,"

$$d(B, C) = a, \qquad d(C, A) = b, \qquad d(A, B) = c,$$
$$\varphi(\overrightarrow{AB}, \overrightarrow{AC}) = \alpha, \quad \varphi(\overrightarrow{BC}, \overrightarrow{BA}) = \beta, \quad \varphi(\overrightarrow{CA}, \overrightarrow{CB}) = \gamma,$$

where, by the definition on p. 347,

(1) $$0 < \alpha < \pi, \quad 0 < \beta < \pi, \quad 0 < \gamma < \pi.$$

The purpose of trigonometry is to deduce relations among the "elements" of the triangle ABC. Setting

$$\overrightarrow{BC} = \mathfrak{a}, \quad \overrightarrow{CA} = \mathfrak{b}, \quad \overrightarrow{AB} = \mathfrak{c},$$

we have (see Fig. 40)

(2) $$\mathfrak{a} + \mathfrak{b} + \mathfrak{c} = \mathfrak{o}$$

and $\alpha = \varphi(\mathfrak{c}, -\mathfrak{b})$, $\beta = \varphi(\mathfrak{a}, -\mathfrak{c})$, $\gamma = \varphi(\mathfrak{b}, -\mathfrak{a})$.
Taking the scalar product of (2) with $\mathfrak{a}^0$ gives

$$\mathfrak{a} \cdot \mathfrak{a}^0 + \mathfrak{b} \cdot \mathfrak{a}^0 + \mathfrak{c} \cdot \mathfrak{a}^0 = \mathfrak{a} \cdot \mathfrak{a}^0 - \mathfrak{b} \cdot (-\mathfrak{a}^0) - \mathfrak{a}^0 \cdot (-\mathfrak{c})$$
$$= a - b \cos \gamma - c \cos \beta = 0$$

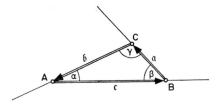

Fig. 40

and two corresponding equations are obtained by cyclic permutation. Thus we have the projection theorems

(P)
$$a = b \cdot \cos \gamma + c \cdot \cos \beta,$$
$$b = c \cdot \cos \alpha + a \cdot \cos \gamma,$$
$$c = a \cdot \cos \beta + b \cdot \cos \alpha,$$

which imply the inequalities

(3)
$$a < b + c, \quad b < c + a, \quad c < a + b;$$

of course, these inequalities could also have been obtained from the triangle inequality (4) for vectors (cf. p. 346).

From (P) we deduce the sine theorem and the angle-sum theorem, as follows. Eliminating $\cos \gamma$ from the first two equations gives

$$a^2 - b^2 = c(a \cos \beta - b \cos \alpha),$$

which together with the third equation implies

$$a^2 - b^2 = a^2 \cos^2\beta - b^2 \cos^2\alpha,$$

or

$$a^2 \sin^2\beta - b^2 \sin^2\alpha = (a \sin \beta - b \sin \alpha)(a \sin \beta + b \sin \alpha) = 0.$$

Since by (1) the second factor is certainly positive, we must have

$$a \sin \beta = b \sin \alpha,$$

so that by division and cyclic permutation we obtain the *law of sines*

(4)
$$\frac{a}{\sin \alpha} = \frac{b}{\sin \beta} = \frac{c}{\sin \gamma} = k > 0.$$

The geometric significance of the constant k follows at once; namely, $k = 2r$, where r is the radius of the circumcircle of ABC.

The equations (4) could have been proved very simply by means of the alternating product, but we have not taken this approach because the orientation of the plane (needed in the definition of an alternating product) is not necessary for trigonometry and would obscure the interdependence of various trigonometric theorems.

Substituting $a = k \sin \alpha$, $b = k \sin \beta$ and $c = k \sin \gamma$ in (P), we obtain three equations containing only the angles

$$\sin \alpha = \sin \beta \cos \gamma + \cos \beta \sin \gamma = \sin (\beta + \gamma),$$
$$\sin \beta = \sin \gamma \cos \alpha + \cos \gamma \sin \alpha = \sin (\gamma + \alpha),$$
$$\sin \gamma = \sin \alpha \cos \beta + \cos \alpha \sin \beta = \sin (\alpha + \beta),$$

which, in view of the restrictions on α, β, and γ, imply

$$\beta + \gamma < \pi, \quad \gamma + \alpha < \pi, \quad \alpha + \beta < \pi$$

and therefore the *angle-sum theorem*

(5) $\alpha + \beta + \gamma = \pi.$

For if (5) were false, we would have $\alpha = \beta + \gamma$, $\beta = \gamma + \alpha$, $\gamma = \alpha + \beta$, so that $\alpha + \beta + \gamma = 0$, which is impossible, since all three angles are positive.

Let us denote by (S) the system of theorems consisting of the law of sines (4) and the angle-sum theorem (5)

(S) $\dfrac{a}{\sin \alpha} = \dfrac{b}{\sin \beta} = \dfrac{c}{\sin \gamma} = k > 0,$

$$\alpha + \beta + \gamma = \pi.$$

Here we have deduced (S) from (P). We now show that conversely (P) follows from (S). First of all, (S) implies that

$$c = k \sin \gamma = k \sin (\alpha + \beta) = k \sin \alpha \cos \beta + k \cos \alpha \sin \beta.$$

Since also $k \sin \alpha = a$ and $k \sin \beta = b$, we see that $a \cos \beta + b \cos \alpha = c$, which is the third equation of (P), and the other two equations are proved in the same way.

By eliminating two of the cosines in (P) it is easy to obtain the system of cosine theorems (see (K) below) but here we prefer to obtain (K) directly from (2)

$$-\mathfrak{c} = \mathfrak{a} + \mathfrak{b},$$
$$\mathfrak{c} \cdot \mathfrak{c} = \mathfrak{a} \cdot \mathfrak{a} + \mathfrak{b} \cdot \mathfrak{b} + 2\mathfrak{a} \cdot \mathfrak{b} = \mathfrak{a} \cdot \mathfrak{a} + \mathfrak{b} \cdot \mathfrak{b} - 2\mathfrak{b} \cdot (-\mathfrak{a}),$$
$$c^2 = a^2 + b^2 - 2ab \cos \gamma,$$

and similarly for the other two cosine theorems, so that altogether

$$a^2 = b^2 + c^2 - 2bc \cos \alpha,$$
(K) $b^2 = c^2 + a^2 - 2ca \cos \beta,$
$$c^2 = a^2 + b^2 - 2ab \cos \gamma.$$

As a particular case of (K) we obtain, say from the third equation, the Pythagorean theorem

$$c^2 = a^2 + b^2 \quad \textit{if and only} \quad \gamma = \frac{\pi}{2}.$$

Again (K) implies (P), as is seen by adding pairs of equations in (K). Thus the three systems (P), (S) and (K) are equivalent, each of them being a set of necessary conditions for the six elements of the triangle

$$(P)$$
$$\swarrow \qquad \searrow$$
$$(S) \iff (K)$$

On the other hand, each of these sets of conditions is also sufficient for the existence of a triangle ABC with the elements $a, b, c, \alpha, \beta, \gamma$, as we shall prove. We first choose an orthonormal basis $\{\mathfrak{e}_1, \mathfrak{e}_2\}$ in $\mathfrak{V}_2$ and set

$$\mathfrak{c} = c \cdot \mathfrak{e}_1,$$
$$\mathfrak{b} = -b(\cos \alpha \mathfrak{e}_1 + \sin \alpha \mathfrak{e}_2),$$
$$\mathfrak{a} = a(-\cos \beta \mathfrak{e}_1 + \sin \beta \mathfrak{e}_2).$$

Then from (P) and (S) we have $\mathfrak{a} + \mathfrak{b} + \mathfrak{c} = \mathfrak{o}$, so that, after choice of an arbitrary point A, we obtain the point B by attaching $\mathfrak{c}$ to A, and obtain the point C by attaching $-\mathfrak{b}$, to A, as the reader should verify.

Each of the sine and cosine theorems contains four triangle elements. Thus if we are given three elements (at least one being the length of a side) that are consistent with (S) and thus also with (K) and (P), the other elements can be calculated from the above formulas. The various special cases once formed an extended field of study, but today there is little interest in them. Trigonometry in a modern school is based on very few formulas, say only the systems (K) and (S), in order that the fundamental interrelations may stand out as clearly as possible. Readers interesed in further details may refer to [4], [7], and [12].

9.2. *Spherical Triangles*

Spherical trigonometry, which investigates the properties of triangles on a sphere, has also been important since ancient times. Here we shall merely deduce a few basic formulas.

Without loss of generality, we may write the equation of the sphere in the form

$$(1) \qquad \qquad \mathfrak{x} \cdot \mathfrak{x} = 1,$$

so that the three unit vectors $\mathfrak{a}, \mathfrak{b}, \mathfrak{c}$ (which we shall assume to be linearly independent) determine three points A, B, C on its surface. In the plane, three noncollinear points uniquely determine the sides of a triangle (as line segments) and its interior, but for spherical triangles this statement is no longer true, as two points A, B determine two "segments" joining them, namely the two arcs of great circles with A and B as endpoints. Thus the points A, B, C determine eight distinct triangles, or even, in the oriented case, sixteen (*Möbius concept of a triangle*).

But if in each case we choose the shorter of the two segments, we arrive at the *Euler concept of a triangle*, in which the vertices uniquely determine the "triangle." In order to describe this idea mathematically we provide the sphere with a suitable *metric* (or *distance*) d, by setting

(2) $$d(A,\ B) = \varphi(\mathfrak{a}, \mathfrak{b}) = \text{arc cos } \mathfrak{a} \cdot \mathfrak{b}$$

for any two points A, B on it, and then showing that in fact d has the properties of a metric (p. 350).

For we see at once that $d(A, B) \geq 0$, with $d(A, B) = 0$ if and only if $A = B$, and that $d(A, B) = d(B, A)$ for all A, B. The triangle inequality will be proved below.

For three points A, B, C on the sphere (1), the distances

$$a = d(B,\ C),\quad b = d(C,\ A),\quad c = d(A,\ B)$$

are thereby defined, and we now call ABC an *Euler spherical triangle* with "sides" defined by these distances. The triangle is already determined by the three unit vectors $\mathfrak{a}$, $\mathfrak{b}$, $\mathfrak{c}$, and for the lengths of the sides a, b, c we have

(3) $\cos a = \mathfrak{b} \cdot \mathfrak{c}$, $\quad \cos b = \mathfrak{c} \cdot \mathfrak{a}$, $\quad \cos c = \mathfrak{a} \cdot \mathfrak{b}$, $\quad$ with $\quad 0 < a, b, c < \pi$.

Any two vertices of the triangle, taken together with the center of the sphere, determine a plane, and the "angles" of the Euler spherical triangle are usually defined by means of the planes. In order to obtain a unique determination of the magnitudes of the angles, we choose the normal vectors for the planes as follows.

(4) $$\mathfrak{a}^* = \frac{\mathfrak{b} \times \mathfrak{c}}{|\mathfrak{b} \times \mathfrak{c}|}, \quad \mathfrak{b}^* = \frac{\mathfrak{c} \times \mathfrak{a}}{|\mathfrak{c} \times \mathfrak{a}|}, \quad \mathfrak{c}^* = \frac{\mathfrak{a} \times \mathfrak{b}}{|\mathfrak{a} \times \mathfrak{b}|}.$$

These unit vectors are the position vectors of three points A^*, B^*, C^* on the sphere, and the Euler spherical triangle formed by them is called the *polar triangle* of ABC. At the same time, the lengths a^*, b^*, c^* of the sides of the triangle $A^*B^*C^*$ measure the angles between the vectors of the pairs $(\mathfrak{b}^*, \mathfrak{c}^*), (\mathfrak{c}^*, \mathfrak{a}^*), (\mathfrak{a}^*, \mathfrak{b}^*)$ and thus also the angles between the corresponding planes. The lengths a^*, b^*, c^*, satisfying the equations

(5) $\quad \cos a^* = \mathfrak{b}^* \cdot \mathfrak{c}^*$, $\quad \cos b^* = \mathfrak{c}^* \cdot \mathfrak{a}^*$, $\quad \cos c^* = \mathfrak{a}^* \cdot \mathfrak{b}^*$

with $0 < a^*, b^*, c^* < \pi$,

measure "exterior angles" of the original triangle (Fig. 41); and the "inte-

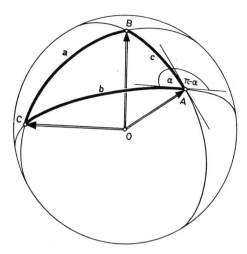

Fig. 41. Spherical triangle

rior angles" of the spherical triangle ABC are then measured by

$$\alpha = \pi - a^*, \quad \beta = \pi - b^*, \quad \gamma = \pi - c^*.$$

It is easy to show by calculation that the polar triangle of a polar triangle is again the original triangle, provided $(\mathfrak{a} \times \mathfrak{b}) \cdot \mathfrak{c} > 0$:

(6) $$\mathfrak{a} = \frac{\mathfrak{b}^* \times \mathfrak{c}^*}{|\mathfrak{b}^* \times \mathfrak{c}^*|}, \quad \mathfrak{b} = \frac{\mathfrak{c}^* \times \mathfrak{a}^*}{|\mathfrak{c}^* \times \mathfrak{a}^*|}, \quad \mathfrak{c} = \frac{\mathfrak{a}^* \times \mathfrak{b}^*}{|\mathfrak{a}^* \times \mathfrak{b}^*|}$$

(but for $(\mathfrak{a} \times \mathfrak{b}) \cdot \mathfrak{c} < 0$, the original triangle is now reflected in the center of the sphere). Thus lengths and exterior angles are dual to each other in spherical trigonometry.

In our definition the polar triangle seems to depend on the orientation of the space, since we have made use of the vector product, but the equations (5) are the same in the other orientation.

The system of equations (4) and (6) now give us various equations for the triangle elements a, b, c, a^*, b^*, c^*; for example, calculating $\cos a = \mathfrak{b} \cdot \mathfrak{c}$ from (6) we obtain

$$\cos a = \mathfrak{b} \cdot \mathfrak{c} = \frac{(\mathfrak{c}^* \times \mathfrak{a}^*) \cdot (\mathfrak{a}^* \times \mathfrak{b}^*)}{|\mathfrak{c}^* \times \mathfrak{a}^*| \cdot |\mathfrak{a}^* \times \mathfrak{b}^*|}.$$

By rule 5') on p. 364 it follows that

$$\cos a = \frac{(\mathfrak{a}^* \cdot \mathfrak{c}^*)(\mathfrak{a}^* \cdot \mathfrak{b}^*) - (\mathfrak{b}^* \cdot \mathfrak{c}^*)(\mathfrak{a}^* \cdot \mathfrak{a}^*)}{|\mathfrak{c}^* \times \mathfrak{a}^*| \cdot |\mathfrak{a}^* \times \mathfrak{b}^*|},$$

$$\cos a = \frac{\cos b^* \cos c^* - \cos a^*}{\sin b^* \sin c^*}.$$

Taking interior angles and permuting cyclically, we obtain the *cosine theorems for the angles*

$$\cos \alpha = -\cos \beta \cos \gamma + \sin \beta \sin \gamma \cos a,$$

$$\cos \beta = -\cos \gamma \cos \alpha + \sin \gamma \sin \alpha \cos b,$$

$$\cos \gamma = -\cos \alpha \cos \beta + \sin \alpha \sin \beta \cos c,$$

and similarly for the *cosine theorems for the sides*

$$\cos a = \cos b \cos c + \sin b \sin c \cos \alpha,$$

$$\cos b = \cos c \cos a + \sin c \sin a \cos \beta,$$

$$\cos c = \cos a \cos b + \sin a \sin b \cos \gamma.$$

It is now easy to show that our distance d (p. 370) satisfies the *triangle inequality*. Since $\cos \alpha > -1$, the first cosine theorem for sides gives

$$(7) \qquad\qquad \cos a > \cos (b + c)$$

and thus

$$a < b + c.$$

For if we had $a \geq b + c$, then both a and $b + c$ would lie in the interval $[0, \pi]$ with the result that $\cos a \leq \cos (b + c)$, in contradiction to (7). Thus if the three points A, B, C are not on a plane through the center O, we have the triangle inequality in stronger form

$$d(B, C) < d(C, A) + d(A, B).$$

Omitting the assumption that ABC is a nondegenerate triangle, we still have

$$(8) \qquad\qquad d(B, C) \leq d(C, A) + d(A, B)$$

as is proved in exactly the same way. Thus we have completed the proof that the sphere (1) with the distance function d is a metric space.

From inequality (7) we now wish to draw a further conclusion; for the lengths of the sides a, b, c we have the so-called "umbrella theorem"

$$(9) \qquad\qquad a + b + c < 2\pi.$$

Again it is simplest to give an indirect proof. The inequality $a + b + c \geq 2\pi$ would imply

$$a \geq 2\pi - (b + c)$$

so that both a and $2\pi - (b + c)$ would lie in the interval $[0, \pi]$. But then it would follow that, in contradiction to (7),

$$\cos a \leq \cos (2\pi - (b + c)) = \cos (b + c)$$

Since (9) also holds for the polar triangle $A^*B^*C^*$,

$$a^* + b^* + c^* < 2\pi,$$

the interior angles α, β, γ of the original triangle must satisfy

(10) $$\alpha + \beta + \gamma > \pi.$$

Thus *the sum of the angles in a spherical triangle is always greater* than π.

For the interpretation of the positive number $\varepsilon = (\alpha + \beta + \gamma) - \pi$, the so-called "spherical excess" of ABC, as a *surface area*, see II14, §3.4. The cosine theorems for angles and sides have been obtained above by calculating the scalar products $b \cdot c$, $c \cdot a$, $a \cdot b$, $b^* \cdot c^*$, $c^* \cdot a^*$, $a^* \cdot b^*$, but we can also form the nine "mixed" scalar products $b^* \cdot c$, ... However, six of these vanish, leaving only $a \cdot a^*$, $b \cdot b^*$, $c \cdot c^*$. From (4) and (6) we have

$$a \cdot a^* = \frac{V}{\sin a} = \frac{V^*}{\sin a^*} = \frac{V^*}{\sin \alpha},$$

$$b \cdot b^* = \frac{V}{\sin b} = \frac{V^*}{\sin b^*} = \frac{V^*}{\sin \beta},$$

$$c \cdot c^* = \frac{V}{\sin c} = \frac{V^*}{\sin c^*} = \frac{V^*}{\sin \gamma}$$

where $D(a, b, c)$ is denoted by V and $D(a^*, b^*, c^*)$ by V^*. We thus obtain the sine theorems

$$\frac{\sin a}{\sin \alpha} = \frac{\sin b}{\sin \beta} = \frac{\sin c}{\sin \gamma} = \frac{V}{V^*},$$

where V and V^* can be interpreted as the volumes of the parallelepipeds spanned by a, b, c and a^*, b^*, c^*, and can be expressed in terms of the elements of the triangle

$$V = \sin b \sin c \sin \alpha = \sin c \sin a \sin \beta = \sin a \sin b \sin \gamma,$$

$$V^* = \sin \beta \sin \gamma \sin a = \sin \gamma \sin \alpha \sin b = \sin \alpha \sin \beta \sin c.$$

For the proof we calculate, say

$$V = D(a, b, c) = (a \times b) \cdot c = \frac{(a \times b) \cdot (a^* \times b^*)}{|a^* \times b^*|}$$

and similarly for V^*.

Furthermore, the Lagrange identity for V^2 gives

$$V^2 = \begin{vmatrix} 1 & \cos c & \cos b \\ \cos c & 1 & \cos a \\ \cos b & \cos a & 1 \end{vmatrix}$$

with a similar result for V^{*2}. Following von Staudt we also call V and V^* the "vertex sines" of the corresponding parallelepipeds.

For further details the reader is referred to the literature (cf. [4], [7], [12]).

10. Isometric Mappings

10.1 Orthogonal Group; Euclidean Group

A linear mapping A of a Euclidean vector space $\mathfrak{B}_n$ into itself is said to be *orthogonal* (or *isometric*) if

(1) $$Ax \cdot Ay = x \cdot y \quad \text{for all} \quad x, y \in \mathfrak{B}_n.$$

In particular, such a mapping preserves the lengths of vectors, and the images of two orthogonal vectors are again orthogonal. Since $Ax = o$ implies that $x = o$, an orthogonal mapping is always invertible, and since by (1) scalar products are preserved, the orthogonal mappings are called *automorphisms* of Euclidean $\mathfrak{B}_n$.

The definition (1) can also be formulated without explicit mention of the vectors x, y by means of the *adjoint* mapping; i.e., corresponding to an arbitrary linear mapping $B: \mathfrak{B}_n \to \mathfrak{B}_n$ there is exactly one linear mapping $B^*: \mathfrak{B}_n \to \mathfrak{B}_n$ such that for all $x, y \in \mathfrak{B}_n$,

$$Bx \cdot y = x \cdot B^*y$$

From the Lagrange identity, applied to $Bx_1, \ldots, Bx_n, y_1, \ldots, y_n$, and to $x_1, \ldots, x_n, B^*y_1, \ldots, B^*y_n$ it follows that

$$\det B^* = \det B,$$

so that in terms of the adjoint mapping A^* we can now write (1) as

$$x \cdot (A^* \circ Ay) = x \cdot y \quad \text{for all} \quad x, y \in \mathfrak{B}_n.$$

Since A is invertible, the orthogonal mappings are characterized by

(2) $$A^* \circ A = A \circ A^* = I.$$

Furthermore, $(\det A)^2 = 1$, i.e.,

(3) $$\det A = 1 \quad \text{or} \quad \det A = -1$$

so that orthogonal mappings preserve the volume (unsigned) of a parallelepiped.

If after choice of an orthonormal basis in $\mathfrak{B}_n$ the mapping A corresponds to the matrix (a_{ik}), then A^* corresponds to the transposed matrix, and (2) gives the *orthogonality relations*

$$\sum_{j=1}^{n} a_{ij}a_{kj} = \sum_{j=1}^{n} a_{ji}a_{jk} = \begin{cases} 0 & \text{for} \quad i \neq k, \\ 1 & \text{for} \quad i = k. \end{cases}$$

By (1) the set of all orthogonal mappings of a Euclidean vector space $\mathfrak{B}_n$ (under the usual composition of mappings) forms a group, called the *orthogonal group* of $\mathfrak{B}_n$. The subgroup of orthogonal mappings with determinant 1 forms a normal subgroup of index 2, the subgroup of *orientation-preserving* or *proper* orthogonal mappings.

Then the automorphisms of a *Euclidean point space* $\mathfrak{E}_n$ consist of those affinities of $\mathfrak{E}_n$ whose corresponding linear mappings are orthogonal, and are therefore represented algebraically by

(4) $\qquad \mathfrak{y} = A\mathfrak{x} + a \quad \text{with} \quad A^* \circ A = A \circ A^* = I.$

Since these automorphisms preserve distances, they are also called *isometries*, or *rigid mappings*.

Conversely, every distance-preserving mapping of $\mathfrak{E}_n$ into itself can be shown to be an automorphism in the above sense, so that in the corresponding vector space it must correspond to an orthogonal *linear* mapping.

Every isometric mapping α of $\mathfrak{E}_n$ into itself also leaves invariant the magnitudes of angles (cf. p. 350). If $\mathfrak{E}_n$ is oriented, the orientation remains unchanged only under the *orientation-preserving* isometries (also called *motions* or *direct isometries*). They are characterized by det $A = 1$.

The group of all isometric mappings of $\mathfrak{E}_n$ into itself is called the *Euclidean group* (*isometric group*), denoted by $\mathbf{B}_n$. Figures equivalent to each other with respect to $\mathbf{B}_n$ (cf. p. 310) are said to be *congruent*. Group $\mathbf{B}_n^+$ of orientation-preserving isometries, also called the *group of motions,* is a normal subgroup of $\mathbf{B}_n$ with index 2. Its coset consists of the *opposite* isometries.

Of interest also is the so-called *equiform* group, generated by $\mathbf{B}_n$ and the dilations. Figures equivalent to each other under the equiform group are said to be *similar.*

In order to survey the various types of isometric mappings we here distinguish, as in §4, the case where α has (at least) one fixed point, and the case where α has no fixed point.

10.2 *Isometries with Fixed Point*

So after choosing one of the fixed points as origin we are faced with the problem of classifying the automorphisms of the Euclidean vector space. Here the situation is much simpler than for general linear mappings. Namely,

For every orthogonal mapping A of a Euclidean vector space $\mathfrak{B}_n$ onto itself there exists a representation of $\mathfrak{B}_n$ as a direct sum

$$\mathfrak{B}_n = \mathfrak{U}_1 \oplus \cdots \oplus \mathfrak{U}_r$$

with the following properties:

1) *Each of the subspaces* $\mathfrak{U}_j$ *is invariant under* A.
2) *The subspaces* $\mathfrak{U}_j$ *are pairwise orthogonal.*
3) *Each of the subspaces* $\mathfrak{U}_j$ *is of dimension* 1 *or* 2.

Thus the whole space $\mathfrak{E}_n$ is spanned by a set of mutually orthogonal planes and lines through the fixed point such that under the given orthogonal mapping each of these lines and planes is mapped onto itself, and each of these submappings is also an isometry.

An orthogonal mapping on the line can be only the identity or reflection in the fixed point, so that only $+1$ or -1 can be an eigenvalue.

To investigate orthogonal mappings on the plane $\mathfrak{B}_2$, we first choose an arbitrary orthonormal basis $\{e_1, e_2\}$ whereupon, for the numbers a, b, c, d in the equations for the mapping

$$A e_1 = a e_1 + b e_2,$$
$$A e_2 = c e_1 + d e_2$$

we obtain the "orthogonality conditions" (cf. p. 374)

$$a^2 + b^2 = c^2 + d^2 = 1, \quad ac + bd = 0.$$

It follows that either $c = -b, d = a$ or else $c = b, d = -a$. In the second case, there exist two mutually orthogonal eigenvectors in $\mathfrak{B}_2$, which can therefore be represented as the direct sum of two invariant one-dimensional subspaces (a case we have already dealt with). Thus we need only consider the first case

$$
\begin{aligned}
A e_1 &= a e_1 + b e_2 \\
A e_2 &= -b e_1 + a e_2
\end{aligned}
\quad \text{with} \quad a^2 + b^2 = 1.
$$

Since for given a, b there is exactly one real number ω in the interval $0 \leq \omega < 2\pi$ such that $a = \cos \omega$ and $b = \sin \omega$, the mapping A can also be described by

(4) $\qquad \begin{aligned} A e_1 &= \cos \omega e_1 + \sin \omega e_2 \\ A e_2 &= -\sin \omega e_1 + \cos \omega e_2 \end{aligned} \quad \text{with} \quad 0 \leq \omega < 2\pi.$

(For $\omega = 0$ or $\omega = \pi$, the space $\mathfrak{B}_2$ is again the direct sum of one-dimensional invariant subspaces.) Here A is an orientation-preserving mapping and the geometric interpretation of ω is given by the equations, valid for every unit vector,

$$\mathfrak{x} \cdot A\mathfrak{x} = \cos \omega,$$
$$[\mathfrak{x}, A\mathfrak{x}] = [e_1, e_2] \sin \omega.$$

In other words if the orientation is determined by e_1, e_2, i.e., $[e_1, e_2] = 1$, then for *every* unit vector $\mathfrak{x}$ and its image $A\mathfrak{x}$ the number ω gives the oriented angle between the ordered pair $(\mathfrak{x}, A\mathfrak{x})$.

Mappings of the type (4) are therefore called *rotations*, and the set of all rotations of $\mathfrak{B}_2$ is the *rotation group*. The rotations can be mapped one-to-one onto the interval $0 \leq \omega < 2\pi$.

$$(5) \qquad\qquad \omega \mapsto A(\omega),$$

where $A(\omega)$ is described by (4).

Under addition mod 2π, the interval $0 \leq \omega < 2\pi$ is a commutative group isomorphic, by (5), to the rotation group, as follows from the addition theorems and

$$(6) \qquad\qquad \omega_1 + \omega_2 (\text{mod } 2\pi) \mapsto A(\omega) \circ A(\omega_2).$$

In particular, the rotation group is therefore commutative.

If the above groups are now considered as topological groups in the obvious way (cf. II16), the isomorphism (5) is *continuous* and is in fact the only continuous isomorphism between angles and their images under rotation, since an arbitrary isomorphism is represented by a one-to-one mapping f of the interval $0 \leq \omega < 2\pi$ onto itself, satisfying the condition

$$f(\omega_1 + \omega_2(\text{mod } 2\pi)) = f(\omega_1) + f(\omega_2) \text{ mod } 2\pi.$$

Thus, if f is continuous, it must follow for "sufficiently small" ω_1 and ω_2 that

$$f(\omega_1 + \omega_2) = f(\omega_1) + f(\omega_2),$$

an equation which implies (the proof will not be given here) that $f(\omega) = c\omega$, and since $0 \leq \omega < 2\pi$ is mapped onto itself, we must have $c = 1$.

So we have answered the question whether, or under what assumptions, the above angle measure is the only possible one. If instead of $[0, 2\pi]$ we start from any interval $[0, 2\pi c]$, only the normalization is changed, i.e., the definition (5) on p. 347 must be changed to

$$\varphi(\mathfrak{x}, \mathfrak{y}) = c \cdot \text{arc } \cos(\mathfrak{x}^0 \cdot \mathfrak{y}^0).$$

Example. In the Euclidean plane are two types of isometries with fixed point (we write only the corresponding orthogonal mappings of $\mathfrak{B}_2$)

$$1) \qquad \begin{array}{l} A\mathfrak{e}_1 = a\mathfrak{e}_1 + b\mathfrak{e}_2 \\ A\mathfrak{e}_2 = b\mathfrak{e}_1 - a\mathfrak{e}_2 \end{array} \quad \text{with } a^2 + b^2 = 1.$$

Here A has two mutually orthogonal eigenvectors $\mathfrak{e}_1, \mathfrak{e}_2$ with the corresponding eigenvalues 1 and -1. Referred to the basis $\{\mathfrak{e}_1, \mathfrak{e}_2\}$, the equations for A are

$$\begin{array}{l} A\mathfrak{e}_1 = \mathfrak{e}_1, \\ A\mathfrak{e}_2 = -\mathfrak{e}_2. \end{array}$$

The corresponding isometries have exactly one pointwise fixed line through O, namely the *axis of reflection,* and all lines perpendicular to it are fixed lines. The isometry α is then called an *orthogonal reflection.*

The other type of isometry in the plane is given by

$$
\text{2)} \qquad
\begin{aligned}
A\mathfrak{e}_1 &= \cos \omega \mathfrak{e}_1 + \sin \omega \mathfrak{e}_2 \\
A\mathfrak{e}_2 &= -\sin \omega \mathfrak{e}_1 + \cos \omega \mathfrak{e}_2
\end{aligned}
\qquad 0 \leq \omega < 2\pi.
$$

For $\omega = 0$ this isometry is the identity, and for $\omega = \pi$ it is the point reflection in O. For all other ω the mapping A has no eigenvectors, i.e., α has no fixed lines. Then α is called a *rotation* about O, since all oriented angles $(\mathfrak{x}, A\mathfrak{x})$ have the same oriented angle measure ω (angle of rotation).

In three-dimensional Euclidean space there are six types of isometry with fixed point, which the reader may investigate for himself.

10.3   *Isometries without Fixed Point*

By §4.4 the orthogonal mapping A corresponding to an isometry without fixed points

$$
\mathfrak{x} \mapsto A\mathfrak{x} + \mathfrak{a}
$$

must have the eigenvalue 1. Thus (cf. §10.2, p. 375) there is a representation of $\mathfrak{V}_n$ as the direct sum

$$
\mathfrak{V}_n = \mathfrak{U}' \oplus \mathfrak{U}''
$$

of two mutually orthogonal, invariant (under A) subspaces such that in $\mathfrak{U}'$ the mapping A is the identity, and in $\mathfrak{U}''$ no vector remains fixed under A.

By a suitable choice of origin (cf. the discussion of affinities without fixed point) we can arrange that $\mathfrak{a}$ belongs to $\mathfrak{U}'$, but no further simplification is possible, since A maps $\mathfrak{U}'$ identically onto itself.

So we can give a complete description of isometries without fixed point. The space $\mathfrak{E}_n$ can be spanned by two mutually orthogonal subspaces $\mathfrak{A}$ and $\mathfrak{B}$ with generating vector spaces $\mathfrak{U}'$ and $\mathfrak{U}''$ such that the isometry may be considered as a "rotation" (taking $\mathfrak{B}$ into itself without fixed halflines) followed by a "translation" taking $\mathfrak{A}$ into itself. The isometries without fixed points are generalized *screw motions.*

As important special cases let us mention

1)   The *translations* $\mathfrak{x} \mapsto \mathfrak{x} + \mathfrak{a}$; here $\mathfrak{U}' = \mathfrak{V}_n$.

2)   The *glide reflections*; here the space $\mathfrak{B}$ is mapped onto O, i.e., A acts on $\mathfrak{U}''$ like $(-I)$.

Thus in three-dimensional space there are two types of glide reflections which, with a suitable choice of basis, are described by

$$
\mathfrak{x} \mapsto A\mathfrak{x} + \mathfrak{a} \quad \text{with} \quad
\begin{aligned}
A\mathfrak{e}_1 &= \mathfrak{e}_1 \\
A\mathfrak{e}_2 &= \mathfrak{e}_2 \\
A\mathfrak{e}_3 &= -\mathfrak{e}_3
\end{aligned}
\quad \text{and} \quad \mathfrak{a} = a\mathfrak{e}_1 \ (\dim \mathfrak{U}'' = 1)
$$

and

$$A e_1 = e_1$$
$$x \mapsto A x + a \quad \text{with} \quad A e_2 = -e_2 \quad \text{and} \quad a = a e_1 \ (\dim \mathfrak{U}'' = 2).$$
$$A e_3 = -e_3$$

II. Quadrics in Euclidean Geometry

11.1. *Symmetric Bilinear Forms in the Euclidean Vector Space*

We start from the same definition of quadrics as in §5 and make use of the results of that section. The classification problem now reads: When are two quadrics *congruent*, i.e., when does there exist an isometric mapping of $\mathfrak{C}_n$ taking the one quadric into the other? In order to answer this question we shall make use, as in affine geometry, of systems of coordinates $\{O, e_1, \ldots, e_n\}$ adapted to the given figure, where the basis $\{e_1, \ldots, e_n\}$ of $\mathfrak{B}_n$ is orthonormalized as far as possible.

Thus we must determine whether in the set of all conjugate bases (i.e., bases of vectors mutually conjugate with respect to the bilinear form f) there exist any orthogonal bases. This is in fact the case, as is most simply proved by the following argument from analysis. For all nonzero vectors $x \in \mathfrak{B}_n$ the Rayleigh quotient

$$F(x) = \frac{f(x, x)}{x \cdot x}$$

defines a real-valued continuous function F. Since the values of this function are obviously independent of the lengths of its argument vectors, we may consider F on the unit sphere $|x| = 1$. On this compact set the continuous function F must attain a maximum, say for $x = e_1$, so that for all $\mathfrak{y} \neq o$

$$F(\mathfrak{y}) \leqq F(e_1).$$

From this inequality we shall deduce that the set of vectors conjugate to e_1 coincides with the set of vectors orthogonal to e_1. Here we must, of course, assume that $f(e_1, e_1) \neq 0$, as may always be done since we could also start from a minimum of F, and the maximum and minimum could coincide only for the zero form, which is here excluded.

Suppose first that $f(e_1, x) = 0$ and $x \neq o$. Setting $\mathfrak{y} = e_1 + tx$ gives for all t with $e_1 + tx \neq o$

$$\frac{f(e_1 + tx, e_1 + tx)}{(e_1 + tx) \cdot (e_1 + tx)} = \frac{f(e_1, e_1) + t^2 f(x, x)}{1 + 2t(e_1 \cdot x) + t^2(x \cdot x)} \leqq f(e_1, e_1)$$

or equivalently,

$$0 \leqq 2t(e_1 \cdot x) f(e_1, e_1) + t^2((x \cdot x) f(e_1, e_1) - f(x, x))$$

which means that $e_1 \cdot x = 0$ (for if we had both $e_1 \cdot x \neq 0$ and $f(e_1, e_1) \neq 0$, the right side of this inequality would be negative for t sufficiently small and of suitable sign). In fact, since $(n - 1)$-dimensional subspaces are described by $f(e_1, x) = 0$ and $e_1 \cdot x = 0$, the two conditions $e_1 \cdot x = 0$ and $f(e_1, e_1) \neq 0$, are equivalent to each other.

If we now begin the construction of a conjugate basis with a (unit) vector e_1 of this sort and successively choose its other vectors in the same way, we will obtain a basis of $\mathfrak{B}_n$ which is both conjugate and orthonormal.

For every symmetric bilinear form f over a Euclidean vector space V_n there is an orthonormal basis $e_1, \ldots, e_n$ such that

$$f(e_i, e_k) = 0 \qquad \text{for} \quad i \neq k,$$

$$f(e_i, e_i) = \frac{1}{a_i^2} \qquad \text{for} \quad i = 1, \ldots, p,$$

$$f(e_i, e_i) = -\frac{1}{a_i^2} \qquad \text{for} \quad i = p + 1, \ldots, r,$$

$$f(e_i, e_i) = 0 \qquad \text{for} \quad i = r + 1, \ldots, n$$

with positive and negative scalars written in the usual way. With respect to this orthonormal basis the quadratic form corresponding to f will be

$$f(x, x) = \sum_{i=1}^{p} \frac{x_i^2}{a_i^2} - \sum_{i=p+1}^{r} \frac{x_i^2}{a_i^2}$$

where our construction makes it clear that not only the natural numbers p and r but also the system of positive scalars $a_1, \ldots, a_r$ are characteristic features of the bilinear form f.

11.2. *Quadrics with Center*

The preceding discussion shows that for a central quadric $\mathfrak{Q}$ that is not a hypercone, there is always an orthonormal system $\{0, e_1, \ldots, e_n\}$ in which the equation of $\mathfrak{Q}$ becomes

$$f(x, x) = \sum_{i=1}^{p} \frac{x_i^2}{a_i^2} - \sum_{i=p+1}^{r} \frac{x_i^2}{a_i^2} = 1.$$

The lines through O in the directions $e_1, \ldots, e_n$, are called the *principal axes* of the quadric, and the scalars a_i, with $i = 1, \ldots, p$ (and for that matter, with $i = p + 1, \ldots, r$, as well) are the (lengths of the) *principal semiaxes*.

For the hypercone the scalars a_i have no immediate geometric significance, since the equation $f(x, x) = 0$ is determined only up to a scalar factor. If say $a_r \neq 0$, only the quotients $a_1/a_r, \ldots, a_{r-1}/a_r$ are characteristic of the hypercone. The reader should consider their geometric significance.

Examples. In the Euclidean plane there are the following four classes of proper conic sections with center:

1) $\dfrac{x_1^2}{a_1^2} + \dfrac{x_2^2}{a_2^2} = 1$ ellipse (circle for $a_1 = a_2$),

2) $\dfrac{x_1^2}{a_1^2} - \dfrac{x_2^2}{a_2^2} = 1$ hyperbola,

3) $\dfrac{x_1^2}{a_1^2} = 1$ pair of parallel lines,

4) $\dfrac{x_1^2}{a_1^2} - \dfrac{x_2^2}{a_2^2} = 0$ pair of intersecting lines.

The focal properties of a conic section are conveniently studied by choosing one such property as the original definition and seeing how far it agrees with the algebraic definition of conic sections given here.

Similarly, in Euclidean three-dimensional space there are eight types of proper central quadrics:

1) $\dfrac{x_1^2}{a_1^2} + \dfrac{x_2^2}{a_2^2} + \dfrac{x_3^2}{a_3^2} = 1$ ellipsoid,

2) $\dfrac{x_1^2}{a_1^2} + \dfrac{x_2^2}{a_2^2} - \dfrac{x_3^2}{a_3^2} = 1$ one-sheeted hyperboloid,

3) $\dfrac{x_1^2}{a_1^2} - \dfrac{x_2^2}{a_2^2} - \dfrac{x_3^2}{a_3^2} = 1$ two-sheeted hyperboloid,

4) $\dfrac{x_1^2}{a_1^2} + \dfrac{x_2^2}{a_2^2} = 1$ elliptic cylinder,

5) $\dfrac{x_1^2}{a_1^2} - \dfrac{x_2^2}{a_2^2} = 1$ hyperbolic cylinder,

6) $\dfrac{x_1^2}{a_1^2} = 1$ pair of parallel planes,

7) $\dfrac{x_1^2}{a_1^2} + \dfrac{x_2^2}{a_2^2} - \dfrac{x_3^2}{a_3^2} = 0$ cone,

8) $\dfrac{x_1^2}{a_1^2} - \dfrac{x_2^2}{a_2^2} = 0$ pair of intersecting planes.

11.3. *Quadrics without Center*

In our study of affine geometry we have seen that for a noncentral quadric $\mathfrak{Q}$ there is at least one (unit) vector (call it e_n) in the null space $\mathfrak{U}_f$ but not in

$\mathfrak{U} = \{x \mid l(x) = 0\}$, and that in constructing the most convenient coordinate system we may choose the origin arbitrarily on $\mathfrak{Q}$.

But now, in order to construct the most convenient orthonormal coordinate system, we attempt to choose the origin in such a way that the tangential hyperplane at that point is orthogonal to the vector $\mathfrak{e}_n$. If the equation of $\mathfrak{Q}$ is $f(x, x) + 2l(x) = 0$, the tangential hyperplane at an arbitrary point O' with $\mathfrak{p} = \overrightarrow{OO'}$ has the equation

$$f(\mathfrak{p}, x) + l(x) + l(\mathfrak{p}) = 0.$$

So we must try to determine $\mathfrak{p}$ in such a way that the linear forms $x \mapsto f(\mathfrak{p}, x) + l(x)$ and $x \mapsto (\mathfrak{e}_n \cdot x)$ have the same solutions, i.e.,

$$a(\mathfrak{e}_n \cdot x) = f(\mathfrak{p}, x) + l(x)$$

for some scalar a and all $x \in \mathfrak{B}_n$.

To prove the existence of such a vector p, we start from the fact that, by p. 331, there is a basis $\{\mathfrak{e}_1, \ldots, \mathfrak{e}_n\}$, conjugate with respect to f, with $l(\mathfrak{e}_1) = \cdots = l(\mathfrak{e}_{n-1}) = 0$ and $l(\mathfrak{e}_n) \neq 0$. For $\mathfrak{p} = p_1 \mathfrak{e}_1 + \cdots + p_n \mathfrak{e}_n$ we then have the equations

$$p_j f(\mathfrak{e}_j, \mathfrak{e}_j) = a(\mathfrak{e}_n \cdot \mathfrak{e}_j) - l(\mathfrak{e}_j) \quad (j = 1, \ldots, n).$$

Since $\mathfrak{e}_n \in \mathfrak{U}_f$, the right side vanishes and therefore $a = l(\mathfrak{e}_n)$. If f is of the maximal rank $r = n - 1$ (paraboloid), the $p_1, \ldots, p_{n-1}$ are uniquely determined, and finally, the scalar p_n is determined by the condition $f(\mathfrak{p}, \mathfrak{p}) + 2l(\mathfrak{p}) = 0$, since $l(\mathfrak{e}_n) \neq 0$.

Thus for $r = n - 1$ there is exactly one point O' at which the tangential hyperplane is orthogonal to $\mathfrak{e}_n$. This point O' is called the *vertex* of the paraboloid.

If Q is a parabolic cylinder, so that $r < n - 1$, then $\mathfrak{U}_f \cap \mathfrak{U} = \mathfrak{Z}$ is a nonzero subspace of dimension $d - 1 = n - r - 1$, with the following geometrical significance. Since $f(x, x) + 2l(x) = 0$ is satisfied by all $\mathfrak{z} \in \mathfrak{Z}$ and also by all $x = \mathfrak{p} + \mathfrak{z}$ with $\mathfrak{z} \in \mathfrak{Z}$ and $f(\mathfrak{p}, \mathfrak{p}) + 2l(\mathfrak{p}) = 0$, through every point of $\mathfrak{Q}$ there is a subspace contained in $\mathfrak{Q}$ and generated by $\mathfrak{Z}$.

In affine geometry the vector $\mathfrak{e}_n \in \mathfrak{U}_f$ had to satisfy only the condition $l(\mathfrak{e}_n) \neq 0$, but now we also choose $\mathfrak{e}_n$ orthogonal to $\mathfrak{Z}$. Then the intersection of $\mathfrak{Q}$ with a subspace $\mathfrak{B}$ (of dimension $r + 1$) orthogonal to $\mathfrak{Z}$, is a noncentral quadric $\mathfrak{Q}'$, in $\mathfrak{B}$ of maximal rank. The corresponding null space is one-dimensional, being generated by $\mathfrak{e}_n$. The paraboloid $\mathfrak{Q}'$ in $\mathfrak{B}$ has exactly one vertex O', and the hyperplane tangent at O' to the parabolic cylinder is orthogonal to $\mathfrak{e}_n$. In this case also we call O' a *vertex*. All points of the subspace through O' spanned by $\mathfrak{Z}$ are vertices of $\mathfrak{Q}$.

If a vertex is taken for origin, then $\mathfrak{e}_n$ is orthogonal to $\mathfrak{U}$, and there exists in

$\mathfrak{U}$ a basis $e_1, \ldots, e_{n-1}$ which (with respect to f) is both conjugate and orthonormal, so that, including the unit vector e_n, we obtain an orthonormal basis for $\mathfrak{B}_n$. Referred to this basis, the equation of $\mathfrak{Q}$ becomes

$$2x_n = \sum_{i=1}^{p} \frac{x_i^2}{a_i^2} - \sum_{i=p+1}^{r} \frac{x_i^2}{a_i^2} \qquad (\text{with } r \leqq n - 1).$$

Examples. In the Euclidean plane there is only one class of non-central conics, namely *parabolas* with the equation

$$2x_2 = \frac{x_1^2}{a_1^2},$$

but in three-dimensional Euclidean (or affine) space there are three types:

1) $2x_3 = \dfrac{x_1^2}{a_1^2} + \dfrac{x_2^2}{a_2^2}$ elliptic paraboloid,

2) $2x_3 = \dfrac{x_1^2}{a_1^2} - \dfrac{x_2^2}{a_2^2}$ hyperbolic paraboloid,

3) $2x_3 = \dfrac{x_1^2}{a_1^2}$ parabolic cylinder.

This classification enables us to answer the question on p. 379 about the congruence of two quadrics as follows:

Two quadrics are congruent if and only if they belong to the same type and have the same scalars.

Bibliography

[1] BIRKHOFF, G.–MACLANE, S.: A survey of modern algebra, 3rd ed., New York, London 1966.
[2] BOURBAKI, N.: Elements of mathematics, 2 Vols., Reading, Mass., 1966.
[3] DIEUDONNÉ, J.: Algèbre linéaire et géométrie élémentaire, Paris 1964.
[4] FLADT, K.: Elementarmathematik vom höheren Standpunkt aus, Part III, Elementargeometrie II, Stuttgart 1959.
[5] GREUB, W. H.: Linear algebra, 3rd ed., Berlin 1967.
[6] HALMOS, P. R.: Finite dimensional vector spaces, 2nd ed., Princeton 1958.
[7] HESSENBERG, G.: Ebene und sphärische Trigonometrie, Sammlung Goschen Vol. 99, 5th ed., revised by H. Kneser, Berlin 1957.
[8] KOWALSKY, H. J.: Lineare Algebra, Berlin 1963.
[9] KUIPER, N. H.: Linear algebra and geometry, 2nd ed., Amsterdam 1965.
[10] LAUGWITZ, D.: Die Geometrien von H. Minkowski, Der Mathematik unterricht, Vol. 4 (1958) p. 27–42.
[11] LENZ, H.: Grundlagen der Elementarmathematik, 2. Aufl., Berlin 1967.
[12] PAULI, L.–Post, M.: Trigonométrie, Lausanne 1946.

[13] PICKERT, G.: Analytische Geometrie, 6th ed., Leipzig 1967.

[14] —: Axiomatische Begründung der ebenen euklidischen Geometrie in vektorieller Darstellung, Math. Phys. Sem. Ber. X (1963) S. 65–85.

[15] SPERNER, E.: Einführung in die Analytische Geometrie und Algebra, Part I, 5th ed., Göttingen 1961; Part II, 4th ed., Göttingen 1961.

[16] TIETZ, H.: Lineare Geometrie, Münster 1967.

[17] WEYL, H.: Space, time, matter. Translated from the German by Henry L. Brose, Methuen & Co., London, 1922.

[18] —: Mathematische Analyse des Raumproblems, Berlin 1923 (Reprint Darmstadt 1963).

[19] WILKER, P.: Algebraische Lösung des Helmholtzschen Raumproblems für reelle Vektorräume, Math. Ann. 139 (1959/60), p. 433–445.

From Projective to Euclidean Geometry

1. Coordinates in Projective Space

1.1. *Coordinate Vectors*[1]

The most easily visualized way of extending the affine plane (II9, §2) to the projective is as follows. We imagine a plane **E** in three-dimensional affine space and a point O (Fig. 1) not on **E**. Then to the point X on **E** we assign all the points $\neq 0$ on the line OX. In this way the lines through O not parallel to **E** represent the totality of points of **E**, while the lines g through O parallel to **E** remain for the improper points (often called "points at infinity"), i.e., each line g is assigned to the improper point common to the parallels to g in E. Then planes through O correspond to lines in the projective plane to which **E** has thus been extended and, in particular, the plane through O parallel to **E** corresponds to the improper line.

Since points of the (three-dimensional) space can also be determined by

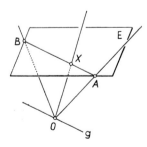

Fig. 1

[1]Except when otherwise stated, the coordinate domain (*domain of scalars*) will be taken to be the real field.

385

their position vectors with respect to O, we have thereby set up a (many-to-one) correspondence between the vectors $\neq 0$ and the points of the projective plane. Here the vectors are called the *coordinate vectors* (CV) of the points, and two vectors $\mathfrak{x}, \mathfrak{x}'$ ($\neq 0$) are the CV of the same point if and only if they are linearly dependent, i.e., if there exists[2] a scalar c with $\mathfrak{x}' = \mathfrak{x}c$. If $\mathfrak{a}, \mathfrak{b}$ are the CV of distinct points A, B, then to the points X of the line[3] joining A, B there correspond as CV all the

(1) $$\mathfrak{x} = \mathfrak{a}s + \mathfrak{b}t \quad (s = t = 0 \text{ excluded});$$

for these are precisely the position vectors of the points $\neq 0$ in the plane OAB. It follows that *three points lie on a line if and only if their* CV *are linearly dependent.*

In order to carry out the same procedure for three-dimensional space, we must imagine this space as a hyperplane in affine four-dimensional space.[4] Of course, such a space cannot be visualized, but the correspondence between points and CV can nevertheless be defined exactly as before. If $\mathfrak{a}, \mathfrak{b}, \mathfrak{c}$ are the CV of three noncollinear points A, B, C, the $\mathfrak{a}r + \mathfrak{b}s + \mathfrak{c}t$ ($r = s = t = 0$ excluded) are the CV of all the points of the plane ABC (in projective space).

From the discussion at the beginning of the present section, it is clear that the affine plane can be obtained from the projective plane by omitting an arbitrarily chosen line (and also its points), where we may choose any position for the improper line that happens to be especially convenient for a proof.

The projective spaces of arbitrary dimension can be treated in exactly the same way. For simplicity, we deal here only with three-dimensional space, but many of our results can be extended to arbitrary projective spaces. In the notation of II9, §1 the one-dimensional subspaces of a $\mathfrak{B}_{n+1}$ may be taken as the points of the n-dimensional projective space, and the m-dimensional linear projective subspaces ($0 \leq m \leq n$) then correspond precisely to the $(m + 1)$-dimensional subspaces of $\mathfrak{B}_{n+1}$, where the sum (see II9, §1) corresponds to the union of the spaces, the intersection corresponds to their intersection, and the zero-dimensional subspace of $\mathfrak{B}_{n+1}$ (consisting only of the zero vector) corresponds to the empty (projective) subspace, to which we therefore assign the dimension -1.

---

[2]Cf. p. 295, II9, §1.2

[3]In general the union of the (linear projective) subspaces (i.e., points, lines, planes) II, II′, i.e., the smallest subspace containing both of them, will be denoted by IIII′; the intersection of II, II′, i.e., the largest subspace contained in both of them, will be denoted by II ∩ II′, and $P \in$ II will mean that P lies in II.

[4]Of course, the improper points and lines, as well as the improper plane, can be introduced without recourse to a four-dimensional space (cf. II1, No. 24, or II3, §3.1.) But here again we wish to have a representation of the points, (including the improper points) by means of vectors.

Since all the CV $\mathfrak{x}$ of the points of a plane, taken together with the zero-vector, form a three-dimensional vector space, they can be described[5] in terms of a linear form ($\neq 0$) by an equation

$$(2) \qquad\qquad \langle \mathfrak{u}, \mathfrak{r} \rangle = 0,$$

where $\mathfrak{u}$ is determined only up to a scalar factor, so that two equations (2) represent the same plane if and only if their linear forms are linearly dependent.

Instead of defining a line g by (1), which is a so-called parametric representation, we can also define it as the intersection of two planes, by means of a pair of equations $\langle \mathfrak{u}, \mathfrak{r} \rangle = 0$, $\langle \mathfrak{v}, \mathfrak{r} \rangle = 0$ with linearly independent linear forms. Then it is clear that

$$(3) \qquad u\langle \mathfrak{u}, \mathfrak{r} \rangle + v\langle \mathfrak{v}, \mathfrak{r} \rangle = 0 \quad (u = v = 0 \text{ excluded})$$

likewise defines a plane through g. But in this way we obtain every plane through g, since such a plane is uniquely defined by g and a point not on g; i.e., replacement of $\mathfrak{x}$ in (3) by the CV of such a point determines the u, v up to a common factor. Thus (3) describes the *pencil of planes with the carrier g.* In plane geometry the pencil of lines through a point is obtained in the same way.

1.2. Cross Ratios

If (in the notation of Fig. 1) we take $\mathfrak{a}_0 = \overrightarrow{OA}$, $\mathfrak{b}_0 = \overrightarrow{OB}$, $\mathfrak{r}_0 = \overrightarrow{OX}$ as the CV of A, B, X then for $X \neq B$ there exists a scalar u (since $\overrightarrow{AX} = \mathfrak{x}_0 - \mathfrak{a}_0$ and $\overrightarrow{BX} = \mathfrak{x}_0 - \mathfrak{b}_0$) with $(\mathfrak{x}_0 - \mathfrak{b}_0)u = \mathfrak{x}_0 - \mathfrak{a}_0$, so that

$$(4) \qquad\qquad \mathfrak{r}_0 = (\mathfrak{a}_0 - \mathfrak{b}_0 u)(1 - u)^{-1}.$$

In this case we say that X *divides* A, B *in the ratio* $u = (ABX)$.[6] In (4) we may obviously replace $\mathfrak{a}_0$, $\mathfrak{b}_0$, $\mathfrak{x}_0$ by any CV of A, B, X, which means that this ratio is not a concept of projective geometry. But the situation is different if we consider a further point $X' \neq A$, B with $u' = (ABX')$, and form the quotient u/u'. If $\mathfrak{a}$, $\mathfrak{b}$, $\mathfrak{x}$, $\mathfrak{x}'$ are now any CV of A, B, X, X' with

$$(5) \qquad\qquad \mathfrak{r} = \mathfrak{a}s + \mathfrak{b}t, \quad \mathfrak{r}' = \mathfrak{a}s' + \mathfrak{b}t'$$

and $\mathfrak{a} = \mathfrak{a}_0 a$, $\mathfrak{b} = \mathfrak{b}_0 b$, then $\mathfrak{x} = \mathfrak{a}_0 a s + \mathfrak{b}_0 b t$ is a CV of X but is not necessarily $= \overrightarrow{OX}$. To obtain $\overrightarrow{OX}$ we must multiply $\mathfrak{x}$ by $(as + bt)^{-1}$ (since the sum of the coefficients of $\mathfrak{a}_0$ and $\mathfrak{b}_0$ is now equal to 1, the representation (4) is

---

[5]Cf. II9, §5.1 and IB3, §2.6.
[6]Note that by this definition $(ABM) = -1$ for the midpoint M of A and B; often $-u$ is taken instead of u.

possible). Thus

$$\mathfrak{r}_0 = (\mathfrak{a}_0 as + \mathfrak{b}_0 bt)(as + bt)^{-1},$$

and comparison with the corresponding equation (4) shows that $u = -bt/as$, $u' = -bt'/as'$ and therefore that $u/u' = s't/st'$.[7] So this quotient can already be calculated from the equations (5). In general, assuming (5) and $X \neq B$, $X' \neq A$, the quotient $s't/st'$, which is shown by the above calculation to depend only on the points A, B, X, X', is called the *cross ratio* $(ABXX')$.[8] It is easy to calculate (for four distinct points) that

(6) $(BACD) = (ABDC) = (ABCD)^{-1}$, $(DBCA) = 1 - (ABCD)$,

and then to determine the effect of an arbitrary permutation of A, B, C, D on the value of the CR; in particular, $(CDAB) = 1 - (BDAC) = 1 - (DBCA) = (ABCD)$.

If in the above notation we take X' as the improper point U of AB, then (cf. Fig. 1) $\mathfrak{x}'$ must be a multiple $\overrightarrow{AB} = \mathfrak{b}_0 - \mathfrak{a}_0$, which means that in (5), with $\mathfrak{a}_0 = \mathfrak{a}$, $\mathfrak{b}_0 = \mathfrak{b}$, we have $s' = -t'$ and therefore $(ABXU) = (ABX)$. In particular, $(ABXU) = -1$ if and only if X is the midpoint of A and B. In general, if $(ABCD) = -1$, the points (C, D) are called the *harmonic conjugates* of (A, B), and since $(ABCD) = (CDAB)$, the points (A, B) are also the harmonic conjugates of (C, D); here the point D is called the *fourth harmonic point* for A, B, C. The affine theorem—that in a parallelogram the intersection of the diagonals is the midpoint of each diagonal—then becomes the *theorem of the complete quadrangle* (Fig. 2). Each pair of diagonal points E, F is harmonic to the pair G, H of points of intersection of EF with the sides not through E or F. The *vertices* of the complete quadrangle are A, B, C, D and the *sides* are the six lines joining these four points in pairs. This theorem is obtained from the affine theorem on

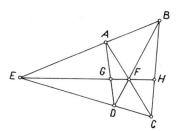

Fig. 2

---

[7]Here for the first time we require the commutative law for multiplication. Up to now the domain of scalars could have been any skew field, but from now on it must be a field, for example, the field of rational, real, or complex numbers.

[8]Abbreviated CR in the discussion below.

parallelograms by taking BC as the improper line and A, F, D, E as the vertices of the parallelogram.

If four planes **A**, **B**, **Γ**, **Γ′** through the line g with the equations (3) $\langle \mathfrak{u}, \mathfrak{r} \rangle = 0$, $\langle \mathfrak{v}, \mathfrak{r} \rangle = 0$ and

(7) $$u'\langle \mathfrak{u}, \mathfrak{r} \rangle + v'\langle \mathfrak{v}, \mathfrak{r} \rangle = 0$$

are cut by a line (skew to g) at the points A, B, C, C', with $\mathfrak{a}$, $\mathfrak{b}$, $\mathfrak{c}$, $\mathfrak{c}'$ for their CV, then by (1)

$$\mathfrak{c} = \mathfrak{a}s + \mathfrak{b}t, \quad \mathfrak{c}' = \mathfrak{a}s' + \mathfrak{b}t'$$

and by substituting these values in (3) and (7) we can express the CR $(ABCC')$ by means of the u, v, u', v' alone. So this value

$$(ABCC') = u'v/uv'$$

is called the cross ratio (**ABΓΓ′**) of the four planes. *Four planes through a line g have the same CR as the points in which they are cut by a line skew to g.*

In the same way we can define the CR of four coplanar lines through a point, and then four planes through a line g will have the same CR as the four lines in which they are cut by a plane not containing g.

The concept of "harmonic conjugates" is defined for planes and lines in exactly the same way as for points.

1.3.   *Homogeneous Coordinates*

If in the four-dimensional space of the CV we introduce a basis (II9, §1.3) $\{\mathfrak{e}_0, \mathfrak{e}_1, \mathfrak{e}_2, \mathfrak{e}_3\}$, we thereby assign to each CV $\mathfrak{x} = \sum_{i=0}^{3} \mathfrak{e}_i x_i$ a quadruple (x_0, x_1, x_2, x_3) of scalars, not all zero, which are called the homogeneous coordinates (hC) of the point X corresponding to the CV $\mathfrak{x}$. Obviously these coordinates are determined by X only up to a common nonzero factor. With $\langle \mathfrak{u}, \mathfrak{e}_i \rangle = u_i$ as the coefficients of u with respects to the basis $\{\mathfrak{e}_0, \mathfrak{e}_1, \mathfrak{e}_2, \mathfrak{e}_3\}$, the equation (2) for a plane takes the form

$$\sum_{i=0}^{3} u_i x_i = 0.$$

The u_i, which like the hC of a point are determined only up to a common factor ($\neq 0$), are called the *plane coordinates* of the plane.

The points E_i ($i = 0, 1, 2, 3$) with the CV $\mathfrak{e}_i$ are called the *reference points* for the coordinate system, and the point E with the CV $\mathfrak{e} = \sum_{i=0}^{3} \mathfrak{e}_i$ is its *unit point*. Since the $\mathfrak{e}_i$ are linearly independent, the reference points are not coplanar and can therefore be considered as the vertices of a tetrahedron. Since $\mathfrak{e}$ is not linearly dependent on any three of the $\mathfrak{e}_i$, it follows that E does not lie on any side of the tetrahedron. These are the only conditions that must be satisfied by the reference points and the unit point. Obviously the

CV of four noncoplanar points form a basis for the space of the CV, so that these points can be taken as the reference points; and then the CV $\mathfrak{e}$ of a point E not on any side of the reference tetrahedron is a linear combination (with nonvanishing coefficients) of the basis vectors, so that when these vectors are multiplied by suitable nonzero scalars the linear form $\mathfrak{e}$ becomes the sum of the new basis vectors and E becomes the new unit point.

If E_0 is the only proper point among the reference points and if, in the notation of §1.1, we take $\overrightarrow{OE_0} = \mathfrak{e}_0$, then since the $\mathfrak{e}_1, \mathfrak{e}_2, \mathfrak{e}_3$ form a basis for the affine (three-dimensional) space, it follows that for each CV $\mathfrak{x}$ of a proper point X there must exist scalars c, x_i^* $(i = 1, 2, 3)$ with

$$\mathfrak{x}c - \mathfrak{e}_0 = \sum_{i=1}^{3} \mathfrak{e}_i x_i^*,$$

so that $x_0 c = 1$, $x_i c = x_i^*$ $(i = 1, 2, 3)$, and therefore

(8) $x_i^* = x_i x_0^{-1}$ $(i = 1, 2, 3)$.

Here the x_i^* are the affine (or nonhomogeneous) coordinates of X in the affine coordinate system with origin E_0 and basis vectors $\{\mathfrak{e}_1, \mathfrak{e}_2, \mathfrak{e}_3\}$ (cf. II9, §§2.5 and 4.2), since in fact $\overrightarrow{E_0 X} = \mathfrak{x}c - \mathfrak{e}_0$. The equation of the improper plane is obviously $x_0 = 0$.

But if all the reference points are proper and if we set $\overrightarrow{OE_i} = \mathfrak{e}_i$ $(i = 0, 1, 2, 3)$, then the $\mathfrak{e}_i - \mathfrak{e}_0$ $(i = 1, 2, 3)$ form a basis for the affine space, so that for a proper point X there exist scalars c, c_1, c_2, c_3 with

$$\overrightarrow{E_0 X} = \mathfrak{x}c - \mathfrak{e}_0 = \sum_{i=1}^{3} (\mathfrak{e}_i - \mathfrak{e}_0)c_i.$$

But this means that $1 - x_0 c = \sum_{i=1}^{3} c_i$, $x_i c = c_i$, so that

$$\overrightarrow{E_0 X} = (\sum_{i=0}^{3} \overrightarrow{E_0 E_i} x_i)(\sum_{i=0}^{3} x_i)^{-1}.$$

Thus the improper plane has the equation $\sum_{i=0}^{3} x_i = 0$, and the proper point with the hC x_i $(i = 0, 1, 2, 3)$ is the center of gravity[9] of masses x_i situated at the base points E_i. So in this case we also speak of *barycentric coordinates* (cf. II9, §6.3). The fact that the unit point is now simply the geometric center of gravity of the reference points must not cause us to forget that in general, as was pointed out above, the unit point is not determined by the reference points. With barycentric coordinates the determination of the unit point lies in our setting $\overrightarrow{OE_i} = \mathfrak{e}_i$.

With the help of the CR we can now show that the reference points E_i and

---

[9]Since the x_i can also be negative, it would perhaps be better to speak of point charges in a homogeneous electric field.

the unit point E uniquely determine a coordinate system. For let X be a point with CV $\mathfrak{x}$ and hC x_i ($i = 0, 1, 2, 3$), so that we wish to prove that the x_i are determined, up to a common factor, by the points X, E, E_i ($i = 0, 1, 2, 3$). For this purpose we make use of the identity $\mathfrak{x} = \mathfrak{e}x_0 + \sum_{i=1}^{3} \mathfrak{e}_i(x_i - x_0)$, where $\mathfrak{e} = \sum_{i=0}^{3} \mathfrak{e}_i$ is a CV of E, and we now make the assumption $x_i \neq 0$. Then

$$\mathfrak{x} - \mathfrak{e}_3(x_3 - x_0) = \mathfrak{e}x_0 + \mathfrak{e}_1(x_1 - x_0) + \mathfrak{e}_2(x_2 - x_0)$$

is a CV for the intersection S of the line XE_3 with the plane $E_1E_2E_3$. The intersection X' of XE_3 with the plane $E_0E_1E_2$ obviously has $\mathfrak{x} - \mathfrak{e}_3x_3$ as a CV, so that comparison with (5) leads to the equation

$$(9) \qquad\qquad (XE_3SX') = 1 - x_0x_3^{-1}.$$

The corresponding argument for X' leads to

$$(10) \qquad\qquad (X'E_2S'X'') = 1 - x_0x_2^{-1}$$

with $S' = X'E_2 \cap E_1E'$, $X'' = X'E_2 \cap E_0E_1$, where $E' = E_3E \cap E_0E_1E_2$ and therefore has $\sum_{i=0}^{2} \mathfrak{e}_i$ as a CV. Since $\mathfrak{e}_0x_0 + \mathfrak{e}_1x_1$ is a CV of X'' and $\mathfrak{e}_0 + \mathfrak{e}_1$ is a CV of $E'' = E'E_2 \cap E_0E_1$, we finally have

$$(11) \qquad\qquad (E_0E_1E''X'') = x_0x_1^{-1}.$$

The equations (9), (10), and (11) now show that the x_i are determined, up to a common factor, by the points X, E, E_i ($i = 0, 1, 2, 3$). For $x_3 = 0$, x_0, $x_1 \neq 0$ we begin the proof with $X' = X$, and for $x_3 = x_2 = 0$, $x_0, x_1 \neq 0$ with $X'' = X$, while for $x_3 = x_2 = x_1 = 0$ there is nothing to prove, in view of the fact that $X = E_0$. Since re-indexing allows us to reduce all possible cases to these three, the proof is complete.

1.4. *Real and Complex Geometry*

In order to remain within the range of our powers of visualization, it is necessary (as is in fact implied by the axioms of Euclidean geometry) to take the real numbers as scalars, a choice which then involves distinguishing various special cases for nonlinear configurations; for example, although two lines in the projective plane always have at least one point in common, this is no longer the case for a line and a conic section. Thus it is natural to introduce the complex numbers as scalars. We then speak of *complex geometry, complex space,* the *complex plane,* and the *complex line*, and use the adjective *real* when we wish to confine our attention to real scalars.

Real projective space can always be imagined as imbedded in complex projective space in the following way. Let us choose a coordinate system in the complex space and also in the real space, and then with each point of the real space let us identify that point of the complex space which has the same homogeneous coordinates.[10] Of course, this imbedding depends on the choice of the two coordinate systems, but in what follows we shall consider that it has been made once for all. Every coordinate system of the real space is then a coordinate system for the complex space as well, and it is precisely these coordinate systems of the complex space that we shall call *real coordinate systems*.

A point in the complex space is called *real* if it is in the (imbedded) real space. Obviously a point is real if and only if it has real coordinates in some (and therefore in every) real coordinate system. For lines and planes the corresponding definition is somewhat more involved; a *real line* (*plane*) is one which contains a line (plane) of the real space. Every line (plane) of the real space lies in exactly one line (plane) of the complex space, since it can be represented as the line (plane) incident to two (three) points, and in the complex plane, as well as in the real, these points determine a line (plane). So we see at once that a line (plane) is real if and only if it can be represented as a line (plane) joining real points.

Let us note that every real line (plane) of the complex space must also contain nonreal points; for example, in a real coordinate system, the line E_0E_1 contains the point[11] with coordinates $(1, i, 0, 0)$, and this point is nonreal, since there is no complex number s for which $1s$ and is are both real. In order to establish the existence of nonreal lines, it is only necessary to choose a real point P and then a nonreal point Q on a real line g not through P; then PQ is not real, since it has only one point (the nonreal point Q) in common with the line g; in fact, apart from P it contains no real point, since otherwise it would necessarily be real, since it would join real points. It is easy to prove that in a real plane (of the complex space) every nonreal line contains exactly one real point. A line without real points, which therefore does not lie in any real plane, can be obtained by joining the points with coordinates $(1, i, 0, 0), (0, 0, 1, i)$ in a real coordinate system; for there are no complex numbers s, t, not both zero, such that all four numbers s, is, t, it are real.

Since the coefficients of the equation of a plane can be represented as a subdeterminant of a matrix formed from the hC of three noncollinear points on the plane, the coefficients of the equation of a real plane in a real coordinate system can always be taken to be real. Conversely, a linear equation with real coefficients in the real coordinate system represents a real plane, since it already represents a plane if the scalars are restricted to be real.

The distinct real planes obviously have a real line as their intersection. Conversely, through a given real line it is always possible to draw two

---

[10]Of course, it also has other, nonreal hC, since in the complex space we may multiply the hC by an arbitrary nonzero complex factor.

[11]Here, as usual, i denotes the "imaginary unit" for which $i^2 = -1$.

distinct real planes with the given line as their intersection; for we only need join the line to two real points not coplanar with the line.

For many geometric theorems it is by no means necessary to make use of all the properties of the real or the complex numbers; often it is enough to know only that the scalars form a field (in the sense of algebra see IB5, §1.10). Thus we can develop a geometry in which the scalars are the elements of an arbitrary field (cf. II3, §3.3 and II9). But if we wish to consider metric geometry, this field must satisfy certain conditions (see II4, II9, §7, and Pickert [4], p. 167). On the other hand, many parts of projective geometry remain valid if we dispense with the commutative law for the multiplication of scalars, i.e., if a skew field (IB3, §1.1) is taken as the domain of scalars (see, e.g., II3, §2.1, and Pickert [4]).

2. Collineations

2.1. *Central Collineations*

If a translation (II9, §4.4) in the affine plane takes the point P into P' ($\neq P$), it is a well-known fact that the image X' of an arbitrary point X not on PP' is the intersection of the parallel to PP' through X with the parallel to PX through P'. If we now extend the affine plane to the projective plane by adjoining the improper line a and its points, then the lines PP', XX', and also the lines PX, $P'X'$, pass through a point of a. Thus, with $Z = PP' \cap a$ as the improper point of PP', we obtain

$$(12) \qquad X' = ZX \cap (PX \cap a)P'.$$

Since a translation takes every pencil of parallel lines into itself, it is natural to extend the concept to the projective plane by prescribing that every improper point remain fixed, and this prescription is clearly unavoidable if we wish to have a *collineation* (i.e., a one-to-one collinearity-preserving mapping of all points in the projective plane) which is a translation of the proper points.

Now the collineation that we have just constructed from a translation obviously has the following special properties: every point on a remains fixed and also every line through Z.[12] We then speak of a *central collineation*[13] with *center* Z and *axis* a. If such a transformation takes P into P', then (12) must always hold if $Z \neq P$, P' and P, $P' \notin a$, and also $X \notin ZP$. If for the points Y on ZP we write (12) with X, X' instead of P, P' and Y, Y' instead of X, X', we see that a central collineation is completely determined[14] by its center Z, its axis a and its effect on a point P with

---

[12]That is, such a line, regarded as a set of points, is mapped into itself.

[13]Often also called a *perspectivity*, but this name is used here for another concept. In II3, §3.2 the central collineations are also called *perspective collineations*.

[14]Of course, for the image P' of P we must have $P' \in ZP$, $P' \in a$, $P' \neq Z$.

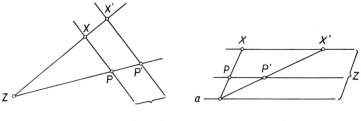

Fig. 3   Fig. 4

$Z \neq P \notin a$. If the center lies on the axis and latter is regarded as the improper line of an affine plane, the central collineations are precisely the translations from which we began.

For a dilation (II9, §4.3) in the affine plane it is clear (see Fig. 3) that (12) again holds except that now $Z = PP' \cap XX'$ is a proper point. The extension of a dilation to a projective plane is obtained in the same was as for translation; namely, if the center does not lie on the axis, and if the axis is regarded as the improper line, then the dilations are given by the central collineations.

From the known existence of translations and dilations it now follows that for distinct collinear points Z, P, P' and a line a not through P or P', there always exists a central collineation, and by the above argument exactly one, with center Z and axis a taking P into P'.

To obtain a mapping of the affine plane from a central collineation it is not necessary to take the improper line as the axis, as we have done up to now. We must only take care that the improper line is transformed into itself and thereby[15] every proper point into a proper point. If the improper line is not the axis, it must pass through the center, i.e., the center must be an improper point. Thus, according to whether the center lies on the axis or not, we obtain the mappings in Figs. 4 and 5. The former are called *shears* and the latter *axial affinities* (cf. II9, §4.3).

If the axis and the center are both proper, then for the construction of X' from X it is often convenient to make use of the *vanishing line f*, namely the image of the improper line under the transformation. Since the point $f \cap a$ must remain fixed, it is improper, so that $f \parallel a$. The correctness of Fig. 6 follows immediately from (12). If f lies exactly halfway between Z and a (of course, Z must not lie on a), then by §1.2 the pair (SP, SP') is harmonically conjugate to the pair (SY, SZ) and therefore (X, X') to (Y, Z), so that in this case the relationship between point and image is symmetric; in other words, the central collineation is *involutory,* i.e., it coincides with its inverse. In Fig. 5, where Z is the improper point of the line perpendicular to a, the involutory central collineation is a reflection in the line a (for this concept

---

[15]See the proof at the beginning of §2.2

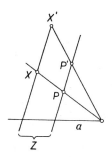

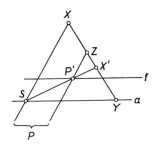

Fig. 5   Fig. 6

see §5.3), and in Fig. 3 the involutory central collineation is a *reflection in the point Z*. As is shown by the translations, if the center lies on the axis, there are no involutory central collineations.[16]

The concept of a central collineation can be defined for a projective space of arbitrary dimension (≥ 3) in the same way as in the plane, if a hyperplane, i.e., a plane in three-dimensional space, is taken as the axis.

2.2. *Projective Collineations*

As in §2.1, a one-to-one mapping onto itself of the set of all points in a projective space of arbitrary dimension is called a *collineation* if collinear points are taken into collinear points. The collineations of a projective space form a group, if the product $\kappa\kappa'$ of the collineations κ, κ' is defined by

$$\kappa\kappa'(X) = \kappa(\kappa'(X)) \quad \text{for all points } X.$$

For then $\kappa\kappa'$ is again a collineation, and the identical mapping ε defined by $\varepsilon(X) = X$ (for all X) is obviously a collineation with the properties of the neutral element of a group; namely, $\varepsilon\kappa = \kappa = \kappa\varepsilon$ for all collineations κ. Moreover, multiplication (successive application) of mappings is well known to the associative (IB2, §1.2.5). Thus we only need to prove that with κ the *inverse mapping* (or *inverse*) κ^{-1}, with $\kappa\kappa^{-1} = \varepsilon = \kappa^{-1}\kappa$, is again a collineation, i.e., that collinearity of distinct images A', B', C' implies collinearity of their preimages A, B, C. But if $D' \notin A'B'$ is the image of D and therefore $AB \neq CD$, and if we write $E = AB \cap CD$, then $E = C$, so that $C \in AB$, as desired, since otherwise $D' \in E'C' = A'B'$.

Since under a collineation κ a set of points is collinear, as we have just seen, if and only if its image points are collinear, the transformation κ maps the set of lines one-to-one onto itself, if to each line g we assign the line $\kappa(g)$ containing the points $\kappa(X)$ with $X \in g$. The set of points on the plane

---

[16]If instead of the real (or the complex) field we take an arbitrary field (or even a skew field) for the domain of scalars (see §1.4), this statement holds only if the characteristic of the field (IB5, §1.11) is $\neq 2$; for characteristic 2 every translation is involutory.

E $= ABC$ through the noncollinear points A, B, C then consists of the points lying on a line XC with $X \in AB$. Thus their images fill up the plane $\kappa(A)\kappa(B)\kappa(C)$, which we shall denote by $\kappa(\mathbf{E})$. Examination of the mapping κ^{-1} shows that in this way κ also maps the set of planes one-to-one onto itself. Thus (for three dimensions) we can consider every collineation as an incidence-preserving one-to-one mapping of the set of all points, lines, and planes, under which points go into points, lines into lines and planes into planes.

Now the central collineations have the property that the CR of points remains unchanged, as is easily seen if the points lie on a line not through Z; for then by §1.2 their CR is equal to the CR of the lines joining them with Z and by (12) this is also the CR of their image points. But if the points X_i ($i = 1, 2, 3, 4$) lie on a line $\neq PP'$ through Z, then by §1.2 they have the same CR as the points $PX_i \cap a$ and by the equation $(PX_i \cap a)P' = X_i'P'$, implied by (12), the points $PX_i \cap a$ have the same CR as the image points X_i.

Collineations which in this way leave the CR unchanged are called *projective collineations*.[17] Obviously these collineations form a group, the *projective group*. Moreover, this group is generated[18] by the central collineations, i.e., every projective collineation is a product of central collineations, which means that the projective group is the smallest group of collineations containing all the central collineations.

Since by (1) and (5) the collinearity, and also the CR, of a set of points is determined by the relationship between their hC, we can obtain a projective collineation κ of the three-dimensional[19] space by choosing two coordinate systems and assigning to every point X the point X' which in the second coordinate system has the same hC as X in the first system. In particular, the mapping κ will take the reference points E_i and the unit point E of the first system into the reference points E_i' and the unit point E' of the second system. Since by §1.3 the hC are determined by the CR up to a common factor, the mapping κ is uniquely determined by the conditions

$$\kappa(E_i) = E_i' \ (i = 0, 1, 2, 3), \quad \kappa(E) = E'.$$

Noting further that by §1.3 the reference points and unit point of a coordinate system are subject to the single condition that no four of them are coplanar, we have the following theorem:

In three-dimensional projective space, given five noncoplanar points A_i

---

[17]Also called *projective mappings*, especially for dimension 1. In §2.4 we show that nonprojective collineations also exist.

[18]For the proof see Baer [1], p. 66, Theorem 2, and Pickert [4], p. 286. Cf. II3, §3.2.

[19]This choice for the number of dimensions is unimportant. As far as the results in §2.2–2.3 are concerned, it is only the theorem on the generation of the projective group by the central collineations that requires the restriction to dimensions ≥ 2; all the other results (with suitable changes of language) are valid for any dimension, and in particular for the projective line.

$(i = 1, \ldots, 5)$, *and five noncoplanar points* A'_i $(i = 1, \ldots, 5)$, *there exists exactly one projective collineation taking* A'_i *into* A_i $(i = 1, \ldots, 5)$.

This theorem, generalized to spaces of arbitrary dimension, is called the *fundamental theorem of projective geometry.*[20] In the plane it refers to four points, no three of which are collinear, and in a line to three distinct points.[21]

If c_i and c'_i $(i = 0, 1, 2, 3)$ are the basis vectors (in the sense of §1.3) for the above coordinate systems, we have the following equations (cf. IB3, §1.5)

$$(13) \qquad c'_k = \sum_{i=0}^{3} c_i a_{ik} \qquad (k = 0, 1, 2, 3)$$

with a regular matrix $A = (a_{ik})_{i,k=0,1,2,3}$. If X has the hC x_i in the first system, then $\sum_{k=0}^{3} c'_k x_k$ is a CV for its image X'. Thus the equations

$$(14) \qquad \sum_{k=0}^{3} c'_k x_k = \sum_{i=0}^{3} c_i x'_i$$

determine the hC x'_i of X' in the first system. Since equations (13) and (14), and the fact that the c_i form a basis, show that

$$(15) \qquad x'_i = \sum_{k=1}^{3} a_{ik} x_k \qquad (i = 0, 1, 2, 3).$$

On the other hand, since the regular matrix A can be chosen arbitrarily, all projective collineations are given by the one-to-one linear mappings of the CV space (cf. IB3, §2). Since by (13) a_{ik} $(i = 0, 1, 2, 3)$ are the hC of E'_k, they are determined, in a given coordinate system, by the projective collineation κ, up to a common factor c_k. Moreover, since $\sum_{k=0}^{3} c'_k$ must be a CV for E', the c_k are also determined up to a common factor; in other words, *a projective collineation determines the coefficients of its coordinate representation* (15) *up to a common factor.*

A projective collineation of complex space is called *real* if it takes every real point into a real point, i.e., if in a real coordinate system it is represented by a real matrix.

2.3. *Affine Mappings*

If we distinguish one plane as the improper plane and wish thereby to proceed from projective to affine three-dimensional space, we must of course restrict ourselves to projective collineations that take proper points into proper points, i.e., (see §2.2) that leave the improper plane fixed. These mappings are called *affine mappings*[22] or *affinities*. Obviously they form a group, the *affine group*.

---

[20]The name is often used for the one-dimensional case alone.

[21]Namely, three points such that no two of them "lie in a point."

[22]The name is often used for mappings of this kind that are not one-to-one; see, e.g., Pickert [4], p. 72. Cf. also the detailed discussion of affine mappings in II9, §4.

Since affine mappings take every improper point into an improper point, they transform parallel lines (intersecting in an improper point) into parallel lines and leave ratios of division of segments unchanged (see §1.2). These properties are sufficient to characterize affine mappings among the one-to-one mappings of the affine space onto itself.[23]

In terms of the relationship in (8) between nonhomogeneous and homogeneous coordinates, the equations (15) represent an affine mapping if and only if $x_0 = 0$ implies $x'_0 = 0$, so that $a_{0k} = 0$ for $k = 1, 2, 3$. If we set $x_0 = x'_0 = 1$ for the proper points, as is obviously possible, equations (15) become

$$(16) \qquad x'_i = \sum_{k=1}^{3} a_{ik}x_k + a_{i0} \qquad (i = 1, 2, 3),$$

where now, in view of (8), the x_k, x'_i are also the affine coordinates of the original point and its image. Here, of course, the matrix $A_0 = (a_{ik})_{i,k=1,2,3}$ must be regular, and conversely, under this condition, the equations (16) represent an affine mapping, since we can then regain (15) with a regular matrix A. Since (16) can be supplemented by $x'_0 = x_0$ to (15), an affine mapping uniquely determines the matrix A_0 in a given coordinate system.

2.4. General Collineations

If the collineation κ takes the reference points E_i ($i = 0, 1, 2, 3$) and the unit point E into the points E'_i, E', then by §2.2 there is exactly one projective collineation κ_0 that will produce the same result, and then the mapping $\kappa' = \kappa_0^{-1}\kappa$ leaves the points E_i, E fixed. Consequently, since $\kappa = \kappa_0\kappa'$ and since we already know the coordinate representations of the projective collineations κ_0, it is only necessary, in order to determine the coordinates for all collineations, to study those collineations κ' for which the reference points and the unit point of the coordinate system remain fixed.

If the plane spanned by the points E_0, E_2, E_3 is taken as the improper plane, then κ' takes the proper planes through E_2E_3 again into proper planes through E_2E_3. For a proper point X with affine coordinates x_i ($i = 1, 2, 3$) the point $E_2E_3X \cap E_0E_1$ is on the first axis and its first coordinate is x_1 (see Fig. 7). Thus the first coordinate x'_1 of the point $\kappa'(X)$ depends only on x_1, and the corresponding remark is obviously true for every coordinate. Consequently, there exist one-to-one mappings f_1, f_2, f_3 of the domain of scalars onto itself such that $x'_i = f_i(x_i)$ ($i = 1, 2, 3$) is an affine coordinate representation of κ'. But now the line E_0E left fixed by κ' has the equations $x_1 = x_2 = x_3$. Thus $x_1 = x_2 = x_3$ implies $x'_1 = x'_2 = x'_3$, so that $f_1 = f_2 = f_3$. As a result, the representation can be written in terms of a

[23]See, e.g., Pickert [4], p. 72 and 338.

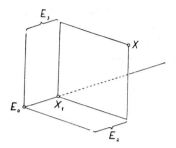

Fig. 7

single mapping f in the form

(17) $$x'_i = f(x_i) \qquad (i = 1, 2, 3).$$

Let us now examine the line with equations $x_3 = 0, x_2 = ax_1 + b$, which lies in the plane $E_0 E_1 E_2$ but does not pass through E_2. It must be taken by κ' into a line with the same property, i.e., there must exist scalars a', b' such that $x_2 = ax_1 + b$ implies $x'_2 = a'x'_1 + b'$. Thus

(18) $$f(ax_1 + b) = a'f(x_1) + b'.$$

Since E_0, E are fixed points, it follows that $f(0) = 0, f(1) = 1$. So with $x_1 = 0$ we have $b' = f(b)$ from (18). Parallel lines go into parallel lines, i.e., a' depends only on a and not on b. But if we set $b = 0$ and $x_1 = 1$, then (18) gives $a' = f(a)$. So (18) goes into $f(ax_1 + b) = f(a)f(x_1) + f(b)$, from which, setting $a = 1, x_1 = x, b = y$, and again $a = x, x_1 = y, b = 0$, we obtain

(19) $$f(x + y) = f(x) + f(y), \quad f(xy) = f(x)f(y) \quad \text{for all } x, y.$$

From the second of these equations and from (17), we can derive a representation of κ' in hC by letting i take the value 0 as well; for then all we need to show is that multiplication of the x_i by a common factor merely multiplies the x'_i likewise by a common factor. Furthermore, it is easy to show that we thereby obtain a collineation with the fixed points E_i, E, provided f is an *automorphism* (IB1, §4.6) of the domain of scalars, i.e., a one-to-one mapping for which (19) holds. Since $\kappa = \kappa_0 \kappa'$, with κ_0 represented by (15), all collineations can be written

(20) $$x'_i = \sum_{k=1}^{3} a_{ik} f(x_k) \qquad (i = 0, 1, 2, 3).$$

Here the coefficients a_{ik} form a regular matrix and f is an automorphism of the domain of scalars. If f is the identical automorphism, $f(x) = x$ for all x, then (20) is identical with (15), i.e., we obtain the projective collineations. Under an arbitrary automorphism f it is easy to show from (20) that all CR undergo the f-transformation.

For example, the field of complex numbers has a nonidentical auto-morphism, namely, passage to complex conjugates, but the real field has no such nonidentical automorphism (IB1, §4.6). Thus in real geometry all collineations are projective, but not in complex geometry.

The above results are valid for every projective space of dimension ≥ 2, but on the projective line, of course, they cannot hold in their present form. However, they are valid under the additional condition that the mapping takes harmonic pairs of points into harmonic pairs.[24]

3. Projectivities

3.1. *Generation of Projective Collineations by Projections*

Let **E**, **E**′ be two distinct planes and S, S' two distinct points not on **E** or **E**′. If we project **E** from S onto **E**′ and then project **E**′ from S' back onto E, i.e., if to each point X in **E** we assign the point $X' = (XS \cap \mathbf{E}') \, S' \cap \mathbf{E}$, it is easy to see that we have thereby produced a collineation of the plane **E**; and in fact even a central collineation, since every point on the line $\mathbf{E} \cap \mathbf{E}'$ and every line through $SS' \cap \mathbf{E}$ is obviously left unchanged. The sketch in Fig. 8 (the plane intersection of a three-dimensional figure) shows that every central collineation (with center Z and axis a) can be represented in this way. Since a projective collineation is the product of central collineations (see §2.2), every projective collineation can threfore be produced by a sequence of projections. Conversely, from the results in §1.2 on the CR of points and lines, it is easy to see that every sequence of projections leading back to the original plane is a projective collineation.

These statements can easily be generalized to higher dimensions, but they fail to hold in the one-dimensional case. Here we turn to the fundamental theorem of projective geometry (see §2.2), by which every projective

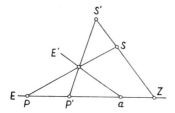

Fig. 8

[24]The fact that under this condition f must be an automorphism in the one-dimensio-nal case also is proved in detail, e.g., in Blaschke [2], p. 45, who thereby provides another form of our proof for spaces of dimension ≥ 2, since then the additional condition is automatically satisfied by every collineation, in view of the quadrangle construction for harmonic pairs of points (see §1.2).

mapping of a line with three fixed points must be the identity. If a one-to-one mapping of a range of points (set of points on a line) into another range of points is defined to be a *projective mapping* if it preserves the CR, we have the following theorem.

Every projective mapping of a range of points is the result of at most three successive projections.

For a first projection will ensure, if necessary, that the set of images is distinct from the original range of points, and then a second projection, if necessary, produces the case of a projective mapping (onto a distinct range of points) with a fixed point. But such a projective mapping is always a projection from a point; for if A is the fixed point and if two other points B, C are taken into B', C', then a projection from $S = BB' \cap CC'$ back onto BC produces a projective mapping with the three fixed points A, B, C, and is therefore the identity.

3.2. *Projectivities*

In (three-dimensional) projective space seven kinds of *primitive geometric forms* are distinguished; namely, *range of points* (line),[25] *pencil of lines* (point and plane), *pencil of planes* (line), *plane of points* (plane), *bundle of lines* (point), *plane of lines* (plane), *bundle of planes* (point). As is indicated by the names, we are dealing here, respectively, with the set of all points, lines, or planes incident with a fixed element of the type indicated by the word in parentheses,[26] the *carrier* of the primitive form. The first three primitive forms are *one-dimensional,* and the others *two-dimensional.*

We now consider one-to-one mappings of one primitive form onto another such that preimage and image are always incident to each other and the mapping is thereby uniquely determined. It is easy to see that such mappings are possible only between two one-dimensional primitive forms of different kinds or else between the first two, or the last two, two-dimensional primitive forms. Each of these mappings, and any sequence of two of them, is called a *perspectivity.*[27] The successive application of two mappings to form such a product is permissible here, of course, only if the first mapping takes the primitive form A onto a primitive form B that is then taken by the second mapping onto C, in which case the sequence of mappings is called a *perspectivity of A onto C from the carrier of B.* In particular, the pro-

---

[25]If lines are introduced as sets of points, this concept is identical with "line." Our notation ∈ for incidence of point and line arose in this way but does not imply that the lines have been so introduced. Since the set of point on a conic section, for example, is also called a range of points, we should really speak here of "ranges of points on a line."

[26]Only for the pencil of lines, namely, all lines through a point and on a plane, are there two such elements; in plane geometry we usually think of a pencil of lines as a primitive geometric form and then only the point is called the carrier.

[27]This name is often used for the central collineations.

jections in §3.1 are therefore perspectivities. A perspectivity of an entirely different kind is obtained from three pairwise skew lines g, g', h if to each point $X \in g$ we assign the point $Xh \cap g'$, i.e., if we first carry out the perspectivity of the range of points on g onto the pencil of planes on h and then the perspectivity from this pencil of planes onto the range of points on g'; such a mapping is called an *axial perspectivity*.

From the results on CR in §1.2 it is clear that every perspectivity leaves the CR of any four elements unchanged. Conversely, any one-to-one mapping between one-dimensional primitive forms that leaves all CR unchanged can be extended, by adjunction of perspectivities if necessary, to a mapping with the same property between ranges of points. But by §3.1 such a mapping can be regarded as a sequence of perspectivities. So if a *projectivity*[28] is defined as a sequence of perspectivities, we have the following theorem.

The projectivities of one-dimensional primitive forms are precisely the one-to-one mappings between such forms that leave all CR unchanged.

From the results in §3.1 it also follows that a *projectivity of one range of points onto another is a perspectivity from a point if and only if the projectivity has a fixed point.* So in this case the carriers must have a point in common, i.e., they must lie in a plane, and if they do not, we obtain the theorem that *a projectivity between ranges of points with skew carriers is an axial perspectivity.* For by the foregoing argument, the projectivity of the range of points on g onto the range of points on g', where g, g' are skew lines, is the result of two successive perspectivities from points S, S', which means (see Fig. 9) that the given projectivity is precisely the perspectivity from SS'.

3.3. Involutions

Particularly important among the projectivities of a one-dimensional primitive form onto itself are the *involutions*, i.e., the nonidentical mappings

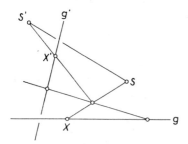

Fig. 9

---

[28]The projective collineations are often called projectivities, but the two concepts should be kept apart, in spite of the close relationship between them, with its unfortunate confusion in terminology.

identical with their inverse, or more specifically, the point involutions, line involutions, or plane involutions according to the type of primitive form. Since every involution can be reduced by a perspectivity to a point involution,[29] the following concepts and theorems, introduced here only for point involutions, are valid for the other two types as well.

The pair of points consisting of a point and its image under an involution is called a *conjugate pair* (*of points*) *of the involution*. Thus if (A, B) is a conjugate pair, so also is (B, A), so that these pairs do not need to be distinguished. Fixed points of an involution are also called *double points*. We now prove the following theorem.

Given four collinear points A, A', B, B' with $A \neq B$, B' and $A' \neq B$, B', there exists exactly one involution with the conjugate pairs (A, A') and (B, B').

If $A = A'$, $B = B'$, we choose a coordinate system in the line in such a way that $E_0 = A$, $E_1 = B$. Then if E_1 is taken as an improper point, the desired involution can be described as an affine mapping with the fixed point E_0. By (16) the affine coordinates x, x' of the point X and its image X' satisfy the equation $x' = ax$ with a fixed scalar $a \neq 0$, which will describe an involution if and only if $a \neq 1 = a^2$, i.e., $a = -1$; i.e., if *the pair X, X' is harmonically conjugate to the pair of double points.*[30] By §1.2 we thus have the following relation for A, B, X, X': *The points A, B are diagonal points of a complete quadrangle in which X, X' lie on the sides not through A, B.*

If we now imagine the coincident points A, A' and B, B' as moving apart from each other, it is natural to try to prove the theorem by means of a complete quadrangle in which two of the diagonal points no longer necessarily lie on the carrier g of the given range of points, and for $A \neq A'$ this idea leads in fact to the desired result. For let us draw lines a, b, $b' \neq g$, through the three points A, B, B', respectively, and set $S = a \cap b$, $S' = a \cap b'$ (see Fig. 10), and then let us project the line g[31] from S onto b', and b' from A' onto b, and finally b from S' back onto g. It is easy to see that the resulting projectivity π takes the points A, A', B, B' into A', A, B', B, respectively, and by the fundamental theorem of projective geometry (see §2.2) it is the only projectivity taking A, A', B into A', A, B'. So the quadrangle used to determine $X' = \pi(X)$ (with the vertices S, S', $b \cap S'X'$, $B' \cap SX$) can be chosen arbitrarily except that three of its sides meeting at a vertex must pass through A, B, X and the other three must pass through A', B', X'. Since A can be interchanged with A' and B with B', we see that $\pi(X') = X$, i.e., that π is in fact an involution. The proof thus completed also shows that

A projectivity of a range of points onto itself which interchanges two distinct points is an involution.

---

[29]By simply drawing a line to intersect the pencil of lines or the pencil of planes.
[30]Except, of course, for the cases $A = X = X'$ and $B = X = X'$.
[31]For brevity, we write g instead of "the range of points on g."

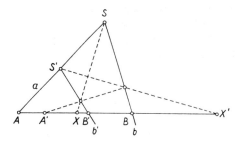

Fig. 10

Another proof of this theorem is obtained by choosing the two inter-changed points as reference points E_0, E_1 of a coordinate system. Then the coordinate representation (15) must be of the form $x_0' = a_0 x_1$, $x_1' = a_1 x_0$ $(a_0, a_1 \neq 0)$. For the affine coordinates $x = x_1/x_0$, $x' = x_1'/x_0'$ we thus have the following equation for points $\neq E_0$, E_1, with $a = a_1/a_0$,

(21) $xx' = a$,

which shows not only that the given projectivity is an involution but also:
An involution has either two double points or none.

The first case occurs if and only if a has a square root. Thus if we allow all the complex numbers to be scalars, every involution has two double points. But if the scalars are restricted to the real numbers, the involution (21) has two double points if $a > 0$ and no double points if $a < 0$. In the first case the involution is called *hyperbolic,* and in the second *elliptic.*[32] The elliptic involutions can serve as real representatives of their pairs of double points existing only in complex space, and in this way the real space can be extended to the complex without the use of coordinates (von Staudt).

If in the first theorem (of §3.3) the pairs (A, A'), (B, B') are regarded as pairs of double points of two involutions, the theorem shows that for the complex field of scalars there exists exactly one pair of conjugate points common to both these involutions, namely, the pair of double points of the involution with the conjugate pairs (A, A'), (B, B'). For we see from the second theorem of §3.3 that the conjugate pair (C, C') has the desired property if and only if it is harmonically conjugate both to (A, A') and to (B, B'), from which it follows by §1.2 that each of the pairs (A, A'), (B, B') is harmonically conjugate to (C, C') and is thus a pair of conjugate points in the involution with the double points C, C'. With the real field as scalars we shall now also prove that

---

[32]Since the diameter involutions of the hyperbola and the ellipse (see §6.2) are of these kinds.

Two distinct involutions on a line, at least one of which is elliptic[33] *have exactly one (real) pair of conjugate points in common.*

For with a suitable choice of coordinate system the elliptic involution can be represented by (21) with $a < 0$, and for the other involution we may make use of (15), where the matrix A can differ from its inverse only by a scalar factor, so that $a_{00} = -a_{11}$. Passage to affine coordinates gives

$$(22) \qquad x' = \frac{a_{10} - a_{00}x}{a_{00} + a_{01}x},$$

or in combination with (21), for the case $a_{00} \neq 0$,

$$x^2 + a_{00}^{-1}(aa_{01} - a_{10})x + a = 0,$$

and since $a < 0$ this equation has real roots. But for $a_{00} = 0$ the equation (22) is of the same form as (21), which means that the two involutions have the pair of points (E_0, E_1) in common.

4. Polarities

4.1. *Correlations*

In §1.3 we have seen that the planes of three- dimensional projective space can be represented by quadruples of scalars in the same way as the points. This fact suggests that for a given coordinate system we should consider the mapping which to each point with the hC a_i ($i = 0, 1, 2, 3$) assigns the plane with the equation $\sum_{i=0}^{3} a_i x_i = 0$. If for brevity we let a' represent the linear form whose coefficients are the coordinates of the vector a, the point with CV a then corresponds to the plane with equation $\langle a', x \rangle = 0$. The points of the line AB, where A, B have the CV a, b, correspond to the planes with equations $\langle sa' + tb', x \rangle = 0$, i.e., to the planes of a pencil, which means that the line AB corresponds to a line, namely, the carrier of the pencil, and the totality of lines is mapped one-to-one onto itself. The points of the plane spanned by the noncollinear points A, B, C with CV a, b, c correspond to the planes with equations $\langle ra' + sb' + tc', x \rangle = 0$. Since the a', b', c' are linearly independent, this system of equations has a one-dimensional family of solutions (see IB3, §2.4) and all of its planes pass through one point D, whose hC are obviously the plane coordinates of the plane ABC. Thus the totality of planes is mapped onto the totality of points (of little importance to us here is the fact that this mapping is precisely the inverse of the mapping of points onto planes from which we began). Altogether, we now have a one-to-one mapping δ of the totality of points, lines, and planes onto the totality of planes, lines, and points, in which points go into planes, lines into lines, and

---

[33]This condition is sufficient but not necessary.

planes into points, where[34]

(23) $P \subset g$ if and only if $\delta(g) \subset \delta(P)$;

(24) $P \subset \mathbf{E}$ if and only if $\delta(\mathbf{E}) \subset \delta(P)$.

A mapping with these properties is called a *correlation*[35] of the space; the special correlation corresponding in the above manner to each coordinate system will be called δ_0. From (23), (24) it also follows that

(24′) $g \subset \mathbf{E}$ if   and only if $\delta(\mathbf{E}) \subset \delta(g)$,

since $g \subset \mathbf{E}$ means that $X \subset \mathbf{E}$ for all $X \subset g$, i.e., by (23), (24), that every plane ($= \delta(X)$) through $\delta(g)$ also contains the point $\delta(\mathbf{E})$, which will be the case if and only if $\delta(\mathbf{E}) \subset \delta(g)$. Recalling further that δ takes distinct elements into distinct elements, that the intersection of two planes is the only line common to both planes, and that a line joining two points is the only line containing both the points, we see from (23), (24) that

(24″)
$$\delta(PP') = \delta(P) \cap \delta(P'), \quad \text{if} \quad P \neq P';$$
$$\delta(\mathbf{E} \cap \mathbf{E}') = \delta(\mathbf{E})\delta(\mathbf{E}'), \quad \text{if} \quad \mathbf{E} \neq \mathbf{E}'.$$

A correlation δ is already determined by its effect on the points, since $\delta(g)$ is the line common to all planes $\delta(X)$ with $X \subset g$, and $\delta(\mathbf{E})$ is the point common to all planes $\delta(X)$ with $X \subset \mathbf{E}$. We now wish to show that every one-to-one mapping of the totality of points onto the totality of planes is a correlation if it satisfies the condition that three points are collinear if and only if the corresponding planes are collinear. For if such a mapping is followed by δ_0^{-1}, the result is obviously a collineation, which by §2.2 can be regarded as an incidence-preserving mapping κ of the set of all points, lines, and planes, and therefore $\delta_0\kappa$ is the desired correlation. Thus every correlation is of the form $\delta_0\kappa$ and conversely, for every collineation κ, the mapping $\delta_0\kappa$ is obviously a correlation. Thus (20) shows that, given a regular matrix $(a_{ik})_{i,k=0,1,2,3}$ and an automorphism f of the domain of scalars, every correlation can be written in the form

(25) $$u_i = \sum_{k=0}^{3} a_{ik} f(x_k) \qquad (i = 0, 1, 2, 3),$$

where the u_i are the plane coordinates of the plane corresponding to the point with the hC x_i, and the matrix a_{ik} is arbitrary to the extent described in §2.2.

---

[34]Here we have used $\subset$, instead of $\in$, to indicate incidence of a point with a line or a plane, in order to emphasize the analogy with incidences of lines and planes, occurring below. In general, $\in$ denotes inclusion of an element in a set, and $\subset$ inclusion of a (proper) subset.
[35]Often called a *duality*, the name "correlation" being reserved for the mapping of the points only in the duality.

To define a correlation in the plane it is only necessary to replace "plane" by "line," whereupon (24) becomes (23) and (24′) is deleted. In an arbitrary n-dimensional space the correlation δ_0 is obtained from the one-to-one relation proved in linear algebra for the subspaces of a vector space and the subspaces of the dual space of linear forms;[36] namely, the $(m + 1)$-dimensional vector space $\mathfrak{V}$ of CV of the points of a projective m-dimensional subspace corresponds one-to-one to the $(n - m)$-dimensional space $\mathfrak{L}$ of linear forms in such a way that $\mathfrak{x}$ lies in $\mathfrak{V}$ if and only if $\langle \mathfrak{u}, \mathfrak{x} \rangle = 0$ for all $\mathfrak{u} \in \mathfrak{L}$. Thus a correlation relates every $(n - m - 1)$-dimensional projective subspace to an m-dimensional projective subspace.[37]

A correlation δ is called *projective* if the CR of any four collinear points A_i is equal to the CR of the image points $\delta(A_i)$. Since δ_0 is easily seen to be projective, it is clear that δ is projective if and only if the collineation κ with $\delta = \delta_0 \kappa$ is projective. So, corresponding to (25), the projective collineations are represented by

$$(26) \qquad u_i = \sum_{k=1}^{3} a_{ik} x_k \qquad (i = 0, \mathfrak{n}, 2, 3).$$

The projective correlation δ leaves invariant not only the CR of collinear points, but also those of the lines and planes of a pencil. For if g_i $(i = 1, 2, 3, 4)$ are four lines passing through the point P and lying in the plane **E**, we need only intersect them with a line g not through P, since then the CR of the lines g_i is equal to that of the lines $g \cap g_i$, which in turn is equal to the CR of the planes $\delta(g \cap g_i)$ through $\delta(g)$. But the latter (see the end of §1.2) is equal to the CR of the lines $\delta(g \cap g_i) \cap \delta(P)$, and these lines are in fact the $\delta(g_i)$, since $(g \cap g_i)P = g_i$ by (24″). The case of four planes of a pencil is dealt with similarly but more simply.

The inverse of a correlation is obviously again a correlation, the product of a correlation with a collineation is a correlation, and the product of two correlations is a collineation. Thus the correlations alone (with successive application as multiplication) do not yet form a group, but the correlations and collineations together do form a group, and the corresponding statement holds for projective correlations.

4.2.   *Principle of Duality*

The projective correlation δ of a plane provides us with an important transference principle, which we first illustrate by an example. If π is a projectivity of one pencil of lines onto another in the plane, this projectivity is obviously transformed by δ into a projectivity $\pi' = \delta \pi \delta^{-1}$ of one range of points onto another. Then π' is a perspectivity if and only if it has a fixed

---

[36]Easily derived from IB3, §§2.4 and 2.6; see also Pickert [4], p. 59.
[37]For details see, e.g., Pickert [4], p. 293.

point (see §3.2), and also, in view of (23), if and only if π is a perspectivity. Since the preimage under δ of a fixed point of π' is obviously a fixed line of π, the theorem on projectivities of ranges of points at once enables us to assert its *plane dualization*

A projectivity of one pencil of lines onto another in the same plane is a perspectivity (from a line) if and only if it has a fixed line.

Similarly, if δ is a projective correlation of space, we have the *space dualization*

A projectivity of one pencil of planes onto another is a perspectivity from a plane if and only if it has a fixed plane.

Finally, the space dualization of the plane dualization reads

A projectivity of one pencil of lines onto another through the same point is a perspectivity from a line[38] *if and only if it has a fixed line.*

It is clear that the last of these theorems can also be obtained from the original one by joining the points and lines occurring in it to a point not on the plane. If we apply this procedure to the plane dualization, we have precisely the space dualization, but with the restriction that the carriers of the plane pencils must not pass through one point.[39] The space dualization of the plane dualization is also called the *bundle dualization* of the (similarly restricted) space dualization. Both theorems deal only with the lines and planes of a bundle, and one arises from the other by interchange of "plane" and "line."

To obtain the dualizations of the original theorem, we need only consider the effect of a correlation on the primitive geometric forms; in the plane a range of points and a pencil of lines are *dual* to each other, and in space there are four sets of dual forms, namely, a range of points and a pencil of planes, a plane of points and a bundle of planes, a plane of lines and a bundle of lines, and finally, a pencil of lines is a *self-dual* concept.[40]

Of course this transference of theorems from forms of one kind to forms of another is possible only for theorems on points, lines, and planes that involve only properties remaining unchanged under a projective correlation. Among such properties we a ready have discussed 1) incidence and 2) cross ratio. In view of (23), (24), and (24'), incidence must be regarded as a symmetric relation, i.e., the statement "P and g are incident"[41] means not

---

[38]Here the primitive form for the perspectivity is the pencil of planes with this line as carrier.

[39]This distinction arises from the fact that in formulating the space dualization we considered the original theorem as a space theorem, so that the carriers of the ranges of points were not assumed to be lines in a plane.

[40]But the carrier point and the carrier plane interchange roles under the duality.

[41]Of course, this statement means the same as "g and P are incident."

only "g passes through P" but also "P lies on g." A theorem stated in terms of concepts of this sort is called a *projective theorem* and its *space dualization* is the theorem arising from it by interchange of the words "point" and "plane." Then, exactly as in the forgoing example, by applying the transformation δ to the assumptions for the dualized theorem, we obtain the principle of *space duality*

The space dualization of a valid projective theorem of space geometry is also valid.

The corresponding principle for plane geometry is obtained by simply replacing "space" by "plane," and interpreting *plane dualization* as interchange of "point" and "line," the proof being exactly analogous to the three-dimensional case. In general, dualization in n-dimensional geometry consists of replacing an "m-dimensional subspace" by an "$(n - m - 1)$-dimensional subspace."

One must not suppose that the principle of space duality in the form given here, but omitting the CR-concept, can be justified by the simple remark that the axioms of incidence (see II3, §4.1) are dualized into consequences of themselves. This fact justifies only the following, weaker principle of duality; *an incidence theorem holding in every (three-dimensional) projective space (all skew fields being allowed as domains of scalars) provides a space dualization valid also in every (three-dimensional) projective space.* In fact it is possible to specify a projective space (by using a suitably chosen skew field) in which the duality principle is not valid when restricted to this single space.[42] Since the corresponding statement holds for the projective plane, the duality principle in II3, §3.1 necessarily includes the restrictive condition "the proof of which involves only Axioms P1–P3."

4.3. *Polarities*

The plane plays the same role in three-dimensional space as the line in two-dimensional and the point in one-dimensional space (i.e., on the line). In this sense the definition of a projective correlation in one-dimensional space becomes identical with that of a projective mapping, and as natural analogies for the point involutions discussed in §3.3 we thus have, in two-dimensional and three-dimensional space, the involutory projective correlations.

The projective correlation δ whose effect on points is described by (26) is involutory if and only if it coincides with its inverse δ^{-1}, so that we first wish to determine for δ^{-1} its representation

$$v_i = \sum_{k=0}^{3} b_{ik}y_k \qquad (i = 0, 1, 2, 3)$$

[42]H. Kneser, Jber. dtsch. Math. Ver. 45, p. 77 and 78 (1935); cf. also Pickert [4], p. 298. In these cases the commutative law of multiplication cannot hold in the domain of scalars.

corresponding to (26). Thus the plane with coordinates v_i is taken by δ into the point with $hC\, y_k$; in other words, in view of (24), it follows from (26) and $\sum_{k=0}^{3} v_k x_k = 0$ that $\sum_{i=0}^{3} u_i y_i = 0$, which will certainly be the case we set $v_k = \sum_{i=0}^{3} a_{ik} y_i$ $(k = 0, 1, 2, 3)$. Since the coefficient matrix in (26) is uniquely determined up to a scalar factor, the $(b_{ik})_{i,k=0,1,2,3}$ can differ only by a scalar factor from the transpose $A^T = (a_{ki})_{i,k=0,1,2,3}$ of $A = (a_{ik})_{i,k=0,1,2,3}$. Thus δ is involutory if and only if there exists a scalar c with $A^T = cA$. Transposing a second time shows that $A^{TT} = A = cA^T$ and therefore $A = c^2 A$, so that c can have only the values 1 or -1; in other words, A must be either *symmetric* $(A^T = A)$ or *skew symmetric* $(A^T = -A)$.

These two cases can be distinguished geometrically in a very simple way. The skew symmetry $A^T = -A$ means[43] that $\sum_{i,k=0}^{3} a_{ik} x_i x_k = 0$ for all x_i, or, in view of (26), that $P \in \delta(P)$ for all points P. A projective correlation with this property is called a *null system*.[44]

In the symmetric case, i.e., with a point P for which $P \notin \delta(P)$, the mapping δ is called a *polarity*, where $\delta(P)$ is the *polar* of P and P is the *pole* of $\delta(P)$. Points X, Y with $X \in \delta(Y)$, or alternatively (in view of (24) and $\delta = \delta^{-1}$) with $Y \in \delta(X)$, are said to be *conjugate* to each other, a relation which from (26) is expressed by the equation $\sum_{i,k=0}^{3} a_{ik} x_k y_i = 0$ and is at once seen to determine δ uniquely. The same notation is used in two-dimensional space as well.

If for a polarity δ we choose as E_3 a point not conjugate to itself, we can then take the other reference points E_0, E_1, E_2 in $\delta(E_3)$. For the a_{ik} in (26) this obviously means that $a_{i3} = 0$ $(i = 0, 1, 2)$, so that the matrix $(a_{ik})_{i,k=0,1,2}$ is also regular. Since the points X, $Y \in \delta(E_3)$ are conjugate if and only if $\sum_{i,k=0}^{2} a_{ik} x_k y_i = 0$, the conjugacy relation for δ in $\delta(E_3)$ also provides a polarity. Thus we may take E_2 to be non-selfconjugate and E_0, $E_1 \in \delta(E_2)$. Then $a_{2i} = 0$ $(i = 0, 1)$. Finally, on the line $E_0 E_1$ we have an involution,[45] so that we can take E_0, E_1 to to be conjugate to each other, i.e., $a_{01} = 0$. Since the matrix A is symmetric, the equations (26) now have the form

$$(27) \qquad\qquad u_i = a_i x_i \qquad (i = 0, 1, 2, 3)$$

with $a_i \neq 0$. If the a_i are squares, say $a_i = c_i^2$, as will certainly be the case if the domain of scalars is the complex field, we can simplify further by means of the coordinate transformation $x_i' = c_i x_i$. Since the plane equation is invariant, the new plane coordinates are $u_i' = u_i/c_i$, so that (27) becomes $u_i' = x_i'$; i.e., in complex space, every polarity is a correlation δ_0 (see §4.1), relative to a suitably chosen coordinate system.

---

[43]Here we are assuming that the domain of scalars, in addition to being a field, satisfies $1 + 1 \neq 0$, so that $a_{ii} = 0$.

[44]For the details see, e.g., Pickert [4] and Weiss [5].

[45]Namely, the one-to-one point mapping $X \rightarrow \delta(X) \cap E_0 E_1$, which leaves the CR unchanged and has the property that its point pairs are precisely the pairs of conjugate points on the line $E_0 E_1$. Moreover, with $u_0 + u_1 x' = 0$, $x' = x_1/x_0$, $x_2 = 0$, $x_3 = 0$, we could actually derive from (26) an equation of the form (22), though with different indices for the coefficients.

4.4. *Polar Triangles*

In §3.3 we have seen that any two involutions of the real line, at least one of which has no double points, must have a pair of conjugate points in common. The feature of a polarity in the plane that corresponds to a point pair for an involution on the line is a set of three non-collinear, pairwise conjugate points, such a set being called the *polar triangle* of the polarity. The points are its vertices and the lines joining them are its sides. Obviously every side of the polar triangle is the polar of the vertex not on it. The following theorem corresponds to the above statement about involutions.

Two polarities in the real plane, at least one of which has no selfconjugate point, have a polar triangle in common.[46]

For the proof we express the polarities in the form corresponding to (26) by $u_i = \sum_{k=0}^2 a_{ik}x_k$ and $u_i = \sum_{k=0}^2 b_{ik}x_k$, respectively, whereupon, in order to find a corresponding polar triangle, we require a point with the same polar in both polarities. A point with the hC x_i will have this property if and only if there exists a scalar $c \neq 0$ such that $\sum_{k=0}^2 (a_{ik} - cb_{ik})x_k = 0$ ($i = 0, 1, 2$). So we must solve this system of equations nontrivially for the x_k, which will be possible if and only if its coefficient matrix has determinant zero. But this requirement leads to a cubic equation in c, which will always have a (real) root, and the root will certainly be $\neq 0$, since the a_{ik} form a regular matrix. Thus we have found a point P with the same polar p in both polarities. Since P is not selfconjugate in one of the two polarities, it follows that $P \notin p$. As discussed in §4.3, the polarities induce involutions in p, one of which will not have a double point. Consequently there exists a pair of conjugate points Q, R common to the involutions; i.e., P, Q, R are pairwise conjugate.

If we take the vertices of a common polar triangle as reference points, the argument leading to (27) shows that the two polarities can be represented in the form $u_i = a_i x_i$ and $u_i = b_i x_i$. If the second one has no selfconjugate point, which means that $\sum_{i=0}^2 b_i x_i^2 = 0$ implies $x_0 = x_1 = x_2 = 0$, then $b_i > 0$, and by the argument at the end of §4.3 we can set $b_0 = b_1 = b_2 = 1$, for a suitable choice of unit point.

4.5. *Degenerate Polarities*

In a polarity given by (26) two points X, Y with the hC x_i and y_k are conjugate if and only if

$$(28) \qquad \sum_{i,k=0}^3 a_{ik}x_i y_k = 0.$$

---

[46] The theorem holds in three-dimensional space if "polar triangle" is replaced by "polar tetranedron," and with the corresponding change of wording it holds in spaces of arbitrary dimension. For dimensions greater than 1, the assumption that at least one of the polarities has no selfconjugate point is sufficient but not necessary.

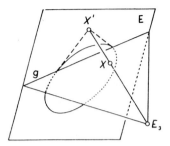

Fig. 11

But how can this conjugacy be described if the matrix A, instead of being regular, is now only required not to be the zero matrix? If A is of rank 3, we know (IB3, §2.4) that there is exactly one point Y with $\sum_{k=0}^{3} a_{ik}y_k = 0$ ($i = 0$, 1, 2, 3), i.e., exactly one point conjugate to all the points X. If we take this point as the reference point E_3, we must have $a_{i3} = 0$ ($=a_{3k}$), so that in (28) the indices need to run only up to 2. Then since $(a_{ik})_{i,k=0,1,2}$ is regular, there will be a polarity δ in the plane $\mathbf{E} = E_0E_1E_2$, and the equation (28) for the points X, Y (with X, $Y \neq E_3$) will mean that $E_3X \cap \mathbf{E}$ and $E_3Y \cap \mathbf{E}$ are conjugate. Thus the set of points conjugate to X ($\neq E_3$) in the sense of (28) coincides with the set of all points on the plane E_3g, where $g = \delta(X')$, $X' = E_3X \cap E$ (Fig. 11). In particular, the set of selfconjugate points is simply the totality of points on the lines joining E_3 to the selfconjugate points of δ. This configuration is called a *cone*,[47] and the set of selfconjugate points of the plane polarity δ (the ellipse in Fig. 11) is called a *conic section* (cf. also §6.1).

If A is of rank 2, then E_3 and E_2 can be taken conjugate to every point (IB3, §2.4) and we obtain an involution on the line E_0E_1, taking the place of the polarity δ. The selfconjugate points thus fill a *pair of planes*, namely, the two planes joining the double points of the involution with the line E_2E_3. Finally, for rank 1 the points conjugate to all points fill a plane, in which we can choose E_1, E_2, E_3. Since (28) then becomes $a_{00}x_0y_0 = 0$ with $a_{00} \neq 0$, the selfconjugate points are precisely the points of the plane $E_1E_2E_3$. A plane considered as the set of selfconjugate points for a degenerate polarity of rank 1 is called a *double plane*. But considered without reference to the degenerate polarity, i.e., merely as a set of points, a double plane is in no way different from a plane.

The conjugacy (28) with nonregular symmetric matrix A is called a *degenerate polarity*. Many writers use the words *non-degenerate polarity* to refer to the correlations that we have called polarities, and the word "polarity" to include both concepts.

----

[47]In the projective sense; in affine geometry a distinction is made between *cones* and *cylinders*, depending on whether the *vertex* E_3 is proper or improper.

5. Euclidean Geometry

5.1. *Orthogonality*

In order to introduce the concept "perpendicular" (orthogonal) in the real affine plane, let us consider what properties our intuition naturally requires of such a concept. If to each line in a pencil of lines we assign the line perpendicular to it, the result is a one-to-one involutory mapping of the pencil onto itself without fixed line; moreover, all the essential geometric properties of lines must be preserved, since this correspondence can be realized by rotation. So we set up the requirement that the mapping in question must be an elliptic involution of lines, which is also called a *right-angle involution*.[48] For another pencil of lines this involution must be obtainable by a parallel displacement, in order that transference to parallel lines will not destroy orthogonality. In other words, the elliptic point involution obtained by intersecting the pencil of lines with the improper line must be independent of the given pencil of lines. So we define the concept "perpendicular" in the real affine plane as follows.

The (proper) lines are said to be perpendicular to each other if their improper points are a pair of conjugate points in a preassigned elliptic involution, called the *absolute involution,* on the improper line.

In space the procedure is exactly the same. In the improper plane there is given a polarity without selfconjugate points, called the *absolute polarity,* and two (proper) lines are said to be perpendicular to each other if their improper points are conjugate. This definition includes the possibility that skew lines are perpendicular to each other.

Since the points conjugate to a given point fill up a line, we have the result (in agreement with our intuition) that *the lines perpendicular to a given line g through a given point P lie in a plane,* called the *plane through P perpendicular to g.*

For "perpendicular" we shall use the symbol $\perp$, and if $\mathfrak{x} = \overrightarrow{AB}$, $\mathfrak{y} = \overrightarrow{CD}$, $AB \perp CD$ we shall also write $\mathfrak{x} \perp \mathfrak{y}$, as is permissible, since $\overrightarrow{AB} = \overrightarrow{A'B'}$ implies $AB \parallel A'B'$ and thus also $A'B' \perp CD$. Since by §1.1 the $\mathfrak{X} = AB$ can be taken as the CV of the improper point X on the line AB in the improper plane, the coordinates x_1, x_2, x_3 of $\mathfrak{x}$ with respect to a basis $\{\mathfrak{e}_1, \mathfrak{e}_2, \mathfrak{e}_3\}$ of the affine space are also hC of X in the improper plane, provided the improper points of the affine coordinate axes are taken as reference points and the improper point on the line $x_1 = x_2 = x_3$ is taken as the unit point. Then by §4.5 there exists a symmetric matrix $A = (a_{ik})_{i,k=1,2,3}$, and thus a symmet-

---

[48]The pair of elements common to this involution and to the line involution with the same carrier and with the double lines g, h consists of the pair of angle-bisectors of g and h.

ric bilinear[49] form F, defined by

$$F(\sum_{i=1}^{3} c_i x_i, \sum_{k=1}^{3} c_k x_k) = \sum_{i,k=1}^{3} a_{ik} x_i y_k$$

such that $F(\mathfrak{x}, \mathfrak{y}) = 0$ is equivalent to $\mathfrak{x} \perp \mathfrak{y}$. Of course, the concept of orthogonality defines F only up to a scalar factor. By the argument at the end of §4.3 we can choose the basis in such a way that A is the unit matrix, so that $F(\mathfrak{x}, \mathfrak{y}) = \sum_{i=1}^{3} x_i y_i$ and in particular $c_1 \perp c_2 \perp c_3 \perp c_1$; i.e., *it is always possible to choose a basis consisting of mutually orthogonal vectors.*

The affine mappings that take mutually perpendicular lines into mutually perpendicular lines are called *similarities;* obviously they form a subgroup of the affine group, which is called the *equiform group* or *principal group.* They can also be characterized as those projective collineations that take the absolute polarity (in the plane, the absolute involution) into itself. For complex scalars the absolute involution is determined by its pair of double points, which are called the *absolute circular points.*[50] Similarly, the absolute polarity is determined by the conic section consisting of its selfconjugate points (see §6.1), which is called the *absolute conic section.*[51] A real plane cuts the absolute conic section in two points, namely, the absolute circular points of the plane. The proper lines through one of the absolute circular points, or though a point of the absolute conic section, i.e., the lines perpendicular to themselves, are called *isotropic.*[52] If we wish to avoid such lines in the complex plane as well, it is only necessary to replace the absolute polarity by a suitable nonprojective involutory correlation, namely, to take the unit matrix as the coefficient matrix in (25) and to define f as passage to complex conjugates, so that $F(\mathfrak{x}, \mathfrak{y})$ is replaced by $\sum_{i=1}^{3} \bar{x}_i y_i$ and therefore, exactly as in the real plane, $\mathfrak{x} \perp \mathfrak{x}$ implies $\mathfrak{x} = 0$. In other words, we introduce a *unitary* or *hermitian metric.*

5.2. Length

On parallel lines we can easily compare lengths by means of translations; for after choice of a vector $e \neq 0$ as *unit vector* the *length* or *magnitude* $|\mathfrak{x}|$ of the vector $\mathfrak{x} = ex$ (x = real number) can be defined as the nonnegative number $|x|$. In order to extend the definition of length to vectors $\mathfrak{x}$ linearly independent of e, we recall the theorem from elementary geometry that the diagonals of a parallelogram are perpendicular to each other if and only if the parallelogram is a rhombus, so that length is to be defined in such a way

---

[49]For this concept see II9, §§5.1 and 7.1; the bilinear form F is said to be symmetric if $F(x, y) = F(y, x)$ for all vectors x, y.

[50]Since every circle passes through them; see §6.3.

[51]Or also the *absolute sphere-circle*, since all spheres pass through it.

[52]That is "equal under turning," since an isotropic line in the plane remains fixed under any rotation about its real point.

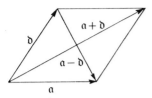

Fig. 12

that the two vectors $\mathfrak{a}$ and $\mathfrak{b}$ are equally long (Fig. 12) if and only if $\mathfrak{a} - \mathfrak{b} \perp \mathfrak{a} + \mathfrak{b}$. In the notation of §5.1, $\mathfrak{a} - \mathfrak{b} \perp \mathfrak{a} + \mathfrak{b}$ means that $F(\mathfrak{a} - \mathfrak{b}, \mathfrak{a} + \mathfrak{b}) = 0$, i.e., $F(\mathfrak{a}, \mathfrak{a}) = F(\mathfrak{b}, \mathfrak{b})$, since F is linear and symmetric. Thus it is natural to define the length of $\mathfrak{x}$ by means of $F(\mathfrak{x}, \mathfrak{x})$. Since the absolute polarity determines F only up to a factor, we can arrange that $F(\mathfrak{e}, \mathfrak{e}) = 1$ and therefore $F(\mathfrak{x}, \mathfrak{x}) \geqq 0$ for all $\mathfrak{x}$; for by §5.1 we can choose a basis such that the matrix of F is a multiple of the unit matrix, which means that $F(\mathfrak{x}, \mathfrak{x})$ has the same sign for all $\mathfrak{x} \neq 0$. After this normalization, by which F is now completely determined, let us write $\mathfrak{a} \cdot \mathfrak{b}$ instead of $F(\mathfrak{a}, \mathfrak{b})$ and speak of the *inner* (or *scalar*) product of $\mathfrak{a}$, $\mathfrak{b}$ in view of the fact that F now has the properties of scalar multiplication discussed in II9, §7.1; $\sqrt{\mathfrak{x} \cdot \mathfrak{x}}$ is the *length* or *magnitude* $|\mathfrak{x}|$ of the vector $\mathfrak{x}$. Since $\mathfrak{e}x \cdot \mathfrak{e}x = x^2$, this definition means, in particular, that $|\mathfrak{e}x| = |x|$ and therefore agrees with the definition at the beginning of this paragraph. If $\mathfrak{x} = \overrightarrow{AB}$ we shall also write $d(AB)$ in place of $|\mathfrak{x}|$. Vectors $\mathfrak{x}$ with $|\mathfrak{x}| = 1$ are called *unit vectors*.

The function which to the vector $\mathfrak{x}$ assigns the square of its length $\mathfrak{x} \cdot \mathfrak{x}$ $(= \mathfrak{x}^2)$ is called the *fundamental metric form*. It already determines the inner product, and therefore the absolute polarity, since

(29) $$(\mathfrak{a} + \mathfrak{b})^2 = \mathfrak{a}^2 + \mathfrak{b}^2 + 2\mathfrak{a} \cdot \mathfrak{b},$$

and it includes the *Pythagorean theorem* that *the lines CA, CB are perpendcular to each other if and only if $d(A, B)^2 = d(C, A)^2 + d(C, B)^2$*. Since "perpendicularity" is thus described in terms of "length," an affine mapping that leaves length invariant is also a similarity. Moreover, in view of (29), such a mapping leaves all inner products invariant.

By §5.1 there exists a basis $\{\mathfrak{e}_1, \mathfrak{e}_2, \mathfrak{e}_3\}$ with $F(\mathfrak{x}, \mathfrak{x}) = \sum_{i=1}^{3} x_i y_i$, so that $\mathfrak{e}_1^2 = \mathfrak{e}_2^2 = \mathfrak{e}_3^2 = 1$, $\mathfrak{e}_1 \cdot \mathfrak{e}_2 = \mathfrak{e}_2 \cdot \mathfrak{e}_3 = \mathfrak{e}_3 \cdot \mathfrak{e}_1 = 0$. Such a basis is said to be *orthonormal* and a coordinate system formed from it is called *Cartesian*.

5.3. *Isometries*

An affine mapping that leaves all lengths unchanged is called an isometry (or a *rigid mapping*).[53] The isometries obviously form a group, which by §5.2 is a subgroup of the equiform group.

[53]Cf. the discussion of isometries in III1, §10.

A similarity can be represented as the product of an isometry and a dilation. For by §5.2 equality of length of the vectors $\mathfrak{a}$, $\mathfrak{b}$ means the same as orthogonality of $\mathfrak{a} + \mathfrak{b}$, $\mathfrak{a} - \mathfrak{b}$ and thus remains unchanged under a similarity transformation. Then if the similarity takes the unit vector $\mathfrak{e}$ into the vector $\mathfrak{e}'$, the two vectors $\mathfrak{x}$ and $e|\mathfrak{x}|$ of equal length are taken into the vectors $\mathfrak{x}'$, $\mathfrak{e}'|\mathfrak{x}|$, which are likewise of equal length; in other words $|\mathfrak{x}'| = |\mathfrak{e}'|\,|\mathfrak{x}|$, so that all lengths are multiplied by a fixed factor.

For a given orthonormal basis $\{\mathfrak{e}_1, \mathfrak{e}_2, \mathfrak{e}_3\}$ an affine mapping is isometric if and only if the images $\mathfrak{e}'_i$ of the $\mathfrak{e}_i$ again form an orthonormal basis, since in an isometry $\mathfrak{e}'_i \cdot \mathfrak{e}'_k \cdot = \mathfrak{e}_i \cdot \mathfrak{e}_k$, which implies conversely that $\mathfrak{x}'^2 = \mathfrak{x}^2$ for the image $\mathfrak{x}' = \sum_{i=1}^{3} \mathfrak{e}_i x_i$ of the vector $\mathfrak{x} = \sum_{i=1}^{3} \mathfrak{e}_i x_i$; and then $\mathfrak{e}'_k = \sum_{i=1}^{3} \mathfrak{e}_i a_{ik}$ in the representation (16). Thus the $\mathfrak{e}_k$ form an orthonormal basis if and only if

$$\sum_{i=1}^{3} a_{ik} a_{il} = 0 \quad \text{for} \quad k \neq l, \quad \text{and} = 1 \quad \text{for} \quad k = l;$$

i.e., the matrix $A_0^T A_0$ is the unit matrix. *So in a Cartesian coordinate system the equations* (16) *represent an isometry if and only if the coefficient matrix* $(a_{ik})_{i=1,2,3}$ *is orthogonal.*

The determinant of an orthogonal matrix is equal to 1 or -1. Since $\det A_0$ is independent of the coordinate system (see II9, §6, or IB3, §3.7) there exist two kinds of isometries, namely, those with $\det A_0 = 1$ and those with $\det A_0 = -1$. The first are called *direct isometries* (or *motions*) and the second are called *opposite isometries*.[54] Direct isometries leave the orientation (cf. II9, §§6.2 and 7.4) of a trilateral unchanged, and opposite isometries change it. Translations ($A_0 = $ unit matrix) are direct isometries, *reflections* (in a plane in the three-dimensional case, and about a line in the two-dimensional case) are opposite isometries, since, e.g., reflection in the plane with equation $x_1 = 0$ is represented in a Cartesian coordinate system by $x'_1 = -x_1$, $x'_2 = x_2$, $x'_3 = x_3$.[55] Obviously the direct isometries form a group and, together with their products by a fixed opposite isometry, provide all the isometries, whereas the product of two opposite isometries is a direct isometry.

Let us note (see II5, §1) that every isometry is a product of reflections, which in turn are involutory similarities. Conversely, every involutory similarity is an isometry, since the (positive) factor by which all lengths are multiplied must be a square root of 1, and must therefore be $+1$. Thus the group of isometries is the subgroup of the equiform group generated by its involutory elements. This characterization of isometries, which makes no use of the concept of length, is valid for arbitrary dimensions.[56]

---

[54]For this nomenclature, cf. II4, §2, footnote.
[55]The corresponding statement holds for reflection in a line in the plane if we merely omit the third coordinate axis.
[56]See, e.g., Pickert [4], p. 351.

A direct isometry with a fixed point is called a *rotation* (around this point). For such a rotation (with E_0 as a fixed point) we may assume $a_{i0} = 0$ in (16). With E as the unit matrix we obtain

$$\det(A_0 - E) = \det(A_0 - A_0^T A_0) = \det(E - A_0^T) \cdot \det A_0$$

$$= \det(E - A_0^T) = \det(E - A_0)^T = \det(E - A_0) = -\det(A_0 - E),$$

and therefore $\det(A_0 - E) = 0$, so that the set of fixed points characterized by $x_i = x_i (i = 1, 2, 3)$ contains a whole range of points. Thus the coordinate system can be so chosen that the points with $x_1 = x_2 = 0$ remain fixed, which means that $a_{13} = a_{23} = 0, a_{33} = 1$. Since $e_1', e_2' \perp e_3' = e_3$, we also have $a_{31} = a_{32} = 0$, so that $x_3' = x_3$ and the equations

$$(30) \qquad x_1' = a_{11}x_1 + a_{12}x_2, \quad x_2' = a_{21}x_1 + a_{22}x_2$$

represent a rotation in every plane perpendicular to the third coordinate axis, the *axis of rotation*. In this way a rotation in space is completely reduced to a plane rotation.

Apart from translations and rotations, the only other direct isometries are the *screw displacements*, consisting of rotations followed by translations in the direction of the axis of rotation.[57] Correspondingly, in the plane the only direct isometries are the rotations and the translations, a statement easily proved in the following way. From a plane rotation (30) we obviously obtain an arbitrary direct isometry by adding a_1 and a_2. If this isometry is not a translation, in other words if the matrix of the a_{ik} is not the unit matrix, then the linear homogeneous system obtained from (30) by setting $x_i' = x_i$ has a nonzero determinant, as is easily seen from the discussion of the a_{ik} in §5.4. So the corresponding nonhomogeneous system (with absolute terms a_1, a_2) has a solution (see IB3, §3.6), and therefore the isometry has a fixed point, which means that it is a rotation.

5.4. Angles[58]

If in the representation (30) of a plane rotation about the coordinate origin we set $a_{21} = b, a_{22} = a$, the orthogonality of the coefficient matrix means that

$$a_{11}^2 + a_{12}^2 = 1 = a^2 + b^2, \quad a_{11}b + a_{12}a = 0, \quad a_{11}a - a_{12}b = 1$$

or (equivalently) $a_{11} = a, a_{12} = -b, a^2 + b^2 = 1$, and if we now admit all complex numbers, then (30), restricted to real points,[59] can be written

---

[57]See II9, §10, and also, e.g., Pickert [4], p. 188.

[58]Cf. the somewhat different treatment of angles in II9, §§7.1 and 8.3. If we wish to consider angles as geometric figures, we should here speak of "angle measure."

[59]In the complex field (31) is also valid but must be completed by the equation obtained from it by taking complex conjugates.

(under the sole condition[60] that $|a + ib| = 1$) in the form

(31) $$(x_1' + ix_2') = (a + ib)(x_1 + ix_2),$$

which shows that the multiplication of two rotations consists simply of multiplying together the corresponding factors $a + ib$. If we now look for a real number that will characterize the rotation and will behave additively when rotations are combined, it is natural to think of the familiar theorem in analysis that $a + ib = e^{i\varphi}$, i.e., $a = \cos \varphi$, $b = \sin \varphi$.[61]

Then φ is defined only up to integer multiples of 2π, and can thus be uniquely determined, for example, by the requirement $0 \leq \varphi < 2\pi$, which has the disadvantage that the additivity is lost. The angle φ, called the *angle of rotation*, enables us to write (30) in the form

(32) $\quad x_1' = x_1 \cos \varphi - x_2 \sin \varphi, \quad x_2' = x_1 = x_1 \sin \varphi + x_2 \cos \varphi.$

To what extent does φ depend on the coordinate system? Since $2 \cos \varphi$, being the sum of the diagonal coefficients in (32), is invariant (see IB3, §3.7), we need only investigate the sign of $\sin \varphi$ for $\sin \varphi \neq 0$, where the equation

$$x_1 x_2' - x_1' x_2 = (x_1^2 + x_2^2) \sin \varphi$$

shows that the ordered pair consisting of a vector and its image has the same orientation as (e_1, e_2), or the opposite orientation (II9, §6.2) according as $\sin \varphi > 0$ or < 0. So if we restrict ourselves to Cartesian coordinate systems with the same orientation, the value of φ does not depend on the coordinate system.

After choice of orientation, the angle of rotation taking the unit vector e_1 into the unit vector e is called the *oriented angle* between e_1 and e and, if $a|a|^{-1} = e_1$, $b|b|^{-1} = e$, it is also called the oriented angle between a and b. In order to show that this angle is uniquely determined, we need only take e_1 as the first basis vector, and the uniquely determined e_2 as the second, of an orthonormal basis with the given orientation, whereupon a, b are uniquely determined by $e = e_1 a + e_2 b$. In particular, $e_1 \cdot e = a = \cos \varphi$, so that

(33) $$\cos \varphi = \frac{a \cdot b}{|a| \, |b|}.$$

The angle φ uniquely determined by this equation under the restriction $0 \leq \varphi < \pi$ is called the *unoriented angle* between a and b. Since the equation (33) does not depend on the orientation, it is also convenient for defining an angle in space. Obviously the unoriented angle remains unchanged under any isometry. When a, b are replaced by ac, bc (with real c), equation (33)

---

[60] $|a + ib| = \sqrt{a^2 + b^2}$ is called the absolute value of the complex number $a + ib$.
[61] We here take the point of view that the functions cos, sin are defined analytically, say by power series.

Fig. 13   Fig. 14

remains unchanged ($c^2 = |c|^2$), which means that unoriented angles are not changed by similarity transformations (cf. §5.3).

The oriented or unoriented angle between two oriented lines is simply defined as the corresponding angle between the orienting vectors. If both lines are unoriented, there are four possibilities for orienting them, which give rise, however, to at most two values for the oriented angle and two values for the unoriented (Figs. 13 and 14). The oriented angle is restricted by $0 \le \varphi < 2\pi$, and for unoriented lines it is convenient to define the angle between them as the smaller of the two possible values. (Cf. the discussion of these four kinds of angle in III, No. 12.)

The oriented angle φ between unoriented lines can also be characterized as follows. Let us choose the coordinate system in such a way that e_1 and $e_1 \cos \varphi + e_2 \sin \varphi$ are vectors determining the two lines g, h (in this order). Then the hC of the improper points G, H on g, h are $(0, 1, 0)$ and $(0, \cos \varphi, \sin \varphi)$, and those of the absolute circular points K, K' are $(0, 1, -i), (0, 1, i)$. By §1.2 we then have[62]

(34) $$(KK'GH) = e^{2i\varphi},$$

and this equation, named after Laguerre, determines the angle up to integer multiples of π, which is exactly what we need (cf. Fig. 13). Choice of the other orientation, e.g., replacement of e_2 by $-e_2$, would mean interchange of the two absolute circular points in (34).

6. Conic Sections

6.1. *Polars and Tangents*

As mentioned above in §4.5, the set of selfconjugate points of a plane polarity is called a *conic section,* and the same name is often given to the corresponding sets of points for a degenerate polarity. It is easy to show (as was done above in §4.5 for three-dimensional space) that such a *degenerate conic section* is either (for rank 2) a *pair of lines*[63] or (for rank 1) a single

---

[62]Take $s = t = 1$, $s' = e^{i\varphi}$, $t' = e^{-i\varphi}$ in (5).
[63]More precisely, the set of points on two lines.

line,[64] but in what follows we shall pay no attention, in general, to these degenerate cases. We now allow the coordinates to be complex numbers but consider only *real polarities,* i.e., polarities in which every real point has a real polar. Then for a real coordinate system these polarities have a real matrix, every real line has a real pole, and restriction to the real plane produces a polarity.

Now let a conic section be given as the set κ of selfconjugate points of a polarity δ. On a line g with $\delta(g) \notin g$ the polarity δ produces an involution $X \to \delta(X) \cap g$ (cf. §4.3), so that (by §3.3) the conic section κ, if we allow complex scalars, is intersected by g in exactly two points, the double points of the involution. But if $\delta(g) \in g$, then no point of g, except $\delta(g)$ itself, can lie on κ, since such a point, being conjugate to itself and to $\delta(g)$, must be the pole of g and therefore $= \delta(g)$. Now $\delta(g) \in g$ means that g is the polar of a point of κ, namely the polar of $\delta(g)$. Lines with this property are called *tangents* of κ, so that δ takes κ precisely into the set of its tangents. The pole of a tangent is called its *point of contact.* As we have just shown, a given line is a tangent of κ if and only if it has only one point in common with κ,[65] and this point is then the point of contact of the tangent. So through a point $P \in \kappa$ there passes only one tangent, namely its polar, which is then said to be the tangent *to κ at P.* Since the polar of a point Y consists of the points conjugate to Y, its equation is given by (28). Thus we have obtained the equation of the tangent at Y, if Y lies on κ.

In order to show that the concepts introduced[66] here without limit processes coincide with those customary in differential geometry (see II14, §1), we take $E_0 \in \kappa$ and E_0, E_1 conjugate to each other, i.e., $E_0 E_1$ is tangent to κ at E_0, which means that $a_{00} = a_{01} = 0$ in the coefficient matrix of δ, so that the nonhomogeneous coordinates x_1, x_2 of a point $X \in \kappa$ satisfy the equation $a_{11}x_1^2 + 2a_{12}x_1x_2 + a_{22}x_2^2 + 2a_{02}x_2 = 0$. With $t = x_2/x_1$, $x_1 \neq 0$, we thus obtain the parameter representation

$$x_1 = -2a_{02}t(a_{11} + 2a_{12}t + a_{22}t^2)^{-1}, \quad x_2 = x_1 t,$$

which does in fact satisfy the equations

$$\lim_{t \to 0} x_1 = 0 = \lim_{t \to 0} x_2, \quad \left. \left(\frac{dx_2}{dt} \middle/ \frac{dx_1}{dt} \right) \right|_{t=1} = 0.$$

---

[64]Often also called a *double line;* but it is to be noted that this double line, regarded as a set of points, is in no way distinguished from a line; the special name merely indicates that we regard it as the set of selfconjugate points of a degenerate polarity.

[65]It is easy to see that we do not need complex scalars here. But we must note that the statement does not always hold in the affine plane, since every parallel to an asymptote (distinct from the asymptote itself) cuts a hyperbola in only one point.

[66]This definition without limit processes is possible, in a different way, for all algebraic curves; in defining a tangent there is no need to restrict oneself to such scalar domains as the real or the complex field.

Now let P be a point not on κ and let $p = \delta P$. If the line g through P is not a tangent, it has exactly two points D, D' in common with κ (in the complex field). Since P, $g \cap p$ is a point pair in the involution with double points D, D', the two points P, $g \cap p$ are harmonically conjugate to D, D' (see §3.3). But if g is a tangent, its point of contact, being conjugate to P, lies on p; and, conversely, the line joining P to a point Q lying both on κ and on p is the polar of Q and is therefore a tangent. *The polar p of P cuts κ in the points of contact of the tangents through P.* But if P lies on κ, then $\delta(P)$, namely the tangent at P, is the only line through P that has no second point in common with κ. Thus (in the complex plane), the polarity δ is completely determined by the conic section κ.

By applying δ to the involution induced by δ on a line not tangent to κ, we at once obtain the following result. If for a point P not on κ we assign to every line x through P the line $\delta(x)P$, the result is a line involution whose double lines are the tangents through P. For brevity, this involution is called the *line involution at P* (with respect to κ) and the lines x, $\delta(x)P$ are said to be *conjugate* to each other.

6.2. *Center Properties*

Let us now choose a line u as the improper line and thereby transfer our attention to the affine plane. For a conic section κ it is then of basic importance to distinguish two separate cases. If u is a tangent, κ is called a *parabola,* and otherwise the pole of u is a proper point M, called the *center* of κ, a name which comes from the fact that by §6.1 the points D, D' of κ lying on a line g through M are harmonically conjugate to M, $g \cap u$, so that M is the midpoint, or center, of the pair D, D'. The lines through M are called *diameters* of κ and the line involution in M with respect to κ is called a *diameter involution*. In the case of real scalars we must then ask whether a diameter involution has double lines or not. In the first case, i.e., when the diameter involution is hyperbolic, the conic section κ is called a *hyperbola,*[67] and in the second case, when the diameter involution is elliptic, κ is called an *ellipse* (but only if real points exist). The double lines for a hyperbola are called *asymptotes.* If κ has no real point, it is called an *imaginary conic.*

We now apply the results of §6.1 to the case $P \in u$ (see Fig. 15). Then $p = \delta(P)$ passes through $M = \delta(u)$ and is therefore a diameter, and PM is the diameter conjugate to p. The tangents of the two points $\in \kappa \cap g$ pass through P and are therefore parallel. For every other line g through P the point $g \cap p$ is the midpoint of the pair of points $\in \kappa \cap g$. The tangents at these points again pass through the pole of g, and therefore through a point of p.

---

[67]These names, and also the name "parabola," arose in antiquity from the study of the vertex equation (40) by "application of areas."

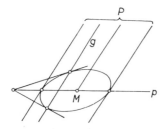

Fig. 15

In the parabola we must take for the point M, not a center but the point of contact of u. The straight lines $\neq u$ through this point are called *diameters*, and every diameter p obviously intersects κ in only one proper point. The relations indicated in Fig. 15 remain valid for the tangent at this point and the lines g parallel to it.

6.3.   *Axes*

Since "perpendicularity" has been introduced in §5.1 by means of an involution, it follows from §3.3 that for every conic section κ with center M there exists at least one pair of mutually perpendicular, conjugate diameters. If these diameters are chosen as coordinate axes, so that E_1, E_2 are their improper points and E_0 is the point M, then (independently even of the fact that $E_0E_1 \perp E_0E_2$) the three points E_i are pairwise conjugate, which means that $a_{01} = a_{02} = a_{12} = 0$ in the equation for κ, and thus $a_{00}, a_{11}, a_{22} \neq 0$. By suitable choice of the unit point we can ensure that the coordinate system is Cartesian, and the improper points of the asymptotes will then be determined as the solutions $\in \kappa \cap E_1E_2$ of the equation

$$(35) \qquad\qquad a_{11}x_1^2 + a_{22}x_2^2 = 0.$$

Thus a_{11}, a_{22} have the same or different signs according as κ is an ellipse (or an imaginary conic) or a hyperbola. Setting $x_0 = 1$, so that x_1, x_2 become nonhomogeneous coordinates, we can then, in all three cases, write the equation of κ as follows (in the third case it may be necessary to interchange E_1, E_2)

$$(36) \qquad \begin{aligned} x_1^2 a_1^{-2} + x_2^2 a_2^{-2} &= 1 \qquad \text{(ellipse)} \\ x_1^2 a_1^{-2} + x_2^2 a_2^{-2} &= -1 \qquad \text{(imaginary conic)} \\ x_1^2 a_1^{-2} - x_2^2 a_2^{-2} &= 1 \qquad \text{(hyperbola)} \end{aligned}$$

If κ has two pairs of mutually perpendicular, conjugate diameters, the diameter involution (see §3.3) coincides with the right-angle involution, i.e., conjugate diameters are always perpendicular to each other. This case holds if and only if (35) represents the absolute circular points. Since by §5.1 these points satisfy the equation $x_1^2 + x_2^2 = 0$, we obviously have $a_{11} = a_{22}$, so that (36) can be written in the form

(37) $$x_1^2 + x_2^2 = r^2.$$

If $r > 0$ is real, (37) describes the set of points at distance r from M, i.e., the *circle* with radius r. This name is also used for the more general case in which r^2 need only be real $\neq 0$, so that r can also be a pure imaginary number. *The circles are precisely those conic sections that pass through the absolute circular points.*

In a conic section that is not a circle the uniquely determined, mutually perpendicular, conjugate diameters are called its *axes*[68] and the points of the conic lying on the axes are called *vertices*. Since the values of a_1, a_2 (> 0) in (36) are uniquely determined[69] by the center and vertices, they are invariant under an isometry of the plane. Since any two Cartesian coordinate systems can be transformed into each other by an isometry, it follows that two central conics with the same a_1, a_2, differ only in their position on the plane.

In the parabola let M' denote the point corresponding in the absolute involution to the point of tangency M of u, so that every line through M' is perpendicular to every diameter. Since u is a tangent, there is only one proper tangent through M'. Its point of contact is called the *vertex* of the parabola, and the diameter through it, in other words, the polar of M', is called the *axis*. Then there exists a Cartesian coordinate system with $E_2 = M'$, $E_1 = M$, and the vertex as E_0. Since E_0, E_1 are self-conjugate and also conjugate to E_2, the equation of the conic becomes

(38) $$2a_{01}x_0x_1 + a_{22}x_2^2 = 0,$$

from which, by setting $x_0 = 1$ and $p = -a_{01}/a_{22}$ we obtain the affine equation of the parabola

(39) $$x_2^2 = 2px_1,$$

where it can clearly be arranged that $p > 0$. The fact that p remains unchanged under isometries and therefore characterizes the parabola (apart from its position on the plane) is most easily proved by considering the focus (see §6.4).

If the origin of the coordinate system (E_0) is shifted to a vertex on E_0E_1 $(-a_1, 0)$ for the ellipse and to $(a_1, 0)$ for the hyperbola, the equation takes

---

[68] Often also called *principal axes*.

[69] In the case of the ellipse and the imaginary conic section we must also make some such requirement as $a_1 \geq a_2$, since here the two axes are not distinguished, as they are for the hyperbola, by the fact theat one of the vertices is real and the other is imaginary. In the ellipse the a_1, a_2 are simply the distances from the vertices to the center, and in the hyperbola the same remark holds for a_1.

the form

$$(40) \qquad x_2^2 = 2px_1 + (\varepsilon^2 - 1)x_1^2,$$

involving the *parameter* $p = a_2^2/a_1$ and the *eccentricity* $\varepsilon = \sqrt{a_1^2 \pm a_2^2}/a_1$ (upper sign for the hyperbola, lower for the ellipse). For $\varepsilon = 1$, equation (39) for the parabola is then contained in the general *vertex equation* (40); $\varepsilon = 0$ characterizes the circles, $\varepsilon < 1$ the ellipses, and $\varepsilon > 1$ the hyperbolas.

From (36) and (39) we see that two conic sections of the same kind can be transformed into each other by a real affine mapping. Since by §6.2 the four kinds are characterized by affine concepts, it follows that two conic sections of different kinds cannot be transformed into each other by such a mapping. Of course, if arbitrary complex affine mappings are allowed, there remains only the distinction between parabolas on the one hand and the central conics on the other. In the projective plane even this distinction disappears, since the line u no longer plays a distinguished role. Two conic sections can be transformed into each other by a real projective collineation except in the case that exactly one of them contains real points. Finally, if we allow nonreal projective mappings, this sole remaining distinction also disappears. All conic sections can be transformed into one another by projective collineations.[70]

6.4. *Foci*

The proper point P is called a *focus* of the conic κ if it does not lie on κ and the line involution at P with respect to κ is the right-angle involution, i.e., for every line g through P the line joining P to the pole of g is perpendicular to g. Since an involution is determined by its double elements, P is a focus if and only if the tangents to κ through P are the isotropic lines through the absolute circular points K and K'. Thus all the foci are obtained by considering the intersections with each other of proper tangents through K and K'. For a parabola there is only one proper tangent through each of these two points, so that there is exactly one focus. For a circle, which passes through K and K', there is only one tangent of any kind for each point, and therefore only one focus, namely $\delta(K) \cap \delta(K') = \delta(KK') = \delta(u)$, i.e., the center of the circle. Conversely, by the results in §6.3 for a diameter involution, the center is a focus only for a circle. In every other case there are four foci, no three of them collinear. We now show that these four points lie on the two axes, so that there are exactly two of them on each axis. With B as focus and M as center we have $B\delta(MB) \perp MB$; but $M = \delta(u)$ implies $\delta(MB) \in u$, so that $B\delta(MB) \parallel M\delta(MB)$ and therefore $M\delta(MB) \perp MB$. Since $M\delta(MB)$ is the diameter conjugate to MB, these two lines are the axes of the conic. For the parabola ($M \in u$) the line $B\delta(MB)$ is parallel to the tangent at the proper point of the parabola on MB, so that again MB is an axis.

[70]As could already have been seen from §4.3.

In order to determine from (36) and (39) the position of the foci on the x_1-axis, we need only find those points $(c, 0)$ through which the isotropic lines $y = \pm i(x - c)$ are tangent to κ, i.e., have only one point in common with κ. A simple calculation shows that $c = p/2$ for the parabola and $c = \pm a_1 \varepsilon$ for the ellipse and the hyperbola. The other foci, which exist for the ellipse and the hyperbola, lie on the other axis and are not real, since their x_2-values are $\pm i a_1 \varepsilon$.

6.5. *Projective Generation of the Conics*

Given a conic π on the projective plane let us choose the coordinate system as in §6.3 (for the parabola), i.e., E_0, $E_1 \in \kappa$ and $E_0 E_2$, $E_1 E_2$ tangent to κ. Then the equation of κ has the form (38) and, since κ cannot consist solely of points $\in E_0 E_1$, $E_1 E_2$, $E_2 E_0$, we can still choose $E \in \kappa$ in such a way that (38) becomes

(41) $$x_2^2 = x_0 x_1.$$

Now the lines of the pencil at E_0 described by

(42) $$s x_1 - t x_2 = 0 \quad (s = t = 0 \text{ excluded})$$

intersect κ at points with the coordinates

(43) $$x_0 = s^2, \, x_1 = t^2, \, x_2 = st,$$

so that this *rational parametric representation* of π corresponds to the one-to-one mapping onto κ of the pencil of lines through E_0 that assigns the line (42) to the point (43) lying on it. But the same representation also assigns the line

(44) $$s x_2 - t x_0 = 0$$

to the point (43) lying on it and therefore corresponds to a one-to-one mapping onto κ of the pencil of lines through E_1. Since the CR are determined by the pair of parameter values s, t of the line, the correspondence between the lines (44) and the lines (42) is a projectivity of the one pencil of lines onto the other. But this projectivity can also be described by saying that the corresponding lines intersect on κ, and since all points of κ are obtained in this way and $E_0 E_1$ is not a fixed line, we obtain the following theorem.

Every conic section can be obtained as the set of points of section of corresponding lines of two distinct pencils of lines that are mapped projectively, but not perspectively, onto each other.[71]

The fact that every such projectivity generates a conic is proved as follows. For E_0, E_1 let us take the carriers of the pencils, for E_2 the intersection of the

[71]In §6.7 we will describe the connection between this theorem and the theorem on angles at the circumference of a circle.

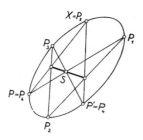

Fig. 16

image and preimage of the line E_0E_1, and for E the intersection with its own image of an arbitrarily chosen line $\neq E_0E_1$, E_0E_2 through E_0. The correspondence of (44) with (42) then actually maps E_0E_1, E_0E_2, E_0E onto the correct image lines and is therefore identical (see §2.2) with the given projectivity; and finally, as the intersection of (42) and (44) we obtain (43).

If E_0, E_1 are now chosen as two arbitrary points P, P' on κ, the projectivity is determined by three further points P_1, P_2, $P_3 \in \kappa$, since it must take PP_i into $P'P_i$ ($i = 1, 2, 3$) and by the argument dual to that of §3.1 it is the product of the perspectivity from P_2P_3 onto the line pencil at S (see Fig. 16) followed by the perspectivity of this pencil of lines from P_1P_2, so that every point $X \in \kappa$ is obtained by the construction of Fig. 16. With $P = P_6$, $P' = P_4$, $X = P_5$ we thus obtain *Pascal's theorem; For six distinct points* P_i ($i = 1, \ldots, 6$) *on a conic the points of intersection*

$$P_1P_2 \cap P_4P_5, \quad P_2P_3 \cap P_5P_6, \quad P_3P_4 \cap P_6P_1$$

lie on a line.

The dual of this theorem, in which the points are replaced by tangents, is called Brianchon's theorem. Coincident points P_i in Pascal's theorem, and coincident tangents in Brianchon's, are to be regarded as a pair consisting of tangent and point of contact, i.e., the tangent[72] replaces the line joining the points, and the point of contact replaces the intersection of the tangents.

From the above result it follows at once that for five points P, P_i ($i = 1$, 2, 3, 4) on κ the CR of the lines PP_i ($i = 1, 2, 3, 4$) is independent of P, and can therefore be defined as the CR of four points P_i on κ. In view of (42), this CR is determined from the parametric representation (43) exactly as though the s, t were the homogeneous coordinates of points on a line. A CR-preserving mapping of κ onto itself is called a *projectivity on κ*.[73] Any projective collineation of the entire plane that maps κ onto itself obviously

---

[72]For the proof of this degenerate case of the theorem there is no need of any limit process, since our definition of a tangent does not involve any such process; see, e.g., Pickert [4], p. 312 and 313.

[73]Many authors also use this name for a CR-preserving one-to-one mapping of κ onto a range of points, a pencil of lines, or another conic.

induces a projectivity on κ.[74] In particular, (41) shows that this is true for the central collineation with center E_2 and axis E_0E_1 represented by $x_0' = x_0$, $x_1' = x_1$, $x_2' = -x_2$. Since E_2 may be taken as an arbitrary point $\notin \kappa$ (and then E_0E_1 as its polar), the intersection of κ with any pencil of lines with carrier not on κ induces an involutory projectivity, or more briefly an *involution*, on κ with fixed points on the polar of the carrier of the pencil. Since an involution is determined by its double points, all involutions on κ are obtained in this way. So we have arrived at the *transference principle of Hesse; the points $P \notin \kappa$ correspond one-to-one to the involutions on κ, and in turn these involutions correspond, after projection from a point $\in \kappa$ onto a line, to all the involutions on the line.*

6.6. Pencils of Conics

Clearly, a conic passes through the four points E_0, E_1, E_2, E if and only if the coefficients of its equation satisfy the conditions $a_{00} = a_{11} = a_{22} = 0$, $a_{01} + a_{12} + a_{20} = 0$. Thus its equation can be written in the form

$$(45) \qquad ux_0(x_1 - x_2) + vx_2(x_1 - x_0) = 0.$$

Here the factors by which u and v are multiplied are simply the products of the left sides of the equations of the lines E_1E_2, E_0E and E_0E_1, E_2E, respectively, and this rule enables us to write down (45) immediately, even when the coordinate system has not been adapted as above to the four points. The conic passing through a given fifth point is then determined by substituting the coordinates of the point into (45), thereby determining u, v up to a common factor ($u = v = 0$ excluded).

Equation (45) is a special case of the more general

$$(46) \qquad u \sum_{i,k=0}^{2} a_{ik}x_ix_k + v \sum_{i,k=0}^{2} b_{ik}x_ix_k = 0,$$

where the only conditions on the u, v are that $u = v = 0$ is excluded and that for no pair u, v do we have $ua_{ik} + vb_{ik} = 0$ for all i, $k = 0, 1, 2$. Then for every choice of u, v equation (46) represents either a conic or a degenerate conic, and their totality is called a *pencil of conics*, which is said to be *generated* by the (possibly degenerate) conics with the equations $\sum_{i,k=0}^{2} a_{ik}x_ix_k = 0$, $\sum_{i,k=0}^{2} b_{ik}x_ix_k = 0$. Not every pencil of conics consists, like the one represented by (45), of all the conics and degenerate conics through four points.[75] The degenerate conics in (46) are obtained by setting the determinant of the matix $(ua_{ik} + vb_{ik})_{i,k=0,1,2}$ equal to 0. This homogeneous cubic equation in u, v (if not satisfied by all u, v) provides up to three distinct

---

[74]Conversely, every projectivity on κ can be extended in exactly one way to a projective collineation of the entire plane; see, e.g., Pickert [4], p. 330.

[75]See, e.g., Blaschke [2], p. 89, and Keller [3], p. 77.

values for the ratio of u to v, and in any case gives at least one such value (for real a_{ik}, b_{ik} certainly at least one real value). For the pencil (45) we see at once that the three degenerate conics are the three pairs of lines joining the four points in pairs; they are the three pairs of opposite sides[76] of the complete quadrangle with the four points as vertices. It is very convenient to use one of the pairs of lines in (46) to determine the points of intersection of the generating conics, since it is easy to see that this pair of lines has exactly these four points of intersection in common with one of the two conics.

Every line g having no point in common with all the conics of a pencil, cuts the conics of the pencil in point pairs of an involution.[77]
For if we denote by s, t the first two hC of one of the points of intersection on g, and set $a_{00} = a_0, a_{11} = a_1, 2a_{01} = a_2, b_{00} = b_0, b_{11} = b_1, 2b_{01} = b_2$, we have

$$u(a_0 s^2 + a_1 t^2 + a_2 st) + v(b_0 s^2 + b_1 t^2 + b_2 st) = 0,$$

i.e., the point (43) lies on the line

$$(47) \qquad u \sum_{i=0}^{2} a_i x_i + v \sum_{i=0}^{2} b_i x_i = 0.$$

The carrier of the pencil of lines (47) does not lie on the conic (43), since otherwise, with a suitable choice of s, t, the equation (47) would be satisfied for all u, v, so that one point of g would lie on all the conics in the pencil. Thus (47) produces an involution on (43). But since (43) obviously produces a projective mapping of g onto the conic, the theorem is thereby proved.

This theorem is very convenient for the problem of finding a conic through four given points (no three of which are collinear) tangent to a given line g not containing any of the four points. Namely, the pairs of lines of the pencil of conics determined by the four points provide three pairs of conjugate points of the involution, and a tangent conic must then pass through one of its double points.[78]

If we wish to determine double points of an involution from two given pairs of conjugate points, we need only assign a point P to the involution in accordance with the Hesse transference principle for a conic κ, where it is clear that two pairs of points of the involution are quite sufficient to determine P. Then the polar of P intersects κ in two points corresponding to the double points of the involution. If we take a circle for κ, the points of inter-

---

[76]Namely, sides with no vertex in common.

[77]Applying this theorem to the degenerate conics of the pencil (45), we obtain the relation already proved in §3.3 between an involution and a complete quadrangle.

[78]It is not necessarily true that a tangent conic will pass through a double point, since the carrier of one of the pairs of lines can lie on g and will then be a double point of the involution.

section of the polar with κ can be constructed by elementary geometry as the points of contact of the tangents through P. Another procedure for the construction of the double points of an involution will be given in the following section.

6.7. *Pencils of Circles*

A pencil of conics containing two circles κ, κ' consists entirely of circles (apart from its lines or pairs of lines), since the κ, κ' have the absolute circular points in common, and these points therefore lie on every conic in the pencil. In this case we speak of a *pencil of circles*. With the abbreviations

$$\kappa(X) = (x_1 - a_1)^2 + (x_2 - a_2)^2 - r^2,$$

$$\kappa'(X) = (x_1 - a_1')^2 + (x_2 - a_2')^2 - r'^2$$

for the left sides of the equations of κ, κ' in a Cartesian coordinate system, the equation of any circle of the pencil becomes

(48) $u\kappa(X) + v\kappa'(X) = 0.$

If κ, κ' have the same center M, then (48) under the condition $u + v \neq 0$ obviously represents all the circles with center M and the pair of isotropic lines through M, and the improper line is a further degenerate conic of the pencil. In the following discussion this case is excluded.

For $u + v = 0$, (48) becomes the equation of a line p, which from the projective point of vew is to be paired with the improper line u. The number $\kappa(X)$ $(= d(X, M)^2 - r^2$, where M is the center of κ) is called the *power* of X with respect to κ, and p is called the *radical axis* of κ and κ', where for κ, κ' we may, of course, take any two circles in the pencil; *a point lies on p if and only if it has the same power with respect to κ and κ'*.

If the radical axis cuts a pencil of circles at two points, (48) shows that at the same two points it must also cut κ, κ', and therefore every circle in the pencil. So the only other possibility is that p is tangent to every circle in the pencil at one and the same point. The two points G, G' thus determined on p (they may coincide) are called the *base points* of the pencil. Since the pencil is now determined by G, G' and the absolute circular points K, K', it must contain all the circles through G, G'.[79] The pairs of lines GK, $G'K'$ and GK', $G'K$ are now seen to be the further pairs of lines belonging to the pencil. If we set $N = GK \cap G'K'$, $N' = GK' \cap G'K$, the theorem on the complete quadrangle (see Fig. 17) shows, for $N \neq N'$, that $NN' \cap u$, $p \cap u$ are har-

---

[79]This statement also holds for $G = G'$ if by "passing through G, G''' we mean tangency of p at G; for then, corresponding to (45), the equation of a conic passing through E_0, E_1, E and touching $E_0 E_1$ can be written in the form $u x_1 (x_0 - x_2) + v x_2 (x_1 - x_2) = 0$.

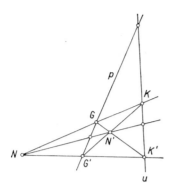

Fig. 17

monic to K, K' and thus $NN' \perp p$.[80] Since $NN' \cap p, p \cap u$ are also harmonically conjugate to G, G', we see that NN' is the polar of $p \cap u$ with respect to every pencil in the circle, so that the midpoint of such a circle, since it is the pole of u, must lie on NN'.[81] Thus NN' is called the *central line* of the pencil. If $G = G'$, the line NN' is to be replaced by the perpendicular to p at G ($= G' = N = N'$). Since with n_1, n_2 as coordinates of N the pair of isotropic lines NK, NK' has the equation

$$(x_2 - n_2 + i(x_1 - n_1))(x_2 - n_2 - i(x_1 - n_1)) = 0,$$

or

$$(x_2 - n_2)^2 + (x_1 - n_1)^2 = 0,$$

such a pair of lines, or also the point N, is called a *null circle*. Of course, in the case $G \neq G'$, not all four points G, G', N, N' can be real. In fact, one can show that, if the base points are nonreal, the points N, N' are real and conversely.[82]

We now set ourselves the task of finding those circles in a given pencil that are tangent to a given line g. By §6.6, this means that we must construct the double points D_1, D_2 of an involution with the pairs of conjugate points P, P' and Q, Q' (Fig. 18); for in a circle every diameter is perpendicular to its conjugate diameter and therefore (see §6.2) perpendicular to the tangents at its endpoints, so that the centers M_1, M_2 of the touching circles lie on the perpendiculars to g at D_1, D_2. Since GG' and the improper line together constitute a pair of lines in the pencil, it follows that $O = GG' \cap g$ and the

[80]Of course, this result could also be obtained by direct calculation from (48).

[81]The line NN' is even the right bisector of G, G'. The fact that the center of every circle in the pencil must lie on this line, follows at once from the results of §6.2 if we note that here any two conjugate diameters are perpendicular to each other.

[82]See, e.g., Pickert [4], p. 224.

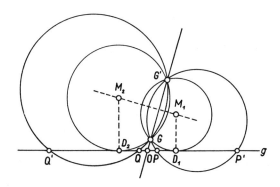

Fig. 18

improper point U on g form a pair of conjugate points of the involution. Thus we may apply (21) with $O = E_0$, $U = E_1$. Then D_1, D_2 are real if and only if O does not lie between P, P' and thus also not between Q, Q'; and in this case we have the further result $d(O, D_1)^2 = d(O, D_2)^2 = d(O, P) \cdot d(O, P') = d(O, Q) \cdot d(O, Q')$ ($=$ the power of O with respect to any circle in the pencil).[83]

To construct the square root of a product of segments, as we have just now had to do, it is customary to use the theorem of Thales on the angle at the circumference of a semicircle, which is, of course, a special case of the theorem that the angle at the circumference is equal to the angle between chord and tangent. The latter theorem is subsumed under the present projective concepts in the following way. For a circle κ, we consider the projectivity at the beginning of §6.5 for the pencil of lines through E_0 onto the pencil through E_1, where E_0, E_1 are proper points. By the fundamental theorem of projective geometry, this projectivity is also characterized by the fact that it takes $E_0 K$ into $E_1 K$, $E_0 K'$ into $E_1 K'$ and $E_0 E_1$ into $E_1 E_2$. But any projectivity with these properties is the product of a rotation about E_0 and a translation by $\overrightarrow{E_0 E_1}$, so that, for every point X on the circle, the oriented angle between $E_0 X$, $E_1 X$ ($=$ angle at the circumference) is in fact equal to the oriented angle between $E_0 E_1$, $E_1 E_2$ ($=$ angle between chord and tangent).

7. Quadric Surfaces[84]

7.1. *Introduction*

A set of self-conjugate points of a polarity or of a degenerate polarity in

---

[83]If D_1, D_2 are not real, we thus obtain $-d(O, P) \cdot d(O, P') =$ the power of O. Taken together, the two cases are equivalent to the theorem that the product of segments of chords is constant.

[84]Also called quadrics, for brevity.

three-dimensional space is called a *quadric surface* or a *degenerate quadric surface*.[85] As in §6.1, we now admit all complex numbers as coordinates but employ only real polarities. If the plane **E** does not contain its pole $\delta(\mathbf{E})$ with respect to the quadric **K**, then no point of **E** is conjugate to all the points of this plane, so that δ induces a polarity in **E** (cf. also §4.3); in other words, **K** ∩ **E** is a conic. On the other hand, if $\delta(\mathbf{E}) \in \mathbf{E}$, then **E** is called a *tangent plane* and $\delta(\mathbf{E})$ is its *point of contact;* and in this case **K** ∩ **E** is a degenerate conic, namely a pair of lines; for δ induces a degenerate polarity on **E**, which must be of rank 2, since otherwise every point of a line would be conjugate to all points of **E** and would therefore be $= \delta(\mathbf{E})$. The line g through a point $P \in \mathbf{K}$ then contains no further point of **K** if and only if it lies in the tangent plane $\delta(P)$ but not in **K**. For on the one hand, $P \neq Q \in \mathbf{K}$, $\delta(P)$ implies $P \in \delta(Q)$, $Q \in \delta(Q)$, which means $PQ \subset \delta(Q)$, $PQ \subset \delta(P)$ and therefore $PQ = \delta(Q) \cap \delta(P) = \delta(PQ)$, so that $X \in PQ$ always implies $PQ \subset \delta(X)$, i.e., every point of PQ is selfconjugate and thus lies on **K**; and on the other hand, $g \subset \delta(Q)$ does not hold for any point Q of the line g not in the plane $\delta(P)$, since for $Q \neq P$ the point Q is not conjugate to P, and therefore δ induces on g the involution $X \rightarrow g \cap \delta(X)$, which must have another double point in addition to P. The lines in the tangent plane through the point of contact are called *tangents*.

Let us now transfer our attention to the affine space by distinguishing a plane **Y** as the improper plane. If $\delta(\mathbf{Y}) \notin \mathbf{Y}$, $M = \delta(\mathbf{Y})$ is called the *center* of **K**, and it is easy to show that this point has the properties corresponding to those of the center of a conic. We now take the center for E_0, so that $a_{01} = a_{02} = a_{03} = 0$ in the equation of the quadric. By §4.4 the conic **K** ∩ **Y** and the absolute conic have a polar triangle in common, whose vertices we take for E_1, E_2, E_3. Then $a_{12} = a_{23} = a_{31} = 0$, and by a suitable choice of the unit point the coordinate system with the axes $E_0 E_i$ becomes Cartesian. Because the rank of the corresponding matrix is 4, we have $a_{ii} \neq 0$, so that with $x_0 = 1$, i.e., with x_1, x_2, x_3 as nonhomogeneous coordinates, the equation of the quadric becomes

$$(49) \qquad \sum_{i=1}^{3} a_i x_i^2 = 1 \qquad (a_i \neq 0).$$

Since replacement of the right side by 0 produces the equation of **K** ∩ **Y** (in hC), we see that as far as real affine mappings are concerned (§6.3) we may distinguish, according to the signs of the a_i, the following four kinds of quadric surfaces with a center: *ellipsoid* (+ + +), *hyperboloid of one sheet* (+ + −), *hyperboloid of two sheets* (+ − −), and *imaginary quadric*

---

[85]Often the corresponding configuration in n-dimensional space are called quadrics, and often also the degenerate quadrics are called quadrics, as in II9, §5, where, however, the quadrics are considered only in the *affine* plane, so that for purposes of clsssification only the affine mappings are available.

$(-\ -\ -)$. Here the ellipsoids are distinguished from the imaginary quadric surfaces by the presence of real points, and the one-sheeted hyperboloids from the two-sheeted by the presence of real lines. Since $\sum_{i=1}^{3} x_i^2 = 0$ is the equation of the absolute conic, **K** passes through this conic if and only if $a_1 = a_2 = a_3$, whereupon **K** is called a *sphere;* for $a_1 > 0$ this sphere obviously consists of the points X with $d(M, X) = 1/\sqrt{a_1}$.

If $\delta(\mathbf{Y}) \in \mathbf{Y}$, let $\mathbf{K} \cap \mathbf{Y}$ consist of the lines u, u', and set $\delta(\mathbf{Y}) = u \cap u' = E_3$. Then by §3.3 the line involution at E_3 with respect to the absolute conic has a real pair of conjugate lines g, g', which also belongs to the involution with the double lines u, u' and is therefore harmonic to u, u'. On g and g' we now choose as E_1, E_2 the points conjugate to E_3 in the absolute polarity, so that for arbitrary choice of E_0 ($\in \mathbf{Y}$) and suitable choice of E we obtain a Cartesian coordinate system, where E_1, E_2 are conjugate to each other with respect to **K** and also conjugate to E_3. The points conjugate to E_1, E_2 (with respect to **K**) now fill up the line $\delta(E_1 E_2)$ through E_3, and this line does not lie on **Y**, since $\delta(E_1 E_2) \subset \mathbf{Y}$ would imply $E_3 = \delta(\mathbf{Y}) \in E_1 E_2$. Thus $\delta(E_1 E_2)$ contains a further point of **K**, which we may choose as E_0. In this way we have arranged that $a_{00} = a_{33} = a_{13} = a_{23} = a_{01} = a_{02}$ $a_{12} = 0$ and thus $a_{03} \neq 0$, so that after division by a_{03} the equation of **K** with $x_0 = 1$ can be written

(50) $$a_1 x_1^2 + a_2 x_2^2 = 2x_3.$$

As with (49) above, we may now distinguish the following two types, with respect to real affine mappings: *elliptic paraboloid* $(+\ +)$ and *hyperbolic paraboloid* $(+\ -)$; in the first case $\mathbf{K} \cap \mathbf{Y}$ consists of nonreal lines and in the second of real lines.

For the degenerate quadric surfaces of rank 3, i.e., for the cones, §4.5 shows that we need only consider the intersection with an (arbitrarily chosen) plane not through the vertex of the cone. If the vertex is a proper point, we may choose this plane as **Y**. Since the affine mappings of the space induce the projective collineations in **Y**, the discussion in §6.3 shows that for proper cones we can distinguish only the following two cases, with respect to real affine mappings: cones with real generators and cones without real generators.[86] If the vertex is improper, we speak of a *cylinder*. In this case the intersecting real plane must be chosen $\neq \mathbf{Y}$, so that, according to the nature of its intersection, we may distinguish the following types: *elliptic cylinder, hyperbolic cylinder, cylinder without real generators, and parabolic cylinder.*

How this classification of quadric surfaces is altered by considering projective space, or nonreal mappings, is easily determined in the same way as at the end of §6.

---

[86] By the *generators* of a cone we mean the lines joining the vertex to the other points of the cone; thus the generators lie entirely on the cone.

7.2. *Reguli*

Through each point P of a quadric **K** there pass (by §7.1) exactly two lines[87] lying entirely in **K**, namely, the two lines in which **K** intersects the tangent plane at P. For $E_0 \in$ **K** we let g_1, g_2 denote these two line. Then any point of **K** not on g_1, g_2 may be chosen as E_3. The plane $E_3 g_i$ cannot intersect **K** in a conic, since a conic does not contain any line. Thus $E_3 g_i \cap$ **K** is a pair of straight lines or, in other words, through E_3 there passes a line $g_i' \subset$ **K** meeting g_i. Since $E_3 \notin g_1 g_2$ and $E_0 \notin g_i'$, it follows that $g_1' \neq g_2'$, so that these must be the two lines $\subset$ **K** through E_3. We may now set $E_i = g_i \cap g_i'$ $(i = 1, 2)$ and take $E \in$ **K**. Since $E_0 E_i, E_3 E_i \subset$ **K** $(i = 1, 2)$, the coefficients in the equation for the quadric must satisfy $a_{00} = a_{33} = a_{ii} = a_{0i} = a_{3i} = 0$ $(i = 1, 2)$, so that in view of $E \in$ **K** the equation of **K** becomes

$$(51) \qquad\qquad x_0 x_3 = x_1 x_2.$$

We see that for every pair of values of s, t and s', t', excluding $s = t = 0$ and $s' = t' = 0$, the equations

$$(52) \qquad\qquad s x_0 - t x_1 = 0, \quad s x_2 - t x_3 = 0,$$

as well as

$$(53) \qquad\qquad s' x_0 - t' x_2 = 0, \quad s' x_1 - t' x_3 = 0,$$

represent a line lying on **K**. Since the coefficient matrix of the four equation (52), (53) is of rank 3, the two equations (52), (53) have exactly one point in common, so that, in particular, they do not coincide. It is clear that exactly one line (52) passes through each point of **K**, and exactly one line (53). Thus (52), (53) represent all the lines on **K**. Since two lines of the same type are easily seen to be skew to each other, each of the two types includes exactly those lines that lie on **K** and intersect a fixed line on **K**, so that the distinction between the two kinds is independent of the coordinate system. The totality of lines of one type is called a *regulus* on **K**. The equation (52) also shows that a regulus can be generated as the intersection of two projectively related pencils of planes with skew carriers, namely $E_2 E_3$ and $E_0 E_1$, a result which corresponds exactly to the projective generation of the conics (see §6.5).

7.3. *Circular Intersections*

A real plane **E** not tangent to the quadric **K** cuts K in a circle if and only if its absolute circular points lie on **K**, i.e., if its improper line $u =$ **E** $\cap$ **Y** meets the absolute conic[88] κ_Y in points lying **K** $\cap$ **Y**. Thus if **K** $\cap$ **Y** $= \kappa_Y$,

---

[87]However, it is only for the hyperboloid of one sheet and for the hyperbolic paraboloid that these lines are real.

[88]Since the absolute circular points lie on this conic (see §5.1).

so that **K** is a sphere (see §7.1), then every real nontangent plane cuts **K** in a circle. But if **K** is not a sphere, the intersection will be a circle only if u passes through two points of intersection of $\kappa_\mathbf{Y}$ with **K** $\cap$ **Y**. From (49), (50) it is easy to calculate that there can be at most two real lines with this property. Thus the real planes intersecting **K** in circles fall into two (possibly coincident) parallel pencils. The pencils coincide if and only if in (49) two of the a_i are equal or if $a_1 = a_2$ in (50), and then we are dealing with a *surface of rotation*. If **K** $\cap$ **Y** is a pair of real lines, the carrier u for each of the two pencils of planes must be one of the lines of this pair. But then no plane through u can cut **K** in a conic and therefore every such plane must be a tangent plane. Consequently, the hyperbolic paraboloid has no circular intersections (in real planes). When the same argument is applied to cones and cylinders, we see that only the hyperbolic and parabolic cylinders have no circular intersections (in real planes).

7.4. *Stereographic Projection*

Let P be a point of the sphere **K** and let **E** be a plane parallel to (and distinct from) the plane tangent to **K** at P. Then by $X \to PX \cap$ **E** the points on the sphere $\neq P$ are mapped one-to-one onto the (proper) points of **E**, a mapping which is called *stereographic projection* (cf. also II12). It is clear that the circles on the sphere that pass through P are mapped one-to-one onto the totality of (proper) lines of **E**. We now wish to show that the circles on the sphere not through P correspond one-to-one to the circles on **E**. Since the tangent plane at P is parallel to **E**, the two lines f, f' in which it cuts **K** pass through the two circular points of **E**. Every circle κ on **K** not through P cuts f, and also f', in a point $\neq P$, so that the stereographic image of κ passes through the absolute circular points of **E**. Since this image is also a conic, in view of the fact that it is the intersection of a cone, it must therefore be a circle. On the other hand, any circle on **E** is determined by three noncollinear points and thus, by projecting these points onto **K**, we can at once find the circle on the sphere into which the given circle is mapped. In this way we have also shown that pencils of circles in **E** correspond to sets of circles on **K** intersected by pencils of planes.

We now wish to show that our mapping preserves angles. The angle between two circles at one of their points of intersection is defined as the angle between their tangents at this point. The tangents, at a point Q, to circles on the sphere **K** have only one point in common with **K** and therefore lie in the tangent plane Γ at Q. Under projection from P they go into tangents of the image circles.[89] By §7.2, the two lines $\Gamma \cap$ **K** can be denoted by g, g' in such a way that g is cut by f and g' by f' (see Fig. 19). Thus if Q' is the image point

---

[89]For a circle lying on the sphere and passing through P such a tangent naturally goes into the image line of the circle.

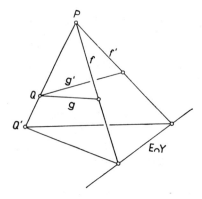

Fig. 19

of Q, the lines g, g' go into the lines joining Q' to the absolute circular points of **E**. Since g, g' are the lines joining Q to the absolute circular points of **Γ** and since projection of the plane **Γ** from P onto the plane **E** preserves the CR, we see by (34) that, as desired, the magnitude of angles is left unchanged.

Bibliography

[1] BAER, R.: Linear algebra and projective geometry. New York 1952.
[2] BLASCHKE, W.: Projektive Geometrie. 3rd ed., Basel and Stuttgart 1954.
[3] KELLER, O. H.: Analytische Geometrie and lineare Algebra. 2nd ed., Berlin 1963.
[4] PICKERT, G.: Analytische Geometrie. 6th ed., Leipzig 1967.
[5] WEISS, E. A.: Liniengeometrie and Kinematik. Leipzig 1935.

CHAPTER II

Algebraic Geometry

Introduction and Historical Outline

The mathematics of Greek antiquity dealt not only with conics, as in the well-known work of Apollonius, but also with various higher curves, most of them introduced as geometric loci for the solution of problems like doubling the cube or trisecting an angle. Such curves are the cissoid of Diocles, the conchoid of Nicomedes, or the spiric curves of Perseus, which are now called special algebraic curves of third and fourth degree. But transcendental curves, like the spiral of Archimedes, are also to be found in the mathematics of ancient Greece.[1]

Of course, this distinction between algebraic and transcendental curves according to whether they can or cannot be defined by an algebraic equation in Cartesian coordinates could not be made by the ancient Greeks, since they lacked the necessary algebraic equipment. Consequently, it was impossible for them to discuss the higher curves in any systematic way. For curves of the third degree this was first done by Newton in the year 1704, and subsequent authors, particularly Cramer in 1750 and Plücker in 1839, made many discoveries in the rich, varied field of higher plane curves. Algebraic surfaces of the second degree had been investigated since the 17th and 18th centuries (Fermat and Euler) but only in the 19th century was work begun on surfaces of higher degree, from which arose the great, far-reaching mathematical discipline of modern algebraic geometry.

Algebraic geometry may be understood as a continuation of the branch of mathematics which in schools and colleges is called analytic geometry. For the most part, analytic geometry is confined to the study of geometric figures with linear or quadratic equations, i.e., lines, planes, conics, etc., but alge-

---

[1]These and many other curves are discussed in the book of Hermann Schmidt, intended for secondary schools, Ausgewählte höhere Kurven, Wiesbaden, 1949.

braic geometry, properly so-called, considers figures whose equations are of arbitrarily high degree. As a result, algebraic geometry deals with numerous special questions about the form, and in particular the projective classification, of higher algebraic curves and surfaces, questions which are naturally much more complicated and comprehensive than the corresponding questions for conics, so that a higher point of view has gradually been adopted. The whole discipline is now centered on the study of curves, surfaces, and sets of points in higher dimensions, not with respect to their invariance under projective transformations, but under the more general birational transformations, so that the resulting equivalence classes are now much larger. In this chapter we shall explain the essential features of birational transformations by means of simple examples.

Even in the 19th century various branches of mathematics, such as projective geometry, the theory of functions of a complex variable, and topology had a combined influence on algebraic geometry. After a golden age in Germany, associated with the names of Brill and Noether, further progress, since the beginning of the 20th century, was made chiefly in Italy, where the leading names are Segre, Enriques, Castelnuovo, and Severi. Then modern algebra and topology produced a renaissance in the whole discipline, in which the main emphasis was placed on deepening the foundations, though with much other progress.

We will now try to give an approximate idea of the subject by means of the simplest algebraic curves and surfaces and the simple mappings already familiar to students in secondary school.

I. Remarks on Plane Algebraic Curves

1.1. *The Conic as an Algebraic Curve*

After the straight lines, the simplest algebraic curves on the plane are the conics. In real or complex homogeneous projective coordinates the equation of a conic section is

$$(1) \qquad f = \sum_{i,j=0}^{2} a_{ij} x_i x_j = 0$$

with real or complex constants a_{ij} (cf. II10, §§6 and 4.5). Since the left side of (1) is an entire rational function of second degree, the conics are called algebraic curves of *order* 2. In general, by an algebraic curve of order k we mean the locus of all points on the plane whose homogeneous coordinates satisfy the equation obtained by setting a homogeneous, entire rational function of degree k (IB4, §2.3) equal to zero. Those points on the curve in which the first partial derivatives with respect to the three homogeneous coordinates all vanish are called *singular* points. Thus the singular points of the conic (1) must be nontrivial solutions of the homogeneous linear system of equations

(2) $$\sum_{j=0}^{2} a_{ij}x_j = 0 \qquad (i = 0, 1, 2)$$

which exist only if

(3) $$|(a_{ij})_{i,j=0,1,2}| = 0,$$

i.e., if the determinant of the equation for the conic vanishes. But then, as is well known, the conic consists of a pair of lines G, H. If G and H are distinct, the solution of the system (2) consists of the coordinates of one point, the intersection of G and H (the matrix (a_{ij}) is then of rank 2); but if G and H coincide, we speak of a double line, and then every point on such a line satisfies (2) and is therefore singular (the matrix (a_{ij}) is now of rank 1). Thus the nondegenerate conics are algebraic curves without singular points.

A further property of the conics as algebraic curves is that if (u_0, u_1, u_2) are homogeneous *line coordinates*[2] on the plane and if the conic (1) is not degenerate, then the coordinates of its tangents satisfy the condition

(4) $$\sum_{i,j=0}^{2} A_{ij}u_iu_j = 0,$$

where (A_{ij}) is the inverse of the matrix (a_{ij}). This fact is also expressed by saying that the conic is an algebraic curve of class 2. Here by the class of a curve we mean the degree of its algebraic equation in the line coordinates of its tangents. Conversely, the conics are the only plane algebraic curves that are both of order 2 and of class 2.

Finally, in the sense of algebraic geometry, the most important property of the conics is that they admit a so-called *rational parametrization,* which is most conveniently obtained by transforming to a coordinate system in which the given nondegenerate conic K^2 has the form

(5) $$x_1^2 - x_0x_2 = 0.$$

If K^2 is not imaginary, this form is easily obtained by a real transformation of coordinates and we can also arrange that two given real points A and B on K^2 have the coordinates $(1, 0, 0)$ and $(0, 0, 1)$. Then every point P on K^2 is determined by the intersection of the two lines

(6) $$sx_1 + tx_2 = 0 \qquad \text{and} \qquad sx_0 + tx_1 = 0,$$

one each from the pencils through the base points A and B (Steiner's projective generation of K^2). From (6) it follows at once that the coordinates of the point P are

(7) $$x_0 = t^2, \quad x_1 = -st, \quad x_2 = s^2.$$

---

[2]Cf. II10, §1.1, and II3, §3.3.

If we now regard (s, t) as homogeneous coordinates of the points of a line K^1, we see at once that to each point of the line K^1 there corresponds a point on the conic K^2. Conversely, for each point $P(x_0, x_1, x_2)$ of K^2 it follows from (7) and (5) that $s:t = -x_1:x_0 = -x_2:x_1$. Consequently, the points of K^2 can be put into one-to-one *birational* (i.e., rational in both directions) correspondence with the points on K^1. In algebraic geometry these facts are expressed by saying that the conic admits a rational parametrization, or that it is a rational curve, or also that it is of genus 0.

1.2. *Example of a Curve of Third Order with a Double Point*

After the conic sections, the simplest plane algebraic curves are the so-called cubics or curves of third order. Here too we can find examples from the secondary school; e.g., the rational function

(1)
$$y = \frac{ax^2 + 2bx + c}{px^2 + 2qx + r}$$

is almost always discussed in elementary calculus. If we assume $q^2 - pr > 0$ and $b^2 - ac > 0$, the function (1) has two real poles and two real zeros, i.e., there exist two points at which the curve defined by (1) cuts the x-axis and the curve has two asymptotes parallel to the y-axis. The shape of the curve is illustrated in Fig. 1. But the same curve can also be discussed in the projective plane. For this purpose we replace x and y in (1) by the quotients of the homogeneous coordinates x_1/x_0 and x_2/x_0 and then multiply out all the denominators. In the projective plane the equation of the curve becomes

(2) $(px_1^2 + 2qx_0x_1 + rx_0^2)x_2 - (ax_1^2 + 2bx_0x_1 + cx_0^2)x_0 = 0,$

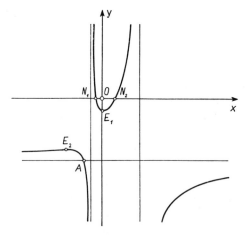

Fig. 1. $y = \dfrac{-5x^2 + 2x + 3}{x^2 - 2x - 3}$

from which we deduce the following facts.

a)   The curve (2) is of order 3, i.e., a cubic.

b)   The three partial derivatives of the entire rational function (2) all vanish at the point with coordinates $(0, 1, 0)$. Thus by the definition in §1 the cubic has a singular point at $(0, 1, 0)$, i.e., at the point at infinity on the y-axis, but unlike the conics, the cubic does not thereby degenerate into curves of lower degree (say a line and a conic, or three lines). At the point at infinity $(1, 0, 0)$ on the x-axis, however, which also lies on the curve, the derivatives of (2) do not all vanish, i.e., the point at infinity is a so-called simple or ordinary point. If by a projective transformation we bring the three points $(1, 0, 0)$, $(0, 1, 0)$ and $(0, 0, 1)$ into the finite part of the plane, the transformed curve has the shape indicated in Fig. 2. In particular, the singular point $(0, 1, 0)$ is called a double point, since, as can be seen both intuitively and by calculation, there pass through it two so-called branches of the curve, with the tangents

$$(3) \quad px_1 + (q - \sqrt{q^2 - rp})x_0 = 0 \quad \text{and} \quad px_1 + (q + \sqrt{q^2 - rp})x_0 = 0.$$

The lines (3), which under the assumption $p \neq 0$ and $q^2 - rp \neq 0$ are certainly distinct from each other, are seen to be the two asymptotes of the curve parallel to the y-axis. For if we return to nonhomogeneous coordinates $(x_0 = 1)$, the product of the left sides of (3) coincides, up to the factor p, with the denominator of the function (1).

c)   The second derivative y'' of the function (1) has a polynomial of third degree in its denominator, which means that there will be three points, in general not always real, in which y'' vanishes; these are the familiar points of

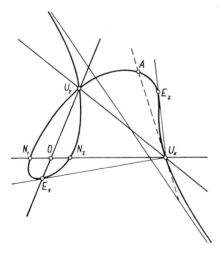

Fig. 2. Projective image of the curve in Fig. 1.

inflection. They can be described as ordinary points at which the tangent (for cubics) has no point in common with the curve except its point of contact. For real coefficients in (2) the three points of inflection are all real or else one of them is real and the other two are conjugate complex. It is clear that there are no points of inflection on the conics; they are first encountered on the cubics.

d)   The condition satisfied by the line coordinates of the tangents of our curve will not be calculated here. It is an algebraic equation of fourth degree, i.e., the class of our cubic is 4. Thus the order and the class need not coincide for higher curves, as they do for conics. In §1.6 we will give the formulas expressing the connection, discovered by Plücker, between order and class of an algebraic curve. The class k of a curve is dual to its order and can be defined geometrically as the greatest number of distinct (not always real) tangents that can be drawn to the curve from an arbitrary point. A simple argument from continuity shows that from an ordinary point P (not a point of inflection) on the curve we can draw $k - 2$ tangents $(k > 2)$ to a curve of class k, in addition to the tangent at P. If for our curve (2) we choose for P the point at infinity $(1, 0, 0)$, which we have already determined to be simple, we see that the line at infinity $x_0 = 0$ is a tangent at P, which is unlike a tangent at a point of inflection in that it cuts the curve again, i.e., at the point $(1, 0, 0)$. But the differential calculus shows that apart from $x_0 = 0$ there are two further tangents that pass through P, i.e., are parallel to the x-axis. From this point of view the class $k = 4$ becomes understandable.

1.3.   Neil Parabola

A curve of third order often drawn in secondary school is the so-called Neil parabola (Fig. 3) with the equation

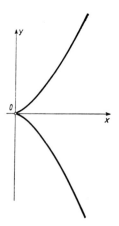

Fig. 3. $y^2 - x^3 = 0$

(1) $y^2 - x^3 = 0$, or in homogeneous coordinates, $x_0 x_2^2 - x_1^3 = 0$.

The origin, with homogeneous coordinates $(0, 0, 1)$, is seen to be singular. An easy calculation, left to the reader, shows that a line $ux_1 + vx_2 + wx_0 = 0$ is tangent to the curve (1) if and only if its coordinates (u, v, w), satisfy the condition

(2) $4u^3 + 27v^2 w = 0.$

Thus, unlike the curve in the preceding section, the Neil parabola is both of order 3 and of class 3, a property which is due to the fact that the singularity of the Neil parabola is a so-called cusp and not a double point. In §1.4 we will give a more detailed discussion of cubics with singular points.

1.4.   Cubics with Singularities as Rational Curves

We now prove the following important theorem. All cubics with a singular point have a rational parametrization.[3]

Proof. Let there be given a cubic which in homogeneous coordinates satisfies an equation of type

(1) $a_0 x_0^3 + a_1 x_1 x_0^2 + a_2 x_2 x_0^2 + b_0 x_1^2 x_0 + 2b_1 x_1 x_2 x_0$

$$+ b_2 x_2^2 x_0 + c_0 x_1^3 + c_1 x_1^2 x_2 + c_2 x_1 x_2^2 + c_3 x_2^3 = 0$$

and has $P = (1, 0, 0)$ as a singular point. Since all the first partial derivatives vanish at P, it follows that $a_0 = a_1 = a_2 = 0$. Then the line through P

(2) $G_2 x_1 - G_1 x_2 = 0,$ i.e., $x_1 = G_1 t,$ $x_2 = G_2 t,$ $x_0 = s,$

will intersect the curve at one further point, in general distinct from P, with coordinates

$$x_1 = G_1(b_0 G_1^2 + 2b_1 G_1 G_2 + b_2 G_2),$$

(3) $x_2 = G_2(b_0 G_1^2 + 2b_1 G_1 G_2 + b_2 G_2),$

$$x_0 = -(c_0 G_1^3 + c_1 G_1^2 G_2 + c_2 G_1 G_2^2 + c_3 G_2^3).$$

If we now let G_1, G_2 run through all pairs of numbers, not both zero, and interpret these pairs as coordinates of the points on an auxiliary line G, we have the desired parametrization. To each point of the curve distinct from P there must correspond a line of the pencil (2), i.e., a point of G, so that (3) is the desired, so-called rational parametrization of the curve.

If $D = b_0 b_2 - b_1^2$ is not zero, there are two lines in the pencil (2) that have only the point P in common with the curve, and then P is called a double point, which is said to be real or isolated, depending on whether $D < 0$ or > 0. These lines, which are real only for $D < 0$, are called double-point

---

[3]This property is expressed more concisely by saying that such cubics are rational.

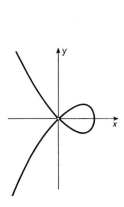

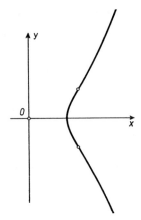

Fig. 4. $y^2 - x^2 + x^3 = 0$ Fig. 5. $y^2 + x^2 - x^3 = 0$

tangents. See the examples $y^2 + x^2(x - 1) = 0$ and $y^2 - x^2(x - 1) = 0$ in Figs. 4 and 5.

For $D = 0$ the singular point is called a cusp; here there is only one line of the pencil that has no further point in common with the curve, the so-called cuspidal tangent. In the birational mapping (3) of the points of the cubic onto the points of the auxiliary line it must then be noted that the double point corresponds to two distinct (real or conjugate complex) points on G.

1.5. *Cubics without Singular Points*

In general, unlike the examples considered up to now, a cubic has neither double points nor cusps. The general case is well illustrated by the curves with an equation in Cartesian coordinates of the form

(1) $$y^2 = a_0x^3 + a_1x^2 + a_2x + a_3.$$

It is easy to calculate that singular points of the curves (1) can lie only on the line $y = 0$ and certainly do not exist if the polynomial $a_0x^3 + a_1x^2 + a_2x + a_3$ has no multiple roots. Examples of cubics without singular points are given by the two curves (Figs. 6 and 7)

(2) $$y^2 - x(x + 1)(x - 1) = 0 \quad \text{and} \quad y^2 - x(x^2 + 1) = 0.$$

The first of these curves has two separate real parts, and the second only one. Let us also note that for all curves of the form (1) the point at infinity on the y-axis is a point of inflection. The second curve in (2) has two other real points of inflection, symmetric to the x-axis, i.e., collinear with the point of inflection at infinity. It can be shown that a cubic without singular points has

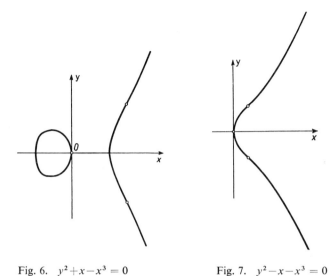

Fig. 6. $y^2 + x - x^3 = 0$ Fig. 7. $y^2 - x - x^3 = 0$

nine points of inflection altogether, not more than three of which can be real, and that, in contrast to the cubics with singular points, no cubic curve without singular points admits a rational parametrization.

1.6.   *Examples of Higher Curves, Genus and Plücker Formulas*

Curves of order higher than 3 can have the double points and cusps that occur for cubics, but in addition to these so-called ordinary singularities they can have higher singularities. As an example of a quartic, i.e., a curve of fourth degree, with a so-called threefold point at the origin of Cartesian coordinates, we mention the easily drawn curve whose equation in polar coordinates is $r = \sin 3t$, or in Cartesian coordinates (Fig. 8)

$$(1) \qquad (x^2 + y^2)^2 - 3x^2y + y^3 = 0.$$

By the method of §1.4 we can show that this curve has a rational parametrization. An example of a curve of fourth degree without singular points (even in the complex case) is the quartic

$$(2) \qquad x^4 + y^4 - 1 = 0.$$

In the examples above we have spoken about the order and class of algebraic curves, and have said that these numbers could be defined as the degrees of the polynomial equations satisfied, respectively, by the coordinates of the points of the curve and by the line coordinates of its tangents, which means, of course, that they are projective invariants. However, they

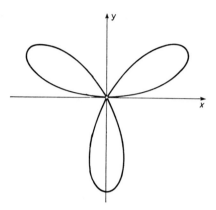

Fig. 8. $(x^2+y^2)^2 - 3x^2y + y^3 = 0$

are not invariants from the more comprehensive point of view of higher
algebraic geometry. From this point of view, all rational curves, e.g.,
straight lines, conic sections, singular cubics, etc., are equivalent to one
another, since they can all be mapped birationally onto the line. But even in
the early days of the subject, another important integer was introduced,
namely the so-called *genus p*, invariant under the (much more inclusive)
birational transformations of algebraic geometry. If K is a plane algebraic
curve of order n without singularities, its genus p is defined as the number

(3) $$p = \tfrac{1}{2}(n - 1)(n - 2),$$

so that, e.g., the cubics without singular points, are all of genus 1. If a curve
has singularities, the above number $_{n-1}C_2$ is decreased according to the
number and type of the singularities. For example, if a curve of order n has
h singularities, all of ordinary type (i.e., double points or cusps), its genus is

(4) $$p = \tfrac{1}{2}(n - 1)(n - 2) - h.$$

Thus all singular cubics are of genus $p = 0$, and the quartic (1) is also of
genus $p = 0$, since its threefold point can easily be visualized as the coales-
ence of three simple points. For more complicated higher singularities the
definition of genus is also more complicated. The quartic (2) without singular
points is of genus $p = 3$.

   Finally, let us mention without proof some further formulas due to
Plücker. If a plane algebraic curve is of order n in class K and if it has only
ordinary singularities, namely d double points and s cusps, and if w is the
number of points of inflection on the curve, then

(5) $$k = n(n - 1) - 2d - 3s, \quad w = 3n(n - 2) - 6d - 8s.$$

2. Algebraic Curves in Space

2.1. *The Standard Rational Curve in P_3*

Even though we may be interested only in the geometry of algebraic curves, it is still useful not to confine our attention to the plane but to consider space curves as well, or better still curves in a projective space P_n of arbitrarily high dimension. Nevertheless, the very definition of an algebraic curve in space cannot be given without profound algebraic preparation, which will be discussed to some extent in §2.3. However, the simplest nonplanar algebraic curve, the so-called standard rational third-degree curve K^3 in P_3, can be defined at once by parametric equations; namely, in a suitable coordinate system,

$$(1) \qquad x_0 = s^3, \quad x_1 = s^2t, \quad x_2 = st^2, \quad x_3 = t^3,$$

where s, t run through all pairs of numbers except $(0, 0)$. Clearly, K^3 generalizes the straight line and the conic section K^2 in the plane (cf. §1.1 (7)) and can itself easily be generalized to the space P_n.

The equations (1) assign the points of K^3 one-to-one to the points of a parameter line G with homogeneous coordinates (t, s). Since K^3, like the conics and the rational cubics, thereby admits a rational parametrization, it is called a rational space curve or a curve of genus 0.

From (1) it also follows that for all points of K^3

$$(2) \qquad x_0x_2 - x_1^2 = 0, \quad x_1x_3 - x_2^2 = 0, \quad x_0x_3 - x_1x_2 = 0;$$

in other words, K^3 lies on the three quadric surfaces with equations (2), and there are no other points common to all three surfaces. Thus the curve can be just as well defined by the equations (2) (i.e., as the intersection of the three quadric surfaces) as by the parametrization (1). The first two quadric surfaces Q and Q' in (2) are obviously cones with their vertices at the points $(0, 0, 0, 1)$ and $(1, 0, 0, 0)$, respectively. These cones have the common generator $x_1 = x_2 = 0$, which in turn cuts K^3 at the points with parameter values $(1, 0)$ and $(0, 1)$, i.e., the vertices of the cones, and is therefore a chord of the curve. The third quadric surface Q'' in (2) is nondegenerate and does not contain this generator. Conversely, two quadratic cones Q and Q' with a generator g in common can be shown to intersect each other (except in special cases) in a curve K^3 with g as a chord. It follows at once from (1) that, in general, a plane has at most three points in common with K^3, which is therefore said to be a space curve of third order.

In real affine space we must particularly consider the behavior of K^3 with respect to a plane F_2 distinguished as the plane at infinity in the projective space P_3. Depending on whether the three points of intersection of the curve with the plane F_2 are real and distinct, only one is real, or they are not all distinct, it was once customary to distinguish various affine types, which in analogy with the conic sections were given such names as cubic hyperbola and the like. Let us only note that obviously this classification cannot contain any type corresponding to the ellipse, whose real points all lie in the finite part of the plane.

2.2.   Plane Rational Cubics as Projections of the Curve K^3

If we now investigate what happens to the space curve K^3 under projection onto the plane P_2, we will obtain the familiar types of rational plane cubics. Let us first assume that the center Z_0, from which we wish to project, lies on the curve K^3, and for simplicity let us take it to be the point $(0, 0, 0, 1)$. But this point is also the vertex of the first cone Q in §2.1 (2), so that the projecting rays joining Z_0 to the various points of K^3 are generators of the cone Q, and the limiting position of these projecting rays is the line $x_0 = x_1 = 0$, i.e., the tangent to K^3 at the point Z_0. Thus the result of projecting the curve K^3 from Z_0 onto a plane (say $x_3 = 0$) not through the center is a conic section.

Now let us assume that the center of projection Z_0 does not lie on K^3. Then we must distinguish the two cases that Z_0 lies, or does not lie, on a tangent to K^3.

a)   Let Z_0 be the point $(0, 0, 1, 0)$. Certainly this point does not lie on the curve K^3, but it does lie on the line $x_0 = x_1 = 0$, which is easily seen to be tangent to K^3 at the point $(0, 0, 0, 1)$. Projection of K^3 from the center Z_0 onto the plane $x_2 = 0$ (which does not contain Z_0) produces a plane curve whose parametric representation is obviously obtained from (1) in §2.1 by simply omitting the x_2-coordinate, and is therefore

$$(1) \qquad\qquad x_0 = s^3, \quad x_1 = s^2 t, \quad x_3 = t^3$$

in the plane $x_2 = 0$, i.e., the Neil parabola $x_0^2 x_3 - x_1^3 = 0$. The vertex $(0, 0, 0, 1)$ of this parabola is the point of intersection of the tangent $x_0 = x_1 = 0$ of the curve K^3 with the plane of projection $x_2 = 0$.

b)   Now let Z_0 be the point $(1, 0, 0, 1)$, so that it lies on the chord joining the points $(1, 0, 0, 0)$ and $(0, 0, 0, 1)$ of the curve K^3. The transformation of coordinates

$$(2) \qquad y_0 = x_0 - x_3, \quad y_1 = x_1, \quad y_2 = x_2, \quad y_3 = x_0 + x_3$$

brings the center Z_0 to $(0, 0, 0, 1)$, which is more convenient. The parametric representation of K^3 in the new coordinates is

$$(3) \qquad y_0 = s^3 - t^3, \quad y_1 = s^2 t, \quad y_2 = st^2, \quad y_3 = s^3 + t^3.$$

Projection onto the plane $y_3 = 0$ then produces a plane curve with the

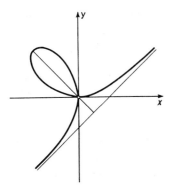

Fig. 9. $x^3 - y^3 - axy = 0$ (folium of Descartes)

parametric equations

(4) $y_0 = s^3 - t^3, \quad y_1 = s^2 t, \quad y_2 = st^2.$

By eliminating the parameter we obtain the equation

(5) $y_1^3 - y_2^3 - y_0 y_1 y_2 = 0,$

or in nonhomogeneous coordinates

$$x^3 - y^3 - xy = 0.$$

This curve is called the *folium of Descartes;* at the origin of the x, y-coordinates it has a double point with the coordinate axes as tangents (see Fig. 9), and every other cubic with a double point can be obtained in the same way by projecting K^3 onto some plane. Of greatest interest here is the fact that the double point is the projection of two distinct points on the curve K^3, since the chord from $(1, 0, 0, 0)$ to $(0, 0, 0, 1)$, containing the center Z_0, is one of the projecting rays. We now see more clearly why in §1.4 it was necessary, when mapping the plane cubic curve birationally onto the line G, to assign two distinct points of G to the double point of the cubic.

Curves which, like K^3 and the plane singular cubics, can be projected into each other are not essentially distinct in the sense of algebraic geometry. Up to now we have mentioned only rational curves; the simplest example of a nonrational space curve will be discussed in §2.3.

2.3.   *Space Curves of Fourth Order, First Species*

In §2.1 the standard rational curve K^3, defined in the first place by its parametric representation, was subsequently recognized as the remaining part of the intersection of two cones with a line in common. Thus the next simplest type of algebraic space curves will be the intersection of two quadric surfaces (in particular, cones) that have no line in common and do not have the other special property that their intersection consists of two distinct conic sections. The particular case of the intersection of a sphere and a cylinder is called a space curve of fourth order, first species. The term species refers to a classification not further explained here, the phrase "first species" indicating that there exists another type of space curves of fourth order, namely of second species, that cannot be obtained as the intersection of two quadric surfaces.

By the order of a curve we mean, in general, the greatest number of distinct, not necessarily real points that the curve can have in common with a plane E (cf. the order 3 of K^3), and then our curve C, defined as the intersection of two quadric surfaces Q and Q', is of order 4, since the two conics K and K' in which the quadric surfaces cut the plane E can have at most four points in common with each other. If the equations of Q and Q' are

(1) $$\sum_{i,j=0}^{3} a_{ij}x_ix_j = 0 \quad \text{and} \quad \sum_{i,j=0}^{3} b_{ij}x_ix_j = 0,$$

all the quadric surfaces of the pencil

(2) $$s\sum a_{ij}x_ix_j + t\sum b_{ij}x_ix_j = 0$$

will pass through C, so that instead of Q and Q' we could have chosen any two other quadric surfaces $\bar{Q}$ and $\bar{Q}'$ of this pencil as a base for all the quadric surfaces (2), which means that C is also defined by the intersection of $\bar{Q}$ and $\bar{Q}'$. In particular, it is always convenient to choose cones for this purpose. Now it is well known that among the quadric surfaces of the pencil (2) the cones are defined by the values of (s, t) satisfying the condition

(3) $$|(sa_{ij} + tb_{ij})_{i,j=0,1,2,3}| = 0.$$

As a homogeneous polynomial of fourth degree in s, t the equation (3) has at most four roots, which in special cases may coincide in various ways. Let us choose the pencil in such a way that the roots are distinct. Then there exist four distinct cones through our curve C, and intersection of any two of them already determines C.

2.4. *Plane Curves of Fourth Order as Projections of the Space Curve of Fourth-Order First Species*

An algebraic space curve C does not differ essentially, in the sense of higher algebraic geometry, from a plane curve C' arising from it by projection, provided that the projection sets up a correspondence (one-to-one with at most finitely many exceptions) between the points of C and those of C', as was the case, for example, with the projections in §2.2 of the curve K^3 onto the plane. We now wish to project the curve of fourth order in §2.3 from a center Z_0 onto the plane. Let us first note the very special case that Z_0 is the vertex of one of the four cones Q in §2.3. Here the projector rays coincide with the generators of Q, and C is obviously projected into a conic section K. But in general there will be two points of intersection between an arbitrary projector and an arbitrary quadric (distinct from Q) in the pencil (2) (in §2.3), so that the projector will meet the curve C in two points. Thus the projectors do not set up a one-to-one correspondence between points of C and K, but rather a one-to-two correspondence, so that we cannot say that the curve C and the conic K (i.e., also the straight line) are birationally equivalent. But we can say that C is projected into a doubly covered conic section.

Now let Z_0 be a point in general position, by which we mean that it does not lie on any of the four cones. Then there exists a well-defined, nondegenerate quadric $\bar{Q}$ in the pencil §2.3 (2) through Z_0. We now define C as the intersection of $\bar{Q}$ with another quadric $\bar{Q}'$ of the pencil §2.3 (2). As a nondegenerate quadric, $\bar{Q}$ contains two (not always real) lines g^I and g^{II} through Z_0. Let g^I have the points A, A' in common with Q', and g^{II} the points B, B' in common with $\bar{Q}$, $\bar{Q}'$. Then all four points A, A', B, B' belong to C, since they lie both on $\bar{Q}$ and $\bar{Q}'$. But this means that g^I and g^{II} are chords of C, and no other chord of C can pass through Z_0, since such a chord, having three points in common with $\bar{Q}$, must lie entirely on $\bar{Q}$, whereas g^I and g^{II} are the only such lines through Z_0. By a projection of the curve C from Z_0 onto a plane E' we then obtain a plane curve C' which is cut by any line G' of the plane E' in four points at most, namely the projections of the four points common to C and the plane through Z_0 and G'. Thus C' is a plane curve of fourth order with two double points at the intersections of the lines g^I and g^{II} with the plane E'. Further singularities, as we have seen, cannot arise from the projection, and if the curve C has no other special features, then C' has only these two double points and thus by formula (4) in §1.6 its genus is $p = 1$. Then C and C' are birationally mapped onto each other by the projection, so that to C also we assign the genus $p = 1$ and say that the complete curve of intersection of two quadric surfaces in general position is elliptic, as curves of genus 1 are also called.

As a simple example let us mention the curve of intersection of two circular cones with parallel axes. By projection parallel to the axes of the cones this curve becomes the so-called *oval of Descartes*, an elliptic curve (Fig. 10) with two complex double points (cf. H. Schmidt [4], p. 143).

2.5. *General Remarks on the Geometry of Algebraic Curves*

Algebraic space curves are defined, with complete generality, as the intersection of algebraic surfaces, i.e., they are sets of points in P_3 (more generally, in P_n) defined by the vanishing of finitely many homogeneous polyno-

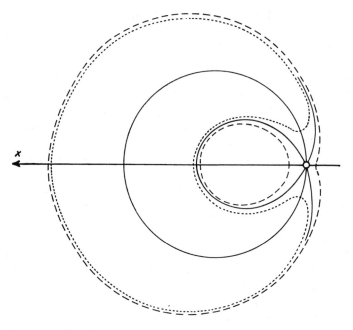

Fig. 10. $(x^2+y^2-5x)^2-4(x^2+y^2)+a = 0, a = 3, 0, -3$

mials in the projective coordinates. Through the given curve there always pass infinitely many surfaces, i.e., there exist infinitely many polynomials vanishing at the points of the curve. These polynomials form a so-called polynomial ideal (cf. IB5, §3). A so-called irreducible space curve occurs only when this polynomial ideal is a so-called prime ideal with a residue class field. From the purely algebraic point of view, the fundamental problem of algebraic geometry consists in characterizing the function fields defined in this way. All curves birationally equivalent to one another, e.g., the rational curves, or the fourth-order, first-species space curve and its plane projection, define isomorphic fields. In classical algebraic geometry, an algebraic curve, or a surface or figure of higher dimension, is considered as the set of its complex points; i.e., in algebraic language, the ground field of constants is the complex field. But in modern algebraic geometry, it is customary, to an increasing extent, to allow more general ground fields, which means, of course, that we get farther away from intuition. But intuition comes into its own again when we turn our attention to the intriguing special problems about the possible shapes of algebraic curves in the real. As an example, consider the remark in §1.5 that in the real a plane elliptic cubic can consist of one branch or (as can be shown) of at most two branches.

The birational equivalence of two curves is a one-to-one correspondence between their complex points which is distorted, but only at finitely many points, by the fact that finitely many points of one curve may correspond to one point of the other, as in some of the above examples. All rational curves are equivalent to one another in this sense, but for nonrational curves the condition that they have the same genus is only necessary, but not sufficient for birational equivalence. The genus of a space curve is most simply defined as the genus of the general projection of the curve.

3. Algebraic Surfaces and Surface Transformations

3.1. *Quadric Surfaces as Rational Surfaces*

After curves, the next objects of interest in algebraic geometry are the algebraic surfaces, among which the planes and the quadric surfaces are already familiar from elementary analytic geometry (cf. II9, §5, and II10, §7). An algebraic surface of kth order in the complex projective space P_3 is defined, in analogy with the algebraic curves in P_2, by the vanishing of an entire rational homogeneous function, irreducible in the complex field,

(1) $$f(x_0, x_1, x_2, x_3) = 0$$

of degree k. Among these surfaces we then define the rational ones, corresponding to the rational curves of the preceding section. Thus a surface is said to be rational if it admits a rational parametrization, i.e., if there exist four entire rational functions (of two parameters in the nonhomogeneous case and three in the homogeneous case) satisfying (1) identically. Interpreting the parameters as point coordinates of a plane P_2 we can also say that a rational surface is a surface representable as the rational image of a plane, where it may happen that more than one point of the plane must be assigned to the same point of the surface, so that the mapping is no longer one-to-one; and even if the mapping is one-to-one in general, it may fail on certain curves to be well defined in one direction on the other. Let us first study this possibility for a quadric surface.

Let there be given a nondegenerate quadric Q in P_3. Then it is well known that in the complex case we can always introduce projective coordinates in which Q has the equation

(2) $$x_0 x_3 - x_1 x_2 = 0.$$

Further, we see at once that the following rational functions satisfy (2) identically:

(3) $$x_0 = uv, \quad x_1 = uw, \quad x_2 = vw, \quad x_3 = w^2.$$

We now interpret (u, v, w) as homogeneous coordinates on a plane P_2' and thereby obtain a point P on Q uniquely assigned to a point P' in P_2' provided that the coordinates (u, v, w) of P', when substituted in (3), do not annul all the x_i. Certainly all points P' not on the line $w = 0$ are of this kind; and conversely for every P on Q but not on the plane $x_3 = 0$, i.e., not on either of the lines of intersection

$$(4) \qquad g^{\mathrm{I}}(x_1 = x_3 = 0) \quad \text{and} \quad g^{\mathrm{II}}(x_2 = x_3 = 0)$$

of this plane with Q, we can find the corresponding point P' on P_2. On the other hand, all points on $w = 0$ distinct from the two points $U(v = w = 0)$ and $V(u = w = 0)$, correspond to the single point $(1, 0, 0, 0)$ on Q. The points U and V themselves have no pre-image on Q, since they annul all the x_i in (3). But by the following passage to the limit we see that all points on the lines g^{I} and g^{II} are to be assigned to U and V, respectively. To begin with, the points

$$(5) \qquad (h, a, ha) \text{ on } P_2 \text{ and } (1, h, a, ah) \text{ on } Q \text{ for } a \neq 0, h \neq 0$$

are certainly assigned to each other. But under the passage to the limit $h \to 0$ the point (h, a, ha) goes into V and the point $(1, h, a, ha)$ into the point $(1, 0, a, 0)$ on the line g^{I}, i.e., the whole of g^{I} must be assigned to the point V, and analogously the whole of g^{II} to the point U.

Thus we have shown that the so-called birational correspondence (3) between the surfaces Q and P_2 is not in every case one-to-one, but that certain points of the one surface correspond to a whole curve on the other. This state of affairs, which is encountered again and again in surface mappings in algebraic geometry, can already be observed in the birational transformations between two planes, to which we now turn.

3.2. Plane Cremona Transformations

It is well known that projective mappings between two projective spaces P_n and P'_n, or from one P_n onto itself, are without exception one-to-one. But for $n \geq 2$, many other mappings between P_n and P'_n are also one-to-one, at least in general. Such mappings are called *Cremona transformations* in honor of a great Italian geometer of the past century; here we restrict ourselves to the case $n = 2$, for which we have the following general definition.

Let there be given two planes P_2 and P'_2 with coordinates (x_i) and (x'_i) and also two sets of formulas

$$(1) \qquad x'_i = f_i(x_0, x_1, x_2) \quad (i = 0, 1, 2),$$

$$(2) \qquad x_j = f'_j(x'_0, x'_1, x'_2) \quad (j = 0, 1, 2),$$

where in each set the f_i and the f'_j are polynomials of the same degree. Then (1) and (2) define a Cremona transformation between the planes P_2 and P'_2 if

$$(3) \qquad f'_j(f_i(x_k)) = H(x_0, x_1, x_2)x_j \quad (j = 0, 1, 2)$$

for some polynomial H. If g is the degree of the functions f_i, it can be shown that the degree of the f'_j is also g, and we speak of a Cremona transformation of gth degree (for $g = 1$ it is a projective mapping, with H a constant). In general, the lines on one plane correspond to a certain system of curves of gth degree on the other plane, but for $g > 1$ there always exist points in the one plane whose images in the other plane are not uniquely determined. Here we will discuss this phenomenon only for the simplest type of nonprojective Cremona transformation, namely the quadratic transformation defined by

$$
\begin{aligned}
x'_0 &= x_1 x_2, & x'_1 &= x_0 x_2, & x'_2 &= x_0 x_1; \\
x_0 &= x'_1 x'_2, & x_1 &= x'_0 x'_2, & x_2 &= x'_0 x'_1,
\end{aligned}
\tag{4}
$$

so that now the function H in (3) is $x_0 x_1 x_2$. From (4) it follows for the points of a conic

$$
u x_1 x_2 + v x_0 x_2 + w x_0 x_1 = 0
\tag{5}
$$

in the plane P_2 that their images in the plane P'_2 are the points of the line

$$
u x'_0 + v x'_1 + w x'_2 = 0.
\tag{6}
$$

The conics that can be written in the form (5) are said to form a *homaloidal net*; clearly they all pass through the three fundamental points $A(0, 0, 1)$, $B(0, 1, 0)$ and $C(1, 0, 0)$. But in contrast to the projective transformations we must now note the following facts.

a) The three vertices A, B, C of the coordinate system in the plane P_2 have no image point, at least so far, in the plane P'_2, since for these points the x'_i assigned to them by (4) all vanish.

b) The points distinct from A, B, C on the three coordinate lines $x_i = 0$ all have the same image; e.g. $(1, 0, 0)$ is the image of all the points $(0, b, c)$.

c) Since the formulas (4) are symmetric in the primed and unprimed coordinates, we see again that all the points (except $(0, 0, 1)$ and $(0, 1, 0)$) on the coordinate line $x'_0 = 0$ are assigned to the same point C in P_2, so that the whole line $x'_0 = 0$ may be said to be the image of the point C, and the corresponding statement holds for A and B.

Up to now we have considered the planes P_2 and P'_2 as distinct, but of course we can let them coincide and then consider the various types of quadratic transformations of a plane onto itself. Two elementary examples are considered in the next two sections.

3.3. Isogonal Correspondence as an Example of a Quadratic Transformation in Elementary Geometry

In a plane with a Euclidean metric (II10, §5), let there be given a triangle ABC and a point X not on any of its sides. By reflection of the three lines XA, XB and XC in the angle bisectors at the vertices A, B, and C, we obtain three lines which again can easily be shown to pass through one point X', and then a simple calculation shows that this correspondence between the points X and X' defines a quadratic mapping of the plane onto itself, which is obviously involutory, since X' is assigned in turn to X. Furthermore, if we now let X lie on a side of the triangle, but not at a vertex, the point X' is still determined. For example, if X lies on AB, then $X' = C$ etc. But the images of the vertices themselves are not determined, since to each of them we are compelled to assign as images all the points of the opposite side, a situation similar to the quadratic transformation between two distinct planes discussed in the preceding section.

3.4. Inversion in the Plane as a Cremona Transformation

Another important example of a quadratic transformation of the plane onto itself is the inversion in a circle (II7, §2.4) familiar from secondary school mathematics. If the equation of the circle in Cartesian coordinates is $x^2 + y^2 = 1$, the inversion is defined by the well-known formulas

$$(1) \qquad\qquad x' = \frac{x}{x^2 + y^2}, \quad y' = \frac{y}{x^2 + y^2}.$$

In the homogeneous coordinates obtained by replacing x and y by x_1/x_0, x_2/x_0, the formulas (1) become

$$(2) \qquad\qquad x_0' = x_1^2 + x_2^2, \quad x_1' = x_0 x_1, \quad x_2' = x_0 x_2.$$

The inverse formulas are the same, with interchange of primed and unprimed coordinates, so that, like the example of the preceding section, the transformation is involutory. The conics of the form

$$(3) \qquad\qquad r(x_1^2 + x_2^2) + s x_0 x_1 + t x_0 x_2 = 0$$

now constitute the homaloidal net corresponding to the net of straight lines, where (3) obviously represents all circles through the origin O (including all the lines through O, and also, to be more precise, the line at infinity). But if the real plane is extended to the complex, these circles can be interpreted as the conics through the two so-called circular points

(4) $I(x_0 = 0, x_1 = 1, x_2 = i)$ and $J(x_0 = 0, x_1 = 1, x_2 = -i)$.

Thus the net (3) consists of all conics through O, I, J and is therefore analogous (apart from questions of reality) to the system of all conics through three real points considered in §3.2. The triangle OIJ, formed by the line at infinity IJ and the two complex so-called isotropic lines OI and OJ (conjugate to each other), plays the role of the fundamental triangle in §3.2. In the real plane, the only phenomenon in the present case analogous to the exceptions to one-to-oneness in §3.2 (namely that each vertex of the triangle corresponded to the whole opposite line) is that the point O corresponds to the whole line at infinity. But inversion in a circle is usually considered, not in the real projective plane but in the Gaussian or conformal plane, completed by a single point at infinity, in which it is one-to-one without exception.

3.5 Linear Families of Curves in the Plane. Rational Surfaces

One of the most important concepts of plane algebraic geometry is that of a linear family of curves, by which we mean the totality of all curves of fixed degree k whose equations can be written in the form

$$(1) \quad s_0 f_0(x_0, x_1, x_2) + s_1 f_1(x_0, x_1, x_2) + \cdots + s_d f_d(x_0, x_1, x_2) = 0,$$

where $f_i = 0$ are the equations of certain fixed curves of degree k and the s_i are arbitrary constants. If the f_i are linearly independent, i.e., if no one of the $d + 1$ curves $f_i = 0$ is a linear combination of the others, we say that (1) defines a linear ∞^d-system, and for $d = 1$ and 2 we speak of pencils and nets. For example, pencils of circles and pencils of conics are well known. The homaloidal nets of conics, by which we have defined quadratic transformations in §3.2, are another special case. An important example of a ∞^3-system of conics results from examining the images, in the mapping of §3.1, of the points on the plane intersections of the quadric. By §3.1 (3) these points constitute all the conic sections of the linear ∞^3-system

$$(2) \qquad s_0 uv + s_1 uw + s_2 vw + s_3 w^2 = 0.$$

This system consists of all conics through the points $U(v = w = 0)$ and $V(u = w = 0)$, which are therefore called its base points, and the whole geometry on the quadric surface can be studied from it. The system (2) is said to be simple, i.e., not all of its curves through an arbitrary point of the plane automatically pass through another point of the plane. Such simple ∞^3-systems always define rational surfaces, their parametric representation being given by

$$(3) \qquad x_i = f_i(u, v, w) \quad (i = 0, 1, 2, 3),$$

where $\sum_{i=0}^{3} s_i f_i = 0$ are the curves of the system.

Another example of a net of curves of third degree is provided by the ∞^3-system consisting of all cubics through six given points (not all on one conic) on the plane P_2. The rational surface with the corresponding parametrization is the general surface of third degree, as was proved by Clebsch in the 19th century.

3.6. Ruled Surfaces and Cones as Further Examples for the Theory of Algebraic Surfaces

Up to now, in analogy with the rational curves, we have been dealing with rational surfaces, as being the simplest class of surfaces, and we can now proceed, as before, to define birationally related classes of surfaces that are no longer rational. But the theory of such surfaces, which has been developed chiefly by the Italians in the present century, is essentially more complicated than that of algebraic curves. We shall be satisfied here with pointing out a special type of algebraic surfaces encountered in a natural way in projective geometry, namely the ruled surfaces (reguli). An algebraic ruled surface is an algebraic surface that can also be interpreted as the set of all points on a one-parameter family of lines. Familiar examples are the well-known nondegenerate (complex) quadric surfaces, on which there are even two such one-parameter families of lines. All other ruled surfaces carry only one such family of so-called generating lines. A particularly simple case arises when all these lines pass through a fixed point S. Then the ruled surface is called a cone with the vertex S. In a system in which S has the coordinates $(0, 0, 0, 1)$, the equation of a cone is

$$(1) \qquad\qquad f(x_0, x_1, x_2) = 0,$$

i.e., an equation in which x_3 is missing. For it is clear that this equation is satisfied by the coordinates of the points on all lines joining points of S to points on the curve with equation $f(x_0, x_1, x_2) = 0$ in the plane $x_3 = 0$. The well-known quadratic cones, i.e., quadric surfaces with matrix of rank 3, are also rational. For in a coordinate system in which such a cone K has the equation

$$(2) \qquad\qquad x_2^2 - x_0 x_1 = 0,$$

the parametric representation of the cone is

$$(3) \qquad\qquad x_0 = u^2, \quad x_1 = w^2, \quad x_2 = uw, \quad x_3 = uv.$$

The linear system corresponding by §3.5 to the plane intersections of this cone is

$$(4) \qquad\qquad s_0 u^2 + s_1 w^2 + s_2 uw + s_3 uv = 0,$$

which obviously consists of all conics tangent to the line $u = 0$ at the point $(0, 1, 0)$. Clearly it is a limiting case of the system (2) in §3.5 corresponding to the general quadric surface.

If (1) is the equation of a cubic (without singularities) in the plane $x_3 = 0$ and if, as above, we regard (1) as the equation of a cone in P_3 (called a cubic cone), it turns out that this cone is nonrational. A cubic cone of this sort, namely a so-called elliptic cone, is the simplest example of a nonrational surface, since it can be shown that every other surface of third degree is rational, a rather remarkable fact, since in general the plane curves of third order are not rational. For more varied examples of nonrational surfaces it is necessary to consider surfaces of the fourth degree.

Bibliography

[1] BURAU, W.: Algebraische Kurven und Flächen I: Algebraische Kurven der Ebene, Sammlung Göschen Bd. 435 (1961); Algebraische Kurven und Flächen II, Algebraische Flächen 3. Grades und Raumkurven 3. u. 4. Grades, Sammlung Göschen Bd. 436/436a (1962).

[2] FLADT, K.: Analytische Geometrie spezieller ebener Kurven. Frankfurt/M. 1962.

[3] HAUSER, W., BURAU, W.: Integrale algebraischer Funktion und ebene algebraische Kurven. Berlin 1958 (VEB Deutscher Verlag der Wissenschaften).

[4] SCHMIDT, H.: Ausgewählte höhere Kurven. Wiesbaden 1949.

[5] SEVERI, F.: Vorlesungen über algebraische Geometrie. Leipzig 1921 (translated into German by Löffler.)

[6] VAN DER WAERDEN, B. L.: Einführung in die algebraische Geometrie. Berlin 1939.

[7] WIELEITNER, H.: Theorie der ebenen algebraischen Kurven. Sammlung Schubert. Leipzig 1905.

Erlanger Program and Higher Geometry

1. Geometries and Their Groups. The Erlanger Program

A geometry consists (cf. II1, No. 20) of certain sets of objects and relations among them, with certain properties (the axioms and theorems). Then the corresponding relation-preserving mappings, or automorphisms, are said to belong to the geometry and to form its *automorphism group.*

Conversely, it may happen that starting from a *group* we can form a *geometry* whose automorphism group is identical with the given group, as was particularly emphasized by Felix Klein in his Erlanger Program ([5]).

In the following sections these two points of view will be discussed in greater detail.

1.1. *Geometries and Their Automorphism Groups*

Various geometries have been studied in the preceding chapters, e.g., the Euclidean, equiform, affine, and projective geometries in II9 and 10, and in II3–6; these and the non-Euclidean geometries have been constructed from the point of view of the foundations of geometry.

Every geometry of this sort proceeds, as stated above, from certain objects and relations, which define the corresponding relation-preserving mappings, namely the *automorphisms* of the geometry, e.g., the rigid mappings (isometries) of Euclidean geometry, the similarities of equiform geometry, and the (nonsingular) affine and projective mappings of affine and projective geometry. For example, the nonsingular affine mappings transform an affine space one-to-one onto itself with preservation of the corresponding relations (collinearity, parallelism, ratios of division of line segments).

The automorphisms of any geometry form a group, the *automorphism group,* under the operation of composition of mappings, since the product of two automorphisms is again an automorphisms, the identical mapping is the neutral element, every automorphism has its inverse, and the associative law holds for all mappings (cf. IB2, §1.2.5).

These group properties enable us to define equivalence of figures, where by a figure we mean, as always in this chapter, a set of points of the geometry in question, and two figures F_1, F_2 are said to be equivalent, with respect to the given group of automorphisms, if F_1 can be mapped onto F_2 by one of the automorphisms. This relation is symmetric since if $F_1 \sim F_2$, so that $\alpha(F_1) = F_2$ for some automorphism α, then $\alpha^{-1}(F_2) = F_1$, since with α the group also contains α^{-1}, so that $F_2 \sim F_1$; and the reader can easily verify that the relation is also reflexive and transitive, and is therefore a general equivalence relation in the sense of IA, §8.5. Consequently, the figures are classified into sets of equivalent figures differing, from the point of view of the given geometry, only in position, as for example in the classification of quadrics in Euclidean and affine geometry (II9, §§5 and 11).

If, as is often the case in what follows, a certain figure F is distinguished as the *absolute figure,* then the automorphisms leaving F "fixed" (i.e., mapping F onto itself as a whole, not necessarily pointwise) obviously form a subgroup U of the automorphism group G. The class of figures equivalent to F_1 with respect to U is contained in the equivalence class of F_1 with respect to G (in general, as a proper subclass).

1.2. *The Euclidean, Equiform, and Affine Groups as Subgroups of the Projective Group*

After pointing out the importance of a given automorphism group for its corresponding geometry, let us now discuss the interrelations among various groups of automorphisms; for example, the Euclidean, equiform, and affine groups may be considered as subgroups of the projective group, as was done in II10. Let us briefly recapitulate the properties of this hierarchy of groups and its effect on the corresponding (real) geometries, with particular attention to the properties left invariant by each of the given subgroups.

The *projective* mappings leave collinearity and cross ratios invariant.

The subgroup of projective mappings leaving invariant a distinguished line in P_2 (or plane in P_3) is the *affine* group of the plane (or space). The absolute figure is the "improper" line (or plane) and parallelism and ratios of division of parallel segments are invariant.

If a pair of conjugate complex improper points, i.e., the pair of "absolute circular points" (or in space a circle on the "absolute sphere") is also fixed, we obtain the subgroup of similarities, also called the *equiform* group. In the geometry of similarities, angles and therefore orthogonality are invariant. A nondegenerate conic section through the circular points is called a "circle," with the analogous definition for "spheres." Then nonparallel segments can also be compared and the ratio of their lengths remains invariant. On the other hand, the length of a segment cannot yet be defined, since any two segments are equivalent under the equiform group, so that in the geometry of similarities there is no unit of length. Felix Klein referred to similarity geometry as *elementary geometry* and to its group as the *principal group*.

Finally, the subgroup of similarities mapping a prescribed circle (or sphere) K onto some circle (or sphere) in the set M of all circles (or spheres) equivalent to K under a translation is called the group of rigid, or isometric, mappings and belongs to *Euclidean* geometry. Its most important subgroup consists of orientation-preserving isometries, or motions.

Summing up, we see that this hierarchy of groups leads to a *hierarchy* of geometries; descent to a smaller subgroup corresponds to ascent to an "over-geometry," richer in concepts, theorems, and invariants. In general, a figure F gains additional properties and its equivalence class becomes smaller. Let us illustrate by starting with a circle K of radius 1 in the Euclidean plane E_2 and looking for the equivalence class of K in each of our geometries. One after another, in ascending order of subgroups, we obtain, in Euclidean geometry all circles of radius 1, in equiform geometry all circles, in affine geometry all ellipses, and in projective geometry all conic sections. The reader should consider the corresponding classes of figures for a square.

1.3. *Klein's Erlanger Program*

In view of the important role of automorphism groups in the corresponding geometries, it is natural to adopt the second of the above two points of view, i.e., starting from a given group to construct, in a sense to be made more precise below, a geometry as the "invariant theory" of the given group.

It is the great merit of Felix Klein to have recognized the importance of group theory for geometry and to have formulated it, in a systematic way, in 1872 in his *Erlanger Program* "Recent trends in geometric research" ([5]), which for many years provided the framework for geometric studies and gave them a special impetus.[1] The following sections outline its basic ideas, with some critical remarks.

1.3.1. *Geometry as the "invariant theory" of a group.* To quote from the Erlanger Program [5], §1, "Let there be given a manifold, and in it a group of transformations; it is our task to investigate those properties of a figure belonging to the manifold that are not changed by the transformations of the group."[2]

By a "manifold" Klein means a "numerical manifold," i.e., a set[3] whose

---

[1]Cf., e.g., G. Fano, Enzykl. Math. Wiss. III *A B* 4 b.

[2]Here the words "invariant theory" are not to be taken too narrowly, i.e., not merely in the sense of the algebraic invariant theory of that time.

[3]For the concept of a manifold in modern topology cf. II16, §2.

elements are coordinate n-tuples, usually of real or complex numbers. Geometric objects, like points, lines, circles, quadrics, etc., are first defined by coordinates and coordinate-equations, and then the transformations, also defined by equations, act on these objects. Klein could never quite free himself from this coordinate point of view,[4] although it is no longer adopted today, as a result of the rapid development, toward the end of the century, of the foundations of geometry. Nowadays we introduce a geometry without use of coordinates, invariantly, so to speak, from the very beginning, as a set of objects with relations, a geometry based on axioms formulated either in geometric language (foundational or "synthetic" point of view, cf., II3, 4 and 6) or else in terms of the vector space over a number field ("analytic" point of view, cf., II9 and 10).

Replacing "manifold" by "set," we thus interpret Klein's principle as follows. Let there be given a set M. The one-to-one mappings of M onto itself, under the ordinary operation of successive application of transformations, form the symmetric group S_M of M (cf., IB2, §1(2). Every subgroup T of S_M is called a *transformation group* of M, and we investigate those properties of figures that are left invariant under all transformations of a given transformation group T. In contrast to §§1.1 and 1.2, M has no geometric properties, at least not at first, and these properties are imprinted, so to speak, on M by the group T.

This point of view can be made more precise if, in the spirit of the analytic approach, we distinguish between the (abstractly defined) group G and the transformation group T, where T is a homomorphic image $D(G)$ of G, the homomorphism $D:G \rightarrow S_M$ being called a representation of G in M. Then the question arises of determining the group that corresponds to a given geometry, in particular, to a given classical geometry, a question discussed in detail in the following chapter (II13).

In this interpretation, Klein's principle has proved equally fruitful for geometry and for group theory, e.g., in the group-theoretic construction of absolute geometry (II5) or of plane elliptic and hyperbolic geometry (II13, §2), or in the Helmholtz-Lie space problem (cf. II13, §3 and II9, §7.2).

1.3.2.   *The transference principle.*   Let there be given a geometry with set of objects M and automorphism group G, and also a second set M'. Let the mapping $\Theta: M \rightarrow M'$ map M one-to-one onto M'. Then Θ also effects a "transference" of the group G onto a certain group G' of transformations of M'.

For to every automorphism $\alpha \in G$ we can assign a mapping $\alpha' = \Theta\alpha\Theta^{-1}$ (cf. Fig. 1) of M' into itself. Moreover, the mapping $\alpha \rightarrow \alpha' = \pi(\alpha)$ of G into the set of mappings of M' into itself is a homomorphism, since

$$\pi(\alpha)\pi(\beta) = \Theta\alpha\Theta^{-1}\Theta\beta\Theta^{-1} = \Theta\alpha\beta\Theta^{-1} = \pi(\alpha\beta),$$

[4]H. Freudenthal, Math. -Phys. Semesterber. 7, pp. 2–25 (1961).

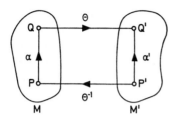

Fig. 1

so that the set $\pi(G) = G'$, as the homomorphic image of a group, is itself also a group. By transferring G in this way, we obtain the group G' as a group of transformations of M'.[5]

We also see that $\pi(\alpha) = \pi(\beta)$, i.e., $\Theta\alpha\Theta^{-1} = \Theta\beta\Theta^{-1}$ implies $\alpha = \beta$. So the groups G, G' are isomorphic (cf. IB2, §4), and then the *geometries* of M and M' are also said to be *isomorphic*,[6] and the geometry of M' is said to be obtained from that of M by the *transference* (Θ, π), where Θ maps the set M one-to-one onto M' and π maps the group G one-to-one onto G'.

Following Hesse (cf. II10, §6.5), Felix Klein formulated this situation, though in a less precise way, as a *transference* principle for geometries, allowing us to provide isomorphic geometries for sets of geometric objects of extremely varied character, a principle which, in the following sections, will be seen to be extremely fruitful. For example, by stereographic projection we can map the group of Möbius circle geometry isomorphically onto the subgroup of projective mappings of the space P_3 leaving invariant a fixed quadric of spherical type, whereby the Möbius geometry is *linearized,* i.e., can be dealt with by linear transformations, and similar statements will be proved below for the circle geometries of Laguerre and Lie.

1.3.3.   *The hierarchy principle.*   It is now obvious how the discussion in §1.2 can be generalized to obtain a *general hierarchy principle* for geometries. To a subgroup U of G there corresponds an "over-geometry," richer and more inclusive than the geometry corresponding to G.

The relationship of Euclidean, equiform, and affine geometry to projective geometry has been discussed above in §1.2. But many other geometries can be arranged in similar hierarchies. For example, certain subgroups of the projective group are isomorphic to sphere and line geometries, and to the classical non-Euclidean geometries.

But in the Erlanger Program Klein also points to other mappings, whose groups include the projective group, e.g., the birational transformations

---

[5]If Θ is not invertible, this transfer will be possible only under certain compatibility conditions on Θ and G.

[6]Cf. the treatment of "similar" representations in II13, §1.12.

(II11), the topological (II16) and the contact transformations,[7] for which geometric information is given in [6]. But here an important restriction must be made. As automorphisms we may consider only those birational or topological mappings that take the *whole* space into itself, and not those that refer only to an imbedded curve or surface.

Again, a great deal of modern geometric research, say in differential or algebraic geometry, does not fit satisfactorily into Klein's scheme of hierarchies. Here his expectations have not been fulfilled; the hierarchy principle of the Erlanger Program has been confined essentially to projective geometry.

1.4. *Further Examples for the Erlanger Program. Higher Geometry*

But for projective geometry and its subgroups the Erlanger principles of hierarchy and transference have always been important. A projective subgroup is distinguished by the (nonsingular) projective mappings of projective space P_n that map an absolute figure onto itself; for example, the affine group and its subgroups were obtained in §1.2 by distinguishing a hyperplane in P_n.

In the following sections other important examples will be discussed from this uniform point of view; the *absolute figure* will in each case be a *quadric* in the real space P_n, whereby we shall obtain the classical non-Euclidean geometries and, by suitable transferences, the circle, sphere, and line geometries of Möbius, Laguerre, Lie, and Plücker, geometries which for our purposes we shall include under the general name of "higher geometry." Since all the groups of these geometries are isomorphic to projective subgroups, we thereby give effect, at least to that extent, to the saying of Cayley that "projective geometry is all geometry."[8]

2. Möbius Geometry

The circle geometries of Möbius, Laguerre, and Lie all deal with circles in the real Euclidean plane E_2, but start from different fundamental elements, i.e., *objects of the first kind;* Möbius geometry from points (*M*-points) and Laguerre geometry from oriented lines (spears). In both these geometries the circles are regarded as sets of objects of the first kind, i.e., as sets of points and spears, respectively. More precisely, the *objects of the second kind* are the *M*-circles in Möbius geometry and the cycles (oriented circles) in Laguerre geometry. In Lie geometry, on the other hand, the circles are them-

---

[7]Cf. Lie-Scheffers, Geometrie der Berührungstransformationen. Leipzig 1896.

[8]A. Cayley, Sixth memoir upon quantics. Coll. math. papers II, p. 500. At that time Cayley still wrote "Descriptive geometry."

selves the objects of the first kind; there is only one kind of object, the $\mathfrak{L}$ circles.

Historically, these geometries were developed as geometries of the circle in the Euclidean plane, and were only later brought into relationship with the geometries of subgroups. Here, in view of our interest in the Erlanger Program, they will be obtained by transference from the geometries of projective subgroups.

The analogous sphere geometries (in space) of Möbius, Laguerre, and Lie will be mentioned occasionally.

2.1. *Stereographic Projection. The Möbius Plane*

In P_3 let the absolute figure Q_M be a quadric of spherical type, and let G_M be the projective subgroup leaving Q_M fixed. The particular choice of quadric Q_M is of no importance, since Q_M can be taken by a nonsingular mapping into any other quadric Q_M' of spherical type, when (by §1.3.2) the corresponding subgroups G_M, G_M' will be isomorphic. In a suitably chosen projective coordinate system Q_M will have the equation

$$(1) \qquad\qquad (\mathfrak{x}, \mathfrak{x}) = -x_0^2 + x_1^2 + x_2^2 + x_3^2 = 0$$

with $(\mathfrak{x}, \mathfrak{x})$ as an abbreviation for $-x_0^2 + x_1^2 + x_2^2 + x_3^2$. Here $\mathfrak{x}$ is a vector in the real four-dimensional vector space V_4 which, together with its multiples $\rho\mathfrak{x} \neq \mathfrak{o}$ represents a point X in P_3 (cf. II10, §1.1), lying on Q_M if and only if (1) holds. The bilinear form (cf. II9, §5.1) corresponding to $(\mathfrak{x}, \mathfrak{x})$ is

$$(2) \qquad\qquad (\mathfrak{x}, \mathfrak{y}) = -x_0 y_0 + x_1 y_1 + x_2 y_2 + x_3 y_3;$$

with $x_0 = 0$ as the improper plane and $x = x_1/x_0, y = x_2/x_0, z = x_3/x_0$ as nonhomogeneous coordinates, equation (1) can be written in the form $x^2 + y^2 + z^2 = 1$, and Q_M can be interpreted as the unit sphere in Euclidean space E_3.

The fact that planes intersect Q_M in circles, which remain circles under the mappings in G_M, and that *stereographic projection* is circle-preserving (see the proof below, or II10, §7.4), naturally suggests a relationship between projective geometry and the circle geometry of the plane, which we now proceed to set up.

By projection (Fig. 2) from a point N on the sphere (which we may call the North Pole) onto an image plane at N (which we may take to be the equatorial plane $z = 0$, since by §1.3.2 any other choice would amount to an automorphism of the geometry) we assign to every point $P \neq N$ on the sphere exactly one point P' in E_2. From Fig. 2 we see that[9]

$$1 : \left(1 - \frac{x_3}{x_0}\right) = r' : r = x : \frac{x_1}{x_0} = y : \frac{x_2}{x_0},$$

[9]For the exceptional points with $x_1 = 0$ and $x_2 = 0$ the result (3) holds equally well.

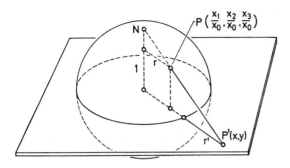

Fig. 2

so that for $P \in Q_M$, $P \neq N$,

(3)
$$x = \frac{x_1}{x_0 - x_3}, \quad y = \frac{x_2}{x_0 - x_3}.$$

Thus from (1) we obtain the inverse formulas

(4)
$$\begin{cases} \rho x_0 = x^2 + y^2 + 1 \\ \rho x_1 = 2x \\ \rho x_2 = 2y \\ \rho x_3 = x^2 + y^2 - 1, \end{cases}$$

where the x_i, determined only up to a common factor ρ, are sometimes called the *tetracyclic coordinates* of $P'(x, y)$.

The deleted sphere, i.e., without its North Pole, is mapped by (3) one-to-one onto the proper points of the equatorial plane E_2, and now since we wish to assign an image to N also, we extend E_2, not by its infinitely many improper points in P_3, but by *one* point ∞, so as to form the *Möbius plane* $M_2 = E_2 \cup \{\infty\}$ (as is done in the theory of functions of a complex variable, see III7), whose elements, both proper and improper (i.e., including the point ∞) will be called *M-points*. Then the stereographic projection (3), together with

(5)
$$N \leftrightarrow \infty$$

forms a *one-to-one mapping* Θ_M of Q_M onto M_2.

The most important properties of stereographic projection are 1) it *preserves circles* and 2) it *preserves angles,* as we shall now prove.

1)   A circle on Q_M is the intersection of Q_M with a plane, whose pole U with respect to Q_M may be represented by a vector $\mathfrak{u}$ with coordinates (u_0, u_1, u_2, u_3). Then by (2) and (1) the equation of the circle is

(6) $(\mathfrak{u}, \mathfrak{x}) = -u_0 x_0 + u_1 x_1 + u_2 x_2 + u_3 x_3 = 0, \quad (\mathfrak{x}, \mathfrak{x}) = 0$

and by (4) its image under Θ_M is

(7) $(u_0 - u_3)(x^2 + y^2) - 2u_1 x - 2u_2 y + (u_0 + u_3) = 0.$

For $u_0 - u_3 = 0$ this equation represents a line, and for $u_0 - u_3 \neq 0$ comparison with the equation of a circle

(8) $x^2 + y^2 - 2ax - 2by + (a^2 + b^2 - r^2) = 0$

shows that (7) represents a circle with radius $r > 0$ in M_2, provided $-u_0^2 + u_1^2 + u_2^2 + u_3^2 > 0$. But this condition is satisfied, since the pole U lies outside Q_M, so that $(\mathfrak{u}, \mathfrak{u}) > 0$.

Conversely, the circles (8) and the lines in M_2 can be represented in the form (7) (with the stated conditions on u_i) and are therefore taken by Θ_M^{-1} into circles (6) on Q_M, the lines being taken (since $u_0 - u_3 = 0$) into circles through N. Also, it is clear from (6) that in tetracyclic coordinates the equation of a line or circle in M_2 is linear (*linearization*).

If circles (with $r > 0$) and lines are included together under the concept of M-circles, Θ_M is circle-preserving, since it maps the circles on Q_M one-to-one onto the M-circles in M_2.

2)   The proof that Θ_M is *angle-preserving* is as follows. The angle between two M-circles is defined as the angle between their tangents t_1, t_2, and these tangents are taken (Fig. 3) into circles k_1, k_2 through N. The angle between the circles at N is equal to the angle between t_1 and t_2 (since t_1', t_2' are parallel to t_1, t_2) and therefore, by symmetry, also equal to the angle between t_1' and t_2'.

2.2.   The Möbius Group

The stereographic projection $\Theta_M : Q_M \to M_2$ now allows us, in the sense of §1.3.2, to transfer the group G_M (i.e., the projective subgroup leaving the sphere Q_M invariant) onto the Möbius plane M_2. Before making this trans-

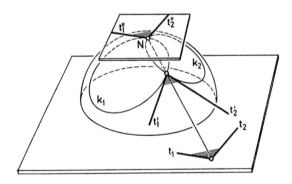

Fig. 3

ference, we can first extend Θ_M so as to include the points outside Q_M. To each point outside Q_M we assign the circle in which its polar plane intersects Q_M, and then also assign to it the corresponding M-circle in M_2, and conversely. This extended mapping[10] will be denoted by $\hat{\Theta}_M$

$$\hat{\Theta}_M : \begin{cases} \text{points on } Q_M & \to M\text{-points} \\ \text{points outside } Q_M & \to M\text{-circles} \end{cases}$$

Let us now give the mapping equations for $\hat{\Theta}_M$. Comparing (7) and (8) we find for *circles* $(u_0 - u_3 \neq 0)$

(9)
$$\begin{cases} \rho u_0 = a^2 + b^2 - r^2 + 1 \\ \rho u_1 = 2a \\ \rho u_2 = 2b \\ \rho u_3 = a^2 + b^2 - r^2 - 1 \end{cases}$$

and similarly, for lines $(u_0 - u_3 = 0)$, comparison of (7) with

(10) $a_1 x + a_2 y + a_3 = 0$

gives

(11) $\rho u_0 = a_3, \quad \rho u_1 = -a_1, \quad \rho u_2 = -a_2, \quad \rho u_3 = a_3$

and the equations (9) and (11) can, of course, be inverted. The u_i in (9) and (11) are sometimes called *tetracyclic coordinates of the M-circle*. The result (9) also holds for proper points, since (9) goes over into (4) if $r = 0$.

By §1.3.2 the transference $(\hat{\Theta}_M, \hat{\pi}_M)$ maps the group G_M, sometimes called the hyperbolic group of three-dimensional space, onto an isomorphic group G'_M, called the *Möbius group*. Its elements, namely the *Möbius mappings*, take M-points into M-points and M-circles into M-circles and are therefore *circular transformations*.[11] (But circles and lines may be interchanged.) In this way we have defined the *Möbius geometry* on M_2.

Since Q_M is invariant under G_M, polarity with respect to Q_M (cf. II10,§4) is also invariant, which means, since the equation $(\mathfrak{x}, \mathfrak{x}) = 0$ is invariant, that the *condition for polarity* of two points $U, V \in P_3$ (i.e., the condition that each of them lies on the polar of the other) is

(12) $(\mathfrak{u}, \mathfrak{v}) = 0.$

In M_2 equation (12) means that the corresponding M-circles u, v are *orthogonal*, since Θ_M preserves angles (see Fig. 4). For the invariance of angles between arbitrary M-circles cf. §2.5.

We also note that if u is an M-circle and X is an M-point, the polarity condition $(\mathfrak{u}, \mathfrak{x}) = 0$ means that X lies on u (*incidence*), since X corresponds

---

[10]To the interior points of the sphere we may, if we wish, assign the imaginary circles.

[11]Conversely, every circular transformation (i.e., every M-circle-preserving invertible mapping of M_2 onto itself) is a Möbius mapping. For the proof see §2.4.

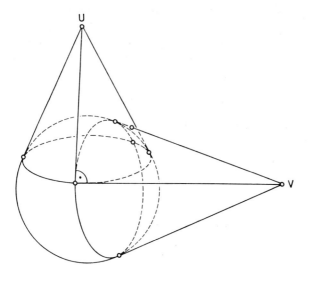

Fig. 4

to the point on the sphere polar to U and therefore lies on the circle $\Theta_M^{-1}(u)$ corresponding to u.

2.3.   The Möbius Inversions

The most important Möbius mappings are the *M-inversions*. In order to define them, we begin (Fig. 5) with a point U outside the sphere Q_M and its polar plane $(u, x) = 0$. We consider the *projective reflection* σ_U corresponding to U, by which we mean the central collineation (II10, §2.1) with center U and with the polar plane as "axis," under which, for every X and its image

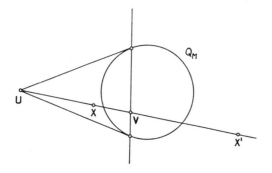

Fig. 5

X', the cross ratio (U, V, X, X') has the value -1, since V lies in the polar plane. In particular, σ_U is involutory, i.e., $\sigma_U\sigma_U$ is the identical mapping.

It is easy to show that this projective reflection σ_U is given by

(13)
$$\mathfrak{x}' = \mathfrak{x} - 2\frac{(\mathfrak{u}, \mathfrak{x})}{(\mathfrak{u}, \mathfrak{u})}\mathfrak{u}$$

and belongs to the group G_M, since σ_U leaves Q_M fixed.

(13')
$$(\mathfrak{x}', \mathfrak{x}') = (\mathfrak{x}, \mathfrak{x}) - 4\frac{(\mathfrak{u}, \mathfrak{x})}{(\mathfrak{u}, \mathfrak{u})}(\mathfrak{u}, \mathfrak{x}) + 4\frac{(\mathfrak{u}, \mathfrak{x})^2}{(\mathfrak{u}, \mathfrak{u})^2}(\mathfrak{u}, \mathfrak{u}) = (\mathfrak{x}, \mathfrak{x}).$$

We now ask: Into what sort of Möbius mapping $\sigma'_U = \Theta_M\sigma_U\Theta_M^{-1}$ is σ_U taken by our transference principle? Let u be the M-circle belonging to U, and let X be an arbitrary M-point (Fig. 6). We choose an M-circle w through X orthogonal to u, whereupon, by §2.3 we have $(\mathfrak{w}, \mathfrak{x}) = 0$, because of the incidence, and $(\mathfrak{w}, \mathfrak{u}) = 0$ because of the orthogonality. Then the image $X' = \sigma_U(X)$ of X can be determined from σ_U by going back to the sphere Q_M, i.e., if $\mathfrak{x}$ represents point $\Theta_M^{-1}(X)$ on the sphere, its image under σ_U is given by (13). Now

$$(\mathfrak{w}, \mathfrak{x}') = (\mathfrak{w}, \mathfrak{x}) - 2\frac{(\mathfrak{u}, \mathfrak{x})}{(\mathfrak{u}, \mathfrak{u})}(\mathfrak{w}, \mathfrak{u}) = 0$$

so that X' lies on every w through X orthogonal to u, and we have the following two cases.

a) If u is a circle (Fig. 6a), then X' lies on the diameter MX and by the secant-tangent theorem (for $X \neq M$)

(14)
$$d(M, X)\cdot d(M, X') = r^2,$$

so that we are dealing with the well-known *inversion in the circle u* (cf. II7, §2.4), which to every $X \neq M$ assigns the image X' determined by (14) on the ray MX. The point M is taken into ∞.

b) If u is a line (Fig. 6b), X' lies on the perpendicular to u through X, so that we have the Euclidean *line reflection* in u.

Both of these types of mappings will be called *M-inversions*.

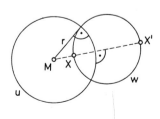

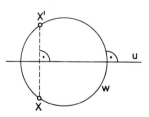

Fig. 6a   Fig. 6b

It is an important fact that the *M-inversions* generate the *whole Möbius group*, i.e., every *M*-mapping is the product of *M*-inversions, as we shall show in §2.4.

2.4. *Linear Transformantions of the z-Plane*

Our Möbius plane M_2 can also be regarded as the z-plane ($z = x + iy$) extended by the point ∞. In the elementary theory of functions of a complex variable, the group H^+ of (fractional) linear transformations of the z-plane

(15) $$z' = \frac{az + b}{cz + d}; \quad a, b, c, d \text{ complex}; \quad ad - bc \neq 0$$

plays an important role (cf. III5). For $c \neq 0$ the transformation (15) is the product of three transformations

(I) $$z \rightarrow z_1 = cz + d,$$

(II) $$z_1 \rightarrow z_2 = \frac{1}{z_1},$$

(III) $$z_2 \rightarrow z' = \frac{bc - ad}{c} z_2 + \frac{a}{c},$$

where (I) and (III) are (entire) linear (and therefore either translations or rotation-dilations) and (II) is the product of an *M*-inversion in the unit circle and in the real axis. If $c = 0$, then (15) is entire linear.

It is well known that translations and rotations are products of line reflections, and that dilations are products of inversions in circles.

$$\left. \begin{array}{c} z \rightarrow z_1 = \dfrac{1}{\bar{z}} \\[2ex] z_1 \rightarrow z' = \dfrac{r^2}{\bar{z}_1} \end{array} \right\} z' = r^2 z,$$

where r is real and $\bar{z}$ is the complex conjugate of z. Thus (15), as a product of *M*-inversions, is a Möbius mapping, so that $H^+ \subset G'_M$.

If the Möbius plane is oriented, it is clear that (I), (II), (III), and thus also (15), are orientation-preserving, so that H^+ is the group of linear *orientation-preserving (sense-preserving)* transformations. In addition to (15), we may also consider the transformations

($\overline{15}$) $$z' = \frac{a\bar{z} + b}{c\bar{z} + d}; \quad a, b, c, d \text{ complex}; \quad ad - bc \neq 0,$$

where ($\overline{15}$) arises from (15) by the line-reflection $z \rightarrow \bar{z}$ and therefore *reverses sense* (i.e., is not orientation-preserving). The set H^- of transformations ($\overline{15}$) is not in itself a group, but forms a group when taken together with

H^+. Calling the extended group H, we again have

(16) $H \subset G'_M$.

But also conversely,

(17) $G'_M \subset H$,

so that every Möbius mapping is given either by (15) or by $(\overline{15})$.

For let $\alpha \in G'_M$ and $\alpha(\infty) = z_0$. Then in the Möbius mapping β with $z' = 1/(z - z_0)$ we have $\beta(z_0) = \infty$, so that $\beta\alpha \in G'_M$ leaves the improper point fixed and takes lines into lines. In the real plane $E_2 = M_2/\{\infty\}$, which for the moment we may take to be projectively extended, $\beta\alpha$ is therefore an invertible, line-preserving mapping of the projective plane onto itself, and is therefore a projective mapping (a projective collineation, cf. II10, §2.4). Since proper points remain proper, $\beta\alpha$ is an affine mapping, and finally, since it is circle-preserving, it is a similarity transformation $z \to az + b$ or $z \to a\bar{z} + b$. Consequently α is of the form $z \to z_0 + 1/(az + b)$ or $z \to z_0 + 1/(a\bar{z} + b)$.

Let us also remark that, as can be proved in the same way, if α is a circular transformation (cf. §2.2) then it is of the form (15) or $(\overline{15})$, and is therefore a Möbius mapping.

From (16) and (17) it follows that $G'_M = H$, so that *the Möbius group is identical with the group H*. At the same time it is clear that the M-inversions generate the Möbius group G'_M, since they generate H.

2.5.   *Invariants of Möbius Geometry. Pencils and Bundles of Circles*

Let us mention some other basic facts in Möbius geometry. An important invariant is the angle φ between two M-circles u, v, or more precisely, the expression (18) for $\cos^2\varphi$. For if u, v are represented by the points U, $V \in P_3$ with vectors $\mathfrak{u}$, $\mathfrak{v}$, then

(18) $$\cos^2\varphi = \frac{(\mathfrak{u}, \mathfrak{v})^2}{(\mathfrak{u}, \mathfrak{u})(\mathfrak{v}, \mathfrak{v})},$$

as is proved by calculating the right-hand side (from (9) and (11)) and showing by elementary geometry that the result is equal to $\cos^2\varphi$. But (18) is an invariant, since the right-hand side does not change under renormalizations $\mathfrak{u} \to \lambda\mathfrak{u}$, $\mathfrak{v} \to \mu\mathfrak{v}$ nor under Möbius mappings (by (13) it is invariant under the generating M-inversions).

In §6.1 we shall see that φ can also be interpreted as the elliptic distance between U and V, provided the line UV has no point in common with the quadric Q_M, and similarly we can show that the *cross ratio* of four M-points on an M-circle is invariant.

Let us now consider briefly the *linear sets of circles* in M_2, defined as the images of projective subspaces (lines and planes) of P_3.

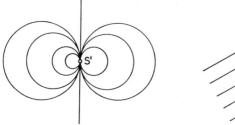

Fig. 7a. *S'* proper   Fig. 7b. *S'* = ∞

From a *line g* in P_3 we obtain by $\hat{\Theta}_M$ a *pencil of circles* in M_2, for which there are three types, depending on the position of g with respect to the sphere Q_M.

If g is tangent to the sphere at a point S, we obtain a *degenerate pencil* (Fig. 7a, b). If g is not tangent to the sphere, let H be the reciprocal polar of g (intersection of planes polar to points on g) with respect to Q_M. One of the two lines cuts Q_M in two points, and the other is a "nonintersector" of Q_M. Because of the polar relationship of g and h, the corresponding pencils of circles (Fig. 8a, b) in M_2 are orthogonal, one of them having two (real) base points, and the other not. The corresponding *bundle of circles* in M_2 obtained from a *plane* in P_3 will not be discussed here.

We conclude this section with a typical *problem of Möbius geometry*. Given three *M*-circles u, v, w not in a pencil, determine the *M*-circles orthogonal to all three u, v, w.

Solution. If U, V, W representing the *M*-circles in P_3 span the whole plane ε, the only possible solution is the *M*-circles corresponding to the pole S of ε with respect to Q_M. So there is a common orthogonal *M*-circle (which

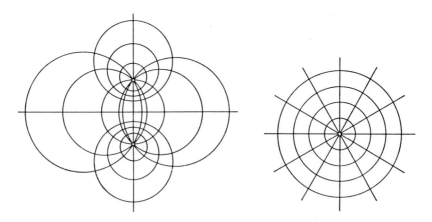

Fig. 8a.   Both base points are proper   Fig. 8b.   One base point is improper

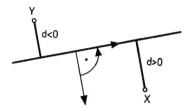

Fig. 9

is then uniquely determined) if and only if ε cuts the quadric Q_M, and then s corresponds to the circle of intersection.

3. Laguerre Geometry

3.1. *The Laguerre Plane and the Quadric Q_L*

In Möbius geometry the basic elements were points, but Laguerre based his treatment of circle geometry on oriented lines, which he called directions, halflines, or *spears*.

By a *spear* we mean an oriented line in the oriented plane E_2 (cf. II9, §7.4). The orientation of the line is defined by distinguishing one of the two half-spaces (by means of a vector normal to it), which means, since the plane is oriented, that a sense of rotation is also given (Fig. 9). Let the equation of the spear in Hesse normal form be

(1) $$a_1 x + a_2 y + a_3 = 0, \quad a_1^2 + a_2^2 = 1$$

with (a_1, a_2) the normal unit vector, and with a_3 as the signed distance from the origin O to the spear. The set of spears is called the *Laguerre plane L_2*.

Circles are also oriented, by distinguishing the interior or the exterior, and are thus provided with a sense of passage around the circumference. The two possible orientations of a circle will be indicated here by the sign of the radius r, namely $r > 0$ or $r < 0$ (Fig. 10a, b). An oriented circle will be regarded as the set of its tangent spears (all oriented in the same sense at

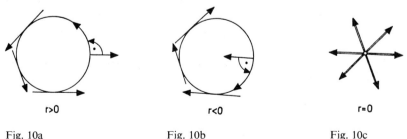

Fig. 10a   Fig. 10b   Fig. 10c

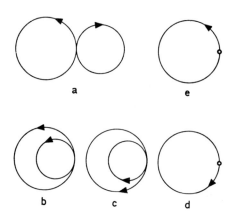

Fig. 11

their points of contact) and a point (a "zero cycle" without orientation) will be regarded as the set of spears incident with it. Points[12] and oriented circles will both be called *cycles*. The spears belonging to a cycle are said to be tangent to it (Fig. 10a, b, c), and two cycles are tangent to each other if they have exactly one spear in common (Fig. 11).

Our present task is to relate the Laguerre plane L_2 in a suitable way to a quadric in P_3. Since $-1 + a_1^2 + a_2^2 = 0$, it is natural to set

(2) $$a_1 = \frac{x_1}{x_0}, \quad a_2 = \frac{x_2}{x_0}, \quad a_3 = \frac{x_3}{x_1}.$$

Then for the above equation we obtain

(3) $$\langle \mathfrak{x}, \mathfrak{x} \rangle = -x_0^2 + x_1^2 + x_2^2 = 0,$$

where $\langle \mathfrak{x}, \mathfrak{x} \rangle$ again denotes the quadratic form. Here (3) represents a cone Q_L in P_3 with the vertex $Z(0, 0, 0, 1)$, where $x_0 \neq 0$ for all "proper" points $\neq Z$ on the cone. Since (2) can be solved to give

(4) $$\rho x_0 = 1, \quad \rho x_1 = a_1, \quad \rho x_2 = a_2, \quad \rho x_3 = a_3,$$

it therefore provides a one-to-one mapping Θ_L of $Q_L \backslash \{Z\}$ onto the set L_2 of spears, and it is this mapping that we shall use in §3.2 for our transference. In contradistinction to Möbius geometry, the exceptional point Z has no image.

As in §2.1, we again consider intersections of Q_L with planes not passing through Z, and thus obtain the nondegenerate conic sections with equations

(5) $$\langle \mathfrak{x}, \mathfrak{x} \rangle = 0, \quad u_0 x_0 + u_1 x_1 + u_2 x_2 + u_3 x_3 = 0 \quad (u_3 \neq 0).$$

---

[12]In Laguerre geometry we consider only proper points.

Points on the conic sections (5) are taken by Θ_L into spears which by (4) satisfy the equation

(6) $$u_0 + u_1 a_1 + u_2 a_2 + u_3 a_3 = 0.$$

Since equation (6) states that a spear (1) has the fixed distance $-u_0/u_3$ from the fixed point $(a = u_1/u_3, b = u_2/u_3)$, all image spears (6) are tangent to the cycle with center (a, b) and radius $r = u_0/u_3$ (cf. Fig. 10). More precisely, *to every nondegenerate conic section on Q_L the mapping Θ_L assigns a cycle in L_2, and conversely,* so that we let $\hat{\Theta}_L$ denote the correspondence between cycles and planes not through Z. For $\hat{\Theta}_L$ and the inverse mapping we have

(7) $$r = \frac{u_0}{u_3}, \quad a = \frac{u_1}{u_3}, \quad b = \frac{u_2}{u_3},$$

(7') $$\rho u_0 = r, \quad \rho u_1 = a, \quad \rho u_2 = b, \quad \rho u_3 = 1.$$

As in Möbius geometry, the cycles of Laguerre geometry can again be represented linearly in P_3, namely by (5) (*linearization*).

Tangency between cycle and spear corresponds on Q_L to incidence of conic section and point, and tangency of two cycles corresponds to tangency of two conic sections. Hence it is tangency that is important here; orthogonality and angles play no role.

3.2.   *The Laguerre Group. L-inversions*

We now take the cone Q_L as the absolute figure in P_3, and let G_L be the projective subgroup leaving Q_L fixed (as a whole). The mapping Θ_L in §3.1 now leads to a *transference* by which G_L goes over into the isomorphic group G'_L called the *Laguerre group*. Laguerre mappings take spears into spears and cycles into cycles (but in general they interchange circles and points). Thus we are no longer dealing, in general, with point transformations but always with "contact transformations" since, by the remark at the end of §3.1, contact (tangency) is invariant.

Let $\alpha' \in G'_L$ be a Laguerre mapping, and let $\alpha \in G_L$ be the corresponding projective mapping of P_3. Under the mapping α the *equation* $\langle x, x \rangle = 0$ of Q_L is taken into itself, since α leaves Q_L fixed. However, the left side of the equation, i.e., the quadratic *form* $\langle x, x \rangle$ will not, in general, remain unchanged but will be multiplied by a factor $\lambda \neq 1$. Those Laguerre mappings α' for which the corresponding mappings α in a normed representation[13] leave even the form $\langle x, x \rangle$ itself unchanged obviously constitute a semigroup, the restricted Laguerre group G''_L, the solution here being like simi-

---

[13]The mapping α can be represented as $\eta = Ax$, where A is a linear mapping of the V_4 corresponding to P_3 with det $A \neq 0$; and α is also represented by $\eta = \bar{A}x$ with $\bar{A} = \rho A$, $\rho \neq 0$. Now ρ can be so chosen that det $\bar{A} = \rho^4$ det $A = \pm 1$, in which case we speak of *normed representation* for α.

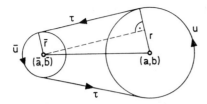

Fig. 12

larity geometry in projective space, where the subgroup of isometries leaves the form $x^2 + y^2 + z^2$ fixed, whereas a nonisometric similitude takes it into $\rho^2(x^2 + y^2 + z^2)$. Since nonisometric similitudes exist only for degenerate forms (cf. II9, §5.1), they do not exist at all in Möbius or Lie geometry, for example, or in hyperbolic or elliptic geometry (cf. §7).

Unlike Möbius geometry, the geometry of Laguerre has no concept of angle. Instead, we have the *tangent distance* τ between two cycles. From Fig. 12 we see that

(8) $$\tau^2 = (a - \bar{a})^2 + (b - \bar{b})^2 - (r - \bar{r})^2,$$

which in view of (7) and (3) can be put in the form

(9) $$\tau^2 = \langle \mathfrak{v}, \mathfrak{u} \rangle \quad \text{with} \quad \mathfrak{v} = \mathfrak{u} - \bar{\mathfrak{u}}, u_3 = \bar{u}_3 = 1.$$

It is easy to show that G''_L leaves all tangent distances unchanged, whereas G'_L multiplies them by a "similarity factor." The reader should compare the corresponding property of distance in equiform and Euclidean geometry.

The most important Laguerre mappings are the *L-inversions,* defined as in §2.3. If U is a point $\notin Q_L$ in P_3 and σ_U denotes projective reflection in U and in the polar plane of U, then σ_U is involutory and thus has an equation like the one in §2.3 (13)

(10) $$\mathfrak{x}' = \mathfrak{x} - 2\frac{\langle \mathfrak{u}, \mathfrak{x} \rangle}{\langle \mathfrak{u}, \mathfrak{u} \rangle}\mathfrak{u},$$

and the corresponding Laguerre mapping $\sigma'_U = \Theta_L \sigma_U \Theta_L^{-1}$ is called an *L-inversion.* We note without proof (cf. [2], §37) that the *L*-inversions themselves generate the restricted group G''_L and taken together with the *L*-similitudes they generate the full group G'_L. A Euclidean interpretation of the *L*-inversions will not be given here.

Instead, let us give some examples. 1) For $\mathfrak{u} = \mathfrak{b}_0 = (1, 0, 0, 0)$ we obtain from (10) the mapping equations

(11) $$x'_0 = -x_0, \quad x'_1 = x_1, \quad x'_2 = x_2, \quad x'_3 = x_3.$$

From (4), (5), and (7) it follows that all points, lines, and circles remain fixed, but the orientation of all spears and cycles is changed (*change of direction*).

2) If we perform two *L*-inversions with $\mathfrak{u}$ and $\mathfrak{u} + \lambda \mathfrak{z}, \mathfrak{z} = \mathfrak{b}_3 = (0, 0, 0, 1)$ one

after the other, we obtain the mapping

(12) $x' = x + 2\lambda\langle v, x\rangle b_3$ with $v = \dfrac{u}{\langle u, u\rangle},$

and thus

(12′) $x_0' = x_0,$ $x_1' = x_1,$ $x_2' = x_2,$ $x_3' = x_3 + 2\lambda(-v_0 x_0 + v_1 x_1 + v_2 x_2).$

From (5) and (7) it follows that

(13) $\bar{r} = r + 2\lambda v_0,$ $\bar{a} = a - 2\lambda v_1,$ $\bar{b} = b - 2\lambda v_2.$

For $v_0 = 0$ this is a *translation*, and for $v_1 = v_2 = 0$ it is a so-called *dilation*, under which the radii of all cycles are changed by the same amount, the centers of the cycles remaining fixed.

3.3.   *Pencils and Bundles of Cycles*

We shall now give a brief discussion of linear sets of cycles, i.e., pencils and bundles of cycles obtained by $\hat{\Theta}_L$ from pencils and bundles of planes in P_3.

A *pencil of planes* in P_3 has a carrier line g. If g is "improper," i.e., if g goes through the vertex Z of the cone, then a pencil of planes intersects the cone in two generators, one generator, or no generator at all. Such a generator has the equation $x + \lambda 3$ with $\langle x, x\rangle = 0$ and $3 = (0, 0, 0, 1)$ and therefore corresponds by (2) to a pencil of parallel like-oriented spears (*pencil of spears*). If g is "proper," $Z \notin g$, then the planes of the pencil not through Z intersect the cone in nondegenerate conic sections, corresponding in L_2 to the cycles of a *pencil of cycles*. Such pencils are of the following three types.

a)   If g is a tangent, then all the conic sections have exactly one point S in common. All cycles are tangent to one another, and have the tangent spear s in common, so that we have an *isotropic* pencil of cycles (called an oriented "line element," cf. Fig. 13a).

b)   If g is a secant cutting Q_L in S and T, we have a *spacelike*[14] pencil of cycles, namely all cycles tangent to the two spears s and t. They have a common exterior center of similarity, with an arbitrary real dilation factor (Fig. 13b).

c)   If g does not intersect the cone, we have a *timelike* pencil of cycles, namely (we omit the proof here) a pencil with an interior center of similarity (Fig. 13c).

A *bundle of planes* with "proper" carrier point $S \neq Z$ corresponds to a *bundle of cycles*, which can also, depending on the position of S with respect to Q_L, be isotropic, spacelike, or timelike. The reader should vizualize these possibilities. For example, for $S = B_0$ we obtain a spacelike bundle of cycles consisting precisely of the points of E_2.

---

[14]The terminology is taken from the special theory of relativity, cf. §3.4.

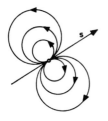

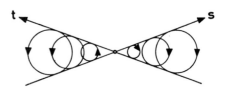

Fig. 13a   Fig. 13b

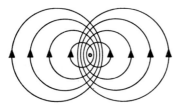

Fig. 13c

We have now developed the basic features of Laguerre geometry. A typical problem is the well-known *tangency problem of Apollonius,* namely to determine the cycles tangent to three given circles. However, we will discuss this problem later in the more general setting of Lie geometry (§4).

3.4. *Cyclographic Transference of Laguerre Geometry. Laguerre Geometry in Space and in Special Relativity*

We have developed Laguerre geometry by taking as the absolute figure a cone in P_3, which clearly showed the relationship with Möbius geometry, since in both cases the absolute figures were quadrics (though of different rank). Another, more customary, transference is based on cyclography and makes use of the model dual to ours.

Our cone Q_L in P_3 (as a quadratic configuration of lines through a point) corresponds dually to a conic section C (as a quadratic configuration of lines in a plane). Let this plane be the improper one and let C be the absolute figure with the equations

(14) $x_0 = 0, \quad x_1^2 + x_2^2 - x_3^2 = 0.$

If (14) is projected from a proper point, the result is a cone, which in the nonhomogeneous coordinates $x = x_1/x_0$, $y = x_2/x_0$, $z = x_3/x_0$ has the equation

(15) $(x - a)^2 + (y - b)^2 - (z - r)^2 = 0,$

where $S(a, b, r)$ is the vertex, since (15) cuts the improper plane at C. Thus all cones (15) are called *isotropic.*

Furthermore, the cone (15) cuts the plane $z = 0$ in

(16) $$(x - a)^2 + (y - b)^2 - r^2 = 0,$$

so that in a Euclidean interpretation we have the circular cone with half-angle $\pi/4$, base circle (16), and vertex with signed z-coordinate r (cf. Fig. 14). This one-to-one mapping $\hat{\Theta}$ (of E_3 onto the set of cycles at $z = 0$), which to every point $S(a, b, r)$ assigns the oriented circle (16), is called a *cyclographic mapping*. The planes tangent to the isotropic cone, i.e., the *isotropic planes* inclined at an angle $\pi/4$ to $z = 0$, obviously correspond to the *spears*, again under a one-to-one mapping Θ (Fig. 14).

Then the transference is clear. The projective subgroup G leaving C fixed corresponds to an isomorphic group G' of transformations of the set of spears and cycles. By duality, we have $G \simeq G_L$ and therefore $G' \simeq G'_L$, so that we have again obtained the Laguerre geometry. In a cyclographic transference G obviously consists of the affine mappings whose corresponding linear mappings (II9, §4) leave the equation $x^2 + y^2 - z^2 = 0$ unchanged, a fact which makes clear the relationship to similarity geometry in space with the corresponding quadratic form $x^2 + y^2 + z^2$.

Let us here make a few remarks about the next higher dimension. In *Laguerre geometry of space* the form $x^2 + y^2 + z^2 - w^2$ plays the corresponding role. If we write $w = ct$, the restricted space Laguerre group $\mathfrak{G}'$ (and also the group $\mathfrak{G}$ corresponding to it cyclographically in four-dimensional space) leaves the form

(17) $$x^2 + y^2 + z^2 - c^2 t^2$$

unchanged and is therefore isomorphic to the *Lorentz group* of special relativity. In fact, the Lorentz group is identical with the group $\mathfrak{G}$, which explains the terminology "timelike" and "spacelike" in §3.3. In the sense of Laguerre geometry, a line g in four-dimensional space is timelike, spacelike, or isotropic if, as a "worldline" of a uniform motion, it lies inside,

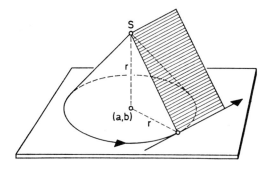

Fig. 14. For $r > 0$

outside, or on a lightcone, in other words, if its velocity vector is smaller than, greater than, or equal to the velocity of light c.

4. Lie Geometry

4.1.   *The Lie Plane and the Quadric Q_ϱ*

Lie geometry includes Möbius and Laguerre geometry in a more general framework. Its basic elements are cycles (oriented circles and proper points), spears (oriented lines), and the improper point ∞. The set of these $\mathfrak{L}$-circles is called the *Lie plane* $\mathfrak{L}_3$.[15]

We start from the mapping $\hat\Theta_M$ of Möbius geometry, which refers to nonoriented circles and lines, so that in the mapping equations §2.2 (9) (and generally in the whole of Möbius geometry, which is therefore said to be part of the "elementary" geometry of circles) the parameter r occurred only in the form r^2. If we wish to include orientation, as in Laguerre geometry, i.e., to pass to the "higher" geometry of circles, it is desirable to let the signed radius r be one of the coordinates.

I.   As an extension of §2.2 (9), the equations of *cycles* then become

$$(1') \qquad \begin{cases} \rho x_0 = a^2 + b^2 - r^2 + 1 \\ \rho x_1 = 2a \\ \rho x_2 = 2b \\ \rho x_3 = a^2 + b^2 - r^2 - 1 \\ \rho x_4 = 2r \end{cases} \qquad x_0 - x_3 \neq 0$$

where the vector x represents a point X in the real space P_4 lying on a quadric Q_ϱ with the equation

$$(*) \qquad [x, x] = -x_0^2 + x_1^2 + x_2^2 + x_3^2 - x_4^2 = 0.$$

Equation $(1')$ can be solved for a, b, r thus

$$(1) \qquad a = \frac{x_1}{x_0 - x_3}, \quad b = \frac{x_2}{x_0 - x_3}, \quad r = \frac{x_4}{x_0 - x_3}.$$

II.   In view of §2.2 (11) and §3.1 (1) it is natural to write the equations of spears in the form

$$(2') \qquad \begin{cases} \rho x_0 = a_3 \\ \rho x_1 = -a_1 \\ \rho x_2 = -a_2 \\ \rho x_3 = a_3 \\ \rho x_4 = 1 \end{cases} \qquad a_1^2 + a_2^2 = 1, \quad x_0 - x_3 = 0, \quad x_3 \neq 0.$$

[15]In this notation the set of $\mathfrak{L}$-circles depends on three parameters.

Here again (*) holds and we have

(2) $$a_1 = -\frac{x_1}{x_4}, \quad a_2 = \frac{x_2}{x_4}, \quad a_3 = \frac{x_3}{x_4}.$$

III. The only remaining point on Q_ϱ is given by $x_0 - x_3 = 0$, $x_4 = 0$ (*), or in other words by $(\rho, 0, 0, \rho, 0)$, so that the only remaining $\mathfrak{L}$-circle, the *improper point*, we set

(3) $$\infty \leftrightarrow (\rho, 0, 0, \rho, 0).$$

In summary, *equations* (1), (2), (3) *represent a one-to-one mapping Θ_ϱ of Q_ϱ onto $\mathfrak{L}_3$*, which in §4.2 will provide a transference. The $x_0, \ldots, x_4$ are also called *pentacyclic coordinates* of the $\mathfrak{L}$-circle.

What is the significance, for two $\mathfrak{L}$-circles y, z, of the *polarity condition* $[\mathfrak{y}, \mathfrak{z}] = 0$? We shall say that y, z are tangent to each other, or *in contact* with each other (in the sense of Lie geometry), if and only if $[\mathfrak{y}, \mathfrak{z}] = 0$. Then for the case, say, that y, z are cycles, (1') shows that

$$[\mathfrak{y}, \mathfrak{z}] = -y_0 z_0 + y_1 z_1 + y_2 z_2 + y_3 z_3 - y_4 z_4$$
$$= -2\{(a - \bar{a})^2 + (b - \bar{b})^2 - (r - \bar{r})^2\},$$

so that $[\mathfrak{y}, \mathfrak{z}] = 0$ if and only if the circles are tangent to each other with the same orientation, or else the point y is incident with the circle z. The reader should convince himself that eight of the ten possible types of tangent figures involve tangents in the sense of Laguerre geometry (§3.1, Figs. 10 and 11). The only new possibilities are tangency of two parallel like-oriented spears and tangency of a spear with ∞. In P_4, the equation $[\mathfrak{y}, \mathfrak{z}] = 0$ means that if the points Y and Z lie on Q_ϱ, then so does the whole line YZ.

4.2. The Lie Group, $\mathfrak{L}$-Pencils of Circles, and $\mathfrak{L}$-Bundles of Circles

If quadric Q_ϱ is taken as the absolute figure in P_4 with corresponding subgroup G_ϱ, our mapping Θ_ϱ provides a transference $(\Theta_\varrho, \pi_\varrho)$, taking Q_ϱ into the Lie plane $\mathfrak{L}_3$ and G_ϱ into the Lie group G'_ϱ. The Lie mappings take $\mathfrak{L}$-circles into $\mathfrak{L}$-circles, so that in general they are not point transformations, but contact transformations, since with Q_ϱ the polarity condition is also invariant.

In analogy with §§2 and 3, we now introduce the $\mathfrak{L}$-*inversions*, which are again involutory and generate G'_ϱ (cf. [2], §49). They include the familiar reflections of earlier geometry and enable us, for example, to prove that any two $\mathfrak{L}$-circles y, z can be transformed into each other. For if Y, $Z \in Q_\varrho$ represent these $\mathfrak{L}$-circles and if the line YZ does not lie entirely on Q_ϱ, then the projective reflection in any point U on YZ ($U \neq Y$, Z) interchanges the two points; and if the line YZ lies on Q_ϱ, we can choose a suitable auxiliary point on Q_ϱ and carry out two successive projective reflections. Similarly,

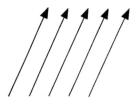

Fig. 15.   Timelike pencil

any two points inside[16] Q_ϱ or any two outside, are on equal footing. These
remarks facilitate the classification and geometric interpretation of figures
in Lie geometry, for example, pencils and bundles.

Pencils and *bundles* of $\mathfrak{L}$-circles are again defined by means of projective
subspaces in P_4.

A *line* in P_4 lying entirely on Q_ϱ corresponds to an "isotropic" pencil of
(pairwise tangent by §4.1) $\mathfrak{L}$-circles. Thus we have either Fig. 13a or Fig. 15.

Let us now consider a *plane* ε in P_4 intersecting Q_ϱ in a (real one-dimensional) conic section C. Its total-polar space is a line g, which may be a secant
or a nonintersector of Q_ϱ. If g is a secant and cuts Q_ϱ at Y and Z, then C
provides all $\mathfrak{L}$-circles tangent to two fixed $\mathfrak{L}$-circles y, z. Thus we have a
"spacelike" pencil of $\mathfrak{L}$-circles (cf. §3.3). If g has no point in common with
Q_ϱ, we have a "timelike" pencil.

A three-dimensional *hyperplane* in P_4 defines an $\mathfrak{L}$-bundle of circles. Here
too the interrelations and terminology are similar to those of Laguerre
geometry, and we shall not discuss them further.

4.3.   *Subsuming of Möbius and Laguerre Geometries under Lie Geometry*

In the present section we shall distinguish subgroups of the Lie group,
entirely in spirit of the Erlanger hierarchy principle, by adding to the absolute figure a point S in P_4 (and thereby also its polar hyperplane s with
respect to Q_ϱ). In this way we shall be able to subsume the Möbius and
Laguerre geometries under the Lie geometry.

We first of all distinguish a point S in P_4 *inside* Q_ϱ, say the base point B_4
of the projective coordinate system with the coordinates $(0,0,0,0,1)$. By the
remark in §4.2 and by §1.3.2, the choice of any other interior point would
merely lead to an isomorphic geometry. In view of (*), the equation of the
polar hyperplane s of B_4 is $x_4 = 0$. For the intersection $s \cap Q_\varrho$ we have

$$(4) \qquad x_4 = 0, \qquad -x_0^2 + x_1^2 + x_2^2 + x_3^2 = 0,$$

which, as in §2.1 (1), is the equation of a quadric Q_M of spherical type in the
three-dimensional subspace s. We now ask: How is the group G_M leaving

[16]A point X lies "inside" Q_ϱ if $[x, x] < 0$, and "outside" if $[x, x] > 0$.

Q_M fixed in the subspace s related to the Lie group $G_\mathfrak{Q}$ leaving s and therefore $s \cap Q_\mathfrak{Q} = Q_M$ fixed in P_4?

In any case, an element $\alpha \in H$ induces an $\bar{\alpha} \in G_M$, as is easily seen if we consider α only on s. Conversely, however, to every $\bar{\alpha} \in G_M$ there correspond exactly two $\alpha_1, \alpha_2 \in H$ with $\alpha_2 = \sigma_S \alpha_1$, differing from each other by the projective reflection σ_S in $S = B_4$ (the proof will not be given here), where σ_S corresponds in $\mathfrak{Q}_3$ to a "change of direction" (cf. §3.2, Example 1). Thus it is the factor group $H/\{\iota, \sigma_S\}$ and not H itself, that is isomorphic to the Möbius group.

For $\mathfrak{Q}_3$ this means that the subset of $\mathfrak{Q}$-circles determined by $x_4 = 0$, namely, the set of M-points (as is seen from (1') and (3)), is mapped onto itself, and also the set of oriented M-circles. So we are dealing here with a Möbius geometry that includes orientation and with Möbius mappings that include change of direction. The ordinary Möbius geometry is obtained by identifying each oriented M-circle with the oppositely oriented circle, i.e., by neglecting the orientation.

Similarly, if we distinguish a point S on $Q_\mathfrak{Q}$, say $(\rho, 0, 0, \rho, 0)$, its polar hyperplane s intersects $Q_\mathfrak{Q}$ in a cone Q_L with the vertex S

$$(5) \qquad x_0 - x_3 = 0, \quad x_1^2 + x_2^2 - x_4^2 = 0.$$

The points on the cone distinct from S correspond by (1)–(3) to the set of spears, which is therefore the distinguished set, so that in this case (cf. §3.1(3)) we are back to Laguerre geometry. The details will not be given here.

4.4. The Apollonius Contact Problem

Let us conclude this section with a problem typical of Lie geometry, namely the *contact problem of Apollonius*. Given three $\mathfrak{Q}$-circles u, v, w, no two of which are tangent to each other, it is required to find the $\mathfrak{Q}$-circles tangent to all three of the given circles. The solution is as follows. By hypothesis, the corresponding points U, V, W in P_4 determine a plane ε, which intersects $Q_\mathfrak{Q}$ in a (one-dimensional, real) conic section. The desired $\mathfrak{Q}$-circles correspond to points that are polar to U, V, W and therefore lie on the total-polar line g of ε. By §4.2, the line g is either a secant or a non-intersector of $Q_\mathfrak{Q}$, so that there are either two solutions or no solution. (Since the set of three circles can be oriented in eight ways, there are at most eight nonoriented solutions for the nonoriented circles.)

It is very easy now to give a *constructive* solution. If ∞ is one of the u, v, w, say $u = \infty$, then v, w are cycles, and the solutions are spears; but the spears common to two cycles are easily found. If the three given elements do not include ∞ but at least one of them is a cycle, this cycle can be taken into ∞; for a dilation will take it into a point, which can then be taken into

∞ by an M-inversion. If u, v, w are all spears (i.e., do not include any cycle) the solution is elementary.

5. Plücker Geometry

The analogous Lie geometry of $\mathfrak{L}$-spheres in space is related in a remarkable way to Plücker line-geometry, as we shall now briefly show. The basic elements of this geometry are the lines in P_3.

5.1. *Line Geometry and the Plücker Quadric* Q_P

Let Γ_4 denote the set[17] of lines in the real P_3. A line g is determined by two of its points R, S. If $\mathfrak{r}$, $\mathfrak{s}$ are the corresponding vectors in V_4 and r_i, s_i ($i = 0, 1, 2, 3$) are its components with respect to a basis $\mathfrak{b}_i$, we form the subdeterminants of the matrix

$$(1) \qquad \begin{pmatrix} r_0 & r_1 & r_2 & r_3 \\ s_0 & s_1 & s_2 & s_3 \end{pmatrix}$$

namely

$$(2) \qquad g_{ik} = \begin{vmatrix} r_i & r_k \\ s_i & s_k \end{vmatrix}, \quad g_{ik} = -g_{ki}, g_{ii} = 0, \quad i, k = 0, 1, 2, 3.$$

The set of six "essential" subdeterminants is now regarded as a six-vector $\mathfrak{G}$, which is then assigned to the line g

$$(3) \qquad \mathfrak{G} = \mathfrak{r} \wedge \mathfrak{s} = (g_{01}, g_{02}, g_{03}, g_{23}, g_{31}, g_{12}).$$

We also call $\mathfrak{G} = \mathfrak{r} \wedge \mathfrak{s}$ the outer product of $\mathfrak{r}$ and $\mathfrak{s}$ (cf. in general IB3, §3.3, and for the case E cf. II9, §8.3). By (2) and (3) the outer product is
(a) alternating $\mathfrak{r} \wedge \mathfrak{s} = -\mathfrak{s} \wedge \mathfrak{r}$,
(b) bilinear[18] $\mathfrak{r} \wedge (\lambda \mathfrak{s} + \mu \mathfrak{t}) = \lambda (\mathfrak{r} \wedge \mathfrak{s}) + \mu (\mathfrak{r} \wedge \mathfrak{t})$,
(r) regular. If $\mathfrak{b}_0$, $\mathfrak{b}_1$, $\mathfrak{b}_2$, $\mathfrak{b}_3$ are linearly independent in V_4,
then $\mathfrak{b}_0 \wedge \mathfrak{b}_1$, $\mathfrak{b}_0 \wedge \mathfrak{b}_2$, $\mathfrak{b}_0 \wedge \mathfrak{b}_3$, $\mathfrak{b}_2 \wedge \mathfrak{b}_3$, $\mathfrak{b}_3 \wedge \mathfrak{b}_1$, $\mathfrak{b}_1 \wedge \mathfrak{b}_2$ are linearly independent in V_6.
Conversely, these properties can be used to define the outer product: let there be given an alternating, bilinear, regular mapping of $V_4 \times V_4$ into V_6, in which the ordered pair $(\mathfrak{r}, \mathfrak{s})$ has the image $\mathfrak{r} \wedge \mathfrak{s}$. Then for a choice of basis as described in (r), the image $\mathfrak{r} \wedge \mathfrak{s}$ has the coordinates (3), as can be shown from (a) and (b).

If instead of R, S we choose two other points on g, say X, Y with $\mathfrak{x} = \lambda \mathfrak{r} + \mu \mathfrak{s}$, $\mathfrak{y} = \varrho \mathfrak{r} + \sigma \mathfrak{s}$, $\lambda \sigma - \mu \varrho \neq 0$, then from (a) and (b) we see that $\mathfrak{x} \wedge \mathfrak{y} = (\lambda \mathfrak{r} + \mu \mathfrak{s}) \wedge (\varrho \mathfrak{r} + \sigma \mathfrak{s}) = (\lambda \sigma - \mu \varrho)(\mathfrak{r} \wedge \mathfrak{s})$, so that the outer

---

[17] Since this set depends on four parameters.
[18] It follows from (a) and (b) that the outer product is linear in the first factor. From (a) it follows that $\mathfrak{r} \wedge \mathfrak{r} = \mathfrak{o}$.

products $\mathfrak{x} \wedge \mathfrak{y}$ and $\mathfrak{r} \wedge \mathfrak{s}$ are proportional, and if then, as is customary, we consider the one-dimensional subspaces of V_6 as points of a projective P_5, every line g in P_3 corresponds to exactly one point $\underline{G}$ in P_5. The point $\underline{G}$ lies on a quadric Q_P in P_5, since expansion of the determinant $D(\mathfrak{r}, \mathfrak{s}, \mathfrak{r}, \mathfrak{s})$ by the two first rows gives

(4)
$$\begin{vmatrix} r_0 & r_1 & r_2 & r_3 \\ s_0 & s_1 & s_2 & s_3 \\ r_0 & r_1 & r_2 & r_3 \\ s_0 & s_1 & s_2 & s_3 \end{vmatrix} = 2g_{01}g_{23} + 2g_{02}g_{31} + 2g_{03}g_{12} = \{\mathfrak{G}, \mathfrak{G}\} = 0.$$

Then Q_P is called the *Plücker quadric* and we again write $\{\mathfrak{G}, \mathfrak{G}\}$ as an abbreviation for the quadric form in (4).

Thus we have defined a mapping $\Theta_P : \Gamma_4 \to Q_P$, which maps Γ_4 one-to-one onto Q_P and is therefore suitable for a transference (§5.2).

The fact that θ_P maps the set Γ_4 onto Q_P is proved as follows. If $\underline{G}$ is a point of Q_P with $\mathfrak{G} = (g_{01}, g_{02}, g_{03}, g_{23}, g_{31}, g_{12}) \neq \mathfrak{o}$ and if say $g_{01} \neq 0$, let us choose $\mathfrak{r} = \sum_k g_{0k}\mathfrak{b}_k$, $\mathfrak{s} = \sum_k g_{1k}\mathfrak{b}_k$. We then find $\mathfrak{r} \wedge \mathfrak{s} = g_{01}\mathfrak{G}$, so that the line $g = RS$ is the preimage corresponding to $\underline{G}$. If $g_{01} = 0$, we proceed in the same way with a $g_{ik} \neq 0$.

The mapping θ_P is also *one-to-one*. For let g, h be two lines whose six-vectors represent the same point on Q_P, $\mathfrak{H} = \lambda\mathfrak{G}$. As we shall show below, the polarity relationship $\{\mathfrak{G}, \mathfrak{H}\} = \lambda\{\mathfrak{G}, \mathfrak{G}\} = 0$ means that g and h have a point R in common. Then we can set $\mathfrak{G} = \mathfrak{r} \wedge \mathfrak{s}$, $\mathfrak{H} = \mathfrak{r} \wedge \mathfrak{t}$, and obtain $\mathfrak{H} - \lambda\mathfrak{G} = \mathfrak{r} \wedge (\mathfrak{t} - \lambda\mathfrak{s}) = \mathfrak{o}$. Thus in view of the property of regularity $(\mathfrak{r})$, the vectors $\mathfrak{r}$ and $\mathfrak{t} - \lambda\mathfrak{s}$ are linearly dependent, which means that $\mathfrak{t} - \lambda\mathfrak{s} = \mu\mathfrak{r}$, so that g and h coincide.

The g_{ik} are called Plücker *line coordinates*, and we can also introduce dual line coordinates. Since a line g can be represented by two planes u, v, we obtain a corresponding outer product $\bar{\mathfrak{G}} = \mathfrak{u} \wedge \mathfrak{v}$, related to (3) by $\bar{\mathfrak{G}} = \rho(g_{23}, g_{31}, g_{12}, g_{01}, g_{02}, g_{03})$.

What is the significance of the *polarity condition*

(5) $\quad \{\mathfrak{G}, \mathfrak{H}\} = g_{01}h_{23} + g_{02}h_{31} + g_{03}h_{12} + g_{23}h_{01} + g_{31}h_{02} + g_{12}h_{03} = 0$

for two lines g, h?

In P_3 the condition (5) *means that g and h either intersect each other or are identical*. For let $\mathfrak{G} = \mathfrak{r} \wedge \mathfrak{s}$, $\mathfrak{H} = \mathfrak{p} \wedge \mathfrak{q}$. Then expansion of the determinant $D(\mathfrak{r}, \mathfrak{s}, \mathfrak{p}, \mathfrak{q})$ by the first two rows (cf. (4) above) gives

$$\begin{vmatrix} r_0 & r_1 & r_2 & r_3 \\ s_0 & s_1 & s_2 & s_3 \\ p_0 & p_1 & p_2 & p_3 \\ q_0 & q_1 & q_2 & q_3 \end{vmatrix} = \{\mathfrak{G}, \mathfrak{H}\}.$$

Thus $\{\mathfrak{G}, \mathfrak{H}\} = 0$ if and only if the points R, S, P, Q are coplanar, i.e., if g, h either intersect of coincide.

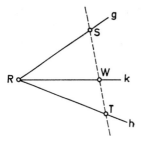

Fig. 16

In P_5 the condition (5) for distinct points $\underline{G}$, $\underline{H}$ means that with $\underline{G}$, $\underline{H}$ the line $\underline{G\,H}$ lies entirely on Q_P. The points of $\underline{G\,H}$ then correspond in $\overline{P}_3$ to the lines of a *pencil;* for let R be the point of intersection (which must exist by (5)) of g and h; then for a $\underline{K} \in \underline{GH}$,

$$\mathfrak{K} = \lambda\mathfrak{G} + \mu\mathfrak{H} = \lambda(\mathfrak{r} \wedge \mathfrak{s}) + \mu(\mathfrak{r} \wedge \mathfrak{t}) = \mathfrak{r} \wedge (\lambda\mathfrak{s} + \mu\mathfrak{t}) = \mathfrak{r} \wedge \mathfrak{w},$$

so that the line $k = RW$ with $W \in ST$ corresponds to $\underline{K}$; and conversely, to a line k in the pencil there corresponds a point $\underline{K} \in \underline{GH}$ (Fig. 16).

On Q_P there also exist (two-dimensional) planes, as can be seen as follows. The set of lines P_3 through a fixed point (*bundle of lines*) or the set of lines in a fixed plane (*field of lines*) corresponds to such a plane on Q_P, as the reader can easily prove by analogy with the above discussion for pencils. On the other hand, there cannot be any three-dimensional subspaces on Q_P.

5.2. The Plücker Group

In P_5 we take the Plücker quadric Q_P as the absolute figure and let G_P be the projective subgroup that leaves Q_P fixed. The mapping Θ_P from §5.1 then determines a transference from Q_P to Γ_4 and from G_P to the *Plücker group* G_P'. Thus the Plücker mappings are line-preserving mappings of P_3.

We know that the projective mappings (collineations) and the correlations of P_3 are line-preserving (II10, §§2 and 4) and it is easy to show, on the basis of their mapping equations, that collineations and correlations actually lead to Plücker mappings. Conversely, every Plücker mapping corresponds to a collineation or correlation of P_3.

Let us sketch the proof: α' maps Γ_4 one-to-one onto itself, and under this mapping bundles go either into bundles or into fields (since by §5.1 the bundles and fields in P_5 are represented by planes in Q_P), and planes in Q_P go into planes in Q_P.

Case 1. There exists a bundle b that goes into a bundle $\alpha'(b)$. Then every bundle c goes into a bundle $\alpha'(c)$. For if $\alpha'(c)$ went into a field, then the intersection $\alpha'(b) \cap \alpha'(c)$ would either be empty or would be a pencil, whereas in fact $b \cap c$ is a line. Thus points (as carriers of bundles) go into points. But a one-to-one line-preserving point mapping of the real P_3 is a collineation (II10, §5.5).

Case 2. All bundles go into fields; but then points go into planes, so that we have a correlation.

The result is that the Plücker group consists of the collineations and correlations of P_3. Plücker line-geometry is simply the projective geometry of P_3, but from the new point of view that lines are considered as the basic elements.

5.3. Linear Sets of Lines. Lie Line-Sphere-Transformations

The study of linear sets of lines has been very fruitful in line-geometry.

A hyperplane U_4 in P_5 (more precisely: its intersection with Q_P) corresponds in P_3 to a *linear complex* (or *wreath*). The hyperplane U_4 touches or cuts Q_P according to whether the pole $\underline{S}$ of U_4 with respect to Q_P lies on Q_P or not. In the first case, $\{\mathfrak{S}, \mathfrak{S}\} = 0$ and $\{\mathfrak{S}, \mathfrak{X}\} = 0$ for all $\underline{X} \in U_4$, so that all lines in the complex in P_3 intersect a fixed line s, and the complex is called *special*. In the second case, the complex is closely related to the projective reflection σ_S in the point $\underline{S}$. For it is clear that σ'_S leaves U_4 point-wise fixed, so that σ'_S leaves the complex linewise fixed. The Plücker mapping σ'_S is called the null system of the complex and σ'_S is a correlation. The null systems generate the whole Plücker group G'_P; for further details see II10, §4.3 and [11]. A three-dimensional subspace U_3 of P_5 corresponds in P_3 to a *linear congruence*. The line g that is total-polar to U_3 can be either a generator, a tangent, a nonintersector, or a secant of Q_P, and we speak correspondingly of a special, parabolic, elliptic, or hyperbolic linear congruence. In the last case, for example, all points of U_3 are polar to the two points of intersection of g with Q_P, so that all lines in a hyperbolic congruence meet two real "guidelines."

A plane U_2 in P_5 can have various positions with respect to Q_P. If it intersects Q_P in a (real one-dimensional) conic section, then the latter corresponds in P_3 to one of the two families of lines in a ruled quadric (of the one-sheeted hyperboloid type).

These statements show that line-geometric problems (and their applications in kinematics and mechanics) are made much clearer by a five-dimensional discussion.

We close with a remark on the Lie *line-sphere-transformation*. Under the transformation of coordinates

$$g'_{01} = y_0 + y_3, \quad g_{23} = -y_0 + y_3,$$

$$g'_{02} = y_1 + y_4, \quad g_{31} = y_1 - y_4,$$

$$g'_{03} = y_2 + y_5, \quad g_{12} = y_2 - y_5,$$

the equation (4) for Q_P (without the factor 2) is transformed into

$$(6) \qquad -y_0^2 + y_1^2 + y_2^2 + y_3^2 - y_4^2 - y_5^2 = 0$$

where the equation (6) is of rank 6 and index 3 (cf. II9, §5.2). We now compare (6) with the equation for the absolute quadric Q_ϱ of the Lie sphere-geometry

$$(7) \qquad -x_0^2 + x_1^2 + x_2^2 + x_3^2 + x_4^2 - x_5^2 = 0$$

of rank 6 and index 4. The so-called line-sphere transformation of Lie

$$(8) \qquad \begin{cases} y_k = x_k, & k = 0, 1, 2, 3, 5 \\ y_4 = ix_4, & i^2 = -1 \end{cases}$$

takes Q_P into Q_ϱ. This (complex) transformation, which takes lines into spheres and intersections of lines into contacts of spheres, can be applied in a remarkable way to the geometry of surfaces, where asymptotic lines and lines of curvature of a surface (cf. II14, §2) can be transformed into each other. For details see [2], [6], and Lie-Scheffers, Geometric der Berührungs-transformationen, Leipzig 1896.

6. Projective Determination of Measure. The Geometries of Cayley-Klein

We now wish to show that the classical non-Euclidean geometries can also be subsumed under projective geometry, and again we shall confine our attention to the two-dimensional case. Here too we distinguish an absolute figure Q, namely a conic section C in P_2, thereby determining a projective subgroup as the group of the geometry corresponding to C. No transference will be needed here (but cf. §7). According to the choice of C we will obtain the (classical) elliptic or hyperbolic geometry of the plane, and also the equiform geometry and other degenerate geometries.

This method of establishing the classical geometries is due to A. Cayley (1821–1895) and F. Klein (1849–1925), and we shall now group all the geometries thus obtained under the name of Cayley-Klein geometries. Cayley was the first to note that Euclidean geometry can also be regarded as an "over-geometry" if as the absolute figure we take the pair of circular points, and he used other absolute conic sections as well, but it was the great contribution of Felix Klein that he recognized these geometries as models of the non-Euclidean geometries, much discussed at that time on epistemological grounds, and in this way gave a decisive turn to the discussion. Thus when C. von Staudt had shown how to establish projective geometry independently of Euclidean, the hyperbolic and elliptic geometries of Gauss, Bolyai, Lobachevski, and Riemann thereby became "autonomous."

In this section we first investigate the various ways of setting up a projective measure on the line, in a pencil of lines, and then in the plane.

6.1. Hyperbolic and Elliptic Measure on the Line

On the real projective line P_1 we choose a quadric Q, i.e., a pair of points A, B given by

$$(1) \qquad F(\mathfrak{x}, \mathfrak{x}) = F(x_0\mathfrak{b}_0 + x_1\mathfrak{b}_1, x_0\mathfrak{b}_0 + x_1\mathfrak{b}_1)$$
$$= x_0^2 F(\mathfrak{b}_0, \mathfrak{b}_0) + 2x_0x_1 F(\mathfrak{b}_0, \mathfrak{b}_1) + x_1^2 F(\mathfrak{b}_1, \mathfrak{b}_1) = 0.$$

Here F is a nonzero symmetric bilinear form, and the nonzero vector $\mathfrak{x}$ ($\neq \mathfrak{o}$) in the two-dimensional real vector space V_2 with the basis $\mathfrak{b}_0, \mathfrak{b}_1$ represents a point $X \in P_1$. The absolute points A, B can be real and distinct, conjugate complex, or real and coincident, corresponding to the *hyperbolic, elliptic,* or *parabolic* case, respectively.

Disregarding the parabolic case for the moment (cf. §6.2), we have $A \neq B$, which means that the cross ratio $(XYAB)$, defined for any two nonabsolute points X, Y, is a projective invariant on which we can base our projective measure of length. For all nonabsolute X, Y, Z we have

$$(2) \qquad (XYAB) \cdot (YZAB) = (XZAB) \neq 0.$$

But we wish our measure of length to be additive, not multiplicative, and therefore we make use of the (natural) logarithm of the cross ratio to set up the following definition of *distance between two nonabsolute* points X, Y

$$(3) \qquad d(X, Y) = \frac{1}{2k} \ln (XYAB), \quad k = \text{const} \neq 0.$$

We now wish to investigate the properties of this definition of a distance function, and in particular, its relationship to the real numbers. We first note that $(XYAB) = 1$ for all $X = Y$, so that $d(X, Y) = 0$; and for $X \neq Y$ we can represent the absolute points A, B by

$$(4) \qquad\qquad \mathfrak{a} = \mathfrak{x} + a\mathfrak{y}, \quad \mathfrak{b} = \mathfrak{x} + b\mathfrak{y}$$

and find[19]

$$(5) \qquad\qquad (XYAB) = \frac{a}{b}.$$

I. In the *hyperbolic* case $(XYAB) = a/b < 0$ or > 0 according as the point pairs X, Y and A, B separate each other or not. Thus if X, Y lie in the same segment (of the two determined by A, B on the line P_1) we may take the real logarithm ln in (3). If X tends to A or B, then $|d(X, Y)|$ increases

---

[19]The reader should verify these statements for real and for conjugate complex A, B (see the definition of cross ratio in II10, §1.2).

beyond all bounds (by (4) and (5)) so that distances from the absolute segments will be regarded as undefined; in other words, we restrict ourselves to "accessible" points. Then a measure of length has been defined in each of the two segments, and we can speak of the *hyperbolic measure of length*. The constant $k = k_h$ in (3) can be taken to be real, and is often set equal to 1.

The hyperbolic distance function has the following properties.
(I) d_h is real, and $d_h(X, Y) = 0$ if and only if $X = Y$.
(II) $d_h(Y, X) = -d_h (X, Y)$ [since $(YXAB) = (XYAB)^{-1}$].
(III) $d_h(X, Y) + d_h(Y, Z) = d_h(X, Z)$ [from (2)].
(IV)   The hyperbolic length of a segment is infinite.
Let us also note that by (II) the distances are signed, so that each of the two segment is oriented by the hyperbolic measure of length. If a projective mapping of P_1 onto itself interchanges A and B, then it reverses the orientation.

From (II) we see that d_h does not have the properties of a distance function for metric space (cf. II16, §9), but the function δ_h with $\delta_h(X, Y) = |d_h(X, Y)|$ obviously does have these properties. Thus the situation is quite analogous to the Euclidean measure of length; $d_h(X, Y)$ corresponds to the coordinate difference $y - x$, whereas $\delta_h(X, Y)$ corresponds to the absolute value $|y - x|$.

II.   In the *elliptic* case, A, B are conjugate complex. (Here the reader should compare the remarks in II10, §1.4 on the imbedding of real projective geometry into complex.) If we set $a = |a|e^{i\varphi}$, $b = \bar{a}$, then (5) becomes

(5′) $(XYAB) = e^{2i\varphi},$

which is a complex number with absolute value 1. Then in (3) we must use the logarithm function defined in the complex plane by $\ln z = \ln |z| + i \arg z + 2m\pi i$ (m an integer). Thus from (5′) we have

$$\ln (XYAB) = 2i(\varphi + m\pi), \quad m \text{ an integer.}$$

The definition (3) now defines an *elliptic measure of length*. The distances are real if for the measure constant we set $k = ik_e$ (k_e real). In what follows we shall always take $k_e = 1$.

The properties of the elliptic distance function d_e are somewhat different from those of d_h. (I) d_e is real, but is defined only up to an integer multiple of π, so that $X = Y$ if and only if $d_e(X, Y) = m\pi$. (II) and (III) are valid only up to integer multiples of π. It is easy to show that the point Y traverses the line P_1 precisely once if $d_e(X_0, Y)$ traverses the interval $(-\pi/2, \pi/2]$, say, where X_0 is a fixed point. With this restriction on d_e, the line P_1 is given an orientation, which is reversed if A and B are interchanged. (IV) The elliptic length of the line P_1 is finite, namely, by the above, equal to π. Here too it is not d_e but δ_e with $\delta_e(X, Y) = |d_e(X, Y)|$, $-\pi/2 < d_e(X, Y) \leq \pi/2$, that has the properties of a metric distance function.

The various properties of the elliptic distance function on P_1 can easily be visualized (Fig. 17). In a projective plane P_2, let P_1 be chosen as the improper line u. Then the projective coordinate system can be so chosen that the points A, B on u have the coordinates $(0, 1, \pm i)$ and can therefore be interpreted as the absolute

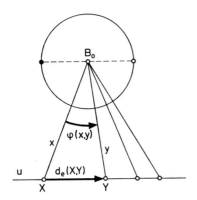

Fig. 17

circular points. If we now project u from the base point $B_0(1, 0, 0)$, we obtain a one-to-one mapping of u onto the pencil of lines with center B_0. Then (5') is identical with the Laguerre angle formula II10, §5.4 (34), so that the distance $d_e(X, Y)$ is equal to the angle φ between the corresponding lines x, y while δ_e is the nonoriented acute (or right) angle between the two lines. Finally, we can map u onto a circle with center at B_0, and identify diametrically opposite points, in order to obtain a one-to-one mapping, or else we can at once restrict ourselves to a semicircle (closed at one end).

The hyperbolic and elliptic distance functions d can be expressed in terms of the form F. By (3), (5) we have

$$e^{2kd} = (XYAB) = \frac{a}{b},$$

and thus

(6) $$(e^{kd} + e^{-kd})^2 = e^{2kd} + e^{-2kd} + 2 = \frac{(a+b)^2}{ab}.$$

Then by (4) and (1) a, b are solutions of the quadratic equation

(7) $$F(\mathfrak{x} + \lambda\mathfrak{y}, \mathfrak{x} + \lambda\mathfrak{y}) = F(\mathfrak{x}, \mathfrak{x}) + 2\lambda F(\mathfrak{x}, \mathfrak{y}) + \lambda^2 F(\mathfrak{y}, \mathfrak{y}) = 0.$$

Expressing $a + b$, ab in terms of the coefficients of (7) we obtain from (6)

(8) $$\left.\begin{array}{r}\cosh^2 k_h d_h(X, Y) \\ \cos^2 k_e d_e(X, Y)\end{array}\right\} = \frac{F^2(\mathfrak{x}, \mathfrak{y})}{F(\mathfrak{x}, \mathfrak{x})F(\mathfrak{y}, \mathfrak{y})}$$

and then calculate

(9) $$\left.\begin{array}{r}\sinh^2 k_h d_h(X, Y) \\ \sin^2 k_e d_e(X, Y)\end{array}\right\} = \varepsilon\frac{F(\mathfrak{x}, \mathfrak{x})F(\mathfrak{y}, \mathfrak{y}) - F^2(\mathfrak{x}, \mathfrak{y})}{F(\mathfrak{x}, \mathfrak{x})F(\mathfrak{y}, \mathfrak{y})}, \quad \varepsilon = \begin{cases}\varepsilon_h = -1, \\ \varepsilon_e = +1.\end{cases}$$

The important equations (8) and (9) will often be used below.

6.2.   *Parabolic Measure on the Line*

In the parabolic case $A = B$ definition (3) leads in every case to the distance zero between distinct points X, Y, and this "trivial" parabolic measurement of length does, in fact, have applications in various geometries (cf. §6.4). But one can also introduce a nontrivial parabolic length. For this purpose we proceed as follows. If in P_1 we introduce a projective coordinate system B_0, B_1, E, in which the base point B_1 coincides with $A = B$, a point X will be represented by $\mathfrak{x} = x_0 \mathfrak{b}_0 + x_1 \mathfrak{b}_1$ or, in the nonhomogeneous projective coordinate $\xi = x_1/x_0$, by $\mathfrak{x} = \mathfrak{b}_0 + \xi \mathfrak{b}_1$. For another "admissible" coordinate system B_0', $B_1' = B_1$, E' with

(10) $$\mathfrak{b}_0 = \mathfrak{b}_0' + p\mathfrak{b}_1', \quad \mathfrak{b}_1 = q\mathfrak{b}_1', \quad q \neq 0$$

we then have $\mathfrak{x} = \mathfrak{b}_0' + \xi'\mathfrak{b}_1'$ with

(11) $$\xi' = p + q\xi.$$

For the coordinate difference of nonabsolute points X, $Y \neq B_1$, we therefore have

(12) $$\eta' - \xi' = q(\eta - \xi).$$

With a distinguished coordinate ξ of this kind we now define the *parabolic distance* of X, Y as

(13) $$d_p(X, Y) = k_p(\eta - \xi), \quad k_p \text{ real} \neq 0.$$

Here d_p is fixed only for $q = 1$, since under a coordinate transformation (11), or in exactly the same way under a "similarity mapping" with $q \neq 1$, the distance is multiplied by a factor independent of X, Y.

The properties (I)–(IV) of d_h hold for d_p also.

Remark.   The parabolic measurement of length can be obtained by a *passage to the limit* from the hyperbolic or elliptic. The absolute points A, B can be represented (e.g., for the hyperbolic case) by prescribing k and a suitable choice of coordinate system as follows.

(14) $$F(\mathfrak{x}, \mathfrak{x}) = x_0^2 - k^2 x_1^2 = 0.$$

In order to determine d_h from the first equation (9) we now calculate

(15) $$F^2(\mathfrak{x}, \mathfrak{y}) - F(\mathfrak{x}, \mathfrak{x})F(\mathfrak{y}, \mathfrak{y}) = k^2(x_0 y_1 - x_1 y_0)^2$$

and for $k_h = k$ we obtain from (9)

(16) $$\frac{1}{k^2}\sinh^2 kd(X, Y) = \frac{(x_0 y_1 - x_1 y_0)^2}{(x_0 - k^2 x_1^2)(y_0^2 - k^2 y_1^2)}.$$

Under passage to the limit $k \to 0$ both of the points A, B go, by (14), into

$B_1(0, 1)$, and (16) goes into

(17) $$d^2(X, Y) = \left(\frac{y_1}{y_0} - \frac{x_1}{x_0}\right)^2 = (\eta - \xi)^2.$$

Thus we obtain d_p up to the (in any case undetermined) similarity factor.

6.3. *Projective Measurement of Angles in a Pencil of Lines*

The discussion in §§6.1 and 6.2 can also be dualized, to give the analogous measurement of angles in a pencil of lines. We first distinguish a pair of absolute lines a, b by setting equal to zero a quadratic form in the line vectors

(18) $$G(\mathfrak{u}, \mathfrak{u}) = 0$$

and then, in the hyperbolic and elliptic cases we define, in analogy with (3), the angle between two lines u, v by

(19) $$w(u, v) = \frac{1}{2k} \ln (u, v, a, b).$$

Then, in analogy with (8) and (9), we have the following equations

(20)
$$\left.\begin{array}{c} \cosh^2 k_h w_h(u, v) \\ \cos^2 k_e w_e(u, v) \end{array}\right\} = \frac{G^2(\mathfrak{u}, \mathfrak{v})}{G(\mathfrak{u}, \mathfrak{u})G(\mathfrak{v}, \mathfrak{v})},$$

$$\left.\begin{array}{c} \sinh^2 k_h w_h(u, v) \\ \sin^2 k_e w_e(u, v) \end{array}\right\} = \varepsilon \frac{G(\mathfrak{u}, \mathfrak{u})G(\mathfrak{v}, \mathfrak{v}) - G^2(\mathfrak{u}, \mathfrak{v})}{G(\mathfrak{u}, \mathfrak{u})G(\mathfrak{v}, \mathfrak{v})}, \quad \varepsilon = \left\{\begin{array}{c} -1 \\ +1 \end{array}\right..$$

The properties of the angle function are the same as those of the corresponding distance function.

In particular, by comparison as in §6.1 with the Laguerre formula for angles, we see that for $k_e = 1$ the *elliptic measurement of angles* coincides with the usual Euclidean measurement. The complete elliptic angle in a pencil of nonoriented lines is thus equal to π. Two lines u, v are said to be orthogonal if and only if $\cos w_e = 0$, or from (20) if $G(\mathfrak{u}, \mathfrak{v}) = 0$, which means that u, v are polar with respect to a, b, so that the pairs of lines u, v and a, b separate each other harmonically. Similarly, we speak of orthogonal points X, Z on a line if $\cos d_e = 0$.

6.4. *The Cayley-Klein Geometries*

Following Cayley and Klein, we now wish to define projective measures in the plane P_2.

If C is a conic section, then every line g not lying entirely on C intersects C in two points A, B, which may be real and distinct, conjugate complex, or real and coincident. Thus the conic section C, when taken as the absolute figure, defines a projective measurement of length on all nongenerators. We take the constant k_h to be the same (say $= 1$) on all lines with hyperbolic measure, and similarly on all lines with elliptic measure. For the lines with parabolic measure it is clear from §6.2 that we cannot at once do this, and that a particular discussion is necessary, if we wish to avoid the trivial measure (all distances equal to zero).

Similarly, in order to obtain a measurement of angles in P_2, instead of starting from "point conic" C, we start from a "line conic" C^* and proceed dually.

We can get an over-all view of the conic sections at our disposal here by considering the (projective) classification of conic sections (cf. II10, §6 of quadratic forms in II9, §5.2). This classification produces five types of point conics (Fig. 18), namely, a) imaginary, b) real, nondegenerate, c) conjugate complex lines, d) pair of real lines, and e) double line.

Dually, we have the corresponding five types of line conics (Fig. 19), namely a*) imaginary b*) real, non-degenerate, c*) conjugate complex points, d*) pair of real points, and e*) double point. (Here a "point" is always a set of lines, namely the corresponding pencil.)

If we now choose a *point conic* a as the absolute figure, we thereby determine a line conic a*, namely the set of tangents of *a*. Similarly, starting from b, we obtain the combination bb*. If we take as the absolute figure a pair of lines c and d (corresponding to certain projective mappings of P_2) with the

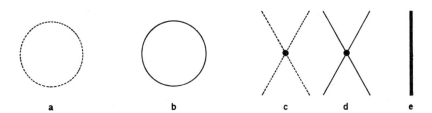

a b c d e

Fig. 18.   Dotted lines indicate nonreal clements

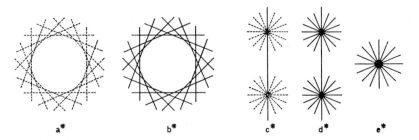

a* b* c* d* e*

Fig. 19.   Dotted lines indicate nonreal elements. Real points of intersection are filled in; nonreal points are not filled in.

intersection S, the pencil of lines of type e* determined by S is, of course, fixed as a whole, so that we obtain the cases ce* and de*. Only the double line e fails to provide a line conic, without the adjunction of further absolute elements.

Dually, we can start from a *line conic* and obtain the combinations a*a, b*b, c*e, and d*e.

Thus we find combined point and line conics of the six types aa*, bb*, c*e, d*e, ce*, de*. As absolute figures, they determine in each case "over-geometries" of the plane projective geometry, which we shall call *Cayley-Klein geometries.*[20]

Let us also note that for aa* and bb* the absolute conic, and consequently the corresponding quadratic forms F, G, are not degenerate, whereas in the other cases they are always degenerate. Thus we also speak of *nondegenerate* or *degenerate* Cayley-Klein geometries. In the discussion of Laguerre geometry above (§3.2) we have noted that for the degenerate forms there are also nonisometric "similarity" mappings. For the non-degenerate geometries, on the other hand, only isometric (rigid) mappings are admissible; if desired they can be subdivided into direct isometries (motions) and indirect isometries.

We now give a brief description of the Cayley-Klein geometries.

Case aa.* The absolute figure is an imaginary conic section C; for all lines and pencils in P_2 we have an elliptic measure. We are dealing here with *elliptic geometry.*

Every two lines intersect, so that there are no parallels. The length of a line and the complete angle in a pencil is equal to π. There exist pairs of orthogonal points and pairs of orthogonal lines. All perpendiculars to a line g pass through its pole G with respect to C, and the corresponding dual statement also holds. There exist polar triangles, in which, therefore, any two sides are orthogonal to each other. (For elliptic geometry see also II3, 5, and 6, also §7 of the present chapter.)

Case bb.* The absolute figure is a non-degenerate conic C. In the exterior of C the measurement of length is not uniform, i.e., it is hyperbolic, parabolic, or elliptic according to the position of the line, so that lines are not "freely movable." In order to arrive at a uniform measurement of length, as in the other Cayley-Klein geometries, we restrict ourselves to the *interior* of C and thereby obtain the *hyperbolic geometry.*

Measurement of lengths is now everywhere hyperbolic, and measurement of angles for lines intersecting in the interior is everywhere elliptic. The perpendicular to a line g through a point R is uniquely determined (Fig. 20).

There are also nonintersecting lines. For a given line h we can find infinitely many nonintersecting lines through a point $\notin h$, and among them

---

[20]We could also prescribe a point conic C and a line conic C^*, independently of each other, as the absolute figure (C, C^*) and study the corresponding geometry. But here we only mention this possibility.

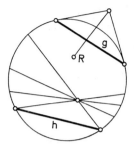

Fig. 20

there will be exactly two ("parallels") which intersect h on C (Fig. 20). For lines intersecting outside C we can measure the angles hyperbolically. If the point of intersection lies on C, we take the trivial measurement of angles (dispensing with any parabolic measure) as a continuation of the elliptic measurement of angles. Thus the angle between parallels is equal to zero. (For hyperbolic geometry, see also II3, 5, and 6, and also §8 of the present chapter.)

Case c*e. The absolute figure is a double line (chosen as the improper line) together with a conjugate complex pair of points and, by §1.2, the corresponding projective subgroup is the equiform group, so that we obtain the usual *equiform geometry*.

The measurement of angles for intersecting lines is everywhere elliptic, and the measure constant k_e can be uniformly set equal to 1. The angle between parallel lines is again measured in the trivial parabolic way, and thus is set equal to zero.

The measurement of length is parabolic on all lines. By §6.2, the measure factor k_p depends on the choice of coordinates on each individual line, so that a priori it will be different on different lines. Let us now give a brief discussion of the way in which, with a uniform choice of $k_p = $ const, we can bring the parabolic measurement of length into conformity with the comparison of lengths in the equiform geometry.

After a suitable choice of coordinate system in P_2 let $x_0^2 = 0$ be the absolute double line and let

(21) $x_0 = 0, \quad F(x, x) = x_1^2 + x_2^2 = 0$

be the pair of absolute points (circular points). We now agree on the following "normalization." Proper points X are represented by vectors x with $x_0 = 1$, and (real) improper points T by vectors t with $F(t, t) = t_1^2 + t_2^2 = 1$. This normalization is meaningful, since in any other "admissible" coordinate system, in which the circular points are represented by (21), it remains unchanged up to a constant factor (independent of X or T).

In our equiform geometry a circle K is determined (see §1.2) by the circular points, and its center is the pole of the line at infinity with respect to K. The equation of K becomes

$$\left(\frac{x_1}{x_0} - \frac{m_1}{m_0}\right)^2 + \left(\frac{x_2}{x_0} - \frac{m_2}{m_0}\right)^2 - r^2 = 0,$$

and it is established that all points on K have the same distance from M.

A line MT cuts K in Y_1, Y_2 with $\mathfrak{y}_j = \mathfrak{m} + \xi_j \mathfrak{t}$, $\xi_j^2(t_1^2 + t_2^2) - r^2 = \xi_j^2 - r^2 = 0$, so that $|\xi_j| = r$ for variable T. On the other hand, the normalization adopted means that for the parabolic distance we have $d_p(M, Y_j) = k_p \xi_j$, if we take $\mathfrak{b}_0 = \mathfrak{m}$ and $\mathfrak{b}_1 = \mathfrak{t}$ as basis vectors on MT. The desired conformity of parabolic measurement of length with equiform comparison of lengths will thus be attained if for all lines through M (and all lines in general) we set up $k_p = \text{const}$. For example, for $k_p = 1$ we now have

$$d_p^2(M, Y) = \left(\frac{y_1}{y_0} - \frac{m_1}{m_0}\right)^2 + \left(\frac{y_2}{y_0} - \frac{m_2}{m_0}\right)^2,$$

which is equal to the Euclidean distance up to the usual similarity factor.

Note.   Equiform geometry can also be obtained by *passage to the limit* from hyperbolic (or elliptic) geometry, if we proceed in exact analogy with the passage to the limit in §6.2. Here we start from the hyperbolic geometry with

$$(23) \qquad \begin{cases} F(\mathfrak{x}, \mathfrak{x}) = x_0^2 - k^2(x_1^2 + x_2^2) = 0 \\ G(\mathfrak{u}, \mathfrak{u}) = -k^2 u_0^2 + u_1^2 + u_2^2 = 0 \end{cases}$$

and equation (9) for $\sinh^2 k_h d_h(X, Y)$ where $k_h = k$. For $k \to 0$ the equations (23) go into the absolute figure for the equiform geometry

$$(24) \qquad F(\mathfrak{x}, \mathfrak{x}) = x_0^2 = 0, \quad G(\mathfrak{u}, \mathfrak{u}) = u_1^2 + u_2^2 = 0$$

and equation (9) for the hyperbolic distance goes into

$$d^2(X, Y) = \left(\frac{y_1}{y_0} - \frac{x_1}{x_0}\right)^2 + \left(\frac{y_2}{y_0} - \frac{x_2}{x_0}\right)^2$$

which corresponds to (22).

Case d*e. The absolute figure is a double line (chosen as the improper line) with a real pair of points. This geometry is called *pseudo-equiform*. For intersecting lines, the measurement of angles is hyperbolic, and for parallel lines, the angle is again set equal to zero. The measurement of length is parabolic. As in the equiform geometry, we can also determine k_p uniformly for all lines, with the exception of the "isotropic" lines through the two distinguished improper points. For the distance (cf. (22)), we find

$$(25) \qquad d_p^2(X, Y) = \pm\left(\frac{y_1}{y_0} - \frac{x_1}{x_0}\right)^2 \mp \left(\frac{y_2}{y_0} - \frac{x_2}{x_0}\right)^2$$

up to the usual factor. For $d_p = \text{const}$, equation (25) gives the pseudo-

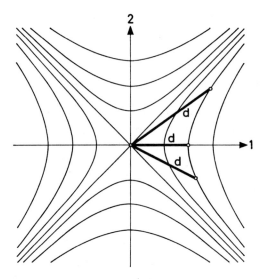

Fig. 21

equiform "circles," i.e., hyperbolas (see Fig. 21 with $x_1 = x_2 = 0$), which for the upper sign in (25) lie in the right and left quadrants, and for the lower sign, in the upper and lower quadrants. These hyperbolas are the orbits of the *pseudorotations*

(26) $$\begin{cases} y'_1 = y_1 \cosh t + y_2 \sinh t, \\ y'_2 = y_1 \sinh t + y_2 \cosh t, \end{cases} \quad y_0 = y'_0 = 1,$$

which correspond to the Euclidean rotations. On the isotropic lines $\lambda y_0 + y_1 \pm y_2 = 0$ the distance between two points is equal to zero, by (25). We shall pay no attention to the nontrivial parabolic measure of length.

The distance (25) occurs in the special theory of relativity. There the square of the distance is defined by the form $x^2 + y^2 + z^2 - c^2 t^2$ of §3.4 (17), and for $y = z = 0$ we arrive at (25). The pseudorotations (26) take the form, with $\cosh t = 1/\sqrt{1 - \beta^2}$, $\sinh t = -\beta/\sqrt{1 - \beta^2}$, $\beta = v/c$ and with $y_1 = x, y_2 = ct$

$$x' = \frac{x - \beta ct}{\sqrt{1 - \beta^2}}, \quad ct' = \frac{-\beta x + ct}{\sqrt{1 - \beta^2}}$$

and are therefore lines in the (two-dimensional) Lorentz transformations of the special theory of relativity.

Note. The pseudo-equiform geometry can also be obtained by a *passage to the limit* if we start from

$$(28) \qquad \begin{cases} F(\mathfrak{x}, \mathfrak{x}) = x_0^2 \mp k^2(x_1^2 - x_2^2) = 0, \\ G(\mathfrak{u}, \mathfrak{u}) = \mp k^2 u_0^2 + u_1^2 - u_2^2 = 0, \end{cases}$$

then for $k \to 0$ the equations (28) go into

$$(29) \qquad F(\mathfrak{x}, \mathfrak{x}) = x_0^2 = 0, \quad G(\mathfrak{u}, \mathfrak{u}) = u_1^2 - u_2^2 = 0$$

and (9) goes into (25).

Cases ce* and de* are the dual counterparts of the equiform geometry (c*e) and the pseudo-equiform geometry (d*e). Their absolute figure is a double pencil with conjugate complex lines or a pair of real lines. Thus we speak of *central-elliptic* or *central-hyperbolic geometry*. The theorems in these geometries arise from dualization of the theorems of equiform and pseudo-equiform geometry, so that we do not need to discuss them further.

7. Subsuming of Equiform and Elliptic Geometry under Möbius Geometry

The three classical geometries, equiform, elliptic, and hyperbolic, can be brought into relationship with Möbius geometry in a uniform and simple way. Namely, their groups are obtained as subgroups of the Möbius group if to the absolute quadric Q_M we adjoin a fixed point, lying on Q_M, inside Q_M, or outside Q_M, respectively. In the present section we discuss this principle for equiform and elliptic geometries, and in §8 for hyperbolic geometry. By suitable transferences we will obtain further important models for elliptic and hyperbolic geometry and will then discuss the *isometries* in these geometries.

7.1. Subsuming of Equiform Geometry under Möbius Geometry

Starting from §2 we now let Q_M be the sphere in §2.1 (1) corresponding to Möbius geometry. We distinguish a point on Q_M and consider the subgroup of all Möbius mappings that leave this point fixed. By §1.3.2 the particular choice of this point makes no difference, since the corresponding subgroups are all isomorphic. So we choose the North Pole N and call the corresponding subgroup U_{eq}. Through transference by the stereographic projection Θ_M into the Möbius plane M_2 we obtain the subgroup U'_{eq}, isomorphic to U_{eq}, that leaves ∞ fixed. Among all the Möbius mappings (see §2.4 (15) and $(\overline{15})$) precisely those of the form

$$(1) \qquad z \to z' = pz + q, \quad z \to z' = p\bar{z} + q \quad (p \neq 0)$$

are contained in U'_{eq}. But (1) represents the similarity mappings of equiform geometry, as the reader can easily show, say by separation of real and imaginary parts. Thus equiform geometry is recognized as an "over-geometry" of Möbius geometry.

7.2. *Subsuming of Elliptic Geometry under Möbius Geometry*

This time we choose an interior point of Q_M. Here too the particular choice is unimportant (cf. §7.1), and we take the center O (1, 0, 0, 0) of the sphere as the absolute point. Then the subgroup U_e of G_M distinguished in this way also leaves fixed the polar plane ω of O (equation $x_0 = 0$), and the circle of intersection of this plane with the sphere (equations $x_0 = 0$, $x_1^2 + x_2^2 + x_3^2 = 0$). Thus U_e consists, by §1.2, of the Euclidean isometries, both direct and opposite, leaving the sphere Q_M fixed (*"sphere isometries."*)

Since the sphere isometries leave fixed the (imaginary) circle in which the plane ω intersects the sphere, they act on ω like *elliptic isometries* with this circle as the absolute figure (§6.4). So in each case U_e induces in ω a subgroup of the elliptic group G_e of ω. In fact, U_e induces the whole group G_e, since for every elliptic mapping η we find exactly two isometries α_1, α_2 that induce η.

Proof.   Let us set $O = B_0 = B_0'$ and then choose four points B_1, B_2, B_3, B_4 in ω, no three of which are collinear. Let their images under η be $B_j' = \eta(B_j)$. Let a point of intersection of OB_4 with Q_M be called E, and similarly let E' be a point of intersection of OB_4' with Q_M. Now B_0, B_1, B_2, B_3, E and B_0', B_1', B_2', B_3', E' are in general position, so that by the fundamental theorem of projective geometry (II10, §2.2) there exists exactly one projective mapping α_1 taking the points B_j, E into B_j', E' ($j = 0, \ldots, 3$). This transformation α_1 leaves O and ω fixed, and also maps Q_M onto itself. (For α_1 leaves the circle on the sphere fixed and is therefore a similarity mapping, and since $E' \in Q_M$, we see that in fact Q_M remains fixed.)[21] Thus α_1 is one of the sphere isometries in U_e. The proof also shows that there is only one further α_2, and that it is obtained if, instead of E', we choose the second point of intersection E'' of OB_4' with Q_M. Moreover, $\alpha_2 = \sigma_0\alpha_1$, where σ_0 is the projective reflection in O and ω.

Since $\alpha_2 = \sigma_0\alpha_1$ and since σ_0 reverses sense, exactly one of the α_j is a sense-preserving sphere isometry (rotation of the sphere). Thus the *elliptic group G_e* of ω is isomorphic to the group U_e^+ of sphere rotations, which means that *elliptic geometry is subsumed under Möbius geometry.*

The relationship between η and α_j (as a mapping of Q_M onto itself) is obtained by the projection of the elliptic improper plane ω from the fixed point O. This projection provides us at once with further models of elliptic geometry.

I.   *The bundle model.* "Points" in the bundle model are the lines through O, and "lines" are the planes through O. The set of ω-points and ω-lines is mapped one-to-one onto the set of "points" and "lines" of the model,

---

[21]In projective terminology, Q_M belongs to the pencil of quadrics that is spanned by the plane ω (counted twice) and the cone with O as vertex and the given circle as base. This pencil, which is fixed under α_1, also contains $\alpha_1(Q_M)$ and, since $E' \in Q_M$, it follows that $\alpha_1(Q_M) = Q_M$. In this form the above proof is also applicable to the analogous cases (cf. §4.3 and 8.1).

respectively. The group G_e of the Cayley-Klein model in ω is transferred to the isomorphic group U_e^+ acting on the bundle.

The great advantage of this model is that *projective measure* is so easy to visualize. Length in ω goes into length in the model and by the Laguerre angle formula elliptic distances of points of ω are simply equal to the Euclidean angles between the corresponding lines (Fig. 22). In the same way, elliptic angles between lines in ω are equal to Euclidean angles between the corresponding planes. This simple Euclidean model gives us a clear view of the whole of plane elliptic geometry.

II.   *The sphere model.* The "sphere model" is obtained by considering the intersection of the bundle O with the sphere Q_M. A "point" is a pair of diametrically opposite points, and a "line" is a great circle on Q_M (Fig. 22). The corresponding group is the group U_e^+ of sphere rotations operating on Q_M. Elliptic distances are measured by Euclidean distances along great circles on Q_M, where the identification of diametrically opposite points means that, in agreement with §6.1 (II) (elliptic case), the distance between two "points" is not only s but also $\pi - s$. Elliptic angles are measured by the Euclidean angles between great circles.

The sphere model is important in *elliptic trigonometry*. Since elliptic triangles can always be represented as Euler triangles lying entirely on a

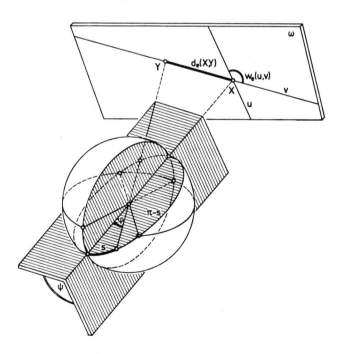

Fig. 22

hemisphere, the theorems of spherical trigonometry can be applied to them.[22] Let us also mention the well-known fact that the sum of the three angles $\varphi_1 + \varphi_2 + \varphi_3$ in an elliptic triangle is always $> \pi$.

Differential geometry in the elliptic plane is also best studied in the sphere model. In particular, it is clear at once that the elliptic plane is a Riemannian space with constant positive Gaussian curvature k_e^2 (cf. II14, §2.5), and conversely, it is proved in differential geometry (cf. [10]) that every closed surface (without singularities) of constant positive Gaussian curvature is a sphere.

III. *The Poincaré model.* Finally, let us transfer the sphere model into the Möbius plane M_2, by means of the stereographic projection Θ_M. A "point" of this Poincaré model[23] is a pair of points $\{z, -1/\bar{z}\}$ (Fig. 23), since it is easy to calculate from §2.1 (4) that the points (x, y) and $(-x/(x^2 + y^2), -y/(x^2 + y^2))$ correspond to diametrically opposite points on a sphere. The "lines" are now images of the great circles on the sphere, and are thus (in addition to the unit circle itself) precisely the M-circles cutting the unit circle in diametrically opposite points.

Since Θ_M is angle-preserving, measurement of angles here is Euclidean, and we see at once that the *sum of the angles* in a triangle is $> \pi$; for if by a rotation of the sphere we bring one of the vertices to the South Pole and then project onto M_2 (Fig. 24), two of the sides become diameters. The reader should convince himself that the sum of the angles can be arbitrarily close to 2π.

Measurement of length is not transferred so easily. The elliptic distance of two "points" z_1, z_2 can be defined, up to a constant, by $\ln(z_1, z_2, w_1, w_2)$, where w_1, w_2 are the points of intersection of the "lines" through z_1, z_2 with the imaginary circle $z\bar{z} + 1 = 0$. By §2.1 (4) this circle corresponds under Θ_M to the circle $x_0 = 0$, $x_1^2 + x_2^2 + x_3^2 = 0$ on the sphere. Thus among all Möbius mappings the *isometries* of the Poincaré model are distinguished by the fact that they leave fixed the equation $z\bar{z} + 1 = 0$. If we start from the

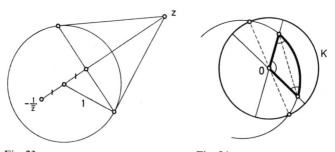

Fig. 23   Fig. 24

---

[22]Cf. II9, §9.2, where the basic concepts of spherical trigonometry are described in detail.

[23]We introduce this name in analogy with hyperbolic geometry (§8.1).

general Möbius mapping (§2.4 (15)) and determine those mappings for which $z'\bar{z}' + 1 = \rho(z\bar{z} + 1)$, we arrive, after some calculation, at the following representation of the general elliptic geometry

(2) $$z \to z' = \frac{az + b}{-\bar{b}z + \bar{a}}.$$

The corresponding equation obtained from §2.4 $(\overline{15})$ with $\bar{z}$ on the right side can be brought into the form (2), since we are identifying z and $-1/\bar{z}$; for by what has been said above, there are no orientation-reversing elliptic isometries.

7.3. *Motions in Elliptic Geometry*

We now give a brief survey of elliptic motions, which is most easily done by means of the sphere model. In it an elliptic motion is a sphere rotation, which in general has exactly one fixed "point," as can also be shown by calculation from (2). Elliptic "circles," i.e., curves at a fixed distance from a "point" are circles on the sphere, and they can easily be transferred to other models, say as pencils of conics in the elliptic plane at infinity.

Rotations through the angle π (half-rotations) are the only motions (excepting the identity) with more than one fixed "point." They can also be called *elliptic reflections*, i.e., either as elliptic "point" reflections (in the "point" $\{G, G'\}$, cf. Fig. 25) or as "line" reflections (in the "line" g polar to $\{G, G'\}$). Their fixed "points" are the "points" on g, and the "pole" of g. If we start from a projective reflection $\sigma \neq \sigma_0$ in the group U_e of P_3, i.e., its center Z is improper and its plane ε passes through O (since O is to remain fixed), then σ acts on Q_M as a Euclidean reflection in the plane ε, so that we can also characterize the elliptic reflections in this way.

Every elliptic motion is the product of two reflections, since every sphere rotation is the product of two Euclidean plane reflections.

The sphere rotations can also be represented in an elegant way by means of *quaternions* (cf. IB8, §3), whereby we also obtain relationships with

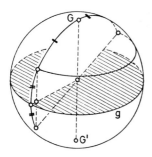

Fig. 25

elliptic geometry in space, which we shall not discuss further here (Cf. W. Blaschke, Kinematik und Quaternionen, Berlin 1960).

8. Subsuming of Hyperbolic Geometry under Möbius Geometry

8.1. *The Hyperbolic Group as a Subgroup of the Möbius Group. Further Models of Hyperbolic Geometry*

Let us again choose a fixed absolute point, this time outside Q_M, which without loss of generality may be taken as the improper point Z $(0, 0, 0, 1)$ on the z-axis. Then the subgroup U_h of the group G_M leaving Z fixed also leaves fixed the polar plane ε of Z with equation $x_3 = 0$ (the equatorial plane) and in particular the unit circle K in it. A mapping of U_h thus induces in ε a hyperbolic isometry with respect to K. Conversely, every such isometry η is induced by exactly two mappings, α_1, α_2 in U_h, with $\alpha_2 = \sigma_z \alpha_1$, where σ_z is the projective reflection in Z and ε, i.e., the Euclidean plane reflection in ε (as can be seen in the same way as before, in §7.2). Exactly one of the two α_j leaves the lower hemisphere Q_M^0 fixed, and these mappings with fixed Q_M^0 form a subgroup U_h^0 isomorphic to the hyperbolic group G_h of ε. Thus *hyperbolic geometry is subsumed under Möbius geometry*.

In the same way as for elliptic geometry in §7.2, the orthogonal projection ζ of the plane ε from the center Z onto the lower hemisphere Q_M^0, and also the stereographic projection Θ_M, will provide us with further models of hyperbolic geometry.

I. *The sphere model.* The orthogonal projection ζ maps the circle K and its interior K_i one-to-one onto the lower hemisphere Q_M^0, thereby transferring the group G_h to U_h^0 and the Cayley-Klein geometry to the hemisphere. "Points" of this sphere model are the points of Q_M^0 (the points on K are improper), and "lines" are the ε-orthogonal semicircles (Fig. 26).

In this sphere model also length is defined by means of cross ratios. Up to a constant the distance between two points X, Y is equal to $\ln(X\ Y\ A\ B)$, where A, B are the improper points of the "line" XY. It is an important fact that the elliptic measurement of angles of the Cayley-Klein model is transferred into Euclidean measurement of angles on Q_M^0.

For in the first place the angle $w_e(g, h)$ of the Cayley-Klein model is equal to the elliptic distance between the poles G, H (Fig. 27):

$$w_e(g, h) = \frac{1}{2i} \ln (ght_1t_2) = \frac{1}{2i} \ln(GHT_1T_2) = d_e(G, H).$$

Here t_1, t_2 are the tangents from R to K, and T_1, T_2 are their points of contact with K. But the elliptic distance $d_e(G, H)$ is equal (see §2.5) to the Euclidean angle φ between the orthogonal circles $\zeta(g)$, $\zeta(h)$.

II. *The Poincaré model.* We now transfer the sphere model by stere-

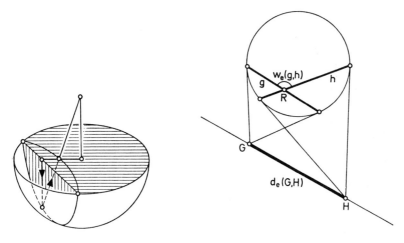

Fig. 26   Fig. 27

ographic projection Θ_M back to K_i (and K). But in contradistinction to the Cayley-Klein model, the "lines" are now arcs of circles intersecting K at right angles, whereupon we speak of the Poincaré circle model of hyperbolic geometry. (There is also the Poincaré halfplane model, arising from the circle model if by a linear transformation, say $z \to z' = i(1 - z)/(1 + z)$, we take K_i into the upper z-halfplane.)

The measurement of length is transferred immediately from the sphere model to the Poincaré circle model, since cross ratios of four points on a Q_M-circle or M-circle remain, by §2.5, invariant under Θ_M. The angles between intersecting "lines" are measured as in Euclidean geometry, since Θ_M is angle-preserving, and it is easy to show, for example, that in every triangle the *hyperbolic angle-sum* $\varphi_1 + \varphi_2 + \varphi_3 < \pi$.

For if one of the vertices of the triangle already lies at the center O of K (Fig. 28), then two of the sides are diameters of K, since they are K-orthogonal arcs of circles, and the third side is a K-orthogonal circle such that O lies outside of it, so that we obtain the desired result immediately by comparison with the Euclidean angle-sum. If no vertex lies at O, we can bring a vertex R to O by a hyperbolic isometry, as can be seen very easily in the sphere model; for if $\Theta_M^{-1}(R) \neq \Theta_M^{-1}(O)$, let the line joining the $\Theta_M^{-1}(R)$ and $\Theta_M^{-1}(O)$ intersect ε at T. The projective reflection σ_T leaves Q_M and Z fixed, and is thus a hyperbolic isometry for the sphere model, which takes $\Theta_M^{-1}(R)$ into $\Theta_M^{-1}(O)$, as desired, and the corresponding mapping of the Poincaré model then takes R into O.

The sum of the angles can obviously be made as small as desired, and for "asymptotic" triangles can even become equal to zero (Fig. 29 a, b).

Hyperbolic isometries in the Poincaré model consist of those Möbius mappings that leave fixed the absolute circle K with equation $z\bar{z} - 1 = 0$.

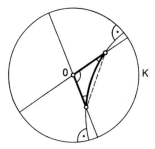

Fig. 28

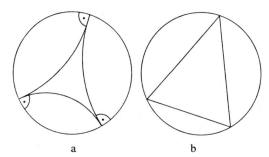

Fig. 29. Asymptotic triangle: a, in the Poincaré model; b, in the Cayley-Klein model

If we start from §2.4 (15) and ($\overline{15}$) and determine the mappings with $z'\bar{z}' - 1 = \rho(z\bar{z} - 1)$, we obtain as the general hyperbolic isometry

$$(3) \quad z \to z' = \frac{az + b}{\bar{b}z + \bar{a}} \quad \text{and} \quad (3') \quad z \to z' = \frac{a\bar{z} + b}{\bar{b}\bar{z} + \bar{a}}, \quad |a| > |b|.$$

By (3) and (3'), the image of $z = 0$ is $z' = b/\bar{a}$ and therefore lies in the interior K_i of K if and only if the above condition $|a| > |b|$ is satisfied. Here, in contradistinction to elliptic geometry, it is clearly possible to have opposite isometries also, for example, for $a = 1, b = 0$.

8.2.   Direct and Opposite Isometries in Hyperbolic Geometry

By definition hyperbolic *isometries* are the mappings that belong to the group of the given model. Again, the most important ones are the hyperbolic *reflections* (more precisely, *line reflections*), by which we mean, in the sphere model, the projective reflections $\sigma_T \in U_h^0$ about a center T lying outside K in the equatorial plane ε or about the ε-orthogonal polar plane of T with respect to Q_M. (For it is precisely for such T that the lower hemisphere Q_M^0 is taken into itself.) Similarly, in the Cayley-Klein model, a hyperbolic reflection is a projective reflection about T and its polar K, and in the

Poincaré model it is an M-inversion in a K-orthogonal circle. As M-inversions the hyperbolic reflections are sense-reversing, and are therefore opposite isometries.

We now wish to determine the *fixed points* of a hyperbolic isometry ($\neq \iota$) and for this purpose we make use of the representation (3), (3′) in the Poincaré model. With $z = z'$, we first obtain for a direct isometry

$$(4) \qquad \bar{b}z^2 + (\bar{a} - a)z - b = 0.$$

There are exactly two solutions z_1, z_2 with $|z_1| \cdot |z_2| = |b/\bar{b}| = 1,$[24] and we are interested only in those for which $|z| \leq 1$. Three cases are to be distinguished.

(I) $|z_1| < 1, \quad |z_2| > 1,$
(II) $|z_1| = |z_2| = 1, \quad z_1 = z_2,$
(III) $|z_1| = |z_2| = 1, \quad z_1 \neq z_2.$

We now discuss these cases for the Cayley-Klein model, keeping in mind that for a fixed point its polar with respect to K (and the points of intersection of this polar with K) are also fixed.

I.   There exists exactly one fixed point F inside K (as well as two conjugate complex fixed points on K). The motion is called a *rotation* with center F. The orbits for the continuous group of all rotations about F are called "circles" and they belong to the pencil of conics (no proof will be given here; cf. Fig. 30, I). The projective reflection in the interior point F and its polar f is a special rotation, called a half-rotation or also a *point reflection,* and as in Euclidean geometry it is sense-preserving.

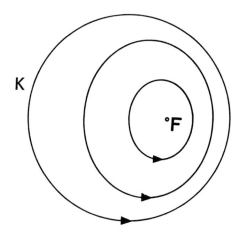

Fig. 30, I

<hr />

[24]The case $b = 0$, so that $\bar{z}_1 = 0$, $z_2 = \infty$ is easily dealt with.

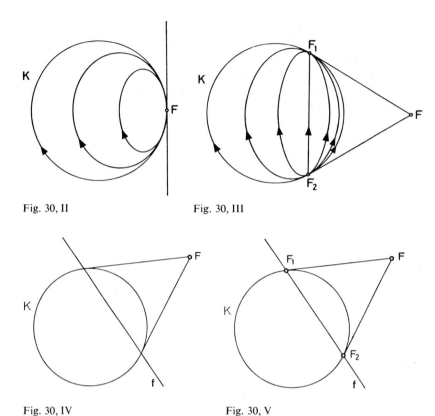

Fig. 30, II   Fig. 30, III

Fig. 30, IV   Fig. 30, V

II.   There exists exactly one fixed point F, and F lies on K; this is a *limit rotation* with center F (Fig. 30, II). The orbits for the continuous rotation group about F are called "limit circles" (horocycles).

III.   There exist exactly two fixed points F_1, F_2 on K and a fixed point outside K; this is a *shift* along $f = F_1 F_2$ with center F (Fig. 30, III). The orbits of the continuous rotation group are called "distance curves" (hypercycles); their hyperbolic distance from f is constant (we do not give the proof), so that the mapping corresponds to Euclidean translation.

For the fixed points of an *opposite isometry* it follows from (3') that

(4') $\bar{b}z\bar{z} + \bar{a}z - a\bar{z} - b = 0.$

If on the basis of §2.1 (4) we recalculate (4') for the sphere model, we find that either all the points of an ε-orthogonal circle (IV), or only the two points of intersection with K (V) remain fixed. We discuss the two cases IV and V again for the Cayley-Klein model.

IV.   All points on a line f and its pole F remain fixed; this is a *(line) reflection* in f with the center F (Fig. 30, IV).

V. There exist exactly two fixed points F_1, F_2 on K and a fixed point F outside K (Fig. 30, V). This opposite isometry is called a *shift reflection* τ with center F, since τ is the product of a reflection σ in f and a shift along f; the product $\tau\sigma = \alpha$ is sense-preserving and leaves F_1, F_2, F fixed, so that it is a shift. Thus $\tau = \alpha\sigma$ as asserted.

We now also wish to show: *Every hyperbolic motion is the product of two reflections in lines through the center F of α.*

Proof. Let σ be the reflection in a given line s through F. The product $\alpha\sigma$ is sense-reversing (of type IV or type V) and leaves F fixed. If we show that $\alpha\sigma$ is a reflection σ' in a line s' through F, the assertion is proved, since $\alpha\sigma = \sigma'$, $\alpha = \sigma'\sigma$.

I) α is a rotation. Then F lies in K_i, so that $\alpha\sigma$ is necessarily of type IV, since in type V there are no fixed points in the interior.

III) Here α is a shift; it is true that $\alpha\sigma$ leaves F fixed, but it exchanges F_1 and F_2, so that it cannot be of type V, but can only be of type IV.

II) α is a limit rotation. Then by the fundamental theorem of projective geometry α is determined by four points of general position and their images under α. We choose $B_0 = F$, B_1, B_2 on K and E as points of intersection of the tangents in B_1, B_2. Let the images be $\alpha(F) = F$, $\alpha(B_1) = B_1'$, $\alpha(B_2) = B_2'$. Then $S = B_1 B_2' \cap B_2 B_1'$ and does not lie on K. Let $\sigma_S(F) = F^*$. Furthermore $T = B_1' B_2' \cap FF^*$ exists and $T \notin K$. The mapping $\sigma_T\sigma_S$ takes F, B_1, B_2 into F, B_1', B_2', and therefore $\sigma_T\sigma_S = \alpha$. We then see that $F = F^*$, since otherwise α would not have had two fixed points on K. Thus S, T lie on the tangent at F and are therefore exterior points, so that σ_S and σ_T are hyperbolic reflections (cf. Fig. 31).

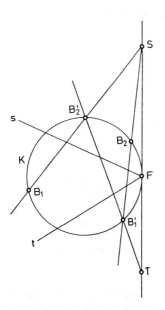

Fig. 31

Correspondingly: *Every opposite isometry is the product of three reflections* (cf. V).

8.3. *Hyperbolic Trigonometry*

Two triangles are said to be hyperbolically *congruent* if they can be taken into each other by a hyperbolic isometry. The basis for the theory of congruence of triangles in hyperbolic geometry is the free mobility of "flags"; any flag, consisting of a point R, an oriented line g incident with R, and a halfplane corresponding to g, can be taken into any other flag (R^*, g^*, η^*). (For, as in §8.1, II, let us map the point R onto R^* and then the flag (R, g, η) onto (R^*, g', η'), and then by reflection map g' into g^* with preservation of orientation, and finally, if necessary, interchange the halfplanes by reflection in g^*.) Then we have at once the following *congruence theorem*. Two triangles in the hyperbolic plane are congruent if they agree in two sides and the included angle.

We shall not discuss the other congruence theorems but merely remark that, in contrast to Euclidean geometry, two triangles are congruent if their three angles are equal, which shows that there are no hyperbolic similarity mappings.

On the basis of projective measure we can now also develop *hyperbolic trigonometry,* in extensive analogy with Euclidean and elliptic (spherical) trigonometry.[25] As an example let us prove the so-called law of sines.

Law of sines for hyperbolic trigonometry. In the triangle $R_1 R_2 R_3$ with sides $a_j = |d(R_k, R_l)|$ and angles

$$\alpha_j = |w(R_j R_k, R_j R_l)|, \quad j, k, l = 1, 2, 3; 2, 3, 1; 3, 1, 2$$

we have

(5) $\sin \alpha_1 : \sin \alpha_2 : \sin \alpha_3 = \sinh a_1 : \sinh a_2 : \sinh a_3.$

Proof.   The points R_j can be represented[26] by vectors $\mathfrak{r}_j$, and the sides $R_k R_l$ by line vectors $\mathfrak{r}_k \wedge \mathfrak{r}_l = \mathfrak{a}_j$.
From §6.1 (9) we have

(6) $\sinh^2 a_1 : \sinh^2 a_2 = \dfrac{F^2(\mathfrak{r}_2, \mathfrak{r}_3) - F(\mathfrak{r}_2, \mathfrak{r}_2)F(\mathfrak{r}_3, \mathfrak{r}_3)}{F^2(\mathfrak{r}_3, \mathfrak{r}_1) - F(\mathfrak{r}_3, \mathfrak{r}_3)F(\mathfrak{r}_1, \mathfrak{r}_1)} \cdot \dfrac{F(\mathfrak{r}_1, \mathfrak{r}_1)}{F(\mathfrak{r}_2, \mathfrak{r}_2)}.$

Let the absolute circle K be represented by $F(\mathfrak{x}, \mathfrak{x}) = -x_0^2 + x_1^2 + x_2^2 = 0$ and, as a line conic, by $G(\mathfrak{u}, \mathfrak{u}) = -u_0^2 + u_1^2 + u_2^2 = 0$. We calculate that

(7) $F^2(\mathfrak{r}_k, \mathfrak{r}_l) - F(\mathfrak{r}_k, \mathfrak{r}_k)F(\mathfrak{r}_l, \mathfrak{r}_l) = G(\mathfrak{r}_k \wedge \mathfrak{r}_l, \mathfrak{r}_k \wedge \mathfrak{r}_l) = G(\mathfrak{u}_j, \mathfrak{u}_j)$

---

[25]Cf. II9, §9.
[26]For the outer product, cf. §5.1. But here, in view of the fact that $\mathfrak{x}, \mathfrak{y} \in V_3$, an outer product $\mathfrak{x} \wedge \mathfrak{y}$ has only *three* components, namely the coefficients (up to a common factor) of the equation for the line XY.

and it follows from (6) that

$$\text{(8)} \qquad \sinh^2 a_1 : \sinh^2 a_2 = \frac{G(\mathfrak{a}_1, \mathfrak{a}_1) \cdot F(\mathfrak{r}_1, \mathfrak{r}_1)}{G(\mathfrak{a}_2, \mathfrak{a}_2) \cdot F(\mathfrak{r}_2, \mathfrak{r}_2)}.$$

In view of the relation corresponding to (7)

$$\text{(7')} \quad G^2(\mathfrak{a}_k, \mathfrak{a}_l) - G(\mathfrak{a}_k, \mathfrak{a}_k)G(\mathfrak{a}_l, \mathfrak{a}_l) = F(\mathfrak{a}_k \wedge \mathfrak{a}_l, \mathfrak{a}_k \wedge \mathfrak{a}_l) = D^2(\mathfrak{r}_1, \mathfrak{r}_2, \mathfrak{r}_3)F(\mathfrak{r}_j, \mathfrak{r}_j)^2{}^7$$

and §6.3 (20), we find from (8) that

$$\text{(8')} \qquad \sin^2 \alpha_1 : \sin^2 \alpha_2 = \frac{F(\mathfrak{r}_1, \mathfrak{r}_1) \cdot G(\mathfrak{a}_1, \mathfrak{a}_1)}{F(\mathfrak{r}_2, \mathfrak{r}_2) \cdot G(\mathfrak{a}_2, \mathfrak{a}_2)}.$$

The desired assertion follows from (8), (8') by cyclic permutation.

Similarly, we can derive the two *cosine theorems*

$$\text{(9)} \qquad \cosh a_1 = \cosh a_2 \cdot \cosh a_3 - \sinh a_2 \cdot \sinh a_3 \cdot \cos \alpha_1$$

$$\text{(10)} \qquad \cos \alpha_1 = -\cos \alpha_2 \cdot \cos \alpha_3 + \sin \alpha_2 \cdot \sin \alpha_3 \cdot \cos a_1,$$

and the corresponding formulas obtained by cyclic permutation. We leave the proofs to the reader.

These sine and cosine theorems form the basis of hyperbolic trigonometry. For further details the reader is referred to the literature ([8], [9]).

8.4. *Differential-Geometric Properties of the Hyperbolic Plane*

Starting from the expression §6.1 (9) for the hyperbolic distance d_h (with $k_h = 1$) and making use of the relation §8.3 (7), which holds in general for the forms $F(\mathfrak{x}, \mathfrak{x}) = -x_0^2 + x_1^2 + x_2^2$ and $G(\mathfrak{u}, \mathfrak{u}) = -u_0^2 + u_1^2 + u_2^2$, we obtain

$$\text{(11)} \quad \sinh^2 d_h(R, X) = \frac{F^2(\mathfrak{r}, \mathfrak{x}) - F(\mathfrak{r}, \mathfrak{r})F(\mathfrak{x}, \mathfrak{x})}{F(\mathfrak{r}, \mathfrak{r})F(\mathfrak{x}, \mathfrak{x})} = \frac{G(\mathfrak{r} \wedge \mathfrak{x}, \mathfrak{r} \wedge \mathfrak{x})}{F(\mathfrak{r}, \mathfrak{r})F(\mathfrak{x}, \mathfrak{x})}.$$

If the distances $d_h(R, X)$ are measured on the line RS, with

$$\text{(12)} \qquad \mathfrak{x} = \mathfrak{r} + t\mathfrak{s},$$

then d_h is a function of t with $d_h(t) \to 0$ for $t \to 0$. Now on the one hand[28]

$$\text{(13)} \qquad \lim_{t \to 0} \frac{1}{t} \sinh d_h(t) = \dot{d}_h(0),$$

and on the other hand, by (11), (12)

$$\text{(14)} \quad \lim_{t \to 0} \frac{1}{t^2} \sinh^2 d_h(t) = \lim_{t \to 0} \frac{1}{t^2} \frac{G(\mathfrak{r} \wedge t\mathfrak{s}, \mathfrak{r} \wedge t\mathfrak{s})}{F(\mathfrak{r}, \mathfrak{r})F(\mathfrak{r} + t\mathfrak{s}, \mathfrak{r} + t\mathfrak{s})} = \frac{G(\mathfrak{r} \wedge \mathfrak{s}, \mathfrak{r} \wedge \mathfrak{s})}{F^2(\mathfrak{r}, \mathfrak{r})}$$

---

[27] $D(\mathfrak{r}_1, \mathfrak{r}_2, \mathfrak{r}_3)$ is the determinant of $\mathfrak{r}_j$.

[28] For simplicity we assume the existence of the derivatives $d_h(0)$, but they could be obtained from (11).

with $\mathfrak{r} = \mathfrak{x}(0)$, $\mathfrak{s} = \dot{\mathfrak{x}}(0)$, it now follows from (13), (14) that

$$(15) \qquad d_h^2(0) = \frac{G(\mathfrak{x} \wedge \dot{\mathfrak{x}}, \mathfrak{x} \wedge \dot{\mathfrak{x}})}{F^2(\mathfrak{x}, \mathfrak{x})}.$$

If we now use nonhomogeneous (say Cartesian) coordinates $x_0 = 1$, $x_1 = x$, $x_2 = y$ for $\mathfrak{x}$ and write s_h for d_h, then from (15)

$$(16) \qquad ds_h^2 = \frac{dx^2 + dy^2 - (y\,dx - x\,dy)^2}{(x^2 + y^2 - 1)^2}.$$

This expression, called the hyperbolic line element ds_h^2 (referred to Cartesian coordinates), has been derived here for (straight) lines but can also be used for measurement of length on arbitrary curves. The line element (16) makes the hyperbolic plane into a two-dimensional Riemannian space, whose Gauss curvature $K = -1$ is already determined (see the Theorema egregium (II14, §2.7)) by the ds_h^2. After somewhat laborious calculation, omitted here, we find that $K = -1$, or if the requirement $k_h = 1$ is omitted, $K = -k_h^2$. Thus the hyperbolic plane with (16) has *constant negative Gauss curvature*.

On the other hand, we know from differential geometry (cf. [10], §25) that all sections of surfaces with constant $K = -1$ are developable (isometrically onto one another) and are therefore *models of hyperbolic geometry* (but only locally, since in the large in E_3 all such surfaces have singular points, in essential contrast to the case $K = 1$) cf. §7.3).

The simplest surface with $K = -1$ in three-dimensional space, and thus the simplest "differential-geometric" model of the hyperbolic plane, is the *pseudosphere,* the surface of rotation of a tractrix with the tangent segment 1. Cf. II14, §§2.3 and 2.5, where the tractrix is defined and it is proved that $K = -1$ for its surface of revolution.

With these remarks on differential-geometric properties of the hyperbolic plane we bring this chapter to a close, since it will already be clear to the reader that large parts of geometry can be subsumed under Klein's Erlanger Program, and therefore in particular, under projective geometry. But here at the end we wish to emphasize again what was stated at the beginning (§1), namely that more recent geometric research has developed largely outside this framework.

Bibliography

[1] BIEBERBACH, L.: Einführung in die höhere Geometrie, Leipzig 1933.
[2] BLASCHKE, W.: Vorlesungen über Differentialgeometrie III, Berlin 1929.
[3] COXETER, H. S. M.: Non-Euclidean Geometry, 4th ed., Toronto 1961.
[4] HEFFTER, L.: Grundlagen und analytischer Aufbau der projektiven, euklidischen, nichteuklidischen Geometrie, 3d ed., Stuttgart 1958.

[5] KLEIN, F.: Vergleichende Betrachtungen über neuere geometrische Forschungen. Erlangen 1872 und Ges. math. Abhandlungen 1, Berlin 1921.

[6] —: Vorlesungen über höhere Geometrie. 3d ed. 1968.

[7] —: Vorlesungen über nichteuklidische Geometrie, Berlin 1928.

[8] LENZ, H.: Nichteuklidische Geometrie, Mannheim 1967.

[9] PERRON, O.: Nichteuklidische Elementargeometrie der Ebene, Stuttgart 1962.

[10] STRUBECKER, K.: Differentialgeometrie III. 2. ed., Berlin 1969.

[11] WEISS, E. A.: Einführung in die Liniengeometrie und Kinematik, Leipzig 1935.

CHAPTER 13

Group Theory and Geometry

In II1, Nos. 20–24, we saw that to a given geometric system there corresponds a group, namely the group of isomorphisms, i.e., of mappings of the system onto itself that preserve the relations of the system. Klein's Erlanger Program (cf. II12) makes use of the fact that certain geometries are characterized by the corresponding group and can be built up on that basis.

1.1.   It is expedient to make a careful distinction between a group G and its representations $D(G)$. The group G can be given by an abstract multiplication table satisfying certain conditions (abstract group), or else by a set of one-to-one mappings of a set X onto itself, where the product fg of two mappings f and g is the mapping obtained by performing first g and then f (transformation group). Of course, in the second case G must contain the identity mapping, the product of every two mappings and the inverse of every mapping; on the other hand, it is not necessary to require the associative law for a transformation group, since it necessarily holds for mappings (see also IB2, §1.2.5). A transformation group can also be considered as an abstract group, if we disregard the fact that its elements are transformations and keep in mind only its abstract multiplication table.

In place of the word "transformations" we could also use "permutations," but that is not customary if X has infinitely many elements, and particularly if it has a topology.

We say that a set X has a topology if something like the limit concept, or the neighborhood concept, of ordinary space is defined in it. In their greatest generality, topological concepts are introduced axiomatically.[1] In this chapter they play only a subordinate role but we have mentioned them here for completeness. For convenience we give the following summary.

[1]Cf. here and below II16, §§10 and 9, and also III1.

In ordinary Euclidean space the concept of distance has the following properties:

The distance function ρ is a nonnegative real function of pairs of points;

$$\rho(x, y) = 0 \quad \text{if an only if} \quad x = y;$$
$$\rho(x, y) = \rho(y, x);$$
$$\rho(x, z) \leqq \rho(x, y) + \rho(y, z).$$

A set X for which such a function has been defined in any way is called a *metric space.*

The metric induces a topology, i.e., in X we define $\lim x_n = x$ if and only if

$$\lim \rho(x_n, x) = 0.$$

This concept of limit has the basic properties in common with the limit in ordinary Euclidean space. In terms of the metric, or of the topology, one can define *continuity* of functions or mappings in the usual way (ε, δ-definition or limit definition). One-to-one mappings that are continuous in both directions are called *topological*, because they preserve the topology. Mappings that preserve the metric are called *isometric.*

A set M in the space R is said to be *closed* if with any convergent sequence in R it also contains the limit of the sequence.

Most similar to ordinary space are the *n-dimensional* manifolds, i.e., metric spaces in which every point has a neighborhood that can be mapped topologically onto an n-dimensional sphere; for example, the sets of points in $(n + k)$-dimensional Euclidean space with the equations $\varphi_1(x_1, \ldots, x_{n+k}) = 0, \ldots,$ $\varphi_k(x_1, \ldots, x_{n+k}) = 0$, where $\varphi_1, \ldots, \varphi_k$ are sufficiently often differentiable and the rank of the Jacobian matrix is equal to k.

Just as the plane can be constructed from two lines as x- and y-axes, so from two spaces R_1, R_2 we can construct the product space $R_1 \times R_2$, namely the set of ordered pairs

$$(x_1, x_2) \quad \text{with} \quad x_1 \in R_1, x_2 \in R_2,$$

where we may set, for example,

$$\rho((x_1, x_2), (y_1, y_2)) = \rho(x_1, y_1) + \rho(x_2, y_2),$$

and then

$$\lim (x_1^{(n)}, x_2^{(n)}) = (x_1, x_2)$$

if and only if

$$\lim x_1^{(n)} = x_1 \quad \text{and} \quad \lim x_2^{(n)} = x_2.$$

Then we can also define continuity of a function of several variables. The function (mapping) f with two arguments x_1, x_2, which run through R_1 and R_2, respectively, is regarded as a function of the pair (x_1, x_2), i.e., as a function defined in $R_1 \times R_2$.

1.2. In the ordinary Euclidean geometry we distinguish, with Hilbert,[2] connective relations (e.g. incidence and parallelism), congruence relations,

---

[2]Cf. also II6.

and topological relations (including order relations like Hilbert's three-place relation of betweenness, and the relations derived from it, like neighborhood and limit). In ordinary real or complex projective geometry there are not only connective relations but also topological relations (which in complex projective geometry cannot be reduced to order relations). If we wish to speak of the automorphism group of a geometry, we must state which relations are to be preserved. In real projective geometry it makes no difference whether, in addition to invariance of connective relations, we also demand invariance of topological relations, since in that geometry all mappings of the space onto itself that preserve the connective relations (collineations) also preserve the topology (are continuous), although for complex projective geometry this statement no longer holds. But if we are interested only in topological relations, then for real projective space the automorphism group consists of all topological mappings of the space onto itself, which is a much larger set than the set of collineations. More generally, the mappings of any space onto itself that preserve its topology are simply the topological mappings, i.e., the one-to-one mappings continuous in both directions.

1.3.   If the set X in §1.1 has a topology and if the transformation group G consists of topological mappings of X onto itself, we can attempt to impose a natural topology on G also, which will be induced by the topology in X. In general, this can be done in many ways, say by defining that for $f_n, f \in G$

$$\lim f_n = f$$

means

$$\lim f_n x = fx \quad \text{for every} \quad x \in X;$$

or more sharply that

$$\lim f_n = f$$

means

$$\lim f_n x_n = fx$$

for all x_n, x with

$$\lim x_n = x.$$

With either definition it follows that in G

$$\lim f_n = f \quad \text{and} \quad \lim g_n = g$$

implies

$$\lim f_n g_n = fg.$$

Under additional hypotheses on the topology in X, which we shall not discuss here, it follows that in G

$$\lim f_n = f$$

implies

$$\lim f_n^{-1} = f^{-1}.$$

These remarks make it natural to define the concept of a *topological group* (cf. III1, §2.3) i.e., an abstract group G which also has a topology such that the group-theoretic structure and the topological are consistent with each other, namely,

$$\lim a_n = a, \quad \lim b_n = b$$

implies

$$\lim a_n b_n = ab, \quad \lim a_n^{-1} = a^{-1}$$

for all $a_n, a, b_n, b \in G$. These requirements can be summarized as follows:

$$ab^{-1} \text{ is continuous in } a \text{ and } b \text{ together.}$$

1.4. By a representation $D(G)$ of the (abstract) group G in X we mean a homomorphism of A into the group of one-to-one mappings of X onto itself (cf. IB2, §15.1). In other words

To every $a \in G$ there corresponds a one-to-one mapping f_a of X onto itself such that

$$f_a f_b x = f_{ab} x \quad \text{for all} \quad a, b \in G, x \in X,$$

or more concisely,

$$f_a f_b = f_{ab} \quad \text{for all} \quad a, b \in G.$$

If G is a topological group, we shall choose a topological space for X and shall further require that

$$\lim f_{a_n} x_n = f_a x$$

for all a_n, a, x_n, x with

$$\lim a_n = a, \quad \lim x_n = x,$$

in other words, we require the continuity of $f_a x$ in a and x together. We then speak of *continuous representation*.

Note the following:

1) If e is the unit element of G, then f_e is the identical mapping of X onto itself, i.e., $f_e x = x$ for all $x \in X$, since $f_e y = f_{ee} y = f_e f_e y$, so that $y = f_e^{-1} x$ implies $x = f_e f_e^{-1} x = f_e f_e f_e^{-1} x = f_e x$.

2) $f_{a^{-1}} = f_a^{-1}$, since $f_a f_{a^{-1}} x = f_{aa^{-1}} x = f_e x = x$. Thus $f_{a^{-1}} = f_a^{-1} x$ for all $x \in X$, i.e., $f_{a^{-1}} = f_a^{-1}$.

1.5.   But the converse, namely if f_a is the identity, then a is the unit of G, does not necessarily hold. But if this converse does hold, i.e., if f_a is the identity only for $a = e$, the representation is said to be *faithful*. Then we also have: If $f_a = f_b$, then $a = b$, since $f_a = f_b$ implies that $f_b^{-1} f_a =$ identity $= f_{b^{-1}} f_a = f_{b^{-1}a}$, so that $b^{-1} a = e$, and $a = b$.

1.6.   The representation is said to be *transitive* if for every pair $x, y \in X$ there exists an $a \in G$ such that $f_a x = y$.

(It is sufficient to assume this property for a fixed $x_0 \in X$ and all $y \in X$, since there then exists an a with $f_a x_0 = x$ and a b with $f_b x_0 = y$, and therefore a c (namely $c = ba^{-1}$) with $f_c x = y$ (i.e., $f_c x = f_{ba^{-1}} x = f_b f_a^{-1} x = f_b x_0 = y$.)

1.7.   Let us choose x_0 arbitrarily in X. The set of $a \in G$ for which $f_a x_0 = x_0$ (i.e., the set such that f_a leaves the element x_0 fixed) is a subgroup of G, called the *stability* group[3] of x_0.

For in fact if $f_a x_0 = x_0$ and $f_b x_0 = x_0$, then $f_{b^{-1}a} x_0 = f_b^{-1} f_a x_0 = x_0$, so that with a, b the stability group also includes $b^{-1}a$.

If the representation is continuous, the stability group H of x_0 is closed, i.e., with every convergent sequence in G, the subgroup H also includes the limit of the sequence; for let $a_n \in H$ and $\lim a_n = a$. Then $f_{a_n} x_0 = x_0$, so that $\lim f_{a_n} x_0 = x_0$. Thus $f_a x_0 = x_0$, so that also $\lim a_n \in H$.

1.8.   Let the representation of G be transitive, and let H be the stability group of x_0. What can we say about the set of the f_a that take x_0 into a given $x_1 \in X$?

In any case there is an $a \in G$ such that

$$f_a x_0 = x_1.$$

Now let

$$f_b x_0 = x_1$$

for a certain $b \in G$. Then

$$f_{a^{-1}b} x_0 = f_a^{-1} f_b x_0 = f_a^{-1} x_1 = x_0,$$

so that

$$a^{-1} b \in H,$$

---

[3]"Fixed group" in the terminology of IB2, §15.

and
$$b \in aH.$$

Conversely, if
$$b \in aH,$$

then
$$b = ah \quad \text{for a certain} \quad h \in H,$$

so that
$$f_b x_0 = f_{ah} x_0 = f_a f_h x_0 = f_a x_0.$$

Thus if we divide G into classes such that a and b are in the same class if $f_a x_0 = f_b x_0$, we obtain a partition of G into left-cosets of the stability group H of x_0.

For a transitive representation of G in X the elements of X correspond one-to-one to the left cosets of the stability group H of x_0 in the sense that the coset aH contains the point $f_{aH} x_0$ of X.

(By $f_M x$ we shall mean the set of all $f_a x$ with $a \in M$; thus in the preceding case $f_{aH} x$ contains exactly one point.)

1.9.   Can every subgroup H in a given group G occur as a stability group for a suitable representation of G? The answer is *yes*, and the results of §1.8 show how the desired stability group must be constructed.

For a given G and H we must find the set X and the representation of G in X. For X we choose the set of left cosets uH of H in G, a set also denoted by G/H.[4] If we define[5]

$$g_a uH = auH \quad \text{for all} \quad a, u \in G,$$

then g_a is in fact a one-to-one mapping of X onto itself, since

$$g_a uH = g_a vH \quad \text{for certain} \quad a, u, v \in G$$

implies
$$auH = avH,$$

and thus
$$uH = vH;$$

and for given uH and vH there exists an a, say vu^{-1}, such that $g_a uH = vH$.

Furthermore,
$$g_a g_b uH = abuH = g_{ab} uH,$$

[4]The notation G/H here, in contrast to IB2, §6, does not (necessarily) stand for a factor group.
[5]For reasons that will become clear later we here choose the letter g instead of f.

so that we do have a representation.

If for x_0 we take the subgroup H itself (regarded as an element of G/H), then the stability group of X_0 consists of the $a \in G$ with

$$aH = H,$$

i.e., with

$$a \in H,$$

so that, as desired, H is the stability group.

1.10. If we require the representation to be continuous, then a necessary and sufficient condition for H to occur as a stability group (cf. §1.7) is that H be closed.

Let us sketch the proof. For the X defined in §1.9, i.e., for G/H, we must define the topology. This we do as follows. If

$$\lim a_n = a,$$

then we define

$$\lim a_n H = aH.$$

(In other words, if U is a neighborhood of a in G, we let the set of all uH with $u \in U$ be a neighborhood of aH.) We can then show that this produces an acceptable topology in G/H. The continuity of $g_a x$ in a and x is now expressed thus:

From $\lim a_n = a$ and $\lim u_n = u$ it follows that $\lim a_n u_n H = auH$, which is in fact true, by §1.3 and by definition.

1.11. Two questions now arise. If as in §1.8 we start from a transitive representation of G in X and choose an $x_0 \in X$, we obtain a stability group H of x_0, and if for this H we construct, as in §1.9, the representation of G in G/H, then G/H is identified in a natural way with X (see §1.8). Then are we also to identify in some way the two representations (the one we started from and the one we have constructed), and if so in what sense? Furthermore, what happens if instead of x_0 we start from some other element of X?

1.12. In order to answer these questions, we make the following definitions.

If G is represented in X by mappings f_a and in Y by mappings g_a, the two representations are said to be *similar* if X can be mapped onto Y by a suitable one-to-one mapping θ in such a way that f_a and g_a become identical with each other when $x \in X$ is identified with the corresponding element $\theta x \in Y$; more precisely (see Fig. 1), if

$$g_a \theta x = \theta f_a x \quad \text{for all} \quad x \in X, a \in G.$$

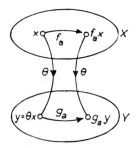

Fig. 1

The property of stabirity can also be written as follows: There exists a θ such that

$$\theta^{-1}g_a\theta = f_a \quad \text{for all} \quad a \in G.$$

It is easy to show that similarity is an equivalence relation in the sense of IA, §8.5.

In the case of topological groups it will also be natural to require that θ be topological (continuous in both directions).

1.13.   In this sense the transitive representation of G from which we started in §1.8 is similar to the representation of G in G/H obtained in §1.9 with H as the stability group of x_0. For if, as at the end of §1.8, we set

$$\theta x = uH$$

if and only if

$$f_{uH}x_0 = x,$$

then

(1) $$\theta f_{uH}x_0 = uH \quad \text{for all} \quad u \in G,$$

so that also

$$\theta f_{auH}x_0 = auH \quad \text{for all} \quad a, u \in G,$$

and therefore

(2) $$\theta f_a f_{uH}x_0 = auH.$$

But

(3) $$auH = g_a uH = g_a \theta f_{uH}x_0$$

by the definition of g in §1.9 and from (1), so that from (2) and (3)

$$\theta f_a f_{uH}x_0 = g_a \theta f_{uH}x_0,$$

and therefore, since $f_{uH}x_0$ runs through all X for $u \in G$,

$$\theta f_a = g_a \theta,$$

as desired.

Thus *a transitive representation of the group G is determined up to similarity if one of its stability groups H is known; for then it is similar to the representation of G in G/H.*

In topological groups this result no longer necessarily holds, if the similarity mapping θ is required to be continuous in both directions (see §1.12 at the end). But the exceptions have a certain pathological character.[6]

1.14.   We now let G be transitively represented as in §1.8 and ask: How are the stability groups H_0 and H_1 of x_0, $x_1 \in X$ related to each other?

In view of the transitivity there exists a $c \in G$ with

$$f_c x_0 = x_1.$$

Now let $a \in H_1$, so that

$$f_a x_1 = x_1.$$

Then

$$f_{c^{-1}ac} x_0 = f_c^{-1} f_a f_c x_0 = f_c^{-1} f_a x_1 = f_c^{-1} x_1 = x_0,$$

so that

$$c^{-1}ac \in H_0,$$

and conversely it follows from $c^{-1}ac \in H_0$ that $a \in H_1$. Thus

$$H_1 = cH_0c^{-1}$$

In a transitive representation of G the stability groups of the various points are conjugate to one another in G.

Moreover, every subgroup in G that is conjugate to H actually occurs as a stability group, since cH_0c^{-1} is the stability group of $f_c x_0$.

By §§1.8, 1.9, and 1.13 it is clear that the representations of G in G/H_0 and in G/H_1 must be similar to each other, as can also be proved directly. For let

$$f_a uH_0 = auH_0,$$
$$g_a uH_1 = auH_1,$$
$$H_1 = cH_0c^{-1}.$$

Now set

$$\theta uH_0 = uH_0c^{-1},$$

[6]H. Freudenthal, Annals of Math. 37 (1936), pp. 46–56.

which is also

$$= uc^{-1}H_1,$$

and is therefore an element of G/H_1. Thus θ maps G/H_0 (one-to-one) onto G/H_1. Now

$$\theta f_a u H_0 = \theta a u H_0 = a u H_0 c^{-1} = g_a u H_0 c^{-1} = g_a \theta u H_0,$$

so that in fact

$$\theta f_a = g_a \theta.$$

1.15.   When is the representation of G in $X = G/H$ faithful? The kernel of the homomorphism

$$a \to f_a$$

consists by definition (see IB2, §10.1) of the a for which f_a is the identity, leaving all the elements of X fixed. The kernel is the intersection of the stability groups of all $x \in X$, and is therefore the intersection of all cHc^{-1} (where H is one of these stability groups and c runs through the whole of G). But this intersection is the maximum normal subgroup of G in H.

The representation of G in G/H is faithful if and only if H contains no normal subgroup of G (except the group consisting of the unit element alone).

1.16.   A transitive representation is called *asystatic* if it never happens that precisely those f_a leaving a certain x fixed also leave a certain y fixed, or in other words if the stability groups of distinct $x \in X$ are distinct.

Now suppose that the representation of G in G/H is *not* asystatic. Then there exists a $c \in G$, $c \notin H$ with $cHc^{-1} = H$. Let H' be the smallest subgroup of G containing H and c. Certainly H' contains all $c^i H$ (where i is an integer), but the union of these $c^i H$ is already a group, since $cH = Hc$ and therefore $c^i Hc^j H = c^i c^j H$. But H is a normal subgroup in H', since for $h \in H$

$$(c^i h)H(c^i h)^{-1} = c^i hHh^{-1}c^{-i} = c^i Hc^{-i} = c^i c^{-i}H = H,$$

which gives the following result. *If H is not a normal subgroup of a strictly larger subgroup of G, then G is asystatically represented in G/H.*

The converse is also true. *If G is asystatically represented in G/H, then H is not a normal subgroup of any strictly larger subgroup of G.*

For if H is a normal subgroup of $H' \subset G$ and if $c \in H'$, $c \notin H$, then $cHc^{-1} = H$, so that the representation is not asystatic.

1.17.   The representation of an arbitrary group G in terms of left multiplication is an important one. If we take $X = G$ and set

$$\lambda_a x = ax \quad \text{for all} \quad a, x \in G,$$

it is clear that λ_a are one-to-one mappings of X onto itself. Furthermore

$$\lambda_a\lambda_b x = \lambda_a(bx) = abx = \lambda_{ab}x,$$

so that we actually have a representation, which moreover is transitive, since

$$\lambda_{xy^{-1}}x = y.$$

For each x_0, the corresponding stability group consists of the unit element alone, since $ax_0 = x_0$ only if $a = e$.

Another important (in general, nontransitive) representation of the group G is the *adjoint* representation, formed as follows. If we set $X = G$ and

$$\varphi_a x = axa^{-1} \quad \text{for all} \quad a, x \in G,$$

it is clear that the φ_a are one-to-one mappings of X onto itself, and that

$$\varphi_a\varphi_b x = a(\varphi_b x)a^{-1} = abxb^{-1}a^{-1} = (ab)x(ab)^{-1} = \varphi_{ab}x,$$

so that we have a representation.

Moreover, the φ_a are ("inner") automorphisms of G

$$\varphi_a(xy) = axya^{-1} = axa^{-1} \cdot aya^{-1} = \varphi_a(x)\varphi_a(y).$$

The unit element e remains invariant under all φ_a, so that if G consists of more than one element, the adjoint representation is certainly not transitive.

The $\varphi_a x$ for fixed x form a class of conjugate elements of G. On such a class the representation is transitive.

The stability groups of an x in the adjoint representation consists of the a with $axa^{-1} = x$ (the normalizer of x, cf. IB2, §13).

1.18.   For an asystatic transformation group, X may be regarded not only as the set G/H of cosets of H in G, but also as the set Y of subgroups xHx^{-1} conjugate to H in G under transformation by the φ_a (see §1.17), i.e.,

$$\varphi_a(xHx^{-1}) = axHx^{-1}a^{-1} = (ax)H(ax)^{-1}.$$

The stability group of the element H in Y consists of the a with $aHa^{-1} = H$, i.e., in the asystatic case (by §1.16) of H itself.

2.   Examples from Geometry

2.1.   Let P be the (real) projective line and $G(P)$ the projective group of P, i.e., the group of fractional linear transformations

(1)
$$\xi \rightarrow \frac{\alpha\xi + \beta}{\gamma\xi + \delta}$$

with real α, β, γ, δ and

$$\begin{vmatrix} \alpha & \beta \\ \gamma & \delta \end{vmatrix} \neq 0.$$

The transformations with

$$\begin{vmatrix} \alpha & \beta \\ \gamma & \delta \end{vmatrix} > 0$$

form a subgroup, $G^+(P)$.

Both of these transformation groups are transitive and asystatic.

If we now omit the point ∞ from P, we obtain the affine line P'. The isotropy group of ∞ under $G(P)$ is the affine group $G(P')$. This group (now regarded as a group of transformations of $X = P'$) is also transitive and asystatic. The isotropy group of the point 0 under $G(P')$ is isomorphic to the multiplicative group of the real numbers $\neq 0$.

The fractional linear transformations (1) remain meaningful for complex ξ, so that, if we wish, we may consider their extensions to the complex projective line P_{co}.

If ξ, η are two points of P or two conjugate complex points of P_{co}, let

$$G(P; \xi; \eta) \quad \text{and} \quad G^+(P; \xi; \eta)$$

consist of the transformations of $G(P)$ and $G^+(P)$ leaving ξ and η fixed, and let

$$G(P; \xi, \eta) \quad \text{and} \quad G^+(P; \xi, \eta)$$

be the transformations in $G(P)$ and $G^+(P)$ that either leave ξ and η fixed or interchange them. We now wish to give a meaning to $G(P; \xi; \eta)$ and $G^+(P; \xi; \eta)$ in the case that $\xi = \eta$, namely as the group of transformations that have $\xi = \eta$ as a two-fold fixed point.[7]

If ξ, η are real, we may assume $\xi = \infty$, $\eta = 0$, without loss of generality; similarly, for the conjugate complex case, we may set $\xi = i$, $\eta = -i$; and for a two-fold root $\xi = \eta = \infty$.

The group $G(P; \infty; 0)$ is already known, i.e., it is the multiplicative group of the real numbers, and $G^+(P; \infty; 0)$ is the multiplicative group of the positive numbers. If we take P to be the line at infinity of the Euclidean plane, and $\pm i$ as the isotropic points (II10, §5.1), then $G(P; i, -i)$ is generated by the direct isometries of the Euclidean plane, and $G(P; i, -i)$ by the direct and opposite isometries. Let us also note that $G^+(P; i, -i) = G^+(P; i; -i) = G(P; i; -i)$. Finally, $G(P; \infty, \infty) = G^+(P; \infty, \infty)$ is the additive group of the real numbers.

---

[7]The point ξ is a two-fold fixed point if ξ is a double root of the quadratic equation obtained by setting (1) equal to zero (so that the discriminant of the equation must vanish). The point ∞ is a two-fold fixed point if and only if $\gamma = 0$ and $\alpha = \delta$.

2.2.   Let us now construct representations of

$$G = G(P) \quad \text{in} \quad G/H$$

with

1. $H = G(P; i, -i)$,
2. $H = G(P; \infty, 0)$,
3. $H = G(P; \infty, \infty)$.

We regard P as the line at infinity of the Euclidean plane E', extended to the projective plane E by the adjunction of P. In E we take a real conic K (say a circle in E'), with interior K_i and exterior K_e. By the projection, call it π, from a point of K, we can map P onto K (the center of projection corresponds to the tangent to K at that point). Then π transfers the group G acting on P to a group Γ acting on K; namely, $g \in G$ corresponds to $\pi g \pi^{-1} \in \Gamma$, so that G is mapped onto Γ by the similarity π (i.e., similarity in the sense of §1.12).

The group Γ is called the projective group of K. Let us also consider the projective group $G(E)$ of the plane E, and in it the subgroup $G(E; K)$ of transformations leaving K fixed as a whole. Then the following statement holds:

The restriction of $G(E; K)$ to K is precisely Γ; and conversely, every transformation in Γ can be extended in exactly one way to a transformation in $G(E; K)$ (we omit the proof).

$G(E; K)$ is not transitive, but if the set X on which $G(E; K)$ is to act is restricted to

$$K, K_i, K_e$$

respectively, we obtain three transitive representations of G, namely

$$G(E; K|K), \quad G(E; K|K_i), \quad G(E; K|K_e),$$

Since $G(E; K|K)$ is precisely Γ, it is not essentially different from $G(P)$. The interior K_i of the conic K is the Cayley model of the hyperbolic plane and $G(E; K|K_i)$ is its group of congruences (direct and opposite isometries). The exterior K_e is sometimes called the pseudo-hyperbolic plane. But if to each point of K_e we assign its polar with respect to K, then K_e can be interpreted as the set of lines of the hyperbolic plane. Thus we can say that K_i is also a model of the hyperbolic plane of points and K_e of the hyperbolic plane of lines.

In $G(E; K|K_i)$ let us determine the stability group of the center m of K. This subgroup leaves K fixed as a whole, leaves m fixed, and is projective, so that it either leaves tangents from m to K fixed or else interchanges them, and similarly for the points of contact, i.e., the stability points. Thus it is essentially identical with $G(P; i, -i)$ (i.e., is related to it by the similarity π). Thus problem 1 is solved, in the following sense.

$G(P)/G(P; i, -i)$, with the group $G(P)$ acting on it, *is* essentially *the hyperbolic plane of points* (with the corresponding group).

Similarly, in $G(E; K|K_e)$ we determine the stability group of a point $m \in K_e$. Then there are two real points of K (the points of contact of the tangents from m) which it either leaves fixed or else interchanges with each other, and for a suitable choice of m these will be the points 0 and ∞. Thus the stability group is essentially $G(P; \infty, 0)$ and we have solved problem 2, as follows.

$G(P)/G(P; \infty, 0)$ *is* essentially *the hyperbolic plane of lines*.

In both cases the representations are asystatic. Moreover, the above relations are topological, a fact which we mention here without proof.

2.3. The complex projective line P_{co} is precisely the plane of the theory of functions of a complex variable (see IB8, §1). If we represent it in the usual way (as the Gauss plane extended to the Riemann sphere by means of a point ∞), the real axis (including ∞) is the real projective line, which we shall again call P, denoting the upper and lower halfplanes by P^+ and P^-. If complex ξ and complex coefficients are admitted in §2.1 (1), we obtain the projective group $G(P_{co})$ of P_{co}, and if only complex ξ is admitted, while the $\alpha, \beta, \gamma, \delta$ remain real, we obtain a subgroup $G_{re}(P_{co})$ of $G(P_{co})$. The transformations with positive determinant form a subgroup $G_{re}^+(P_{co})$ of $G_{re}(P_{co})$.

Again, $G_{re}(P_{co})$ and $G_{re}^+(P_{co})$ are representations of $G(P)$ and $G^+(P)$, respectively. These representations are not transitive, but if $G_{re}^+(P_{co})$ is restricted to

$$P, P^+, P^-$$

respectively, it becomes a transitive transformation group, namely

$$G_{re}^+(P_{co}|P), \ G_{re}^+(P_{co}|P^+), \ G_{re}^+(P_{co}|P^-).$$

The first of these groups is identical with $G^+(P)$.

Transformations in $G_{re}(P_{co})$ with negative determinant (for example, $\xi \to -\xi$) interchange P^+ and P^-. If we consider $G_{re}(P_{co})$ on the set Q of (unordered) pairs of conjugate complex points $(\xi, \bar{\xi})$ with $\xi \neq \bar{\xi}$, a representation of $G(P)$ which is also called $G(Q)$, the isotropy group of the pair $\pm i \in Q$, e.g., is essentially $G(P; i, -i)$. Thus

Q *is a model of the hyperbolic plane of points with $G(Q)$ as corresponding group.*

Instead of Q we can also take P^+, by letting ξ replace the pair $(\xi, \bar{\xi})$, $\xi \in P^+$. The result is the Poincaré model of the hyperbolic plane of points, although we should note that the group of the hyperbolic plane is no longer described by §2.1 (1). To get such a description, the image of a $\xi \in P^+$ which happens to lie in P^- still has to be replaced by its conjugate. We thus obtian

$$\xi \to \frac{\alpha\xi + \beta}{\gamma\xi + \delta} \quad \text{for} \quad \begin{vmatrix} \alpha & \beta \\ \gamma & \delta \end{vmatrix} > 0,$$

$$\xi \to \frac{\alpha\bar{\xi} + \beta}{\gamma\bar{\xi} + \delta} \quad \text{for} \quad \begin{vmatrix} \alpha & \beta \\ \gamma & \delta \end{vmatrix} < 0$$

for the direct and opposite isometries of the hyperbolic plane of points in the Poincaré model.

2.4.   Let us turn to problem 3 of §2.2. To the transformation

(2) $$\xi \to f(\xi) = \frac{\alpha\xi + \beta}{\gamma\xi + \delta}$$

we assign a plane transformation $\tilde{f}$, in which the point (ξ, η) with $\eta \neq 0$ is taken into the point (ξ^*, η^*) with

(3) $$\xi^* = f(\xi), \quad \eta^* = \frac{df}{d\xi}(\xi)\eta.$$

Since the chain rule of differential calculus shows that $\widetilde{fg} = \tilde{f}\tilde{g}$, we are in fact dealing here with a representation, seeing that η is transformed under $\tilde{f}$ like the differential $d\xi$, so that the point (ξ, η) can be interpreted as a line element $(\xi, d\xi)$ and $\tilde{f}$ as the contact transformation corresponding to f.

In order to obtain the stability group say of the point $\xi = 0, \eta = 1$, we must solve both of the equations

$$f(0) = 0, \quad \frac{df}{d\xi}(0) = 1$$

which leads to $\beta = 0, \alpha = \delta$; i.e., the f that have 0 as a two-fold fixed point. So the isotropy group is essentially the same as in problem 3 (with 0 instead of ∞). Thus

$G(P)/G(P; \infty, \infty)$ *is* essentially *the set of line elements of the projective line with the transformation group induced by the projective group.*

The representation is not asystatic, since with $\xi = 0, \eta = 1$ all line elements "$\xi = 0, \eta$ arbitrary" remain fixed.

The analytic representation (3) is not altogether satisfactory, since it has singularities arising from the denominator of (2). We shall not discuss the way in which these singularities can be removed, but shall give another model of $G(P)/G(P; \infty, \infty)$.

In P_{co} let us consider the set C consisting of circles tangent to the real axis p (the "horocycles" of hyperbolic geometry, so far as they lie in P^+) and of lines parallel to P. The transformations in $G_{re}(P_{co})$ take circles into circles, and thus define a transformation group on C, which we shall call $G(C)$. It is clear that $G(C)$ is transitive.

Let c_0 be the line $\alpha + i$ (α a real variable), and let c_0 (regarded as a circle) be tangent to P at ∞. Then the stability group of c_0 consists precisely of the transformations (2) that have ∞ as a two-fold fixed point. Thus

C is essentially $G(P)/G(P; \infty, \infty)$ with the corresponding group.

With c_0 all parallel lines remain fixed, a fact which also shows the non-asystatic character of the representation.

By means of tetracyclic coordinates the set C can be interpreted as a surface in space, whereby we are led to the following model. Consider a sphere S in space and a cylinder C' tangent to S, excluding its points of contact. The projective transformations of space leaving S and C' fixed (each of them as a whole), again generate on C' a transformation group similar to $G(P)/G(P; \infty, \infty)$ (we do not give the proof).

2.5. Let F be the skew field of quaternions (see IB8, §3). The $a \in F$ with $|a| = 1$ form a group G with respect to multiplication. Let us consider the adjoint representation (see §1.17)

$$\varphi_a x = axa^{-1}.$$

The φ_a are orthogonal transformations of R_4 (since $|\varphi_a x| = |axa^{-1}| = |a||x||a|^{-1} = |x|$) and leave the real axis invariant. Thus they also leave the orthogonal complement R_3 ($=$ the set of x with scalar part zero) invariant. Moreover, the sphere $|x| = 1$ in R_4, and thus also its intersection with R_3, namely the 2-sphere S_2, are left invariant. The representation of G by means of φ_a on $X = S_2$, i.e., the group of rotations of R_3 about 0 (see IB8, §3.2), is transitive, but it is not faithful, since $\varphi_a = \varphi_{-a}$. The isotropy group of an x_0 consists of all φ_a with $a = \cos \tau + x_0 \sin \tau$, and is isomorphic to the group of rotations of the circle; $\pm x_0$ have the same isotropy group; the representation is not asystatic. If for all $x \in X$ we identify x with $-x$, we obtain the elliptic plane with its group.

2.6. Let G_n be the projective group of n-dimensional projective space P_n. Let $V_{n,k}$ be the set of k-dimensional projective subspaces of P_n ($0 \leq k \leq n$). On $V_{n,k}$ the group G_n induces a group of transformations, which we shall call $\Gamma_{n,k}$; this group is transitive and asystatic and for $k < n$ it is a faithful representation of G_n.

The set of transformations in G_n leaving a $P \in V_{n,k}$ invariant as a whole will be called $G_n(P)$. It is precisely the stability group of P under $\Gamma_{n,k}$. For each k ($0 \leq k \leq n$) we choose one $P_k^0 \in V_{n,k}$ so that $P_k^0 \subset P_l^0$ for $k \leq l$. We can also consider $V_{n,k}$ as the set of cosets $uG_n(P_k^0)$ (u variable in G_n). The inclusion relation $P_k \subset P_l$ between arbitrary elements of $V_{n,k}$ and $V_{n,l}$, respectively, is easily seen to be equivalent, for the corresponding cosets

$uG_n(P_k^0)$ and $vG_n(P_l^0)$, to the statement: $uG_n(P_k^0) \cap vG_n(P_l^0)$ is not empty.

In view of the asystasy we can also interpret $V_{n,k}$ as a set of subgroups conjugate to $G_n(P_k^0)$ in G_n. In this interpretation we can give another group-theoretic characterization of the inclusion relation between elements of distinct $V_{n,k}$, a procedure which leads to interesting theories.[8]

3. The Space Problems

3.1. The discovery of non-Euclidean geometry (1800 up to 1830) and the Habilitation Lecture of Riemann (1854) long remained unknown, or at least unregarded, so that the first study of the foundations of geometry that attracted much attention was Helmholtz's "On the facts that lie at the basis of geometry," dated 1868, i.e., before the work of Klein (say from 1873 on), Pasch (1882), and Hilbert (1899). Up to the time of Hilbert, the phrase "foundations of geometry" was generally understood to refer to work in the direction indicated by Helmholtz.

Helmholtz undertook to characterize space, among all possible manifolds, by a requirement of homogeneity, namely free mobility of rigid bodies. His interpretation was essentially group-theoretic, although the word "group" does not occur in his work, and perhaps he was not even acquainted with the concept; but his method of formulating the space problem, though quite precise for his time, was soon seen to require revision. We give it here in the latest revision, due in preliminary form to Kolmogorov (1938) and completed in an essential way by J. Tits (1952)[9]

Let R be a connected metric space that is locally compact, and let there exist for any two (sufficiently small) congruent triangles in R an isometry of R that takes the one into the other. Then R is a real Euclidean, elliptic, spherical, or hyperbolic space (of some dimension).

By an isometry of a metric space we here mean a distance-preserving mapping of the space onto itself, and by a triangle we mean an ordered system of three points. As rigid bodies in the sense of Helmholtz we therefore need only to postulate triangles. If we require rigidity only for pairs of points, we obtain further non-Euclidean geometries, with complex numbers, quaternions, or octaves (Cayley numbers) as their coefficient domains.

3.2. Sophus Lie (1890) noted the incompleteness of Helmholtz's proof. Instead of the "free mobility of rigid bodies" Helmholtz had required the much more incisive "free mobility in the infinitesimal" (i.e., of the tangential

---

[8]See J. Tits, Mém. Acad. Roy. Belg. 29 (1955), p. 3, and Bull. Soc. Math. Belg. 8 (1956), pp. 48–81.

[9]See H. Freudenthal, Math. Zeitschrift 63 (1956), pp. 374–405, with the literature given there, and H. Freudenthal, Nieuw Archief voor Wiskunde 5 (4) (1957), pp. 105–142.

vector spaces). Accordingly Lie formulated a new simpler problem, which today, unfortunately, is often called *the* Helmholtz problem:

By a total flag in a sufficiently often differentiable n-dimensional manifold R we mean a system consisting of a point x in R, a direction $u^{(1)}$ through x, a two-dimensional direction $u^{(2)}$ through $u^{(1)}$, and so forth up to an $(n - 1)$-dimensional direction $u^{(n-1)}$ through $u^{(n-2)}$.

If there is a transformation group G in R such that for any two total flags there is exactly one transformation of G taking the one into the other, then R is one of the above-mentioned Euclidean or non-Euclidean spaces with the corresponding group of motions.

In recent times many important refinements have been made.[10] A much weaker problem arises if to the assumptions we adjoin the invariance of a Riemann metric. The problem was handled in an incidental way by Riemann (1854); to call it the Helmholtz problem is quite inappropriate.

3.3.   In his Erlanger Program (1872) F. Klein emphasized the necessity of a group-theoretic treatment of geometry. He was concerned not with the foundations of geometry (in his work neither the geometries nor their groups are characterized axiomatically) but with certain special geometries and groups of historical importance. E. Cartan, who took up the ideas of Klein again in an imposing series of articles,[11] gave a group-theoretic characterization of those Riemann spaces in which "geodesic reflection" is an isometry, and related them to the semisimple Lie groups. The first two sections of the present chapter are elementary examples of the spirit of Cartan's investigations. In recent times, investigations of this sort have been continued, in the sense of the foundations of geometry, by J. Tits.[12]

[10]See J. Tits, Mém. Acad. Roy. Belg. 29 (3), 1955.

[11]See here for example E. Cartan, Enseignement math. 26 (1927), 200–225.

[12]*Loc. cit.* See also H. Freudenthal, *Lie groups in the foundations of geometry*, Advances in Math. 1 (1964), fasc. 2.

Differential Geometry of Curves and Surfaces[1]

I. Curves

1.1. *Concept and Representation of a Curve, Transformation of Parameters*

In differential geometry curves and surfaces are investigated by means of the differential and integral calculus, and must therefore be described in a way that makes them available to the calculus, namely, in terms of functions. If we introduce an orthogonal coordinate system in which each point is determined by its three coordinates x^i, $i = 1, 2, 3$, or alternatively by the vector $\overrightarrow{OX} = \mathfrak{x} = (x^i)$, then by a curve we mean a set of points which we may imagine as being traversed in a certain time ($t_1 \leq t \leq t_2$); in other words, a curve is determined by the fact that the coordinates x^i depend on a "parameter" t

$$(1) \qquad\qquad x^i = \xi^i(t), \quad i = 1, 2, 3.$$

These three functions ξ^i represent a curve in the sense that certain statements (but not all, see below) about the functions ξ^i may be regarded as statements about the curve.

We shall now attempt to replace this description of a curve by a definition that makes no use of intuition, although our intuition should always serve as a guideline. Thus the above situation for a curve can be described by saying that to every element t of a segment of the real line there is assigned a triple of real numbers. If these triples are called points of three-dimensional number space, then (1) is a mapping of a segment of the number line into three-dimensional number space. For the time being, we may say that a curve *is* such a mapping.

For the purposes of *differential* geometry we must require that the func-

---

[1]The authors are deeply indebted to H. Freudenthal for numerous suggestions, some of them quite extensive.

tions ξ^i be sufficiently often differentiable, usually three times, and we must require that the mapping be one-to-one, at least "locally," i.e., in a neighborhood of a given point, which means, in view of our differentiability conditions, that the derivatives $d\xi^i/dt$ do not all vanish at the same point.

This definition does not yet correspond altogether to our intuitive idea of a curve, since we do not care when or how rapidly it is traversed, i.e., in abstract terms we allow transformations of parameter; namely, if τ is a one-to-one, sufficiently often differentiable (i.e., *admissible*) mapping of the segment (u_1, u_2) onto the segment (t_1, t_2), and if for all $u \in (u_1, u_2)$

(2) $$\xi^i(\tau(u)) = \eta^i(u),$$

then the triple of functions ξ^i represents the *same* curve as the triple η^i. Thus a curve is a class of triples of functions that can be taken into one another by suitable parameter transformations. Any given triple of functions is one of many possible *representations* of the curve, and a statement about the functions ξ^i is a statement about the curve if and only if it remains unchanged under all admissible transformations of parameter.

For example, consider the direction of a tangent to the curve. Taking for granted (IB3) that our triples of numbers can be regarded as vectors, and relying again for the moment on intuition, we see that

$$\left(\frac{\xi^i(t + h) - \xi^i(t)}{h} \right)$$

is a vector in the direction of a chord, and

$$\left(\lim_{h \to 0} \frac{\xi^i(t + h) - \xi^i(t)}{h} \right) = \left(\frac{d\xi^i(t)}{dt} \right)$$

is a vector in the direction of the corresponding tangent, a remark which serves only as motivation for considering the vector $d\xi^i/dt$. Under a transformation of parameter (2), we have

$$\frac{d\eta^i(u)}{du} = \frac{d\xi^i(t)}{dt} \bigg|_{t = \tau(u)} \frac{d\tau}{du},$$

i.e., all three coordinates of the vector are multiplied by the same factor, in which case we say that the vectors have the same *position*. If two vectors differ only by a *positive* factor, we say that they have the same *direction*.

The position of a vector $d\xi^i/dt$ is invariant under admissible transformations of parameter, and is therefore a property of the *curve;* we may *define* it as the *position of the tangent* to the curve.

Here a distinction may be made. Under our assumptions the function $d\tau/du$ is nonzero and continuous, so that the function τ is monotonic. But for monotone increasing τ ($d\tau/du > 0$) the direction of the vector $d\xi^i/dt$ also remains invariant. A statement about the functions ξ^i that is invariant only for monotone increasing τ will be said to refer to the *directed* (or *oriented*) curve. Thus a directed curve is a class of triples of functions derivable from one another by admissible transformations of parameter with monotonically increasing τ. The class of triples of functions representing undirected curves breaks up into two classes, representing two oppositely directed curves.

The coordinates of the points on the curve are denoted by x^i and their dependence on the parameter t or u by $x^i(t)$ and $x^i(u)$, although in general the mapping $t \to x^i$ is different from $u \to x^i$. Furthermore, we shall write $\mathfrak{x}(t)$ instead of $(x^i(t))$ and $\dot{\mathfrak{x}}(t)$ instead of $(d\xi^i(t)/dt)$, the dot denoting differentiation with respect to the parameter in question, and we then combine the three Taylor's series for the $\xi^i(t)$ into

$$\mathfrak{x}(t + h) = \mathfrak{x}(t) + h\dot{\mathfrak{x}}(t) + \cdots + \frac{h^n}{n!}\mathfrak{x}^{(n)}(t) + o(h^n),$$

where $o(h^n)$ goes to 0 faster than h^n, i.e., $\lim_{h\to 0}(o(h)/h^n) = 0$.

The tangent to the curve $\mathfrak{x}(t)$ at the point corresponding to the parameter value t has the equation

$$\mathfrak{y}(v) = \mathfrak{x}(t) + v\dot{\mathfrak{x}}(t).$$

1.2. *Transformation of Coordinates*

The values of the functions $x^i = \xi^i(t)$ can be regarded as homogeneous or as nonhomogeneous coordinates (in the above discussion, when a triple of coordinates, and not a one-dimensional vector space, was called a point, we were thinking of nonhomogeneous coordinates) and we are interested only in statements about curves (and later about surfaces) that are invariant under certain groups of coordinate transformations (cf. the Erlanger Program). Depending on the group in question, we distinguish projective, affine, and Euclidean differential geometry, but here we shall discuss only the Euclidean case, referring the reader to W. Blaschke, Vorlesungen über Differentialgeometrie II. Affine Differentialgeometrie, Berlin 1923; and E. Salkowski, Affine Differentialgeometrie. Berlin, Leipzig 1934 for affine differential geometry. For projective geometry he is referred to G. Bol, Projektive Differentialgeometric, Göttingen, Part I 1950, Part II 1954, Part III 1967.

Euclidean differential geometry deals with invariants under (proper) orthogonal transformations of coordinates, i.e., under translations and rotations. Differences, and therefore derivatives, of position vectors are obviously invariant under translations $\mathfrak{x} \to \mathfrak{y} = \mathfrak{x} + \mathfrak{a}$. The inner product of two vectors, and thus the length (absolute value) of a vector and the angle between two vectors are invariant under rotations. So whenever we are dealing with the inner products of differences or derivatives of (position-) vectors, the desired invariance is guaranteed.

1.3. *Arclength*

In order to define the arclength of a curve $x(t)$, $(a \leqq t \leqq b)$ we subdivide the segment (a, b) of the real number line by means of the intermediate points

$$(3): a = t_0 < t_1 < \cdots < t_{n-1} < t_n = b,$$

and then join the corresponding points on the curve by line segments. The length of the resulting polygonal line is given by

$$L(3) = \sum_{v=1}^{n} |x(t_v) - x(t_{v-1})|.$$

If these lengths $L(3)$ remain bounded for all possible subdivisions 3 and if L is their least upper bound, the curve is said to be rectifiable and L is its length (arclength).

A sufficient condition for rectifiability is that the $x(t)$, i.e., the coordinate functions $x^i = \xi^i(t)$, are continuously differentiable.

Proof: The differentiability of $x(t)$ can be expressed as

$$x(t + h) - x(t) = h \cdot (\dot{x}(t) + r(t, h)),$$

where $r(t, h) \to 0$ with $h \to 0$. So we must show that for all t in the closed interval (a, b) the vector r approaches zero *uniformly* if x is continuous in (a, b).

It is obviously enough to prove the assertion for each coordinate. Setting $x' = \xi(t)$, $r' = \eta(t, h)$, we have only to show that in

* $$\xi(t + h) - \xi(t) = h \cdot \dot{\xi}(t) + h \cdot \eta(t, h)$$

the factor $\eta(t, h)$ approaches zero uniformly, as follows at once from the fact that by * and the mean-value theorem for the differential calculus

$$\eta(t, h) = \dot{\xi}(t + \vartheta h) - \dot{\xi}(t), \quad (0 = \vartheta \leqq 1),$$

and in the closed interval $\dot{\xi}$ is continuous and therefore uniformly continuous.

Thus for every $\varepsilon > 0$ there exists a $\delta > 0$ independent of t such that for all t in (a, b):

$$|r(t, h)| < \varepsilon \quad \text{for} \quad |h| < \delta.$$

If we now choose the subdivision 3 in such a way that $|t_v - t_{v-1}| < \delta$ for all v and then note that

$$|\dot{x}| - |r| \leqq |\dot{x} + r| \leqq |\dot{x}| + |r|,$$

we obtain

$$\sum_{v=1}^{n} (t_v - t_{v-1}) |\dot{x}| - (b - a)\varepsilon \leqq L(3) \leqq \sum_{v=1}^{n} (t_v - t_{v-1}) |\dot{x}| + (b - a)\varepsilon,$$

so that by the definition of the integral

$$L = \int_a^b |\dot{x}(t)|dt,$$

where L is independent of the choice of parameter, since if we set $t = \varphi(\tau)$ with $d\varphi/d\tau > 0$, $x(t) = x(\varphi(\tau)) = \mathfrak{y}(\tau)$, we have

$$\int |\dot{\mathfrak{y}}|d\tau = \int |\dot{x}|\frac{d\varphi}{d\tau}d\tau = \int |\dot{x}|dt.$$

As a function of t the arc-length (measured from t_0) is given by

$$s(t) = \int_{t_0}^t |\dot{x}(t)|\ dt.$$

For $t < t_0$ the arclength is negative. From

$$\frac{ds(t)}{dt} = |\dot{x}(t)|$$

it follows that for the special choice of arclength s as parameter,

$$|x'(s)| = 1.$$

Example. $x(t) = \rho(e_1 \sin t + e_2 \cos t)$, with mutually orthogonal vectors e_1, e_2, defines a circle with center 0 and radius ρ in the plane spanned by e_1 and e_2, and $\dot{x}(t) = (e_1 \cos t - e_2 \sin t)$. Since $|\dot{x}(t)| = \rho$, arclength is given by $s = \rho t$, and with this parameter the equation of the circle takes the form

$$x^*(s) = \rho\left(e_1 \sin\frac{s}{\rho} + e_2 \cos\frac{s}{\rho}\right).$$

1.4. The Moving Trihedral

For a curve $x = x(s)$ with arclength s as parameter, it follows from $x'(s)x'(s) = 1$ by differentiation with respect to s that

$$x''(s)\cdot x'(s) = 0,$$

so that $x''(s)$ is perpendicular to $x'(s)$.

Here $x'(s)$ is the *tangent vector* to the curve $x = x(s)$; for plane curves the vector $x''(s)$ determines the direction of the normal. For space curves x'' determines the direction of the principal normal, and the vector $x''(s)/|x''(s)|$ of unit length is the *principal normal*. The plane spanned by the tangent vector $t = x'(s)$ and the principal normal vector $\mathfrak{h} = x''(s)/|x''(s)|, (x''(s) \neq 0)$ is called the *osculating plane* of the curve.

The osculating plane is the one into which the curve fits most closely, i.e., it is the limit position of the plane determined by three "neighboring" points $x(s), x(s + h), x(s + 2h)$ on the curve. In order to prove this statement, we consider the two vectors in this plane

$$x(s + h) - x(s) = hx'(s) + \frac{h^2}{2}x''(s) + o(h^2),$$

$$x(s + 2h) - x(s) = 2hx'(s) + 2h^2x''(s) + o(h^2).$$

The position of the plane is determined by the vector

$$\frac{1}{h^3}(x(s + h) - x(s)) \times (x(s + 2h) - x(s)) = (x' \times x'') + o(h)$$

orthogonal to it. For $h \to 0$ the limit value of this vector $x' \times x''$, which is orthogonal to the osculating plane.

The unit vector in this direction

$$\mathfrak{b} = \mathfrak{t} \times \mathfrak{h}$$

is called the *binormal*. The three mutually orthogonal unit vectors $\mathfrak{t}$, $\mathfrak{h}$, $\mathfrak{b}$ form the *moving trihedral* of the curve, and the geometrical properties of the curve can be studied by investigating the motion of the trihedral as its base point moves along the curve (see §1.6).

We have here assumed $x'' \neq 0$. If $x''(s_0) = 0$ at some point, the direction of the tangent is stationary at that point, which means that only the "normal plane" orthogonal to x' is still determined, but not the vectors $\mathfrak{h}$ and $\mathfrak{b}$ in it.

The equality $x''(s) = 0$ holds for all s if and only if $x' = \mathfrak{a} = \text{const}$, $x = \mathfrak{a}s + \mathfrak{c}$, i.e., if the curve is a straight line. It is clear that no particular direction is distinguished in the plane orthogonal to a straight line.

1.5. Curvature

The curvature of a curve may be understood as the change in direction of the tangent for a given change in arclength. To put this idea into a form suitable for calculation, we picture $\mathfrak{t}(s)$ as marked off from a fixed point, so that its endpoint describes a curve on the unit sphere, called the *tangent representation* of the given curve $x(s)$. Then the *curvature* of the curve $x(s)$ is defined by $\kappa(s) = d\theta/ds$, where θ is arclength of the tangent representation (and s arclength of the curve). From the equation $ds(t)/dt = |\dot{x}(t)|$ in §1.3 we see that replacement of x by $\mathfrak{t}$, and therefore s by θ and t by s, gives

$$\kappa(s) = \frac{d\theta}{ds} = |\mathfrak{t}'(s)| = |x''(s)|.$$

The equality $\kappa(s) = x''(s) = 0$ for all s is characteristic of straight lines.

If the curve is a circle with a center $\mathfrak{m}$ and radius ρ, i.e.,

$$x^* = \mathfrak{m} + \rho(\mathfrak{e}_1 \sin (s/\rho) + \mathfrak{e}_2 \cos (s/\rho)),$$

then $\kappa = |x^{*''}| = 1/\rho$.

In general, the curvature of a curve is the reciprocal of the radius of the circle that conforms most closely to the curve at the given point. For the proof we refer the curve and the circle to their arclengths as parameter, starting $(s = 0)$ at the point in question. Expanding the curve and the circle in power series in s

$$x(s) = x(0) + sx'(0) + \frac{s^2}{2}x''(0) + o(s^2)$$

$$x^*(s) = m + \rho e_2 + se_1 - \frac{s^2}{2}\frac{1}{\rho}e_2 + o(s^2),$$

we see that if they are to agree up to expressions vanishing faster than s^2, we must have

$$m + \rho e_2 = x(0),$$

$$e_1 = x'(0),$$

$$-\frac{1}{\rho}e_2 = x''(0),$$

i.e., for $x''(0) \neq 0$,

$$e_1 = x'(0) = t, \quad e_2 = -\frac{x''(0)}{|x''(0)|} = -\mathfrak{h},$$

$$\frac{1}{\rho} = |x''(0)| = \kappa,$$

(which is the desired relation)

$$m = x(0) - \rho e_2 = x + \rho\mathfrak{h},$$

where ρ is called the *radius of curvature,* and m is the *center of curvature* (lying on the principal normal at a distance ρ from the point on the curve).

Thus we can also *define* κ as the reciprocal of the radius of curvature, but then we must consider the case $x'' = 0$ separately.

Remark.   We have written κ for the curvature and τ for the torsion, whereas some authors write $1/\rho$ and $1/\tau$. In our notation the singularities occur at $\kappa = \infty$ or $\tau = \infty$, and $\kappa = 0$ or $\tau = 0$ represent points at which the tangent line or the osculating plane are merely stationary.

1.6.   Frenet Formulas

Let us investigate how the moving trihedral t, $\mathfrak{h}$, b turns as it moves along the curve $x(s)$ by expressing the derivatives of these three vectors with respect to arclength as linear combinations of the vectors themselves. To do this we first write

$$t' = a_{11}t + a_{12}\mathfrak{h} + a_{13}b,$$

$$\mathfrak{h}' = a_{21}t + a_{22}\mathfrak{h} + a_{23}b,$$

$$b' = a_{31}t + a_{32}\mathfrak{h} + a_{33}b,$$

with undetermined coefficients a_{ik}. From $tt = \mathfrak{h}\mathfrak{h} = bb = 1$ it follows that $tt' = \mathfrak{h}\mathfrak{h}' = bb' = 0$, so that $a_{11} = a_{22} = a_{33} = 0$. From $t\mathfrak{h} = 0$ we have $t\mathfrak{h}' = -t'\mathfrak{h}$, so that $a_{12} = -a_{21}$. Similarly, $a_{13} = -a_{31}$, $a_{23} = -a_{32}$. Also $\mathfrak{x}'' = t' = \kappa\mathfrak{h}$, so that $a_{12} = \kappa$, $a_{13} = 0$. If we now set $a_{23} = \tau$, we obtain

$$t' = \qquad \kappa\mathfrak{h}$$

$$\mathfrak{h}' = -\kappa t \qquad + \tau b$$

$$b' = \qquad -\tau\mathfrak{h}.$$

These formulas are due to Frenet (1847).

The τ here is called torsion of the curve. (As noted above, many authors call $1/\tau$ the torsion.) If $\tau = 0$ at a point of the curve, then $b' = 0$, so that the osculating plane is stationary. If $\tau = 0$ for all s, then $b = $ const and, since

$$(\mathfrak{x}b)' = tb = 0$$

it follows that $\mathfrak{x}b = $ const, i.e., the projection of $\mathfrak{x}(s)$ onto the constant unit vector b is constant, so that the curve $\mathfrak{x}(s)$ lies in a plane.

If we introduce the Darboux rotation vector

$$\mathfrak{d} = \tau t + \kappa b$$

the Frenet formulas take the simple form

$$t' = \mathfrak{d} \times t, \quad \mathfrak{h}' = \mathfrak{d} \times \mathfrak{h}, \quad b' = \mathfrak{d} \times b.$$

For the kinematic interpretation of these Frenet formulas see, e.g., W. Blaschke [3].

We now wish to express τ in terms of $\mathfrak{x}$ and its derivatives. From $b' = -\tau\mathfrak{h}$ we have $\tau = -b'\mathfrak{h}$, where the b' is calculated from $b = t \times \mathfrak{h} = (1/\kappa)(\mathfrak{x}' \times \mathfrak{x}'')$. Here we must have $\kappa \neq 0$, i.e., we must assume that $\mathfrak{x}'' \neq 0$ (see §1.4.).

$$b' = \left(\frac{1}{\kappa}\right)' (\mathfrak{x}' \times \mathfrak{x}'') + \frac{1}{\kappa}(\mathfrak{x}' \times \mathfrak{x}''')$$

$$\tau = -\frac{1}{\kappa}b'\mathfrak{x}'' = \frac{[\mathfrak{x}', \mathfrak{x}'', \mathfrak{x}''']}{\mathfrak{x}''^2}.$$

From this rational representation of the torsion, we see that its sign depends on the orientation of the coordinate axes. On the other hand, the sign for the curvature κ can be chosen at will.

For curves in an oriented plane it is customary to set

$$\mathfrak{h}(s) = \pm \frac{x''(s)}{|x''(s)|},$$

and then to choose the sign in such a way that the bilateral $\mathfrak{t}$, $\mathfrak{h}$ corresponds to the orientation of the plane. Similarly, the sign for κ should be so chosen that the first Frenet formula is valid.

If the plane is oriented in the usual way (counterclockwise) the curvature of the unit circle when traversed counterclockwise is equal to $+1$.

1.7.   *Natural Equations*

The importance of the Frenet formulas is shown by the following theorem. *If the invariants κ and τ are given as functions of s, the curve is uniquely determined up to motions.* The proof is based essentially on integration of the Frenet formulas, which represent a system of ordinary differential equations. Let us start from the series expansion

$$x(s) = x(0) + sx'(0) + \frac{s^2}{2}x''(0) + \frac{s^3}{3!}x'''(0) + \cdots.$$

Here we have

$$x' = \mathfrak{t},$$

$$x'' = \mathfrak{t}' = \kappa\mathfrak{h},$$

$$x''' = -\kappa^2\mathfrak{t} + \kappa'\mathfrak{h} + \kappa\tau\mathfrak{b}$$

and so forth.

From the Frenet formulas it follows that all the derivatives of x can be written in terms of κ, τ and their derivatives, and $\mathfrak{t}$, $\mathfrak{h}$, $\mathfrak{b}$. Thus if κ, τ are functions of s, and if $x(0)$, $\mathfrak{t}(0)$, $\mathfrak{h}(0)$, $\mathfrak{b}(0)$ are given, then $x(s)$ is uniquely determined. Of course, our proof is valid only for arbitrarily often differentiable functions κ, τ and also only if the above series converges. (For more general proofs we refer to the textbooks on differential geometry.)

Since x, $\mathfrak{t}$, $\mathfrak{h}$, $\mathfrak{b}$ can be given any prescribed values or positions for $s = 0$ (i.e., by a motion), the equations $\kappa = k(s)$, $\tau = w(s)$ are called the *natural equations* of the curve.

2.   Surfaces

2.1.   *Concept and Representation*

A surface (more precisely, a segment of a surface) is defined as a mapping from a domain in the number plane into three-dimensional number space, and is therefore represented by triples of functions of two variables. We write

$$\mathfrak{x} = (x^i) = (\xi^i(u^1, u^2)) \quad \text{or also} \quad \mathfrak{x}(u^1, u^2)$$

$$\mathfrak{x}_1 = \frac{\partial \mathfrak{x}}{\partial u^1} = \left(\frac{\partial \xi^i}{\partial u^1}\right), \quad \mathfrak{x}_2 = \frac{\partial \mathfrak{x}}{\partial u^2} = \left(\frac{\partial \xi^i}{\partial u^2}\right),$$

and require that the ξ^i be sufficiently often differentiable, and that

(1) $$\mathfrak{x}_1 \times \mathfrak{x}_2 \neq 0,$$

which means that the mapping is locally one-to-one.

Two triples of functions $\xi^i(u^1, u^2)$ and $\eta^i(v^1, v^2)$ represent the same surface if there exist sufficiently differentiable functions φ^1, φ^2 for which the functional determinant (Jacobian) does not vanish

$$D = \frac{\partial(\phi^1, \phi^2)}{\partial(v^1, v^2)} \neq 0,$$

and for which

$$\xi^i(u^1, u^2) = \xi^i(\phi^1(v^1, v^2), \phi^2(v^1, v^2)) = \eta^i(v^1, v^2).$$

A statement about the functions ξ^i is called a *statement about the surface* if it is invariant under these "admissible" transformations of the parameter. If it is invariant only under admissible transformations with $D > 0$, it is called a statement about the *oriented surface*.

2.2. Tangent Plane. Normal

The equations $u^1 = \psi^1(t), u^2 = \psi^2(t)$, instead of which we shall also write $u^1 = u^1(t), u^2 = u^2(t)$, define a curve $\mathfrak{x}(u^1(t), u^2(t)) = \mathfrak{x}(t)$ on the surface. The curves $u^2 = \text{const}$ and $u^1 = \text{const}$ are called the parameter curves, and the vectors $\mathfrak{x}_1$ and $\mathfrak{x}_2$ are tangent to them, respectively. The tangent vector for an arbitrary curve on the surface is

$$\frac{d\mathfrak{x}}{dt} = \mathfrak{x}_1 \frac{du^1}{dt} + \mathfrak{x}_2 \frac{du^2}{dt}$$

or more concisely,

$$\frac{d\mathfrak{x}}{dt} = \mathfrak{x}_i \frac{du^i}{dt},$$

with the usual summation convention for an index occurring both as superscript and as subscript in a product. Thus all the tangent vectors to the surface at a fixed point lie in a plane spanned by the vectors $\mathfrak{x}_1, \mathfrak{x}_2$. This plane, called the *tangent plane,* is oriented by the bilateral $\mathfrak{x}_1$, $\mathfrak{x}_2$ (in this order). The vector $\mathfrak{x}_1 \times \mathfrak{x}_2$ is perpendicular to it and

$$\mathfrak{n} = \frac{\mathfrak{x}_1 \times \mathfrak{x}_2}{|\mathfrak{x}_1 \times \mathfrak{x}_2|}$$

is called the *normal* to the surface. For a parameter transformation with $D > 0$, this vector remains unchanged, and for a transformation with $D < 0$ it changes sign, i.e., direction.

Here $\mathfrak{x}_1$, $\mathfrak{x}_2$ appear as basis vectors in the tangent plane. A parameter transformation $u^i = \varphi^i(v^1, v^2)$ produces a linear homogeneous transformation of the coordinate system in the tangent plane; for if

$$\mathfrak{y}(v^1, v^2) = \mathfrak{x}(\varphi^1(v^1, v^2), \varphi^2(v^1, v^2)),$$

then

$$(3) \qquad \frac{\partial \mathfrak{y}}{\partial v^l} = \frac{\partial \mathfrak{x}}{\partial u^k} \frac{\partial \varphi^k}{\partial v^l}.$$

In comparison with IB3, §1.5 we now write $\mathfrak{x}_{l'}$ instead of $\partial \mathfrak{y}/\partial v^l$ and $t_{l'}^k$ instead of $\partial \varphi^k / \partial v^l$

$$(3') \qquad \mathfrak{x}_{l'} = t_{l'}^k \mathfrak{x}_k,$$

where it is to be noted that the prime on the subscript indicates a different system of coordinates or parameters.

2.3. *Arclength. First Fundamental Form*

The arclength of the curve $\mathfrak{x}(u^i(t))$ is calculated as follows (see §1.3)

$$(4) \qquad \left(\frac{ds}{dt}\right)^2 = \left(\frac{d\mathfrak{x}}{dt}\right)^2 = \left(\sum_i \mathfrak{x}_i \frac{du^i}{dt}\right)^2 = \sum_{i,k} \mathfrak{x}_i \mathfrak{x}_k \frac{du^i}{dt}\frac{du^k}{dt},$$

where the summation convention would allow us to omit the symbol $\sum$. It is customary to write $\mathfrak{x}_i \mathfrak{x}_k = g_{ik}$ (so that $g_{12} = g_{21}$) and with Gauss

$$g_{11} = E, \quad g_{12} = F, \quad g_{22} = G,$$

and it is also convenient to write a vector in the tangent plane in the concise form

$$(5.1) \qquad d\mathfrak{x} = \mathfrak{x}_1 du^1 + \mathfrak{x}_2 du^2 = \mathfrak{x}_i du^i,$$

and then a second one as

$$(5.2) \qquad \delta\mathfrak{x} = \mathfrak{x}_i \delta u^i.$$

Here $\mathfrak{x}_1$, $\mathfrak{x}_2$ are the basis vectors in an (oblique) coordinate system in the tangent plane, and du^1, du^2 and δu^1, δu^2 are the coordinates of the given vectors.

Since we shall now make systematic use of the concept of a tensor introduced in IB3, §2.6, let us give a brief account of it here,[2] restricting our attention to the two-dimensional case. In a plane let there be given an affine coordinate system with the basis vectors $\mathfrak{e}_1$, $\mathfrak{e}_2$, so that a vector is represented by

(a) $\qquad\qquad \mathfrak{x} = \mathfrak{e}_1 x^1 + \mathfrak{e}_2 x^2 = \mathfrak{e}_i x^i \quad$ (summation over i).

Now let

(b) $\qquad\qquad\qquad \mathfrak{e}_{l'} = \mathfrak{e}_i t^i_{l'},$

i.e.,

$$\mathfrak{e}_{1'} = \mathfrak{e}_1 t^1_{1'} + \mathfrak{e}_2 t^2_{1'}, \quad \mathfrak{e}_{2'} = \mathfrak{e}_1 t^1_{2'} + \mathfrak{e}_2 t^2_{2'}$$

define a transformation of basis, where the determinant of the matrix $(t^i_{l'})$ is assumed to be $\neq 0$. In terms of the new basis the vector $\mathfrak{x}$ is represented by

(c) $\qquad\qquad\qquad \mathfrak{x} = \mathfrak{e}_{l'} x^{l'},$

where the new coordinates are connected with the old ones by

(d) $\qquad\qquad\qquad x^i = t^i_{l'} x^{l'},$

as may be seen by substituting (b) into (c) and comparing with (a). If $(t^{m'}_i)$ is defined (note that the prime is now on the superscript) as the matrix inverse to $(t^i_{l'})$, then

(e) $\qquad\qquad t^i_{l'} t^{m'}_i = \delta^{m'}_{l'} = \begin{cases} 1 & \text{for } m' = l', \\ 0 & \text{for } m' \neq l'. \end{cases}$

Thus from (d) we obtain

(d') $\qquad\qquad\qquad x^{m'} = t^{m'}_i x^i.$

A system of magnitudes with the transformation equations (d), (d') is called a contravariant tensor of first order (more precisely, these magnitudes are the coordinates of the tensor in the given coordinate system) i.e., a contravariant vector. On the other hand, the transformation equations

$$y_i = t^{l'}_i y_{l'}$$

---

[2] For literature see A. Duschek and A. Hochrainer, Grundzüge der Tensorrechnung in analytischer Darstellung. Wien. 1. Teil: Tensoralgebra, 1946. (2. Teil: Tensoranalysis. 1950. 3. Teil: Anwendungen in Physik und Technik. 1955.).—J. A. Schouten: Ricci-Calculus. 2. Aufl. Berlin 1954. —P. K. Raschewski: Riemannsche Geometrie und Tensoranalysis. Berlin 1959 (Russian original 1953).

determine a covariant tensor of first order, i.e., a covariant vector. Systems of magnitudes which, under a transformation of basis (b), are transformed according to the equations

(f₀) $$g_{ik} = t_i^{l'} t_k^{m'} g_{l'm'},$$

(f⁰) $$c^{ik} = t_{l'}^i t_{m'}^k c^{l'm'}$$

constitute a covariant or contravariant tensor of second order.

The great convenience of this notation is explained by the fact that behavior under a transformation of coordinates is indicated by the position of the indices. Expressions of the form $x_i y^i$ or $g_{ik} du^i du^k$ represent invariants, as can easily be calculated from (d) and (f), and here the summation convention is essential; in fact, it was adopted precisely for the reason that it leads in this way to invariants.

In our case a parameter transformation on the surface $u^i = \varphi^i(u^{k'})$ produces a linear transformation of basis in every tangential plane. In place of (b) we now have

$$x_{l'} = t_{l'}^i x_i \quad \text{with} \quad t_{l'}^i = \frac{\partial \varphi^i}{\partial u^{l'}}.$$

The coordinates of the vectors dx are transformed according to

$$du^i = \frac{\partial \varphi^i}{\partial u^{l'}} du^{l'} = t_{l'}^i \cdot du^{l'},$$

and the $g_{ik} = x_i x_k$ according to

$$g_{l'm'} = x_{l'm'} = t_{l'}^i \cdot t_m^k \cdot g_{ik}.$$

Thus the g_{ik} are the coordinates of a covariant tensor of second order. By (4), this tensor determines length in the tangent plane, or more generally, it determines the inner product of vectors in the tangent plane

(6) $$dx\delta x = g_{ik} du^i \delta u^k.$$

In particular, the angle α between dx and δx is then defined by

$$\cos \alpha = \frac{dx\delta x}{|dx||\delta x|} = \frac{g_{ik} du^i \delta u^k}{\sqrt{g_{ik} du^i du^k} \sqrt{g_{ik} \delta u^i \delta u^k}},$$

The tensor with coordinates g_{ik} is therefore called the *metric fundamental tensor*, and the quadratic form $ds^2 = g_{ik} du^i du^k$ determined by it is the *first fundamental form of surface theory*.

When the fundamental tensor is used to measure lengths in the tangent plane we can pass directly to the *intrinsic geometry of the surface*, where we pay no attention to the imbedding of the surface in the Euclidean plane but simply take the fundamental tensor as given. In the actual world of physics we are precisely in this position in one more dimension. For if we take the (admittedly doubtful) point of view that our actual space *has* a well-determined metric structure, which it is our task to discover, we do not know whether this three-dimensional space is Euclidean or "curved," being perhaps imbedded in a higher-dimensional Euclidean space. However, we can measure lengths in our space, say by means of rigid bodies, whereupon the transference of the measurement of length from one point in space to another becomes a special problem, corresponding, in the two-dimensional case of the present chapter, to the situation that at every point of our surface a fundamental tensor is given, determining the metric "locally," i.e., in a plane that best approximates the surface at the point in question.

As an example, let us calculate the fundamental tensor for a surface of revolution, obtained by rotating the meridian curve $z = f(r)$ about the z-axis, so that it has the equation

$$\mathbf{x} = (r \cos \varphi, r \sin \varphi, f(r)).$$

Then

$$\mathbf{x}_1 = \mathbf{x}_r = (\cos \varphi, \sin \varphi, f')$$

$$\mathbf{x}_2 = \mathbf{x}_\varphi = (-r \sin \varphi, r \cos \varphi, 0),$$

$$g_{11} = 1 + f'^2, \quad g_{12} = g_{21} = 0, \quad g_{22} = r^2.$$

As a particular meridian curve let us take the tractrix, i.e., the curve described say by a pocket watch if we drag it by its chain along a line l with the endpoint of the chain moving along the z-axis. The equation of such a curve (see Fig. 1) is

$$\sqrt{h^2 + r^2} = l = \text{const.}$$

Now $h = rf'$, so that

$$l = r\sqrt{1 + f'^2},$$

$$f' = \sqrt{l^2 - r^2}/r.$$

Thus we obtain $g_{11} = l^2/r^2$.

2.4. Surface Area

The area of a surface segment can be calculated from the fundamental tensor alone, but in order to define it, we shall start from the space in which the surface is imbedded. We think of the segment of the surface as being divided up into "infinitesimal parallelograms" with sides parallel to the parameter curves. The area of such a parallelogram (cf. II9, §8.3) is

$$\sqrt{(\mathbf{x}_1 du^1 \times \mathbf{x}_2 du^2)^2} = \sqrt{(\mathbf{x}_1 \times \mathbf{x}_2)^2} du^1 du^2,$$

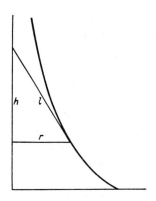

Fig. 1

and by II9, §8.3 (7)

$$(6')\qquad (\mathfrak{x}_1 \times \mathfrak{x}_2)^2 = \begin{vmatrix} \mathfrak{x}_1\mathfrak{x}_1 & \mathfrak{x}_1\mathfrak{x}_2 \\ \mathfrak{x}_2\mathfrak{x}_1 & \mathfrak{x}_2\mathfrak{x}_2 \end{vmatrix} = \det (g_{ik}),$$

a determinant which we shall denote by g,

$$g = \det (g_{ik}) = g_{11}g_{22} - g_{12}g_{21} = EG - F^2.$$

So for the area of an infinitesimal parallelogram we obtain

$$\sqrt{g}\,du^1 du^2,$$

which is in fact determined by the fundamental tensor alone. The area of the surface *is defined* by integration of this expression.

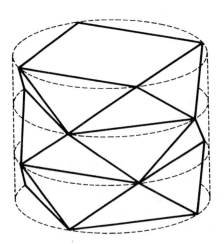

Fig. 2

The definition of surface area analogous to the definition of arclength, i.e., by means of inscribed polyhedra, is possible only under certain restrictions, as is shown by the following example, the "crushed cylinder" of H. A. Schwarz.[3] A right circular cylinder of radius r and height h is divided, by $k - 1$ planes parallel to the base, into k cylinders of equal height. In each of the corresponding circles a regular n-gon ($n > 2$) is inscribed in such a way that in two adjacent circles the vertices of the one n-gon are directly above the centers of the corresponding chords in the lower circle. If the vertices are joined so as to form triangles (see Fig. 2), there will be $2kn$ of these triangles, the length of a side will be $2r \sin(\pi/n)$ and the altitude will be

$$\sqrt{r^2(1 - \cos(\pi/n))^2 + h^2/k^2} = \sqrt{4r^2 \sin^4(\pi/2n) + h^2/k^2},$$

so that the sum of all the triangular surfaces is

$$J = 2knr \sin(\pi/n) \sqrt{4r^2 \sin(\pi/2n) + h^2/k^2}$$

$$= 2\pi r \frac{\sin (\pi/n)}{\pi/n} \sqrt{\frac{1}{4}\pi^4 r^2 \left(\frac{\sin (\pi/2n)}{\pi/2n}\right)^4 \left(\frac{k}{n^2}\right)^2 + h^2}.$$

If we now let k and n go simultaneously to ∞, in such a way that k/n^2 approaches the limit q, then J will have the limit

$$2\pi r \sqrt{\tfrac{1}{4}\pi^4 r^2 q^2 + h^2},$$

which is equal to the area of the surface enclosing the cylinder only if $q = 0$.[4]

Instead of discussing the restrictions necessary for a definition of surface area by inscribed polyhedra, we have defined the area of a surface by approximation with segments of the tangent plane, a definition

$$J = \iint \sqrt{g}\, du^1 du^2,$$

which satisfies the condition that J is to be invariant under transformations of parameter, since under a transformation $u^i = \varphi^i(u^{l'})$ with $t^i_{l'} = \partial\varphi^i/\partial u^{l'}$ the g_{ik} become

$$g_{l'm'} = t^i_{l'} \cdot g_{ik} t^k_{m'}.$$

If the corresponding matrices are denoted by $\mathfrak{G}'$, $\mathfrak{T}$, $\mathfrak{G}$ and the transpose of $\mathfrak{T}$ by $\mathfrak{T}^T$, the transformation for the g_{ik} becomes

$$\mathfrak{G}' = \mathfrak{T}\mathfrak{G}\mathfrak{T}^T,$$

so that, by the rule for the multiplication of determinants (IB3, §3.4)

$$g' = g|\mathfrak{T}|^2,$$

---

[3] Ges. Abh. 2, pp. 309–311.
[4] Cf. Mangoldt-Knopp, Einführung in die höhere Mathematik, Vol. 3, 11. Ed. of 1958, p. 388, where further historical information is given.

where $|\mathfrak{T}|$ is the functional determinant of the transformation, from which follows the invariance of J.

A definition of surface area of another kind was given by Minkowski.[5] If a sphere of radius h is described about each point of the surface F as center, the set of points of space lying either inside or on the boundary of one of these spheres defines the region of space consisting of points at a distance $\leq h$ from the surface F. If $V(2h)$ is the volume of this region, we may consider $\lim_{h \to 0}(V(2h)/2h)$ (provided that $V(2h)$ exists and that the limit also exists) as the area of the surface F.

This definition of Minkowski requires no differentiability of any kind. If we apply it to a segment of surface $\mathfrak{x}(u^1, u^2)$, where u^1, u^2 range over a domain G of the u^1, u^2-plane, then (assuming differentiability of $\mathfrak{x}$ and existence of the normal) we can replace the region of space at a distance $\leq h$ by the region obtained by marking off a fixed length h on both sides of the normal to the surface (Fig. 3). This region is described by

$$\mathfrak{y}(u^1, u^2, k) = \mathfrak{x}(u^1, u^2) + k\mathfrak{n}(u^1, u^2),$$
$$u^1, u^2 \in G, \quad -h \leq k \leq h.$$

Now if $\mathfrak{y}$ has the Cartesian coordinates y^1, y^2, y^3, then

$$V(2h) = \int \int \int dy^1 dy^2 dy^3 = \int \int_{u^1, u^2 \in G} \int_{k=-h}^{h} [\mathfrak{y}_1, \mathfrak{y}_2, \mathfrak{y}_k] du^1 du^2 dk,$$

with

$$\mathfrak{y}_1 = \mathfrak{x}_1 + k\mathfrak{n}_1, \quad \mathfrak{y}_2 = \mathfrak{x}_2 + k\mathfrak{n}_2, \quad \mathfrak{y}_k = \mathfrak{n},$$

so that

$$V(2h) = \int \int_{u^1, u^2 \in G} \int_{k=-h}^{h} [\mathfrak{x}_1, \mathfrak{x}_2, \mathfrak{n}] du^1 du^2 dk + \int \int \int k \cdot (\ldots) du^1 du^2 dk.$$

In the second summand we are interested only in the fact that after integration with respect to k it contains the factor h^2, while the first sum becomes

$$2h \int \int_G [\mathfrak{x}_1, \mathfrak{x}_2, \mathfrak{n}] du^1 du^2.$$

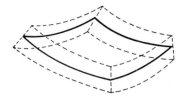

Fig. 3

[5]Ges. Abh. Bd. 2, p. 122.

Thus

$$\lim_{h \to 0} \frac{V(2h)}{2h} = \iint_G [\mathbf{x}_1, \mathbf{x}_2, \mathbf{n}] du^1 du^2.$$

Substitution of $\mathbf{n} = \dfrac{\mathbf{x}_1 \times \mathbf{x}_2}{\sqrt{(\mathbf{x}_1 \times \mathbf{x}_2)^2}}$ gives $\displaystyle\iint_G \sqrt{(\mathbf{x}_1 \times \mathbf{x}_2)^2}\, du^1 du^2.$

A more systematic discussion of the intrinsic geometry of a surface will be given in §3. For the time being we regard the surface as imbedded in Euclidean space.

2.5. *Spherical Representation. Second Fundamental Form. Gaussian Curvature*

To a point $\mathbf{x}$ on the surface the normal vector $\mathbf{n}$ assigns a point on the unit sphere (since $|\mathbf{n}| = 1$). This assignment is called the *normal mapping*, and the resulting image is the *normal* or *spherical representation* of the surface.

If $\mathbf{x}$ moves on the surface along a curve $\mathbf{x}(t)$, the normal $\mathbf{n}$ moves with velocity

$$\frac{d\mathbf{n}}{dt} = \mathbf{n}_1 \frac{du^1}{dt} + \mathbf{n}_2 \frac{du^2}{dt} = \mathbf{n}_i \frac{du^i}{dt}.$$

Since $\mathbf{n}^2 = 1$, it follows that $\mathbf{n}\mathbf{n}_i = 0$, so that $\mathbf{n}_1$ and $\mathbf{n}_2$ are orthogonal to $\mathbf{n}$, i.e., they are parallel to the tangent plane, and can thus be expressed linearly in $\mathbf{x}_1, \mathbf{x}_2$

(7) $$\mathbf{n}_i = -b_i^k \mathbf{x}_k.$$

The mapping that takes $d\mathbf{n}$ into $d\mathbf{x}$ is then a linear mapping of the tangent plane into itself, whose matrix we shall call $-\mathfrak{Z}$

$$d\mathbf{n} = -\mathfrak{Z}d\mathbf{x}, \quad \mathfrak{Z} = (b_i^k).$$

To calculate the b_i^k we multiply (7) by $\mathbf{x}_l$

(8) $$\mathbf{n}_i \mathbf{x}_l = -b_i^k \mathbf{x}_k \mathbf{x}_l = -b_i^k g_{kl}.$$

Solution of this system of equations gives the b_i^k in terms of the $\mathbf{n}_i$, the $\mathbf{x}_l$ and the g_{kl}, the latter again being given in terms of the $\mathbf{x}_l$.

We also define

(9) $$b_{il} = b_i^k g_{kl}$$

and for an arbitrary tensor the "lowering" of an index by the fundamental tensor is defined in the same way. The opposite operation of raising an index is obtained as follows. If we define the matrix (g^{lm}) inverse to (g_{ik}) by

$g_{ik}g^{kl} = \delta_i^l$, it is easy to verify that

$$g^{11} = g_{22}/g, \quad g^{22} = g_{11}/g, \quad g^{12} = g^{21} = -g_{12}/g.$$

Then

(10) $$b_i^k = b_{il}g^{lk},$$

i.e., if we write $\mathfrak{B}$ for the matrix (b_{ik}) and $\mathfrak{G}$ for (g_{ik}),

(11) $$\mathfrak{Z} = \mathfrak{B}\mathfrak{G}^{-1} \quad \text{or} \quad \mathfrak{Z}\mathfrak{G} = \mathfrak{B}.$$

From (8) and (9)

$$b_{ik} = -\mathfrak{n}_i \mathfrak{x}_k.$$

From $\mathfrak{n}\mathfrak{x}_k = 0$ it follows that

(12) $$b_{ik} = -\mathfrak{n}_i \mathfrak{x}_k = \mathfrak{n}\mathfrak{x}_{ki}$$

and therefore

$$b_{ik} = b_{ki}.$$

Thus $\mathfrak{B}$ is a symmetric matrix, whereas $\mathfrak{Z}$ is not always symmetric. (The reader should calculate b_1^2 and b_2^1.)

By the rule of multiplication of determinants it follows from (11) that

(13) $$\det \mathfrak{Z} = \det (b_i^k) = \frac{b}{g},$$

where we have written b for $\det (b_{ik})$. It is customary to set $b_{11} = L, b_{12} = b_{21} = M, b_{22} = N$, whereupon (13) takes the form

$$\det \mathfrak{Z} = \frac{LN - M^2}{EG - F^2}.$$

This determinant is the factor by which the surface element is multiplied in the normal mapping. The ratio of the two areas, namely on the normal representation and on the surface itself,

(14) $$\det \mathfrak{Z} = K = \frac{b}{g}$$

is called the *Gauss curvature* of the surface. Compare the definition in §1.5 of the curvature of a plane curve as the ratio of arclength on the normal representation and on the curve itself.

The quadratic form determined by $\mathfrak{B} = (b_{ik})$ is called the *second fundamental form*. Its geometric interpretation is as follows. Differentiation with respect to arclength on the surface gives

$$\frac{d^2\mathfrak{x}}{ds^2} = \mathfrak{x}_{ik}\frac{du^i}{ds}\frac{du^k}{ds} + \mathfrak{x}_i\frac{d^2u^i}{ds^2},$$

and thus by (12)

$$\mathfrak{n}\mathfrak{x}'' = \frac{b_{ik}du^i du^k}{ds^2} = \frac{b_{ik}du^i du^k}{g_{ik}du^i du^k} = \frac{\mathrm{II}}{\mathrm{I}}.$$

So the ratio of the two fundamental forms is equal to the cosine of the angle between the normal to the surface and the principal normal to the curve on the surface, multiplied by the curvature of the curve.

From this geometric interpretation we see that *the second fundamental formula is invariant under parameter transformations* just like the first one, defining arclength. The same result could, of course, have been obtained by calculation from $\mathrm{II} = -\mathfrak{n}_i \mathfrak{x}_k du^i du^k$.

If the second fundamental form vanishes along a curve on the surface, the osculating plane of the curve coincides with the tangent plane. Such curves are called osculating curves, or *asymptotic lines*. At each point on the surface there exist (in general) two asymptotic directions, which can be real and distinct, conjugate complex, or coincident.

For the surface of revolution of the tractrix (see the end of §2.3),

$$\mathfrak{n} = \frac{\mathfrak{x}_r \times \mathfrak{x}_\varphi}{|\mathfrak{x}_r \times \mathfrak{x}_\varphi|} = \frac{1}{l}(-\sqrt{l^2 - r^2}\cos\varphi, -\sqrt{l^2 - r^2}\sin\varphi, r),$$

$$b_{11} = -\frac{l}{r\sqrt{l^2 - r^2}}, \quad b_{12} = b_{21} = 0, \quad b_{22} = \frac{r}{l}\sqrt{l^2 - r^2},$$

$$b = -1, \quad K = b/g = -1/l^2,$$

i.e., this surface is of constant negative Gauss curvature, with $K = -1$ for $l = 1$, which means that hyperbolic geometry, with the geodesic lines for "straight lines" (cf. II6, §6.2, and II12, §8.4), is valid on it.

2.6. *Principal Curvatures*

We now ask for the eigenvectors of the mapping $\mathfrak{Z}$, i.e., the vectors $\mathfrak{v} \neq 0$ in the tangent plane such that

$$(15) \qquad \qquad \mathfrak{Z}\mathfrak{v} = \lambda\mathfrak{v}.$$

By IB3, §3.7, they correspond to the eigenvalues of the matrix $\mathfrak{Z} - \lambda\mathfrak{E}$, i.e., to the solutions of

$$(16) \qquad \qquad |\mathfrak{Z} - \lambda\mathfrak{E}| = 0 \quad (\mathfrak{E} = \text{unit matrix}).$$

In order to show that here too the eigenvalues are always real, we must make a slight change in the proof given in IB3, §3.7, since now the matrix $\mathfrak{Z}$ is not always symmetric.

Since (1) and (6′) imply $g = |\mathfrak{G}| \neq 0$, the solutions of (16) are also solutions of

(17) $$|\mathfrak{J}\mathfrak{G} - \lambda\mathfrak{E}\mathfrak{G}| = |\mathfrak{B} - \lambda\mathfrak{G}| = 0,$$

and from (15)

(18) $$\mathfrak{B}\mathfrak{v} = \lambda\mathfrak{G}\mathfrak{v},$$

so that, since $\mathfrak{B}$ and $\mathfrak{G}$ are symmetric and real, the proof in IB3, §3.7, will now carry over in the following way. Let λ be a complex eigenvalue with $\mathfrak{v}$ as a corresponding eigenvector. Then, together with (18), we have

(19) $$\mathfrak{B}\bar{\mathfrak{v}} = \bar{\lambda}\mathfrak{G}\bar{\mathfrak{v}}.$$

Multiplying (18) by $\bar{\mathfrak{v}}^T$, and (19) by $\mathfrak{v}^T$

$$\bar{\mathfrak{v}}^T\mathfrak{B}\mathfrak{v} = \lambda\bar{\mathfrak{v}}^T\mathfrak{G}\mathfrak{v},$$

$$\mathfrak{v}^T\mathfrak{B}\bar{\mathfrak{v}} = \bar{\lambda}\mathfrak{v}^T\mathfrak{G}\bar{\mathfrak{v}},$$

taking transposes of the matrices in the second of these equations and subtracting the result from the first gives

(20) $$(\lambda - \bar{\lambda})\bar{\mathfrak{v}}^T\mathfrak{G}\mathfrak{v} = 0.$$

If we now assume that the Hermitian form determined by $\mathfrak{G}$ is positive definite (as is natural, since it represents the square of arclength) it follows from (20) that $\lambda = \bar{\lambda}$, so that λ is real. Since (17) is a quadratic equation, $\mathfrak{J}$ has two real eigenvalues; if these eigenvalues coincide, the corresponding point on the surface is said to be *umbilical*.

A sphere consists entirely of umbilical points. The fact that a closed convex surface has at least two umbilical points was very troublesome to prove.

The directions given by the eigenvectors are called *directions of principal curvature,* and curves lying in the direction of principal curvature at each point are *lines of curvature.* On each surface there exists a two-parameter family of lines of curvature, which for many investigations may conveniently be taken as parameter lines.

Let us now intersect the surface at a given point with a normal plane through the eigenvector $d\mathfrak{x}$, i.e., with a plane through $d\mathfrak{x}$ and $\mathfrak{n}$, so that $\pm\mathfrak{n}$ is the principal normal $\mathfrak{h}$ of the resulting plane curve. The Frenet equation

$$\frac{d\mathfrak{n}}{ds} = -\kappa\frac{d\mathfrak{x}}{ds}$$

together with $d\mathfrak{n} = -\mathfrak{J}d\mathfrak{x} = -\lambda d\mathfrak{x}$ shows that

(21) $$\lambda = \pm\kappa,$$

with plus or minus sign according to whether or not the principal normal is in the same direction as the normal to the surface.

The normal sections corresponding to the two eigenvalues of λ are called *principal sections*. (When the λ are equal, i.e., at an umbilical point, every section is a principal section.) The curvatures of the plane curves on principal sections are called *principal curvatures;* by (21) they are precisely the eigenvalues λ. From the characteristic equation (16) we see that the product of the principal curvatures is equal to $\det \mathfrak{Z} = K$, and that their sum is equal to the trace of the matrix $\mathfrak{Z}$, i.e. (assuming that principal normal and surface normal have the same direction)

$$\kappa_1 + \kappa_2 = b_1^1 + b_2^2 = 2H,$$

where H is called the *mean curvature* of the surface.

2.7. Determination of the Surface by its Fundamental Forms

The two fundamental forms

$$\mathrm{I} = g_{ik}du^i du^k, \quad \mathrm{II} = b_{ik}du^i du^k$$

are invariants of the surface. In analogy with the theory of curves (where the invariants are κ and τ) we now ask whether the surface is thereby uniquely determined up to motions, i.e., by the values of g_{ik} and b_{ik} as functions of u^1, u^2. Since we have not chosen invariant parameters, the g_{ik}, b_{ik} are to be regarded not as scalars but as tensors, i.e., with the above rule of transformation for other parameters (§2.3).

Let us now choose the vectors $\mathfrak{x}_1$, $\mathfrak{x}_2$, $\mathfrak{n}$ as a moving trihedral on the surface. Although they are not all unit vectors (only $\mathfrak{n}$ is always a unit vector), and not always mutually orthogonal ($\mathfrak{n}$ is orthogonal to $\mathfrak{x}_1$ and $\mathfrak{x}_2$), they are always linearly independent, and their derivatives can be represented as linear combinations of the vectors themselves. Some of these equations have already been set up, namely

$$(7) \qquad \frac{\partial \mathfrak{n}}{\partial u^i} = \mathfrak{n}_i = -b_i^k \mathfrak{x}_k,$$

which are called the *Weingarten equations for the derivatives*.

The rest of them we first write in a purely formal way, i.e., with arbitrary symbols for the coefficients

$$\mathfrak{x}_{ik} = \Gamma_{ik}^l \mathfrak{x}_l + c_{ik}\mathfrak{n}$$

and then calculate the coefficients. Inner multiplication with $\mathfrak{n}$ gives, from (12)

$$(22) \qquad \mathfrak{x}_{ik}\mathfrak{n} = c_{ik} = b_{ik},$$

and in order to calculate the Γ_{ik}^l we form the inner products

$$\mathfrak{x}_{ik}\mathfrak{x}_m = \Gamma_{ik}^l \mathfrak{x}_l \mathfrak{x}_m = \Gamma_{ik}^l g_{lm} = \Gamma_{ik,m},$$

where the last equality sign means that the $\Gamma_{ik,m}$ are thereby defined. Then we obtain the Γ^l_{ik} by raising the index

$$\Gamma^l_{ik} = \mathfrak{x}_{ik}\mathfrak{x}_m g^{ml}.$$

In order to calculate $\mathfrak{x}_{ik}\mathfrak{x}_m$, we differentiate $g_{im} = \mathfrak{x}_i\mathfrak{x}_m$ with respect to u^k

$$(23) \qquad \frac{\partial g_{im}}{\partial u^k} = \mathfrak{x}_{ik}\mathfrak{x}_m + \mathfrak{x}_i\mathfrak{x}_{mk} = \Gamma_{ik,m} + \Gamma_{mk,i},$$

permute the indices i, k, m cyclically, add the first two of the resulting equations and subtract the third, thereby obtaining

$$(24) \qquad \Gamma_{ik,m} = \mathfrak{x}_{ik}\mathfrak{x}_m = \frac{1}{2}\left(\frac{\partial g_{im}}{\partial u^k} + \frac{\partial g_{mk}}{\partial u^i} - \frac{\partial g_{ki}}{\partial u^m}\right),$$

which means that the $\Gamma_{ik,m}$ and Γ^l_{ik} have now been calculated in terms of the g_{ik} and their derivatives alone. Retaining the symbol Γ^l_{ik} instead of the complicated expression for it, we arrive at the *Gauss equations for the derivatives*

$$(25) \qquad \mathfrak{x}_{ik} = \Gamma^l_{ik}\mathfrak{x}_l + b_{ik}\mathfrak{n}.$$

The Γ^l_{ik} and the $\Gamma_{ik,m}$ are not the coordinates of tensor, since they do not behave appropriately under transformations. They are the Christoffel three-index symbols, for which it is also customary to write $\Gamma^l_{ik} = \{^{ik}_l\}$, $\Gamma_{ik,m} = [^{ik}_m]$.

The Gauss and Weingarten equations for the derivatives correspond to the Frenet equations in the theory of curves, so that if we ask whether the tensors g_{ik} and b_{ik} determine a surface up to motions, we must integrate this system of equations for the derivatives. We are dealing here with a system of partial differential equations for the trihedral vectors, which has a uniquely determined solution, with prescribed initial trihedral, if the integrability conditions

$$(26.1) \qquad \frac{\partial \mathfrak{x}_{ik}}{\partial u^j} = \frac{\partial \mathfrak{x}_{ij}}{\partial u^k},$$

$$(26.2) \qquad \frac{\partial \mathfrak{n}_k}{\partial u^j} = \frac{\partial \mathfrak{n}_j}{\partial u^k}$$

are satisfied. The necessity of these equations is clear (i.e., they are satisfied for any given surface); let us state here, without proof, that they are also sufficient.

Let us write out part of (26.1) in detail. If for this purpose we differentiate the Gauss equations (25) and then use these and the Weingarten equations (7), we obtain

$$\frac{\partial \mathfrak{x}_{ik}}{\partial u^j} = \left[\frac{\partial}{\partial u^j} \Gamma_{ik}^m + \Gamma_{ik}^l \Gamma_{lj}^m - b_{ik} b_j^m \right] \mathfrak{x}_m + \left[\Gamma_{ik}^l b_{lj} + \frac{\partial b_{ik}}{\partial u^j} \right] \mathfrak{n},$$

$$\frac{\partial \mathfrak{x}_{ij}}{\partial u^k} = \left[\frac{\partial}{\partial u^k} \Gamma_{ij}^m + \Gamma_{ij}^l \Gamma_{lk}^m - b_{ij} b_k^m \right] \mathfrak{x}_m + \left[\Gamma_{ij}^l b_{lk} + \frac{\partial b_{ij}}{\partial u_k} \right] \mathfrak{n}.$$

Since $\mathfrak{x}_1$, $\mathfrak{x}_2$, $\mathfrak{n}$ are linearly independent, it follows from (26.1) that

(27) $\qquad \dfrac{\partial}{\partial u^j} \Gamma_{ik}^m - \dfrac{\partial}{\partial u^k} \Gamma_{ij}^m + \Gamma_{ik}^l \Gamma_{lj}^m - \Gamma_{ij}^l \Gamma_{lk}^m = b_{ik} b_j^m - b_{ij} b_k^m.$

The left side, which will be denoted here by $r^m_{.ijk}$, is a tensor of fourth order (as is easily calculated from the right side), called the *Riemann curvature tensor*. Of essential importance is the fact that it can be calculated from the g_{ik} and their derivatives alone, in other words without use of the b_{ik}, so that it is part of the intrinsic geometry of the surface. With

$$g_{nm} r^m_{.ijk} = r_{nijk}$$

we obtain

(28) $\qquad\qquad r_{nijk} = b_{ik} b_{jn} - b_{ij} b_{kn},$

so that we have the following symmetry properties

$$r_{nijk} = -r_{injk}, \text{ and therefore } = 0 \text{ for } i = n,$$
$$r_{nijk} = -r_{nikj}, \text{ and therefore } = 0 \text{ for } j = k.$$

The only nonzero coordinates are

$$r_{1212} = -r_{2112} = r_{2121} = -r_{1221} = b_{11} b_{22} - b_{12} b_{21} = b.$$

Thus *the curvature tensor (in the two-dimensional case) is already determined by one of these coordinates.* The same statement also holds for $r^m_{.ijk}$.

Equation (27) contains the Theorema egregium of Gauss; namely, *b and therefore also the Gaussian curvature $K = b/g$ depends only on the g_{ik} and their derivatives,* and thus only on the fundamental tensor, not on the imbedding of the surface in space.

The further conditions that follow from (26) will not be discussed here.

It is also an interesting question whether the g_{ik} alone determine the surface. Of course, they do not determine it altogether, since any two surfaces that can be mapped onto each other by a length-preserving mapping (e.g., the plane and a cylinder) have the same first fundamental form, and thus also the same K. But if we add subsidiary conditions, e.g., that the surface is to be closed and convex (cf. II15), it turns out that the surface is determined, up to motions, by the g_{ik} alone, i.e., that it is a rigid surface. The developments following from this question and the results obtained up to now, are given in Efimov: Flächenverbiegung im Grossen. With supplements by K. P. Grotemeyer. Berlin 1957.

2.8. *Linear Differential Forms*

The theory of surfaces can also be built up from linear differential forms instead of the two quadratic fundamental forms I and II. For this purpose we assign to each point $x(u, v)$ on the surface a trihedral consisting of the orthogonal unit vectors, $\mathfrak{v}_1, \mathfrak{v}_2, \mathfrak{v}_3$. (In this section only the parameters will be denoted by u, v instead of u^1, u^2, so as not to have too many indices.) Ordinarily we shall take $\mathfrak{v}_3 = \mathfrak{n}$ to be the normal vector of the surface, and we may think of $\mathfrak{v}_1, \mathfrak{v}_2$ as the unit tangent vectors to the lines of curvature, but for the moment we shall not make any definite choice.

We obtain equations for the derivatives by representing the differentials $dx, d\mathfrak{v}_1, d\mathfrak{v}_2, d\mathfrak{v}_3$ as linear combinations of $\mathfrak{v}_1, \mathfrak{v}_2, \mathfrak{v}_3$

$$(29) \qquad\qquad dx = \sum_i \omega_i \mathfrak{v}_i,$$

$$(30) \qquad\qquad d\mathfrak{v}_i = \sum_k \omega_{ik} \mathfrak{v}_k, \quad i = 1, 2, 3,$$

where ω_1, for example, is calculated from

$$dx = x_u du + x_v dv = \sum_i \omega_i \mathfrak{v}_i$$

by inner multiplication with $\mathfrak{v}_1$

$$\omega_1 = (x_u \mathfrak{v}_1)du + (x_v \mathfrak{v}_1)dv.$$

An expression of this kind is called a *linear differential* form or a *linear Pfaffian form*. The ω_i, ω_{ik} are linear Pfaffian forms, with $\omega_{ik} = -\omega_{ki}$.

If $\mathfrak{v}_3 = \mathfrak{n}$, then (29) gives

$$dx = \omega_1 \mathfrak{v}_1 + \omega_2 \mathfrak{v}_2,$$

and for the element of arclength

$$ds^2 = dx^2 = \omega_1^2 + \omega_2^2.$$

The first quadratic fundamental form has thus been represented as a sum of squares of differential forms, a result obtained by replacing the x_u, x_v as basis vectors for the coordinate system in the tangent plane by the orthogonal unit vectors $\mathfrak{v}_1, \mathfrak{v}_2$.

Thus (29) is related to the first fundamental form but (30) corresponds to the Gauss and Weingarten equations for the derivatives (25), (7). The system of partial differential equations (29), (30) is solvable if the Pfaffian forms ω_1, ω_{ik} satisfy certain integrability conditions, and then the surface is determined up to motions, by these forms.

This method of the "moving trihedral," due to E. Cartan, allows us to obtain many of the results of surface theory in a simpler and clearer way but, of course, it demands a knowledge of Pfaffian forms (see, e.g., W. Blaschke [4] and W. Haack [5]).

3.   Geometry on a Surface

3.1.   *Geometry on a Surface as Riemannian Geometry*

In the arguments of this section we shall usually pay no attention to the imbedding of the surface in (three-dimensional Euclidean) space, although this imbedding will be mentioned whenever it seems convenient for purposes of visualization.

Thus we start with a two-dimensional manifold, i.e., a set of points with a one-to-one, continuous mapping onto pairs of real numbers u^1, u^2; in other words, a topological image of a segment of the number plane (cf. II16, §2). At each point let there be given a fundamental tensor $g_{ik}(u^1, u^2)$ and thereby a measurement of length

(1) $$ds^2 = g_{ik}du^i du^k$$

and of angle as in §2.3. A geometry of this sort, in which the element of arclength is defined as the square root of a (positive definite) quadratic form, is called a *Riemannian geometry*. Thus we consider our surface as a two-dimensional Riemannian space. Many of our results will hold equally well for a Riemannian space of arbitrary dimension n.

What sort of questions can be dealt with on this basis? In §2.4 we saw that the area of a segment of a surface is determined by

$$\iint \sqrt{g}\,du^1\,du^2, \quad n\text{-dimensional} \quad \int \cdots \int \sqrt{g}\,du^1 \ldots du^n$$

and we can also ask for the shortest curve joining two points, i.e., for geodesic lines.

3.2.   *Geodesic Lines*

Let two points be given by their coordinates u_0^i, u_1^i ($i = 1, 2$) (in n-dimensional space we need only think of $i = 1, \ldots, n$), and let the desired shortest line between these two points have the equations

$$u^i = u^i(t) \quad \text{with} \quad u^i(t_0) = u_0^i, u^i(t_1) = u_1^i$$

(the same letters are used for the function and for its value). We now seek functions $u^i(t)$ such that the arclength

(2) $$J = \int_{t_0}^{t_1} \sqrt{g_{ik}\dot{u}^i\dot{u}^k}\,dt$$

is a minimum.

The calculus of variations deals with the general problem of finding an extreme value of an integral of the form

(3) $$J = \int_{t_0}^{t_1} f(u^i, \dot{u}^i)\,dt,$$

where $f(u^i, \dot{u}^i)$ is an abbreviation for $f(u^1, \ldots, u^n, \dot{u}^1, \ldots, \dot{u}^n)$, by suitable choice of the functions $u^i(t)$. To solve this problem we imbed the curve $u^i(t)$ in a family of comparison curves

(4) $u^i(t, \varepsilon)$

in such a way that $u^i(t) = u^i(t, 0)$ and

(5) $\dfrac{\partial u^i(t, \varepsilon)}{\partial \varepsilon} = 0 \quad \text{for} \quad t = t_0 \quad \text{and} \quad t = t_1,$

in order that for $t = t_0$ and $t = t_1$ all the comparison curves may pass through the prescribed points. We shall also assume the necessary properties of continuity and differentiability. Then (3) becomes a function of ε

$$J(\varepsilon) = \int_{t_0}^{t_1} f(u^i(t, \varepsilon), \dot{u}^i(t, \varepsilon))\, dt$$

with the familiar necessary condition for $J(\varepsilon)$ to have an extreme value at $\varepsilon = 0$

$$\delta J = \frac{dJ}{d\varepsilon}\bigg|_{\varepsilon=0} = \int_{t_0}^{t_1} \left(\frac{\partial f}{\partial u^i} \frac{\partial u^i}{\partial \varepsilon} + \frac{\partial f}{\partial \dot{u}^i} \frac{\partial \dot{u}^i}{\partial \varepsilon} \right)_{\varepsilon=0} dt = 0.$$

Integrating the second summand by parts and taking (5) into account, we have

$$\delta J = \int_{t_0}^{t_1} \left(\frac{\partial f}{\partial u^i} - \frac{d}{dt}\left(\frac{\partial f}{\partial \dot{u}^i} \right) \right) \frac{\partial u^i}{\partial \varepsilon}\, dt,$$

so that a sufficient condition for $\delta J = 0$ is that

(6) $A_i(t) = \dfrac{\partial f}{\partial u^i} - \dfrac{d}{dt}\left(\dfrac{\partial f}{\partial \dot{u}^i} \right) = 0$

for all i and for all t with $t_0 \leq t \leq t_1$, and (6) is also necessary, as may be seen by special choice of the comparison curves. For example, if we had $A_k(t_x) > 0$ for some i $(i = k)$ and for some value t_x, then by the assumed continuity $A_k(t_x)$ would be greater than zero in an interval (t_a, t_b), and if we then took

$u^i(t, \varepsilon) = u^i(t) + \varepsilon w^i(t),$
$w^i(t) = 0 \quad \text{for} \quad i \neq k \quad \text{and} \quad t_0 \leq t \leq t_1,$
$w^k(t) = 0 \quad \text{outside } (t_a, t_b),$
$ \geq 0 \quad \text{inside } (t_a, t_b) \text{ and}$
$ > 0 \quad \text{inside an interval contained in } (t_a, t_b),$

we would have $\delta J > 0$.

Since $dJ/d\varepsilon = 0$ is necessary but not sufficient, the *Euler differential equa-*

tions (6) are only necessary conditions for the integral J to have an extreme value for the curve $u^i(t)$. Curves satisfying (6) are called *extremals* for the variational problem, and in the theory of surfaces the extremals of the variational problem (2) are called *geodesic lines*. Whether, and under what conditions they are actually shortest lines, is a question requiring further investigation, which will be omitted here.

For $f = \sqrt{g_{ik}\dot{u}^i\dot{u}^k}$ we have

$$\frac{\partial f}{\partial u^j} = \frac{1}{2f}\frac{\partial g_{ik}}{\partial u^j}\dot{u}^i\dot{u}^k, \quad \frac{\partial f}{\partial \dot{u}^j} = \frac{1}{f}g_{ij}\dot{u}^i.$$

Note that in $g_{ik}\dot{u}^i\dot{u}^k$ a given index j occurs both for $i = j$ and $k = j$

$$\frac{d}{dt}\left(\frac{\partial f}{\partial \dot{u}^j}\right) = \frac{1}{f}\left(\frac{\partial g_{ij}}{\partial u^k}\dot{u}^i\dot{u}^k + g_{ij}\ddot{u}^i\right) - \frac{1}{f^2}g_{ij}\dot{u}^i\frac{df}{dt}.$$

So from (6)

$$g_{ij}\ddot{u}^i + \left(\frac{\partial g_{ij}}{\partial u^k} - \frac{1}{2}\frac{\partial g_{ik}}{\partial u^j}\right)\dot{u}^i\dot{u}^k - g_{ij}\dot{u}^i\frac{1}{f}\frac{df}{dt} = 0.$$

Now

$$\left(\frac{\partial g_{ij}}{\partial u^k} - \frac{1}{2}\frac{\partial g_{ik}}{\partial u^j}\right)\dot{u}^i\dot{u}^k = \frac{1}{2}\left(\frac{\partial g_{ij}}{\partial u^k} + \frac{\partial g_{ij}}{\partial u^k} - \frac{\partial g_{ik}}{\partial u^j}\right)\dot{u}^i\dot{u}^k;$$

so that by renaming the summation indices in the middle summand we see (since $g_{kj} = g_{jk}$) that the left side is also equal to

$$\frac{1}{2}\left(\frac{\partial g_{ij}}{\partial u^k} + \frac{\partial g_{ik}}{\partial u^i} - \frac{\partial g_{ik}}{\partial u^j}\right)\dot{u}^i\dot{u}^k = \Gamma_{ik,j}\dot{u}^i\dot{u}^k$$

(by §2 (24)). Multiplication by g^{jl} and summation over j gives

(7) $$\ddot{u}^l + \Gamma^l_{ik}\dot{u}^i\dot{u}^k - \dot{u}^l\frac{1}{f}\frac{df}{dt} = 0.$$

Taking arclength for parameter on the extremal curves, we obtain the differential equations of the geodesic lines in the form

(8) $$\ddot{u}^l + \Gamma^l_{ik}\dot{u}^i\dot{u}^k = 0,$$

where the dots now indicate derivatives with respect to arclength.

In the argument leading to this result there has been no mention of the imbedding of the surface in space. But if this imbedding is taken into account, the points on the surface can also be represented by their vectors $\mathfrak{x}$ from a given origin O, whereupon the problem is to make the integral

$$J = \int_{t_0}^{t_1}\sqrt{\dot{\mathfrak{x}}^2}dt$$

a minimum among the comparison curves lying on the given surface. Let the family of comparison curves be denoted by $x(t, \varepsilon)$, where we shall at once take arclength for the parameter and write s instead of t. (Above we made this simplification only later, because we occasionally wish to use general equation (7).) Then since $\sqrt{\dot{x}^2(s, 0)} = 1$,

$$\delta J = \left.\frac{dJ}{d\varepsilon}\right|_{\varepsilon=0} = \int_{s_0}^{s_1} \dot{x}(s, 0)\frac{\partial \dot{x}}{\partial \varepsilon}(s, 0)\, ds.$$

Since (cf. (5)) $\partial x/\partial \varepsilon$ must vanish for $s = s_0$ and $s = s_1$, integration by parts gives

$$\delta J = \int_{s_0}^{s_1} \ddot{x}(s, 0)\frac{\partial x}{\partial \varepsilon}(s, 0)\, ds$$

and therefore as differential equation for the extremal curves

(9)
$$\ddot{x}\frac{\partial x}{\partial \varepsilon} = 0.$$

Also, since the comparison curves must lie on the given surface, the vector $\partial x/\partial \varepsilon$ is tangent to the surface, which means, since the comparison curves may be chosen arbitrarily, that x is perpendicular to *every* tangent vector.

Since the dots indicate derivatives with respect to arclength, $\ddot{x}$ lies in the same direction as the principal normal to the curve. Thus *a geodesic line is characterized by the fact that its principal normal is identical (up to sign) with the normal to the surface.*

This theorem, which was communicated (without proof) to Leibniz by Johann Bernoulli in 1698 and is thus one of the oldest in surface theory, can be visualized as follows. Let three "neighboring" points O, P, R on the curve be projected onto a plane (Fig. 4) perpendicular both to the osculating plane of the curve and to the tangent plane of the surface, and let the images of Q and R coincide. If the osculating plane is not perpendicular to the tangent plane, we can obviously find a point P' in the latter for which QPR' is shorter than QPR, which means that the original curve is not a geodesic line.

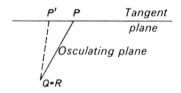

Fig. 4

An intuitive treatment of geodesic lines is given in the booklet by L. A. Lyusternik "Kürzeste Linien" (translated from the Russian) Berlin 1957.

Then (8) follows from (9) and the equations for the derivatives §2, (7) and (25)

$$\dot{\mathfrak{x}} = \mathfrak{x}_i \dot{u}^i,$$

$$\ddot{\mathfrak{x}} = \mathfrak{x}_i \ddot{u}^i + \mathfrak{x}_{ik} \dot{u}^i \dot{u}^k = (\ddot{u}^l + \Gamma^l_{ik} \dot{u}^i \dot{u}^k) \mathfrak{x}_l + b_{ik} \dot{u}^i \dot{u}^k \mathfrak{n},$$

so that the $\ddot{\mathfrak{x}}$ is parallel to $\mathfrak{n}$ if and only if the coefficients of $\mathfrak{x}_l$ vanish.

It was important for us to show that the differential equations for the geodesic lines can be obtained without reference to the imbedding of the surface, since we can now make use of them in spaces whose imbeddability in Euclidean spaces remains undecided.

By the general theory of differential equations (under some regularity conditions, assumed here tacitly) it follows from the second-order system (8) that *through every point there is exactly one geodesic line in each direction.*

3.3. *Levi-Civita Parallel Displacement*

The metric (1) provides a way of measuring length and angle for "infinitesimal" vectors (§2)

(10) $$d\mathfrak{x} = \mathfrak{x}_i du^i,$$

i.e., for vectors in the tangent plane, if we think of a surface in space. But if we wish to disregard the imbedding, we may proceed as follows. The manifold with the given fundamental tensor is replaced at each of its points (and in a "small" neighborhood of each of them) by a Euclidean space in which (1) is the metric referred to an (in general) oblique coordinate system; i.e., the Riemannian space is *locally Euclidean*. Instead of (10) we now write

$$\mathfrak{v} = \mathfrak{x}_i v^i$$

and no longer take any account of the fact that $\mathfrak{x}_1$, $\mathfrak{x}_2$ (the tangent vectors to the parameter curves) are derivatives of the vector $\mathfrak{x}$, but regard them as basis vectors for the coordinate system in the surface. The inner product of two vectors $\mathfrak{v}$, $\mathfrak{w}$ is defined (cf. II9, §7.1) by $\mathfrak{x}_i \mathfrak{x}_k = g_{ik}$, i.e.,

$$\mathfrak{v}\mathfrak{w} = g_{ik} v^i w^k.$$

For vectors in the tangent plane, i.e., in the Riemannian space, we must distinguish between coordinates in the three-dimensional Euclidean imbedding space and coordinates in the plane tangent to the surface. If in a coordinate system $\mathfrak{e}_1, \mathfrak{e}_2, \mathfrak{e}_3$ the vector $\mathfrak{x}_i$ has the coordinates x_i^λ ($\lambda = 1, 2, 3$), then $\mathfrak{v}$ has the coordinates $v^\lambda = x_i^\lambda v^i$ as a three-dimensional vector but as a vector in the tangent plane its coordinates are v^i ($i = 1, 2$). Since we shall not be making any explicit use of the spatial coordinates, the v^1, v^2 will always represent coordinates in the Riemannian space.

But how are the Euclidean spaces, assigned to the individual points of a Riemannian space, related to one another? The question amounts to this: When is a vector with coordinates v^i at the point $P_0(u_0^i)$ to be regarded as the "same" vector as the one with coordinates w^i at the point $P_1(u_1^i)$, or in other words, does there exist some procedure that could reasonably be described as parallel displacement of v^i into w^i? We shall develop such a procedure and shall see that in general the result is dependent on the path along which the vector is displaced, so that in its first form the question was wrongly posed. This concept of parallel displacement is one of the fundamental concepts of Riemannian geometry.

At each point (u^k) of the Riemannian space, let there be given a vector with the coordinates $v^i(u^k)$. In this vector field we consider only the family of vectors along a curve $u^k(t)$: $v^i(u^k(t))$, and seek to give a definition for the statement that the vectors of this family arise from one another by "parallel displacement." As a first attempt let us require that lengths of vectors and the magnitudes of angles between vectors in two families remain invariant, i.e., that the inner product

$$(11) \qquad \frac{d}{dt}(g_{ik}v^iw^k) = \frac{\partial}{\partial u^l}(g_{ik}v^iw^k)\dot{u}^l = 0$$

is not changed. Then

$$\frac{\partial}{\partial u^l}(g_{ik}v^iw^k) = \frac{\partial g_{ik}}{\partial u^l}v^iw^k + g_{ik}\frac{\partial v^i}{\partial u^l}w^k + g_{ik}v^i\frac{\partial w^k}{\partial u^l}.$$

Now by §2 (23)

$$\frac{\partial g_{ik}}{\partial u^l} = \Gamma_{kl,i} + \Gamma_{il,k} = g_{ij}\Gamma_{kl}^j + g_{kj}\Gamma_{il}^j,$$

so that, with re-indexing,

$$(12) \qquad \frac{\partial}{\partial u^l}(g_{ik}v^iw^k) = g_{ij}\left(\frac{\partial w^j}{\partial u^l} + \Gamma_{kl}^jw^k\right)v^i + g_{jk}\left(\frac{\partial v^j}{\partial u^l} + \Gamma_{il}^jv^i\right)w^k.$$

Thus (11) is certainly satisfied if we define the parallel displacement of a vector by

$$(13) \qquad D_lv^j = \frac{\partial v^j}{\partial u^l} + \Gamma_{il}^jv^i = 0.$$

For later use we note that (12) can also be written

$$(14) \qquad \frac{\partial}{\partial u^l}(g_{ik}v^iw^k) = g_{ik}(v^iD_lw^k + w^iD_lv^k).$$

As for visualization, we obtain a transfer of vectors preserving lengths and angles in the following way. Let $E(t)$ be the plane tangent to the surface at the successive points $u_i(t)$, and let us think of this family of planes as being unrolled onto *one* plane E_0. Along the image of the curve $u^i(t)$ in E_0 we let the vector undergo a parallel translation in the ordinary sense, and then transfer the result back to the surface by "rolling up" the family of planes.

If a vector $\mathfrak{v}(t) = \mathfrak{x}_i(t)v^i(t)$ in the plane $E(t)$ is translated to the point $u^i(t + dt)$, it becomes $\mathfrak{v}(t + dt) = \mathfrak{v} + (d\mathfrak{v}/dt)dt$, with

$$\frac{d\mathfrak{v}}{dt} = \mathfrak{x}_i\frac{\partial v^i}{\partial u^l}\dot{u}^l + \mathfrak{x}_{il}\dot{u}^l v^i = \left(\frac{\partial v^j}{du^l} + \Gamma^j_{il}v^i\right)\dot{u}^l\mathfrak{x}_j + b_{ik}\dot{u}^k v^i\mathfrak{n},$$

and rotation of the new vector from the plane $E(t + dt)$ into the plane $E(t)$ amounts in the limit to orthogonal projection, i.e., the normal component of $d\mathfrak{v}/dt$ is discarded. If the resulting vector is to come from $\mathfrak{v}$ by parallel displacement, the remaining part, i.e., the tangential component of $d\mathfrak{v}/dt$, must vanish, which gives us precisely (13).

We see that the definition (13) of parallel displacement cannot be uniquely determined from (11). For we could certainly rotate the tangent plane $E(t + dt)$ further about its point of contact, e.g., about an angle proportional to the arc-length of the curve $u^i(t)$, and this transfer would also preserve lengths and angles.

The expression $D_l v^j = \partial v^j/\partial u^l + \Gamma^j_{il}v^i$ is called the covariant derivative of the vector v^i. It is a tensor of second order, whereas $\partial v^j/\partial u^l$ does not transform like a tensor (cf. IB3, §2.6).

If in $D_l v^j$ we replace the v^j by the vector tangent to the curve $u^j(t)$, i.e., if we set $v^j = \dot{u}^j$, and if then t is arclength, we have

$$\dot{u}^l D_l v^j = \ddot{u}^j + \Gamma^j_{il}\dot{u}^i\dot{u}^l.$$

This expression vanishes if and only if $u^j(t)$ is a geodesic line, which means that the tangent vectors of a geodesic line go into one another by parallel displacement. In this sense a geodesic line is a (straight) line under parallel displacement.

Let us follow the analogy with place curves somewhat further. The vector $D\mathfrak{t}$ with coordinates $D\dot{u}^j = D_l\dot{u}^j \cdot \dot{u}^l$ can be regarded as the covariant derivative of the tangent vector $\mathfrak{t} = (\dot{u}^j)$ with respect to arclength. Also, $D\mathfrak{t}$ is orthogonal to $\mathfrak{t}$, since from $g_{ik}\dot{u}^i\dot{u}^k = 1$ (the dot representing differentiation with respect to arclength) it follows by differentiation (cf. (14)) that

$$\frac{\partial}{\partial u^l}(g_{ik}\dot{u}^i\dot{u}^k)\dot{u}^l = 2g_{ik}\dot{u}^i D_l\dot{u}^k\dot{u}^l = 2g_{ik}\dot{u}^i D\dot{u}^k = 0.$$

The length of the vector $D\mathfrak{t}$

$$\kappa_g = \sqrt{g_{ik}D\dot{u}^i D\dot{u}^k}$$

is called the *geodesic curvature*. The above discussion shows that these magnitudes depend only on the fundamental tensor and not on the imbedding of the surface in space.

Analytically, a parallel displacement amounts to integration of the system of differential equations

(13)
$$\frac{\partial v^j}{\partial u^l} = -\Gamma_{il}^j v^i$$

and will therefore be independent of the path if and only if

$$\frac{\partial^2 v^j}{\partial u^l \partial u^m} = \frac{\partial^2 v^j}{\partial u^m \partial u^l}.$$

(To see this, the reader need only examine the change in the function $v(u^1, u^2)$ under passage around an infinitesimal parallelogram (Fig. 5) with sides parallel to the axes. Moving the vector along the path $P_0 P_1 P_{12}$ gives

$$v + \frac{\partial v}{\partial u^1} du^1 + \frac{\partial v}{\partial u^2} du^2 + \frac{\partial^2 v}{\partial u^1 \partial u^2} du^1 du^2,$$

and along $P_0 P_2 P_{12}$ gives

$$v + \frac{\partial v}{\partial u^2} du^2 + \frac{\partial v}{\partial u^1} du^1 + \frac{\partial^2 v}{\partial u^2 \partial u^1} du^2 du^1.$$

Of course this is only the basic idea of a proof.) Now

$$\frac{\partial^2 v^j}{\partial u^l \partial u^m} = -\frac{\partial}{\partial u^m} \Gamma_{il}^j \cdot v^i - \Gamma_{il}^j \frac{\partial v^i}{\partial u^m} = -\frac{\partial}{\partial u^m} \Gamma_{il}^j \cdot v^i + \Gamma_{il}^j \Gamma_{km}^i v^k,$$

so that

$$\frac{\partial^2 v^j}{\partial u^l \partial u^m} - \frac{\partial^2 v^j}{\partial u^m \partial u^l} = \left(\frac{\partial}{\partial u^l} \Gamma_{km}^j - \frac{\partial}{\partial u^m} \Gamma_{kl}^j + \Gamma_{il}^j \Gamma_{km}^i - \Gamma_{im}^j \Gamma_{kl}^i \right) v^k = r_{klm}^j v^k$$

(cf. §2.7 (27)), so that a parallel translation is independent of the path if and only if the Riemann curvature tensor vanishes. In view of its symmetry properties, this tensor will vanish, in the two-dimensional case, if and only if the Gaussian curvature K is equal to zero. Now surfaces with $K = 0$ are precisely those (we omit the proof) that can be mapped onto the plane with

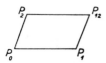

Fig. 5

preservation of lengths. On such surfaces Riemannian geometry is simply Euclidean geometry.

3.4. Sum of the Angles in a Geodesic Triangle

In a passage around a "small" parallelogram the change in a vector undergoing parallel displacement depends on the Gaussian curvature. Thus we may conjecture that in a passage around a finite segment of the surface, the change in the vector will depend on $\iint K\,do$, which is called the total curvature of the segment. So in the special case that the segment is a triangle with geodesic sides, the excess of the triangle, i.e., the difference between π and the sum of its angles, will depend on the total curvature, in the following way.

Since the fundamental magnitudes here are invariant under a change of parameters, and the other magnitudes are tensors with known transformation laws, we may choose any system of parameters convenient for the problem. So we choose geodesic polar coordinates, as follows. Through a point P_0 there exists exactly one geodesic line in each direction. Let these curves be taken for $u^2 = \varphi = \text{const}$, where φ is the angle with a fixed direction. As curves $u^1 = r = \text{const}$ we choose the orthogonal trajectories (Fig. 6) of these lines, which means that the scalar product of the two vectors $dx = (dr, 0)$ and $\delta x = (0, d\varphi)$ vanishes: $g_{12}\, dr\, d\varphi = 0$, i.e.,

$$(15) \qquad\qquad g_{12} = 0.$$

(In more detail: if the fundamental tensor $g_{l'm'}$, given originally with respect to some other coordinate system, has the coordinates g_{ik} when transformed to our present system, then $g_{12} = 0$.)

The fact that the curves $\varphi = \text{const}$ are geodesic lines means that the differential equations (8) must be satisfied for $\dot{u}^2 = \dot{\varphi} = 0$. It follows that

$$(16) \qquad\qquad \Gamma^2_{11} = 0.$$

Since $g_{12} = 0$ and therefore $g^{12} = 0$, it then follows

$$\Gamma^2_{11} = \Gamma_{11,2}g^{22} = 0$$

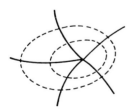

Fig. 6.   Since in the plane of the paper geodesic lines are straight lines, any nontrivial diagram is necessarily false.

and, since $g^{22} \neq 0$ (for otherwise we would have $g = 0$), we see that

(17) $\Gamma_{11,2} = 0,$

and thus, from §2 (24),

(18) $\dfrac{\partial g_{11}}{\partial u^2} = \dfrac{\partial g_{11}}{\partial \varphi} = 0.$

We have not yet chosen the scale for values of r. If arclength from P_0 is taken as parameter on the curve $\varphi = 0$, then $ds^2 = g_{11}dr^2 = dr^2$ on this curve, so that $g_{11} = 1$, a result which by (18) holds on every curve $\varphi = $ const. Thus the curves $r = $ const are obtained by marking off the same arclength from P_0 on all curves $\varphi = $ const; and are therefore called *circles of constant geodesic distance*.

Let us mention without proof that it is only on surfaces of constant Gaussian curvature that these circles are identical with the curves of constant geodesic curvature.

Thus in our geodesic polar coordinate system, the fundamental tensor has the coordinates

(19) $g_{11} = 1, \quad g_{12} = g_{21} = 0, \quad g_{22} = g$

(g_{22} is equal to the determinant g), and the element of arclength is

(20) $ds^2 = dr^2 + gd\varphi^2,$

where we must assume $g > 0$, since the metric fundamental form must be positive definite.

From (20) it follows that the integral

$$\int_0^r ds^2 = \int_0^r (dr^2 + gd\varphi^2)$$

takes its smallest value along the curves $\varphi = $ const, so that the geodesic line is actually the shortest line joining two points, to the extent that our parameter net is admissible, i.e., represents a one-to-one correspondence between the surface and the r, φ-plane; in other words, two geodesic lines from P_0 do not have a second intersection.

Later we shall need the statement

(21) $\lim_{r \to 0} \sqrt{g}/r = 0.$

which is plausible on the ground that for polar coordinates in Euclidean plane geometry

$$ds^2 = dr^2 + r^2 d\varphi^2,$$

and in the limit (see §3.3 above) Riemannian space is Euclidean.

For the proof we may, for example, introduce the so-called "normal coordinates" $r \cos \varphi = u$, $r \sin \varphi = v$, and then show by calculation that the line element remains regular at the point $u = v = 0$ (cf. Blaschke [3], p. 152.)

Setting $\partial g/\partial r = g_r$, $\partial g/\partial \varphi = g_\varphi$, we have from (19), by a straightforward but somewhat laborious calculation

$$(22) \quad \Gamma^1_{11} = \Gamma^2_{11} = \Gamma^1_{12} = 0, \quad \Gamma^2_{12} = \frac{g_r}{2g}, \quad \Gamma^1_{22} = -\frac{1}{2}g_r, \quad \Gamma^2_{22} = \frac{g_\varphi}{2g},$$

$$(23) \qquad\qquad K = r^2_{.121} = -\frac{1}{\sqrt{g}}\frac{\partial^2 \sqrt{g}}{\partial r^2}.$$

We now ask for the sum of the angles of a triangle whose sides are geodesic lines. For this purpose we choose a vertex P_0 of the triangle as origin for a system of geodesic coordinates, and the two adjacent sides as curves $\varphi = 0$ and $\varphi = \gamma = $ const, and let the third side be given by $r = f(\varphi)$. In order to calculate that this side is a geodesic line, we must use the two equations (7) ($l = 1$, $l = 2$), since φ is now the parameter, not arclength. Setting $df/d\varphi = f_\varphi$, we obtain

$$(24) \qquad\qquad f_{\varphi\varphi} = \frac{g_r}{g}f^2_\varphi + \frac{g_\varphi}{2g}f_\varphi + \frac{1}{2}g_r.$$

We are interested in the angle α formed by the third side of the triangle with the geodesic line $\varphi = $ const (see Fig. 7). The tangent vectors are

$$dx = (1, 0), \quad \delta x = (f_\varphi, 1).$$

For the value of $\cos \alpha = dx\,\delta x/\sqrt{dx^2}\sqrt{\delta x^2}$ (scalar products formed with the fundamental tensor), we obtain $\cos \alpha = f_\varphi/\sqrt{f^2_\varphi + g}$, and thus

$$(25) \qquad\qquad f_\varphi = \sqrt{g}\cdot\text{ctg } \alpha.$$

Substitution of this result in (24) gives

$$(26) \qquad\qquad \frac{d\alpha}{d\varphi} = -\frac{\partial}{\partial r}\sqrt{g}.$$

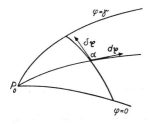

Fig. 7

We now calculate the total curvature, namely the integral

$$\iint_D K\, do = -\int_{\varphi=0}^{\gamma} \int_{r=0}^{f(\varphi)} \frac{\partial^2 \sqrt{g}}{\partial r^2} dr\, d\varphi,$$

taken over the surface D of the triangle, where we make use of the fact that $do = \sqrt{g}\, dr\, d\varphi$. To begin with (from (21))

$$\int_{r=0}^{f(\varphi)} \frac{\partial^2 \sqrt{g}}{\partial r^2} dr = \left[\frac{\partial \sqrt{g}}{\partial r}\right]_0^{f(\varphi)} = \left(\frac{\partial \sqrt{g}}{\partial r}\right)_{f(\varphi)} - 1.$$

So from (26)

$$\iint_D K\, do = \int_{\varphi=0}^{\gamma}\left(1 + \frac{d\alpha}{d\gamma}\right)d\varphi = \gamma + \alpha(\gamma) - \alpha(0).$$

The interior angles of the triangles are $\beta_1 = \pi - \alpha(0)$, $\beta_2 = \alpha(\gamma)$, $\beta_3 = \gamma$. Thus

(27) $$\iint_D K\, do = \beta_1 + \beta_2 + \beta_3 - \pi.$$

So the total curvature of a geodesic triangle is equal to the excess of the triangle, i.e., if $K = \pm 1$, the total curvature is equal in magnitude to the area of the triangle, a theorem which is important in non-Euclidean geometry.

The result (27) can be regarded as a special case of the Gauss-Bonnet integral theorem, which states that if the arbitrary closed curve C has no double points and can be shrunk to a point continuously on the surface, then

(28) $$\iint K\, do + \oint \kappa_g\, ds = 2\pi.$$

If C is a geodesic triangle we need only round off the corners, say by geodesic circles, and pass to the limit. Then $\oint \kappa_g\, ds$ gives the sum of the exterior angles, which is equal to $\pi - \beta_1 + \pi - \beta_2 + \pi - \beta_3$ (cf. Blaschke [3], pp. 163 and 164).

Conversely, we can obtain (28) from (27) by approximating C with a geodesic polygon (cf. Haack [5], p. 120ff.)

Bibliography

[1] BEHNKE, H.: Vorlesungen über Differentialgeometrie. 4th ed., Münster 1958 Multigraphed lecture notes.
[2] BIEBERBACH, L.: Differentialgeometrie. Teubners math. Leitfäden, Vol. 31. Leipzig, Berlin 1932.

[3] BLASCHKE, W.: Vorlesungen über Differentialgeometrie I. 4th ed., Berlin 1945. Springer.
[4] BLASCHKE, W.: Einführung in die Differentialgeometrie. Springer. Berlin Göttingen Heidelberg 1950.
[5] COXETER, H. S. M.: Introduction to geometry (espec. Part IV), John Wiley and Sons, Inc., New York London 1961.
[6] HAACK, W.: Elementare Differentialgeometrie. Birkhäuser, Basel 1955.
[7] KRUPPA, E.: Analytische und konstruktive Differentialgeometrie. Springer, Vienna 1957.
[8] LAUGWITZ, D.: Differentialgeometrie. Teubner, Stuttgart 1960.
[9] STRUBECKER, K.: Differentialgeometrie. Sammlung Göschen. Berlin. 3 Vols.

Convex Figures

I. Properties of General Convex Bodies

A point set in n-dimensional Euclidean space is said to be *convex* if with any two points it contains the line segment joining them. If it is also closed and bounded, it is called a *convex* or *ovoid body*, an *ovoid region*, or simply an *ovoid*. An ovoid region is said to be k-dimensional if it contains points about which we can describe a k-dimensional sphere lying entirely in the region but no points about which we could describe such a $(k + 1)$-sphere.

A 0-dimensional ovoid is a point, a 1-dimensional ovoid is a line segment, and a 2-dimensional ovoid is also called an oval. If the ovoid contains four points not in one plane, then it contains the whole tetrahedron with these points for vertices.

The intersection of a set of ovoids is again an ovoid; for if two points lie in the intersection, and thus in all the ovoids, then the line joining them lies in all the ovoids. The intersection may have a smaller dimension than any of the ovoids in the original set.

By a *parallel body* E_p of an ovoid E we mean the set-theoretic union of all spheres of radius p with centers in E. For any two given ovoids E and F there exists a smallest p with the property that E lies in F_p and F in E_p. This p is called the *distance* between E and F, and is written $d(E, F) = \inf p$. The triangle inequality

$$d(E, F) \leqq d(E, G) + d(G, F)$$

holds for this definition of distance.

If E is an ovoid, then for every $\varepsilon > 0$, there exists a convex polyhedron P such that P lies in E_ε and E lies in P_ε; the distance between E and P is $d(E, P) \leqq \varepsilon$, which means that an ovoid can be *approximated* with arbitrary accuracy by a polyhedron.

In particular, for plane convex regions:

If every n ($\geq$ 4) points of a bounded closed curve K are the vertices of a convex polygon, then the curve is convex.

2. Theorems on Ovals (Two-dimensional Ovoids)

In this section all geometric figures lie in the plane of the given oval E. For the position of a line g with respect to E there are three possibilities.

1.   If g has no point in common with E, it is called a *bounding line* in the direction of the normal $\mathfrak{n}$ pointing away from E.

2.   If g contains at least one boundary point but no interior point of E, it is called a *support line* in the direction $\mathfrak{n}$. An oval has exactly one support line[1] in each direction, and lies entirely in one of the two halfplanes determined by its support line.

3.   Finally, the line g may intersect E, in which case it will contain both boundary points and interior points of E.

(With suitable changes of wording, these three cases hold for the position of a plane with respect to a three-dimensional ovoid.)

A characteristic property of ovals is that they have at least one support line at every boundary point.

Let us give an indirect proof (Fig. 1) that this property could have been used as a definition.

We assume that the region $\mathfrak{B}$ has a support line at every point, but is not convex; i.e., it has points P and Q such that the line joining them contains a point A not in the region. If I is an interior point of $\mathfrak{B}$, the line AI must contain a boundary point R, through which there will then pass a support line with the three interior points P, Q, and I on one side of it, which is impossible.

Two parallel lines of support $g(\mathfrak{n})$ and $g(-\mathfrak{n})$ form a *support strip* contain-

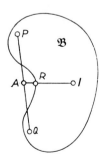

Fig. 1

[1]Assertions of this sort appear evident, but they require proof. See [4], p. 5.

ing the whole of E. The distance between the two lines is called the *breadth* $b(\mathfrak{n})$ of E in the direction $\mathfrak{n}$. Since b is a continuous function of $\mathfrak{n}$, it has at least one minimum, called the *thickness* of E, and at least one maximum, called the *diameter* of E. (Here the concept diameter is not the same as in analytic geometry.) If b is constant for all directions, we speak of *curves of constant breadth*.

Helly's theorem[2] is very remarkable in character: *If k convex figures are so situated in the plane that every three of them have a point in common, then all of them have a point in common.* (If k is infinite, at least one of the figures must be bounded.)

For n dimensions the theorem reads: *If k convex figures are so situated in n-dimensional space that every $(n + 1)$ of them has a point in common, then all of them have a point in common.*

This theorem is the starting point for many related questions, not only in the field of convex figures. Let us mention a theorem proved by L. Danzer[3] (on the basis of a question proposed by H. Hadwiger): *If a set of congruent nonintersecting circles is so situated in the plane that every five of them are met by some straight line, then there exists a straight line meeting all of them.* For problems of this sort H. Hadwiger has introduced the name *Combinatorial Geometry*.

Helly's theorem can also be used to prove theorems that at first sight have very little to do with it; for example: In the interior of a convex figure E with area F there exists a point O such that every line through O divides the figure into two parts neither of which is less than $\frac{4}{9}F$ in area.

For the proof let us consider all halfplanes (to obtain bounded figures, we could take their intersections with a circle containing E) that contain more than $\frac{5}{9}$ the area of E, and then prove that any three of these have a point in common. By Helly's theorem all of them have a point O in common, which is the desired point.[4]

Questions referring to inscribed and circumscribed polygons for ovals are interesting from many points of view. To avoid degenerate cases, we shall assume that the boundary of the oval does not contain a line segment.

Theorem 1: *A square can be inscribed in any oval.*

Idea of the proof. For every chord of E joining boundary points with nonparallel support lines there exists a parallel chord of the same length and therefore an inscribed parallelogram. By continuity the pair of chords can be given a parallel displacement producing a rhombus. If the direction

---

[2] E. Helly: Über Mengen konvexer Körper mit gemeinschaftlichen Punkten. Jahresber. Deutsch. Math. Ver. 32, 175 and 176 (1932); see also [4], p. 13; [6], Chapter 2.

[3] L. Danzer: Über ein Problem aus der kombinatorischem Geometrie. Arch. der Math. 8, 347–351 (1957); see also B. Grünbaum: On Common Transversals. Arch. der Math. 9, 465–467 (1958).

[4] For similar questions see Yaglom [4], p. 16.

of one of the pair of sides is now changed continuously (with preservation of the rhombus property) until it reaches the original position of the other pair, the acute angle of the rhombus has been changed continuously (as must and can be proved) into the obtuse angle, so that there must be an intermediate position in which it is a right angle, and the rhombus is a square.[5]

Theorem 2: *A square can be circumscribed about any oval.*

Idea of the proof. Construct a rectangle consisting of support lines and proceed as in Theorem 1.

Theorem 3: *An affine-regular hexagon (i.e., affinely transformable into a regular hexagon) can be inscribed in any oval.* In other words, *every oval can be affinely transformed into an oval with an inscribed regular hexagon.*

For the proof consider the hexagon formed by the six endpoints of two parallel chords a, c of equal length and an intermediate chord b with $b = 2a = 2c$ (such chords obviously exist). By continuous rotation we can arrange that its three diagonals have a common midpoint, i.e., that it is affine-regular.

Theorem 4: *Infinitely many affine-regular pentagons can be inscribed in any plane convex figure.*[6]

Closely related are questions concerning the smallest or largest inscribed or circumscribed figures for ovals of given diameter or thickness. For example,

Theorem 5 (Gale's theorem):[7] *Any oval of diameter 1 is contained in an equilateral triangle with inscribed circle of diameter 1. This triangle has the smallest area of all triangles containing every oval of diameter 1.*

Theorem 6 (Jung's theorem):[8] *The smallest circle containing every oval of diameter 1 is the circle circumscribed about an equilateral triangle of side 1.*

The three-dimensional analog is: *Every three-dimensional ovoid*[9] *of diameter 1 is contained in a sphere of radius $\frac{1}{4}\sqrt{6}$ i.e., in the sphere circumscribed about a regular tetrahedron of side 1.*

---

[5]For inscribed rectangles see O. Giering: Ausfüllung von Eilinien durch einbeschriebene gleichseitige und gleichwinklige n-Ecke. Elemente d. Math. 16/1961 No. 4.

[6]W. Böhme: Ein Satz über ebene konvexe Figuren, Math.-phys. Semesterber. 6, p. 153 (1958), p. 341.

[7]D. Gale: On inscribing n-dimensional sets in a regular n-simplex. Proc. Amer. Math. Soc. 4, 222–225 (1953). U. Viet: Über die ebene Eibereichen eingeschriebenen Dreiecke. Math.-phys. Semesterber. 4, 57 and 58 (1955).

[8]H. W. E. Jung: Über die kleinste Kugel, die eine räumliche Figur einschließt. J. reine angew. Math. 123, 241–257 (1901); W. Süss: Durchmesser und Umkugel bei mehrdimensionalen Punktmengen. Math. Zeitschr. 40, 315 and 316 (1935).

[9]The diameter of an ovoid K is the maximum distance between two points of K.

Theorem 7 (Blaschke's theorem):[10] *Every bounded convex figure of thickness 1 includes a circle of radius $\frac{1}{3}$.*

The three-dimensional analog here is: *Every three-dimensional convex body of thickness 1* (i.e., smallest distance between pairs of parallel support planes) *contains a sphere of diameter* $1/\sqrt{6}$.

3. Measurements of an Oval[11]

The most important measurements of an oval are its *curvature, circumference,* and *area.* The differential geometric methods of the present section require certain assumptions about continuity and differentiability, which will here be taken for granted. Since the figure is always considered as the whole, we speak of "differential geometry in the large."

Here we give just a few theorems for purposes of illustration.

If the curve is represented by the position vectors $x(s)$ (where s is arclength) issuing from an interior point, then $x'(s) = t$ is the unit vector in the tangent direction, and if φ is the angle between the tangent and the x-axis, the curvature is defined as

$$\frac{d\varphi}{ds} = \frac{i}{\rho}$$

and the radius of curvature is $\dot{s} = ds/d\varphi = \rho$. Then the Frenet formulas for the plane (see II14, §1.6) are

$$\frac{dt}{ds} = t' = \frac{1}{\rho}n; \quad \frac{dn}{ds} = n' = -\frac{1}{\rho}t,$$

where n is the unit vector normal to the curve. It follows that

$$\frac{dt}{d\varphi} = \dot{t} = n; \quad \frac{dn}{d\varphi} = \dot{n} = -t.$$

If we now introduce the distance p from the origin to the tangent at an arbitrary point of the curve

(1)
$$p = -xn,$$

differentiation with respect to φ gives

(2)
$$\dot{p} = -\dot{x}n - x\dot{n},$$

with $\dot{x}n = 0$, since $\dot{x}$ and n are orthogonal. Further differentiation gives

$$\ddot{p} = \rho - p,$$

[10][4], p. 15.
[11]Cf. [3], p. 30 ff.

and thus for the radius of curvature

(3) $$\rho(\varphi) = \ddot{p}(\varphi) + p(\varphi).$$

The circumference of an oval is given by

(4) $$U = \oint ds = \int_0^{2\pi} \rho(\varphi)d\varphi = \int_0^{2\pi} p d\varphi$$

and the area (taking p as the altitude and ds as the base of a surface element) by

(5) $$F = \frac{1}{2} \oint p ds = \frac{1}{2} \int_0^{2\pi} p\rho d\varphi.$$

These results imply the following theorems, among others.

1. *All curves of constant breadth b have circumference πb.*

Proof. From $b(\varphi) = p(\varphi) + p(\varphi + \pi) = \text{const}$ it follows that

$$U = \int_0^{2\pi} p d\varphi = \int_0^{\pi} [p(\varphi) + p(\varphi + \pi)] d\varphi = \int_0^{\pi} b d\varphi = b\pi.$$

2. The *Steiner formulas* for the circumference U_a and the area F_a of a parallel surface E_a.[12]

For a given curve $\mathfrak{x}$, a parallel curve is defined by $\mathfrak{x}_a = \mathfrak{x} - a\mathfrak{n}$, so that on the one hand $\dot{\mathfrak{x}}_a = \dot{\mathfrak{x}} - a\dot{\mathfrak{n}} = (\rho + a)\cdot\mathfrak{t}$, and on the other $\dot{\mathfrak{x}}_a = \rho_a \mathfrak{t}_a$ (definition of ρ_a, $\mathfrak{t}_a$). It follows that $\mathfrak{t}_a = \mathfrak{t}$ (which means that tangents at corresponding points are parallel) and

(6) $$\rho_a = \rho + a.$$

Substitution of this result and $p_a = p + a$ in (4) and (5) gives the Steiner formulas

$$U_a = U + 2\pi a,$$
$$F_a = F + aU + a^2\pi.$$

For polygons these formulas may be visualized as in Fig. 2.

3. A theorem of Blaschke: *On every oval curve (boundary of an oval) with continuously turning tangent there exist at least three pairs of points with parallel tangents (antipodal points) and with equal curvature.*

Let the components of the tangent vector at a point of the curve (x_1, x_2) be $t_1 = x_1' = \cos \varphi$; $t_2 = x_2' = \sin \varphi$. Then since the oval curve is closed,

(7)
$$\oint \cos \varphi \, ds = \int_0^{2\pi} \rho(\varphi) \cos \varphi \, d\varphi = 0,$$
$$\oint \sin \varphi \, ds = \int_0^{2\pi} \rho(\varphi) \sin \varphi \, d\varphi = 0,$$

[12]J. Steiner: Über parallele Flächen. Monatsber. d. Akad. d. Wiss. Berlin 1840. Collected works, Vol. 2, 171–176.

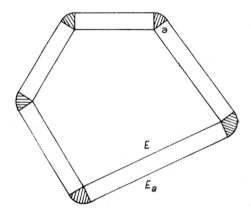

Fig. 2

which gives

$$\int_0^\pi \rho(\varphi) \cos \varphi \, d\varphi + \int_\pi^{2\pi} \rho(\varphi) \cos \varphi \, d\varphi$$

(8)
$$= \int_0^\pi \rho(\varphi) \cos \varphi \, d\varphi - \int_0^\pi \rho(\varphi + \pi) \cos \varphi \, d\varphi$$

$$= \int_0^\pi [\rho(\varphi) - \rho(\varphi + \pi)] \cos \varphi \, d\varphi = 0$$

and also

(9)
$$\int_0^\pi [\rho(\varphi) - \rho(\varphi + \pi)] \sin \varphi \, d\varphi = 0.$$

The function

$$D(\varphi) = \rho(\varphi) - \rho(\varphi + \pi)$$

satisfies the condition

$$D(\phi + \pi) = -D(\phi)$$

and thus has a change of sign for at least one pair of antipodal points, i.e., an oval curve has at least one pair of antipodal points with the same curvature.

But if we assume that $D(\varphi)$ vanishes only say at $\varphi = 0$ and $\varphi = \pi$, then for $0 < \varphi < \pi$ we would have say $D(\varphi) > 0$ and therefore

$$\int_0^\pi D(\varphi) \sin \varphi \, d\varphi > 0$$

in contradiction to equation (9). Thus $D(\varphi)$ must change sign at least once

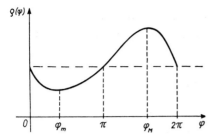

Fig. 3

more, which is possible only if $D(\varphi) = 0$ for at least a third pair of antipodal points.

4. The four-vertex theorem: *Every oval curve has at least four vertices, i.e., four points at which the curvature has a maximum or minimum.* (We will disregard the circle, whose curvature is constant.)

There are numerous proofs of this theorem.[13] We shall here make use of the preceding theorem of Blaschke[14] (Fig. 3). As a continuous function on a closed curve, ρ has at least one maximum (for φ_M) and at least one minimum (for φ_m). If the oval curve had only these two vertices, there would exist only one pair of antipodal points with the same curvature since $\rho(\varphi)$ would then be monotonically increasing from φ_m to φ_M, in contradiction to the preceding theorem. Thus every oval curve must have at least one more pair of vertices. The minimum number of four vertices occurs, for example, in the ellipse.

Let us also mention, without proof, the following theorems about oval curves with continuously turning tangent.[15]

5. *The radius of curvature of an oval curve of length U assumes the values $U/2\pi$ at least four times. At least two of the maxima of the radius of curvature are greater than $U/2\pi$ and at least two of its minima are less* (apart from the circle).

6. *Every curve of constant breadth has at least six vertices.*

7. *Curves of constant breadth b have at least six points at which their radius of curvature takes the same value $b/2$.*

In addition to the usual measurements for an oval, such as its area, circumference, and curvature, others may be introduced. For example, Kowner's theorem deals with deviation from central symmetry.

Kowner's theorem:[16] *The interior of every ovoid E contains a central-*

[13]Barner and Flohr have given a survey of the methods of proof and generalizations of this theorem. Cf. Math. Unterr. 4/58.

[14]Cf. [1], p. 160.

[15]W. Süss: Über Krümmungeseigenschaften im Grossen von Eilinien und Eiflächen. Sitz. -Ber. Her. Heidelb. Akad. d. Wiss. math. -naturw. Kl. 1935, 4. Abh.

[16]Cf. [4], p. 25.

symmetric convex figure Z with area F(Z) not less than $\frac{2}{3}$ of the area F(E) of E. If E is a triangle, not more than $\frac{2}{3}$ of its area can be covered by a figure in its interior

(10)
$$\frac{2}{3} \leq \max\left|\frac{F(Z)}{F(E)}\right| \leq 1.$$

Thus $\max |F(Z)/F(E)|$ measures the *centricity* of ovoid curves; central-symmetric curves have measure 1, and triangles, i.e., the ovoid curves "departing most" from central-symmetry, have measure $\frac{2}{3}$.

An analogous theorem states that an oval region E is contained in a central-symmetric figure S with area not more than twice the area of E. *If E is a triangle, this figure is the best possible,* i.e.,

(11)
$$\frac{1}{2} \leq \max\left|\frac{F(E)}{F(S)}\right| \leq 1.$$

Here also the triangle is a "limit case" among the convex figures.

4. Central Ovoids[17]

A plane convex region is called a *central ovoid* if it contains a point O bisecting every chord through O. Such a region is taken into itself by a rotation through 180° about O.[18]

1. For arbitrary ovals it follows from continuity that every interior point bisects at least one chord, and that at least one point bisects more than one chord; in fact, we shall prove the following theorem.

Theorem: *The center of gravity S of an oval E (covered with material of homogeneous density) bisects at least three chords* (Fig. 4).

Let us choose a system of polar coordinates (r, φ) with S as pole, such that $r(0) = r(\pi)$ and

$$\Delta(\varphi) = r(\varphi) - r(\varphi + \pi)$$

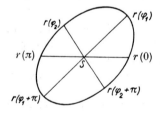

Fig. 4

[17]W. Süss: Über Eibereiche mit Mittelpunkt. Math. -phys. Semesterber. 1, 273–287 (1950).
[18]See II7, §2.3.

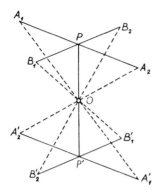

Fig. 5

changes sign at $\varphi = 0$. Since S is the center of gravity, we cannot have $r(\varphi) > r(\varphi + \pi)$ for all $\varphi(0 < \varphi < \pi)$ nor $r(\varphi) < r(\varphi + \pi)$ for all $\varphi(0 < \varphi < \pi)$. If we assume that $\Delta(\varphi) = r(\varphi) - r(\varphi + \pi)$ is negative in a sector following $\varphi = 0$ in the clockwise direction, then it must be positive in a subsequent sector, which means that there must be an intermediate φ_1, $0 < \varphi_1 < \pi$, for which $\Delta(\varphi_1) = 0$, i.e., $r(\varphi_1) = r(\varphi_1 + \pi)$. If we now consider the corresponding vertex changes of sign at $\varphi = \pi$ and $\varphi = 0$, we see that $\Delta(\varphi)$ must have still another change of sign at a value φ_2 with $r(\varphi_2) = r(\varphi_2 + \pi)$ and $\varphi_1 < \varphi_2 < \pi$. Thus S bisects the three chords corresponding to $\varphi = 0$, $\varphi = \varphi_1$, $\varphi = \varphi_2$.

2. For central ovals we now have the following theorem.

Theorem of Brunn:[19] *In a central oval that has no line segments on its boundary no point, except the center itself, bisects more than one chord.*

We give an indirect proof (Fig. 5). Let us assume that there is another point P ($\neq O$) that bisects at least two chords, say $A_1 A_2$ and $B_1 B_2$. Then the image point P' of P has the same property, since it bisects the chords $A_1' A_2'$ and $B_1' B_2'$ arising from rotation through $180°$ about O. The line segments $A_1 A_2$ and $B_1 B_2'$ lie on the same side of the segment PP' and are parallel to it. Then if the boundary points A_1, A_2', B_1, B_2' are distinct, we have a contradiction to the hypothesis that the region is convex, and if two of them coincide (more than two cannot coincide), the others lie on a line segment of the boundary, which is again contrary to hypothesis.

3. Brunn's theorem characterizes the central ovals. Namely, a plane oval with no line segments on its boundary has a center if and only if all its points, excepting the center of gravity, bisect exactly one chord.[20]

4. If an oval has a point O such that all chords through it bisect the area

[19]H. Brunn: Über Ovale und Eiflächen. Diss. München 1887.
[20]U. Viet: Umkehrung eines Satzes von Brunn über Mittelpunktseibereiche. Math. -phys. Semesterber. 5, 141 and 142 (1956).

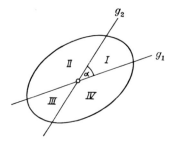

Fig. 6

of the oval, then O is a center. For let us refer the boundary curve to polar coordinates r, φ with the origin at O. Let g_1 and g_2 be two "neighboring" lines through O (see Fig. 6). Then

$$\text{I} + \text{II} = \frac{F}{2},$$

$$\text{III} + \text{II} = \frac{F}{2}$$

imply that

$$\text{I} = \text{III},$$

which means that

$$\frac{1}{2}\int_0^\alpha r^2(\varphi)d\varphi = \frac{1}{2}\int_\pi^{\pi+\alpha} r^2(\varphi)d\varphi.$$

If now on the right-hand side we replace the variable of integration by $\varphi + \pi$ and subtract, we obtain

$$\int_0^\alpha [r^2(\varphi) - r^2(\varphi + \pi)]d\varphi = 0.$$

Differentiation with respect to the upper limit gives

$$r^2(\alpha) - r^2(\alpha + \pi) = 0,$$

and thus

$$r(\alpha) = r(\alpha + \pi), \quad \text{q.e.d.}$$

5. If an oval has a point O such that all chords through it bisect the circumference, then O is a center.

As before, we show that $s(\alpha) = \bar{s}(\alpha)$ (Fig. 7). In polar coordinates the arclength is given by

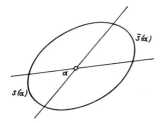

Fig. 7

$$s(\alpha) = \int_0^\alpha \sqrt{r^2(\varphi) + r'^2(\varphi)}\,d\varphi,$$

$$\bar{s}(\alpha) = \int_0^\alpha \sqrt{r^2(\pi + \varphi) + r'^2(\pi + \varphi)}\,d\varphi.$$

Then for all α

$$\bar{s}(\alpha) - s(\alpha) = \int_0^\alpha [\sqrt{r^2(\varphi + \pi) + r'^2(\varphi + \pi)} - \sqrt{r^2(\varphi) + r'^2(\varphi)}]\,d\varphi.$$

Differentiation with respect to the upper limit gives

$$\sqrt{r^2(\alpha + \pi) + r'^2(\alpha + \pi)} - \sqrt{r^2(\alpha) + r'^2(\alpha)} = 0,$$

$$r^2(\alpha + \pi) - r^2(\alpha) = r'^2(\alpha) - r'^2(\alpha + \pi).$$

If we set

$$r(\alpha) + r(\alpha + \pi) = \sigma(\alpha),$$

$$r(\alpha + \pi) - r(\alpha) = \delta(\alpha),$$

the preceding equation can be restated as

$$\sigma(\alpha)\delta(\alpha) + \sigma'(\alpha)\delta'(\alpha) = 0.$$

The function δ cannot be strictly monotonic, and at an extremal point $\delta' = 0$ and therefore $\sigma\delta = 0$. It follows that δ vanishes for every extreme value and therefore vanishes identically.

6. Blaschke's theorem has the following corollary.

If at every point the curvature is the same as at its antipodal point, then the oval has a center.

Proof. Let $\rho(\varphi) = \rho(\varphi + \pi)$, so that by §3 (3)

$$p(\varphi) + \ddot{p}(\varphi) = p(\varphi + \pi) + \ddot{p}(\varphi + \pi).$$

From

$$\delta(\varphi) = p(\varphi + \pi) - p(\varphi)$$

we obtain the differential equation

$$\breve{\delta}(\varphi) + \delta(\varphi) = 0$$

with the general solution $\delta(\varphi) = a \cos \varphi + b \sin \varphi$.

If the origin is so chosen that $p(0) = p(\pi)$ and $p(\pi/2) = p(3\pi/2)$, it follows from $\delta(0) = \delta(\pi/2) = 0$ that $a = b = 0$, which means that $\delta = 0$. Thus in every case $p(\varphi + \pi) = p(\varphi)$, which completes the proof.

7. In conclusion, let us mention a characterization of central ovals due to Asplund and Grünbaum.[21] This criterion differs from the foregoing theorems in that the characteristic property in question does not concern the center itself.

Theorem: *A strictly convex oval K has a center if and only if the follow-ing condition is satisfied. If K_1, K_2, K_3 are three translates (images under translation) of K with a common intersection but no common tangent, there exists a fourth translate K_4 of K passing through the three other points of intersection of K_1, K_2, K_3.*

5. Some Characteristic Properties of the Circle

1. Theorem of Fujiwara-Kubota: *An oval curve that has at most two points in common with any other curve congruent to it is a circle.*

Proof. Let S be the center of gravity of the given oval E. If E is not a circle, some rotation about S will take E into a congruent oval E' which is distinct from E and therefore has exactly two points of intersection with E (since by hypothesis it cannot have more than two such points). If g is the line joining these points of intersection, the part of E inside E' must lie on one side of g and the part of E' inside E must lie on the other, which is in contradiction to the fact that E and E' have the same center of gravity S. Thus E must be a circle.

This theorem carries over to three-dimensional space: *An ovoid is a sphere if its intersection with any other congruent copy of it lies on a plane.*

2. *An oval curve, representable by a differentiable function, is a circle if it touches every circumscribed rectangle at the midpoints of the sides* (Fig. 8).

Proof. Let $\mathfrak{x}$, $\bar{\mathfrak{x}}$ be the positive vectors of a pair of antipodal points P, $\bar{P}$, so that by hypothesis P and $\bar{P}$ are the midpoints of opposite sides of a circumscribed rectangle and the line joining them is perpendicular to these

[21]Asplund and Grünbaum: On the geometry of Minkowski planes. L'Enseignement Mathématique, VI, p. 299 (1960).

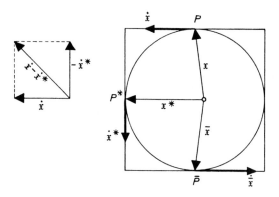

Fig. 8

sides, and therefore to the tangents at x and $\bar{x}$. Then

(1)
$$(x - \bar{x})\dot{x} = (x - \bar{x})\dot{\bar{x}} = 0$$

or

$$(x - \bar{x})(\dot{x} - \dot{\bar{x}}) = 0,$$

so that

$$(x - \bar{x})^2 = \text{const} = a^2,$$

which means that the given oval curve bounds a region of constant breadth, and every circumscribed rectangle is a square of side a.

Now let the points of contact with adjacent sides of the square be P and P^*, with position vectors x and x^*. Then

$$(x - x^*)^2 = \frac{a^2}{2},$$

$$(x - x^*)(\dot{x} - \dot{x}^*) = 0.$$

As we see from Fig. 8 (vector diagram), $x - x^*$ is perpendicular to $\dot{x} - \dot{x}^*$ only if $\dot{x}^2 = \dot{x}^{*2}$; and similarly $\dot{x}^{*2} = \dot{\bar{x}}^2$, so that $\dot{x}^2 = \dot{\bar{x}}^2$. Since $\dot{x}$, $\dot{\bar{x}}$ point in opposite directions, it follows that $\dot{x} + \dot{\bar{x}} = 0$, $(x + \bar{x})/2 = \text{m}$, so that the oval curve of constant breadth has a fixed center m which bisects the chords joining points of contact of parallel tangents. By (1) these chords are perpendicular to the parallel tangents, which completes the proof that the curve is a circle.

3.   The isoperimetric problem.[22] A particularly significant property of the circle is that among all curves of equal perimeter it encloses the greatest

[22]See [1], p. 38 ff., for historical information and references.

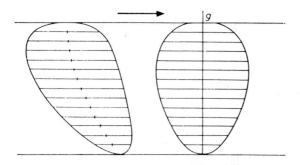

Fig. 9

area; or worded differently, among all curves enclosing the same area, the circle has the smallest perimeter.

The argument given by Steiner,[23] although incomplete because of the lack of existence proofs, is nevertheless extremely interesting, since it involves only very elementary ideas. Let us give an indication of it here (Fig. 9). If the oval is approximated by trapezoids which then undergo a parallel shift toward a line g, perpendicular to the direction of the shift, in such a way that g becomes the axis of symmetry of the region (Steiner symmetrization), the area will remain constant but the perimeter will become smaller (or remain constant if the region was already symmetric in the given direction), as may be seen from the individual trapezoids. If this symmetrization is carried out in every direction, all ovals, except the circle, will decrease in circumference while the area remains constant. Thus the circle is a region of smallest circumference with given area, if such an extreme region exists at all.

Let us now give a proof due to Erhard Schmidt, which is more general in that nonconvex curves are also allowed, but less general in requiring certain differentiability assumptions.

Let the closed (simple, i.e., without double points) curve $\mathfrak{x} = \mathfrak{x}(s)$ of length U (here s is arclength) be enclosed between two parallel tangents (Fig. 10) of arbitrary direction, at a distance $2r$ apart, and let us also draw a circle $\bar{\mathfrak{x}}(s)$ between these tangents, where $\bar{x}(s) = x(s)$ as in the figure (if $\mathfrak{x}(s)$ is not convex, parts of the circle may be traversed more than once, as s runs from 0 to U). Then, for a y-axis parallel to the tangents, the areas of the two regions can be written

(1)
$$F(\mathfrak{x}) = -\int_0^U xy' \, ds,$$

(2)
$$F(\bar{\mathfrak{x}}) = \int_0^U \bar{y}\bar{x}' \, ds = r^2\pi$$

[23]J. Steiner: Einfache Beweise der isoperimetrischen Hauptsätze. Journ. reine angew. Math. 18, 1838. Collected works 2, 75–91.

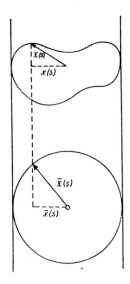

Fig. 10

and, by adding,

(3) $$F(x) + \pi r^2 = \int_0^U (\bar{y}\bar{x}' - xy')ds$$

or since $x = \bar{x}$,

(4) $$F(x) + \pi r^2 = \int_0^U (\bar{y}x' - \bar{x}y')ds,$$

an expression which can be estimated from below and from above.

From below: since the geometric mean is not greater than the arithmetic,

(5) $$2r\sqrt{\pi F} \leq F(x) + \pi r^2;$$

And from above: since the integrand is the scalar product of $\bar{x}^* = \{\bar{y}, -\bar{x}\}$ and $x' = \{x', y'\}$ with $|\bar{x}^*| = r, |x'| = 1$,

(6) $$\int_0^U (\bar{y}x' - \bar{x}y')ds \leq rU.$$

Thus we have derived the isoperimetric inequality·

(7) $$4\pi F \leq U^2.$$

We must now see when the equality holds. In the first inequality (5), it will hold for $r = \sqrt{F/\pi}$, which means, since F/π is independent of the direction of the parallel strip, that $x(s)$ is a region of constant breadth $2r$. In the second inequality (6) the equality holds if x' is parallel to $\bar{x}^*$. But $\bar{x}^*$, being perpendicular to $\bar{x}$, is tangent to the circle, which means that the two curves

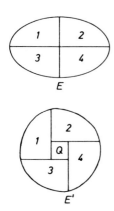

Fig. 11

differ only by a shift in the direction of the y-axis. Since $\bar{x}$ is a circle, the proof is thereby complete.

The isoperimetric inequality is used to prove many estimates concerning area or circumference; for example:

If a and b $(a > b)$ are the semidiameters of an ellipse and U its circumference, then $U > 2\pi(a + b)/2 = \pi(a + b)$.

Proof.[24] Let the ellipse E be slit along the axes and put together again as in Fig. 11 to form the oval curve E'. Then

$$U_E = U_{E'},$$
$$F_{E'} = F_E + F_Q = \pi ab + (a - b)^2.$$

But for E' (which is not a circle) we see from (7) that

$$U_E^2 > 4\pi F_{E'} = 4\pi^2 ab + 4\pi(a - b)^2 > 4\pi^2 ab + \pi^2(a - b)^2 = \pi^2(a + b)^2,$$

i.e., $U > \pi(a + b)$, as desired.

6. Characteristic Properties of the Ellipse

1. In analogy to the Fujiwara-Kubota theorem for a circle in §5 we now prove the following characteristic property of an ellipse.[25]

An oval curve that has not more than four points in common with any other oval curve affinely related to it is an ellipse.

Since the given oval curve E cannot have a line segment on its boundary (for otherwise it would have more than four points in common with a

---

[24]H. Hadwiger, Zur Schätzung des Ellipsenumfangs. Elem. Math. 4, pp. 11 and 12 (1949). The proof given above is due to G. D. Chakerian, On estimating the perimeter of an ellipse, Elem. Math. 20, p. 89 (1965).

[25]W. Süss: Eine charakteristische Eigenschaft der Ellipse. Math. -phys. Semesterber. 4, 54–56 (1955).

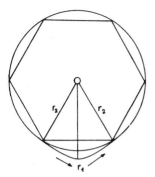

Fig. 12

translation along this line segment), its center of gravity S (by the first theorem in §4) bisects at least three chords AA', BB', and CC'.

Moreover, S is also a center of E, since rotation about S through $180°$ interchanges the endpoints of the three chords, so that the rotated figure E' has at least six points in common with E and therefore, by assumption, coincides with E. So S bisects all chords, and is thus a center. We now let E be mapped by a suitable affine transformation into an oval curve E^+ with an inscribed regular hexagon centered on S (§2, Theorem 3).

Since rotation about S through $60°$ permutes the vertices of the hexagon cyclically, the rotated figure again has six points in common with E^+ and must therefore coincide with it.

We now prove that E^+ must be a circle, which means that the original oval curve E is an ellipse.

Let K be the circle with the same regular inscribed hexagon as E^+, so that K and E^+ coincide at six points, and assume that the radius vector r_1 issuing from S, along an angle bisector from the center of the hexagon, is greater than the radius of the circle r_2 (Fig. 12). Then we also consider an oval curve $\bar{E}$ obtained from E^+ by rotation about S through $30°$. After the rotation, the radius vector increases from r_2 to r_1 in the sector $30° \leq \alpha \leq 60°$, whereas before the rotation it decreased from r_1 to r_2. Thus the two oval curves must intersect at least once in the sector $30° \leq \alpha \leq 60°$, so that altogether they must intersect more than four times and must therefore coincide, by hypothesis; i.e., $r_1 = r_2 = r_0$. Consequently K and E^+ coincide at all the vertices of a regular dodecagon, and thus, by continual bisection, at an everywhere dense set of vertices. But then, by continuity, the oval curve E^+ must coincide altogether with the circle K, so that E is an ellipse.

2. *An oval curve is an ellipse if all its midlines* (loci of midpoints of parallel chords) *are straight lines.*[26]

[26]K. H. Berger: Eilinien mit perspektiv liegenden Tangenten- und Sehnendreiecken. Sitz. -Ber. Heidelb. Akad. d. Wiss. math. -naturw. Kl. 1936. 4. Abh. (especially Section 6). M. Kneser: Eibereiche mit geraden Schwerlinien. Math. -phys. Semesterber. 1, 97 and 98 (1950).

Proof.   An oval curve with the given property (only such curves are considered below) is *affine-symmetric,* where the only difference from ordinary axial symmetry is that the family of parallel lines, i.e., of the lines determining the reflection, is not necessarily orthogonal to the axis. The support lines at the endpoints of a curve of gravity are parallel to the chords determining it, all straight midlines go through a common point, namely the center of gravity S, and the oval curve E is central-symmetric with respect to this point.

Thus the oval curve E has no corners and no line segments in its boundary, since otherwise, by the affine symmetry, every point on the boundary would be a corner, and every part of the boundary in a corresponding parallel strip would be a straight line. The line segment of gravity s_1 determined by the midpoints of an arbitrary family of parallel chords is the midparallel in any parallelogram constructed from two chords of equal length in the family, and the other pair of sides of the parallelogram belong to a second family of parallel chords, whose midline s_2 belongs, in turn, to the first family of chords. Two midlines associated with each other in this way will be called *conjugate* (in analogy with the conjugate diameters of an ellipse). They bisect each other at the center of gravity S (which is also the center) of the oval region. The lines of support at the endpoints of a conjugate pair of midlines form a parallelogram with s_1 and s_2 as midparallels and S as center. On the other hand, the pair of midpoint lines s_1, s_2 are conjugate diameters of a uniquely determined ellipse K inscribed in the same support-line parallelogram as E and with the same points of support. All midlines of K are straight, and like those of E, pass through the common midpoint S.

We now give an indirect proof that E is identical with the ellipse K determined in this way (Fig. 13).

The two ellipses E and K have in common the conjugate pair of midlines s_1 and s_2, the pairs of corresponding endpoints P, P' and Q, Q', and the four support lines $t(P)$, $t(P')$, $t(Q)$, and $t(Q')$ at these points.

If K and E have a finite arc in common, they must coincide altogether, since this common arc could be extended indefinitely by suitable reflection.

We now assume that E and K are not identical with each other. Then starting from the point P, the curves E and K must first separate from each other, but then must certainly come together again at a point C (and its mirror image C' in s_1), the existence of which is guaranteed by the fact that, at the latest, E and K meet again at the endpoints Q, Q' of s_2. (At C and C', the two curves E and K also have a common support line.) Then inside the parallel strip $t(P)$ and CC' it follows from the affine symmetry about s_1 that either the corresponding segment of K lies inside of E or vice versa. So except for the three points C, P, and C' (where they meet and have the same lines of support) the two ellipses E and K have no further points in

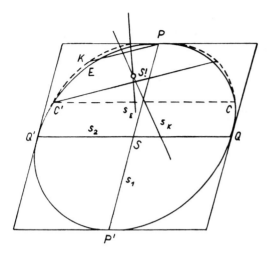

Fig. 13

common in the strip. If starting from C' we draw a chord meeting K and E in the arcs between C and P, and then from P we draw a second chord, parallel to the first one, meeting K and E in the arcs between P and C', the midpoints of these chords will determine a midline with respect to K and E, which must go through S (since it is straight). But we now see, in view of the intertwined order of corresponding midpoints of the curves, that the midlines intersect in the strip formed by the two parallel chords. However, this point cannot be S, since S lies on QQ' and is therefore outside the strip formed by the chords. Thus assumption that the oval curve E is distinct from the ellipse K leads to a contradiction, so that E is necessarily an ellipse.

The assumption that the two triangles formed by the tangents and the corresponding chords are in perspective also leads to the conclusion that the midlines are straight. Thus we have the following theorem.

Theorem:[27] *An oval curve is an ellipse if any two triangles, the one formed by tangents and the other by chords joining the points of contact, are in perspective.*

Bibliography

[1] BLASCHKE, W.: Kreis und Kugel. Leipzig 1916, 2nd ed., Berlin 1956.
[2] BONNESEN, T. und FENCHEL, W.: Theorie der konvexen Körper. Ergebnisse

[27]K. H. Berger, *loc. cit.* 18. T. Kubota: Kennzeichnende Eigenschaften der Ellipse und des Ellipsoids. The Tôhoku Mathematical Journal. Vol. 43, 392 and 393 (1937).

der Math. Bd. 3, Heft 1. 164 S. Berlin 1934. Zusammenfassende Darstellung der Theorie und ihrer Ergebnisse bis 1934. Umfassende Literaturübersicht.

[3] HADWIGER, H.: Altes und Neues über konvexe Körper. 116 S. Basel. Stuttgart 1955. Eine kurze, eingäangige Einführung.

[4] YAGLOM, I. M. and BOLTYANSKY, V. G.: Convex figures. Berlin 1956. Translated form the Russian.

[5] TOTH, L. FEJES: Lagerungen in der Ebene, auf der Kugel und im Raum. Berlin-Göttingen-Heidelberg 1953.

[6] EGGLESTON, H. G.: Convexity. Cambridge 1958.

Aspects of Topology

Introduction

The development of mathematics in recent decades has shown that one of its great supporting pillars is topology. In every individual subject one looks for the topological and algebraic structures involved, since these structures form a unifying core for the most varied branches of mathematics. Thus some account of topology must be given here, although lack of space precludes any systematic or thoroughgoing treatment of the subject.

What is topology? A first answer to this question can be found in geometry. In a space of three or more dimensions, let us consider a one-to-one bicontinuous mapping of a geometric figure (defined as a set of points) and ask for those properties of the figure that remain unchanged under the mapping. The continuity of the mapping can be taken into account most simply by referring the space to a Cartesian coordinate system and describing the mapping by continuous functions of the coordinates, which need to be defined only for the points of the figure. If we imagine the figures, or surfaces, as made of a material that can be stretched but not torn, we may visualize them as being deformed arbitrarily. One-to-one bicontinuous mappings of this kind are said to be *topological,* and two figures that can be mapped onto each other by a topological mapping are *homeomorphic.*

Here one is tempted to think of Klein's Erlanger Program and to regard topology as the theory of invariance under topological transformations, but in topology we are dealing, in general, not with mappings of the whole space, which could then be interpreted as transformations, but only with mappings of imbedded figures. For example, the knotted cloverleaf (Fig. 1) is homeomorphic to the circumference of a circle, since we need only refer both of these closed curves to their arclengths as parameter, and then map them

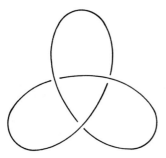

Fig. 1

linearly onto each other. But there is no topological mapping of the whole space under which a cloverleaf loop of this sort is taken into the circumference of a circle. This example shows that although visualizable deformations produce topological mappings, they do not include all such mappings. But the knottedness of the cloverleaf is a topological property only when it is imbedded in three-dimensional space. If we adjoin a fourth coordinate, the loop can be taken into the circumference of a circle by a suitable topological transformation of the whole four-dimensional space (cf. Seifert-Threlfall [6], p. 315). Figures homeomorphic to each other are considered in topology as not essentially distinct, but the analogy with congruence in elementary geometry is not complete. The latter case refers to motions of the whole space taking congruent figures into one another, but topological mappings of homeomorphic figures concern only points of the figure itself. Nevertheless, it may happen, as in the example of the cloverleaf, that topological mappings of the whole space are of interest, and in order to make the distinction, we give the name "intrinsic topological properties" or "topological properties of shape" to such properties as are common to all homeomorphic figures, whereas "topological properties of position" refer to the way in which the figure, imbedded in the space, behaves under topological mappings of the whole space.

But these remarks describe topology only to a certain extent. The first figures to be considered topologically were polyhedra (see Euler's formula for polyhedra, II8, §5), for which it is characteristic that they can be constructed from triangles related to one another in a very definite way. For example, an octahedron and a cube can easily be divided into triangles (triangulated) in such a way that the subdivisions are isomorphic to each other; i.e., there is a one-to-one correspondence between their triangles (Fig. 2) such that triangles with common vertex (common edge) go into triangles with common vertex (common edge). (For the concept of isomorphism see also II8, §4.1.) Since any two corresponding triangles can be mapped affinely onto each other in such a way that the images of common edges coincide, surfaces with isomorphic triangulations are always homeo-

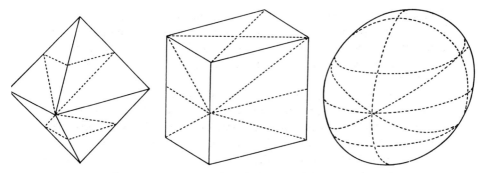

Fig. 2

morphic, under a topological mapping composed of the affine mappings of the triangles. Furthermore, if these triangulations, first of the octahedron and then of the cube, are projected onto the circumscribed sphere from its center, so as to produce curved triangles on its surface, the same scheme of triangles will be obtained in each of the two cases, together with a topological mapping of the sphere onto the octahedron and cube, respectively. These examples illustrate the importance in topology of sets of points admitting subdivision into triangles or, more generally, into simplexes, i.e., points, line segments, triangles, tetrahedra, etc., where the simplexes (straight or curved, and of various dimensions) are joined to one another, if at all, only along common sides of arbitrary dimension. The study of these subdivisions is then based solely on the corresponding framework of simplexes, with various boundary properties. The simplexes are now regarded as new elements forming a so-called complex, and the topology of complexes, the so-called "combinatorial topology," makes no further use of arguments from continuity, but only of algebraic methods. In our first section we shall introduce the most important concepts of combinatorial topology by discussing the relatively simple topology of surfaces.

Let us return to sets of points in Euclidean space. In geometry we are not interested in arbitrary sets but only those with certain properties. For example, everyone has an intuitive idea of a "curve," but we must make a very careful analysis of its set-theoretic topological foundations before we can arrive at an unobjectionable definition. The Jordan curve theorem is intuitively obvious, namely that a plane topological image of the circumference of a circle divides the plane into precisely two regions, for which it forms the common boundary; but L. E. J. Brouwer discovered that in the plane it is also possible to have *three* simply-connected regions, all of them with the same boundary. Such a "boundary curve" is therefore a set of points with a different topological structure from the circumference of the circle, and the question, which sets of points are to be considered as curves, is by no means as trivial as it may seem at first sight. The second section of the chapter will deal with the basic questions of the topology of curves. More generally, the aim of this type of topology, namely the geometry of continuity, is to determine which sets of points are to be considered as geometric figures (including curves), and to investigate them by set-theoretic and combinatorial methods.

In one respect the above discussion of the nature of topology departs altogether from the field of geometry. The study of simplexes as elements of a complex leads to an abstract interpretation, in which these elements, freed from their geometric origin, are merely symbols with no concrete meaning, to be manipulated in a definite way, as we shall see in detail in the first section, so that combinatorial topology can be dealt with axiomatically. Again, the point of view we have taken up to now in set-theoretic topology, namely that it is the study of one-to-one bicontinuous mappings of sets of points in Euclidean space, is unnecessarily special. One can start from arbitrary abstract elements called "points," and then define a topological space by means of axioms. For example, since there are no coordinates in these abstract spaces, continuity can no longer be described as before, and we must define axiomatically what is meant by a "neighborhood of a point." Our third section will deal with the construction of such spaces. As a result of the fact that modern topology takes no account of the individual nature of the elements, but only of their mutual relationships, the concepts of general topology have become a basic part not only of geometry but also of analysis.

Topology of Surfaces

1. Deformation of Surfaces

Among all topological mappings, i.e., one-to-one bicontinuous map-

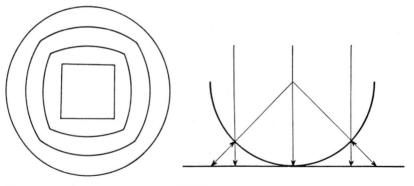

Fig. 3   Fig. 4

pings, the deformations are particularly easy to visualize, where by a deformation we mean a continuous passage from one geometric figure to another, i.e., more precisely, a one-parameter family of topological mappings, depending continuously on the parameter, that takes the one figure into the other. For example, a circle can be deformed into a square by a family of projections (Fig. 3). Such bendings and distortions can very often be described by suitable projections, which provide many examples of homeomorphic, i.e., topologically equivalent, figures. Arbitrary plane closed polygons without multiple points can obviously be deformed into the circumference of a circle, and convex polyhedra into the surface of a sphere. But it is much more difficult, in general, to show that two given figures are not homeomorphic, since deformations are only part of all possible topological mappings. Let us give a few easily visualized examples.

a) The Euclidean plane can be mapped topologically, by central projection, onto the open hemisphere, i.e., a hemisphere without its boundary points, and then the hemisphere can be projected orthogonally onto the open circular disk (Fig. 4).

b) The surface of a sphere from which one point has been removed is homeomorphic to the open circular disk. For by stereographic projection from the excluded point onto a tangential plane the so-called deleted sphere is mapped topologically onto the Euclidean plane (Fig. 5), whereupon it follows from a) (since the composition of two topological mappings is again a topological mapping) that the deleted surface of the sphere is homeomorphic to the open circular disk.

The question whether or not two more complicated surfaces are homeomorphic is easier to decide if by means of cuts they are brought into a normal form, namely the Poincaré fundamental polygon, where it is necessary to state how the cuts are to be fastened together again. For example, an

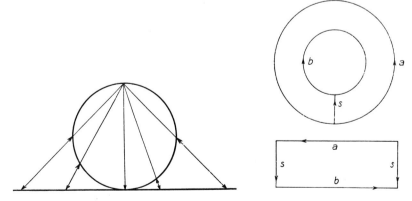

Fig. 5   Fig. 6

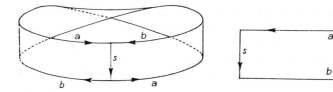

Fig. 7

annulus can be cut and then transformed into a rectangle in which two opposite sides are to be identified, i.e., pasted together (Fig. 6).

The situation can also be described by a formal schema. The sides of the fundamental polygon are oriented and named in such a way that sides to be identified are given the same names and are oriented according to the way in which they are to be fastened together (Fig. 6). If the fundamental polygon is then traversed and the names of the sides are written multiplicatively, one after the other (opposite orientation of a side being denoted by the exponent -1), the resulting formula completely describes the geometric figure. The formula for the annulus is $asbs^{-1}$, and for the Möbius strip, when described by a fundamental rectangle, the formula is $asbs$ (Fig. 7).

From the fact that the annulus and the Möbius strip have different formulas we cannot yet conclude that the surfaces bounded by them are topologically distinct, but this conclusion does follow at once from the fact that the annulus is bounded by two closed bounded curves, and the Möbius strip by only one. Closed surfaces are particularly important, but before dealing with them in general, let us continue our list of examples.

c)   In the theory of functions of a complex variable, the open plane is closed by adjunction of a point at infinity, and can then be transferred by stereographic projection to a sphere, the Riemann number sphere (cf. IB8, §3.2), which can also be described by a fundamental polygon, a polygon of two sides: aa^{-1} (Fig. 8).

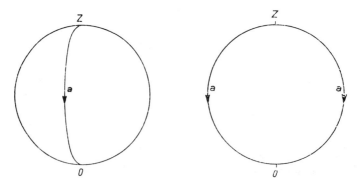

Fig. 8

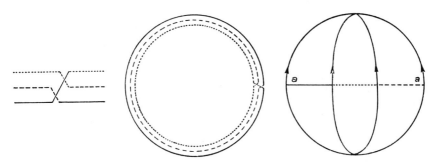

Fig. 9

For we have only to cut the sphere along a meridian and bend the two sides of the semicircle outwards about the hinges O and Z onto the plane.

d) An n-sheeted Riemann surface with two branch points and a branch cut can be mapped by the same stereographic projection onto an n-times covered sphere. The corresponding cut along a meridian then produces the fundamental polygon aa^{-1}, where one sheet of the sphere corresponds to an nth part of the circle (Fig. 9).

e) In analytic geometry the open plane is closed by the adjunction of an infinitely distant line to form the projective plane. If the plane is projected onto a hemisphere as in a), the infinitely distant point in a given direction will correspond to two boundary points of the hemisphere, which must therefore be identified (Fig. 10). A subsequent orthogonal projection then produces a circular disk with identification of diametrically opposite points,

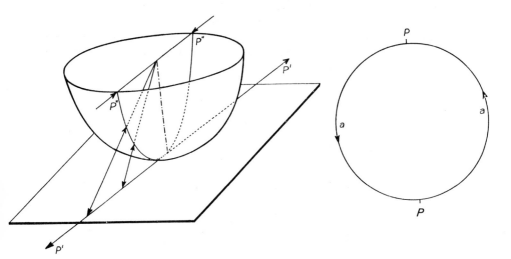

Fig. 10

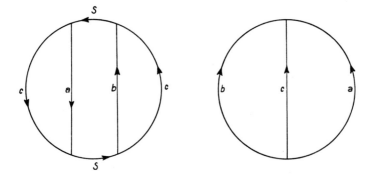

Fig. 11

i.e., with the fundamental polygon *aa*. By removing a segment from each side of the disk, we obtain a Möbius strip, and the segments can then be pasted together along *c* to form a circular disk (Fig. 11). We thus have the important result that if a circular disk and a Möbius strip are pasted together along their boundaries (in three-dimensional Euclidean space, self-intersection will be unavoidable), the result is a closed surface homeomorphic to the projective plane.

f) Identification of both pairs of opposite sides of a fundamental rectangle gives a torus for $aba^{-1}b^{-1}$, and a Klein bottle for $aba^{-1}b$ (Figs. 12 and 13).

The example of the Klein bottle shows that pasting together assigned sides in a given fundamental polygon may produce a surface that cannot be imbedded in three-dimensional Euclidean space without self-intersection and, as we have just seen, the projective plane can also be represented, in a similar but more complicated way, as a closed surface with self-intersection (cf., for example, Hilbert, Cohn-Vossen [4], §47). But these self-intersections

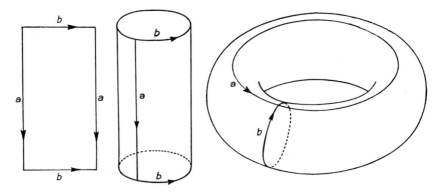

Fig. 12

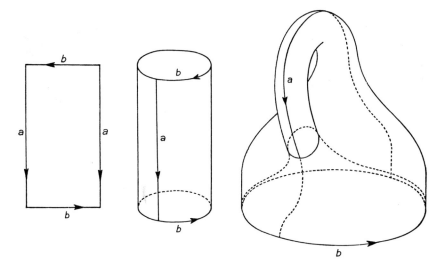

Fig. 13

are of no importance for the topological form of the surface, since we require only a topological image of the fundamental polygon; in three-dimensional space the uniqueness of the mapping can be ensured by counting the points of intersection doubly in a suitable way, and in a four-dimensional space the Klein bottle or the projective plane can be represented without self-intersection. As is clear from its fundamental polygon, the Klein bottle can be divided by a cut into two Möbius strips (in Fig. 14 the second strip is formed by pasting together, along *a*, the triangles that have been cut off). By cutting and subsequent pasting along assigned edges, any fundamental polygon can be put into a different form, e.g., Fig. 15 shows how we can obtain another formula for the Klein bottle, namely *aamm*.

g)   An infinite sequence of closed surfaces of distinct type can be created very simply by cutting circular holes in the surface of a sphere and closing them again by segments of surfaces bounded by a topological circle, the simplest examples of such surfaces being the "handle" and the "cross-cap."

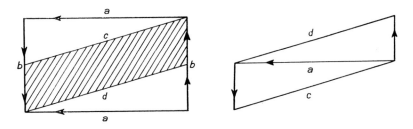

Fig. 14

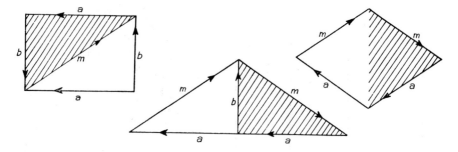

Fig. 15

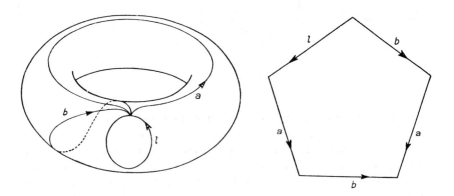

Fig. 16

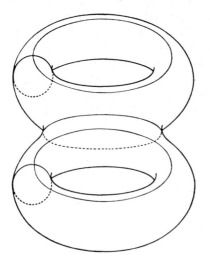

Fig. 17

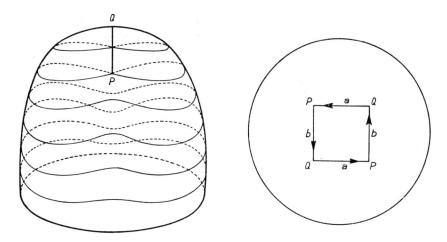

Fig. 18

By a handle we mean the surface of a torus from which a circle has been removed (Fig. 16), so that by cutting a handle along the curves a and b we obtain the fundamental polygon $laba^{-1}b^{-1}$. A sphere with two attached handles is a double ring surface (Fig. 17), and as for the cross-cap, we have already become acquainted with it under the name "Möbius strip." The new name arises from a different representation of it in three-dimensional space with self-intersection (Fig. 18). By cutting along the curve PQ of self-intersection and distorting the surface in a suitable way, we obtain a circular disk with a square hole, with edges assigned to each other as in the diagram. To prove its equivalence with the Möbius strip, we cut this circular disk apart as in Fig. 19 and fasten the parts together along the edges ab.

So in addition to deformations, which can be visualized at once, we have a second method of constructing homeomorphic figures, namely, by cutting a given figure into a polygonal surface and then identifying, or pasting together, certain edges. We now ask what topological classes of surfaces can be obtained in this way. To begin with, difference of the fundamental polygons is no proof of difference of topological class, since fundamental polygons are not uniquely determined. Again, the surface of the torus and the Klein bottle, for example, differ in their position in three-dimensional space; the torus is two-sided, and the Klein bottle one-sided. A bug crawling on the Klein bottle will be able to reach the same point in two antipodal positions, which cannot be done on the two-sided surface of the torus. But this distinction is still not an intrinsic topological property. However, if a small oriented circle is moved around on the Klein bottle, it can be brought into coincidence with itself with the opposite orientation (Fig. 20), whereas the torus does not have this property. A surface on which the orientation of

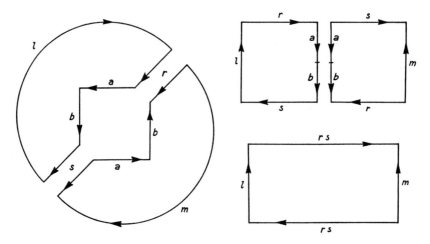

Fig. 19

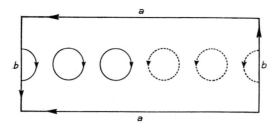

Fig. 20

the neighborhood of a point, as given by a small oriented circle, can be uniquely transferred to every other point by moving the circle is said to be orientable (cf. II8, §4.2). Thus the Klein bottle, the Möbius strip, and therefore the projective plane are *nonorientable*. Since a topological mapping, because of its continuity, always maps a neighborhood of a point onto a neighborhood of the image point, the property of orientability is also transferred, and is therefore a topological invariant. The orientable torus and the nonorientable Klein bottle are not homeomorphic.

2. Two-Dimensional Closed Manifolds

The most interesting of all connected point sets are those in which, from the topological point of view, no point is preferred to any other point. On the sphere and the torus, and in the Euclidean plane, every point has a neighborhood that is the topological image of an open circular disk. The projective plane is also homogeneous in this sense, although here our powers of visualization are not of much help. To get a clear view of its behavior in the neighborhood of improper points we may, for example, consider the topologically equivalent model of the cross-cap closed by a circular disk, but since the endpoints of its curve of self-intersection are singular points on the surface, it is better to go back to the fundamental polygon, from which the equivalence of all the points is immediately evident. In higher dimensions the circular disk is replaced by the full sphere, which has no boundary points.

Certain other sets of elements are also amenable to topological treatment. For example, the set of the lines tangent to a sphere, or the various positions of a mechanical system, or its possible states of motion, have various neighborhood properties that allow us to compare them with the points of an n-dimensional number space and its neighborhoods. The concept of an abstract "topological space" with arbitrary elements, and of its continuous mappings, will be the subject of our third section.

The sets of points such that each point has a neighborhood which is the topological image of the interior of an n-dimensional number-sphere are called n-dimensional manifolds; they are a generalization of the topologically homogeneous surfaces without boundary, some of which we have discussed in §1. In the present state of topology, we cannot decide in every case whether a given set of points is such a manifold or not; so we shall restrict the definition of a manifold by introducing the additional assumption, as in our examples, that the set of points can be divided into subsets which are topological images of n-dimensional polyhedra. Then we can make use of combinatorial methods, some of which are developed below. But even then we will not be able, in more than two dimensions, to classify all the possible topological figures. Only for two-dimensional manifolds can we solve the homeomorphism problem, i.e., decide whether two given manifolds are homeomorphic or not.

Manifolds are called *closed* or *open* according to whether they are made up of finitely or infinitely many polyhedra, so that the name "open circular disk" is justified, since the set of points not including the circumference can only be subdivided into infinitely many polygons. To obtain such a subdivision, it is most convenient to begin by dividing the topologically equivalent Euclidean plane into infinitely many squares, for example, and then to transfer the subdivision onto the circular disk by the projection of §1,a).

The closed two-dimensional manifolds, on the other hand, can be constructed, by definition, from finitely many polygons, which can be chosen arbitrarily under the sole restriction that the total number of sides is even. Any two sides that we wish to regard as edges of a cut (and shall therefore denote by the same letter) we may first map topologically onto each other and then fasten together in the direction determined by their orientation. This pairing of sides of polygons, and also the orientation of the sides, can be chosen arbitrarily, provided care is taken that, starting from any interior

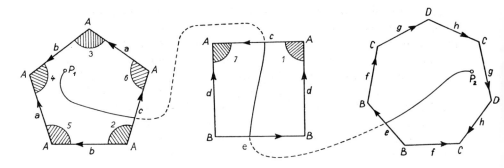

Fig. 21

point of any polygon and proceeding along a path across sides of cuts that
have been fastened together, we can arrive at any interior point of any other
polygon, since otherwise the manifold would not be connected. An example
with three polygons, eight oriented pairs of sides, and a path joining two
interior points is given in Fig. 21, where the correspondences between the
sides makes it clear which vertices are equivalent and are therefore to be
identified. Every interior point of a polygon has a circular neighborhood,
and also every interior point of a side of a polygon, since here each side is
fastened to one other side. But the vertices of the polygons also remain
undistinguished in any way, since for a set of vertices identified with one
another, the corresponding sectors of circles combine to form a full topological circle, as shown in Figs. 21 and 22, for the point A.

We now see why the passage to higher-dimensional manifolds involves
certain difficulties. The polyhedra can no longer be chosen in such an
arbitrary way and the sectors can no longer necessarily be combined to form
full topological spheres.

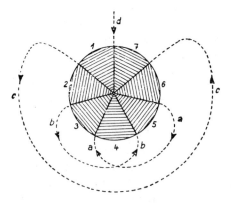

Fig. 22

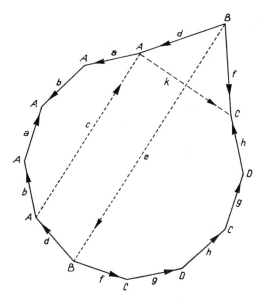

Fig. 23

The system of polygons given in Fig. 21 can be simplified. In the first place, the connectivity allows us to combine all of them into a single polygon (cf. Fig. 23), and further simplifications can be made by introducing new lines of division and fastening the partial segments together again at suitable places. For example, we may decrease the number of equivalent vertices B by cutting the polygon from a point A adjacent to B along a new side k to the other adjacent point C (Fig. 23) and fastening the part cut off (i.e., including the point B) to the side d (Fig. 24). The new polygon contains B only once as vertex and if we pull the sides F together we obtain a simpler fundamental polygon (Fig. 25). By repeated application of this procedure we can remove all the vertices of the polygon not equivalent to A. The result in our present case is shown in Fig. 26. An individual polygon is best described, as before, by a combinatorial schema in which the sides are written multiplicatively one after another with exponents ± 1 depending on whether the side is oriented counterclockwise or clockwise. Our original schema was $caba^{-1}b^{-1}$, $cd^{-1}ed$, $efghghf^{-1}$ (Fig. 21) and the result of our changes up to now is $llbab^{-1}a^{-1}$ (Fig. 26). The steps in the reduction can also be carried out according to purely formal laws involving no visualization.

Comparison with examples in §1 shows that the manifold in Fig. 21 consists of a handle, a sphere with two holes, and a cross-cap. In our preliminary normal form for the fundamental polygon, (with equivalent vertices only) a handle is characterized by two "crossed" pairs of sides $bab^{-1}a^{-1}$ and a cross-cap by a repeated exponent ll (Fig. 26). The presence of a cross-cap

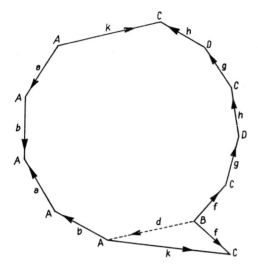

Fig. 24

always implies nonorientability of the manifold (Fig. 26). If the two crossed pairs or the pair of cross-caps are separated by other sides, they can always be brought together again by suitable cuts and fastenings. Thus we obtain three types of fundamental polygons:

α) With p handles and no cross-caps, so that the manifold is orientable. The simplest example is the torus with $p = 1$,

$$(1) \qquad a_1 b_1 a_1^{-1} b_1^{-1} a_2 b_2 a_2^{-1} b_2^{-1} \cdots a_p b_p a_p^{-1} b_p^{-1}.$$

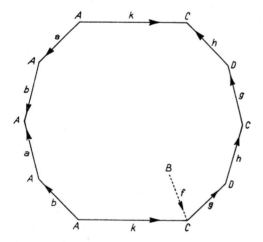

Fig. 25

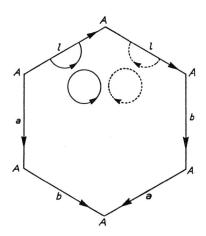

Fig. 26

β) With q cross-caps and no handles, so that the manifold is not orientable. The simplest example is the projective plane with $q = 1$,

(2) $a_1 a_1 a_2 a_2 \cdots a_q a_q.$

γ) With both handles and cross-caps. The manifold is not orientable and can be reduced, as shown above, to case β) if each of the handles is replaced by two cross-caps (Fig. 27):

$$aba^{-1}b^{-1}l^{-1}l^{-1} \rightarrow nnmmp^{-1}p^{-1}.$$

So (1) and (2) may be considered as the final standard forms. The topological invariance of orientability shows that manifolds of type (1) are not homeomorphic to those of type (2), but we have still left open the question whether the numbers p and q have topological significance; in other words, we have not yet completed our enumeration of all closed two-dimensional manifolds under α) and β). If in a polygonal schema the sequence . . . aa^{-1} . . . occurs, it can be eliminated by pulling the sides together, as is shown by Figs. 24 and 25 (with f^{-1} in place of a); only the polygon aa^{-1} itself cannot be further simplified. Thus in addition to the ones given above, there remains the standard form for the surface of the sphere (Fig. 8):

(3) $aa^{-1},$

which is the only standard fundamental polygon without equivalent vertices.

3. Complexes and Polyhedra

We have referred in the introduction to the importance of geometric figures that can be subdivided into particularly simple components, i.e.,

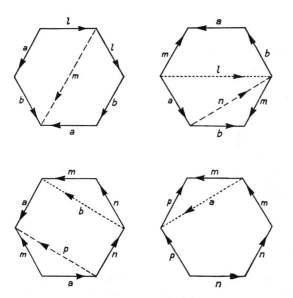

Fig. 27

points, line segments, triangles, tetrahedra, etc., and their topological images. These components are called "simplexes." An n-dimensional simplex in m-dimensional Euclidean space $(0 \leq n \leq m)$ is determined by $(n + 1)$ vertices not lying on an $(n - 1)$-dimensional hyperplane. If arbitrary nonnegative masses are placed at the vertices, their center of gravity is always a point of the simplex, and all points of the simplex can be obtained in this way. So to describe the simplex it is enough to write down the vertices in some order: $P_1 P_2 \cdots P_{n+1}$. Every permutation of the vertices produces the same simplex, but the permutations fall into two classes, even and odd, so that the simplex can be oriented in two ways by its sequence of vertices. An *oriented* simplex is defined by the signed product of its vertices, where even permutations do not change the sign. For example,

$$+ P_1 P_2 P_3 P_4 = - P_2 P_1 P_3 P_4 = + P_3 P_2 P_4 P_1 = + P_4 P_3 P_2 P_1.$$

Every set of $(r + 1)$ vertices of an n-dimensional simplex $(0 \leq r \leq n)$ defines a simplex, called an r-dimensional side of the given simplex. The concept "side" is broader here than in ordinary language, since, for example, the simplex is a side of itself $(r = n)$, and its vertices are zero-dimensional sides $(r = 0)$. A tetrahedron has 1 three-dimensional side, 4 two-dimensional sides, 6 one-dimensional sides, and 4 zero-dimensional sides.

We now consider a finite or at most *countably infinite* set K of simplexes satisfying the following conditions.

1. Every side of a simplex of the set K belongs to the set K.

2. Any two simplexes of the set K have a side in common or else an empty intersection.

3. Every zero-dimensional simplex in K is a side of only finitely many simplexes.

These properties of the set K have been chosen on the ground that basically we are dealing here with decompositions of sets of points. Thus the second condition guarantees that two simplexes link up only along sides without penetrating each other, although one of them may be a side of the other, in view of the extension of the concept of a side. A set K of this sort is called a *complex* or, more precisely, a *simplicial complex*. (Complexes with other components will not be considered here.) Since a complex and its sides can be described solely as a product of its vertices, and since in the conditions for a complex only simplexes occur and not the points of the Euclidean space, we can now adopt an essentially more general, abstract point of view.

We start from a finite or countably infinite set of elements of any kind, denoted by $x_1, x_2, \ldots$ and called "vertices." What other significance these elements may have, whether, for example, they are points in any space, is of no importance. We shall only demand of them that certain subsets go together to form frameworks that we shall call "simplexes" and shall write as products of the vertices occurring in them. The number of vertices, decreased by one, is called the "dimension" of the simplex. As the symbol for a simplex we shall write σ^n, where the superscript denotes the number of dimensions

$$\sigma^n = x_{i_1} x_{i_2} \cdots x_{i_{n+1}}.$$

Every subset of such a framework is itself a framework, and is called a "side" of the given simplex. If the order of the elements is unimportant, we speak of an "absolute simplex," but we can also "orient" the simplexes as above, by considering two orders as equivalent if each is an even permutation of the other. Thus an absolute simplex can be oriented in two ways, which we shall distinguish by plus and minus signs. For example, the absolute simplex $x_1 x_2$ has the two orientations $+x_1 x_2 = -x_2 x_1$ and $-x_1 x_2 = +x_2 x_1$. A set of absolute simplexes satisfying the three conditions for a complex is called an "absolute complex." Similarly, a set of oriented simplexes is called an "oriented complex" if the corresponding absolute simplexes form an absolute complex and each simplex occurs in only one of its two orientations.

Let us now return from abstract simplexes to the sets of points of n-dimensional Euclidean space. The totality of points contained in a finite complex of Euclidean simplexes is called a *polyhedron,* a definition which provides a precise basis for the geometrically intuitive concept of a polyhedron as we have used it above. Here it is not necessary that the so-called principal simplexes, namely, those that are not sides of simplexes of higher

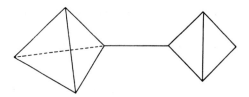

Fig. 29

dimension, all have the same dimension. Figure 28 shows a polyhedron whose complex contains a tetrahedron, two triangles, and a line segment as its principal simplexes.

For infinite simplexes we encounter a difficulty best explained by an example. If the half-open interval $0 < x \leq 1$ is divided into an infinite number of one-dimensional simplexes at the points P_n with $x = 1/n$, $n = 1, 2, 3, \ldots$, then by adjoining the point P_0 with $x = 0$ we obtain an infinite complex with the principal simplexes

$$P_0, P_1P_2, P_2P_3, P_3P_4, \ldots, P_nP_{n+1}, \ldots.$$

But this schema is also obtained by dividing, at the points Q_n with $x = n$, $n = 0, 1, 2, \ldots$, the set of points (Fig. 29) consisting of $x = 0$ and the halfline $1 \leq x < \infty$, since in both cases the complex contains one zero-dimensional principal simplex and infinitely many one-dimensional principal simplexes. Now in the first case the origin completes the half-open unit interval to form the closed unit interval, whereas in the second case Q_0 is isolated. Thus the two sets of points are not homeomorphic, although they correspond to the same complex. So it is clearly inappropriate to regard the closed interval $0 \leq x \leq 1$ both as a polyhedron with one one-dimensional simplex and, as above, as a set of points corresponding to an infinite complex. Thus we restrict the possible relationship between sets of points and their complexes by admitting only such complexes as are *locally finite* with respect to the set of points, i.e., are such that every point of the set has a neighborhood whose points lie in only finitely many simplexes of the complex. Thus the first subdivision in Fig. 29 is excluded, since there are infinitely many Euclidean simplexes in every neighborhood of P_0.

With greater generality than before, we can now define a *polyhedron* as a

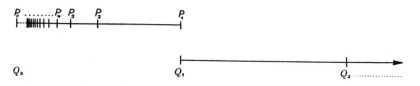

Fig. 29

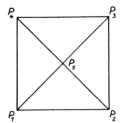
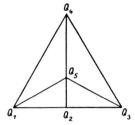

Fig. 30

set of points admitting a locally finite subdivision into a complex of Euclidean simplexes. A half-open line segment is a polyhedron with an infinite complex, and the closed interval is a polyhedron with a finite complex. Since the complex only gives the schema of the polyhedron, two different polyhedra can have the same complex (e.g., the square and the triangle in Fig. 30). If the corresponding principal simplexes are mapped affinely onto each other, these mappings can be combined in a continuous way to produce a topological mapping of the polyhedra, which are therefore homeomorphic to each other; and similarly for infinite complexes, in view of their local finiteness, although this idea will not be developed here.

Finally, the concept of a polyhedron is further generalized by means of topological mappings. Every set of points which is the topological image of a polyhedron in the above sense is also called a polyhedron. Under such a topological mapping the Euclidean simplexes are taken into so-called topological simplexes, while the corresponding complex remains unchanged as an abstract schema of subdivision. Thus all the geometric figures considered up to now are polyhedra since the fundamental polygons of the manifolds, for example, obviously admit simplicial subdivisions into complexes.

A set of points becomes a polyhedron only by admitting subdivision into a complex. But in the above transformations of the fundamental polygons of closed two-dimensional manifolds we have subdivided the same polyhedron in many different ways, by our cutting and pasting. In order to see the connection with our present discussion, we must in each case undertake a simplicial decomposition, i.e., a subdivision into triangles. But not every simplicial decomposition leads to a complex, as can be seen from the projective plane (Fig. 31), where it is true that four triangles are created by three lines not in a pencil (since diametrically opposite boundary points are identified), but none of these triangles is uniquely determined by the vertices P_1, P_2, P_3.

However, it is the task of combinatorial topology to see what topological properties of polyhedra can be read off from the structure of their complexes alone, and as a first step in this direction we now introduce a definition

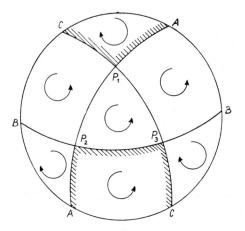

Fig. 31

of equivalence of complexes. Let there be given an absolute complex K, represented by its principal simplexes $x_{i_1} \cdots x_{i_p}$. Then by an "*elementary subdivision* $z_{j_1 j_2, j_0}$" we mean formation of a new complex K' such that every simplex containing the side $x_{j_1} x_{j_2}$ is replaced by two simplexes in which the two vertices x_{j_1} and x_{j_2} are replaced, respectively, by a new vertex x_{j_0}. For example, the complex K in Fig. 32,

$$K: \quad x_1 x_2 x_4 x_5; \quad x_2 x_3 x_4 x_5,$$

with two tetrahedra as its principal simplexes, is replaced, under the elementary subdivision $z_{24,6}$, by

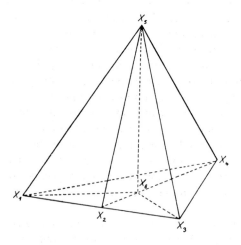

Fig. 32

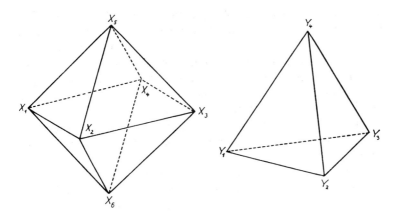

Fig. 33

$$K': \begin{matrix} x_1x_6x_4x_5, \ x_1x_2x_6x_5 \\ x_6x_3x_4x_5, \ x_2x_3x_6x_5. \end{matrix}$$

By the *"reduction $z_{j_1j_2,j_0}^{-1}$"* we mean the inverse process. But if we wish to remove a vertex x_{j_0} from a complex by reduction, certain conditions must be satisfied; x_{j_0} can occur *only* together with x_{j_1} or x_{j_2}, and for every simplex with the side $x_{j_1}x_{j_0}$ there must exist another simplex which is the same except that x_{j_1} is replaced by x_{j_2}. Moreover, $x_{j_1}x_{j_2}$ cannot occur as a side in the complex. Then reduction consists in combining two corresponding simplexes to form one simplex by elimination of the vertex x_{j_0}.

Complexes are said to be *equivalent* if they can be transformed into each other by a sequence of elementary subdivisions and reductions. As an example, let us show that the complex of the octahedron is equivalent to that of the tetrahedron (Fig. 33).

$$K: \ x_1x_2x_5, \ x_2x_3x_5, \ x_3x_4x_5, \ x_1x_4x_5, \ x_1x_2x_6, \ x_2x_3x_6, \ x_3x_4x_6, \ x_1x_4x_6,$$

$$z_{24,6}^{-1}: \ x_1x_2x_5, \ x_2x_3x_5, \ x_3x_4x_5, \ x_1x_4x_5, \ x_1x_2x_4, \ x_2x_3x_4,$$

$$z_{13,4}^{-1}: \ x_1x_2x_5, \ x_2x_3x_5, \ x_1x_3x_5, \ x_1x_2x_3,$$

$$K': \ y_1y_2y_4, \ y_2y_3y_4, \ y_1y_3y_4, \ y_1y_2y_3.$$

Renaming the vertices $x_1 = y_1$, $x_2 = y_2$, $x_3 = y_3$, $x_5 = y_4$, makes no change in the complex, and at once proves the desired equivalence.

The change made in the simplicial decomposition of a polyhedron by a partition or a reduction in its complex is a very simple one, so that the topological properties of a polyhedron indicated by the structure of its complex are invariant under elementary subdivisions and reductions; i.e., two polyhedra are homeomorphic if their complexes are equivalent. For the proof it is only necessary to set up decompositions in the two polyhedra whose absolute complexes are equal or isomorphic (and therefore differ only in the notation for their vertices) since such polyhedra can be mapped topologically onto each other by an affine mapping of their Euclidean simplexes. The converse statement, that two homeomorphic polyhedra must have equivalent complexes, has remained until now a famous conjecture, not yet proved even for finite complexes (fundamental conjecture of combinatorial topology).

But even if the relationship between combinatorial topology and the continuity topology of polyhedra has not yet been fully cleared up, study of the structure of complexes is of great importance; for although it must be admitted that, in general, we are still far from being able to decide on the equivalence or nonequivalence of two given complexes, we do know a number of invariants under elementary subdivision, the simplest among them being the homology groups, to the definition and calculation of which we now turn.

4. Homology Groups

We now start from an arbitrary simplicial complex with arbitrarily oriented simplexes (except that the vertices are always positive) denoted by σ_i^r, where r is the dimension of the simplex and i is a running index. A *chain* $\mathfrak{f}^r$ is a formal finite linear combination of simplexes of dimension r with integer coefficients

$$(4) \qquad \mathfrak{f}^r = \sum_i \lambda_i \sigma_i^r, \quad \lambda_i \text{ an integer,}$$

i.e., a finite set of simplexes of the same dimension, each simplex σ_i^r being counted with a definite multiplicity λ_i. For example, if in an oriented complex we start from a vertex and proceed to another vertex along a path consisting entirely of edges, then a one-dimensional chain will state which edges were traversed, in which direction, and how often. However, positive and negative traversals must be combined, so that edges traversed equally often in both directions occur in the chain with multiplicity zero, i.e., not at all. Generalized in a natural way, this concept leads to a definition of addition of two chains

$$(5) \qquad \mathfrak{f}_1^r + \mathfrak{f}_2^r = (\sum_i \lambda_i^{(1)} \sigma_i^r) + (\sum_i \lambda_i^{(2)} \sigma_i^r) = \sum_i (\lambda_i^{(1)} + \lambda_i^{(2)}) \sigma_i^r.$$

A particularly important example of a chain is the so-called *boundary chain* of a simplex. As suggested by our geometric model, the side opposite the vertex x_{i_1} of $\sigma_i^r = x_{i_1} x_{i_2} \cdots x_{i_{r+1}}$ is defined as the simplex $x_{i_2} \cdots x_{i_{r+1}}$, with the orientation "induced" by this order of the vertices. Each of the other sides, together with its induced orientation, is obtained in the same way from σ_i^r, after a permutation bringing the opposite vertex into first position. The sum of all such oriented sides is called the boundary chain of σ_i^r and is denoted by $\partial\sigma_i^r$

$$(6) \quad \partial\sigma_i^r = x_{i_2} \ldots x_{i_{r+1}} - x_{i_1}x_{i_3} \ldots x_{i_{r+1}} + x_{i_1}x_{i_2}x_{i_4} \ldots x_{i_{r+1}} - \cdots$$
$$+ (-1)^r x_{i_1}x_{i_2} \ldots x_{i_r}.$$

The orientation induced in this way does not necessarily correspond with the orientation arbitrarily prescribed for the complex. For example, in the complex in Fig. 34, where the arrows indicate the prescribed orientation,

$$\sigma_1^2 = x_1x_2x_3;$$
$$\sigma_1^1 = x_3x_2, \quad \sigma_2^1 = x_1x_3, \quad \sigma_3^1 = x_1x_2,$$
$$\partial\sigma_1^2 = x_2x_3 - x_1x_3 + x_1x_2$$
$$= -\sigma_1^1 - \sigma_2^1 + \sigma_3^1,$$

so that only the orientation of σ_3^1 is the same as the induced orientation, as can easily be read off from Fig. 34.

The boundary of an arbitrary chain $\mathfrak{k}^r$ is defined by the requirement that, like differentiation, the operator ∂ be permutable with summation and with integer factors

$$(7) \quad\quad\quad\quad \partial\mathfrak{k}^r = \partial(\sum_i \lambda_i\sigma_i^r) = \sum_i \lambda_i\partial\sigma_i^r.$$

Thus

$$(8) \quad\quad\quad\quad \partial(\mathfrak{k}_1^r + \mathfrak{k}_2^r) = \partial\mathfrak{k}_1^r + \partial\mathfrak{k}_2^r.$$

For *every* zero-dimensional simplex σ_i^0 we correspondingly define

$$(9) \quad\quad\quad\quad \partial\sigma_i^0 = 0.$$

A chain is said to be *closed* or to be a *cycle* if it has zero boundary, so that,

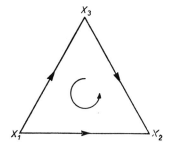

Fig. 34

in particular, zero-dimensional chains are closed. If we calculate the boundary of the boundary chain (6), we have

$$
\begin{aligned}
\partial\partial\sigma_i^r = &+ x_{i_3}x_{i_4} \cdots x_{i_{r+1}} - x_{i_2}x_{i_4} \cdots x_{i_{r+1}} + \cdots \\
&- x_{i_3}x_{i_4} \cdots x_{i_{r+1}} + x_{i_1}x_{i_4} \cdots x_{i_{r+1}} - \cdots \\
&+ x_{i_2}x_{i_4} \cdots x_{i_{r+1}} - x_{i_1}x_{i_4} \cdots x_{i_{r+1}} + \cdots \\
&\cdot \quad \cdot \quad \cdot \quad \cdot \quad \cdot \quad \cdot \quad \cdot \quad \cdot \quad \cdot \quad \cdot \quad \cdot \quad \cdot \quad \cdot \quad \cdot
\end{aligned}
$$

or

$$
\partial\partial\sigma_i^r = 0,
$$

since every $(r - 2)$-dimensional simplex occurs twice with opposite signs, and in general, from (7),

$$
(10) \qquad \partial\partial\mathfrak{t}^r = \sum_i \lambda_i \partial\partial\sigma_i^r = 0.
$$

Thus every boundary chain is a cycle, but not every cycle is a boundary chain. For example, in the annulus complex in Fig. 35 if $\mathfrak{t}^2$ is the chain of all σ_i^2, each counted once positively with the indicated orientation, then

$$
\begin{aligned}
\partial\mathfrak{t}^2 = &\ x_1x_4 + x_4x_3 + x_3x_2 + x_2x_1 \\
&+ x_5x_6 + x_6x_7 + x_7x_8 + x_8x_5.
\end{aligned}
$$

The right side falls into the sum of two one-dimensional cycles (from now on, r-dimensional cycles will be denoted by $\mathfrak{z}^r$)

$$
\begin{aligned}
\mathfrak{z}_1^1 &= x_1x_4 + x_4x_3 + x_3x_2 + x_2x_1, \\
\mathfrak{z}_2^1 &= x_5x_6 + x_6x_7 + x_7x_8 + x_8x_5.
\end{aligned}
$$

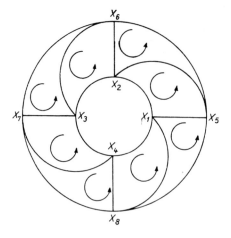

Fig. 35

Each of these cycles is not in itself a boundary, only their sum is a bounding chain. Thus the classification of all cycles into those that bound and those that do not bound is still too coarse.

Now the sum of two cycles again produces a cycle, and for addition the associative law holds

$$(\mathfrak{z}_1^r + \mathfrak{z}_2^r) + \mathfrak{z}_3^r = \mathfrak{z}_1^r + (\mathfrak{z}_2^r + \mathfrak{z}_3^r).$$

Also, there is a "unit element" $\mathfrak{z}_0^r = 0$ and for each cycle $\mathfrak{z}_j^r$ there is an "inverse" $(-\mathfrak{z}_j^r)$. Consequently, the cycles form an Abelian group $\mathfrak{Z}^r$ (for the group-theoretic concepts compare IB2). In view of (8) the bounding chains also have the group property, so that as special cycles they form a *subgroup* $\mathfrak{N}^r$ of $\mathfrak{Z}^r$. Then the totality of all cycles can be divided into cosets with respect to $\mathfrak{N}^r$, i.e., two cycles differing only by a bounding chain are regarded as not essentially different, and are put into the same class. This relationship among cycles is called *homology* (denoted by the symbol $\sim$), and the coset in question is the corresponding *homology* class. Since adjunction of a bounding chain to a cycle makes no change in its homology class, the bounding chains are said to be "zero-homologous." Thus for a bounding chain we have $\mathfrak{z}^r \sim 0$ and for two homologous cycles $\mathfrak{z}_1^r \sim \mathfrak{z}_2^r$ or $\mathfrak{z}_1^r - \mathfrak{z}_2^r \sim 0$. Since in an Abelian group every subgroup is a normal subgroup, we consider the homology classes themselves as elements of a new group, the factor group $\mathfrak{H}^r = \mathfrak{Z}^r/\mathfrak{N}^r$, which is called the *r-dimensional homology group*. Let us calculate it for two examples, the torus and the Klein bottle.

If any two vertices of a given complex can be joined by a path along one-dimensional sides, then the vertices, regarded as chains, are always homologous. For example, for the torus (Fig. 36) we have $x_1 \sim x_6$, since $x_6 - x_1 = \partial(x_1 x_7 + x_7 x_6) \sim 0$. Every zero-dimensional chain is a cycle, and

(11) $$\mathfrak{z}^0 = \sum \lambda_i x_i \sim (\sum \lambda_i) x_1.$$

We can also describe the various homology classes by giving a representative in each class. In the zero-dimensional case, (11) shows that we may take the multiples of a fixed vertex

$$\mathfrak{z}_0^0 = 0,\ \mathfrak{z}_1^0 = x_1,\ \mathfrak{z}_2^0 = -x_1,\ \mathfrak{z}_3^0 = 2x_1,\ \mathfrak{z}_4^0 = -2x_1, \dots.$$

Now if we let $\bar{x}_1$ denote the totality of all cycles homologous to x_1 (cf. the use of an overbar for residue classes of a congruence), the elements of the homology group $\mathfrak{H}^0$ consist of the classes

$$\overline{0},\ \overline{x_1},\ \overline{-x_1},\ \overline{2x_1},\ \overline{-2x_1}, \dots,$$

and the group can be generated by one element $\overline{x_1}$, which is of infinite order. We also write $\mathfrak{H}^0 = \{\overline{x_1}\}$ and call $\mathfrak{H}^0$ a free cyclic group (IB2 §16.1).

Any one-dimensional cycle in Fig. 36 containing "inner" segments, i.e., segments with vertices not on the boundary of the figure, can be replaced

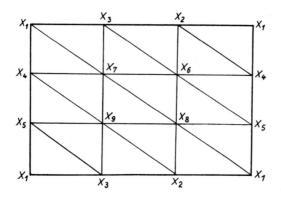

Fig. 36

by a homologous cycle without inner segments. For example, let

$$3_1^1 = x_3x_7 + x_7x_6 + x_6x_5 + x_5x_3,$$
$$3_2^1 = 3_1^1 + \partial x_1x_3x_5 = x_3x_7 + x_7x_6 + x_6x_5 - x_1x_5 + x_1x_3,$$
$$3_3^1 = 3_2^1 + \partial x_3x_6x_7 = x_3x_6 + x_6x_5 + x_5x_1 + x_1x_3,$$
$$3_4^1 = 3_3^1 + \partial x_2x_6x_3 = -x_2x_3 + x_2x_6 + x_6x_5 + x_5x_1 + x_1x_3,$$
$$3_5^1 = 3_4^1 + \partial x_2x_4x_6 = x_3x_2 + x_4x_6 + x_2x_4 + x_6x_5 + x_5x_1 + x_1x_3,$$
$$3_6^1 = 3_5^1 + \partial x_6x_4x_5 = x_3x_2 + x_2x_4 + x_4x_5 + x_5x_1 + x_1x_3,$$
$$3_7^1 = 3_6^1 + \partial x_2x_1x_4 = x_1x_3 + x_3x_2 + x_2x_1 + x_1x_4 + x_4x_5 + x_5x_1 \sim 3_1^1.$$

The procedure consists of adding boundary chains of triangles in such a way that all the inner segments of Fig. 36 are eliminated. If the two cycles corresponding to the edges of the large rectangle are denoted by

$$3_\alpha^1 = x_1x_3 + x_3x_2 + x_2x_1,$$
$$3_\beta^1 = x_1x_4 + x_4x_5 + x_5x_1,$$

then for any one-dimensional cycle 3^1 it follows that

$$3^1 \sim \lambda 3_\alpha^1 + \mu 3_\beta^1.$$

If the fundamental rectangle is folded together to form a torus, the two cycles 3_α^1 and 3_β^1 correspond to a meridian and a circle of latitude. Thus the classes $\bar{3}_\alpha^1$, $\bar{3}_\beta^1$ are generators of $\mathfrak{H}^1$, but we must still determine whether $\lambda 3_\alpha^1 + \mu 3_\beta^1 \sim 0$ always implies $\lambda = \mu = 0$. A two-dimensional chain with only the edges of the large rectangle in its boundary must contain all triangles with equal multiplicity, in order that the common "inner" edges, with suitable orientation, may cancel out. Let

$$\mathfrak{t}^2 = x_1x_4x_7 + x_1x_7x_3 + x_3x_7x_6 + x_3x_6x_2 + \cdots + x_8x_1x_5.$$

We obtain $\partial \mathfrak{t}^2 = 0$, i.e., $\lambda 3_\alpha^1 + \mu 3_\beta^1 \nsim 0$ unless both factors vanish. Thus the elements of the homology group $\mathfrak{H}^1$ are the homology classes $\overline{\lambda 3^1_\alpha + \mu 3^1_\beta}$ $= \lambda 3^1_\alpha + \mu 3^1_\beta$ and $\mathfrak{H}^1$ is a free group with two generators $\mathfrak{H}^1 = \{\bar{3}^1_\alpha, \bar{3}^1_\beta\}$.

The only two-dimensional cycles are the multiples of $\mathfrak{k}^2$ and, since the complex does not contain any three-dimensional simplexes, every homology class consists of a single cycle $\overline{\lambda \mathfrak{k}^2} = \lambda \mathfrak{k}^2$. Again the homology group $\mathfrak{H}^2$ is a free cyclic group $\mathfrak{H}^2 = \{\mathfrak{k}^2\}$.

For comparison let us now calculate the homology groups of the Klein bottle (Fig. 37). As above we have $\mathfrak{H}^0 = \{\overline{x_1}\}$, since the complex is connected, i.e., every vertex can be joined to every other vertex by a broken line. Similarly, $3^1_\alpha \sim \lambda 3^1_\alpha + \mu 3^1_\beta$, but from $\mathfrak{k}^2 = x_1 x_4 x_7 + \cdots + x_8 x_1 x_4$ it follows that $\partial \mathfrak{k}^2 = 2 3^1_\beta$, i.e., $2 3^1_\beta \sim 0$. Thus $\lambda 3^1_\alpha + \mu 3^1_\beta \sim 0$ if and only if $\lambda = 0$ and $\mu \equiv 0 \bmod 2$. The homology class $\bar{3}^1_\beta$ is an element of order 2, since $2 \bar{3}^\beta_1$ is the zero element. All the homology classes are given by $\lambda \bar{3}^1_\alpha$ and $\lambda \bar{3}^1_\alpha + \bar{3}^1_\beta$. The group $\mathfrak{H}^1$ is the direct sum of a free cyclic group and a finite group of order 2. There are no two-dimensional cycles, so that the group $\mathfrak{H}^2$ consists of the zero element alone.

The homology groups in the above examples have a very simple structure. They can be regarded as the direct sum of a free Abelian group, the so-called *Betti group*, and an Abelian group of *finite* order, the so-called *torsion group*. The Betti group consists of the linear combinations of p independent basis elements; p^r is called the *Betti number* of the homology group $\mathfrak{H}^r$ (note that the index r is not an exponent, but indicates dimension). For the annulus we found $p^0 = 1, p^1 = 2, p^2 = 1$, and for the Klein bottle $p^0 = 1, p^1 = 1$, $p^2 = 0$. Thus up to now we have had a torsion group only for the group $\mathfrak{H}^1$ of the Klein bottle, where it is generated by an element of order 2, called a *torsion coefficient*. In group theory it is proved, more generally, that every Abelian group with a finite number of generators is the direct sum of a Betti group and a torsion group (cf. IB2, §9.2), where the torsion group is the direct sum of ρ finite cyclic groups of order $c_1, c_2, \ldots c_\rho$, each of the c_i

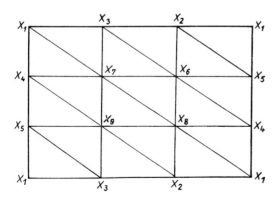

Fig. 37

being a factor of the preceding c_{i-1}. The Betti number p and the torsion coefficients $c_1, \ldots, c_\rho$ are uniquely determined by the Abelian group. Thus for finite complexes the structure of the homology groups is completely determined by the Betti number and the torsion coefficients.

How can we determine the homology groups for more complicated *finite* complexes? For each dimension $r = 0, 1, \ldots, n$, let us choose finitely many chains $\mathfrak{f}_1^r, \mathfrak{f}_2^r, \ldots, \mathfrak{f}_{\alpha^r}^r$ (to be called *fundamental chains*) satisfying the following conditions.

1.   The fundamental chains are linearly independent (cf. II9, §1.2), a condition that is always satisfied if, for example, the chains have no simplex in common.

2.   The boundary of an $(r + 1)$-dimensional fundamental chain is a linear combination of r-dimensional fundamental chains

$$\partial \mathfrak{f}_i^{r+1} = \sum_k \varepsilon_{ik}^r \mathfrak{f}_k^r \quad (\varepsilon_{ik}^r \text{ integers}).$$

3.   Every cycle is homologous to a linear combination of fundamental chains

$$\mathfrak{z}^r \sim \sum_i \lambda_i \mathfrak{f}_i^r.$$

4.   From

$$\sum_i \lambda_i \mathfrak{f}_i^r \sim 0 \quad \text{follows} \quad \sum_i \lambda_i \mathfrak{f}_i^r = \partial \sum_j \mu_j \mathfrak{f}_j^{r+1}.$$

The importance of these conditions lies in the fact that, on the one hand, the lattice of linear combinations of fundamental chains is closed with respect to forming boundaries or passing to homologous linear combinations, and, on the other hand, by condition 3 every homology class contains at least one linear combination of fundamental chains, so that homology classes are included in the lattice. In the last two examples above (the torus and the Klein bottle) we can choose as fundamental chains

$$\mathfrak{f}_1^0 = x_1; \mathfrak{f}_1^1 = \mathfrak{z}_\alpha^1, \mathfrak{f}_2^1 = \mathfrak{z}_\beta^1; \mathfrak{f}_1^2 = \mathfrak{f}^2.$$

By condition 2, all the bounding relations among the fundamental chains are completely described by the coefficients ε_{ik}^r, which may be written in the form of a matrix, the so-called *incidence matrix*,

$$(\varepsilon_{ik}^r) = \mathbf{E}^r = \begin{pmatrix} \varepsilon_{11}^r & \varepsilon_{12}^r & \cdots & \varepsilon_{1\alpha^r}^r \\ \varepsilon_{21}^r & \varepsilon_{22}^r & \cdots & \varepsilon_{2\alpha^r}^r \\ \vdots & \vdots & & \vdots \\ \varepsilon_{\alpha^{r+1}1}^r & \varepsilon_{\alpha^{r+1}2}^r & \cdots & \varepsilon_{\alpha^{r+1}\alpha^r}^r \end{pmatrix}.$$

It is convenient to write this matrix as a double entry table, with the $(r + 1)$-

fundamental chains on the left, and the r-fundamental chains along the top

$\mathbf{E}^r$	$\mathfrak{f}^r_1$	$\mathfrak{f}^r_2$	$\cdots$	$\mathfrak{f}^r_{\alpha^r}$
$\mathfrak{f}^{r+1}_1$	ε^r_{11}	ε^r_{12}	$\cdots$	$\varepsilon^r_{1\alpha^r}$
$\mathfrak{f}^{r+1}_2$	ε^r_{21}	ε^r_{22}	$\cdots$	$\varepsilon^r_{2\alpha^r}$
$\vdots$	$\vdots$	$\vdots$		$\vdots$
$\mathfrak{f}^{r+1}_{\alpha^{r+1}}$	$\varepsilon^r_{\alpha^{r+1}1}$	$\varepsilon^r_{\alpha^{r+1}2}$	$\cdots$	$\varepsilon^r_{\alpha^{r+1}\alpha^r}.$

Since

$$\partial\partial\mathfrak{f}^{r+1}_i = \sum_k \varepsilon^r_{ik}\partial\mathfrak{f}^r_k = \sum_{k,l} \varepsilon^r_{ik}\varepsilon^{r-1}_{kl}\mathfrak{f}^{r-1}_l = 0,$$

it follows from the linear independence of the fundamental chains that

(12) $$\sum_k \varepsilon^r_{lk}\varepsilon^{r-1}_{kl} = 0$$

or in matrix form $\mathbf{E}^r\mathbf{E}^{r-1} = 0$. For the three-dimensional simplex, for example, with the oriented simplexes as fundamental chains in each dimension, we obtain the following incidence matrix:

$\mathfrak{f}^3_1 = x_1x_2x_3x_4,$

$\mathfrak{f}^2_1 = x_2x_3x_4, \quad \mathfrak{f}^2_2 = -x_1x_3x_4, \quad \mathfrak{f}^2_3 = x_1x_2x_4, \quad \mathfrak{f}^2_4 = -x_1x_2x_3,$

$\mathfrak{f}^1_1 = x_1x_2, \quad \mathfrak{f}^1_2 = x_1x_3, \quad \mathfrak{f}^1_3 = x_1x_4, \quad \mathfrak{f}^1_4 = x_2x_3, \quad \mathfrak{f}^1_5 = x_2x_4, \quad \mathfrak{f}^1_6 = x_3x_4,$

$\mathfrak{f}^0_1 = x_1, \quad \mathfrak{f}^0_2 = x_2, \quad \mathfrak{f}^0_3 = x_3, \quad \mathfrak{f}^0_4 = x_4.$

$\mathbf{E}^2$	$\mathfrak{f}^2_1$	$\mathfrak{f}^2_2$	$\mathfrak{f}^2_3$	$\mathfrak{f}^2_4$
$\mathfrak{f}^3_1$	$+1$	$+1$	$+1$	$+1$

$\mathbf{E}^1$	$\mathfrak{f}^1_1$	$\mathfrak{f}^1_2$	$\mathfrak{f}^1_3$	$\mathfrak{f}^1_4$	$\mathfrak{f}^1_5$	$\mathfrak{f}^1_6$
$\mathfrak{f}^2_1$	0	0	0	$+1$	-1	$+1$
$\mathfrak{f}^2_2$	0	-1	$+1$	0	0	-1
$\mathfrak{f}^2_3$	$+1$	0	-1	0	$+1$	0
$\mathfrak{f}^2_4$	-1	$+1$	0	-1	0	0

$\mathbf{E}^0$	$\mathfrak{f}^0_1$	$\mathfrak{f}^0_2$	$\mathfrak{f}^0_3$	$\mathfrak{f}^0_4$
$\mathfrak{f}^1_1$	-1	$+1$	0	0
$\mathfrak{f}^1_2$	-1	0	$+1$	0
$\mathfrak{f}^1_3$	-1	0	0	$+1$
$\mathfrak{f}^1_4$	0	-1	$+1$	0
$\mathfrak{f}^1_5$	0	-1	0	$+1$
$\mathfrak{f}^1_6$	0	0	-1	$+1.$

In *any* complex the oriented simplexes may be chosen as fundamental chains, since they always satisfy the four conditions. But in general, the resulting incidence matrices are very inconvenient, and quite unsuitable for calculation of the homology groups. In the lattice $\mathfrak{G}^r$ of linear combinations of fundamental chains with integer coefficients, the chains $\mathfrak{f}^r_1, \ldots, \mathfrak{f}^r_{\alpha^r}$ form a basis, and we now seek to determine a new basis for $\mathfrak{G}^r$ such that the corresponding incidence matrices take a particularly simple form, a so-called normal form $\mathbf{H}^r$. The change will be carried out step by step in two ways. First we replace $\mathfrak{f}^r_i$ by $\mathfrak{f}^{*r}_i = \mathfrak{f}^r_i + \mathfrak{f}^r_j$ ($i \neq j$), i.e., the matrix $\mathbf{E}^{*r-1}$ is formed from $\mathbf{E}^{r-1}$ by adding the jth row to the ith row, and at the same time $\mathbf{E}^{*r}$ is formed from $\mathbf{E}^r$ by subtraction of the ith column from the jth column. Secondly, we replace $\mathfrak{f}^r_l$ by $\mathfrak{f}^{*r}_l = -\mathfrak{f}^r_l$, i.e., we change the sign in the lth row of $\mathbf{E}^{r-1}$ and in the lth column of $\mathbf{E}^r$. By continued application of these two methods of transforming the basis, and by a suitable interchange of rows or columns, we can arrange that every incidence matrix $\mathbf{H}^r$ has zeros everywhere except in an upper right diagonal, where it has positive numbers, arranged in order of magnitude (Fig. 38).

These numbers are called the *invariant factors* of the matrix, and the normalization process can always be carried out in such a way that every $c^r_{\nu-1}$ divides c^r_ν. Let there be η^r of these positive numbers. Since the above transformations do not change the rank of the matrix, we have $\eta^r = \text{rank}$ $\mathbf{E}^r = \text{rank } \mathbf{H}^r$. The invariance of the factors, i.e., their independence of the choice of steps in the transformations, will not be proved here. For it we refer to Seifert-Threlfall [6], §87, where there is also a proof that the matrices can always be brought into normal form by these elementary transformations.

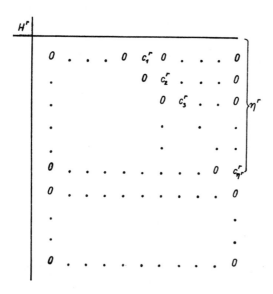

Fig. 38

For three-dimensional simplexes the normal forms of the incidence matrices are

$\mathbf{H}^2$ |

| 0 0 0 1 |

$\mathbf{H}^1$ |

| 0 0 0 1 0 0 |
| 0 0 0 0 1 0 |
| 0 0 0 0 0 1 |
| 0 0 0 0 0 0 |

$\mathbf{H}^0$ |

| 0 1 0 0 |
| 0 0 1 0 |
| 0 0 0 1 |
| 0 0 0 0 |
| 0 0 0 0 |
| 0 0 0 0. |

To see the importance of the normal forms for calculating the homology groups, let us consider $\mathbf{H}^{r-1}$ and $\mathbf{H}^r$ together, since they are related to each other by $\mathbf{H}^r\mathbf{H}^{r-1} = 0$. The square submatrices formed by the invariant factors lead to the classification of matrices given by the dashes in Fig. 39. The line $- \times - \times -$ in $\mathbf{H}^r$ is determined by $\mathbf{H}^{r+1}$, and the line $- \cdot - \cdot -$ in $\mathbf{H}^{r-1}$ by $\mathbf{H}^{r-2}$. To make the distinction clearer, the basis chains for the normal forms in the various sections will be denoted by $u_\lambda^r, v_\mu^r, w_\nu^r$, where the indices take the following values $\lambda = 1, \ldots, \eta^{r-1}; \mu = 1, \ldots, p^r; \nu = 1, \ldots, \eta^r$. The number p^r is calculated from the number of fundamental chains α^r and the ranks η^{r-1}, η^r as follows:

$$(13) \qquad p^r = \alpha^r - \eta^r - \eta^{r-1} \quad (0 < r < n).$$

For $r = 0$, we have $p^0 = \alpha^0 - \eta^0$; and $p^n = \alpha^n - \eta^{n-1}$ for $r = n$. If we set $\eta^n = \eta^{-1} = 0$ by definition, formula (13) includes the cases $r = 0$ and $r = n$. In $\mathbf{H}^0$ the chains u_λ^0 are absent, and in $\mathbf{H}^{n-1}$ the chain w_ν^n is absent.

By condition 3 for fundamental chains, every cycle $\mathfrak{z}^r$ is homologous to a linear combination of the new fundamental chains. But here the chains u_λ^r cannot occur, since they are not closed, and their boundary chains $\partial u_\lambda^r = c_\lambda^{r-1} w_\lambda^{r-1}$ cannot cancel each other out, since the chains w_λ^{r-1} are independent. Consequently,

$$(14) \qquad \mathfrak{z}^r \sim \sum \alpha_\mu v_\mu^r + \sum \beta_\nu w_\nu^r.$$

But by condition 4 this cycle is homologous to zero if and only if it is the boundary of a linear combination of basis chains

$$(15) \qquad \sum \alpha_\mu v_\mu^r + \sum \beta_\nu w_\nu^r = \partial \sum \gamma_\lambda u_\lambda^{r+1}.$$

Here chains v^{r+1}, w^{r+1} do not occur on the right side, since their boundary vanishes. Now $\partial u_\lambda^{r+1} = c_\lambda^r w_\lambda^r$, and therefore, since the basis chains are independent, it follows from (15) that

$$(16) \qquad \alpha_\mu = 0; \quad \beta_\nu = \gamma_\nu c_\nu^r \quad \text{or} \quad \beta_\nu \equiv 0 \bmod c_\nu^r.$$

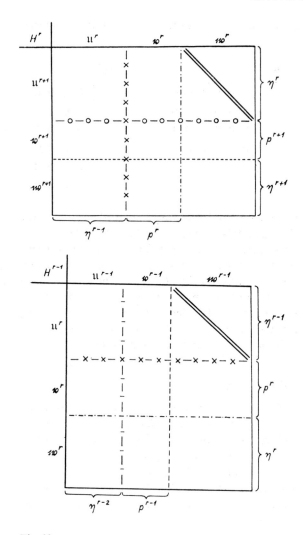

Fig. 39

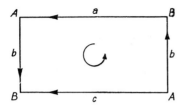

Fig. 40

We can now answer the question which of the cycles generated from v_μ^r, w_ν^r lie in the same homology class. Since their difference must be homologous to zero, it follows from (16) that their coefficients α_μ must be identical, while the coefficients β_ν can differ only by multiples of c_ν^r, i.e., they must belong to the same residue class mod c_ν^r. As representatives of the individual homology classes we thus obtain the cycles

$$(17) \qquad \mathfrak{z}^r = \sum \alpha_\mu v_\mu^r + \sum \beta_\nu w_\nu^r, \quad \begin{cases} \alpha_\mu \text{ arbitrary integers,} \\ 0 \leqq \beta_\nu < c_\nu^r; \, c_\nu^r \neq 1, \end{cases}$$

with the final result that the lattice spanned by the cycles v_μ^r represents the Betti group. The number p^r determined by (13) is the Betti number of the homology group $\mathfrak{H}^r$. The torsion group is generated by the cycles w_ν^r with corresponding invariant factors different from 1 and, since $c_\nu^r w_\nu^r \sim 0$, these factors are identical with the torsion coefficients.

Let us illustrate by calculating the homology groups for some examples.

1. Möbius strip. Since only the fundamental chains are used, the triangulation of the rectangle is not needed. We orient the σ_i^2 in the sense indicated in Fig. 40. As 0-chains we choose $\mathfrak{f}_1^0 \leftrightarrow A$, $\mathfrak{f}_2^0 \leftrightarrow B$, as 1-chains $\mathfrak{f}_1^1 \leftrightarrow a$, $\mathfrak{f}_2^1 \leftrightarrow b$, $\mathfrak{f}_3^1 \leftrightarrow c$, and as 2-chain $\mathfrak{f}_1^2 = \sum_i \sigma_i^2$, namely the sum of all the triangles in the complex. Then the incidence matrices are

$\mathbf{E}^1$	$\mathfrak{f}_1^1$	$\mathfrak{f}_2^1$	$\mathfrak{f}_3^1$
$\mathfrak{f}_1^2$	1	2	-1

$\mathbf{E}^0$	$\mathfrak{f}_1^0$	$\mathfrak{f}_2^0$
$\mathfrak{f}_1^1$	1	-1
$\mathfrak{f}_2^1$	-1	1
$\mathfrak{f}_3^1$	-1	1

$\mathbf{H}^1$
0 0 1

$\mathbf{H}^0$
0 1
0 0
0 0

$$\text{rank}\mathbf{E}^0 = 1, \text{rank}\mathbf{E}^1 = 1,$$
$$p^0 = 2 - 1 = 1, \quad p^1 = 3 - 1 - 1 = 1, \quad p^2 = 1 - 1 = 0.$$

There are no torsion coefficients.

2. Manifold with q cross-caps. The fundamental polygon is

$$a_1 a_1 a_2 a_2 \ldots a_q a_q.$$

The fundamental chains are

$$\mathfrak{k}_1^0 \leftrightarrow A; \mathfrak{k}_1^1 \leftrightarrow a_1, \ldots, \mathfrak{k}_q^1 \leftrightarrow a_q; \mathfrak{k}_1^2 = \sum_i \sigma_i^2.$$

E^1	$\mathfrak{k}_1^1 \cdots \mathfrak{k}_q^1$
$\mathfrak{k}_1^2$	$2 \cdots 2$

E^0	$\mathfrak{k}_1^0$
$\mathfrak{k}_1^1$	0
$\vdots$	$\vdots$
$\mathfrak{k}_q^1$	0

rank$E^0 = 0$,   rank$E^1 = 1$,

$p^0 = 1 - 0 = 1$,

$p^1 = q - 1 - 0 = q - 1$,

$p^2 = 1 - 1 = 0$.

There is one torsion coefficient $c_1^1 = 2$.

3. Manifold with h handles. The fundamental polygon is

$$a_1 b_1 a_1^{-1} b_1^{-1} a_2 b_2 a_2^{-1} b_2^{-1} \cdots a_h b_h a_h^{-1} b_h^{-1}.$$

The fundamental chains are

$$\mathfrak{k}_1^0 \leftrightarrow A; \mathfrak{k}_1^1 \leftrightarrow a_1, \mathfrak{k}_2^1 \leftrightarrow b_1, \ldots, \mathfrak{k}_{2h-1}^1 \leftrightarrow a_k, \mathfrak{k}_{2h}^1 \leftrightarrow b_h; \mathfrak{k}_1^2 = \sum \sigma_i^2.$$

E^1	$\mathfrak{k}_1^1 \cdots \mathfrak{k}_{2h}^1$
$\mathfrak{k}_1^2$	$0 \cdots 0$

E^0	$\mathfrak{k}_1^0$
$\mathfrak{k}_1^1$	0
$\vdots$	$\vdots$
$\mathfrak{k}_{2h}^1$	0

rank$E^0 = 0$,   rank$E^1 = 0$,

$p^0 = 1 - 0 = 1$,

$p^1 = 2h - 0 - 0 = 2h$,

$p^2 = 1 - 0 = 1$.

There are no torsion coefficients.

4. Surface of the sphere. The fundamental polygon is aa^{-1}. The fundamental chains are

$$\mathfrak{k}_1^0 \leftrightarrow A; \mathfrak{k}_2^0 \leftrightarrow B; \mathfrak{k}_1^1 \leftrightarrow a; \mathfrak{k}_1^2 = \sum \sigma_i^2.$$

E^1	$\mathfrak{k}_1^1$
$\mathfrak{k}_1^2$	0

E^0	$\mathfrak{k}_1^0$	$\mathfrak{k}_2^0$
$\mathfrak{k}_1^1$	-1	$+1$

rank$E^0 = 1$,   rank$E^1 = 0$,

$p^0 = 2 - 1 = 1$, $p^1 = 1 - 1 - 0 = 0$,

$p^2 = 1 - 0 = 1$.

There are no torsion coefficients.

5. Octahedral space. The method of constructing two-dimensional manifolds from the fundamental polygon by fastening corresponding sides together is also available in higher dimensions. Here we begin with a Euclidean polyhedron with pairs of opposite faces of the same kind, so that they can be identified. If all the surfaces can be fastened together pairwise, we obtain a closed polyhedron, but the neighborhood of the vertices is not necessarily homeomorphic to a spherical neighborhood. Thus the three-dimensional manifolds cannot be built up in any clear-cut way, and their classification is still an unsolved problem. As an example let us consider the manifold formed by identification of opposite triangles of an octahedron, where the corresponding triangles are twisted with respect to one another

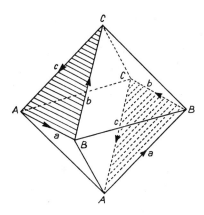

Fig. 41

through an angle of $\pi/3$ (Fig. 41 gives the correspondence for one pair of triangles). A convenient representation is obtained by stereographic projection of the surface of the octahedron from the upper vertex into the plane of symmetry (Fig. 42). By twisting and fastening together all four pairs of opposite sides S_1, S_2, S_3, S_4 we obtain four sides a, b, c, d and a vertex A, whose neighborhood we shall not further investigate, remarking merely that in this way we have obtained a manifold, namely the so-called octa-

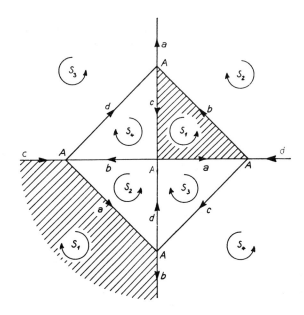

Fig. 42

hedral space. In order to obtain a simplicial complex, we must subdivide the octahedron into tetrahedra σ_i^3, which are then to be oriented coherently, i.e., in such a way that common surfaces inside the octahedron have opposite induced orientations. As fundamental chains for calculating the homology groups it is sufficient to choose the chains corresponding to the vertex A, the edges a, b, c, d, the triangles S_1, S_2, S_3, S_4, and the full octahedron

$$\mathfrak{f}_1^0 \leftrightarrow A; \mathfrak{f}_1^1 \leftrightarrow a, \quad \mathfrak{f}_2^1 \leftrightarrow b, \mathfrak{f}_3^1 \leftrightarrow c, \quad \mathfrak{f}_4^1 \leftrightarrow d; \mathfrak{f}_1^2 \leftrightarrow S_1, \quad \mathfrak{f}_2^2 \leftrightarrow S_2,$$
$$\mathfrak{f}_3^2 \leftrightarrow S_3, \quad \mathfrak{f}_4^2 \leftrightarrow S_4; \mathfrak{f}_1^3 = \sum \sigma_i^3.$$

$\mathbf{E}^2$	$\mathfrak{f}_1^2$	$\mathfrak{f}_2^2$	$\mathfrak{f}_3^2$	$\mathfrak{f}_4^2$
$\mathfrak{f}_1^3$	0	0	0	0

$\mathbf{E}^1$	$\mathfrak{f}_1^1$	$\mathfrak{f}_2^1$	$\mathfrak{f}_3^1$	$\mathfrak{f}_4^1$
$\mathfrak{f}_1^2$	1	1	1	0
$\mathfrak{f}_2^2$	1	1	0	1
$\mathfrak{f}_3^2$	1	0	1	1
$\mathfrak{f}_4^2$	0	1	1	1

$\mathbf{E}^0$	$\mathfrak{f}_1^0$
$\mathfrak{f}_1^1$	0
$\mathfrak{f}_2^1$	0
$\mathfrak{f}_3^1$	0
$\mathfrak{f}_4^1$	0

$\mathbf{H}^1$	$\mathfrak{w}_1^1$	$\mathfrak{w}_2^1$	$\mathfrak{w}_3^1$	$\mathfrak{w}_4^1$
$\mathfrak{u}_1^2$	3	0	0	0
$\mathfrak{u}_2^2$	0	1	0	0
$\mathfrak{u}_3^2$	0	0	1	0
$\mathfrak{u}_4^2$	0	0	0	1

$\eta^0 = 0, \quad \eta^1 = 4, \quad \eta^2 = 0,$
$p^0 = 1 - 0 = 1, \quad p^1 = 4 - 4 - 0 = 0,$
$p^2 = 4 - 0 - 4 = 0, \quad p^3 = 1 - 0 = 1;$
$c_1^1 = 3.$

After bringing the matrices to normal form we find the following relationship between $\mathfrak{f}_i^1$, $\mathfrak{f}_j^2$, $\mathfrak{u}_k^2$, $\mathfrak{w}_l^1$, where the individual steps in the transformation of the matrix $\mathbf{E}^1$ have been applied to the fundamental chains

$$\mathfrak{u}_1^2 = \mathfrak{f}_1^2 - 2\mathfrak{f}_2^2 + \mathfrak{f}_3^2 + \mathfrak{f}_4^2, \quad \mathfrak{w}_1^1 = \mathfrak{f}_3^1,$$
$$\mathfrak{u}_2^2 = \mathfrak{f}_1^2, \quad \mathfrak{w}_2^1 = \mathfrak{f}_1^1 + \mathfrak{f}_2^1 + \mathfrak{f}_3^1,$$
$$\mathfrak{u}_3^2 = \mathfrak{f}_2^2 - \mathfrak{f}_1^2, \quad \mathfrak{w}_3^1 = \mathfrak{f}_4^1 - \mathfrak{f}_3^1,$$
$$\mathfrak{u}_4^2 = \mathfrak{f}_2^2 - \mathfrak{f}_3^2, \quad \mathfrak{w}_4^1 = \mathfrak{f}_2^1 - \mathfrak{f}_3^1,$$
$$3\mathfrak{f}_3^1 \sim 0, \quad \mathfrak{f}_1^1 + \mathfrak{f}_2^1 + \mathfrak{f}_3^1 \sim 0,$$
$$\mathfrak{f}_4^1 - \mathfrak{f}_3^1 \sim 0, \quad \mathfrak{f}_2^1 - \mathfrak{f}_3^1 \sim 0.$$

It follows that $\mathfrak{f}_1^1 \sim \mathfrak{f}_2^1 \sim \mathfrak{f}_3^1 \sim \mathfrak{f}_4^1$ and $3\mathfrak{f}_i^1 \sim 0$; i.e., all edges are homologous to one another and each of them bounds only after being traversed three times.

The above method for practical calculation of the homology group already shows that the special decomposition of the polyhedron into simplexes plays a subordinate role. It is to be expected that the Betti numbers and the torsion coefficients are invariant under elementary subdivision and reduction. This conjecture can be proved (we shall not carry out the details here) by choosing the simplexes themselves as fundamental chains and examining the changes in the incidence matrices under these two processes. So equivalent complexes have the same homology groups, but the converse does not necessarily hold; two complexes with the same homology groups may fail to be equivalent to each other. To be sure, the homology groups of the complex are invariant under elementary subdivision, but they do not characterize the complex completely. Their importance lies rather in the opposite direction. If the homology groups are distinct, we can conclude that the complexes are not equivalent, but so long as the principal conjecture of combinatorial topology remains unproved, we cannot draw any conclusion about whether or not the corresponding polyhedra are *homeomorphic*. It is the great merit of L.E.J. Brouwer that with his method of simplicial approximation of continuous mappings he has created a link between the combinatorial topology of complexes and the (continuity) topology of sets of points. In the topology of polyhedra we ask for properties of the corresponding complexes that remain invariant under all simplicial decompositions of homeomorphic polyhedra. The first proof that the homology groups of a complex are topological invariants of this sort for the corresponding polyhedra, i.e., that two homeomorphic polyhedra always have the same homology groups, was given by Alexander, using the Brouwer method. Although we must be content here with merely stating this fact, we shall return in §11 to a discussion of simplicial approximation, since it is one of the basic concepts, in modern topology, together with topological spaces and complexes.

The topological invariance of homology groups allows us to give a complete classification of closed connected two-dimensional manifolds. We have already shown that every such manifold can be reduced to a sphere with h handles ($h \geq 0$) or with k cross-caps ($k > 0$), and calculation of the homology groups showed that the corresponding pairs of Betti numbers (p^1, p^2) are distinct. Consequently, two manifolds of distinct type are not homeomorphic.

5. Connectivity and the Euler-Poincaré Characteristic

We have seen that two vertices of a complex are homologous if they can be joined by a path consisting entirely of one-dimensional edges. Thus the zeroth Betti number p^0 at once gives the number of pairwise disjoint, connected subcomplexes. But the connectivity relations in higher dimensions will not be determined by the Betti numbers alone. For example, in a triangulation of the projective plane, the triangles are all connected with one another, but since they cannot be coherently oriented, the corresponding chain is not a cycle and therefore $p^2 = 0$.

In any discussion of connectivity it is clear that the orientation of the simplexes is of no importance, but for purposes of calculation of nonoriented simplexes there is no need to introduce any new symbols; we merely need to take the chains mod 2; i.e., the coefficients of the chains are no longer integers, but residue classes mod 2. Then 0 stands for the residue class of the even numbers, and 1 for the class of the odd numbers and, since $+1$ and -1 belong to the same class, there is no difference between $+\sigma^r$ and $-\sigma^r$. Chains mod 2 can be added as before, if we set $1 + 1 = 0$, since the sum of two odd numbers is always an even number. Here too the boundary is to be taken mod 2, so that now, for example, the triangles of the projective plane form a closed chain, which is a cycle mod 2, since exactly two triangles are incident with every edge. We now divide the cycles mod 2 into *connectivity classes* where, in analogy with homology classes, two cycles lie in the same class if they differ by a bounding chain mod 2. These connectivity classes are the elements of a group, the so-called *connectivity group*, the analog of the homology group for oriented complexes. Since $3^r + 3^r = 0$ mod 2, every element is of finite order 2, and for a finite complex the connectivity group is the direct sum of q^r groups of order 2. The number q^r is called the rth *connectivity number* of the complex. To give a geometric interpretation of q^1, we consider a closed cut on the simplicial decomposition of a closed two-dimensional manifold, where by a closed cut we mean a closed path (without double points) along the edges corresponding to a cycle mod 2. If we make more than q^1 closed cuts $3_1, 3_2, \ldots, 3_m$ (for which we shall also assume that no two have a common edge), then since $m > q^1$ the cycles are always homologically dependent, i.e., some linear combination of them is zero-homologous mod 2

$$\varepsilon_1 3_1 + \cdots + \varepsilon_m 3_m = \partial \mathfrak{f}^2.$$

Here the coefficients ε_i have the values 0 or 1, but not all the ε_i vanish. By assumption, the cycles have no common element, so that the chain $\mathfrak{f}^2$ is not the zero chain. On the other hand, since the manifold is closed, $\mathfrak{f}^2$ cannot contain all the triangles of the subdivision. If we now cut the surface along these closed cuts, we will have cut out the part of the surface corresponding to the chain $\mathfrak{f}^2$. For $m \leq q^1$ this conclusion no longer follows. The maximal number of closed cuts which do not divide the surface into pieces can thus be at most equal to q^1. On the other hand, we see from the fundamental polygon that the oriented surfaces have $2h$ closed cuts that do not divide the surface, and the nonoriented surfaces have k such cuts, corresponding to the sides of the polygon (cf. the torus with $2h = 2$, Fig. 43), where h and k are the number of handles and of cross-caps. Since $q^1 = 2h$ and $q^1 = k$ respectively (see below), it follows that the connectivity number q^1 represents precisely the maximal number of closed cuts that can be made without separating the surface into pieces.

The connectivity groups can be calculated in the same way as the homology groups. We only need to reduce to 0 or 1 all the numbers occurring in

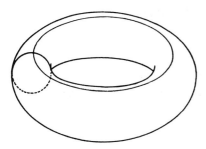

Fig. 43

the incidence matrices mod 2. Thus in the normal form all the torsion coefficients that are even are to be replaced by zero. The connectivity number q^r replacing the Betti number p^r is thus increased by the number of even torsion coefficients of dimension r or $r - 1$; denoting the latter numbers by g^r and g^{r-1}, we have

(18) $$q^r = p^r + g^r + g^{r-1}, \quad 0 < r < n,$$

which also holds for $r = 0$ and $r = n$ if we set $g^{-1} = g^n = 0$. For the orientable closed two-dimensional manifolds we have $p^1 = 2h$ and $q^1 = p^1$, since there are no torsion coefficients. The spheres with k cross-caps have one torsion coefficient of dimension 1, with the value 2, so that here $g^1 = 1$ and $q^1 = p^1 + 1 = (k - 1) + 1 = k$.

The g^r can be eliminated from equations (18). For we need only form the alternating sum

$$\sum_{r=0}^{n} (-1)^r q^r = \sum_{r=0}^{n} (-1)^r p^r + \sum_{r=0}^{n} (-1)^r [g^r + g^{r-1}] = \sum_{r=0}^{n} (-1)^r p^r,$$

which is invariant under elementary subdivision of the complex and is also a topological invariant of the corresponding polyhedron (on account of the topological invariance of the Betti numbers.) It is denoted by χ and is called the *Euler-Poincaré characteristic* of the complex. Its importance lies chiefly in the fact that it can be calculated from the numbers α^r of the fundamental chains, without any need to determine the ranks η^r of the incidence matrices. For from (13), namely $p^r = \alpha^r - \eta^r - \eta^{r-1}$ (with $\eta^{-1} = \eta^n = 0$), we also have $\sum_{r=0}^{n}(-1)^r p^r = \sum_{r=0}^{n}(-1)^r \alpha^r$, and therefore the formula

(19) $$\chi = \sum_{r=0}^{n} (-1)^r q^r = \sum_{r=0}^{n} (-1)^r p^r = \sum_{r=0}^{n} (-1)^r \alpha^r.$$

For a given simplicial decomposition of a Euclidean polyhedron, the chains corresponding to the vertices, edges, and sides of the polyhedron can be taken as the fundamental chains, so that the equation

$$\chi = \sum_{r=0}^{n} (-1)^r \alpha^r$$

is a generalization of the elementary Euler formula for the polyhedron, where α^r is the number of r-dimensional sides. For the tetrahedron, for example,

$$\alpha^0 - \alpha^1 + \alpha^2 = 4 - 6 + 4 = 2,$$

and more generally $\chi = 2 - 2h$ for the orientable connected closed two-dimensional manifolds, and $\chi = 2 - k$ for the nonorientable. Thus each of these manifolds is already determined topologically by its orientability or nonorientability and its Euler-Poincaré characteristic. The topological invariant h or k is also called the *genus* of the surface.

Homology groups can also be considered for a modulus other than 2, or still more generally, we may take elements of an arbitrary Abelian group as coefficients of the chains of simplexes. However, we cannot deal here with this theory of so-called algebraic complexes; it only becomes important in topics that would take us too far afield.

The homology groups do not by any means exhaust our knowledge of topological invariants. We can go much more deeply into the structure of polyhedra by means of the so-called fundamental group. However, this purpose requires much more difficult algebraic methods, so that here we must be content with a brief discussion of the concept.

To each oriented one-dimensional simplex σ_i^1 of a complex we assign a "path" w_i with starting point and endpoint corresponding to the orientation, so that to the simplex $-\sigma_i^1$ the path w_i^{-1} is assigned. Two paths $w_i^{\varepsilon_i}$, $w_j^{\varepsilon_j}$ with $\varepsilon_i = \pm 1$, $\varepsilon_j = \pm 1$, can be combined to form the product path $w_i^{\varepsilon_i} w_j^{\varepsilon_j}$, provided that the endpoint of $w_i^{\varepsilon_i}$ is identical with the starting point $w_{ij}^{\varepsilon_j}$, where it is clear that the order of the factors in a product is essential, since in general the endpoint of $w_j^{\varepsilon_j}$ is not the starting point of $w_i^{\varepsilon_i}$, so that

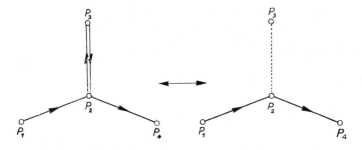

Fig. 44

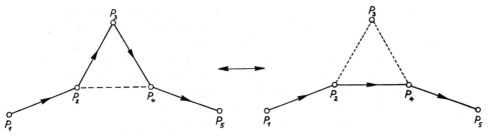

Fig. 45

$w_j^{\varepsilon j} w_i^{\varepsilon i}$ does not necessarily exist; and the product of more than two paths is formed correspondingly. A path $W = w_{i_1}^{\varepsilon i_1} \cdots w_{i_m}^{\varepsilon i m}$ is said to be "closed" if the starting point of $w_{i_1}^{\varepsilon i_1}$ coincides with the endpoint of $w_{i_m}^{\varepsilon i m}$. The product of two closed paths W_1, W_2 with the same starting point is again a closed path $W_{12} = W_1 W_2$ with the same starting point. We now consider modifications of a path, which for simplicity we shall explain from Figs. 44 and 45. If two closed paths with the same starting point can be transformed into each other by deletion or adjunction of segments traversed first in one direction and then in the other (Fig. 44) or else by going across a triangle (Fig. 45), the two paths are said to be "homotopic," and the classes of homotopic paths are the elements of a non-Abelian group (multiplication of classes being defined, as usual, by multiplication of representative paths in them). This group, which is independent of the choice of starting point and is invariant under elementary subdivision, is called the *fundamental group* of the complex and, just as for the homology groups, it can be proved that simplicial decompositions of homeomorphic polyhedra always have the same fundamental group. Consequently, this group is one of the topological invariants of the corresponding polyhedron. It is particularly important for the topological characterization of manifolds of more than two dimensions.

Topology of Curves

6. Curves as Sets of Points

When we compare curves with surfaces or with certain discontinuous infinite sets of points we have the impression that a curve contains fewer points than a surface and more points than a scattered set of points. But a quantitative distinction of this kind is immediately meaningful only for *finite* sets. In order to compare infinite sets, we say that two sets M_1 and M_2 *have the same power* if they can be mapped one-to-one onto each other, i.e., if to each element of M_1 there corresponds exactly one element of M_2 and conversely (see IA, §7.3). The simplest example of an infinite set is the set of all natural numbers, and every infinite set with the same power is said to be *countable*. The set of rational numbers from 0 to 1 is countable, but the set of real numbers in the same interval is not countable, having the so-called power of the continuum. The conjecture that a line segment has "fewer" points than a square, for example, would therefore mean that these two sets of points are of different power. But remarkably, this conjecture is false, since we shall show that the points of the line segment can be mapped one-to-one onto the points of the square.

If x is a real number ($0 < x \leq 1$), let

$$x = 0.\, a_1 a_2 a_3 a_4 \ldots \quad \text{with} \quad a_i = \begin{cases} 0 \\ 1 \end{cases}$$

be its (infinite) dyadic expansion (e.g., for $x = \frac{1}{8}$ we write $x = 0.0001111 \ldots$ $= 0.000\bar{1}$ instead of $x = 0.001$).

Similarly, the p-adic expansion of a real number x in the same segment represents x as a series in powers of $1/p$, with nonnegative integer coefficients,

$$x = \sum_{1}^{\infty} a_v \left(\frac{1}{p}\right)^v \quad \text{with} \quad 0 \leq a_v < p,$$

where for abbreviation (in analogy with the decimal fractions with $p = 10$) we write only the sequence of coefficients

$$x = 0.\, a_1 a_2 a_3 a_4. \ldots$$

For $p = 2$ we speak of the dyadic expansion, and for $p = 3$ of the triadic.

To each expansion we now assign a sequence of natural numbers, stating the distances between the successive 1's in the dyadic representation, e.g.,

$$x = 0.001\,01\,0001\,01\,1\,1\,01\,00001 \ldots \leftrightarrow 3, 2, 4, 2, 1, 1, 2, 5, \ldots.$$
$$\,3\quad 2\quad\ \ 4\quad\ 2\ 1\ 1\ 2\quad\ \ 5$$

Conversely, to each sequence of natural numbers there will correspond a dyadic expansion, and thus a real number x, so that the set of all real numbers $0 < x \leq 1$ has the same power as the set of all sequences of natural numbers. We now decompose each sequence into two sequences by taking the first, third, fifth, . . . term for the first sequence, and the second, fourth, sixth, . . . term for the second

$$x \leftrightarrow 3, 2, \mathbf{4}, 2, \mathbf{1}, 1, \mathbf{2}, 5, \ldots \leftrightarrow \begin{cases} 3, 4, 1, 2, \ldots \leftrightarrow x_1 = 0.0010001101 \ldots, \\ 2, 2, 1, 5, \ldots \leftrightarrow x_2 = 0.0101100001 \ldots. \end{cases}$$

Thus the real numbers $0 < x \leq 1$ are in one-to-one correspondence with the pairs of real numbers $\{x_1, x_2\}$, $0 < x_i \leq 1$. If x is interpreted as a coordinate on the unit interval, and the pair x_1, x_2 as coordinates of a point in the square, the correspondence just described provides a one-to-one mapping of the segment (without its left endpoint $x = 0$) onto the square (without its left and lower edges $x_1 = 0$ and $x_2 = 0$), which means that the powers of the two sets of points are equal. By a suitable partition of the interval and the square we can also show that the *closed* segment and the *closed* square have the same power, but the above simpler example is enough to show that sets of different dimension can certainly have the same power, i.e., that dimension is not an invariant under one-to-one mappings.

Now it might seem that the power of the continuum, taken together with a linear order of its points, would be characteristic for the set of points of an interval. But as a counterexample we may take the set of all irrational points between 0 and 1, which on the one hand contains no subinterval, since between two irrational points there is always a rational point, and on the other hand has the same power as the whole interval $0 \leq x \leq 1$. Another interesting example, which we shall again construct in detail, is the *Cantor discontinuum,* where the point set is considerably more scattered than the irrational points, since every interval (x_1, x_2) with $0 \leq x_1 \leq x \leq x_2 \leq 1$ contains a subinterval entirely free of points of the discontinuum; nevertheless this set too has the same power as the unit interval.

For its construction we first divide the original interval into three equal parts and remove the middle third, not including its endpoints (Fig. 46). The second step then consists of dividing each of the two remaining intervals into three equal parts and again removing the middle part, not including its endpoints. In general, after the nth step, there will be 2^n subdivisions, each of length $\frac{1}{3}^n$, that have not yet been removed, and as the $(n + 1)$st step we divide each of them into three equal parts and remove the open middle intervals. The set of all points on the interval not in any of the deleted subintervals for any natural n is called the discontinuum D. For example, D includes all the endpoints of any of the subintervals. The continued division into three parts corresponds to our expressing the points in a triadic expansion. For each real number x with $0 \leq x \leq 1$

$$x = 0.\, a_1a_2a_3a_4 \ldots, \quad a_i = 0, 1, 2,$$

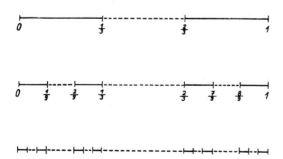

Fig. 46

and the points removed in the first step correspond to $a_1 = 1$, where for the left endpoint $\frac{1}{3}$ we choose the representation $0.0\bar{2}$. The second step removes the numbers with $a_1 \neq 1, a_2 = 1$, where again for the left endpoint we must choose the infinite expansion with no 1, etc. In general, D consists of all those points that have at least one triadic expansion not containing the digit 1. After the first step all points of D are covered by a set of intervals of total length $\frac{2}{3}$, after the second step, this total length has sunk to $\frac{4}{9} = (\frac{2}{3})^2$, and after the nth step to $(\frac{2}{3})^n$. The limit value of the sum of the lengths of the intervals that cover D (the so-called Lebesgue measure of D) is zero. Nevertheless, the set D, paradoxical as this fact may seem, still has the power of the continuum. To prove this, we assign to the endpoints x of the subintervals the following dyadic fractions y'

$$x = 0.\underbrace{00\ldots 0\bar{2}}_{n \text{ places}} \qquad \leftrightarrow y = 0.\underbrace{111\ldots 1}_{n \text{ places}},$$

$$x = 0.a_1 \ldots a_n 2 \qquad \leftrightarrow y = 0.\frac{a_1}{2}\ldots\frac{a_n}{2}01,$$

$$x = 0.a_1 \ldots a_n \underbrace{20\ldots 0\bar{2}}_{m \text{ places}} \leftrightarrow y = 0.\frac{a_1}{2}\ldots\frac{a_n}{2}\underbrace{011\ldots 1}_{m \text{ places}}, \quad \text{with} \quad m \geq 3,$$

while all the other points $x \in D$ correspond to the infinite dyadic expansions with a 1 in place of a 2.

$$x = 0.\, a_1 a_2 \ldots a_n \ldots \leftrightarrow y = 0.\frac{a_1}{2}\frac{a_2}{2}\ldots\frac{a_n}{2}\ldots.$$

If we now adjoin $x = 0 \leftrightarrow y = 0$ and $x = 1 \leftrightarrow y = 1$, we obtain a mapping of the set D onto all points of the segment $0 \leq y \leq 1$.

These examples show that the power of a set of points in itself does not provide any characteristic feature of curves or other geometric figures; some of the neighborhood properties of the points are needed as well. Since for the time being we shall consider only sets of points in the Euclidean space R^n of dimension n we shall mean by an "ε-*neighborhood* $U(p)$ *of a point* p" the totality of points distant from p by less than ε. Then if M is an arbitrary point set in R^n, a point q in R^n ($q \in R^n$) is called a *contact point* of M if every neighborhood $U(q)$ has points in common with M, or in other words, if the intersection of the two sets $U(q) \cap M \neq 0$ is not empty. The intersection of two sets, to be denoted by the symbol $\cap$, consists of all the elements that belong to both sets, and the union of two sets, denoted by $\cup$, consists of all the elements that belong to at least one of the sets (cf. IB, §7.2). Every point $p \in M$ is a contact point of M, since $U(p) \cap M$ contains at least the point p. The totality of all contact points of M forms the so-called *closure* $\bar{M}$ of M.

The set M is always contained in $\overline{M}$, i.e., $M \subset \overline{M}$. A set with $M = \overline{M}$ is said to be *closed*. By the *complement* of M with respect to R^n we mean the set of all points of R^n not contained in M, which we shall write as a difference: $R^n - M$. A set M is said to be *open* if $R^n - M$ is closed. The properties "closed" and "open" are not always mutually exclusive. For example, R^n is closed, $R^n = \overline{R^n}$, so that its complement $R^n - R^n$, the so-called *zero set* 0, containing no points at all, is open. On the other hand, the zero set is also closed: $0 = \overline{0}$, since the set of its contact points is empty, so that its complement R^n is open. Thus the entire space and the zero set are both open and closed. The set of contact points of a set M is to be carefully distinguished from the set of its limit points. The point p is called a *limit point* of M if for every neighborhood $U(p)$ we have $[U(p) - p] \cap M \neq 0$, i.e., every neighborhood contains at least one point of M distinct from p.

These concepts serve to describe the position of a point set M in R^n. For example, M is said to be *dense* in R^n if $\overline{M} = R^n$, and *nowhere dense* in R^n if $R^n - \overline{M}$ is dense in R^n. For example, the Cantor discontinuum D can be shown to be closed and nowhere dense in the unit interval. On the other hand, every point of D is also a limit point of points of D, a property described by saying that D is *dense in itself* [perfect].

Now that "denseness in itself" has been seen to be consistent with a scattered position of the points, let us turn to the requirement that the set shall be connected, defined as follows. Two sets A and B are said to be *disjoint* in R^n if $A \cap B$ is empty; on the other hand, they are said to be *separated* if both $\overline{A} \cap B = 0$ and $A \cap \overline{B} = 0$. For example, the segments $-1 < x < 0$ and $0 < y < 1$ are separated, whereas the adjunction of the origin to either of them would destroy that property, although they would still be disjoint. We now say that a set M is *connected* if it is not the union of two separated (nonempty) sets. An example of a connected set is given in Fig. 47, consisting of two subsets, namely the set of points $(x, \sin \pi/x)$, with $0 < x \leq 1$, and the set of points $(x, 0)$ with $-1 \leq x \leq 0$. For positive x-values the first of these subsets is not closed, since its closure also includes the interval $x = 0$, $-1 \leq y \leq +1$; and since this closure has the origin in common with the second subset, the two subsets are not separated, although they are disjoint and each of them is connected.

In analysis an important role is played by connected open sets of points, which are also called *domains*. Here our interest lies in *connected, closed* sets of points. If such a set is not degenerate, i.e., if it contains more than one point, and if it is bounded, it is called a *continuum*. In a Euclidean space a set of points is said to be *bounded* if the distances between points have a least upper bound. In our third section we shall consider more general spaces, in which the distance between two points is not necessarily defined, so that instead of boundedness we must have some other property. But we shall return to this question later.

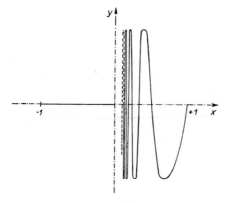

Fig. 47

With the concept of the continuum our first goal has been reached on the way to characterizing curves by the form of their sets of points; for if we wish to be guided by intuition, we must always require connectedness and closedness. Thus the example in Fig. 47 does not represent a curve. On the other hand, boundedness represents a restriction, in comparison with ordinary language, since a parabola, for example, is unbounded and is nevertheless called a curve. Without further discussion of the problems connected with the general idea of a curve, let us restrict ourselves here to the characterization of *arcs of curves,* which in any case must be assumed to be continua. The simplest example of an arc is an interval including its end-points *a* and *b*. If we compare it with a closed square containing *a* and *b*, which is also a continuum, we see that there are infinitely many sub-continua in the square that join *a* to *b* (e.g., T_1, T_2 in Fig. 48), whereas no

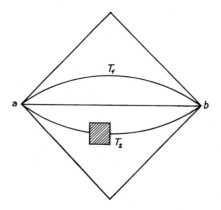

Fig. 48

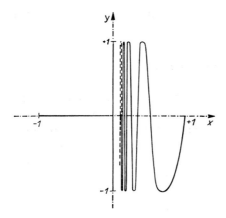

Fig. 49

proper subcontinuum of the interval contains both *a* and *b*, a property called *irreducibility* of the continuum with respect to its points *a*, *b*.

But we have not yet arrived at an acceptable characterization of a curve, as may be seen by extending the example in Fig. 47. Let us close the set there by adjoining the points $(0, y)$ with $-1 \leqq y \leqq +1$, so that we now obtain a continuum (Fig. 49), which is irreducible with respect to $(-1, 0)$ and $(1, 0)$, since every proper subset containing the endpoints is no longer a continuum. Nevertheless, we would scarcely be willing to regard this set as an arc of a curve; in his book on the theory of curves (Teubner, 1932), K. Menger has given a more complicated example of a continuum, irreducible between two points of three-dimensional space, which even contains the surface of a square as a proper subset. Thus we shall narrow the class of point sets under consideration by requiring that the continuum be *irreducibly connected with respect to two points,* by which we mean that no proper subset (which may or may not be a continuum) containing the two points is still connected. In explanation, let us note that the set in Fig. 47 is irreducibly connected with respect to $(-1, 0)$ and $(+1, 0)$ but is not itself a continuum. On the other hand, the example in Fig. 49 is an irreducible continuum with respect to $(-1, 0)$ and $(+1, 0)$, but it is not irreducibly connected since, for example, it contains the set in Fig. 47 as a proper connected subset. Thus neither of the examples is an arc of a curve.

If we now ask whether the definition of an arc of a curve as an irreducibly connected continuum excludes all point sets not ordinarily called arcs, we must make further comparisons with the line segment. For example, the points of a line segment are linearly ordered and this order can be transferred to any image of the segment under a continuous and one-to-one mapping. But since we do not yet know whether all irreducibly connected continua are included among such images or not, we must now turn our attention to continuous images of line segments.

7. Continuous Mappings of a Line Segment

A real function $y = f(x)$ is said to be continuous at the point x_0 if for arbitrarily preassigned ε there exists a δ such that $|f(x) - f(x_0)| < \varepsilon$ for all $|x - x_0| < \delta$. In its domain of definition this function defines a mapping from the x-axis to the y-axis and the continuity means that for every ε-neighborhood $U(y_0)$ on the y-axis with $y_0 = f(x_0)$ there exists at least one δ-neighborhood $U(x_0)$ on the x-axis for which the image $f(U(x_0))$ is contained in $U(y_0)$: $f(U(x_0)) \subset U(y_0)$. Since only the concepts of "neighborhood within a set" and "correspondence of points under a mapping" occur in this formulation, it enables us to define continuity of a mapping of arbitrary sets of points in which neighborhoods are defined. Here as before in this section, we confine ourselves to sets R of points in a Euclidean space R^n. If p is a point of R ($p \in R$), then by a "neighborhood $U(p)$ within the set R" we shall mean the intersection of a Euclidean neighborhood of p with the set R. The property of being a point of contact of a subset M in the set R is preserved under continuous mappings of R, i.e., the image of a point of contact of M is a point of contact of the image set; in other words, if $f(R)$ is a continuous mapping, then $f(\overline{M}) \subset \overline{f(M)}$ for every subset M of R. This invariance of a "point of contact" is characteristic of continuous mappings.

Every set of points that is the continuous image of a closed interval can be traversed continuously, since for this purpose we need only regard the line segment as an interval of time. But it would be wrong to assume that only segments of curves in the ordinary intuitive sense have the property that they can be traversed continuously. In 1890 Peano discovered that the surface of a square and the volume of a cube are also point sets of this kind. In general, Jordan defined curves as continuous images of line segments, but after the results of Peano it is better not to define a curve simply as a set that can be traversed continuously. Under continuous mappings, connected sets are taken into connected sets, but the image of a closed set is not necessarily closed. However, if we also require that the closed set in Euclidean space be bounded, then every continuous image is also closed and bounded. Thus the concept of the continuum is invariant under continuous mappings, and such continuous mappings of a line segment are called *Peano continua*.

But these continua do not necessarily have the property of the line segment of being irreducibly connected, as is shown by the following simple example. Let $\Delta = ABC$ be a right-angled nonisosceles triangle with the right angle at C (Fig. 50). The altitude HC divides Δ into Δ_0 and Δ_1, where Δ_0 denotes the smaller of the two subtriangles. Let H_0 be the foot of the perpendicular from H to the hypotenuse of Δ_0, and similarly for Δ_1 and H_1. Then HH_0 divides

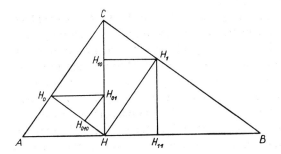

Fig. 50

Δ_0 into Δ_{00} and Δ_{01}, where again the second index 0 is assigned to the smaller subtriangle, and HH_1 divides Δ_1 into Δ_{10} and Δ_{11}. Similarly, the foot of the perpendicular from $H_{v_1 \cdots v_{n-1}}$ to the smaller of the sides of $\Delta_{v_1 \cdots v_{n-1}}$ is denoted by $H_{v_1 \cdots v_{n-1} 0}$ and the foot of the perpendicular to the longer side is $H_{v_1 \cdots v_{n-1} 1}$; and the respective subtriangles are $\Delta_{v_1 \cdots v_{n-1} 0}$ and $\Delta_{v_1 \cdots v_{n-1} 1}$. Every triangle $\Delta_{\mu_1 \cdots \mu_m}$ is properly contained in $\Delta_{\mu_1 \cdots \mu_{m-1}}$ and therefore, by the principle of nested intervals, every sequence of closed triangles $\Delta_{\mu_1}, \Delta_{\mu_1 \mu_2}, \Delta_{\mu_1 \mu_2 \mu_3}, \ldots$ determines exactly one point p of the triangle Δ. These sequences provide a mapping of the points x with $0 \leq x \leq 1$ into the points p of the triangle Δ, if to the subscripts of a given triangle we assign the real number x with these subscripts as its dyadic expansion

$$f: x = 0. \, \mu_1 \mu_2 \mu_3 \ldots \leftrightarrow p = \{\Delta_{\mu_1}, \Delta_{\mu_1 \mu_2}, \Delta_{\mu_1 \mu_2 \mu_3}, \ldots\}.$$

If x has two dyadic expansions, the two representations of x lead to the same point in Δ, for example, $x = 0.1\bar{0} \ldots \rightarrow C$; $x = 0.0\bar{1} \ldots \rightarrow C$, so that the mapping is well defined, although it is not one-to-one, since the point H, for example, is the image of $x_1 = 0.01\bar{0}$ and also of $x_2 = 0.11\bar{0}$, although $x_1 \neq x_2$. On the other hand, for every point p in the triangle Δ there exists at least one sequence of triangles of the above kind, containing p either in the interior or on a side, so that the correspondence f maps the interval $0 \leq x \leq 1$ one-valuedly onto the triangle Δ; and it is easy to see that the mapping is continuous, since in every ε-neighborhood $U(p)$ there is a closed subtriangle $\Delta_{v_1 \cdots v_n}$ containing p whose points are images of all x-values with the same first n dyadic digits. For a vertex p of an acute angle in $\Delta_{v_1 \cdots v_n}$, a suitable neighboring subtriangle must also be taken into account.

The triangle ABC is a reducible continuum, with respect to the points A and B, even though it is the continuous image of an irreducible line segment. So we now restrict the set of Peano continua by the further requirement that the inverse mapping onto the line segment must also exist and be continuous. Such one-to-one bicontinuous mappings are topological, by definition, and the special Peano continua that are homeomorphic to a line segment are called *Jordan arcs*. But even this class of geometric figures contains some very remarkable sets of points. For example, if we let $X(x)$, $Y(x)$ be the plane Cartesian coordinates of the image of x in the continuous mapping from $0 \leqq x \leqq 1$ onto the triangle ABC in Fig. 50 and, following Menger, mark off the value of x as a Z-coordinate $Z(x) = x$ in a third coordinate direction, we obtain a one-to-one bicontinuous mapping of the line segment onto a point set in R^3. This Jordan arc has the property that when it is projected onto the X, Y-plane it completely fills out the triangle, since it then goes into the Peano continuum discussed above. The Jordan arc itself does not fill out any segment of the surface, yet it is irreducibly connected with respect to its endpoints, since otherwise every properly connected subset containing the endpoints would be mapped (because of the one-valuedness and continuity of the inverse mapping) onto a connected proper subset of the line segment that would include the endpoints. But such a subset does not exist.

But even in the Euclidean plane it is possible to give examples of Jordan arcs that are extremely difficult to visualize; for example, the following curve, due to H. von Koch, has no tangent at any point (cf. Mangoldt-Knopp, Einführung in die höhere Mathematik, Vol. 2). The continuum M consisting of the surface of an isosceles triangle with base angles $\pi/6$ is divided into two continua M_0 and M_1 (Fig. 51) by a subdivision of its base into three parts, and elimination of the middle third. The two remaining subtriangles, congruent to each other, are dealt with correspondingly, their

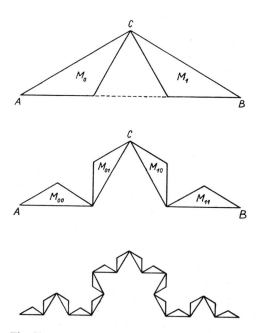

Fig. 51

subdivisions being called M_{00}, M_{01}, M_{10}, M_{11}, where the second index increases in the direction from A to B. Continuation of this procedure leads to a sequence of closed triangles $M_{i_1 i_2 \ldots i_n}$ with $i_v = 0, 1$ $(n \to \infty)$. Then, as before, to each real number in the unit interval $0 \leq x \leq 1$ with the dyadic expansion $x = 0.i_1 i_2 i_3 \ldots$ we assign as image point the point p uniquely determined by the nesting of triangles

$$p = \{M_{i_1}, M_{i_1 i_2}, M_{i_1 i_2 i_3}, \ldots\}.$$

Since this mapping turns out to be topological, the set of image points is a Jordan curve. It consists of the totality of the vertices of all the triangles and their limit points. Since every triangle, no matter how small, is similar to the original triangle and, when it is subdivided, contains the original arrangement of triangles on a smaller scale, the resulting set of points is of a very complicated character.

We have already seen that the topological image of a line segment must be irreducibly connected. But now let us ask conversely: Can every continuum M in R^n irreducibly connected with respect to two points a, b be mapped topologically onto a line segment? We shall first show that the points of M can be "ordered." An arbitrary point p in M, distinct from a, b, is a *separation point* for M, i.e., $M - p$ falls into two separated sets, A', B', with $a \in A'$, $b \in B'$, since otherwise M would not be irreducibly connected. Then we note that the set $A(p) = A' \cup p$ is irreducibly connected with respect to a, p, and that for two distinct separation points p, q in M either $A(p) \subset A(q)$ or $A(q) \subset A(p)$. The correspondence $p \to A(p)$ is one-to-one, and we now supplement it by setting $A(a) = a$, $A(b) = M$.

The inclusion relation among subsets of a set is an example of a more general structure called a *partial order* (or *semi-order*) defined as follows (cf. IA, §8.3). *For the set of elements α, β, γ, $\ldots$ a relation $\leq$ is defined (i.e., for every two elements α, β in the set it is determined whether $\alpha \leq \beta$ or not), such that*

1. $\alpha \leq \alpha$ *for all α (reflexivity),*
2. $\alpha \leq \beta$ *and* $\beta \leq \gamma$ *imply* $\alpha \leq \gamma$ *(transitivity),*
3. $\alpha \leq \beta$ *and* $\beta \leq \alpha$ *imply* $\alpha = \beta$ *(identitivity of Hermes).*

A partial order is called a *total order* if at least one of $\alpha \leq \beta$ and $\beta \leq \alpha$ holds for any two elements α, β. If $\alpha \leq \beta$ with α distinct from β, we shall write $\alpha < \beta$, for brevity. The subsets $A(p)$ enable us to introduce an order into M by setting $p < q$, for distinct p and q, if $A(p) \subset A(q)$. The relation $p < q$ is also described in words by saying that "p comes before q." The real numbers form another example of an ordered set with the order relation "smaller than or equal to." An ordered set may have a first element α_0, in which case $\alpha_0 \leq \alpha$ for α; and similarly α_0 is called a "last element" of the set if $\alpha \leq \alpha_0$ for all α. In order to distinguish among various types of order we consider a decomposition of an ordered set S into two disjoint subsets S_1 and S_2 $(S = S_1 \cup S_2, S_1 \cap S_2 = 0)$ such that $\alpha_1 < \alpha_2$ for any $\alpha_1 \in S_1$ and any

$\alpha_2 \in S_2$. Then there are the following possibilities (Hausdorff): a *jump*, i.e., S_1 has a last element and S_2 has a first element; a *cut*, i.e., either S_1 has a last element but S_2 has no first element or S_2 has a first element but S_1 has no last element; and finally a *gap*, i.e., S_1 has no last element and S_2 has no first element. Ordered sets that have neither jumps nor gaps are said to be *continuously ordered*, a property obviously equivalent to the Dedekind cut axiom (see IB1, §4.3). It is easy to show that the ordered set of points in an irreducibly connected continuum is continuously ordered.

The rational points of the open segment $0 < r < 1$ form a countable set, which can be ordered, for example, in the following way

$$r_1 = \tfrac{1}{2}, r_2 = \tfrac{1}{3}, r_3 = \tfrac{2}{3}, r_4 = \tfrac{1}{4}, r_5 = \tfrac{3}{4}, r_6 = \tfrac{1}{5}, r_7 = \tfrac{2}{5}, r_8 = \tfrac{3}{5}, r_9 = \tfrac{4}{5},$$
$$r_{10} = \tfrac{1}{6}, \ldots .$$

If we denote the set of all r_i by $\{r\}$ and the closed segment by T, then $\{\bar{r}\} = T$, i.e., the set $\{r\}$ is dense in T. A set N containing a countable subset dense in N is said to be *separable*. If a given set M is to be mapped topologically onto the line segment, it must also be separable. If M lies in a Euclidean space R^n, as we are assuming here, it can easily be shown to be separable, but in more general spaces, such as those considered in the third section, this property must be assumed as a supplementary condition.

Thus there exists a countable set $\{p\} = p_1, p_2, p_3, \ldots$, dense in M, from which we shall for the moment exclude the points a, b, between which M is irreducibly connected; and we now define a mapping of $\{p\}$ onto $\{r\}$ as follows.

1. p_1 is assigned to r_1.

2. In the remaining subset $\{r\} - r_1$ we choose the r_i with the smallest index, i.e., in this case r_2. Since $r_2 < r_1$, we may seek in the remaining set $\{p\} - p_1$ the p_k with smallest index that comes before $p_1(p_k < p_1)$ and map p_k onto r_2.

3. From $\{p\} - p_1 - p_k$ we choose the p_j with smallest index. Then p_j may be situated in various ways with respect to p_1, p_k; for example, let us assume $p_k < p_j < p_1$, and we now choose an r_j in the corresponding relationship to r_1, r_2, namely $r_2 < r_j < r_1$, where r_j has the smallest index in $\{r\} - r_1 - r_2$. (In the present case, $r_j = r_7$.)

We then proceed alternately. The choice of smallest index in each case means that in the course of the construction all elements of all the sets are assigned to one another. Since $\{p\}$ and $\{r\}$ are dense in M and neither of them can have a first or a last element, there will always be a further element between any two given elements in the set $\{p\}$, so that the assignment can always be made. If we now adjoin the endpoints $0 \leftrightarrow a$, $1 \leftrightarrow b$, the set $\{p\} \cup a \cup b$ is mapped one-to-one, with preservation of its order, onto the set of rational points in T. The other points in T, namely the irrational ones, can be defined by Dedekind cuts (i.e., decompositions in two subsets) in the set of rational points, and if these decompositions are transferred to the set $\{p\} \cup a \cup b$, it follows from the continuity of the order in M that we have thereby obtained a one-to-one mapping of all the points of M onto T. Is this mapping bicontinuous? The answer might seem to be self-evident, because of the similarity and continuity of the orders in the two sets M and T, but in set-theoretic topology we must be very careful about intuitive conclusions, as the above discussion shows. Consider, for example, the points of the irreducibly connected set M in Fig. 47. A one-to-one mapping of its points $p = (x, \sin \pi/x)$ for $0 \leq x \leq 1$ and $p = (x, 0)$ for $-1 \leq x \leq 0$ onto an x^*-interval $0 \leq x^* \leq 1$ is given by the function $x^* = \frac{1}{2}(x + 1)$, where the points p are ordered in the same way as the points of the line segment, if we define $p_1 < p_2$ by $x_1^* < x_2^*$. But this mapping is not continuous. The set $x^* \subset T$ with $x^* = (3 + 4n)/(2 + 8n)$, $n = 1, 2, 3, \ldots$ has the limit point $x^* = \frac{1}{2}$, but the image set $p = (2/(1 + 4n), 1)$ has no limit point at all in the *interior* of M.

Thus we must still require from the set M that every infinite countable subset of M has at least one limit point in M, since T has this property and does not lose it under topological mapping. Nowadays such a set is said to be compact (Fréchet) or, recently and more precisely, *countably compact,* since the development of topology has led to various other kinds of compactness. In a Euclidean space R^n every bounded closed set M is countably compact, since any countable subset of the bounded set M is bounded and therefore, by the Bolzano-Weierstrass theorem, has at least one limit point in R^n, which must belong to the closed set M. This is the underlying reason why we have considered only continua from the beginning; since they are countably compact, the one-to-one mapping onto a line segment that is determined by the order of their points is necessarily continuous and therefore topological. (The counterexample in Fig. 47 is not a continuum.) A long and complicated path has at last brought us to the result: *Irreducibly connected continua in a Euclidean space are topological images of a line segment* and thus are Jordan arcs. In this way we have given a topological characterization of the Jordan arcs.

Similarly, we could ask for a characterization of the Peano continua, which were defined as continuous images of a line segment and are often called "continuous curves" although, as we have seen, they may fill up an entire segment of a surface. But we must be satisfied here with the above sketch of a proof of the fact, and will now turn to the important question of the imbedding of a Jordan curve in a plane.

8. Jordan Curve Theorem

By a Jordan curve J we mean the topological image of a circle. It is obvious that this set of points is characterized by the property that for any pair of distinct points a, b in J it consists of two Jordan arcs J_1, J_2 from a to b, which are separated if we remove the endpoints a, b. By the earlier definition we can also consider J as a (curvilinear) polyhedron, namely as a one-dimensional sphere. The circle divides the plane R^2 into two disjoint domains and is at the same time their common boundary. Is this property transferred to every Jordan curve? Although the von Koch curve shows that a Jordan curve can be very difficult to visualize, one would still be inclined to believe that not more than two domains can be separated by a boundary curve, and that one of them will lie on the right as we traverse the boundary, and the other on the left. But in the von Koch curve there is no definite direction of traversal at any point, so that statements about direction become meaningless; moreover, boundaries do not need to consist of Jordan arcs. What are we to mean in general by the boundary of a domain? A domain is an open, connected point set G, and the *boundary* of G is defined as the intersection of the closure $\bar{G}$ with the complementary set $R^2 - G$. Every neighborhood of a point on the boundary contains both a point of G and a point of its complement $R^2 - G$. But then in 1910 it was discovered by Brouwer that with this definition of a boundary there can exist three or even more disjoint domains in the plane, all of which have the same boundary.

In order to give at least some idea of the construction of these remarkable sets of points let us consider a simple example of the decomposition of a square into three domains with a common boundary (cf. von Kerékjártó [5]) as in Fig. 52. Each of the three domains G_1, G_2, G_3 is represented by a sequence of closed sets of points which are extended at each step (G_1 is the union of $A_1 \cup A_2 \cup \ldots$, G_2 of $B_1 \cup B_2 \cup \ldots$, and G_3 of $C_1 \cup C_2 \cup \ldots$), so that taken all together they gradually fill out the entire square. The first set in G_1 is the small square A_1, with side $\frac{1}{9}$ of the side of the original square. In the second domain G_2, the first set B_1 is a bent strip, of width $\frac{1}{9}$ of the space between A_i and the left edge of the original square, beginning in the center of this space and continuing up, across and down to the right of A_1 (in the figure B_1 is dotted, but its edge is drawn with a solid line). In the third domain G_3, the first set C_1 begins at the left (with width equal to $\frac{1}{9}$ of the space between B_1 and the original left edge) on the same level as the upper edge of B_1, makes six turns (between A_1, B_1 and the original square) and returns to its original height (in the figure, C_1 is hatched with lines down to the right). Beginning again with G_1, we construct its second set A_2 by the same principle, starting at the far left on the same level as the lower edge of C_1, continuing along a strip through the middle of the free intervening space with $\frac{1}{9}$ of its breadth, over to the far right, and then back around A_1; and finally the space (including A_1) surrounded by the last three segments of A_2 is included in A_2 (in the figure, A_2 is hatched with lines down to the left, $A_1 \subset A_2$). The second part B_2 of G_2 is merely indicated in its various directions by a dashed line, beginning in the upper left corner at the same height as A_2 and

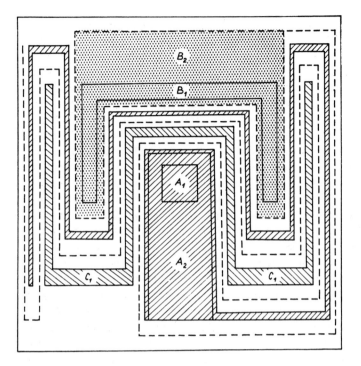

Fig. 52

including B_1 at its other end. The procedure is carried out in cyclic succession for the three domains, beginning alternately either in the lower left corner or in the upper left corner. Every new strip has a breadth equal to $\frac{1}{9}$ of the remaining intermediate space and ends with a circumnavigation of the preceding part of the same domain G_i, which is then included, together with the intervening area in the new part, as before. We thus obtain three disjoint domains $G_1 = \lim_{i \to \infty} A_i$, $G_2 = \lim_{i \to \infty} B_i$, $G_3 = \lim_{i \to \infty} C_i$ which, together with their common boundary, fill up the entire square. In every ε-neighborhood, however small, of any point of the boundary, there lie points of all three domains. For the proof, we need only carry out the construction far enough so that the domain not yet subdivided, which will include all boundary points, does not contain any open circular disk of radius ε. So the boundary of any one of the three domains is at the same time the boundary of the other two and is a closed connected point set.

In order to study the various possibilities for the connectivity of a plane open set M, let us return to the combinatorial methods of the first section. Since any complex for the Euclidean plane will have infinitely many simplexes, let us transfer our attention to the surface S^2 of the sphere obtained by stereographic projection from the plane, with subsequent adjunction of the North Pole N. Then by the projection the set M will be mapped topologically onto a set $M^* \subset S^2 - N$. Now let K be a finite simplicial complex for a sphere S^2 and let H be an open subset of S^2 with finitely many connected components. The triangles in the complex and their subdivisions are to be spherical triangles and, as before, a neighborhood of a point $p \subset S^2$ is the intersection of a three-dimensional neighborhood of p with the sphere S^2. A "chain $\mathfrak{f}^r$ in H" is defined as an integer linear combination of simplexes σ_i^r from K, or from a subdivision complex of K, where the closed sets of points in σ_i^r lie in H. Two "cycles $\mathfrak{z}_1, \mathfrak{z}_2$ in H" are said to be "homologous in H", $\mathfrak{z}_1 \sim \mathfrak{z}_2$, if there exists a chain $\mathfrak{f}$ in H whose boundary $\partial\mathfrak{f} = \mathfrak{z}_1' - \mathfrak{z}_2'$ is the difference of subdivisions of $\mathfrak{z}_1$ and $\mathfrak{z}_2$. Thus the investigation of chains in H involves not only K but all the simplicial subdivisions of K. For example, if we can join two points in K by a spherical finite broken line lying in the interior of H, there will be a chain $\mathfrak{f}^1$ in H corresponding to this broken line, since the original complex can always be subdivided so finely that all points of the broken line correspond to vertices σ_i^0. The set of all cycles can again be classified into homology classes, which form the elements of a homology group, called the *homology group* $\mathfrak{H}^r(H)$ *of the open set H.*

The Betti number p^0 gives the number of connected subsets of H, since every two points of a connected open set G in S^2 can be joined by a spherical broken line in G, e.g., by the line (Fig. 53) joining the midpoints of a finite sequence of (spherical) circular disks U_i with $U_i \cap U_{i+1} \neq 0$ in the interior of G. The Betti number p^1, on the other hand, determines the type of connectivity in the following sense. An open set H is said to be *m-connected* if $p^0 = 1$ and $m = p^1 + 1$. A plane domain G is *m*-connected if its spherical image G^* is *m*-connected. Note that G^* does not contain the North Pole N.

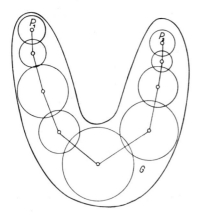

Fig. 53

For example, the exterior of a circle in the plane is 2-connected. The Betti number p^2 of H is different from zero only if $H = S^2$.

If H is the union of two domains H_1 and H_2 on the sphere with nonempty intersection $H_1 \cap H_2 = H_D \neq 0$, let us now investigate the connectivity groups of H and H_D, where we may assume that neither H_1 nor H_2 alone covers the whole sphere S^2. We start from a cycle $\mathfrak{z}_D^r$ in H_D homologous to zero in both domains H_i ($i = 1, 2$): $\mathfrak{z}_D^r = \partial \mathfrak{f}_i^{r+1}$. The chain $\mathfrak{f}_1^{r+1} - \mathfrak{f}_2^{r+1}$ lies in $H = H_1 \cup H_2$ and is a cycle $\mathfrak{z}^{r+1} = \mathfrak{f}_1^{r+1} - \mathfrak{f}_2^{r+1}$ since $\partial \mathfrak{f}_1^{r+1} = \partial \mathfrak{f}_2^{r+1}$. Conversely, to every cycle $\mathfrak{z}^{r+1}$ in H we can assign a cycle $\mathfrak{z}_D^r$ in H_D homologous to zero in both the H_i. For this purpose we need only represent $\mathfrak{z}^{r+1}$ as the sum of two chains $\mathfrak{f}_1^{r+1}$ in H_1 and $-\mathfrak{f}_2^{r+1}$ in H_2, whereupon $\mathfrak{z}_D^r = \partial \mathfrak{f}_1^{r+1} = \partial \mathfrak{f}_2^{r+1}$ in H_D. For this assignment the "Alexander lemma" holds: If $\mathfrak{z}^{r+1} \sim 0$ in H, then $\mathfrak{z}_D^r \sim 0$ in H_D. The proof depends on representing both $\mathfrak{z}^{r+1}$ and $\mathfrak{f}^{r+2}$, where $\mathfrak{z}^{r+1} = \partial \mathfrak{f}^{r+2}$ in H, as a sum of chains in H_i: $\mathfrak{z}^{r+1} = \mathfrak{f}_1^{r+1} - \mathfrak{f}_2^{r+1} = \partial \mathfrak{f}^{r+2} - \partial \mathfrak{f}_2^{r+2}$. The chain $\mathfrak{f}_1^{r+1} - \partial \mathfrak{f}_1^{r+2} = \mathfrak{f}_2^{r+1} - \partial \mathfrak{f}_2^{r+2}$ lies in H_D, and its boundary is $\mathfrak{z}_D^r$.

The Alexander lemma enables us to describe the position of a Jordan arc B on the sphere S^2. If we recall that such an arc can wind around the sphere in an arbitrarily complicated way, and that every subarc can be as non-intuitive as a von Koch curve, we will realize that any statement about the connectivity relations in the open complementary set $S^2 - B$ will not be self-evident. Since B is irreducibly connected, no segment of the surface is completely covered, and $S^2 - B$ is not empty. We choose two points p_1, p_2 in $S^2 - B$, and assign to them (after suitable subdivision of the complex K) the chain $\mathfrak{z}^0 = \sigma_2^0 - \sigma_1^0$, and then ask whether $\mathfrak{z}^0 \sim 0$ in $S^2 - B$. Let $p(x)$, $0 \leq x \leq 1$, be an arbitrary point of B. In the domain $S^2 - p(x)$ the cycle $\mathfrak{z}^0$ is bounding, with $\mathfrak{z}^0 = \partial \mathfrak{f}^1(x)$ in $S^2 - p(x)$, since p_1 can be joined to p_2 by a finite broken line not containing $p(x)$. There always exists a neighborhood $U(p(x))$, disjoint to the broken line, which induces a neighborhood $V(p(x)) \subset U(p(x)) \cap B$ of the point $p(x)$ on the arc. To this neighborhood on the arc there corresponds either an open interval $x' < x < x''$ on the unit line segment, or else a half-open interval if one of the two ends of the arc lies in the given neighborhood (Fig. 54). To every closed subinterval $[x_1 x_2]$, with $x' < x_1 \leq x \leq x_2 < x''$, there corresponds a subarc B_{12} contained in $U(p(x))$. Thus the chain $\mathfrak{f}^1(x)$ also lies in $S^2 - B_{12}$, and $\mathfrak{z}^0 \sim 0$ in $S^2 - B_{12}$. In this way, for a prescribed $\mathfrak{z}^0$, an open interval $(x'x'')$ is assigned to every x. The set of these (open) intervals covers the unit line segment, so that the well-known "Heine-Borel theorem" applies: Every open covering of the unit line segment contains a finite subcovering. Sets for which every open covering contains a finite subcovering are called *compact*. Let us here merely state that in a Euclidean space every countably compact set is compact, and conversely. Thus for a given $\mathfrak{z}^0$ in $S^2 - B$ we can always find a finite subdivision of the unit line segment, i.e., by means of finitely

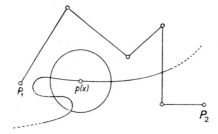

Fig. 54

many points $x_0 = 0, x_1, x_2, \ldots, x_{n-1}, x_n = 1$, such that $\mathfrak{z}^0 \sim 0$ in $S^2 - B_{j,j+1}$, since every interval $[x_j x_{j+1}]$ lies in a suitable open interval $(x'x'')$. The union of the two open sets $S^2 - B_{01}$ and $S^2 - B_{12}$ is the set $S^2 - p(x_1)$ and their intersection is the set $S^2 - B_{01} \cup B_{12} = S^2 - B_{02}$. The cycle $\mathfrak{z}^0$ is zero-homologous in $S^2 - B_{01}$ and $S^2 - B_{12}$, and the corresponding cycle $\mathfrak{z}^1$ in the union $S^2 - p(x_1)$ is likewise zero-homologous, since $S^2 - p(x_1)$ is simply connected, and therefore by the Alexander lemma $\mathfrak{z}^0 \sim 0$ in $S^2 - B_{02}$. We can now repeat the argument by starting from sets $S^2 - B_{02}$ and $S^2 - B_{23}$, and thus obtain $\mathfrak{z}^0 \sim 0$ in $S^2 - B_{03}$. After finitely many steps it follows that $\mathfrak{z}^0 \sim 0$ in $S^2 - B_{0n} = S^2 - B$. Since the proof holds for any two points in $S^2 - B$, all these points can be joined by a broken line in $S^2 - B$, which means that the set $S^2 - B$ is connected. But $S^2 - B$ is also simply connected, as is proved in a similar way. For let us choose a cycle $\mathfrak{z}^1$ in $S^2 - B$ and an arbitrary point $p(x) \in B$. In $S^2 - p(x)$ the cycle $\mathfrak{z}^1$ is zero-homologous, $\mathfrak{z}^1 = \partial \mathfrak{f}^2(x)$, since $S^2 - p(x)$ is simply connected. A neighborhood $U(p(x))$ is then determined which is disjoint to the triangular surfaces corresponding to the chain $\mathfrak{f}^2(x)$, and the rest of the argument is completely analogous. It follows that $\mathfrak{z}^1 \sim 0$ in $S^2 - B$ for all $\mathfrak{z}^1$ in $S^2 - B$, i.e., $S^2 - B$ is simply connected.

The converse of the Alexander lemma, namely if $\mathfrak{z}^{r+1} \sim 0$ in H, then also $\mathfrak{z}_D^r \sim 0$ in H_D, holds only for domains H_i in which every $(r + 1)$-dimensional cycle is bounding. This theorem is proved in the same way as above, by showing under the stated assumptions, that $\mathfrak{z}_D^r \sim 0$ in H_D implies $\mathfrak{z}^{r+1} \sim 0$ in H. As an application let us calculate the connectivity relations in a sphere from which a Jordan curve J has been removed. The curve J can be considered as consisting of two arcs with common endpoints a, b: $J = B_1 \cup B_2$, $B_1 \cap B_2 = a \cup b$. The domains $H_i = S^2 - B_i$ are simply connected, and therefore every cycle $\mathfrak{z}_i^1$ bounds in H_i. Thus the assumptions for the converse of the Alexander lemma are satisfied for the dimension $r = 0$. The union $H_1 \cup H_2$, namely the domain $H = S^2 - a \cup b$, covers the entire sphere with the exception of the two points a, b and is therefore doubly connected, as can be seen at once. Since $p^1 = 1$, the homology group

$\mathfrak{H}^1(H)$ has one free generator. In view of the Alexander lemma and its coverse, this property carries over to the homology group generated by the cycles $\mathfrak{z}_D^0$ in H_D. Every zero-dimensional cycle in H_D that is not bounding in H_i is composed of a suitable cycle $\mathfrak{z}^0$ and a multiple of a fixed chosen vertex σ_j^0 in H_D; i.e., $\mathfrak{z}^0 = \lambda\sigma_j^0 + \mathfrak{z}_D^0$. Thus the homology group $\mathfrak{H}^0(H_D)$ has two free generators, and its Betti number is $p_D^0 = 2$. But now $H_D = (S^2 - B_1) \cap (S^2 - B_2) = S^2 - J$. So our result means that the point set $S^2 - J$ consists of two separated domains G_1 and G_2. The set of limit points of $S^2 - J$ is given by $\overline{(S^2 - J)} \cap J$ and is therefore a subset of J. Every proper closed subset of J lies in a subarc $B \subset J$. But we have shown that a Jordan arc B does not divide the sphere S^2, so that this subset cannot be a proper one, and therefore the Jordan curve J coincides with the whole boundary of $G_1 \cup G_2$. After stereographic projection back onto the plane, we have thus obtained the "Jordan curve theorem": *Every closed Jordan curve divides the plane into two separated domains, for which it is their common boundary.* Consequently, the boundary in the Brouwer example is not a Jordan curve.

The above proof of the Jordan theorem made essential use of the homology properties of the sphere S^2, so that the theorem is not valid for other closed two-dimensional manifolds. For example, on the two-sheeted Riemann surface of the function

$$w = \sqrt{(z - z_1)(z - z_2)(z - z_3)(z - z_4)}$$

with the branch cuts $[z_1 z_2]$, $[z_3 z_4]$, the interior and the exterior of a Jordan curve J in the first sheet, around one of the two cuts, can be joined (Fig. 55) by an arc B crossing the cuts and lying partly in the second sheet. Thus the Riemann surface is not divided by J, but is topologically equivalent to a torus, as can be seen if the cuts in the number spheres corresponding to the two sheets are folded slightly outwards and the spheres are then deformed into cylinders (Fig. 56). On the other hand, there is a generalization of the Jordan theorem in another respect. With the same method of proof it can be shown that the topological image of an n-dimensional sphere in an $(n + 1)$-dimensional sphere S^{n+1} divides this space into two domains with a common boundary.

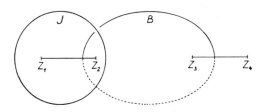

Fig. 55

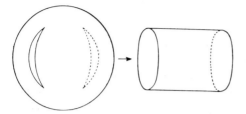

Fig. 56

Other topological theorems can be obtained from the Jordan curve theorem. For example, we have seen that sets of points of different dimension can be mapped onto one another by one-to-one mappings, and also by one-valued continuous mappings. Thus it is not at all self-evident that this possibility does not exist for topological mappings. The topological invariance of the number of dimensions was first proved by Brouwer (1911). We shall here discuss a special case, and for that purpose we start with the full sphere $x^2 + y^2 + z^2 \leq 1$ in the Euclidean R^3. If there existed a topological mapping of this set of points into the Euclidean plane R^2, each of the circular circumferences $x^2 + y^2 + z_0^2 = 1$ with $-1 < z_0 < 1$ would be mapped onto a plane Jordan curve $J(z_0)$. But these image curves cannot all lie outside of one another, for if a point with rational coordinates is chosen inside each $J(z_0)$, the set of these points is countable, whereas the set of Jordan curves $J(z_0)$ has the power of the continuum. So there are noncountably many Jordan curves within one another. Let us choose three of them (J_1, J_2, J_3) such that J_3 lies inside J_2 and J_1 lies outside J_2. The corresponding circles K_1 and K_3 can be joined to each other on the full sphere by a line segment that has no point in common with the third circle K_2. Thus the topological image of this line segment is an arc that joins J_1 with J_3 without crossing J_2. But by the Jordan theorem this is impossible, so that the assumption that the whole sphere can be mapped into the plane leads to a contradiction. Consequently we have proved the Brouwer theorem that the plane cannot be mapped topologically onto a Euclidean space of higher dimension, since the latter always contains three-dimensional full spheres. With the above generalization of the Jordan theorem, this result can be extended to Euclidean spaces of arbitrary dimension.

Topological Spaces

9. Metric Spaces

In the first section we started with triangulations of geometric figures in Euclidean space, but then, in order to bring out more clearly the essential features of the possible relationships among simplexes, we transferred our attention to abstract elements, connected with one another by certain incidence relations, but without any significance in themselves. In this way the applicability of combinatorial methods is greatly extended, since the new elements called "simplexes" can represent objects of the most varied kind. Our second section dealt with the neighborhood properties of sets of points in Euclidean space. But here again it is natural to study these relationships for abstract elements, in order to bring out the essential character of the concept of a space. So in the present section we shall turn from points of Euclidean space to abstract elements of a set S, which we shall also call "points," without thereby implying anything about them. This set S will become a "space R" if for its abstract points we postulate suitable neighborhood properties, as can be done in various ways. To begin with, we may follow the pattern of Euclidean spaces, and require the existence of a real, nonnegative function $d(p, q)$, defined for every pair of elements p, q and called the *distance* between the two points p, q, such that the following axioms are satisfied.

1. $d(p, q) = 0$ if and only if $p = q$ (identitivity of Hermes; cf. IA, §7.2),
2. $d(p, q) = d(q, p)$ (symmetry),
3. for all p, q, r

$$d(p, q) \leq d(p, r) + d(r, q) \quad \text{(triangle inequality)}.$$

A set S with a distance function of this kind is called a *"metric space."* The simplest example of a metric space is the Euclidean space R^n as the set of all n-tuples of real numbers $x = (x_1, x_2, \ldots, x_n)$ with the distance function

$$d(x, y) = \sqrt{\sum_1^n (x_i - y_i)^2}.$$

If the "points" x are the infinite sequences of real numbers $x_1, x_2, \ldots,$ $x_n, \ldots$ such that their squares form a convergent series $\sum_1^\infty x_i^2$, their distance function can be defined by

$$d(x, y) = \sqrt{\sum_1^\infty (x_i - y_i)^2},$$

since it follows from

$$\sqrt{\sum_1^N (x_i - y_i)^2} \leqq \sqrt{\sum_1^N x_i^2} + \sqrt{\sum_1^N y_i^2}$$

for arbitrary N that the series

$$\sum_1^\infty (x_i - y_i)^2$$

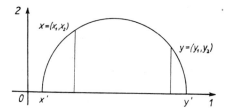

Fig. 57

converges, and it is easily proved that the axioms are satisfied. This infinite-dimensional generalization of the Euclidean spaces is called *"Hilbert space."*

The pairs of real numbers (x_1, x_2) with $x_2 > 0$ can also be metrized in a "non-Euclidean" way. Let x', y' be those points of the 1-axis (Fig. 57) in which a circle through the points $x = (x_1, x_2)$ and $y = (y_1, y_2)$ cuts the 1-axis orthogonally, where $y' < x'$ $(y' > x')$ if $y_1 < x_1$ $(y_1 > x_1)$. As the distance function we define

$$d(x, y) = \frac{1}{2} \log \left(\frac{x_1 - y'}{y_1 - y'} : \frac{x_1 - x'}{y_1 - x'} \right)$$

with $d(x, y) = |\log y_2 - \log x_2|$ for $x_1 = y_1$, the proof of the triangle inequality being now somewhat more troublesome. The metric space thus formed is the Poincaré model of a hyperbolic plane. We see from this example that the point set consisting of the pairs of numbers becomes a space only after a distance function has been defined, and that different functions lead to different spaces (e.g., Euclidean and non-Euclidean). The abstract conception of space is due to Fréchet (1906), who gave the following example of a metric space with applications in functional analysis. The "points" are the real continuous functions in the closed interval $0 \leq t \leq 1$, and the distance between two elements is the maximum of the absolute value of the difference of the two functions, $d(f, g) = \max |f(t) - g(t)|$.

In any metric space we can introduce neighborhoods by defining an ε-neighborhood of p as the set of points of R whose distance from p is less than ε:

$$U_\varepsilon(p) = \{q | d(p, q) < \varepsilon\}.$$

By the distance of a point p from a set $M \subset R$ we then mean the greatest lower bound of the distances $d(p, M) = \inf d(p, q)$ with $q \in M$, and all the set-theoretic topological concepts of the preceding section (contact point, limit point, open, closed, countable, compact, etc.) can be transferred, by means of the ε-neighborhoods, to an arbitrary metric space. The name "contact point of a set M" is then explained by the fact that $p \in \overline{M}$ is equivalent to $d(p, M) = 0$. An arbitrary subset M of a metric space R is in itself a metric space, where the distance function of R is applied only to points in M. To two metric spaces R_1 and R_2 with the functions $d_1(p_1, q_1)$ and $d_2(p_2, q_2)$ we can assign a new metric space R, whose points p are the ordered pairs $p = (p_1, p_2)$ and whose metric $d(p, q)$ is defined by

$$(20) \quad d(p, q) = d((p_1, p_2), (q_1, q_2)) = \sqrt{d_1(p_1, q_1)^2 + d_2(p_2, q_2)^2},$$

where the reader can easily prove that this definition satisfies the axioms. The new metric space R is called the *metric product* of R_1 and R_2. It is obvious that every Euclidean space R^n can be regarded as the metric product of an R^{n-1} with an R^1.

Up to now "neighborhoods" have always been taken to be ε-neighborhoods, i.e., sets of points whose distance from a fixed point is less than ε. But now other subsets will also be taken; in particular, for the point $p = (p_1, p_2)$ the set

$$(21) \qquad\qquad U(p) = \{(q_1, q_2)|q_i \in U_{\varepsilon_i}(p_i)\}.$$

The set of ordered pairs with the neighborhoods defined by (21) is called the *topological product* $R_1 \times R_2$ of the two sets R_1 and R_2. If we compare (Fig. 58) the metric product of two one-dimensional Euclidean spaces with their topological product $R^1 \times R^1$, we see that they differ only in the neighborhoods of their points, which in the first case are open circular disks, and in the second open rectangles.

Before turning in general to the question of how a set of points is made into a space by the definition of neighborhoods, let us extend the Euclidean

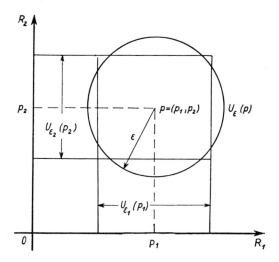

Fig. 58

concept of dimension to certain other metric spaces. For this purpose we consider the space R as made up of pieces and first explain what is meant by a covering of R. A system of subsets M is called a *covering* of a space R if every point $p \in R$ is contained in at least one set M_i. Depending on the properties of the set we speak of "open" or "closed" coverings. The number of subsets may be finite, countably infinite, or of higher power. In a metric space the diameter of each subset, i.e., the least upper bound of all distances between two points in the subset, may be smaller than a fixed ε, in which case we speak of an ε-*covering*. If no point of the space belongs to more than r subsets of a covering, but if at least one point belongs to r subsets, then r is called the *order* of the covering. To every finite or countably infinite system of sets of finite order we can now assign an abstract nonoriented complex with a one-to-one correspondence between its vertices x_i and the sets M_i and between its (nonoriented) simplexes $x_{i_1} x_{i_2} \cdots x_{i_k}$ and the nonempty intersections $M_{i_1} \cap M_{i_2} \cap \cdots \cap M_{i_k} \neq 0$. This complex is called the *nerve* of the covering, the greatest number of vertices of a simplex being identical with the order of the covering.

Since a one-dimensional line segment in R^1 can be decomposed into two disjoint but not separated sets and the nerve of this covering consists of two zero-dimensional simplexes, it is clear that in order to set up a relationship between Euclidean dimension and the dimension of the nerve we must confine ourselves to *closed* coverings, whereupon we see that every covering of a line segment is at least of order two.

Coverings of the plane with congruent domains play a role in ornamentation, e.g., the hexagonal grid of order 3, the rhombic grid of order 6, and a covering of order 4 with circles (Figs. 59–61). Here too the smallest order of a covering is one greater than the Euclidean dimension of the plane. (In the last two examples the order can be reduced to 3 by introducing small closed circular neighborhoods of the nodes (Fig. 62), which provide a first-order covering of their interior points.)

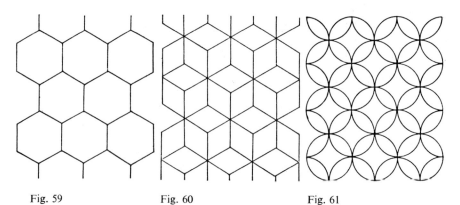

Fig. 59   Fig. 60   Fig. 61

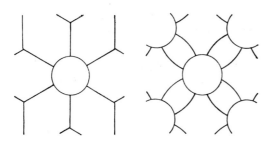

Fig. 62

Let us now investigate the concept of dimension by means of these coverings. Since we will encounter some difficulty for arbitrary sets of points, we first consider countably compact metric spaces (i.e., spaces in which every sequence of points has at least one limit point), which are also called *compacta*. For such a compactum R we define its *dimension* (with Alexandrov) as the smallest nonnegative number $n = \dim R$ for which there exists, for every $\varepsilon > 0$, a finite closed ε-covering of R whose nerve is an n-dimensional complex, or in other words, whose order is $(n + 1)$, and if no such smallest number exists we say that the dimension is infinite. Since it is not difficult to show that this concept of dimension is topologically invariant, more general metric spaces, representable as the union of countably infinitely many compacta, can now be assigned a dimension by means of the "summation theorem in dimension theory," which states that the dimension of the union of countably many n-dimensional closed sets is likewise n-dimensional.

But it still remains to prove that for Euclidean polyhedra the new concept of dimension coincides with the old one. For this purpose it is sufficient, in view of the summation theorem, to consider a Euclidean simplex and to prove the important, easily visualized "Lebesgue paving theorem": every finite closed ε-covering of an n-dimensional simplex, with sufficiently small ε, is at least of order $n + 1$, which was first proved by Brouwer in 1913 (an elementary proof by Sperner is given in Alexandrov [1].) In addition to the method discussed here, by which the dimension of a set of points is reduced to the combinatorial dimension of a complex, there are other methods of assigning a dimension to (essentially arbitrary) sets of points. As an example let us give a set-theoretic definition due to Menger and Uryson, involving the concepts of "neighborhood," "topological space," and "regular space" as explained in the next section. For this purpose we define not only the dimension $\dim R$ of a topological space R but also the *dimension* $\dim_p R$ *at a point* $p \in R$, as follows; $\dim R$ and $\dim_p R$ are functions which to the space R and to each point p assign an integer ≥ -1, or else the value ∞, in such a way that

1. $\dim R = -1$ if and only if R is empty.

2.   If R is not empty, then provided that for all $p \in R$ the values dim R are bounded, dim R is equal to the greatest value of $\dim_p R$, and otherwise dim $R = \infty$.

3. $\dim_p R \leqq n + 1$ if and only if every neighborhood $U(p)$ contains a neighborhood $V(p)$ with boundary B such that dim $B \leqq n$. Here the word boundary means the set of points of contact of $V(p)$ that do not belong to $V(p)$.

The connection between this concept of dimension and our earlier combinatorial concept is given by the following theorem.

The dimension of a regular space R with an at most countable basis is the smallest integer n with the property that for every finite open covering $\mathfrak{U}$ of R there is a finite open covering $\mathfrak{B}$ of order $n + 1$ such that every element of $\mathfrak{B}$ is contained in at least one element of $\mathfrak{U}$.

With this set-theoretic definition of dimension we can give a set-theoretic definition of a curve as a one-dimensional continuum as follows.

A set of points K in a topological space R is called a curve if

1. K is a continuum;

2.   for every point p of R every neighborhood $U(p)$ contains a neighborhood $V(p)$ whose boundary has a discontinuous intersection with K.

For plane curves this definition coincides with the Cantor definition; namely, *a (Cantor) curve is a continuum K in the Euclidean plane none of whose points has a neighborhood contained entirely in K.*

We turn now to the question: How is a set of points to be made into a space if no distance function has been assigned to it in advance?

10.   Topological Spaces

To each point p of an arbitrary abstract point-set S we assign subsets $U(p)$ called "neighborhoods of p." For this system of neighborhoods $\mathfrak{U}$ to make the set of points S into a topological space R the choice of subsets must be restricted by means of axioms. For example, since we naturally think of a point as being in its own neighborhood, we require the following so-called first Hausdorff axiom.

H 1)   *Every point $p \in R$ has at least one neighborhood $U(p)$ and is contained in each of its neighborhoods, $p \in U(p)$.*

With this axiom alone we can define contact point and "closure" of a set $M \subset R$, as before; i.e., p is called a contact point of M if $U(p) \cap M \neq 0$ for all its neighborhoods $U(p)$. The set of all contact points of M is denoted by $\overline{M}$, and is called the *closure* of M, where the correspondence $M \to \overline{M}$ depends in general on the chosen system of neighborhoods. The concepts "closed" and "open" can also be defined as before; i.e., M is closed if $M = \overline{M}$, and open if the complement $R - M$ is closed. But this terminology must not lead us to false conclusions; for example, it is not yet true that the union of two closed sets is necessarily a closed set. So we now restrict the system of neighborhoods by two further Hausdorff axioms.

H 2)   *From every two neighborhoods U(p) and V(p) there exists a neighborhood W(p) contained in the intersection of U(p) and V(p):W(p) ⊂ (U(p) ∩ V(p)).*

The system of all open sets in R can be regarded as a system of neighborhoods $\mathfrak{O}$, if we require that every open set is to be a neighborhood of each of its points. For this system $\mathfrak{O}$ Axioms H 1 and H 2 are satisfied, but the induced formation of closures $M \to \overline{M}$ may be different from the correspondence produced by $\mathfrak{U}$, i.e., the closure of a subset will depend, in general, on whether $\mathfrak{U}$ or $\mathfrak{O}$ is used as a system of neighborhoods. In general, two different systems of neighborhoods of a space are said to be *equivalent* if they produce the same correspondence $M \to \overline{M}$ for all subsets $M \subset R$. This equivalence of $\mathfrak{U}$ and $\mathfrak{O}$ is now enforced by the third Hausdorff axiom.

H 3)   *For every point q in a neighborhood U(p) there exists at least one neighborhood U(q) contained in U(p):U(q) ⊂ U(p).*

These three axioms already enable us to prove the following theorems about open sets.

B 1)   *Every finite or infinite union of open sets is open.*

B 2)   *The intersection of finitely many open sets is open.*

By considering the complementary sets we obtain the corresponding theorems about closed sets; namely, the intersection of finitely or infinitely many closed sets, and the union of finitely many closed sets, is again closed. Since we now have a suitable foundation for the geometry of sets of points, we shall say every set of points S with a system of neighborhoods $\mathfrak{U}$ satisfying H1, 2, 3 is a *topological space R*.

But a topology can also be induced in a point set S by establishing which subsets of S are to be defined as open. For such a system of open sets we must require axiomatically that the Theorems B 1 and B 2 be satisfied, and we must also have a third axiom to the effect that the zero set and the entire set S are open.

B 3)   *The zero set and the set S are open.* Then the set S is made into a topological space by axioms B 1, 2, 3 if the open sets are taken as neighborhoods of their points. This definition lies at the basis of the work of Bourbaki.

A simple example of a topological space containing only three points a, b, c is provided by the requirement that the following sets be open: $0, a, a \cup b, a \cup c, a \cup b \cup c$. We find that $\overline{a} = a \cup b \cup c$, $\overline{b} = b$, $\overline{c} = c$. The point a in this space is not a closed set, since b is a limit point for the subset $M = \{a\}$ consisting of a single point. Such surprising properties of a topological space can be avoided by the so-called first (Fréchet) *separation axiom* T_1.

T_1)   *For any two distinct points p ≠ q there exists a neighborhood U(p) not containing the point q: U(p) ∩ q = 0.*

In a T_1-space, i.e., in a topological space satisfying the T_1-axiom, every point, considered as a subset, is closed $p = \bar{p}$, and it can be shown that every neighborhood $U(p)$ of a limit point p for a subset M contains infinitely many points of M. A sharpening of these separation properties can be obtained by the fourth Hausdorff axiom, the so-called second separation axiom.

T_2) *Every two distinct points* $p \neq q$ *have distinct neighborhoods* $U_1(p)$, $U_2(q)$: $U_1(p) \cap U_2(q) = 0$.

Topological spaces satisfying the T_2-axiom are called *Hausdorff spaces*. We can now specialize the spaces by means of other separation axioms, but we will not carry this procedure further. Let us rather turn to the question how the metric spaces already discussed are related to our present Hausdorff spaces. In these metric spaces the ε-neighborhoods provide a neighborhood system for which it is easy to see that Axioms H 1, 2, 3 and T_2 are satisfied. Thus metric spaces are always Hausdorff spaces. But they have a further property. Since an ε-neighborhood of a closed set M means the totality of points p for which the distance $d(M, p) < \varepsilon$, there always exists a neighborhood $V_{\varepsilon/2}(M)$, $\overline{V_{\varepsilon/2}(M)} \subset U_\varepsilon(M)$, which together with its closure lies entirely in $U_\varepsilon(M)$. This particular property of the system of neighborhoods can be formulated without the concept of distance. To begin with, the concept of "neighborhood of a closed set M" can easily be transferred to more general topological spaces, since for this purpose we need only choose a neighborhood $U(p)$ for every point $p \in M$, and then form the union of all these neighborhoods $U(M) = \bigcup U(p)$ with $p \in M$. A T_1-space in which for every closed set $M \subset R$ and every neighborhood $U(M)$ there exists at least one neighborhood $V(M)$ whose closure $\overline{V(M)} \subset U(M)$ lies in $U(M)$ is called a *normal* space. Thus every metric space is normal. Similarly, a T_1-space in which for every neighborhood $U(p)$ of a point there exists at least one other neighborhood $V(p)$ with $\overline{V(p)} \subset U(p)$ is called a *regular* space. Since every point in a T_1-space is itself a closed set, it follows trivially that every normal space is regular, although the converse is not true. In order to give some idea of how neighborhoods can be used to prove separation properties, we now show that all regular spaces are Hausdorff spaces.

For any two distinct points $p \neq q$ in a regular space the T_1-axiom states that there exists a $U(p)$ with $U(p) \cap q = 0$, and the regularity of the space means that there also exists a $V(p)$ with $\overline{V(p)} \subset U(p)$. Thus the point q lies in the open set $R - \overline{V(p)}$ and $R - \overline{V(p)} \cap V(p) = 0$. Now if every $U(q)$ had a nonempty intersection $U(q) \cap V(p) \neq 0$ with $V(p)$, then q would be a contact point of $V(p)$, i.e., $q \in \overline{V(p)}$, in contradiction to the fact that $q \in R - \overline{V(p)}$. Consequently there exists, as desired, at least one $U(q)$ with $U(q) \cap V(p) = 0$.

The concept of a metric space is still very general since it includes, for example, all sets of points in Euclidean spaces. But on the other hand, in the context of topological spaces, it is very special, since we have seen that a metric space must at least be normal. A topological space is said to be *metrizable* if a distance function can be introduced into it in such a way that the corresponding system of spherical ε-neighborhoods is equivalent to the given system of neighborhoods of the topological space. The problem of finding necessary and sufficient conditions for metrizability has been solved. But here we must be satisfied with a discussion of simple sufficient conditions. Every topological space is characterized by its set of open sets. Since a union of open sets is always open, it is sufficient to start from a subsystem $\mathfrak{B}$ of open sets such that every open set in R can be represented as the union of sets of the system $\mathfrak{B}$. Such a system $\mathfrak{B}$ is called a *basis* for the topological space. As an example of a basis for a Euclidean space, we may take the spherical ε-neighborhoods of all points with rational coordinates, since every irrational number can be represented as the limit of a sequence of rational numbers, and for the same reason we can also confine the ε-values to rational numbers, thereby obtaining a basis with a countable set of elements, or a so-called *countable basis*. A given topological space by no means necessarily contains a countable basis. For example, if in a noncountable set S we define the distance between two distinct points $p \neq q$ by $d(p,q) = 1$, we obtain a metric space, consisting of isolated points, without countable basis. The existence of a countable basis is called the *second Hausdorff axiom of countability*. Topological spaces R with countable basis are separable, i.e., they contain a countable set of points dense in R. Then an important theorem of Uryson states that *a normal space with countable basis is metrizable*. Normality is necessary for metrizability, but a countable basis is only sufficient, as is shown by the simple example of a metric space without countable basis. Uryson has also shown that normal spaces with a countable basis are topological images of sets of points in a Hilbert space, thereby providing a connection between more general topological spaces and a familiar geometric space.

The product of metric spaces as defined in §9 can be carried over to general topological spaces, by extending formula (21). The "points" p of the product space $R_1 \times R_2$ are the ordered pairs of points $p = (p_1, p_2)$ with $p_1 \in R_1$, $p_2 \in R_2$, and the "neighborhoods" $U(p)$ are defined as sets of pairs (q_1, q_2) with $q_i \in U_i(p_i)$. Since Axioms H 1, 2, 3 are satisfied for this system of neighborhoods, the topological product is itself a topological space. For metric products it is then necessary to prove (Fig. 58) that the system of ε-neighborhoods of the metric product space is equivalent to the system of neighborhoods of the topological product. In addition to the Euclidean spaces already mentioned, a simple example is given by the torus as the topological product of two circles.

II. Continuous Mappings of Polyhedra

For continuous mappings of sets of points in a Euclidean space we have already seen that the concept of a contact point is invariant, and for general topological spaces this property can be taken as a definition of the continuity of a mapping; namely, the mapping f of the space R into the space R^* is said to be *continuous* if the image of every contact point of an arbitrary point set $M \subset R$ is a contact point of the image set $f(M) \subset R^*$, i.e., if $f(\overline{M}) \subset \overline{f(M)}$ (cf. §7). A continuous image of a closed set M is not necessarily closed; but if the preimage $f^{-1}(M^*)$ of a subset $M^* \subset R^*$ is defined as the set of all those points of R whose images belong to M^*, then the preimage of a closed set M^* is always closed under a continuous mapping. For the proof we set $M = f^{-1}(M^*)$. Then $M^* = f(M) = \overline{f(M)}$, and by the continuity of f, $f(\overline{M}) \subset \overline{f(M)} = f(M)$ [$p \in \overline{M}$ means $f(p) \in M^*$ so that $p \in M$; therefore $\overline{M} \subset M$]. Thus $\overline{M} = M$, since the closure $\overline{M}$ of M includes M. Similarly, the preimage of an open set under a continuous mapping is open. For topological mappings, since they are one-to-one and bicontinuous, it follows that closed (open) sets are mapped into closed (open) sets. The connectivity of a set is invariant under a continuous mapping. For if a connected set M is decomposed into two disjoint subsets, $M = A \cup B$ with $A \cap B = 0$, it follows that either $\overline{A} \cap B \neq 0$ or $A \cap \overline{B} \neq 0$. Since $f(\overline{A} \cap B) \subset f(\overline{A}) \cap f(B) \subset \overline{f(A)} \cap f(B)$ and $f(A \cap \overline{B}) \subset f(A) \cap \overline{f(B)}$, it follows that $f(A)$ and $f(B)$ are not separated, since either $\overline{f(A)} \cap f(B)$ or $f(A) \cap \overline{f(B)}$ is not empty.

Two continuous mappings f_0, f_1 of a topological space R into a metric space R^* are said to be *homotopic* to each other if f_1 can be transformed into f_0 continuously, i.e., if there exists a one-parameter family of mappings f_t with $0 \leq t \leq 1$, which for $t = 0$ coincides with f_0 and for $t = 1$ coincides with f_1, and is such that all points $p \in R$ differ arbitrarily little from each other for values of the parameter t sufficiently close to each other; or more precisely, for arbitrary preassigned ε there exists a δ such that in the metric image space $d(f_{t_1}(p), f_{t_2}(p)) < \varepsilon$ for $|t_2 - t_1| < \delta$ and all points p. Now as a first principle of classification for continuous mappings we assemble the homotopic mappings into classes, the so-called *homotopy classes*. Here we shall restrict ourselves to the simplest case, in which both R and R^* are finite Euclidean polyhedra. These polyhedra will have simplicial complexes K and K^*, and by a *simplicial mapping* of K into K^* we shall mean an assignment of the vertices x_i of the (abstract) complex K to the vertices x_j^* of K^*, such that to every simplex of K there corresponds a simplex of K^*. An absolute n-dimensional simplex will then be mapped, depending on the number of distinct image vertices, onto an absolute m-dimensional simplex with $m \leq n$. On the other hand, an oriented n-dimensional simplex will either go into an oriented image simplex with the same number of dimensions or else, if at least two image vertices coincide, into the zero-chain. Corresponding to the simplicial mapping of the complexes there exists a continuous mapping of the Euclidean polyhedra onto each other, which is uniquely determined if we map the Euclidean simplexes onto each other affinely. For this purpose we start with the points p_i assigned to the vertices x_i of the complex and describe a point p in the simplex spanned by the linearly independent points $p_{i_1}, \ldots, p_{i_{n+1}}$ by means of its barycentric coordinates λ_v

$$p = \lambda_1 p_{i_1} + \cdots + \lambda_{n+1} p_{i_{n+1}} \quad \text{with} \quad 0 \leq \lambda_v \leq 1, \quad \sum_v \lambda_v = 1.$$

If we now place masses λ_v at the vertices p_{i_v}, the point p is the center of gravity of this distribution of mass. Then the simplicial mapping of K into K^* induces a mapping $p_{j_v}^* = \varphi(p_{i_v})$ of the vertices of the polyhedra p_{i_v} of R into the vertices $p_{j_v}^*$ of R^* and the mapping function φ is extended to an arbitrary simplex point p by setting

$$\varphi(p) = \lambda_1 \varphi(p_{i_1}) + \cdots + \lambda_{n+1} \varphi(p_{i_{n+1}}).$$

This mapping function is linear inside the complex, and piecewise linear and continuous in the whole polyhedron R; it is also called a *simplicial mapping of R into R^** since, just as in our use of the word simplex, there is no danger of misunderstanding.

By the carrier simplex of the point p we mean the simplex in the polyhedron R that enables us to represent p in barycentric coordinates $\lambda_v \neq 0$ such that none of them vanish. The totality of points in R whose carrier simplexes have a common vertex p_i is called the *open star* $St_K(p_i)$ of this vertex with respect to the decomposition complex K. Since such a set is open in the polyhedron R, and since every point belongs to at least one open star of a vertex, the $St_K(p_i)$ form an open covering of the polyhedron. An arbitrary continuous mapping f of the polyhedron R into the polyhedron R^* assigns the open stars $St_K(p_i)$ to certain sets of points $f(St_K(p_i)) \subset R^*$. If a simplicial mapping can be determined in such a way that $f(St_K(p_i))$ lies in the open star of the image vertex $\varphi(p_i)$ with respect to the complex K^*

$$f(St_K(p_i)) \subset St_{K^*}(\varphi(p_i)) \quad \text{for all } i,$$

then φ is called a *simplicial approximation* of f. To see the importance of this concept, let us compare the images $f(p)$ and $\varphi(p)$. If p_i is a vertex of the carrier simplex of p, we have $p \in St_K(p_i)$ and thus $f(p) \subset f(St_K(p_i)) \subset St_{K^*}(\varphi(p_i))$, so that the image point $f(p)$ lies in the intersection of all open stars of the image vertices $\varphi(p_i)$. On the other hand, these vertices determine the affine image of the carrier simplex, so that the points $\varphi(p)$ and $f(p)$ always belong to the same simplex of the polyhedron R^*, namely the carrier simplex of $f(p)$, and in the interior of this simplex the points $\phi(p)$ and $f(p)$ can be joined by a line segment. If we now divide the line segments in the ratio $t:(1 - t)$ and denote the point of division by p_t^*, then for a fixed t with $0 \leq t \leq 1$, the assignment $p \to p_t^*$ is a continuous mapping. Thus for variable t we obtain a family of mappings transforming $f(p)$ for $t = 0$ continuously into $\varphi(p)$ for $t = 1$. Thus the mappings f and φ are homotopic and the Euclidean distance $d(f(p), \varphi(p))$ cannot exceed the maximum of the diameters of the simplexes of R^*.

For every continuous mapping f of R into R^* it can be proved that, for a sufficiently fine subdivision K_i of K, the continuous mapping f can be approximated simplicially by a simplicial mapping φ of K_i into the complex K^*. Since we can also choose K^* such that the maximal diameter of the simplexes of R^* is less than a preassigned ε, the image points $\varphi(p)$ differ arbitrarily little from the image points of the continuous mapping $f(p)$. The existence of a homotopic simplicial approximation to every continuous mapping means that in the study of homotopy classes we can confine our attention to simplicial mappings. Thus the methods of combinatrial topology can be applied to problems in homotopy as well.

In conclusion, we turn to the continuous mapping of a polyhedron onto itself. Here we have $R \equiv R^*$, and we choose K as a subdivision complex of K^*. In a simplicial mapping of K into K^* it may happen that the image of an oriented simplex $\varphi(\sigma_i^r) = \sigma_j^{*r}$ in the subdivision $K^* \to K$ contains the original simplex σ_i^r in the subdivision chain $\sigma_j^{*r} \to \sum_k \varepsilon_{ik}^r \sigma_k^r$, with $\varepsilon_{ik}^r = 0$, ± 1. In this case, $\varepsilon_{ii}^r \neq 0$ and, according to the sign of ε_{ii}^r, σ_i^r is called a positive or negative *fixed simplex*. For example, let K be the two-dimensional complex in Fig. 63 and let K^* coincide with K. A simplicial mapping φ is then given by the following correspondence

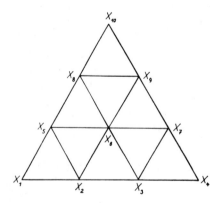

Fig. 63

x_i	1	2	3	4	5	6	7	8	9	10
$\varphi(x_i)$	6	8	6	5	9	9	8	7	6	3

For the oriented simplexes we have

$$\varphi(x_1 x_2 x_5) = x_6 x_8 x_9, \quad \varphi(x_2 x_5 x_6) = 0, \ldots,$$

So that altogether there is only one negative one-dimensional fixed simplex $\varphi(x_6 x_9) = x_9 x_6 = -x_6 x_9$, as is easily verified. As a second example let K be the same complex as before, but let K^* be the original simplex $x_1 x_4 x_{10}$ and let a simplicial mapping be given by

x_i	1	2	3	4	5	6	7	8	9	10
$\varphi(x_i)$	1	1	1	4	10	1	4	10	4	10

Here there are the three zero-dimensional fixed simplexes $\varphi(x_1) = x_1$, $\varphi(x_4) = x_4$, $\varphi(x_{10}) = x_{10}$, and also three positive one-dimensional simplexes $x_3 x_4, x_9 x_{10}, x_1 x_5$, since $\varphi(x_3 x_4) = x_1 x_4 \to x_1 x_2 + x_2 x_3 + x_3 x_4$, etc. Finally, there is a positive two-dimensional fixed simplex $x_6 x_8 x_9$, since

$$\varphi(x_6 x_8 x_9) = x_1 x_{10} x_4 \to x_1 x_5 x_2 + \cdots + x_6 x_8 x_9 + \cdots + x_3 x_7 x_4.$$

By the rth (algebraic) *fixed simplex number* of φ we mean the difference between the number of positive and negative fixed simplexes of dimension r. This number is given by the sum $\sum_i \varepsilon_{ii}^r$, i.e., the trace of the matrix (ε_{ik}^r), and its importance lies in the fact that it can be related to the homology groups in K. The simplicial mapping of K into K^* produces a mapping of the corresponding groups, but since K arises by subdivision from K^*, it follows that $\mathfrak{H}^r(K)$ coincides with $\mathfrak{H}^r(K^*)$, and the mapping in question becomes a mapping of $\mathfrak{H}^r(K)$ into itself (for lack of space we must be satisfied here with this suggestion). A study of this mapping of K into itself shows that, like the Euler-Poincaré characteristic, the alternating sum of the fixed simplex numbers $F_\varphi = \sum_r (-1)^r (\sum_i \varepsilon_{ii}^r)$ has topological significance; for example, under the identical mapping, every simplex is a fixed simplex and the rth fixed simplex number is equal to the number α^r of r-dimensional simplexes, so that in this case the alternating sum of the fixed simplex numbers coincides with the Euler-Poincaré characteristic. In the above examples, $F_\varphi = 1$, and it can be shown that for an arbitrary simplicial mapping φ of a simplex into itself, F_φ retains this value.

The affine mapping of the fixed simplex onto the corresponding more inclusive simplex of the same dimension in the decomposition K^* always contains at least one fixed point. For the proof we again use barycentric coordinates. If $p = \sum_1^r \lambda_i p_i$, with $\sum_1^r \lambda_i = 1$ and $\lambda_i \geq 0$, is the barycentric representation of an arbitrary point of the fixed simplex, and similarly for the image point $p^* = \sum_1^r \lambda_i p_i^*$, then since every vertex p_i belongs to the image simplex, we also have $p_i = \sum_1^r \alpha_{ij} p_j^*$ with $\sum_j \alpha_{ij} = 1$, $\alpha_{ij} \geqq 0$. For a fixed point $p = \varphi(p) = p^*$ the system of equations

$$(22) \qquad \sum_i \lambda_i \alpha_{ij} = \lambda_j$$

must have nonnegative solutions λ_i with $\sum \lambda_i = 1$. Now if the solutions $\lambda_{j_1}, \lambda_{j_2}, \ldots, \lambda_{j_s}$ say were negative, and the remaining magnitudes nonnegative, it would follow from (22) that

$$\sum_{v=1}^s \sum_{i=1}^r \lambda_i \alpha_{ij_v} = \lambda_{j_1} + \cdots + \lambda_{j_s}$$

or

$$\sum_{v=1}^s (\lambda_{j_{s+1}} \alpha_{j_{s+1}j_v} + \cdots + \lambda_{j_r} \alpha_{j_r j_v}) = \sum_{v=1}^s \lambda_{j_v}(1 - \alpha_{j_v j_1} - \alpha_{j_v j_2} - \cdots - \alpha_{j_v j_s})$$

But since $0 \leqq \alpha_{ij} \leqq 1$, $\sum_j \alpha_{ij} = 1$, the left side is nonnegative, and the right side is negative, a contradiction which shows that only solutions of one sign can exist. But these solutions actually do exist, since the rank of the system is smaller than r, in view of the fact that

$$\sum_j (\sum_i \lambda_i \alpha_{ij}) \equiv \sum_j \lambda_j$$

From the number of fixed simplexes of a simplicial mapping we cannot draw any immediate conclusion about the number of fixed points. In the first place, different fixed simplexes can lead to the same fixed point (see the second example, with 7 fixed simplexes but only 4 fixed points), and secondly, if the rank of the system of (22) is lowered, there may be infinitely many fixed points. Thus it is all the more true that the number F_φ does not determine the total number of fixed points. But we have proved the important theorem that $F_\varphi \neq 0$ always implies the existence of at least one fixed point. For if there are any fixed simplexes at all, then certainly there is at least one fixed point (a remark that holds equally well for arbitrary continuous mappings), since it follows from the connection with the homology groups (not given here in detail) that F_φ is invariant under transfer to a homotopic mapping, so that all simplicial approximations of f lead to the same value for F_φ; and since these simplexes may differ arbitrarily little from f, $F_\varphi \neq 0$ also implies that f has at least one fixed point.

If R is an n-dimensional Euclidean simplex Knaster, Kuratowski, and Mazurkiewicz have given an elementary proof of the following theorem on fixed points under an arbitrary continuous mapping. Let $p = \sum_1^{n+1} \lambda_i p_i$ $0 \leqq \lambda_i \leqq 1$, $\sum \lambda_i = 1$ be a point of a simplex and let $p^* = f(p) = \sum_1^{n+1} \mu_i p_i$ with $0 \leqq \mu_i \leqq 1$, $\sum \mu_i = 1$, be its image point. Since f is continuous, the set

M_j of all points $p \in R$ with $\mu_j \leqq \lambda_j$ is closed, and the vertex p_j is contained in M_j. The edge joining $p_{j_1} p_{j_2}$ lies in the union $M_{j_1} \cup M_{j_2}$, since on this edge $\lambda_{j_1} + \lambda_{j_2} = 1$, which implies either $\mu_{j_1} \leqq \lambda_{j_1}$ or $\mu_{j_2} \leqq \lambda_{j_2}$, in view of the fact that $\mu_{j_1} + \mu_{j_2} \leqq 1$, and this conclusion is valid for every r-dimensional side $p_{j_1} \cdots p_{j_{r+1}}$ of R. Since $\lambda_{j_1} + \cdots + \lambda_{j_{r+1}} = 1$ and $\mu_{j_1} + \cdots + \mu_{j_{r+1}} \leqq 1$, the inequality $\mu_{j_\nu} \leqq \lambda_{j_\nu}$, must hold for at least one index j_ν, which means that the side $p_{j_1} \cdots p_{j_{r+1}}$ lies in the set $M_{j_1} \cup M_{j_2} \cup \cdots \cup M_{j_{r+1}}$. For $r = n$ we see that the sets M_j form a closed covering of R, and we now assert that the intersection $M_1 \cap M_2 \cap \cdots \cap M_{n+1}$ is not empty. Let K be a simplicial decomposition complex of the simplex R with the vertices q_i. If q_i is an interior point of the side $p_{j_1} \cdots p_{j_r}$ of the simplex, then q_i must lie in one of the sets from M_{j_1} to M_{j_r}, say in M_{j_ν}, where j_ν is the smallest such index in the sequence $j_1, \ldots, j_r$. The mapping $p_{j_\nu} = \varphi(q_i)$ of q_i onto p_{j_ν} then induces a simplicial mapping φ of the subdivision K onto the undivided simplex, under which the number of fixed n-dimensional simplexes is always odd, by a lemma of Sperner. Sperner's proof of his lemma consists of counting the number of times the $(n - 1)$-dimensional sides of a simplex $q_{i_1} \cdots q_{i_{n+1}}$ have the image $p_1 \cdots p_n$. This number, which we shall denote by $t_{i_1 \ldots i_{n+1}}$ is equal to 1 for a fixed simplex, and for every other simplex it is either 2 or 0. Thus the total number $\sum t_{i_1 \ldots i_{n+1}}$ consists of the number of fixed simplexes F_n and an even number.

This result can be deduced in another way. An $(n - 1)$-dimensional side in the interior of the simplex R occurs twice in the sum, since it is on the boundary of two n-simplexes; but if it lies on the boundary of R, the corresponding summands will be nonzero only on the side $p_1 \ldots p_n$, and their total number will be equal to the number of $(n - 1)$-dimensional fixed simplexes of the corresponding $(n - 1)$-dimensional problem for this side $p_1 \cdots p_n$.

The number of fixed simplexes P is to be clearly distinguished from the algebraic fixed simplex number considered above. Since $\sum t_{i_1 \ldots i_{n+1}} \equiv F_n$ mod 2 and $\sum t_{i_1 \ldots i_{n+1}} \equiv F_{n-1}$ mod 2, we have $F_n \equiv F_{n-1}$ mod 2. For $n = 0$, it follows trivially that $F_0 = 1$, so that F_n is always an odd number. Thus in any subdivision of R, no matter how fine, there exists a fixed simplex whose vertices $q_{k_1}, \ldots, q_{k_{n+1}}$ are mapped onto the vertices $p_1, \ldots, p_{n+1}$ and are therefore contained in the sets $M_1, \ldots, M_{n+1}$. Since the diameter of the simplex $q_{k_1} \cdots q_{k_{n+1}}$ can be made smaller than a preassigned ε, the intersection $M_1 \cap \cdots \cap M_{n+1}$ is not equal to zero. For a point $p \in M_1 \cap \cdots \cap M_{n+1}$, it follows that $\mu_1 \leqq \lambda_1, \mu_2 \leqq \lambda_2, \ldots, \mu_{n+1} \leqq \lambda_{n+1}$, and since $\sum \mu_i = 1$, these relations can hold simultaneously only with equality throughout. Thus p is a fixed point, and we have proved the famous Brouwer theorem that every continuous mapping of a Euclidean simplex into itself contains at least one fixed point.

In the above sections, we have had to confine our attention to the most important concepts of elementary topology, illustrating them by simple examples, and giving some indication of their importance by means of theorems and outlines of the proofs. Our selection has been made with a view to giving the reader some understanding of the problems and encouraging him to undertake a deeper study.

Bibliography

[1] ALEXANDROFF, P.: Einfachste Grundbegriffe der Topologie. Berlin 1932.

[2] ALEXANDROFF, P. and HOPF, H.: Topologie I. Berlin 1935.

[3] HALL, O. W. and SPENCER, G. L.: Elementary Topology. New York 1955.

[4] HILBERT, D. and COHN-VOSSEN, S.: Geometry and the imagination. Translated from the German by P. Nemenyi, Chelsea Publishing Co., New York 1952.

[5] V. KERÉKJÁRTO, B.: Vorlesungen über Topologie I. Berlin 1923.

[6] SEIFERT, H. and THRELFALL, W.: Lehrbuch der Topologie. Leipzig 1934. Reprint New York 1947.

Index